International Symposium on

NEW DEVELOPMENTS IN APPLIED SUPERCONDUCTIVITY

SERIES ON PROGRESS IN HIGH TEMPERATURE SUPERCONDUCTIVITY

International Symposium on

NEW DEVELOPMENTS IN APPLIED SUPERCONDUCTIVITY

Suita, Osaka, Japan   17–19 Oct. 1988

A Supplement of Annual Report of Laboratory for Applied Superconductivity, Osaka University

Editor

Y. Murakami

Lab. for Applied Superconductivity
Faculty of Engineering
Osaka University

World Scientific
Singapore • New Jersey • London • Hong Kong

Published by

World Scientific Publishing Co. Pte. Ltd.

5 Toh Tuck Link, Singapore 596224

USA office: 27 Warren Street, Suite 401-402, Hackensack, NJ 07601

UK office: 57 Shelton Street, Covent Garden, London WC2H 9HE

British Library Cataloguing-in-Publication Data
A catalogue record for this book is available from the British Library.

**INTERNATIONAL SYMPOSIUM ON NEW DEVELOPMENTS IN
APPLIED SUPERCONDUCTIVITY**

ISBN-13 978-9971-5-0816-6
ISBN-10 9971-5-0816-8
ISBN-13 978-9971-5-0834-0 (pbk)
ISBN-10 9971-5-0834-6 (pbk)

Jrganizing Committee

    Katsuhiko FUJII (Chairman)
        Director, LAS, Fac. Engr.

    Kunisuke ASAYAMA
        Dept. Material Physics, Fac. Engr. Science
    Muneyuki DATE
        Dept. Physics, Fac. Science
    Chihiro HAMAGUCHI
        Dept. Electronic. Engr., Fac. Engr.
    Akio HIRAKI
        Dept. Electrical. Engr., Fac. Engr.
    Yoshimasa KYOUGOKU
        Institute for Protein Research
    Akiyoshi MITSUISHI
        Dept. Appl. Phys., Fac. Engr.
    Yoshishige MURAKAMI
        LAS, Fac. Engr.
    Toichi OKADA
        Institute of Scientific and Industrial Research

Editorial Committee
    Yoshishige MURAKAMI(Editor and Chairman)

1. Energy and Systems
    Toshifumi ISE, Kenji MATSUURA, Yasunori MITANI, Yoshishige MURAKAMI, Kiichiro TSUJI, Katsumi YOSHINO
2. Magnet Technology
    Toshihiko KATAOKA, Shigehiro NISHIJIMA, Toichi OKADA, Shigeoki SAJI, Junya YAMAMOTO
3. Materials & Electronics
    Ginya ADACHI, Chihiro HAMAGUCHI, Akio HIRAKI, Toshimichi ITO, Tomoji KAWAI, Takeshi KOBAYASHI, Shinichi NAKASHIMA, Kiyomi SAKAI, Junji SHIRAFUJI

Foreword

The applications of superconductivity have been spectacularly prompted by the overwhelming impact of the discovery of high temperature superconductivity. The developments have been stimulated in the usage of conventional metallic superconductors as well as in the proposals of applications of high temperature superconductors.

Osaka University's International Symposium on "New Developments in Applied Superconductivity" which was co-sponsored by the Ministry of Education, Science and Culture was held October 17-19, 1988.

These proceedings of the Symposium include 27 invited papers and 103 contributed papers, which were intended to overview the developments in these three years after the great discovery. They are edited into four sections. Each section has the invited papers which review the related areas. They are not arranged, however, from a systematic view point in this chaotic situation, but to supply ideas from different fields and stand points.

For instance, in Section 1 New Superconductors and Proposals for Their Applications, Tl compound and Bi compound superconductors are reviewed by the discoverers, which are followed by the review on the recent status of researches of oxide superconductors in China. The film formation and electronic applications of high Tc oxide superconductors are reviewed. The superconducting mechanism is discussed by the comparison of different kinds of substances. The widely spread critical current densities of superconducting oxides are compared from a theoretical view point. The measurements of upper critical fields of high Tc superconductors by high field magnets are lightly touched.

The contributed papers will include the up-to-date developments of researches in oxide superconductors, which are classified along with fabrications, film formations, properties and applications.

In Section 2 Superconducting Magnets and the Related Materials and Section 3 Energy Applications, readers will witness the researches are more or less influenced by the advent of high Tc superconductors.

Section 4 Applications for Accelerators and Measurements, Superconducting Electronics deals with the actual applications of superconductivity. Readers will be impressed by noticing just in these fields there is no competitive method other than the superconductivity.

The symposium was planned at the turning point of superconductivity. We, therefore, did neither intend to edit the presentations systematically, nor select or abandon them along a unique direction, but present them as they are in these chaotic circumstances. We would be very happy if the proceedings could aid readers to have their own ways of thinking of superconductivity.

The Editor

CONTENTS

Section 1   New Superconductors and Proposals for Their Applications

Fabrications of Superconductors

Applications of Superconductors

Section 2   Superconducting Magnets and the Materials

Section 3 Energy Applications

Superconducting Magnet Energy Storage (SMES)

Section 4   Applications for Accelerators and
Measurements, Superconducting Electronics

Welcoming Address

Distinguished guests, ladies and gentlemen:

On behalf of Osaka University, I would like to express a word of welcome to all of you here today at the opening of the International Symposium on New Developments in Applied Superconductivity.

I am very happy to have here with us many fine scientists from all corners of Japan, and I am especially grateful to the distinguished scientists who have traveled far from the Federal Republic of Germany, the People's Republic of China, the Republic of Austria, the United Kingdom and the United States of America to participate in this symposium.

We are also indebted to the Ministry of Education, Science and Culture for the financial support and cooperation to this international symposium.

Osaka University was established in 1931 as the 6th Imperial University of Japan. But we can trace back our origin even further to an old private school called Tekijuku, which was started as early as 1838, exactly 150 years ago, and the building of which is still preserved intact in the heart of Osaka. There, many young boys from all parts of Japan boarded and studied Dutch and Western medicine and science. That was the beginning of the two faculties-Faculty of Medicine and Faculty of Science-which composed Osaka University when we started.

Today, Osaka University encompasses 13 faculties, 3 hospitals, many research institutes and joint-use facilities.

The Laboratory for Applied Superconductivity at Osaka University, which is playing the host at this symposium, was started in 1980 in order to promote research on superconductivity and its engineering applications.

It so happened that my major is Electronic Engineering, and I well understand that the recent discoveries of high-temperature superconductivity will play a key role in the future development of the widespread variety of the technology including electronic industries.

I sincerely hope, therefore, that this symposium will attain its objectives and stimulate interaction among participants.

I also wish to mention to the participants from abroad that Osaka is the center of modern science and industry of west Japan, and our neighboring cities, Kyoto and Nara, had been the capitals of Japan since the dawn of history until the 19th century when the capital moved to Tokyo. I hope that you will find time to visit historic Japan as well as modern science and technology here, and will enjoy the most colorful, and in my opinion, the best season in Japan in and out of the symposium.

I wish you all the success of the symposium and a pleasant stay in Osaka.

Nobuaki Kumagai D. Eng.
President
Osaka University

Opening Remarks

The symposium "New Developments in Applied Superconductivity" held under the co-sponsorship of the Ministry of Education, Science and Culture and Osaka University, is intended to present the most up-to-date information about the applications of and research on superconductivity in the fields of: magnet technology, energy applications, electronics, superconductors and new applications.

Superconducting magnets are essential for modern particle accelerators and are being used in the continuing effort to create fusion magnets. Furthermore, magnet resonance imaging magnets are becoming an indispensable tool for medical diagnosis.

Superconductive magnet energy storage, with its almost 100 percent efficiency, appears to be the future of energy storage.

Josephson effects present us with an insight into the superconducting mechanisms and a macroscopic view of the quantum phenomena of the electron pairs. SQUIDs and SIS junctions are the application of the Josephson effect. They provide us with magnetometers for biomagnetic measurements and with digital devices of unsurpassed combinations of high speed and low power consumption.

However, what really brought superconductivity to the attention of the world was the discovery of high temperature superconductivity by Drs. Bednorz and Muller. The great excitement which was invoked not only in scientists and engineers but also in the general public opened the new age of superconductivity.

Since the discovery of superconductivity by Kamerlingh Onnes 77 years ago, research has been generously supported by both government institutions and private enterprise. Through this support and the hard work of many scientists, research has progressed spectacularly, and with the increased attention brought to superconductivity by the great discovery of high temperature superconductivity, efforts and support have increased even more.

At this time, these resources have to be used to answer some fundamental questions, such as: What are the mechanisms of "novel superconductivity"? Can novel superconductive materials exhibit suitable properties so that they can be used in present conventional superconductive applications or perhaps even as yet unthought of applications? Are there applications that can incorporate both conventional and novel superconductivity? And finally, does a room temperature superconductive material with suitable properties exist? Consequently, as these questions are being answered, in order to determine the direction our results are leading us, communication and cooperation on an international scale is necessary.

In appreciation of this, participants from around the world have been invited here today. Distinguished researchers in superconductivity from the United States, Europe, China and, of course, Japan have gathered here in order to share their knowledge and experience.

Katsuhiko Fujii
Organizing Chairman

Section 1. New Superconductors and Proposals for Their Applications

HIGHEST TEMPERATURE (120 K) TL-BA-CA-CU-O SUPERCONDUCTING SYSTEM

A.M.Hermann and Z.Z.Sheng

Department of Physics, University of Arkansas
Fayetteville, Arkansas 72701, USA

Abstract The highest temperature (120 K) Tl-Ba-Ca-Cu-O superconducting system comprises a number of superconducting compounds. The Tl-Ba-Ca-Cu-O superconductors are easily made. The structure, electronic and magnetic properties are presented. An unusual levitation phenomenon, in which the Tl-Ba-Ca-Cu-O superconductor can be suspended above, below, or to the side of a magnet, is discussed. A new Tl_2O_3-vapor process for fabricating the Tl-Ba-Ca-Cu-O superconductors is described.

Introduction

Discoveries of 30-K La-Ba-Cu-O superconductor [1] and 90-K Y-Ba-Cu-O superconductor [2] have stimulated a worldwide race for new and even higher temperature superconductors. Breakthroughs were made by the discoveries of the 90-K Tl-Ba-Cu-O system [3,4], 110-K Bi-Sr-Ca-Cu-O system [5,6] and 120-K Tl-Ba-Ca-Cu-O system [7-9]. Recently, high temperature superconductivity was also observed in the Tl-Sr-Ca-Cu-O system [10-12], and in the M-Tl-Sr-Ca-Cu-O with M = Pb [13,14] and rare earths [15]. In this paper, we present preparation procedures, structure, and some properties of the 120-K Tl-Ba-Ca-Cu-O superconductors. We discuss an unusual levitation phenomenon of the Tl-Ba-Ca-Cu-O superconductor due to flux pinning [16]. Finally, we present a new Tl_2O_3-vapor-process [17] which allows the highest temperature Tl-Ba-Ca-Cu-O superconductors to be easily made in the forms of complex bulk components, wires and fibers, and thick and thin films, and minimizes problems caused by toxicity and volatility of Tl starting compounds.

Preparation

Tl-Ba-Ca-Cu-O superconductive compounds form easily; there are many ways to make good-quality superconducting samples. One of the typical procedures in preparing the Tl-Ba-Ca-Cu-O samples which we use is the following. Ba-Ca-Cu-oxides are first prepared using the method similar to that we previously developed for preparation of Ba-Cu-oxides [18,19]. Appropriate amounts of $BaCO_3$, CaO (or $CaCO_3$), and CuO are mixed, ground, and heated at 925-950 ^0C in air for 24-48 hour with several intermediate grindings. The resulting uniform black material is ground and served as master material. Appropriate amounts of Tl_2O_3 and Ba-Ca-Cu oxide (depending on the desired stoichiometry) are completely mixed and ground, and pressed into a pellet with a diameter of 7 mm and a thickness of 1-2 mm. The pellet is then put into a tube furnace which had been heated to 880-910 ^0C, and is heated for 2-5 minutes in flowing oxygen, followed by furnace cooling to below 200 ^0C.

Structure

The Tl-Ba-Ca-Cu-O system can form a number of superconducting phases. Two phases, $Tl_2Ba_2Ca_2Cu_3O_{10+x}$ (2223) and $Tl_2Ba_2Ca_1Cu_2O_{8+x}$ (2212), were first identified [20]. The 2223 superconductor has a 3.85 x 3.85 x 36.25 A tetragonal unit cell. The 2212 superconductor has a 3.85 x 3.85 x 29.55 A tetragonal unit cell [20,21]. The 2223 phase is related to 2212 by addition of extra calcium and copper layers. In addition, the superconducting phase in the Ca-free Tl-Ba-Cu-O system is $Tl_2Ba_2CuO_{6+x}$ (2201) [20,22]. Fig. 1

shows schematically the arrangements of metal atom planes in these three Tl-based superconducting phases. The 2201 phase has a zero-resistance temperature of about 80 K, whereas the 2212 and 2223 phases have zero-resistance temperatures 108 K and 125 K, respectively [20-25]. It appears that the addition of each Ca and Cu layer increases the transition temperature about 20 K. If this trend continues linearly, it might be expected that 2234 phase will have a transition temperature at 140-150 K.

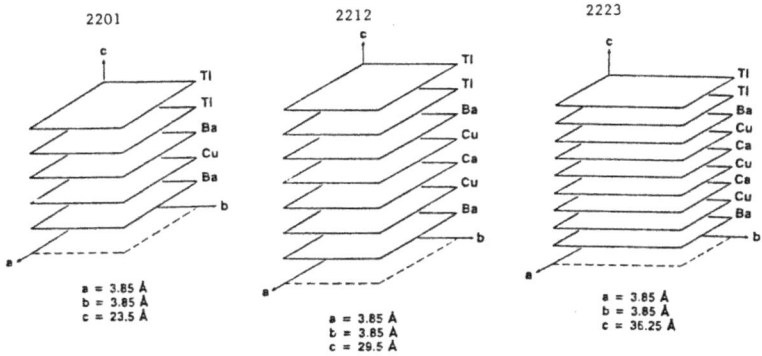

Figure 1   Schematic arrangements of the Tl-based superconducting phases 2201, 2212 and 2223.

A new series of superconducting compounds with a single Tl-O layer, which we denote by $TlBa_2Ca_{n-1}Cu_nO_{2n+2.5}$, was recently also reported [26,27]. The Tl-Ba-Ca-Cu-O superconducting series should be represented using a general formula of $Tl_mBa_2Ca_{n-1}Cu_nO_{1.5m+2n+1}$ with m = 1 and 2, and n = 1, 2, 3, and 4. The Tc of the single Tl-O layer compounds also increases with the number of Cu-Ca layers, and is slightly lower than that of the corresponding double Tl-O layer compounds. Therefore, an increase of Tc might be achieved not only by increasing the number of Ca-Cu layers, but also by increasing the number of Tl-O layers.

Transition Temperature

Fig. 2 shows resistance-temperature variation for a nominal $Tl_{2.2}Ba_2Ca_2Cu_3O_{10.3+X}$ sample. This sample has an onset temperature near 140

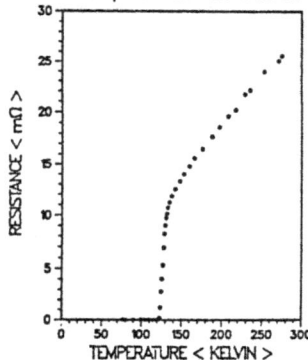

Figure 2   Resistance-temperature dependence of a nominal $Tl_{2.2}Ba_2Ca_2Cu_3O_{10.3+X}$ sample.

K, midpoint of 127 K, and zero resistance temperature at 122 K.

Fig. 3 shows a similar transition temperature for the 2223 phase by 5 kHz AC susceptibility measurements. The onset of the transition is 123 K.

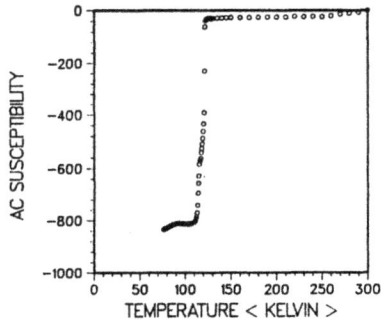

Figure 3   Temperature dependence of AC susceptibility for a nominal $Tl_2Ba_2Ca_2Cu_3O_{10+X}$ sample.

Levitation

Properly prepared Tl-Ba-Ca-Cu-O samples can easily levitate over a magnet. Careful observations have shown that for some Tl-based samples, the force between the sample and the magnet is complicated. In particular, the Tl-Ba-Ca-Cu-O samples specially prepared can be suspended above, below, or to the side of a magnet. Fig. 4a shows two Tl-based superconductors suspended horizontally above and **below** a ring-shaped magnet. A similar magnetic effect

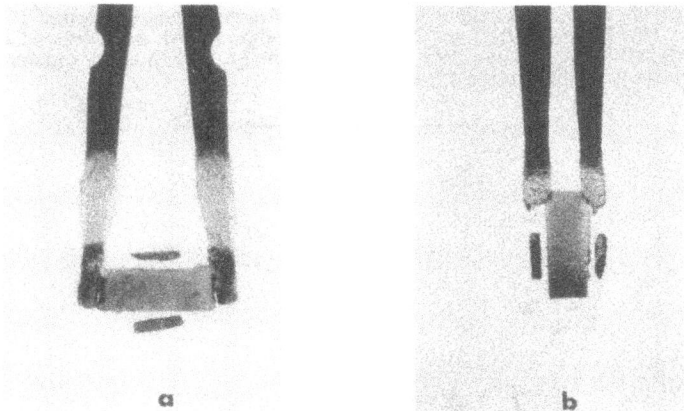

Figure 4   (a) two Tl-based superconductors suspended horizontally above and below a ring shaped magnet, and (b) two Tl-based superconductors suspended vertically near a ring shaped magnet.

showing the coexistence of repulsive and attractive forces was reported for some Y-Ba-Cu-O/AgO samples [28]. Fig. 4b shows two Tl-based superconductors suspended vertically near a ring-shaped magnet. The downward flowing nitrogen vapor is evident in figures 4a and 4b. The levitation beneath or at the side of the magnet clearly involves flux penetration into the non-superconducting regions and corresponding attractive supercurrents which are pinned. Corresponding large residual positive magnetic susceptibility following application of a magnetic field has been observed experimentally for the samples showing unusual levitation [29].

Tl_2O_3-vapor Process

In extensive preparation experiments on Tl-based superconductors, we have found that Tl_2O_3 evaporates above its 717 °C melting point, and the vapor reacts with solid Ba-Ca-Cu-oxides, forming high-quality Tl-Ba-Ca-Cu-O superconductors. This vapor-solid reaction has simplified the making of Tl-Ba-Ca-Cu-O superconductors to the making of Ba-Ca-Cu-oxides, and this minimizes the toxicity problem of Tl starting compounds. In particular, this Tl_2O_3-vapor-process allows the Tl-Ba-Ca-Cu-O superconductors to be easily made in the forms of complex bulk components, wires and fibers, and thick and thin films by fabrication of the precursor Ba-Ca-Cu-oxides and subsequent introduction of Tl [17].

An appropriate amount of the Ba-Ca-Cu-oxide powder was pressed into a pellet, and the pellet was heated at 925-950 °C in flowing oxygen for 5-10 minutes and then was air-cooled. A small platinum boat was put in a quartz boat, and a small amount of Tl_2O_3 (typically 0.1-0.2 gram) was put into the platinum boat. The heated Ba-Ca-Cu-O pellet was put above the platinum boat. The quartz boat with the contents was put into a tube furnace, which had been heated to 900-925 °C, and was heated for about 3 minutes in flowing oxygen followed by furnace-cooling.

After the above heating treatment, the Tl_2O_3 in the platinum boat had completely evaporated, and the Tl_2O_3-vapor-processed Ca-Ba-Cu-O pellet formed a layer of Tl-Ba-Ca-Cu-O superconducting compound(s) on its bottom surface. Figure 5 shows temperature dependences of resistance for some Tl_2O_3-vapor-processed Ba-Ca-Cu-O precursors with the following compositions: (A) $BaCa_3Cu_3O_7$, (B) $Ba_2CaCu_2O_5$, and (C) $Ba_2Ca_2Cu_3O_7$. Their zero resistance temperatures are 110, 96 and 105 K, respectively, which are comparable to those of corresponding sintered samples.

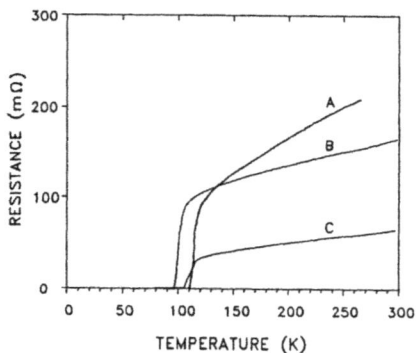

Figure 5   Temperature dependences of resistance for Tl_2O_3-vapor-processed Ba-Ca-Cu-O precursors: (A) $BaCa_3Cu_3O_7$, (B) $Ba_2CaCu_2O_5$, and (C) $Ba_2Ca_2Cu_3O_7$.

In principle, this technique has simplified the preparation of Tl-Ba-Ca-Cu-O to the preparation of Ba-Ca-Cu-O. Therefore, this technique allows Tl-Ba-Ca-Cu-O superconducting components, such as bulk components, wires and fibers, thick and thin films, to be easily made. In particular, Tl_2O_3-vapor-processed molten Ba-Ca-Cu-oxides are also superconducting, and thus this technique allows superconducting components to be easily made in arbitrary shape. Figure 6 shows resistance-temperature dependences for a Tl_2O_3-vapor-processed $Ba_2Ca_2Cu_3O_7$ thick wire (A) and for a Tl_2O_3-vapor-processed $Ba_2Ca_2Cu_3O_7$ thick film (B). The thick film was prepared by first melting $Ba_2Ca_2Cu_3O_7$ on a platinum substrate (heating at 980 ^oC in flowing oxygen for 5 minutes) and then subjecting to Tl_2O_3-vapor-processing. It reached zero resistance at 111 K. It must be emphasized that this technique is of particular importance in making superconducting Tl-Ba-Ca-Cu-O thin films [30,31].

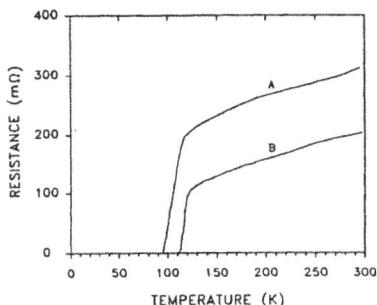

Figure 6   Resistance-temperature dependences for a Tl_2O_3-vapor-processed $Ba_2Ca_2Cu_3O_7$ thick wire (A) and a Tl_2O_3-vapor-processed $Ba_2Ca_2Cu_3O_7$ thick film (B).

Experiments have also shown that some elements, although they can not completely replace any component element in the Tl-Ba-Ca-Cu-O superconducting system, do not influence or influence slightly the superconducting behavior of the Tl-Ba-Ca-Cu-O system. We found that addition of these elements to the Ba-Ca-Cu-O precursors, followed by Tl-vapor-processing, can form various Tl-based superconductors which may satisfy various special requirements for practical applications. Figure 7 shows resistance versus temperature for a Tl_2O_3-vapor-processed $In_2Ba_2Ca_2Cu_3O_{10}$ sample. Note that the $In_2Ba_2Ca_2Cu_3O_{10}$ itself is not superconducting with our preparation conditions.

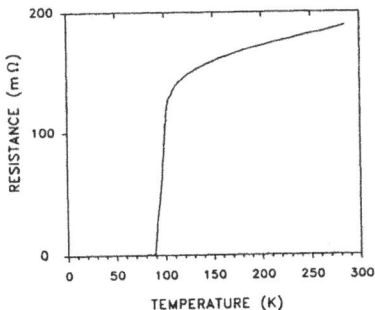

Figure 7   Resistance-temperature dependence for a Tl_2O_3-vapor-processed $In_2Ba_2Ca_2Cu_3O_{10}$.

References

1) J.G.Bednorz and K.A.Muller, Z.Phys.B 64, 189 (1986).
2) M.K.Wu, J.R.Ashburn, C.T.Torng, P.H.Hor, R.L.Meng, L.Gao, Z.J.Huang, Y.Q.Wang, and C.W.Chu, Phys.Rev.Lett. 58, 908 (1987).
3) Z.Z.Sheng and A.M.Hermann, Nature 332, 55 (1988).
4) Z.Z.Sheng, A.M.Hermann, A.El Ali, C.Almason, J.Estrada, T.Datta, and R.J.Matson, Phys.Rev.Lett. 60, 937 (1988).
5) H.Maeda, Y.Tanaka, M.Fukutomi, and T.Asano, Jpn.J.Appl.Phys.Lett. 27, L207 (1988).
6) C.W.Chu, J.Bechtold, L.Gao, P.H.Hor, Z.J.Huang, R.L.Meng, Y.Y.Sun, Y.Q.Wang, and Y.Y.Xue, Phys.Rev.Lett. 60, 941 (1988).
7) Z.Z.Sheng and A.M.Hermann, Nature 332, 138 (1988).
8) Z.Z.Sheng, W.Kiehl, J.Bennett, A.El Ali, D.Marsh, G.D.Mooney, F.Arammash, J.Smith, D.Viar, and A.M.Hermann, Appl.Phys.Lett. 52, 1738 (1988).
9) A.M.Hermann, Z.Z.Sheng, D.C.Vier, S.Schultz, and S.B.Oseroff, Phys.Rev.B 37, 9742 (1988).
10) Z.Z.Sheng, A.M.Hermann, D.C.Vier, S.Schultz, S.B.Oseroff, D.J.George, and R.M.Hazen, Phys.Rev.B (to be published).
11) W.L.Lechter, M.S.Osofsky, R.J.Soulen,Jr., V.M.LeTourneau, E.F.Skelton, S.B.Qadri, W.T.Elam, H.A.Hein, L.Humphreys, C.Skowronek, A.K.Singh, J.V. Gilfrich, L.R.Toth, and S.A.Wolf (submitted).
12) S.Matsuda, S.Takeuchi, A.Soeta, T.Suzuki, K.Aihara, and T.Kamo (submitted).
13) M.A.Subramanian, C.C.Torardi, J.Gopalakrishnan, P.L.Gai, J.C.Calabrese, T.R.Askew, R.B.Flippen, and A.M.Sleight (submitted).
14) Z.Z.Sheng and A.M.Hermann (unpublished).
15) Z.Z.Sheng, L.Sheng, X.Fei, and A.M.Hermann (submitted).
16) W.G.Harter, A.M.Hermann, and Z.Z.Sheng, Appl.Phys.lett. 53, 1119 (1988).
17) Z.Z.Sheng, L.Sheng, H.M.Su, and A.M.Hermann, Appl.Phys.Lett. (accepted).
18) A.M.Hermann and Z.Z.Sheng, Appl.Phys.Lett. 51, 1854 (1987).
19) A.M.Hermann, Z.Z.Sheng, W.Kiehl, D.Marsh, F.Arammash, A.El Ali, G.D.Mooney, L.Sheng, J.A.Woolam and A.Ahmed, Appl.Phys.Comm. 7, 275 (1987).
20) R.M.Hazen, L.W.Finger, R.J.Angel, C.T.Prewitt, N.L.Ross, C.G.Hadidiacos, P.J.Heaney, D.R.Veblen, Z.Z.Sheng, A.El Ali, and A.M.Hermann, Phys. Rev. Lett. 60, 1657 (1988).
21) L.Gao, Z.J.Huang, R.L.Meng, P.H.Hor, J.Bechtold, Y.Y.Sun, C.W.Chu, Z.Z.Sheng, and A.M.Hermann, Nature 332, 623 (1988).
22) C.C.Torardi, M.A.Subramanian, J.C.Calabrese, J.Gopalakrishnan, E,M.McCarron, K.J.Morrissey, T.R.Askew, R.B.Flippen, U.Chowdhry, and A.M.Sleight, Phys.Rev.B 38, 225 (1988).
23) M.A.Subramanian, J.C.Calabrese, C.C.Torardi, J.Gopalakrishnan, T.R.Askew, R.B.Flippen, K.J.Morrissey, U.Chowdhry, and A.M.Sleight, Nature 332, 420 (1988).
24) C.C.Torardi, M.A.Subramanian, J.C.Calabrese, J.Gopalakrishnan, K.J.Morrissey, T.R.Askew, R.B.Flippen, U.Chowdhry, and A.M.Sleight, Science 240, 631 (1988).
25) S.S.P.Parkin, V.Y.Lee, E.M.Engler, A.I.Nazzal, T.C.Huang, G.Gorman, R.Savoy, and R.Beyers, Phys.Rev.Lett. 60, 2539 (1988).
26) Y.Luo, Y.L.Zhang, J.K.Liang, and K.K.Fung (submitted).
27) S.S.P.Parkin, V.Y.Lee, A.I.Nazzal, R.Savoy, R.Beyers, and S.J.La Placa, Phys.Rev.Lett. 61, 750 (1988).
28) P.N.Peters, R.C.Sisk, E.W.Urban, C.Y.Huang, and M.K.Wu, Appl.Phys.lett. 52, 2066 (1988).
29) S.Schultz et al. (unpublished).
30) C.X.Qiu and I.Shih, Appl.Phys.lett. 53, 523 (1988).
31) C.X.Qiu and I.Shih, Appl.Phys.lett. 53, 1122 (1988).

Bi(Pb)-Sr-Ca-Cu-O HIGH-T$_c$ SUPERCONDUCTORS

Hiroshi MAEDA

National Research Institute for Metals, Tsukuba Laboratories,
Sengen, Tsukuba 305, Japan

Abstracts - Bi-Sr-Ca-Cu-O (BSCCO) system has two structurally related superconducting compounds with critical temperature (T$_c$) of 80 K and 110 K called the low-T$_c$ and the high-T$_c$ phases, respectively. The formation of the high-T$_c$ phase is enhanced by the addition of Pb. The Pb-doped material with almost complete high-T$_c$ phase has a T$_c$ of 110 K, and the upper critical fields, H$_{c2}$ of 140 T at 0 K and 60 T at 77 K. With the addition of Pb the critical current density, J$_c$ is also increased to about 400A/cm^2 and alignment of the plate-like grains produces greater than a 2 to 3 fold to increase J$_c$. However, a drastic decrease in J$_c$ is seen in a magnetic field, suggesting the existence of weak superconducting link between grains. When Pb is added an interesting modulated structure in BSCCO system is altered by the introduction of another long modulation wave, which must be related to strain energy reduction.

Introduction

Since the discovery of superconductivity in BSCCO system [1] efforts have been directed towards the identification of the phases responsible. Property measurements and observations by X-ray and electron have shown that the system has at least two superconducting compounds; one with a T$_c$ of about 80 K (low-T$_c$ phase) and the other of about 110 K (high-T$_c$ phase).

The crystal structures of these two superconducting oxides are different from those of (LaSr)$_2$CuO$_4$ (LSCO) and YBa$_2$Cu$_3$O$_7$ (YBCO). However, the oxides still have Cu-O planes believed to produce superconductivity in the copper oxide superconductors. The formation of the high-T$_c$ phase is difficult. The BSCCO sample has a transition at 105 K but complete loss of resistivity is not achieved until about 80 K, the T$_c$ of low-T$_c$ phase, regardless of heat treatment and composition alteration [2]. One of the methods to enhance its formation is the addition of Pb [3]. The Pb-doped compounds has one sharp transition at 110 K, the T$_c$ of the high-T$_c$ phase.

In this report, we describe the preparation, the superconducting properties and the crystal structures of mainly Pb-doped BSCCO compounds.

Crystal Structure

The low-Tc and high-Tc phases can be expressed ideally by the general molecular formula Bi$_2$Sr$_2$Ca$_{n-1}$Cu$_n$O$_x$ (x=2n+4+δ) with n=2 and 3, respectively. But some doubt remains due to the possibility of substitution. Figure 1 shows the ideal structures of the oxides with n=1, 2 and 3. These crystal structures are similar, differing only in the number of Ca-CuO$_2$ slabs inserted between double Bi-O layers. The low-T$_c$ phase has two CuO$_2$ layers and one Ca layer while the high-T$_c$ phase has three CuO$_2$ and two Ca layers. T$_c$ increases with increasing n up to 3.

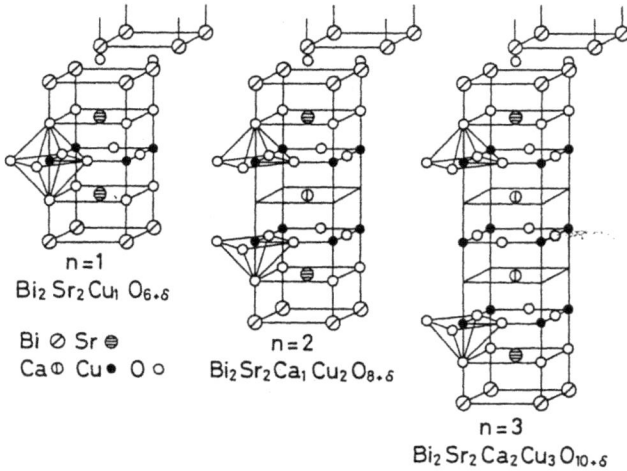

Fig.1   Ideal crystal structures of $Bi_2Sr_2Ca_{n-1}Cu_nO_{(2n+4+\delta)}$

Electron microscope observations show that both phases have an inter-esting structural modulation along the b-axis as shown in Fig. 2.   Both phases have an orthorhombic structure.   The modulated structure is believed

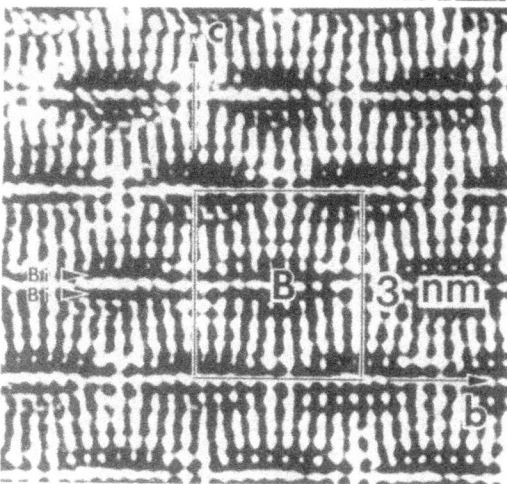

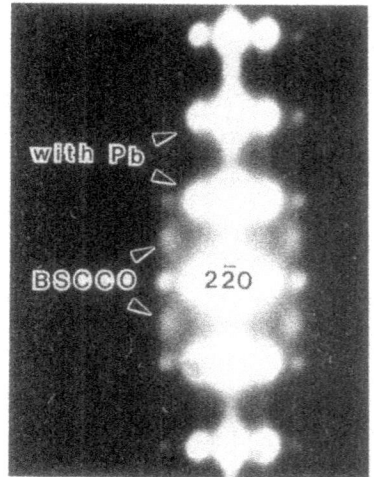

Fig.2   High resolution electron microscope image of the Pb-free low-Tc phase (c-b plane in orthorhombic notation or c- [1$\bar{1}$0] plane in perovskite notation)

Fig.3   Electron diffraction pattern of the Pb-doped sample which has the nominal composition of $Bi_{0.7}Pb_{0.3}Sr_1Ca_1Cu_{1.8}O_x$ and sintering treatment at 845 ℃ for 85hr. (c*- [1$\bar{1}$0]* plane)

to be produced by the periodic displacement of atoms from their ideal lattice positions or the ordering of oxygen vacancies in Bi_2O_2 layers. The wave length of the modulation is incommensurate with the lattice (ideally 2.7 nm). When Pb is added the modulated structure is altered by the introduction of another modulation wave. Figure 3 shows an electron diffraction pattern for Pb-doped sample. Satellite spots (signed "with Pb") due to a modulation are seen along the $[1\bar{1}0]$ direction (b^* direction in orthorhombic notation). From the satellite splitting the modulation wave length is estimated to be about 4.6 nm. Pb addition produces a less regular modulation. Pb is believed to substitute for Bi. The altered modulated structure produced by the introduction of Pb may be due to its valence. Bi has a valence of +3 while the valence of Pb is +2. To retain charge neutrality the oxygen stoichiometry should change which in turn is expected to affect the modulated structure.

Preparation

The low-T_c phase is easily produced using oxide or carbonate powders with the proper cation proportions. While the formation of the high-T_c phase requires meticulous control of 875 °C. The formation reaction proceeds most rapidly when partial melting of the sample occurs. However, a sample with only the high-T_c phase could not be prepared. Very recently, Takano et al. succeeded in preparing samples with almost completely high-T_c phase by the addition of Pb [4]. When Pb is added to the BSCCO system the partial melting temperature is reduced and the temperature range for optimum sintering is expanded.

Figure 4 shows the magnetic susceptibility variation with temperature

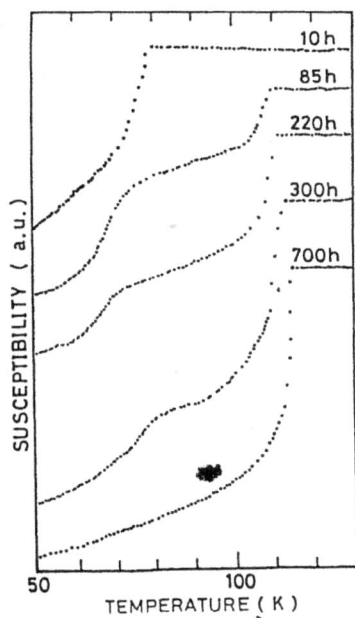

Fig.5 Temperature dependence of resistivity for the same sample sintered for 700hr as in Fig.4.

Fig.4 Temperature dependence of magnetic susceptibility for $Bi_{0.7}Pb_{0.3}$ $Sr_1Ca_1Cu_{1.8}O_x$ powder samples sintered at 845 °C for various times.

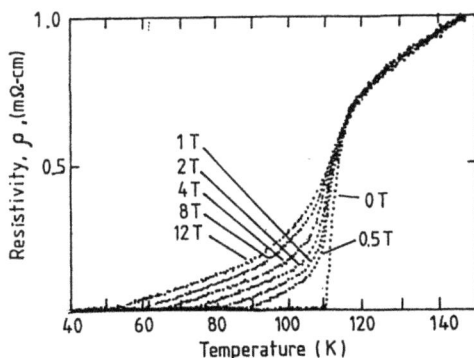

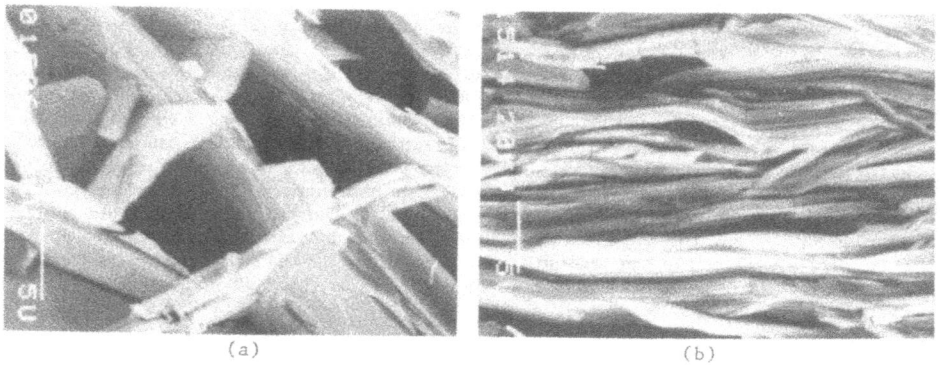

(a)   (b)

Fig.6 SEM micrographs of the fractured surface for the Pb-doped samples of
$Bi_{0.7}Pb_{0.3}Sr_1Ca_1Cu_2O_x$. (a) un-pressed sample; sintered at 845 °C for
200hr, (b) pressed sample; sintered at 845 °C for 100hr, pressed and
sintered for another 100hr at 845 °C.

of Pb-doped samples that received different sintering times at 845 °C.   The
low-T_c phase exists in the sample with a short sintering time and is converted into the high-T_c phase although the exact reaction is not clear.
Finally the almost completely high-T_c phase can be obtained after 700 hr
and the complete zero resistance is achieved at 110 K.   The sintering time
can be reduced with reduced oxygen partial pressure, for example, one-fourth
at 1/13 atm.   The optimum nominal compositional range where the high-T_c
phase is completely formed has been recently reported [4].

The crystalline order of Pb-free high-T_c phase is not good.   Broad X-
ray diffraction peaks and high resolution electron microscope images show
that the plane stacking along the c-axis is not regular due to an intergrowth [5].   When a small amount of Pb is added the intergrowths are completely eliminated.   At present, the origin of the intergrowths in the Pb-
free system is not clear and the removal with a Pb addition is not also understood.

Superconducting Properties

Figure 5 shows resistivity transition curves for the sample sintered
for 700h in Fig.4 at various magnetic fields.   The head of the transition
curve is only slightly influenced by the magnetic field, while the tail of
the curve shifts to lower temperature as the magnetic field is increased.
From the midpoint of the transition curves the upper critical fields, H_{c2},
are estimated to be 60 T at 77 K and 140 T at 0 K (using WHH theory).

The critical current density, J_c in zero field at 77 K of bulk Pb-free
BSCCO is very small (1A/cm²).   With the addition of Pb the J_c is increased
to about 210 A/cm², which is less than the best J_c obtained for YBCO samples
[6].   The low J_c value is partially due to a low density, about 60 % of the
theoretical density and a low preferred orientation, as shown in Fig. 6(a).
A pellet compaction process is an effective method to increase J_c .   For
example, a sample is sintered at 845 °C for 100hr, cooled to room temperature, compacted about one-half in the thickness and finally sintered again
at 845 °C for 100hr in air.   The pressed sample is dense and highly textured
with plate-like grains laminated parallel to the pellet surface as shown in

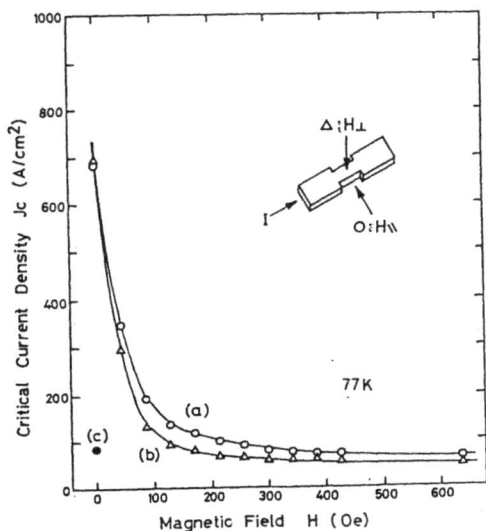

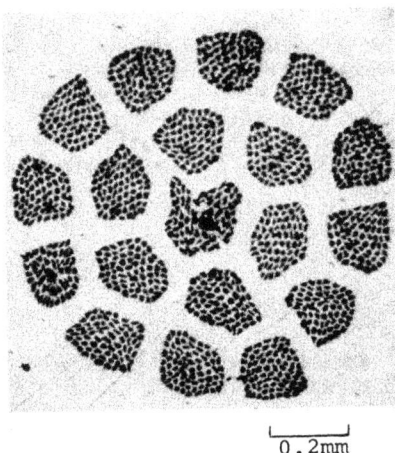

Fig.8 Cross-section of a wire with 1330 Pb-doped BSCCO filaments in a Ag-sheath.

Fig.7 Magnetic field dependence of J_c for the Pb-doped pressed sample with the same composition as in Fig.6. (a)field is parallel and (b) perpendicular to the sample surface. (c) is a J_c of un-pressed sample for the same composition as in Fig.6.

Fig. 6(b). A J_c greater than 1100 A/cm^2 in zero field at 77 K can be easily achieved by this process. However, this J_c value is much smaller than that achieved in polycrystalline thin films (10^6 A/cm^2)[7]. Furthermore, a drastic decrease in J_c is seen in low magnetic fields up to 100 Oe (0.01 T), as shown in Fig.7. This suggests that the superconducting coupling between grains is still very weak in the BSCCO system in a similar manner to that of YBCO.

· Since grains of BSCCO can be easily deformed, not due to plastic deformation but perhaps due to cleavage occurring between double Bi-O layers, it is possible to fabricate an Ag-sheathed wire. Figure 7 shows the cross section of a 1330 multifilamentary wire. After the final heat treatment below 830 C, the oxide filaments are almost composed of the high-T_c phase. Above this temperature a reaction between the oxide and Ag-sheath occurs.

Conclusion

Since the discovery of high-T_c BSCCO superconductors many reports have been presented. However, many problems still remain to be solved. The main problems are the following.
(1) T_c increases with increasing n, the number of CuO_2 layers, up to 3. Will this trend continue with a further increase of n ?, although a lower T_c of 90 K has been obtained in thin films with n=4 [7].
(2) The incommensurate modulate structure appears in BSCCO system. Why and how the highly strained modulated structure is formed ?

(3)   The formation of the high-T_c phase is enhanced with a Pb addition. This is perhaps related to the reduction of strain energy.   The question remains as to which lattice site Pb occupies and if it orders on this site.
(4)   It has been demonstrated that a dense and highly-textured material can easily be produced resulting in a J_c improvement.   However, J_c decreases drastically with increasing magnetic field, showing the existence of super-conducting weak links between grains.   What is the origin of the weak links in BSCCO ?

References

1)   H.Maeda, Y.Tanaka, M.Fukutomi and T.Asano: Jpn.J.Appl.Phys.27(1988)L201.
2)   M.Takano, J.Takada, K.Oda, H.Kitaguchi, Y.Miura, Y.Tomii and H.Mazaki:
     Jpn.J.Appl.Phys.27(1988)L1041.
3) Y.Matsui, H.Maeda, Y.Tanaka and S.Horiuchi: Jpn.J.Appl.Phys.27(1988)L372.
4) S.Koyama, U.Endo and T.Kawai: submitted to Jpn.J.Appl.Phys.
5) S.Ikeda, H.Ichinose, T.Kimura, T.Matsumoto, H.Maeda, Y.Ishida and
   K.Ogawa: Jpn.J.Appl.Phys.27(1988)L999.
6) E.Yanagisawa, D.R.Dietderich, H.Kumakura, K.Togano, H.Maeda and
   K.Takahashi: Jpn.J.Appl.Phys.27(1988)L1460.
7) H.Itozaki, K.Higaki, K.Harada, S.Tanaka, N.Fujimori and S.Yazu: Proc. 1st
   Internl. Symposium on Superconductivity (1988, Nagoya)
8) H.Adachi, S.Kohiki, K.Sekine, T.Mitsuyu and K.Wasa: submitted to Jpn. J.
   Appl.Phys.

SOME RECENT STUDIES ON OXIDE SUPERCONDUCTORS IN CHINA

YANG Qian-sheng

Institute of Physics, Chinese Academy of Sciences
P.O.Box 603, Beijing 100080, China

Abstract-The recent studies in the field of phases, structures and properties on the oxide superconductors in China are briefly reviewed. The fabrication technology on the bulk material and the manufacturing of the thin film and electric devices are also mentioned.

Introduction

The discovery of La-Ba-Cu-O superconductor in late 1986 by Bednorz and Müller (1) prompted an intense study of new oxide superconductivity in a few laboratories including the joint research group of high Tc superconductivity in the Institute of Physics, Chinese Academy of Sciences. A few months later Chu, C.W. et al. (2) and Zhao, Z.X. (3) et al. discovered independently a new superconductor with Tc above LN_2 temperature in Y-Ba-Cu-O system. These important achievements attracted wide interest in the world as well as in China. Soon afterward many interesting progresses were made in China. Most of the work in the last year has been published in the proceeding of "The Beijing International Workshop on High Temperature Superconductivity" (Beijing, China, 1987) and other journals. The main activities and results were also reviewed recently by Yeh, W.J. (4) in the conference of MRS Fall Meeting, Boston(1987).

At the beginning of this year non-rare earth oxide high temperature superconductors (5),(6) were discovered. Since then much studies have been devoted on the structures and properties of these systems. In China, Bi and Tl oxide superconductors were prepared shortly after their discoveries. A series of interesting results have been first discovered or independently obtained as follows. It has been determined that the structure of the nominal composition $BiSrCaCu_2O_y$ is one-dimensional incommensurate (7). Coherent intergrowths in TlBaCaCuO multiphase sample have been found by high resolution electron microscopy(8). A new series of $TlBa_2Ca_{n-1}Cu_n$ Oxides including at least (1201), (1212) and (1234) phases (9) have been found besides the $Tl_2Ba_2Ca_{n-1}Cu_n$ oxides (10). Tentative structure models were proposed (11). Good quality single crystals of $Bi_2Sr_2CaCu_2O_y$ have been grown by self-flux method (12). A set of measurements including anisotropic resistivity (13), thermoelectric power (14), tunneling spectroscopy (15) and modulation structure by precession technique (16) were conducted on the single crystals.

The studies of structure, phase relation and physical properties in rare earth oxide are continuing. The phase diagrams of the ternary oxide system for the rare earth elements La,Y,Gd an Nd, have been determined (17-20). A solid solution in CuO-rich region exists in La and Nd systems. Seven kinds of twin boundaries, with most of which (110) twins have been observed by electron microscope (21). It is possible that twinning would offer effective pinning center for high critical current density. The in-situ observation of phase and structure changes during heating (22), the electron shadow microscopy for direct observation of the distribution of the superconducting phase on the exposed surfaces (23) have been obtained. The effects of doping on the structure and properties have been widely investigated(22).

In order to increase the critical current density of bulk material, several groups in China are engaged in fabrication technology research on the basis of physical and chemical analyses. Fine spherical grains, directional reaction process, laser floating zone melting and second phase doping seem to be effective in improving the bonding between the grains (22). A number of wire fabrication technologies are also in progress (22).

Several methods have been used to prepare the thin films with zero resistance temperature in excess of liquid nitrogen temperature: electron and ion beam evaporation (24), sputtering (25), laser evaporation (22), molecular beam epitaxy and so on. By now the best quality films are obtained with ion beam evaporation and sputtering technique. The primary objective of thin film research has been to achieve high critical current density and stable and reproducible properties for physical studies and for application to electronic devices.

Recently rf SQUID made from bulk material succeeded with noise of $3 \times 10^{-3} \phi_0 / \sqrt{Hz}$. In addition, Josephson frequency mixing microwave resonance and bolometer effect have been tried at several places (22). At the same time, some basic research connected with devices have been carried on(26, 27). The recent progress in the field of superconducting thin film and electronics in China have been briefly reviewed by Cui G.J. in the conference of The International Workshop on Future Electron Devices, High Temperature Superconducting Electon Devices, Miagi-Zao, Japan(28).

In this paper I shall not include all of the recent important works performed in China because of the limited space. Rather, I would like to describe a little more in detail some works which might be interesting.

Bi-Sr-Ca-Cu-O System

The sample of the nominal composition $BiSrCaCu_2O_y$ with Tc above 80K has been studied by electron diffraction and high resolution electron microscopy (7). It has been determined that the crystal structure is one-dimensional incommensurate. The direction of the modulation is almost parallel to the b axis of the average structure. The unit cell parameters are $a = 0.539nm, b = 0.541nm$ and $c = 3.0nm$. The periodicity of modulation is 2.53nm.

The single crystals $Bi_2Sr_2CaCu_2O_y$ with Tc=85K were grown by self-flux method (12). The crystals, with the typical dimensions of 10mm, are stable in the atmosphere and can be easily cleaved after immersed in alcohol. Using such crystals a series of experiments have been made.

Several reciprocal layered photos of the crystals were recorded (16) by X-ray precession camera. About one thousand reflections in total have been observed. After careful examination, an average structure with space group of Pnnn and a real structure with the superspace group of $P^{Pnnn}_{\frac{1}{4},\frac{1}{4},\frac{1}{4}}$ or $P^{Pnnn}_{\frac{1}{4},s,\frac{1}{4}}$ have been found for this one dimensional modulated incommensurate material. As well a basic structure with the space group of Fmmm is deduced from the experimental data.

The resistivity of superconducting single crystal possesses a quasi-two dimensional anisotropy characteristic with a typical metallic behaviour in the a-b planes and a "semiconducting-like" temperature dependence in the c-direction (13). The resistivity characteristic along c axis can be fitted with $A/T + BT$ formula above 110K.

The quasiparticle tunneling spectroscopy of the single crystal has been studied (15). Well-formed gap structures have been observed both with point contact junction and with thin film junction at liquid helium temperature. An excellent fit to the experimental dI/dV curve could be achieved by assuming the density of state of the form $N(E) = Re[E/\sqrt{E^2-\Delta^2}]$, where energy gap Δ is complex and independent of energy. The Δ fitted is 18meV,i.e. $2\Delta/KTc=5.0$. The result is in the strong coupling range.

The thermoelectric power in single crystals has been measured both in a—b plane (Sab) and along c axis (Sc) respectively (14). The temperature dependences are distinctly anisotropic. Sc increases monotonously with temperature and is positive from Tc to room temperature, while Sab increases rapidly near Tc and then decreases with increasing temperature. Sab changes its sign at 250K.

Tl-Ba-Ca-Cu-O System

Two Superconducting phases with Tc at 120K and 90K in the Tl-Ba-Ca-Cu oxide system identified soon after the Tl oxide superconductor had been discovered (10). The 120K superconductor was found to have a composition very close to $Tl_2Ba_2Ca_2Cu_3O_y$ possessing a tetragonal structure with a=0.547nm and c=3.607nm and the 90K superconductor was determined to be the $Tl_2Ba_2Ca_1Cu_2O_y$ compound with a tetragonal structure with a=0.546nm and b=3.006nm respectively by X-ray powder diffraction based on near single phase superconducting materials. These pheses have been also identified by convergent beam electron diffraction in the multi-phase compounds of $TlBaCaCu_2O_y$ with Tc of 114K (8). These phases are tetragonal, with point group 4/mmm. It has been found that c=3.61nm and c=2.97nm phases form coherent intergrowths (8) by high resolution electron microscopy, Both ordered and disordered intergrowths have been observed.

Two new phases $TlBa_2CaCu_2O_y$ and $TlBa_2CuO_y$ have been identified in the multiphase compounds of TlBaCaCuO with a superconducting temperature of 72.5K (9). Both phases are weakly modulated incommensurately in the (100) and (010) directions. Their basic structure are tetragonal, with space group P4/mmm and lattice parameters of a=0.387nm, c=1.29nm and a=0.387nm, c=0.99nm. These two phases and the recently reported $TlBa_2Ca_2Cu_3O_y$ are members of a new structural series $TlBa_2Ca_{n-1}Cu_nO_y$.

Liang et al. proposed a tentative structure model based on the X-ray diffraction data of the Tl oxides. These superconducting phases could be divided into two types according to space group I4/mmm and P4/mmm. The chemical formulas of typy I and type II could be generalized as $TlBa_2Ca_{n-1}Cu_nO_{2n+2.5}$ (n=2,3,4) and $Tl_2Ba_2Ca_{n-1}Cu_nO_{2n+4}$ (n=1,2,3) respectively. The crystal structures of these superconducting phases are somewhat similar to each other. These Tl superconducting phases belong to tetragonal system, with same lattice constant a and different lattice constant c in each type of structure. All cations are distributed at the positions of (O,O,Z) and (1/2, 1/2, Z) alternately along Z axis. The free energies of the formation of the superconducting phases are nearly the same because the crystal structure of these superconducting phases are similar to each other. For this reason, these superconducting phases can coexist. Still more, the two oxygen deficient pseudo-perovskite units adjacent to the same Tl-O layer might be different and the structure with composition modulation alons Z axis is formed.

Structures and Phases in Rare Earth Oxides

The phase diagrams in R_2O_3-BaO-CuO ($R = La,Y,Nd,Gd$) have been completed by means of X-ray diffraction, thermal analysis and superconductivity measurements. In both Nd and La systems there are solid solution $Nd_{1+x}Ba_{2-x}Cu_3O_y$ and $La_{1+x}Ba_{2-x}Cu_3O_y$ but on the contrary, in Y and Gd systems no such solid solution exist. From the view point of crystal chemistry the reason is that the ion radii of Nd^{3+} and La^{3+} are larger than that of Y^{3+} and Gd^{3+} and more close to that of Ba^{2+} and thus Nd^{3+} and La^{3+} can take the sites of Ba^{2+} to form the substitution solid solution.

Twinning occurs commonly in the orthorhombic structure. Twin boundaries are formed during the phase transition from the tetragonal high temperature phase to the orthorhombic phase. Wen S.L. et al. have observed 7 kinds of twin boundaries in $YBa_2Cu_3O_y$ ceramics. The observations show that the (110) twin is the most popular form among them. In the proposed model. the (110) twin boundary is a very thin layer with thickness of 2.7Å and it consists of only bilayer atoms having an orthorhombic unit cell. Although (011) and (032) twin boundaries also consist of a two-atomic layer, their thicknes would be less in comparison with (110) and (301) twins. In spite of the differences in the structure and thickness for a variety of twin boundaries, they are able to keep O_1-Cu_1-O_1 chains to maintain continuity.

It was well known that the complex oxide fabrication technology have led to a wide variety of phases in the sample. It is desirable to establish directly that the regions studied are indeed superconducting. Recently, C.Y. Yang(Institute of Physics, Chinese Academy of Sciences) an J.W. Steeds.(Univ. of Bristol, UK) developed a technique of electron shadow microscopy to observe directly superconducting microregions. Diamagnetism of different granules of $YBa_2Cu_3O_y$ samples has been studied with this method. The microstructures of the sample have also been observed by scanning electron microscopy. The results show that the sample with large, angular (orthorhombic)grains and local texture shows significant Meissner effect, which appears as shadows with big cycloid envelops. However for round particles and grains the Meissner effect is absent in the samples.

Bulk Material Fabrication Technology

Laser heated pedestal growth (LHPG) is a crucible-free floating zone technology for growing oxide crystals from melt. According to the phase equilibrium, it is considerably difficult to grow directly a single crystal of $YBa_2Cu_3O_y$ from melt. The crystal obtained by LHPG may be transformed into single phase $YBa_2Cu_3O_y$ by post heat treatment at $850 \sim 950\,°C$ in air or oxygen for a long time. Many oxide monocrystals of fibers have been successfully grown by his method at the Institute of Metal, Shenyang.LHPG offers several important advantages over other crystal growth methods, for example, eliminating crucible contamination and shaping wire or fiber directly during solidification and rapid growth of crystal due to large thermal gradient.

Recently, the reaction mechanism during the synthesis of superconductor $YBa_2Cu_3O_y$ has been investigated by TG-DTA analysis and X-ray diffraction methods as the General Research Institute for Non-Ferrous Metals. The experimental results indicate that in spite of the starting materials used there are two steps of the reaction for formation of superconductor $YBa_2Cu_3O_y$ The binarr complex oxides $Y_2Cu_2O_5$ and BaCuO are formed, then two binary oxides react each other forming the superconducting phase $YBa_2Cu_3O_y$.

The oxide ·wire was fabricated by extruding the mixtures of the superconducting $YBa_2Cu_3O_y$ powders mixed with 3% fine Ag powder and plasticizer (volume ratio about 35%). The extruded wires with various diameter possess good flexibility. After sintering, the superconducting wire of

0.7mm in diameter with Tc = 93K, Jc(77K) = 466A/cm² was obtained. the tensile strength of the wire is of 75kg/mm², however its plasticity is poor. Adding the trace element Ag with optimum heat treatment is essential to enhance the Jc value of the wire.

Thin Films and Electric Devices

Thin films of YBaCuO were prepared by r.f.magnetron reactive sputtering method. (100)SrTiO₃ single crystal substrate was used. The pressure of the sputtering gas with Ar/O = 5/1 was $2 \sim 5 \times 10^{-2}$ torr. The substrate temperature was $300 \sim 400\,°C$. After sputtering, a post heat treatment was necessary to convert the amorphous structure of the film to a superconducting crystallins form. Annealing was performed in flowing oxygen by heating at $600 \sim 700\,°C$ for about 4 hours and $850 \sim 900\,°C$ for about 3 hours. The critical current densities Jc at 77K of the films with Tc = 88K are 2.5×10^4 A/cm .The films are a-axis preferentially oriented perpendicular to the film surface. This means that most of the Cu-O planes are perpendicular to the film surface. The upper critical field and the critical current appear to be strongly anisotropic. It is obvious that there are also some random weak links in these thin films and these weak links are easily influenced by a magnetic field. Similar results are obtained by reactive ion beam sputtering (24). Thin films with zero resistance temperature in excess of liquid nitrogen temperature have been also fabricated by several other methods: electron beam evaporation, laser evaporation and molecular beam epitaxy.

A rf SQUID from YBaCuO material has been made by the technique of multiloops in parallel at the National Institute of Metrology (29). A flux resolution of 3×10^{-3} $\phi_0/\sqrt{Hz}$ was established by lock mode operation at 77K. A double-hole rf SQUID made from bulk YBaCuO has been also developed in Fudan University (22). It can be operated in lock mode at both 77K and 4.2K.

P.H.Wu et al. at Nanjing University have measured the performance of YBaCuO mixer at 77K. The noise temperature of the mixer is about 1000K. Recently they have observed the harmonic mixing between 36GHz signal and local oscillation which came from the Josephson junction itself at 77K (22).

High Tc film bolometer as an infrared detector (22) has been investigated at the Institute of Electronics, Chinese Academy of Sciences. The sensor made by YBaCuO thin film with Tc = 86K is kept at 80K and current biased at the midpoint of superconducting transition. A equivalent noise power of 5×10^{-8} W/$\sqrt{Hz}$ has been achieved.

On measuring the I—V characteristics of sintered YBaCuO superconductors, a "footlike" structure is developed (27) when the temperature goes down from Tc, strongly reminiscent of the case in superconducting weak links. The energy gap derived from the critical current and the "excess current" is about 18.4 mev at T=0 and with a roughly BCS-type of variation at T>0. The results are indicative of a kind of "cooperative" weak link behavior in sintered superconducting YBaCuO.

Josephson effects have been observed in Nb/YBaCuO point contacts and YBaCuO break junctions. The effect can be easily observed with fresh surface of YBaCuO ceramic obtained in situ at low temperature (26). When the junction, are exposed to microwave field, the I—V characteristics show constant voltage steps, the heights of which follow the Bessel function behavior. The magnetic field dependence of the critical current in break junction at 77K is of the typical Fraunhofer diffraction pattern.

Acknowlegments

I world like to thank Professor Zhao Z.X., Fung K.K., Mai Z.H., Xie S.S., Cui G.J. and Shen J.L. for their helpful discussions and providing valuable informations. Special thanks go to Professors Hong C.S.,Li L.,Gan Z.Z. and Yang Q.Z. for their encouragement and instruction.

The most of the works mentioned above are sponsered by The National Center for Research & Development on Superconductivity.

References

1. Bednorz J.A. and Müller K.A., Z. Phys.B. Condensed Matter 64, 189-193 (1986).
2. Wu M.K., Ashburn R.J., Torng C.J., Hor P.H., Meng R.L., Gao L., Huang Z.J., Wang Y.Q. and Chu P.W., Phys. Rev. Lett. 58,908 (1987)
3. Zhao Z.X., Chen L.Q., Yang Q.S., Huang Y.Z., Chen G.H., Tang R.M., Liu G.R., Cui C.G., Chen L., Wang L.Z., Guo S.Q., Li S.L. and Bi J.Q., Kexue Tongbao (in English) 32 (10), 661 (1987).
4. Yeh W.J., The proceeding of "The MRS Fall Meeting", Boston (1987)
5. Michel C. et al., Z. Phys. B68, 421 (1987).
   Maeda H., Tanaka Y., Fukutomi M. and Asano T., Jpn. J. Appl. Phys. 27, L209 (1988).
6. Shen Z.Z. and Hermann A.H., Nature 332, 138 (1988).
7. Yang D.Y. et al., Supercon. Sci. & Tech. 1,100 (1988).
8. Fung k.k., Zhang Y.L., Xie S.S. and Zhou Y.Q., to be published in Phys. Rev. B. (1988).
9. Huang J.Q., Liang J.K., Zhang Y.L., Xie S.S., Chen X.R. and Zhao Z.X., to be published in Modern Phys. Letts. B. (1988)
   Liu Y., Zhang Y.L., Liang J.K. and Fung, K.K., to be published in J. Phys.C (1988)
10. Mai Z.H., Cui C.G., Xie S.S., Dai D.Y. Chu X., Zhang Y.L., Che G.C., Li S.C., Zhang J.L., Huang W.X. and Cheng X.R., Supercon Sci. Tech. 1 (1988).
11. Liang J.K., Zhang Y.L., Huang J.Q., Xie S.S., Che G.C., Chen X.R., Ni Y.M., Zhen D.N. and Jia S.L., to be published in Physica C (1988).
12. Yan Y.F. et al. Modern Phys. Letts. B2, 571 (1988).
13. Wang J.H., Chen G.H., Chu X., Yan Y.F., Zheng D.N., Mai Z.H., Yang Q.S. and Zhao Z.X., Supercon Sci. Tech. 1,27 (1988)
14. Chen G.H., Wang J.H., Zheng D.N., Yan Y.F., Jia S.L., Yang Q.S., Ni Y.M. and Zhao Z.X., to be published in Modern Phys. Letts. B (1988)
15. Zhao S.P., Tao H.J., Chen Y.F., Yan Y.F. and Yang Q.S., to be published in Solid State Commun. (1988)
   Chen Y.F., Tao H.J., Zhao S.P., Yan Y.F. and Yang Q.S., to be published in Modern Phys. Letts. B (1988).
16. Chu X., Mai Z.H., Yan Y.F., Wang J.H., Zhen D.N., Li C.Z., Yang Q.S., and Zhao Z.X., to be published in Modern Phys. Letts. B (1988)
17. Dong C., Liang J.K., Che G.C., Xie S.S., Zhao Z.X., Yang Q.S., Ni Y.M. and Liu G.R., Phys. Rev. B37, No.7, (1988)
18. Che G.C., Liang J.K., Chen W., Yang Q.S., Chen G.H. and Ni Y.M., J.Less-Common Metals, 138, 137 (1988)
19. Liang J.K., Xu X.T., Rao G.H., Xie S.S., Shao X.Y. and Duan Z.G., J. Phys.D: Appl. Phys. 20, 1324 (1987)

20. Fu S.J., Xie S.S., Liang J.K., Che G.C. and Zhao Z.X., Modern Phys. Letts. B 2, No.9 (1988)
21. Wen S.L., Song X.Y. and Feng J.W., Materials Letts. 6,385 (1988)
22. The proceedings of "The Chinese National Workshop on High Tc Supercondtivity", eds. by Zhou Lian, Wu Xiaozu, Wang Jingrong and Yang Qiansheng, (Baoji, April 1988)
23. Yang C.Y., preprint
    Yang C.Y. and Steeds C.Y, Nature, 331,696 (1988)
24. Zhao B.R., et al., preprint.
25. Gao J., Zhao B.R., Wang H.S., Zhang Y.Z., Lu Y., Zhao Y.Y., Yuan C.W., Xu P. and Li L., preprint
26. Zhao S.P., Tao H.J., Chen Y.F., Che G.C. and Yang Q.S., to be published in Modern Phys. Letts B (1988)
27. Lu Li, Duan H.M. and Zhang D.L., Phys. Rev. B, 37, No.4 (1988)
28. Cui G.J., Extended Abstract of the "5th. International Workshop on Furture Electron Devices, High Temperature Superconducting Electron Devicies", Miyagi-Zao, Japan (2-4) June, (1988),P.19
29. Qiao W.C., private communication.

PROGRESS OF OXIDE SUPERCONDUCTING FILM FORMATION
AND APPLICATION TO ELECTRONICS FIELD

Takeshi KOBAYASHI

Faculty of Engineering Science, Osaka University
1-1 Machikaneyama-cho, Toyonaka, Osaka 560, Japan

Abstract   The epitaxial growth and advanced technology like a hetero-epitaxial growth of $Ln_1Ba_2Cu_3O_y$ (Ln = Y, Er, Nd) systems have been studied. The epitaxy was accomplished at the growth temperature as low as 530 ℃. A challenge to (110)YBaCuO / (110)MgO / (110)YBaCuO double-heterostructure formation made pretty good progress. The as-grown film of rare-earth-element free oxide superconductor BiSrCaCuO exhibited a good morphology, though its Tc_0 remained low (50 K). Using these films, several kinds of two-terminal and three-terminal devices have been fabricated. Among them, 3-terminal one based on the non-equilibrium superconductivity is very promising from view point out only of the electronic application but also of exploring the high T_c superconducting mechanism.

Introduction

Since a discovery of high-T_c oxide superconducting ceramic by Bednorz and Müller[1], there has been a rush of new high-T_c superconducting materials consisting of a modified perovskite structure[2-4]. In these two years, the critical temperature T_c was raised up to 125 K much higher than the liquid nitrogen boiling temperature. According to the imformal communication, some kind of TlBaCaCuO ceramics revealed the critical temperature around 140 K [5]. Very recently, Wu reported his success in getting T_c of 160 K [6].

As a whole, the material prosess required for the high-T_c oxide superconductors are the same as or close to that of the semiconductors. It is quite well known that the single crystal technology greatly helps the physical research and practical use of the semiconductors. The amorphous Si and poly-Si technologies are the exceptions. The defects inside the crystal and the grain-boundary in the polycrystal hamper the electrical and optical characteristics to a large extent. These features are very common to the oxide superconductors. The major reason for this is a reduced carrier density of the high-T_c superconductors, which is the same order of the heavily doped semiconducting materials. Therefore, like a semiconductor engineering, the epitaxial growth is thought to be a key technology to obtain the excellent superconductivity of the high-T_c materials. The epitaxial growth of the oxide superconductors is of great importance in view not only of the application but also of the scientific usage[7]. The latter means that exploring of the high-T_c superconducting mechanism can not be done until the ideal crystallinity of the material comes in our hands.

Nowadays, the high speed and high frequency electronic devices, and the radiant light emitting devices are consisting of highly sophisticated structure. Most of the semiconductor devices are made up with the heterostructure. A typical example is a GaAs / AlGaAs double-heterostructure laser diode working stably at room temperature with fairly long life. The other examples can be found in the Modulation Doped FET which comprises n^+-AlGaAs / GaAs selectively doped heterostructures, and in the Hetero-Bipolar-Transistor. Employing the heterostructure in the device design, the ultimate performance can be drawn in each kind of device. The mature growth technologies such as the CVD, MBE, Plasma Assist Deposition and so on realized the epitaxial heterostructures with the high quality. At present, even the atomic layer epitaxy is not a waking dream. A dream of producing tailor-made crystals has come true.

The concept stated above can be transferred to the superconductivity electronics field, which is expected to open the wide application of the superconducting materials. Until now, only limited device structure has been used in this field. That is a Josephson diode which is consisting of a simple layered structure (superconductor / insulator / superconductor). It does not require the special technique other than the simple deposition of thin films for its fabrication. The simple deposition and oxidation of films have been rather sufficient for the preparation of the conventional superconductor devices. However, because of the poor processing technology, new device which surpasses the familiar Josephson diode could seldom appear in the world. Moreover, the technological poverty hindered us even from changing the research concept. The birth of the high-T_c superconducting materials is necessarily taking us upward from the material processing point of view, and therefore, no one engaged in the electronics field can avoid the epitaxial growth of thin films at present. The oxide superconductor epitaxy should be followed by the hetero-epitaxial growth technique, without which really nice electronic devices like a superconducting transistor can never be brought about [8,9].

In the present work, we first demonstrate the epitaxial growth of an ultra-thin layered film and related interface problems of $Ln_1Ba_2Cu_3O_y$ system, wherein one can see the dramatic degradation of the crystallinity of the grown film with thickness less than 10 nm as long as the $SrTiO_3$ is used as the substrate material. The advanced epitaxial technology is a heterostructure formation of the superconductor / insulator alternation. Our first attemp was done on the (110)YBaCuO / (110)MgO / (110)YBaCuO double-heterostructure. The RHEED (Reflection High Energy Electron-beam Diffraction) observation revealed a feasibility of that kind of epitaxial growth. Next discussion is concerning the as-grown BiSrCaCuO film formation at low temperatures. By using the sputtering deposition, the critical temperature T_{co} of the as-grown film reached 50 K. In this film, the surface morphology was improved so much. Finally, a new high-T_c superconducting transistor is demonstrated which is working on the basis of the enhanced non-equilibium superconductivity. The observed current modulation characteristics was well compared with the analysis obeying the familiar BCS theory.

Thin Film Growth for Heterostructures

Figure 1 shows the successive change of the RHEED pattern from the as-grown YBaCuO film with respect to the grown film thickness. The films were grown on (110) SrTiO$_3$ substrate by the reactive magnetron sputtering. In the thickness range less than 10 nm, the crystal structure deforms from the perovskite. To get more information, the growth experiment was done on (100)SrTiO$_3$ and (100)MgO substrates. As given in Fig.2, the sound crystallinity was obtained only for the film grown on MgO substrate. According to our SIMS depth analysis, an amount of Sr in-diffused from the substrate into the film in the former case. However, it is not known if it is the major cause of the crystal deformation.

Two examples of the multiple hetero-epitaxial growth are shown in Fig.3, where even ultra-thin hetero-epitaxy exhibits a sound RHEED pattern for the film grown on MgO substrate.

A sandwitched structure (110)YBaCuO / (110)MgO / (110)YBaCuO was grown on (110)SrTiO$_3$ substrate by the same technique. The thickness of the bottom and top layers was 150 nm, while the intermediate MgO layer was 50 nm thick. In Fig. 4, we can see RHEED patterns taken after the growth of each layer. The

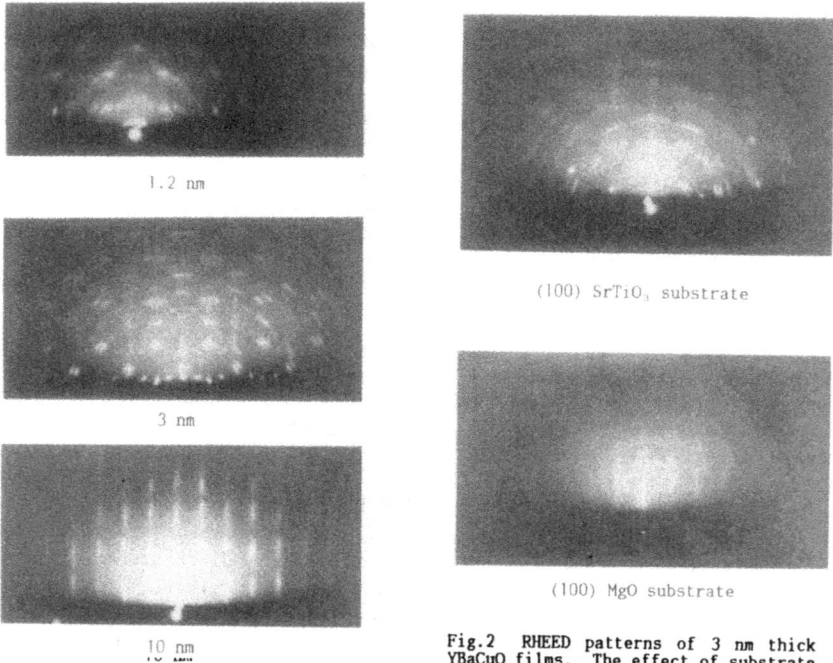

1.2 nm

3 nm

10 nm

(100) SrTiO$_3$ substrate

(100) MgO substrate

Fig.2 RHEED patterns of 3 nm thick YBaCuO films. The effect of substrate material selection is clearly seen.

Fig.1 Film thickness dependence of RHEED pattern from YBaCuO/(110)SrTiO$_3$. RHEED patterns of [1$\bar{1}$0] azimuth.

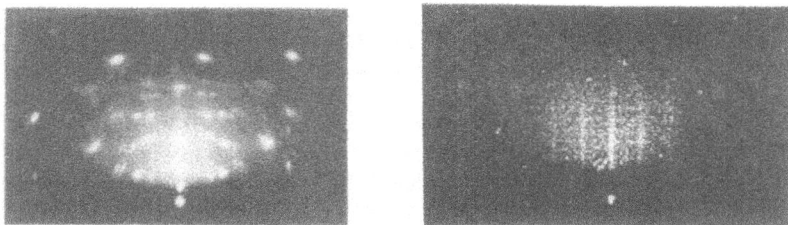

Fig.3   RHEED patterns of high-T$_c$ superconducting layered structure.   Three stucks of YBaCuO(1.2nm)/ErBaCuO(1.2nm) heterostructure grown on (110)SrTiO$_3$ [left] and a single heterostructure of NdBaCuO(3nm)/YBaCuO(3nm) on (100)MgO substrate [right].

successive epitaxial growth has not been established yet for this type of double-heterostructure.   However, the RHEED pattern from the top layer indicates that the epitaxial double-heterostructure is obtained in part of the wafer.

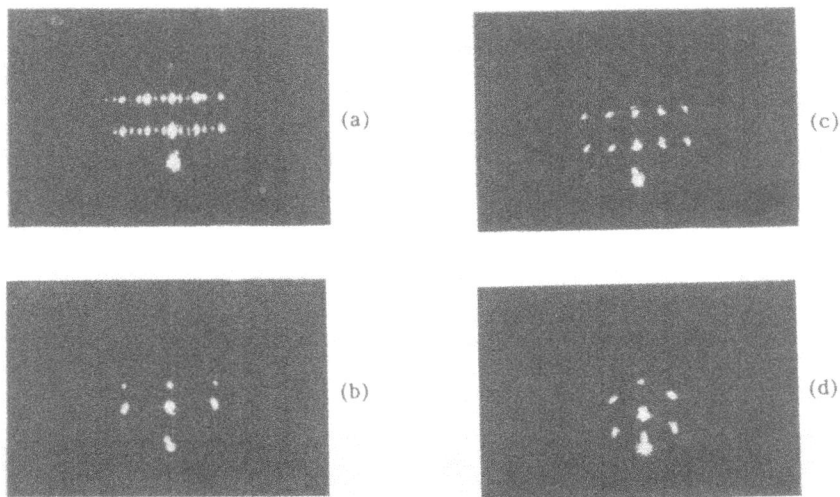

Fig.4   RHEED patterns of (110)YBaCuO/(110)MgO/(110)YBaCuO double-heterostructure. (a) Bottom layer YBaCuO (150 nm) grown on (110)SrTiO$_3$ taken to [110] azimuth. (b) Intermidiate layer MgO (50 nm) successively grown on (a), and (c) Top layer YBaCuO (150 nm) grown on (b).   (d) is the same as (c), but the electron beam goes into [001] azimuth.

As-grown Crystallization of BiSrCaCuO Film

According to findings in YBaCuO thin film growth, as-grown film formation is of great favor as a promising technique to get a mirror-like surface and low temperature crystallization, which are necessary conditions meeting for the electronic application of the oxide superconductors. The BSCCO film was grown on (100)MgO by the reactive rf sputtering at the elevated substrate temperatures. The resistance vs temperature characteristics measured before and after the post anneal (at 830 ℃ in O_2 for 72 hrs) are given in Fig.5, showing the as-grown T_{co} of 50 K. The surface morphologies are shown in Fig.6. One can see a fairly flat underlying film in the as-grown stage. The EPMA analysis tells the small precipitations (~1μm) at the surface are made of Cu-O compound. Through the post annealing, the film changed to aggregation of the high T_c phase BSCCO flakes.

Application to Electronics Field

The exploring of the new superconducting transistor is a central question addressed in the present work. Under support of advanced technologies, we here developed a superconducting transistor, which equipps the quasi-particle injection gate in addition to the source- and drain-electrodes. The schematic view and photo-micrograph of the completed device (five elements are

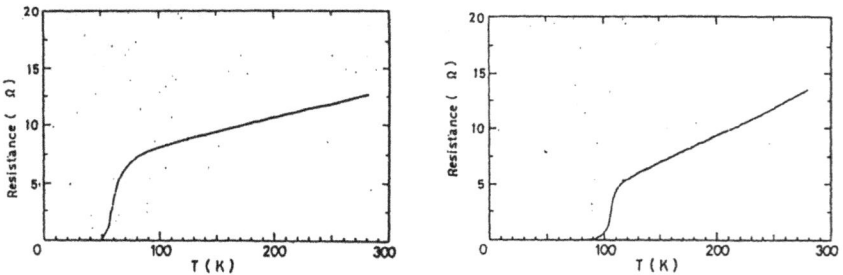

Fig.5 Temerature dependence of the BiPbSrCaCuO thin films. The as-grown [left] and post-annealed [right] films. At present, the highest T_{co} obtained so far on the as-grown film is 50 K.

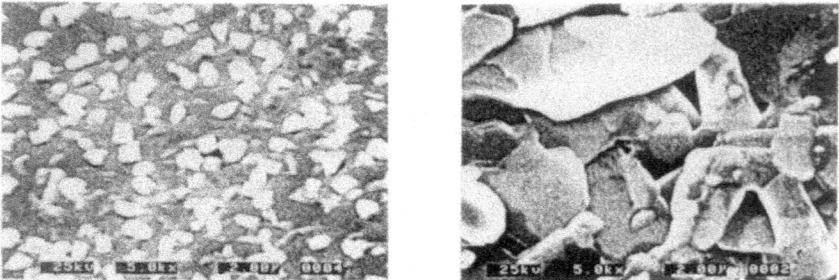

Fig.6 Scanning electron micrograph of the film surface. As-grown film [left] and post-annealed film [right]. The as-grown film surface looks fairly flat. The small precipitations are the micro-crystallization of the excess Cu.

integrated in one chip) are given in Fig.7. The (100)-oriented YBaCuO epitaxial films were used. As clearly seen in Fig.8, the superconducting critical current I_c flowing through the channel without voltage drop was efficiently modulated by the gate current I_g. The analysis based on the μ^*-model by Owen and Scalapino for the non-equilibrium superconductivity well explained our present experiment.

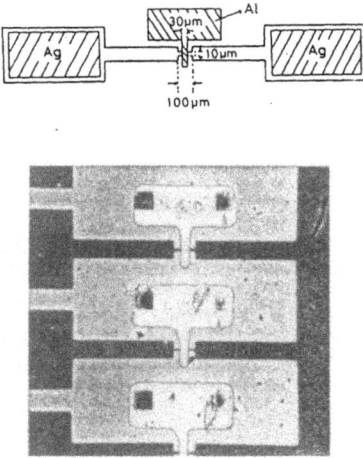

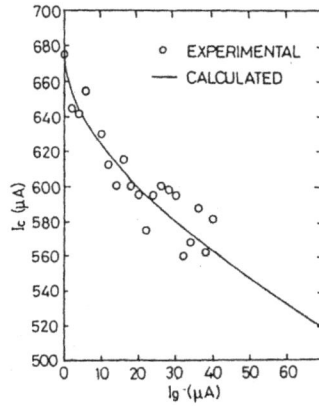

Fig.8 Current modulation characteristics of the new device. I_c denotes the critical current flowing through the channel without voltage-drop. The solid line is a curve-fitting on the basis of BCS non-equilibrium superconductivity.

Fig.7 Schematic top-view and photo micrograph of the three terminal high-T_c superconducting device. Five devices are integrated in one chip.

References

1) J. G. Bednorz and K. A. Müller: Z. Phys. B64 (1986) 189.
2) M. K. Wu, J. R. Ashburn, C. J. Torng, P. H. Hor, R. L. Meng, L. Gao, Z. J. Huang, Y. Q. Wang and C. W. Chu: Phys. Rev. Lett, 58 (1987) 908.
3) H. Maeda, Y. Tanaka, M. Fukutomi and T. Asano: Jpn. J. Appl. Phys. 27 (1988) L209.
4) Z. Z. Cheng, A. M. Hermann, A. El Ali, C. Almasan, J. Estrada, T. Datta and R. J. Matson: Phys. Rev. Lett., 60 (1988) 937.
5) M. A. Subramanian: private communication.
6) P. T. Wu, R. S. Liu, J. M. Liang and L. J. Chen: Int. Supercon. Sympo. (1988, Nagoya) Abstract p.51.
7) Many papers are published dealing with material characterization for the single bulk crystal.
8) T. Kobayashi, H. Sakai and M. Tonouchi: Electron. Lett., 22 (1986) 659.
9) T. Kobayashi, K. Hashimoto, U. Kabasawa and M. Tonouchi: to be published in IEEE Trans. Magn. No.2 (1989).

RECENT ADVANCES IN THE DEPOSITION OF HIGH QUALITY OXIDE SUPERCONDUCTING
FILMS BY PULSED LASER DEPOSITION AT BELLCORE/RUTGERS

T. Venkatesan, C. C. Chang, E. W. Chase, D. M. Hwang, L. Nazar
P. England, C. Rogers, J. M. Tarascon and P. Barboux

Bellcore, Red Bank, NJ 07701

X. D. Wu, A. Inam, B. Dutta[*], M. S. Hegde[**],
J.B. Wachtman, W. L. McLean and M. Croft

Rutgers University, Piscataway, NJ 08854

Abstract - Using pulsed laser deposition technique, high quality Y-Ba-Cu oxide superconducting thin films with critical current densities in excess of 10^6 A/cm^2 at 77 K have been prepared. We have shown that films with state-of-the-art transport properties can be prepared at a substrate holder temperature of 650 C without post annealing. Films as thin as 100 Å on SrTiO$_3$ with 82 K zero transition temperatures (T_{c0}) have been prepared, with T_{c0} greater than 90 K in films of thickness greater than 300 Å.

Introduction

While a number of other techniques have been used to grow high T$_c$ oxide thin films we have used pulsed laser deposition (PLD) technique to prepare these films (1-2). The advantages of PLD are that it is simple, fast and inexpensive (3). But the most important benefit of the technique is that films with composition very close to the target stoichiometry can be easily obtained and the composition is independent of the oxygen pressure in the deposition system.

Experiment

A detailed description of the deposition system has been published elsewhere (1,3). Superconducting or non-superconducting targets of RE$_1$Ba$_2$Cu$_3$O$_x$ were used in the experiments. (100) and (110) SrTiO$_3$, polycrystalline ZrO$_2$, sapphire and single crystal silicon were the substrates used. A Lambda Physik excimer laser (30 ns, 248 nm) was employed. The laser energy density on the target was 1.5 J/cm^2.

[*] Department of Physics, Middlebury College, Middlebury, VT 05753.
[**] Solid State Structural Chemistry Unit, Indian Institute of Science, Bangalore, India.

Results

It was found that there exists an energy density window (1-2 J/cm^2) within which high quality films are obtained(4). For energy density below the lower limit, non-stoichiometric films are obtained. On the other hand, particulates with sizes up to few microns, ejected from the target, were deposited with the films if the energy density was too high. Moreover, films in the central part of the deposited area are stoichiometric while part of the films in the other area are non-stoichiometric (4). There seems to be evidence for no significant cluster emission, but primarily emission of elemental and suboxide species (5). Particles with super thermal energies were also observed.

The films deposited in vacuum at substrate temperatures up to 500 C were insulating (1,3). Post-annealing at high temperatures (850-900 C) in oxygen was needed to have zero resistance temperatures over 80 K. The properties of the films were strongly substrate dependent (6), mainly due to the high temperature processing, with the best results on SrTiO$_3$ and MgO. Thickness dependence for the films, even on SrTiO$_3$, was also observed (7). The superconducting films showed preferred orientations: c axis of the films normal to the surface for the films on (100) SrTiO$_3$, c axis in the plane of the films on (110) SrTiO$_3$, and 'a' axis normal to the surface on sapphire. The orientation dependence was observed in both x-ray diffraction (8) and TEM (9) studies. The crystallite sizes (~ 1-2 μm) in the films on SrTiO$_3$ were much larger than those (~ 50 nm) in the films on sapphire because the lattice match between the high T$_c$ superconductor and SrTiO$_3$ is much better. Since the as-deposited films were disordered, the crystallite nucleation can start anywhere in the film during the high temperature annealing. As a result, the high temperature annealed films were always polycrystalline in nature. The superconducting properties were affected by the grain boundaries (10-11), at which excess Ba, O, and C were observed using a focussed ion beam induced secondary ion imaging technique (12).

As-deposited films, made in a few mTorr oxygen and at substrate temperatures of 650 C, were superconducting at low temperature (~ 30 K). The temperatures quoted were measured on the sample holder since the temperatures measured on the sample surfaces were always lower by 50-150 C and not reproducible. After a low temperature post-annealing at 450 C (compared to 850-900 C) in oxygen, films with zero resistance temperatures as high as 86 K were obtained (13). The transport properties of 2 um wide line were similar to those of a 10 um wide pattern (14). X-ray diffaction and TEM studies showed that the films on SrTiO$_3$ were oriented with c axis normal to the surface (15). The critical current density measured on a films with a zero resistance temperature of 86 K was 1.5x10^5/cm^2 at 82 K, which is high for the films made at such low processing temperatures. A comparison between the films annealed at high temperatures and those made at low temperatures was presented (16). The results showed that the latter films were superior to the former in surface appearance, interface (17), and transport properties.

The low temperature processing made it possible to prepare high T$_c$ superconducting thin films directly on silicon (18). Zero resistance temperature of 67 and 80 K were observed on films deposited directly and with a 50 nm ZrO buffer layer on Si. The low temperature, in-situ deposition suppressed the interface reaction between the films and Si.

Laser deposition techniques have also been used to prepare Bi-Sr-Ca-Cu oxide films (16). Films deposited at room temperature in vacuum have compositions very close to those of the targets. After high temperature annealing in oxygen, 80 K superconducting thin films with a small 110 K phase were obtained on SrTiO$_3$. Films with zero resistance of about 30 K were prepared using in-situ deposition at 600 C. More research is underway.

Moreover, recently we were able to obtain as-deposited Y-Ba-Cu oxide films with zero resistance temperatures of about 90 K on SrTiO$_3$ and over 77 K on sapphire at substrate holder temperatures of about 650 C (19) by increasing the oxygen pressure in the chamber to 100 mTorr. The increased oxygen pressure resulted in an increased oxidation of the elemental species in the laser produced plume (20). The critical current density measured on a film with a zero resistance temperature of 90 K was 4x10^6 A/cm^2 at 77 K, and even in an 0.5 um wide line a critical current density of 2x10^7 A/cm^2 was observed at 4 K (21). These films were found be nearly single crystalline exhibiting an ion channeling minimum yield of 5% (2 Mev He$^+$), implying that over 97% of the atoms are in the right lattice sites (22), to 75% for very thin 100 Å films (Fig. 1). Further in Fig. 2 it is seen that films as thin as 100 Å have a T_{co} of 82 K with the T_{co} rising beyond 90 K for film thickness exceeding 300 Å. There seems to be some correlation between the depressed T_{co} at reduced thickness and the increased disorder in the film as shown in Fig. 3.

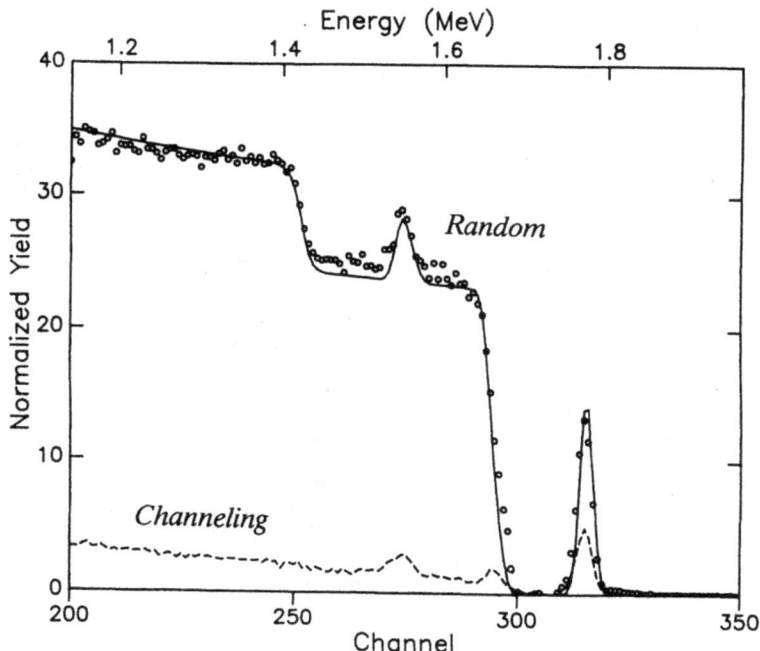

Fig. 1   Random and Channeling Rutherford backscattering spectrum for a Y-Ba-Cu oxide film on (100) SrTiO$_3$. The Solid line is a simulation of 100 Å Y$_1$Ba$_2$Cu$_3$O$_{7-x}$/SrTiO$_3$.

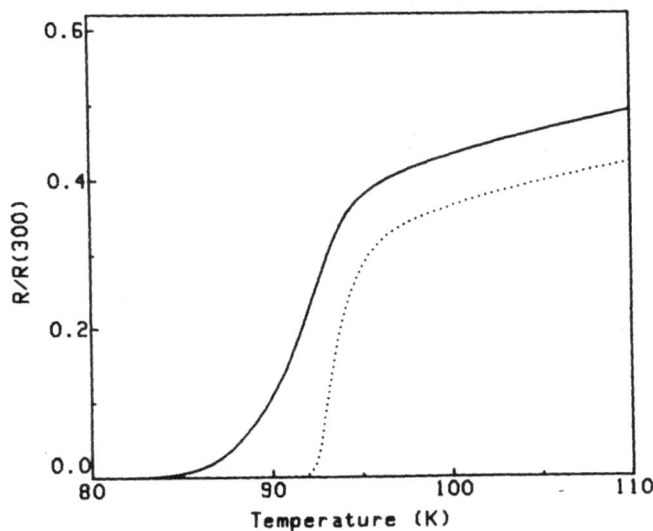

Fig. 2   Resistance vs. temperature for a 100 Å (solid line) and 300 Å (dotted line) thick films on (100) $SrTiO_3$.

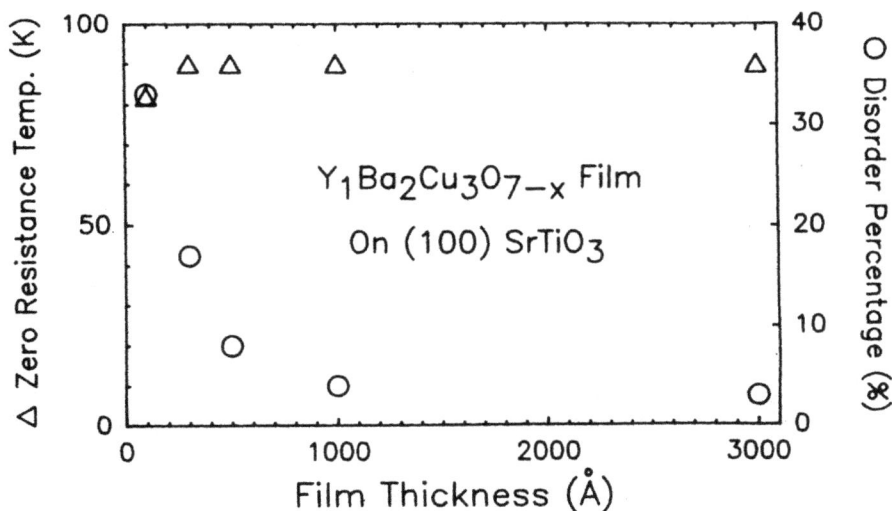

Fig. 3   Zero resistance temperature as a function of film thickness. The disorder percentages, which is obtained by measuring the minimum yield in a Rutherford backscattering and channeling experiment, in the film as a function of thickness is also plotted in the same figure.

Conclusion

The pulsed laser deposition technique is emerging as a reliable technique for the preparation of excellent high T_c oxide thin films. The low temperature processing may enable the synthesis of novel epitaxial superconducting superlattice structures.

Acknowedgments

The authors would like to thank J. M. Rowell, L. H. Greene, B. G. Bagley, P. F. Miceli, R. Krchnavek and R. Levi-setti for their help.

References

1) D. Dijkkamp, T. Venkatesan, X. D. Wu, S. A. Shaheen, N. Jisrawi, Y. H. Min-Lee, W. L. McLean and M. Croft, Appl. Phys. Lett., 51, 619(1987).
2) T. Venkatesan, Solid State Technology, p. 39 (Dec. 1987).
3) X. D. Wu and T. Venkatesan, Chemistry of Oxide Superconductors, ed. C. N. R. Rao (Blackwell Scientific Publications Ltd., Oxford, UK, 1988). p. 175.
4) T. Venkatesan, X. D. Wu, A. Inam and J. B. Wachtman, Appl. Phys. Letts., 52, 1193(1988).
5) T. Venkatesan, X. D. Wu, A. Inam, Y. Jeon, M. Croft, E. W. Chase, C. C. Chang, J. B. Wachtman, R. W. Odom, F. R. di Brozolo and C.A. Magee, Appl. Phys. Lett. (Oct.10, 1988).
6) T. Venkatesan, C. C. Chang, D. Dijkkamp, S. B. Ogale, E. W. Chase, L. A. Farrow, D. M. Hwang, P. F. Miceli, S. A. Schwarz, J. M. Tarascon, X. D. Wu and A. Inam, J. Appl. Phys., 63, 4591(1988).
7) X. D. Wu, D. Dijkkamp, S. B. Ogale, A. Inam, E. W. Chase, P. F. Miceli, C. C. Chang, J. M. Tarascon and T. Venkatesan, Appl. Phys. Lett., 51, 861(1987).
8) P. F. Miceli, T. Venkatesan, X. D. Wu and J. A. Potenza, Thin Film Processing and Characterization of High Temperature Superconductors, (AIP, NY, 1988) p. 150.
9) D. M. Hwang, L. Nazar, T. Venkatesan and X. D. Wu, Appl. Phys. Lett., 52, 1834(1988).
10) S. B. Ogale, D. Dijkkamp, T. Venkatesan, X. D. Wu and A. Inam, Phys. Rev., B36, 7210(1987).
11) P. England, T. Venkatesan, X. D. Wu and A. Inam, Phys. Rev. B (in press).
12) T. Venkatesan, X. D. Wu, A. Inam, E. W. Chase, C. C. Chang, D. M. Hwang, R. Krchnavek, M.S. Hedge, J. B. Wachtman, W. L. McLean, R. Levi-Setti, J. Chabala and Y. L. Wang, J. Materials Education,(in press, 1988).
13) X. D. Wu, A. Inam, T. Venkatesan, C. C. Chang, E. W. Chase, P. Barboux, J. M. Tarascon and B. Wilkens, Appl. Phys. Lett., 52, 754(1988).
14) P. England, T. Venkatesan, T. Cheeks, X. D. Wu, A. Inam and M.S. Hegde(unpublished).
15) D. M. Hwang, C. C. Chang, T. Venkatesan, E. W. Chase, X. D. Wu and A. Inam, (unpublished).

16) X. D. Wu, T. Venkatesan, A. Inam, E. W. Chase, C. C. Chang, Y. Jeon, M. Croft, C.A. Magee, R. W. Odom and F. Radicati de Brozalo, High Tc Superconductivity: Thin Films and Devices, (Proc. of SPIE, 1988).

17) C. C. Chang, X. D. Wu, A. Inam, E. W. Chase and T. Venkatesan, Appl. Phys. Lett., $\underline{53}$, 517(1988).

18) T. Venkatesan, E. W. Chase, X. D. Wu, A. Inam, C. C. Chang and F. K. Shokoohi, Appl. Phys. Lett., $\underline{53}$, 243(1988).

19) A. Inam, M. S. Hegde, X. D. Wu, T. Venkatesan, E. W. Chase, C. C. Chang, P. England, P. F. Miceli, J. M. Tarascon and J. B. Wachtman, Appl. Phys. Lett. $\underline{53}$, 908(1988).

20) X. D. Wu, B. Dutta, M. S. Hedge, A. Inam, T. Venkatesan, E. W. Chase, C. C. Chang and R. Howard, Appl. Phys. Lett. (to be published).

21) C. Rogers et. al., (unpublished).

22) X. D. Wu, A. Inam, M. S. Hedge, T. Venkatesan, C. C. Chang, E. W. Chase, B. Wilkens and J. M. Tarascon, Phys. Rev. B. (in press).

36

STATUS OF HIGH FIELD MAGNETS AND ADVANCED HIGH FIELD SUPERCONDUCTORS

Koshichi NOTO, Kazuo WATANABE and Yoshio MUTO

Institute for Materials Research, Tohoku University,
Katahira 2-1-1 Sendai 980, Japan

Abstract - Recent world wide progress in the hybrid magnet technology has made it available to make many experiments under very strong steady state magnetic fields higher than 30 T. Many invaluable experiments such as critical current densities of advanced superconducting materials, upper critical fields of high T_c oxide superconductors, and so on are now under way. Improvements of high field characteristics in Nb_3Sn conductors have been succeeded recently by the ternary Ti addition in Japan. Significant number of high field magnets which can generate more than 12-15 T have been constructed by using these improved Ti-added Nb_3Sn conductors. Recent progress in advanced high field superconductors such as Nb_3Al, $Nb_3(Al,Ge)$, NbN, $PbMo_6S_8$, and so on which aim higher than 16-20 T is also very remarkable. A very high $J_c = 6.5 \times 10^4$ A/cm^2 was observed recently at 77.3 K and 27 T in a polycrystalline film prepared by a CVD technique.

Introduction

Recently, there are significant progress in the hybrid magnets, in which inner water-cooled magnets are combined with outer superconducting magnets. Following to the generation of 30.1 T at MIT National Magnet Laboratory (1981) [1], 30.7 T (1985), and 31.1 T (1986) at High Field Laboratory for Superconducting Materials (HFLSM), Tohoku University [2-4], 28.5 T was generated at Nijmegen University [5] and a new world record of 31.35 T was generated at High Field Magnet Laboratory in Grenoble [6]. Due to this progress in the hybrid magnet technology in the world, many studies in steady state high magnetic fields have become available.

In HFLSM at Tohoku University, many high field magnets including three hybrid magnets were made open to general use and extensive high field studies such as critical current densities of advanced superconducting materials, upper critical fields of high T_c oxide superconductors, and so on were promoted.

After a brief review on hybrid magnets in the world, the studies on advanced high field superconductors and oxide high T_c superconducting materials held in HFLSM at Tohoku University are overlooked.

Hybrid Magnets

The principle of the hybrid magnet is the combination of an outer superconducting magnet with an inner high power water-cooled magnet. Recently, progress in the technology of this combination is very remarkable [7]. Table 1 is the list of hybrid magnets in the world. As already described in introduction, there are three centers in the world, where steady state fields higher than 30 T are available.

In HFLSM at Tohoku University, total eleven high field magnets [4] shown in Table 2 including three hybrid magnets are open for scientists and engineers in various fields not only from universities and national research institutes but also from various industrial companies. The fields include developmental studies on practical and advanced superconducting materials, researches on high-T_c oxide superconductors, studies on organic superconductors, basic researches on magnetic materials, fractional quantum Hall effect in semiconductors, magneto-optics, and so on.

Table 1. Hybrid magnets in the world.
    COD and CID mean coil outer and inner diameters, respectively.

Location State		SM					WM					HM
		OD mm	COD mm	CID mm	bore mm	field T	COD mm	CID mm	bore mm	power MW	field T	field T
Tohoku Univ. Japan	HM-3	711	435	290	220	8.0	185	38	32	3.1	13.0	20.5
	HM-2	1300	922	420	360	8.0	320	60	52	6.3	16.0	23.2
	HM-1b	1450	1094	430	360	12.0	300	60	52	7.0	17.0	28.1
	HM-1a						300	38	32	7.4	19.6	31.1
MIT USA						7.5	333	60	53		19.6	27.1
			710	400		7.5	330	41	33	8.7	22.9	30.1
Nijmegen Univ. The Netherlands		1360	889	406	356	8.5	333	38	32	5.2	17.0	25.4
			883	420		11.0						28.5
Kurchatov USSR			700	376		6.9	275	48	28	5.6	18.0	25.0
Oxford UK				284	240	6.5	200	53	50		13.5	20.0
Grenoble France + FRG			1087	500		11.0			50	10.0		31.4

38

Table 2. High field magnets installed in HFLSM at Tohoku University.

Magnet	Effective Bore (mm)	Central Field(T)	Power (MW)	Rated Current (A)	Type
16.5 T-SM	56	16.5	——	136	Nb₃Sn tape ·compact DP
13T-SM	50	13	——	450·150	NbTi·fm,Nb₃Sn·fm tape
9T-SM	19x31	9	——	150	Nb₃Sn tape ·split-pair
12T-DRSM	50	12/13.2	——	90/100	Nb3Sn-fm for DR
HM-1	52/32	29/31		1456	Nb₃Sn·fm,NbTi·fm cryostable DP
SM-1	360	12	——		
WM-1a	32	19	7.4	——	Polyhelix
WM-1b	52	17	7.0	——	Polyhelix
HM-2	52	23		1470	NbTi·fm · cryostable DP
SM-2	360	8	——		
WM-2	52	16	6.3	——	Double Bitter
HM-3	32	20		780	NbTi·fm ·compact solenoid
SM-3	220	8	——		
WM-3	32	12.8	3.1	——	Single Bitter
WM-4	32	20	6.8	——	Bitter + Polyhelix
WM-5	82	15	6.7	——	Single Bitter ·
WM-6	62	12.9	4.6	——	Single ·Bitter ·
WM-7	52	14.5	4.5	——	Double Bitter

DP : Double Pancake, fm : fine-multicored, DR : Dilution Refrigerator

High Field Superconducting Magnets

Recently, there are many studies on improvements of practical Nb₃Sn conductors by the addition of a ternary element [8]. In Japan, improvements of high field characteristics in Nb₃Sn conductors [9] have been succeeded by the ternary Ti addition. Significant number of high field magnets which can generate more than 12-15 T have been constructed by employing these improved Ti-alloyed Nb₃Sn conductors [9]. Figure 1 shows the central fields of these magnets plotted against the winding inner diameter. The multipurpose magnet at Karlsruhe named "HOMER" has succeeded very recently in generating more than 20 T [10] by winding one of these conductors in the innermost section of it. The NRIM's magnet which employs a Ti-Nb₃Sn conductor in the middle section, an Al-V₃Ga tape conductor in the highest field section, and can generate 18.1 T is also shown in this figure [11].

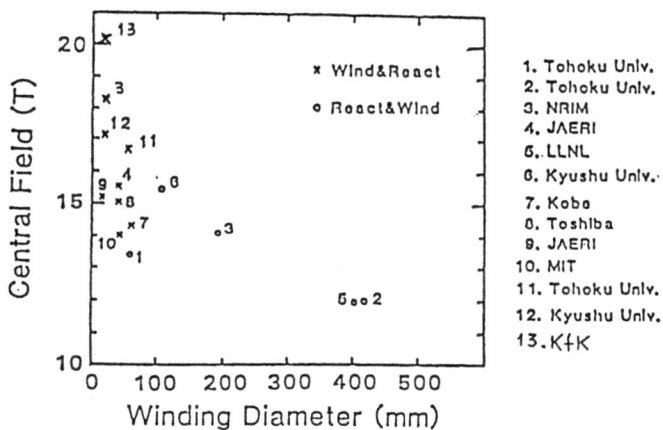

Fig. 1. High field superconducting magnets wound with improved conductors.

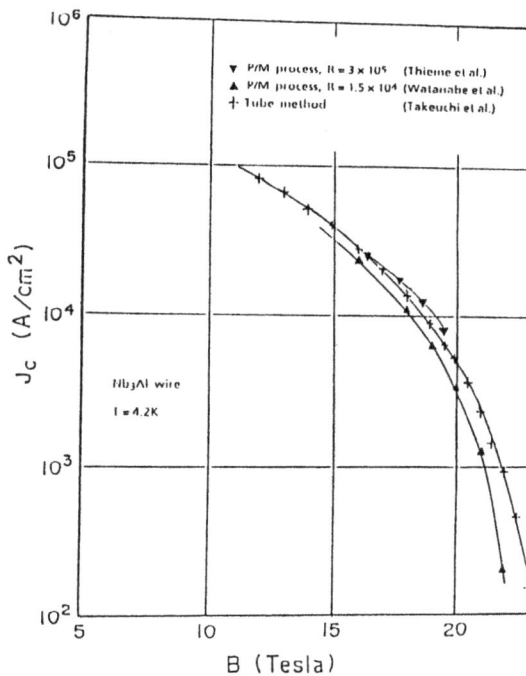

Fig. 2. High field characteristics of Nb_3Al wires.

Advanced High Field Superconducting Materials

There are extensive studies on various advanced high field superconducting materials held in HFLSM, Tohoku University. In these studies, recent progress in Nb_3Al wires and a Chevrel phase $PbMo_6S_8$ wire are briefly described.

Nb_3Al Wires

Following the pioneering study by Thieme et al [12], we also achieved almost the same J_c value in powder metallurgy processed Nb_3Al wires [13,14]. Recently, the NRIM group developed a new process (tube method) for a Nb_3Al multifilamentary wire [15]. High field characteristics of J_c in these Nb_3Al wires are compared in Fig. 2. K. Ikeda and his coworkers also succeeded in developing a Nb_3Al wire by the clad chip method [16].

Chevrel Phase $PbMo_6S_8$ Wires

Since the discovery of very high critical fields of $PbMo_6S_8$ [17], many works have been reported. However, enough high J_c has not yet been obtained for this type of compound. Succeeding to the pioneering work by Seeber et al [18], three groups have been working intensively in Japan to develop the conductor of this material. Very recently, the Mitsubishi group succeeded in preparing Ta- and Nb-sheathed wires with fairly high J_c up to 23 T [19]. Figure 3

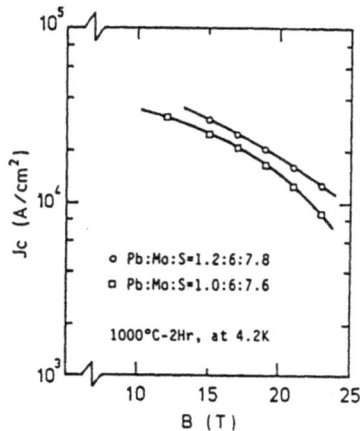

Fig. 3. High field J_c of $PbMo_6S_8$ short samples.

shows the high field characteristics of these samples. Lower and upper curves correspond to the Ta- and Nb-sheathed wires, respectively.

Transport J_c at High Fields in High T_c Oxide Superconductor Films

The discovery of the series of high T_c oxide superconductors since 1986 brought about the feverish activities on this group of materials. According to the studies up to now, although T_c and H_{c2} are high enough, the critical current density, J_c, in sintered pellets or wires of these materials are not high enough. High enough J_c's are reported only in single crystal films [20]. Therefore, studies [21,22] on the cause of the very small J_c in pellets or wires are also important.

Recently, excellent transport critical current properties in high fields up to 27 T are obtained at 77.3 K in a polycrystalline $YBa_2Cu_3O_{7-y}$ film prepared by a chemical vapor deposition technique [23]. Figure 4 shows the results obtained in this sample. As can be seen, J_c is 4.1×10^5, 1.9×10^5, and 6.5×10^4 A/cm^2 at 2, 10, and 27 T, respectively. A peak effect was observed in the field region above 16 T. The upper critical field defined by zero resistivity was estimated to be 35 and 180 T at 77.3 and 0 K, respectively. Satchell et al [24] also pointed out that the material with fairly high J_c under magnetic fields will not necessarily be monocrystalline.

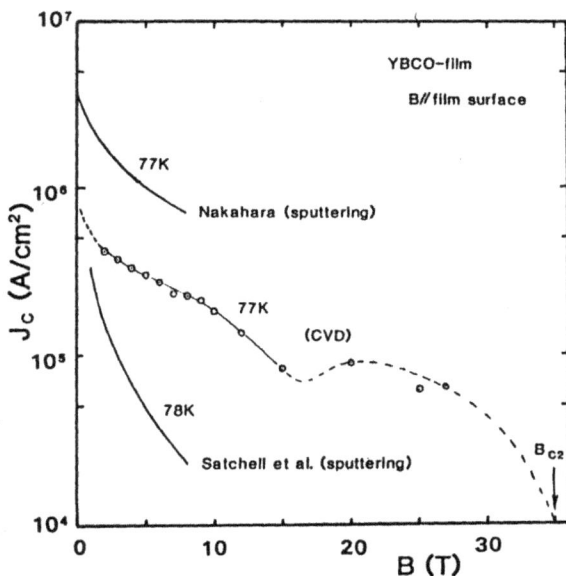

Fig. 4. Magnetic field dependence of transport J_c in high T_c oxide films.

Conclusion

(1) Recent world wide progress in hybrid magnet technology made it available to make various experiments in very high steady state magnetic field higher than 30 T.

(2) Improvement of J_c was succeeded in ternary alloyed A15 conductors, and many high field superconducting magnets up to 12-20 T have been constructed by employing these conductors.

(3) Progress in ordinary advanced high field superconducting materials such as Nb_3Al, Chevrel phase compound, and so on is also remarkable.

(4) There are extensive studies on the newly discovered oxide high T_c superconductors. Very high $J_c=4.1x10^5$, $1.9x10^5$, and $6.5x10^4$ A/cm^2 at 2, 10, and 27 T, respectively, were observed at 77.3 K in a CVD $YBa_2Cu_3O_{7-y}$ film.

References

1) M. J. Leupold et al, Proc. 6th Int. Conf. on Magnet Technology, Bratislava (1977) 400.

2) Y. Nakagawa et al, Proc. 9th Int. Conf. on Magnet Technology, Zurich (1986) 424.

3) K. Noto et al, Adv. Cryog. Engin.-Mat. 34 (1988) 925.

4) Y. Muto et al, Sci. Rep. RITU A33 (1986) 221.

5) M. J. Leupold et al, Proc. 9th Int. Conf. on Magnet Technology, Zurich (1986) 215.

6) Schneider-Muntau et al, Proc. 10th Int. Conf. on Magnet Technology, Zurich (1988)

7) W. Sweet, Physics Today (July, 1988) 61-65.

8) K. Noto, Proc. 9th Int. Conf. on Magnet Technology, Zurich (1986) 199.

9) K. Noto et al, Sci. Rep. RITU A33 (1986) 393.

10) P. Turowski and Th. Schneider, Cryogenics 27 (1987) 403 and R. Flukiger, private communication.

11) K. Tachikawa et al, IEEE Trans. Mag. MAG-23 (1987)

12) C. L. H. Thieme et al, IEEE Trans. Mag. MAG-21 (1985) 756.

13) K. Watanabe et al, IEEE Trans. Mag. MAG-23 (1987) 1428.

14) K. Watanabe et al, IEEE Trans. Mag. MAG-25 (1989) to be published.

15) T. Takeuchi et al, IEEE Trans. Mag. MAG-25 (1989) to be published.

16) K. Ikeda et al, private communication.

17) O. Fischer, Appl. Phys. 16 (1978) 1.

18) B. Seeber et al, IEEE Trans. Mag. MAG-19 (1983) 402.

19) Y. Kubo et al, Proc. MRS Meeting (1988, Tokyo) to be published.

20) T. Nakahara, Proc. ISS-88 (1988, Nagoya) to be published.

21) K. Watanabe et al, Cryogenics (1988) to be published.

22) K. Noto et al, Cryogenics (1988) to be published.

23) K. Watanabe et al, submitted to Appl. Phys. Lett. (1988)

24) J. S. Satchell et al, Nature 334 (1988) 331.

EXPERIMENTAL INVESTIGATIONS FOR NEW SUPERCONDUCTING MECHANISM

Ryozo AOKI, and Testuro NAKAMURA*

Dept. of Physics Kyushu University 33, Fukuoka, 812 Japan
*Research Lab. of Eng. Materials, Tkyo Inst. of Technology,
Yokohama 227, Japan

Abstracts - In order to get some guide to quest for high Tc superconductors
several comments are provided on the basis of experimental investigations
regarding perovskite $SrTiO_{3-x}$/metallic In contact system and the electronic
virtual bound state in narrow gap semiconductor Pb(Tl)Te.

Introduction

    Material search for higher critical temperature superconductor is now
stimulated world wide and various kinds of oxide substances have been
developed recently attaining to 125K. They are much complicated in character
and structure consisting of four or five kinds of elements, and composed of
more than ten layers super-lattices, which enables hardly us to get insight to
the fundamentals of new superconducting mechanism.
    For getting further step-up towards new characteristic superconductors,
it is considered suggestive to broaden our view with relative kinds of substances which have been investigated before and after the 1986 epoch. From
this view points the following comments are provided;

§1. Optimization in layer structures of the oxide superconductors

    Many investigations revealed out the following character on the oxide
supercouductor;
i) The (Cu-O) net-plane is responsible for the superconducting electronic
state.
ii) Its critical temperature Tc is generally provided at 20-40K by the (Cu-O)
mono-layer structure, 80-90K by the double-layers, ∿110K by the triple-layers
∿122K maximum at quadruple-layers and somewhat lower 120K for the quintet-
layers. [1].
iii) The above Tc grade is for the structure with monolayer of (Tl-O) net-
plane. In case of having (Tl-O) double-layer substructure, this Tc grade
pretends one-step shift towards the smaller (Cu-O) layernumbers, and attains
to 125K by the (Cu-O) triple-layers. This may suggests that the (Tl-O) layer
also plays a part of similar role to the (Cu-O) plane.
iv) An assymptotic limit of this multi-layer coupling structure is three
dimensional (Cu-O) network in real perovskite structure such as $LaCuO_3$, where
superconducting state has not ever been realized. This fact and the maximum
Tc character indicate that for obtaining the high Tc, there exists an optimum
condition in the coupling layer numbers [1] or coupling strength as regarding
the dimensionality.

v) Even in certain two-dimensional (Cu-O) netplane structure, Tc does not show monotonical rise-up as the carrier concentration p increases. Once it shows a maximum and then rapidly falls down with further increasing p, as shown by Tokura et al [2] in fig.1 where one finds this character is common for both of LaACuO and YBaCuO systems.

This rapid decrease in Tc is a crucial problem for the promotion of higher Tc investigations, and we have no answer of this problem at present, because the new superconducting mechansism remains unsolved essentially.

To be noted is that similar Tc to carrier concentration characteristics had been also recognized previously in a typical perovskite type superconductor $SrTiO_{3-x}$ system as presented [3] in fig.2 on the basis of the data by Schooly et al [4]. The resemblance among the curves in fig.1 and 2 suggests that the Tc determing mechanism is likely to be ruled commonly, for instance no matter whether magnetic or non-magnetic systems. Really $SrTiO_{3-x}$ has no relation with the magentic spin system, while in $La_{2-x}CuO_{4-y}$ [5] and $YBaCu_{1-x}M_xO_{7-y}$ [9] systems, the [Cu-O] layer shows a strong spin correlation effect.

As for $SrTiO_{3-x}$, this Tc depression mechanism (see fig.2) was discussed by many investigators, and Appel [3] ascribed it to a screening of the Cooper pair attractive potential by the high mobile carrier densities. This screening effect is general for any superconductors. On the other hand, mobile carrier density is necessary for giving rise to the superconducting state. Accordingly Tc shows a maximum at certain appropriate carrier concentrations in these systems.

As for the [Cu-O] layer numbers, both of the superconducting ordering state (ODLRO) and the Coulomb screening effect basically depends on the carrier mobile dimensionality, while in some different styles with each other.

From these view points, a careful optimization in the inter-layer coupling strength and the coupling layer numbers will be required for achieving the highest Tc.

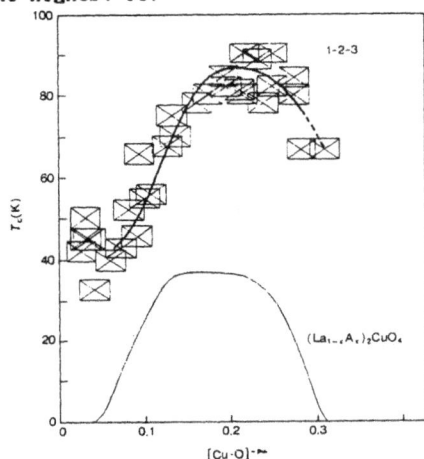

Fig.1 Superconducting Tc dependence of 1-2-3(YBCO) and LaACuO systems on the carrier concentration $[Cu-O]^{+Psh}$ in the (Cu-O) plane; cited from ref 2)

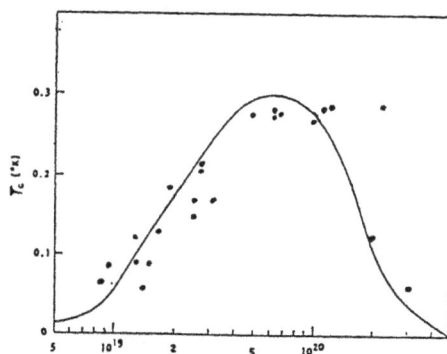

Fig.2 Superconducting Tc of $SrTiO_{3-x}$ as a function of the electron concentration; cited from ref 3)

§.2 Heterogeneous structure and the appropriate coupling

In homogeneous systems such as metals and alloys the superconducting Tc is expressed generally

$$Tc = hw_c \exp[-1/g]$$

For obtaining higher Tc, larger g is required. According to the BCS weak coupling model the parameter g is expressed as g=N(0)·V by the carrier density of states N(0), and the effective pair-attractive potential V. High carrier density increases N(0), but decreases V by the screening effect. Really most of the superconducting elements obey in a rule of N(0)·V=g≃const as shown in fig.3.

In order to avoid this dilemma on N(0) and V, Little and also Ginzburg proposed to consider heterogeneous system which is composed of (I): the highly conducting part with large N(0) by dense carrier, and (II): polarizable part with easily excited state (exciton) for inducing strong pair-attractive inter-action V. Little presented a model [6] of organic molecule with π-electron conducting spine (I) and the polarizable side chain (II), and Ginzburg proposed a multi-layer structure [7] composed of metal film (I) and the dielectric semiconductor (II). The interaction between the free carrier in (I) and the exciton in (II) is considered to take place in the interface boundary region.

We have carried out an investigation [8] of this heterogeneous supercon-ducting model with use of indium metal thin film (I) deposition onto a clean surface of dielectric $SrTiO_3$ (II) single crystal. If the new superconducting interaction takes place, the original Tc of In film would be enhanced. However, little change in Tc of the In specimen was observed on stoichiometric $SrTiO_3$ and it turns out that for those heterogeneous systems composed of quite

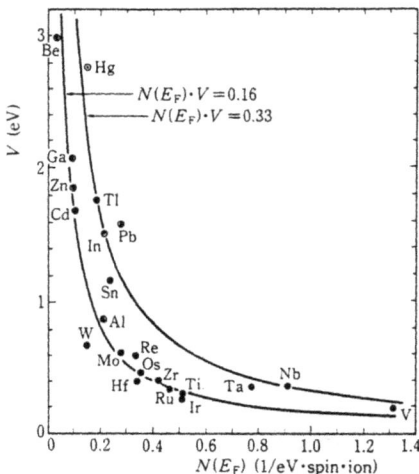

Fig.3 The relation between the superconducting pairing-potential parameter V and the electron density of states N(0) for elements.

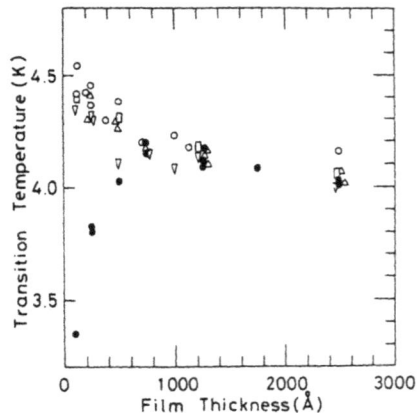

Fig.4 Superconducting transition temperature of indium films on different substrates. ○ on quartz, △ on stoichiometric $SrTiO_3$, ▽ on reduced $SrTiO_{3-x}$ of $n=6.0 \times 10^{17}$ cm^{-3}, □ on reduced $SrTiO_{3-x}$ of $n=1.2 \times 10^{20}$ cm^{-3}, ● on reduced $SrTiO_{3-x}$ of $n=6.4 \times 10^{20}$ cm^{-3}.

different characters (metals and dielectrics), essentially a large barrier potential takes place at the interface and it obstructs the expected interaction between (I) and (II). For obtaining the thinner Schottky barrier and large electron tunneling probability at the interface, the dielectrics must be doped with carrier n by the deoxidation as $SrTiO_{3-x}$. More and more in doping x and at last at $n=6.4 \times 10^{20}$, where the barrier thickness is less than $10^2 A$, the original Tc of In thin film was considerably modified by the interaction as shown in fig.4. However, in that concentration n the dielectrics $SrTiO_{3-x}$ was so highly doped that the excitonic interaction should be quite screened and the heterogeneous character of the system had been smeared out.

In case of the high Tc oxide compounds, the conduction layer part, and other molecular orbital bonding part are consisting of a heterogeneous structure and it is considered that a subtle covalent-ionic bonding sustains an appropriate interaction between the two parts without any severe potential barrier problem.

Really, it has been revealed out by our investigation [9] that in YBCO compound the oxygen atoms, which combines the $(Cu-O)_n$ conduction planes with the $(Cu-O)_m$ chain structure, plays an important role for the mediation of the atom substitution effect on the superconductivity.

In consequence, the viewpoint of heterageneous structure combined with appropriate interaction is considered to be an essential point for the molecular design in the new superconducting compound synthesis.

§3. Semilocalized electronic state and superconductivity

In fig.1, the limit of little carrier-concentration corresponds to a localized electronic state with anti-ferromagnetic spin ordering. As small density of carrier is introduced, the carrier mobility appears and superconductivity takes place on the semi-localized electronic state, and its Tc starts to increase rapidly.

For those oxide superconductor, the carrier is not free-electron-like differently from usual metallic superconductor. It is considered that these semi-localized or heavy-electron-like state may be rather favorable for the high Tc superconductivity. This point had been discussed regarding polaronic semiconductor by Chakraverty, [10] and his remarks stimulated the Bednorz and Muller's pioneer work.

Here we present another investigations [11] which may support this viewpoint. The IV-VI family compound PbTe is a narrow gap semiconductor with considerable dielectric polarizability due to ionic-covalent bonding in NaCl type structure. When it is doped with Tl, semilocalized impurity state takes place closely below top of the valence band. This sharp impurity levels were observed in our tunneling spectrum with two-levels structure, as shown in fig.5.

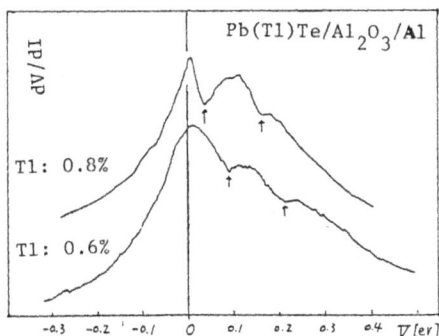

Fig.5 Tunneling impedance spectrum dV/dI versus the bias voltage V of Pb(Tl)Te. The arrow points corresponds to the Tl impurity levels

Another dopant Na providing hole carriers is introduced to this Pb(Tl)Te, and the Fermi level in the valence band is shifted down with Na doping concentration. When it passes through the Tl sharp impurity levels, a resonant scattering of the conduction carrier takes place upon the virtual bound impurity state.

It is interesting that at just of this resonant scattering condition is realized, the superconducting Tc shows a corresponding increase evidently up to about twice, which cannot be understood only from the electronic density of states increase due to the sub % impurity concentrations. The superconductivity and the resonant scattering phenomena in this system were first investigated with sintered samples and reported by Chernik and Lykov [12], and its superconducting mechanism has been discussed in relation of this impurity scattering by Kaidanov et al [13]. We could prepare fine film specimens by a deposition technique with the hot-wall source, and a semi-quantitative correspondence of Tc to the two impurity levels could be observed [11].

From these experimental investigations, it turns out that the semi-localized virtual bound electronic state is evidently correlated with the superconductivity and enhances its Tc. Consequently, this fact may suggest an interesting view point for the new superconducting mechanism in covalent-ionic bonding substances including the current oxide compounds.

References

1) H. Ihara, R. Sugise, T. Shimomura, M. Hirabayashi, N. Terada, M. Jo, K. Hayashi, M. Tokumoto, K. Murata and S. Ohashi; to be published in Proc. Internat. Superconducting Symposium (Tokyo 1988, Springer Verlag).
2) Y. Tokura, J.B. Torrance, T.C. Huang, and A.I. Nazzal; Phys. Rev. B 38 (1988) 7156.
3) J. Appel; Phys. Rev. 180 (1969) 508.
4) J.F. Schooley, W.R. Hosler, E. Ambler, J.H. Becker, M.L. Cohen, and C.S. Koonce Phys. Rev. Letter 14 (1965) 305.
5) R. Aoki, H. Murakami, K. Sakai, T. Nakamura, K. Kawasaki and R. Liang; Presentation in this Symposium CZ-4 (Abstract p.105).
6) W.A. Little; Phys. Rev. 134 (1964) 1416.
7) V.L. Ginzburg; Contemp. Phys. 9 (1968) 355.
8) T. Ohta, R. Aoki, and S. Hayashi; J. Phys. Soc. Jpn. 51 (1982) 1080.
9) R. Aoki, S. Takahashi, H. Murakami, Te. Nakamura, Ta. Nakamura, T. Takagi, and R. Liang; Physica C 156 (1988) 405.
   also presentation in this symposium CZ-12 (Abstract p.113)
10) B.K. Chakraverty; J. de Physique Letters 40 (1979) L97.
11) H. Murakami, T. Migita, Y. Mizomata, Y. Inoue, and R. Aoki; Jpn. J. Appl. Phys. series 1 SC Mat. (1988) 135.
    Also presentation in this symposium CZ-1.
12) I.A. Chernik and S.N. Lykov; Soviet Phys. Solid State 23 (1981) 817.
13) V.I. Kaidanov and Yu. I. Ravich. Soviet Phys. Usp 28 (1985) 31.

CRITICAL CURRENT CHARACTERISTICS IN SUPERCONDUCTING
OXIDES WITH HIGH CRITICAL TEMPERATURES

Teruo MATSUSHITA

Department of Electronics, Kyushu University 36
6-10-1 Hakozaki, Higashi-ku, Fukuoka 812, Japan

Abstract - Theoretical estimates are given for the critical current densities in single-crystalline thin films, polycrystalline bulk materials with oriented textures and those with random textures of high T_c superconductors. In the estimate, it is assumed that twinning planes are causative pinning centers and the percolation theory is used with taking account of depressing effect of the transport current by the grain boundaries. The obtained results are compared with the existing experiments and the possibility of improvement of the characteristics is discussed.

Introduction

The critical current density is a key parameter which determines a potentiality of superconducting high T_c oxides for power application in the future. Large critical current densities reported in single-crystalline thin films [1, 2] guarantees the potentiality of these materials. However, bulk materials which are suitable for mass-producible cables or conductors for large-scale apparatus, can carry only small current densities, although there exist fairly large shielding currents isolated inside the grains [3, 4]. Recently, relatively large critical current densities were reported [5, 6] for specimens with oriented textures.

As reviewed in the above, there are many ranks of critical current densities. Hence, it seems to be necessary to systematically understand these critical current characteristics. In this paper, we theoretically estimate the critical current densities in single-crystalline thin films, polycrystalline bulk materials with oriented textures and those with random textures, where twinning planes are assumed as causative pinning centers. These results are compared with existing experiments. The possibility of improvement of the critical current characteristics in the future is also discussed.

Theory

Elementary pinning force

Causative defects which contribute to the flux pinning in YBaCuO are twinning planes. In this paper, we assume that these planar defects are dominant pinning centers. Most twinning planes are (1 1 0) planes parallel to the c-axis. These flat defects are quite anisotropic and are effective for pinning only when the fluxoids are parallel to them [7]. We suppose that the magnetic field is applied along the c-axis. If we denote the upper critical field in this geometry by $B_{c2\perp}$, it is given by $\phi_0/2\pi\xi_\parallel^2$, where ϕ_0 is the flux quantum and $\xi_\parallel$ is the Ginzburg-Landau coherence length in a basal plane. The dominant pinning interactions between such planar defects and the fluxoids are caused by the electron scattering mechanism [8]. When the Lorentz force is directed normal to the plane, their pinning strength per unit length of the fluxoid is given by [9]

$$\hat{f}_p = (AB_c{}^2\xi_{0\parallel}/\mu_0)[1-(B/B_{c2\perp})],\qquad(1)$$

where A is the constant dependent on the impurity parameter of the material,

B_c is the thermodynamic critical field and $\xi_{0\parallel}$ is the BCS coherence length. The BCS coherence length of oxide superconductors is not clear, but is related with the Ginzburg-Landau coherence length through the impurity parameter. Its minimum value is $1.35\xi_\parallel(0)$ in the clean limit, where $\xi_\parallel(0)$ is the Ginzburg-Landau coherence length at zero temperature. Hereafter, we will use this value for $\xi_{0\parallel}$, whereas this gives the minimum estimate of the elementary pinning force.

J_c in single-crystalline thin film

Here, we treat the case where a magnetic field is applied normal to a single-crystalline thin film with its c-axis normal to a substrate. In this case, the twinning planes are parallel to the fluxoids and Eq.(1) can be used for the elementary pinning force. It is empirically known for ordinary high-field superconductors with planar defects that the critical current density obeys a linear summation of the elementary pinning forces of the defect-fluxoid interaction except in the high-field region:

$$J_c = \zeta \hat{f}_p / Ba_f d, \tag{2}$$

where a_f is a fluxoid spacing, d is a mean pin spacing (a mean spacing of the twinning planes) and ζ is a pinning efficiency smaller than unity. A value of ζ has not yet been derived theoretically for planar defects, but it is empirically known that we can assume $A\zeta \sim 0.2$ from experimental results on ordinary high-field superconductors [10]. Since the twinning planes are not always oriented optimum for the pinning with respect to the direction of the Lorentz force, a factor of $2/\pi$ should be multiplied to the above equation so as to get an expected value of J_c. Thus, we obtain the critical current density in a single-crystalline thin film:

$$J_c = [K/(BB_{c2\perp})^{1/2}][1-(B/B_{c2\perp})] \equiv J_{cs}, \tag{3}$$

$$K = 1.35(\sqrt{3}/\pi^3)^{1/2}(A\zeta B_c^2/\mu_0 d)[1-(T/T_c)^2]^{1/2}. \tag{4}$$

J_c in polycrystalline textured material

Here, we estimate the critical current density in a polycrystalline bulk material with an oriented texture. In most cases, the c-axes of the grains are aligned but the other axes are directed randomly. We assume the case where the magnetic field is applied parallel to the c-axes for such a textured polycrystal. A mean value of the critical current density in the grains is given by Eq.(3).

In polycrystalline materials, grain boundaries depress the transport current. This effect was clarified by Dimos et all.[11] for bicrystalline thin films with a grain boundary parallel to common c-axes. They found that the critical current density j_b depends strongly on the misorientation angle θ of the a-axes between two grains at B=0 as shown by dots in Fig.1. It is to be noted that this observed critical current density is an ideal pin-free characteristic of a weak link and reduces to almost zero at B=2mT [12]. However, the critical current density in textured materials has still a fairly large value even at high fields. This is considered to be caused by the flux pinning at the grain boundaries. We note that, if the pinning potential in the weak-link region prevents the fluxoids from flowing by the Lorentz force, the flow voltage does not appear, resulting in a finite critical current density even at high fields. This critical current density is determined by a depth of the pinning potential, and hence by the condensation energy density in the weak-link region. Recently, Pande [13] found that the superconducting

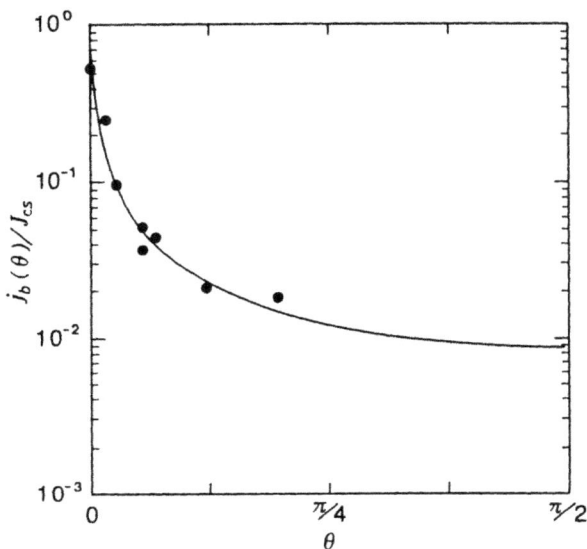

Figure 1. The ratio of the critical current density in grain
boundaries to that in grains versus the misorientation angle
θ in the basal plane [11]. The data for specimens with poor
quality are omitted. The line represents Eq.(5).

order parameter $|\Psi|^2$ is seriously degradated by stresses due to dislocations
at the grain boundary and tried to explain the dependence of the grain
boundary critical current density on the misorientation angle shown in Fig.1.
If this is correct, it is expected that the angular dependence in Fig.1 holds
even for the pinning critical current density at high fields, since the
condensation energy density is proportional to $|\Psi|^2$ at high temperatures. In
the following, we estimate the critical current density based on this
assumption. Normal phases and porosities in the grain boundaries and
unevenness of the boundaries will act as such pinning centers. The observed
angular dependence of the grain boundary critical current density is simply
approximated by

$$j_b(\theta) = J_{cs} \sin\delta/\sin(\theta+\delta),　\qquad (5)$$

as shown by the solid line in Fig.1, where δ is chosen to be 0.87×10^{-2} so as
to get a good fit.
     In polycrystalline materials, the grain boundaries with the current
capacity given by Eq.(5) are expected to distribute randomly in space. That
is, the misorientation angle θ between adjacent two grains will distribute
uniformly between 0 and $\pi/2$. The critical current density in bulk materials
will be approximately obtained by a statistical calculation. For this
purpose, we approximate the critical current density in the present system by
that in an electrical network composed of branches with the same distribution
of the current capacity. We denote a fraction of superconducting grains by
P_m. We define the cumulative distribution function of the current capacity of
the branches $P(J)$, i.e., the fraction of the branches with the current
capacity larger than J:

$$(6)$$

with

$$P(J) = \frac{2}{\pi} P_m \int_0^{\theta_J} d\theta = \frac{2}{\pi} P_m \theta_J, \qquad (7)$$

where we have assumed that the superconducting grains have the same super-conducting properties such as T_c and B_c. Figure 2 represents a plot of $P(J)$ versus J/J_{cs} for various values of P_m. The transport characteristic is seriously affected by the percolation threshold P_c. If Z denotes the number of the branches from one grain in the network, the percolation threshold is given by $P_c = 2/Z$ [14]. For columnar grains with aligned c-axes, it is reasonable to take $Z=6$ and $P_c = 1/3$. The procedure to approximately calculate the critical current density for the system with current capacity distribution is discussed in detail in Ref. 15. According to this procedure, we have

$$J_c = \frac{1}{1-P_c} \int_{P_c}^{P_m} \frac{dP}{P} \int_0^P J(P')dP', \qquad (8)$$

where $J(P)$ is an inverse function of $P(J)$. After a simple calculation, we obtain

$$J_c = \frac{2P_m J_{cs} \sin\delta}{\pi(1-P_c)} \int_{\theta_c}^{\pi/2} \frac{d\theta_J}{\theta_J} \times$$

$$\log\left[\tan\left(\frac{\theta_J + \delta}{2}\right) \Big/ \tan\frac{\delta}{2}\right] \equiv J_{ct}, \qquad (9)$$

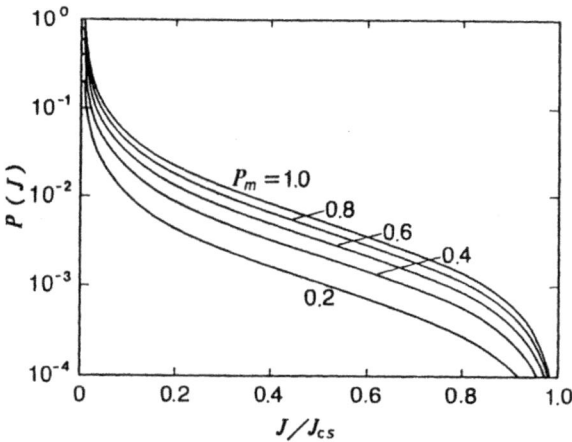

Figure 2. The cumulative distribution function $P(J)$ versus J/J_{cs} for various values of the fraction of superconducting grains P_m.

with

$$\theta_c = \pi P_c / 2P_m. \tag{10}$$

J_c in polycrystalline nontextured material

Main different factors to be taken into account in this case from that for textured materials are:

(i) the superconducting parameters such as B_{c2} are anisotropic with respect to the direction of the magnetic field due to the anisotropy in the coherence length,

(ii) the local critical current density inside grains have also a large anisotropy due to an anisotropic structure of the twinning planes [15], and

(iii) the c-axes are randomly oriented and the depression of the transport current by the grain boundaries will be different from that for textured materials.

The above factors (i) and (ii) associated with the relativity of the direction between the crystalline axis and the magnetic field can be taken into account by repeating the treatment given in Ref.15. However, this treatment is too complicated and such an exact calculation is of no importance because of the ambiguity given in (iii). For simplicity, therefore, we assume that each grain has the same critical current density, which is reasonably given by a mean value, i.e., the intragrain current density [15]:

$$J_g = \frac{K}{4\pi B^{\frac{1}{2}}} \iint_\Omega B_{c2}^{-\frac{1}{2}}(\eta) \left[1 - \frac{B}{B_{c2}(\eta)} \right] \sin\eta \, d\eta \, d\phi, \tag{11}$$

where η is an angle between the magnetic field and the c-axis of the grain and the angular dependence of the upper critical field is given by

$$B_{c2}(\eta) = B_{c2\perp}(\cos^2\eta + \epsilon^2 \sin^2\eta)^{-1/2}. \tag{12}$$

In the above

$$\epsilon = B_{c2\perp} / B_{c2\parallel}, \tag{13}$$

with $B_{c2\parallel}$ denoting the upper critical field in the direction parallel to the basal plane, and ϕ is an angle of the projection of the magnetic field $\vec{B}$ on the basal plane measured from the direction normal to the twinning plane. If the angle of the magnetic field from the twinning plane is represented by ψ, we have

$$\sin\psi = \sin\eta \cos\phi. \tag{14}$$

The area of integration Ω in Eq.(11) is

$$0 \leq |\psi| \leq \Delta\psi, \quad 0 \leq \phi < 2\pi, \tag{15}$$

where $\Delta\psi$ is a width of the angle of effective flux pinning by the twinning planes, which is probably caused by a curvature of the fluxoids and by a deviation of the planar defect from a flat plane.

The depression of the transport current by the grain boundary in randomly textured materials is not clear now. Here, we assume for simplicity that a degree of the depression is similar to the case of textured materials. That is, we use Eq.(5) again. In this case, θ is not the misorientation angle but a parameter representing a distribution of a value of the grain boundary critical current density and J_g is substituted into J_{cs} in Eq.(5). Thus, the critical current density in randomly textured materials J_{cr} can be estimated from Eq.(9), where we take Z=12 and P_c=1/6 for spherical grains.

Results and discussion

We have estimated the critical current density in a single-crystalline thin film, those in bulk materials with oriented and random textures, and the intragrain current density in a bulk material. The obtained results for YBaCuO-type materials at T=77 K are shown in Fig.3. In the estimate, we have used T_c=90 K, $B_{c2\perp}$=10 T, $B_{c2\parallel}$=37 T[16], B_c=0.22 T[9], d=0.1 μm, δ=0.87×10^{-2}, Aζ=0.2 and Δψ=0.35×10^{-1}(2°). Corresponding experimental results are also shown in the figure for comparison.

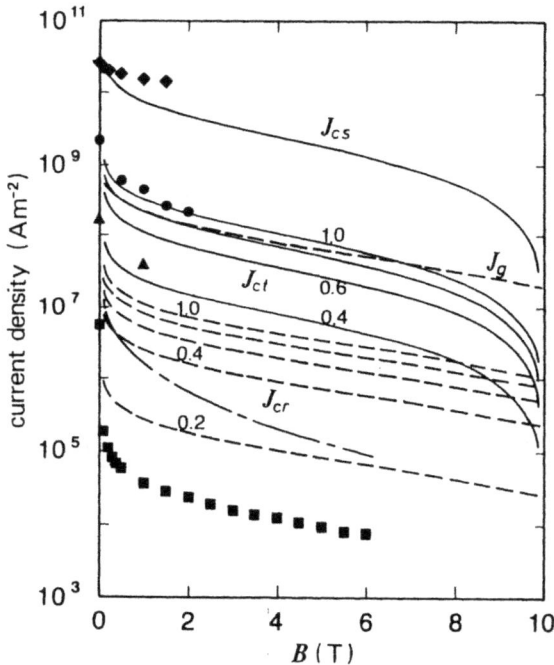

Figure 3. The estimated critical current densities in single-crystalline thin films (J_{cs}), polycrystalline materials with oriented textures (J_{ct}) and sintered materials with random textures (J_{cr}), and the intragrain current density in sintered materials (J_g). The numbers in the figure show the superconducting fraction P_m. The symbols represent the corresponding experimental results of J_c on HoBaCuO thin film (◆)[2], the polycrystal made by the melt-textured growth process (▲)[5], the sintered material (■)[4], and that of J_g on the sintered material (●)[4]. The chained line is the theoretical estimate of J_{cr} for the observed superconducting fraction P_m [17].

As for the critical current density in a single-crystalline thin film and the intragrain current density in a bulk material, the agreement between the present theoretical estimate and the experiments seems to be satisfactory. This suggests that flux pinning mechanism by the twinning planes assumed here works. It is commonly pointed out that point defects may work as effective pinning centers, since the coherence length of this material is very short. In fact, Kupfer et al.[18] showed that the intragrain current density was increased by a factor 10 due to nucleation of point defects by neutron-irradiation. However, the maximum value attained by the point defects is still much smaller than the critical current density in a single-crystalline thin film. The pinning interaction volume of the point defect of the order of ξ^3 is very small and the resultant critical current density is not expected to be large enough. Only the twinning plane is the candidate for practical pinning centers which explains both of the critical current density in a single-crystalline thin film and the intragrain current density in a bulk material. The reason why the intragrain current density in sintered materials is not large is considered that the twinning planes are not oriented in the optimum geometry for the flux pinning.

The superconducting fraction P_m in the specimen made by the melt process [5] is about 0.7 and hence, the measured critical current density is slightly smaller than the present estimate. The critical current density estimated from the dc magnetization measurement on a similar specimen amounts to 1.0×10^8 A/m^2 at B=1 T[19], and the agreement with the present estimate is better.

The critical current density in sintered materials is much lower than the present estimate. The chained line in Fig.3 represents an expected value of the critical current density for the superconducting fraction observed inductively [17]. Experimental results are much smaller even than this value. This implies that the depression of the transport current by the grain boundaries for random textures is much more significant than that for oriented textures. For a grain boundary inclined from the c-axis, the coherence length ξ in the direction of the transport current across the boundary is shorter than ξ in the basal plane for oriented materials. Hence, it is reasonable that the proximity is degradated and the resulting grain boundary critical current density takes a small value.

Here, we briefly discuss the possibility to improve the critical current characteristics of oxide superconductors. The main point is to eliminate or reduce weak-link parts. However, it is difficult to produce single-crystalline long wires for large superconducting magnets. Hence, the weak link will not be able to completely eliminated in polycrystalline wires. However, the experimental result shown in Fig.1 suggests that, if not only the c-axes but also the a-axes are aligned, the weak link can be largely reduced even in polycrystals. In this case it is hopeful to attain the critical current density close to that in single-crystalline thin films. It is necessary to develop a technique to produce materials with highly oriented textures. The next point is to introduce pinning centers stronger than the twinning planes. Pinning potentials of the twinning planes are not deep enough. The idea to use narrow cracks for stronger pinning centers is introduced in Ref.20. It will also be effective to introduce fine precipitates of 211 phase [19]. If these strong pinning centers are successfully introduced, the critical current density higher than that in present single-crystalline thin films will be achieved.

References

1) Y. Enomoto, T. Murakami, M. Suzuki, and K. Moriwaki, "Large Anisotropic Superconducting Critical Current in Epitaxially Grown Ba$_2$YCu$_3$O$_{7-y}$ Thin Film," Jpn. J. Appl. Phys., 26, L1248 (1987).
2) S. Tanaka and H. Itozaki, "High-J$_c$ Superconducting Single Crystalline HoBaCuO Thin Film by Sputtering," Jpn. J. Appl. Phys., 27, L622 (1988).
3) H. Kupfer, I. Apfelstedt, W. Schauer, R. Flukiger, R. Meier-Hirmer, and H. Wuhl, "Critical Current and Upper Critical Field of Sintered and Powdered Superconducting YBa$_2$Cu$_3$O$_7$," Z. Phys. B., 69, 159 (1987).

4) B. Ni, T. Munakata, T. Matsushita, M. Iwakuma, K. Funaki, M. Takeo, and K. Yamafuji, "Ac Inductive Measurement of Intergrain and Intragrain Currents in High T_c Oxide Superconductors," Jpn. J. Appl. Phys., 27, 1658 (1988).

5) S. Jin, T. H. Tiefel, R. C. Sherwood, M. E. Davis, R. B. van Dover, G. W. Kammlott, R. A. Fastnacht, and H. D. Keith, "High Critical Currents in Y-Ba-Cu-O Superconductors," Appl. Phys. Lett., 52, 2074 (1988).

6) M. Murakami, S. Matsuda, K. Sawano, K. Miyamoto, A. Hayashi, M. Morita, K. Doi, H. Teshima, M. Sugiyama, M. Kimura, M. Fujinami, M. Saga, M. Matsuo, and H. Hamada, "Microstructure and Transport Properties of Oxide Superconductors," to be published in Proc. of 1st Inter. Symp. on Super-conductivity, Nagoya, 1988.

7) A. DasGupta, C. C. Koch, D. M. Kroeger, and Y. T. Chou, "Flux Pinning by Grain Boundaries in Niobium Bicrystals," Philos. Mag. B, 38, 367 (1978).

8) See, for example, G. Zerweck, "On Pinning of Superconducting Flux Lines by Grain Boundaries," J. Low Temp. Phys., 42, 1 (1981).

9) T. Matsushita, M. Iwakuma, Y. Sudo, B. Ni, T. Kisu, K. Funaki, M. Takeo, and K. Yamafuji, "Estimate of Attainable Critical Current Density in Superconducting $YBa_2Cu_3O_{7-\delta}$," Jpn. J. Appl. Phys., 26, L1524 (1987).

10) T. Matsushita, A. Kikitsu, H. Sakata, K. Yamafuji, and M. Nagata, "Elementary Pinning Force of Grain Boundaries in Superconducting V_3Ga Tapes," Jpn. J. Appl. Phys., 25, L792 (1986).

11) D. Dimos, P. Chaudhari, J. Mannhart, and F. K. LeGoues, "Orientation Dependence of Grain-Boundary Critical Currents in $YBa_2Cu_3O_{7-\delta}$ Bicrystals," Phys. Rev. Lett., 61, 219 (1988).

12) J. Mannhart, P. Chaudhari, D. Dimos, C. C. Tsuei, and T. R. McGuire, "Critical Currents in [0 0 1] Grains and across Their Tilt Boundaries in $YBa_2Cu_3O_7$ Films," to be published in Cryogenics.

13) C. S. Pande, H. A. Hoff, A. K. Singh, M. S. Osofsky, M. A. Iman, K. Sadananda and L. E. Richards, "Effects of Alloying on Superconducting Properties of $ErBa_2Cu_3O_7$," to be published in IEEE Trans. Mag.

14) S. Kirkpatrick, "Classical Transport in Disordered Media: Scaling and Effective-Medium Theories," Phys. Rev. Lett., 27, 1722 (1971).

15) T. Matsushita, B. Ni, Y. Sudo, M. Iwakuma, K. Funaki, M. Takeo, and K. Yamafuji, "Critical Transport Current Density in Sintered Oxide Super-conductors with High Critical Temperature," Jpn. J. Appl. Phys., 27, 929 (1988).

16) Y. Iye, T. Tamegai, H. Takeya, and H. Takei, "The Anisotropic Upper Critical Field of Single Crystal $YBa_2Cu_3O_x$," Jpn. J. Appl. Phys., 26, L1057 (1987).

17) B. Ni and T. Matsushita, unpublished. The result of P_m is similar to that given in Ref. 4.

18) H. Küpfer, I. Apfelstedt, R. Flükiger, C. Keller, R. Meier-Hirmer, B. Runtsch, A. Turowski, U. Wiech, and T. Wolf, "Intragrain Junctions in $YBa_2Cu_3O_{7-x}$ Ceramics and Single Crystals," to be published in Cryogenics.

19) M. Murakami, M. Morita, K. Miyamoto, and S. Matsuda, "Fabrication of YBaCuO with High Transport J_c," in this proceedings.

20) T. Matsushita, "Strong Flux Pinning by Cracks in Films of Superconducting Oxide," Jpn. J. Appl. Phys., 27, L1712 (1988).

PREPARATION AND SUPERCONDUCTING PROPERTIES OF RARE-EARTH-FREE SUPERCONDUCTORS

Hiromi MUKAIDA, Yuh MATSUTA, Kazuaki SHIKICHI, Kenichi KAWAGUCHI, Masaaki NEMOTO, and Masao NAKAO

Sanyo Tsukuba Research Center
2-1, Koyadai Tsukuba, Ibaraki 305, Japan

Abstract – Tl-based superconductors, $Tl_2Ca_2Ba_2Cu_3O_{10}$ (2223 phase) and $TlCa_3Ba_2Cu_4O_{11}$ (1324 phase), containing Tl-O bilayers or Tl-O monolayers were prepared. The 2223 phase and the 1324 phase exhibited zero resistivity at 114 K and 112 K, respectively. The superconducting transition temperature of the 1324 phase was comparable to that of the 2223 phase. As concerns the formation process, it was found that the 2223 phase was formed at the initial stage of the formation reaction and then the 1324 phase was formed.

Introduction

Since the discovery of the superconducting transition temperatures (Tc's) above 100 K in the Bi-Ca-Sr-Cu-O system [1] and the Tl-Ca-Ba-Cu-O system [2] were reported, superconducting and structural properties of these systems have been investigated [3]. Three related structures containing either Bi-O or Tl-O bilayers have been reported in these systems. These are represented as $Bi_2Ca_{n-1}Sr_2Cu_nO_{2n+4}$ or $Tl_2Ca_{n-1}Ba_2Cu_nO_{2n+4}$ (n=1, 2, 3). The Tc increases with the number of CuO_2 layers.

In the Tl-Ca-Ba-Cu-O system, some ordered intergrowth structures with different phases such as $Tl_2Ba_2CuO_6$ (2021), $Tl_2CaBa_2Cu_2O_8$ (2122), and $Tl_2Ca_2Ba_2Cu_3O_{10}$ were observed by an electron microscope [4]. In addition, it was reported that Tl-O bilayers were substituted occasionally by Tl-O monolayers.

Very recently, a new class of crystal structure represented as $TlCa_{n-1}Ba_2Cu_nO_{2n+3}$ (n=1, 2, 3, 4) with a Tl-O monolayer in the unit cell has been reported [5,6].

In this paper, we report the preparation and the superconducting properties of the 2223 phase, which contains Tl-O bilayers, and the 1324 phase, which contains Tl-O monolayers.

Experiment

Samples were prepared by the conventional solid state reaction method. Tl-Ca-Ba-Cu-O samples with the nominal composition of $Tl_1Ca_3Ba_1Cu_3O_y$ were prepared from powder reagents of 99.9%-pure Tl_2O_3, 99.99%-pure CaO, 99%-pure BaO_2, and 99.99%-pure CuO. After grinding, the mixture was pressed into pellets and wrapped in gold foil. The wrapped pellets were sintered at 875 ℃ for 5 min and then at 855 ℃ for 3 h (sample

A) or sintered at 875 ℃ for 6 h (sample B) in flowing O_2. The samples were heated or cooled at the rate of 30 ℃/min. The temperature dependence of the resistivity was measured by the standard four-probe technique using silver paste contacts. The temperature was determined by a Au+0.07% Fe-Chromel thermocouple. The applied current density was usually 10 mA/cm². X-ray diffraction measurements were carried out in θ-2θ geometry with filtered Cu-Kα radiation (30 kV - 20 mA). Magnetic susceptibility was determined by a superconducting quantum interference device (SQUID) magnetometer.

Results and Discussion

Figure 1 shows powder X-ray diffraction patterns of sample A (a, c) and sample B (b, d). Almost all of the peaks of the pattern in Fig. 1(a) can be indexed as the 2223 phase using the crystal structure data by Torardi et al. [7]. An X-ray diffraction pattern of sample B is shown in

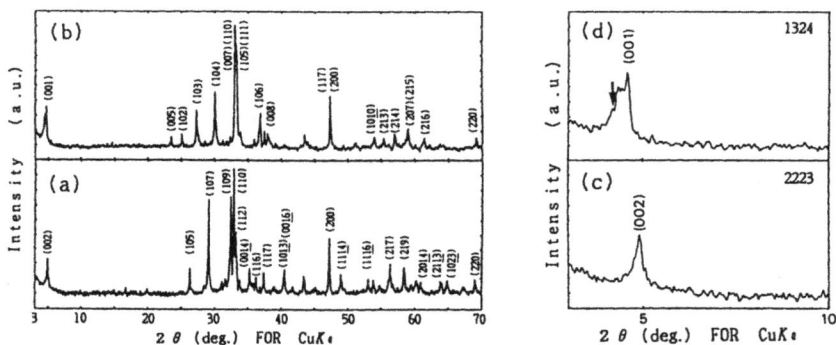

Fig. 1. X-ray diffraction patterns for sample A (a) and sample B (b). Almost all of the peaks in the patterns can be indexed as the 2223 phase (a) or the 1324 phase (b). Peak profiles of the (002) peak for the 2223 phase (c) and the (001) peak for the 1324 phase (d) are shown. A 2θ position of the (002) peak for the 2324 phase is indicated by an arrow in (d).

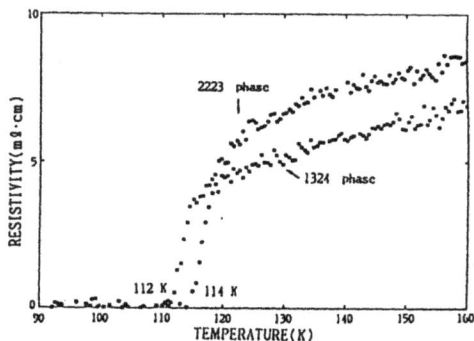

Fig. 2. Temperature dependence of resistivity for the 2223 phase and the 1324 phase. Zero-resistiviy temperature of the 2223 phase is 114 K and that of the 1324 phase is 112 K, respectively.

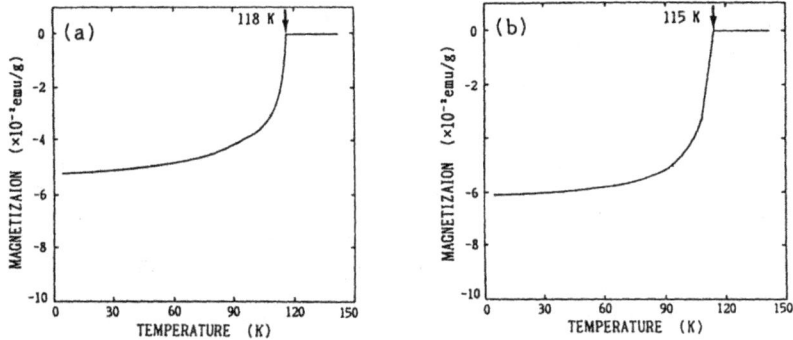

Fig. 3. Temperature dependence of magnetization for the 2223 phase (a) and the 1324 phase (b). The curves were obtained during cooling in a field of 10 Oe.

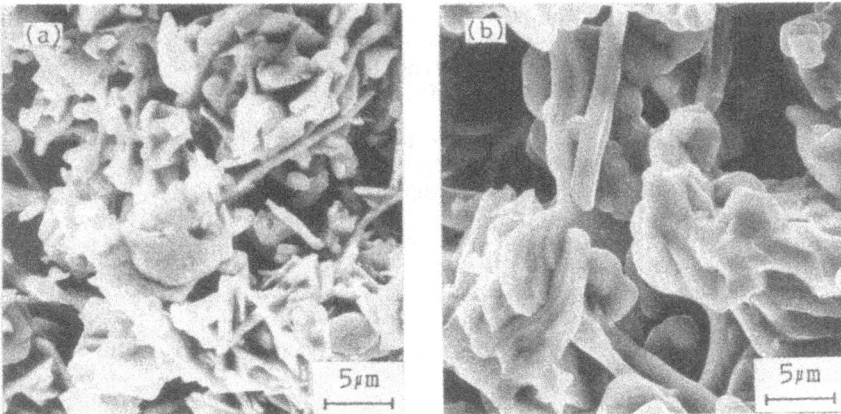

Fig. 4. Scanning electron micrographs of the fraction surfaces of sample A (a) and sample B (b).

Fig. 1(b). In order to index the peaks in this pattern, the intensity calculation was carried out for the structure of the 1324 phase using the structure of $TlCa_2Ba_2Cu_3O_9$ (1223) as a model and making the allowance for additional one CuO_2 layer and one Ca layer. As a result, most of the peaks can be indexed as the 1324 phase with tetragonal cell dimension a=3.849 and c=19.07 Å.

As compared with the composition of the 2223 phase, the nominal composition of the pellets contains excess amounts of Ca and Cu. Thus it is reasonable that the 1324 phase was formed by sintering at 875 ℃ for long duration (6 h). On the other hand, The 2223 phase was formed by sintering at 875 ℃ for short period (5 min). This fact indicates that the 2223 phase was formed at the initial stage of the formation reaction and then the 1324 phase was formed by the addition of Ca and Cu layers.

X-ray diffraction profiles of the (002) peak for the 2223 phase and the (001) peak for the 1324 phase are shown in Figs. 1(c) and 1(d), respectively. The diffraction profile of the (001) peak for the 1324 phase is asymmetrically broadened toward to lower angle (higher d value). This broadening is probably due to the formation of intergrowth and alternating Tl-O monolayer and Tl-O bilayer regions. This alters the c lattice parameter due to varying numer of CuO_2 layers and Tl-O layers. The d value of the (002) peak for $Tl_2Ca_3Ba_2Cu_4O_{12}$ (2324), which contains Tl-O bilayers, is estimated to be ~21 Å. A 2θ position corresponding to the (002) peak for the 2324 phase is indicated by an arrow in Fig. 1(d). If some ordered structure of the 2324 phase exists, the (002) peak for the 2324 phase appears at this position. In fact, the weak peak intensity is observed at this position. This indicates that Tl-O monolayers are occasionally substituted by Tl-O bilayers.

Figure 2 shows the temperature dependence of the resistivity for both the 2223 phase (sample A) and the 1324 phase (sample B). The 2223 phase exhibits zero resistivity at 114 K. On the other hand, the 1324 phase exhibits zero resistivity at 112 K. Although sample B (1324 phase) contains the intergrowth as mentioned earlier, we consider that the Tc of the 1324 phase is comparable to that of the 2223 phase. Comparing the Tc of the 1223 phase (Tc~110 K) reported by Parkin et al.[5], we conclude that the Tc of $TlCa_{n-1}Ba_2Cu_nO_{2n+3}$ (n=1, 2, 3, 4) increases with increasing the number of the CuO_2 layers.

Figure 3 shows the temperature dependence of the magnetization for the 2223 phase (a) and the 1324 phase (b). Samples were cooled in a field of 10 Oe (Meissner effect). The diamagnetism-onset temperature of the 2223 phase is about 118 K and that of the 1324 phase is about 115 K.

Scanning electron micrographs of the fracture surfaces of sample A and sample B are shown in Figs. 4(a) and 4(b), respectively. Porous characteristics are observed for both samples. Because of the long duration of the sintering time, the grain size of sample B is larger than that of the sample A. This porosity must be improved to obtain high critical current density.

Conclusion

It was found that the 1324 phase with four CuO_2 layers separated by Tl-O monolayers was formed after the formation of the 2223 phase.

The 1324 phase and the 2223 phase exhibited superconductivity at 112 K and 114 K, respectively. The Tc of the 1324 phase was comparable to that of the 2223 phase. As compared with the Tc of the 1223 phase, it was found that the Tc of the 1324 phase increased as the number of Cu-O layers increased.

An X-ray diffraction pattern of the 1324 phase indicated that Tl-O monolayers were occasionally substituted by Tl-O bilayers in this sample.

Acknowledgments

We are indebted to S. Fujiwara and M. Harada for help with collecting the scanning electron micrographs. We would like to thank S. Suzuki and A. Mizukami for encouraging this work.

References

1) H. Maeda, Y. Tanaka, M. Fukutomi, and T. Asano, Jpn. J. Appl. Phys. 27, L209 (1988).
2) Z. Z. Sheng, and A. M. Hermann, Nature 332, 138 (1988).
3) K. Takahashi, M. Nakao, D. R. Dietderich, H. Kumakura, and K. Togano, Jpn. J. Appl. Phys. 27, L1457 (1988).
4) S. Iijima, T. Ichihashi, and Y. Kubo, Jpn. J. Appl. Phys. 27, L817 (1988).
5) S. S. P. Parkin, V. Y. Lee, A. I. Nazzal, R. Savoy, and R. Beyers, Phys. Rev. Lett. 61, 750 (1988).
6) H. Ihara, R. Sugise, M. Hirabayashi, N. Terada, M. Jo, K. Hayashi, A. Negishi, M. Tokumoto, Y. Kimura, and T. Shimomura, Nature 334, 510 (1988).
7) C. C. Torardi, M. A. Subramanian, J. C. Calabrese, J. Gopalakrishnan, K. J. Morrissey, T. R. Askew, R. B. Flippen, U. Chowdhry, and A. W. Sleight, Science 240, 631 (1988).

SOLUTION GROWTH OF OXIDE SUPERCONDUCTOR SINGLE CRYSTALS

Tatsuhiko FUJII and Junji SHIRAFUJI

Department of Electrical Engineering, Faculty of Engineering,
Osaka University, 2-1, Yamadaoka Suita, Osaka, 565, Japan

Abstract - Single crystals of oxide superconductor Bi-Sr-Ca-Cu-O have successfully been grown by a slow-cooling method where a slight excess of CuO is added. The single crystals are found to consist of thin platelets.

Introduction

The discovery of high T_c oxide superconductors [1], especially 110K class Bi-Sr-Ca-Cu-O [2] and 120K class Tl-Ba-Ca-Cu-O [3] systems, has led to intense interest in physics and chemistry of these materials promising for various applications. Sintered polycrystalline samples show a high critical temperature as long as simple electrical measurement is made. However, they are unstable in air and have a low critical current density and critical field because of their porous nature of the structure. Moreover, because of highly anisotropic properties of these oxide superconductors, sintered samples are not suitable for studying physical properties and for improving the critical current density and the critical field. Therefore, it is essential to measure thorough physical properties on single crystalline samples of enough size. We have made an attempt to grow single crystals of Bi-Sr-Ca-Cu-O superconductors from solution by slow-cooling method using several kinds of fluxes, for example, Bi_2O_3, CuO and potassium chloride.

Experimental Procedures and Results

Growth experiments of Bi-Sr-Ca-Cu-O system in the case of Bi_2O_3 and CuO fluxes were performed as follows, the reagent grade Bi_2O_3, $SrCO_3$, $CaCO_3$ and CuO powders were weighed to the composition of Bi:Sr:Ca:Cu=2:2:2:3 and mixed thoroughly. Then an excess amount of CuO or Bi_2O_3+CuO was added as fluxes. Because the composition of Bi:Sr:Ca:Cu=2:2:2:3 corresponds to the 110K class superconductor. The starting compositions of solution growth are shown by a to g in Figure 1. Bi-Sr-Ca-Cu-O powder mixtures were placed in alumina crucibles and covered with an alumina lid. The crucible placed in a vertical resistance furnace, heated to 900-1000℃ at a rate of 200-250℃/h, soaked at this temperature for 4-20 hours in air, and then cooled down at a rate of 2-10℃/h to about 750℃, namely below the melting temperature of the mixture. Then the crucible was furnace-cooled to 100℃, and finally taken out of the furnace. It was found experimentally that large crystals were obtained at slower cooling rates.

For typical example, the raw materials of Bi_2O_3, $SrCO_3$, $CaCO_3$ and CuO powders were mixed in the molar ratio of Bi:Sr:Ca:Cu=1:1:1:3. The mixture was charged in an alumina crucible, and heated to 1000℃ at a rate of 250℃/h,

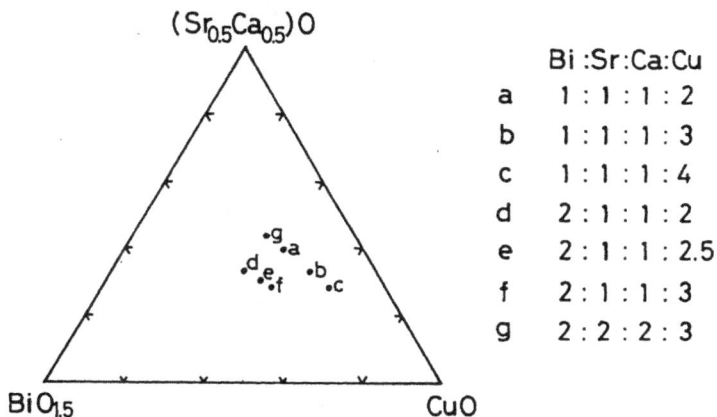

Fig.1 Starting materials of Bi-Sr-Ca-Cu-O system

kept at 1000℃ for 4 hours, cooled to 800℃ at a very slow rate of 2℃/h and cooled rapidly to room temperature at a rate of 200℃/h.

Bi-Sr-Ca-Cu-O crystals were found in the solidified matrix. They were removed mechanically from the matrix. However it was difficult to separate a sizable part of the single crystals from the matrix. Crystal obtained had a metallic luster in color and were thin platelets in shape. An example of the grown crystals is shown in Figure 2.

X-ray Laue pattern of a crystal (the same as in Fig.2) grown in CuO flux is shown in Figure 3. The target was Mo and the accelerating voltage used was 30kV. This Laue pattern shows evidently that the grown crystal is single crystalline with tetragonal structure.

Powder X-ray diffraction measurement was carried out to check the crystal structure and phase in reference to the reports so far [4]. The diffraction pattern using CuKα radiation for the crystal using CuO flux is shown in Figure 4. The accelerating voltage used was 50kV. This pattern indicates that the crystal is composed of $Bi_2Sr_2CaCu_2O_x$ and $Bi_2Sr_2CuO_x$. $Bi_2Sr_2CaCu_2O_x$ is 80K class superconducting phase and $Bi_2Sr_2CuO_x$ is 7K class superconducting or semiconductor phase. 110K class superconducting phase $Bi_2Sr_2Ca_2Cu_3O_x$ dose not exist. The diffraction peaks due to CuO and SrO_2 are also resolved. This is certainly due to a difficulty of separating single crystalline parts from the solidified matrix.

The composition of the crystal was estimated from an electron probe microanalysis (EPMA) or an inductively coupled plasma atomic emission spectroscopy (ICP). EPMA analyses of crystalline materials grown in CuO flux revealed the average composition to be $Bi_{1.0}Sr_{0.8}Ca_{1.2}Cu_{2.9}O_x$. This composition is not close to that of high T_c superconducting phase $Bi_2Sr_2Ca_2Cu_3O_x$, because the sufficient amount of the testing sample of a single phase is hard to be collected from solidified matrix.

CuO and alkaline chloride flux was also tried to grow Bi-Sr-Ca-Cu-O crystals. The powders of the reagent grade Bi_2O_3, $SrCO_3$, $CaCO_3$ and CuO were weighed and mixed in the proper compositions to give the compositions of a,b

64

300 μ m

Fig.2 Photograph of a Bi-Sr-Ca-Cu-O single crystal

Fig.3 X-ray LAUE pattern of Bi-Sr-Ca-Cu-O crystal

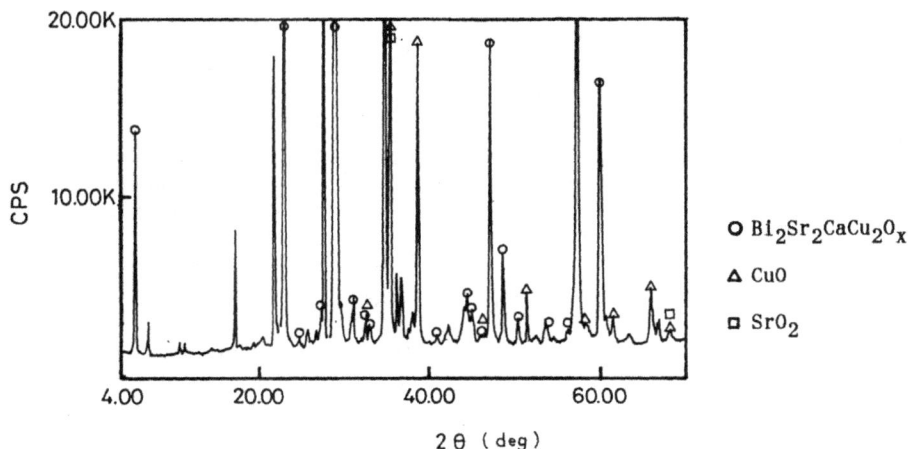

Fig.4 Powder X-ray diffraction pattern of Bi-Sr-Ca-Cu-O system

and g in Fig.1. The powder mixtures were placed in an alumina crucible and
prereacted at about 800 ℃ for several hours in the air. The prereacted
mixtures were ground and mixed again thoroughly with potassium chloride KCl
which acts as a flux. The composition of Bi-Sr-Ca-Cu-O prereacted powder was
20-30wt%. The charge was placed in an alumina crucible and covered with an
alumina lid. The crucible placed in a furnace, heated to 920-960 ℃ at a rate
of 200-250   ℃/h, soaked at this temperature for several hours. Then the
crucible was cooled at a rate of 2-10 ℃/h down to about 700 ℃, namely below
the melting temperature of the salt, followed by furnace cooling to room
temperature.

An example of alkaline chloride flux is shown in more detail in the
following. The raw materials of Bi_2O_3, $SrCO_3$, $CaCO_3$ and CuO powders were
mixed in the molar ratio of Bi:Sr:Ca:Cu=1:1:1:3, followed by firing in air at
800 ℃ for 20 hours. After pulverization of the sintered material, the
powdered Bi-Sr-Ca-Cu-O was mixed with KCl powder at 25wt% of the Bi-Sr-Ca-Cu-
O powder. The charge were placed in a crucible and heated to 950 ℃ at a rate
of 250 ℃/h. After holding at this temperature for 4 hours, it was cooled down
to 750 ℃ at a rate of 2 ℃/h and then rapidly to room temperature at a rate of
200 ℃.

The crystals were embedded on the top of the solidified matrix. They
were separated mechanically from the solidified matrix, and the adhered salt
was easily washed out from the crystals, because potassium chloride (KCl)
dissolves into water very well. The crystals were found in a form of thin
platelet up to 3mm edge or needle up to 7mm in length. It was much easy to
obtain sizable single crystals from the KCl matrix than the case when
Bi_2O_3+CuO flux was used.

ICP analysis of the crystals obtained by KCl flux method showed
unfortunately that most of large size crystals were CuO.

At elevated temperatures as high as 1000 ℃ Bi-Sr-Ca-Cu-O presintered
compound is possibly decomposed into simple binary compounds. More elaborate
experiments are necessary to confirm the advantage of alkaline halide flux
method.

Summary

The solution growth of oxide superconductor Bi-Sr-Ca-Cu-O single crystals in CuO flux has been studied. Although CuO flux method has a problem that grown crystals are embedded in the solidified matrix and are difficult to be separated in a complete form, small size single crystals are obtained. On the other hand, KCl flux seems to give an advantage to remove crystals from the matrix by washing out the attached matrix in water. However, the attempt made to date are not successful. CuO crystals are decomposed from Bi-Sr-Ca-Cu-O charged material.

Acknowledgment

The authors are much indebted to T.Miyatake of Kobe Steel, Ltd. for X-ray diffraction, EPMA and ICP analyses.

Reference

1) J.G.Bednorz and K.A.Müller: Z. Phys. B 64, 189 (1986)
2) H.Maeda, Y.Tanaka, M.Fukutomi, and T.Asano: Jpn. J. Appl. Phys. 27 (1988)
3) Z.Z.Sheng, A.M.Herman, A.El.Ali, C.Almasan, J.Estrada, T.Datta and R.J.Matson: Phys. Rev. Lett. 60, 937 (1988)
4) R.M.Hazen, C.T.Prewitt, R.J.Angel, N.L.Ross, L.W.Finger, C.G.Hadidiacos, D.R.Veblen, P.J.Heaney, P.H.Hor, R.L.Meng, Y.Y.Sun, Y.Q.Wang, Y.Y.Xue, Z.J.Huang, L.Gao, J.Bechtold and C.W.Chu: Phys. Rev. Lett. 60, 1174 (1988)

PROCESSING CONDITION AND TRANSPORT PROPERTIES OF PbBiSrCaCuO COMPOUND

Masato Murakami, Akihiko Hayashi, Masao Kimura, Makoto Saga, and Kenji Doi

R & D Laboratories-I, Nippon Steel Corporation
1618 Ida, Nakahara-ku, Kawasaki 211 Japan

ABSTRACT   High Tc phase in BiSrCaCuO system can be stabilized by Pb addition leading to the improvement in both transport and magnetic critical current density(Jc). However, Jc of Bi system is sensitive to magnetic field compared to Y system presumably due to the absence of twin structure. High Tc phase seems to be formed by the reaction between low Tc phase and $(Sr,Ca)_3Cu_5O_8$. The beneficial effects of Pb addition seem to be the stabilization of high Tc phase.

INTRODUCTION

After the discovery of BiSrCaCuO with Tc exceeding 100K by Maeda et al[1], extensive study has been carried out to enhance critical current density(Jc) of this compound, since Jc is the most crucial property for practical applications. Although some progress has been made[2,3], Jc of Bi compound still remains in lower level than that of Y based oxide superconductors.   This is attributable to its multiphased microstructure[4]. There exists three different superconducting phases in Bi system and their chemical compositions can be described by the following formula $Bi_2Sr_2Ca_{n-1}Cu_nO_x$ (n=1,2,3).   Their Tc's increase with the increase in n and typically 20K, 80K, and 110K, respectively. As expected from chemical formula these three phases have similar structures such that higher Tc phase can be constructed by inserting extra Ca and Cu-O planes into the lower Tc phase. Consequently different superconducting phases tend to coexist in most Bi compounds leading to multiphase structure. It is also notable that a single phase sample cannot be synthesized by starting the stoichiometric composition. Since   Jc is dominated by how the superconducting phases are distributed and depends on the connectivity of 110K superconducting phases, microstructural control is critical to improve Jc.
It has been found that 110K phase preferentially grows at around 880°C in air[1] and long annealing is effective to raise zero resistance temperature above 100K[4].   We have found[5] that additional annealing at 850 °C can promote the formation of high Tc phase. However, the sample was still multiphase and Jc was only 10 A/cm^2 at 77K in zero magnetic field.
Recently Takano et. al[6] found that high Tc phase can be stabilized by substituting a small amount of Bi by Pb, although its effect has not been clarified. In this paper we report our preliminary results on microstructural study and transport measurements of Pb added $BiSrCaCu_2O_y$ compound.

EXPERIMENTAL

Appropriate amounts of mixture of PbO, Bi_2O_3, $SrCO_3$, $CaCO_3$, and CuO were thoroughly ground in an agate mortar and pestle and calcined at 800 C for ten hours followed by pelletizing and sintering at 845°C in air.
Electric resistivity was measured by a standard four probe method. Transport critical current densities were obtained from V-I characteristic curves using 1 μV/cm criterion.   Magnetization measurements were also conducted to estimate intragrain Jc.

Microstructure was observed with a polarized optical microscope and a transmission electron microscope. X-ray diffraction was used to investigate the crystal structure. Chemical compositions of the phases were identified by EDX analysis. Compositional mapping of the sample was performed with a computer assisted electron probe microanalyzer.

RESULTS AND DISCUSSION

Superconducting properties of PbBiSrCaCuO

Figure 1 shows the effect of Pb content on the resistivity versus temperature relationships for $BiSrCaCu_2O_y$ sintered at 845°C for 128 hours. A small amount of Pb addition is effective to improve zero resistance temperature. But an excess addition deteriorates the superconductivity. In the present study, $Pb_{0.2}BiSrCaCu_2O_y$ yielded the best results. Then hereafter we report the results of the sample with this composition.

Figure 2 presents the effects of sintering time on temperature dependence of resistivity for $Pb_{0.2}BiSrCaCu_2O_y$ sintered at 845°C for various periods. As sintering time is increased, zero resistance temperature is raised and reaches 107K with 64 hour sintering. Then, both resistivity and zero resistance temperature deteriorated as sintering time exceeds 128 hours. Figure 3 shows temperature dependence of ac susceptibility for the same samples. The best result is obtained with the sample sintered for 64 hours, which is consistent with resistivity measurement. The data show sharp drops in ac susceptibility at 110 and 95K. However a sharp drop at 95K disappeared after pulverizing the sample, indicating that the peak was due to the weak-link as reported by Mazaki et. al[7].

Critical current density measurements were performed at 4.2K and 77K. Transport Jc values for the sample sintered at 845°C for 32, 64, and 128 hours were 240, 600, 320 A/cm² at 4.2K and 10, 120, and 30 A/cm² at 77K, respectively. We also conducted magnetization measurement to determine intragrain Jc, since the sample contains a significant amount of weak-links.

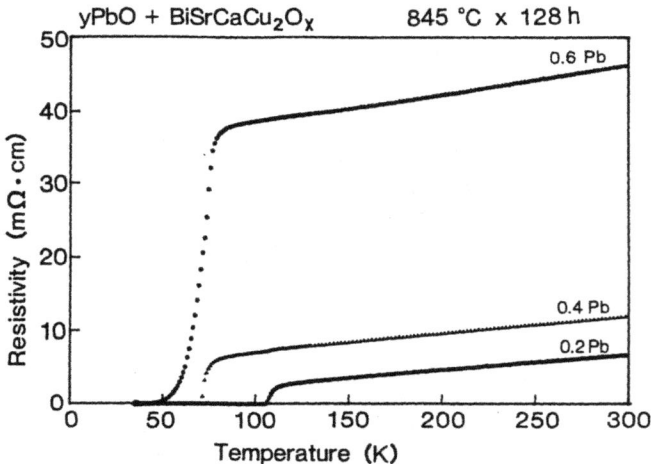

Fig. 1. Effect of Pb content on resistivity versus temperature relationships for PbBiSrCaCuO compound sintered at 845°C for 128 hours.

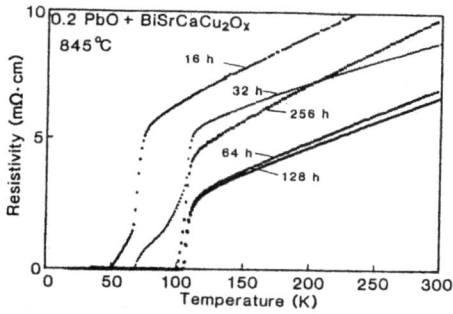

Fig. 2.   Effect of sintering time on resistivity versus temperature for several PbBiSrCaCuO fired at 845 °C.

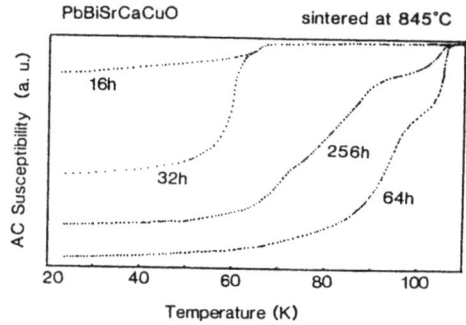

Fig. 3.   Temperature dependence of ac susceptibility for several PbBiSrCaCuO fired at 845 °C.

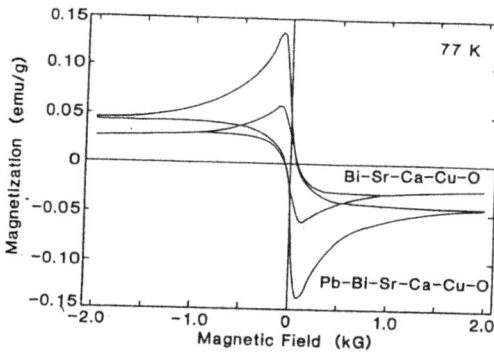

Fig. 4.   Magnetization curves for two Bi compounds.

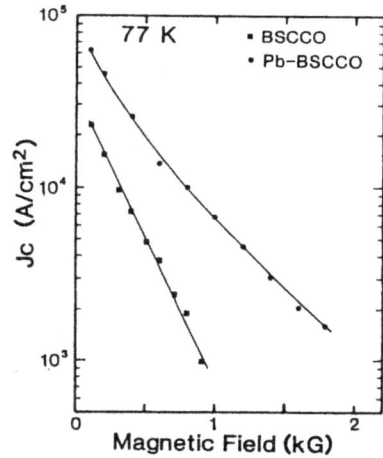

Fig. 5.   Magnetic field dependence of Jc for Bi compounds.

Figure 4 shows magnetization curves for the sample sintered at 845°C for 64 hours. For comparison the data for Pb undoped $BiSrCaCu_2O_y$ with highest Jc is also shown. While no hysteresis is observable in Pb free Bi compound, some hysteresis remained up to 1kG in Pb containing Bi compound, suggesting higher Jc is attainable in magnetic field. If we estimate intragrain Jc from critical state model using the following equation:

$$Jc = 20 \ \Delta M \ / \ d$$

where ΔM is hysteresis between increasing and decreasing field process in emu/cm^3 and d is the sample size in cm. (Here we used grain size as d), magnetic field dependence of Jc for these samples can be estimated as shown in Fig. 5. Compared to Y based oxide superconductors the decrease of Jc in magnetic field is much sharper. This behavior may be understandable by considering the absence of twin structure in Bi compound which is believed to provide pinning centers for Y based oxide superconductors[8].

Microstructure

Figure 6 shows scanning electron micrographs for $Pb_{0.2}BiSrCaCu_2O_y$ at 845 °C for 32 and 64 hours. X-ray diffraction patterns for these two samples are also shown in Fig. 7. The sample sintered for 32 hours are mainly composed of low Tc phase, while volume of high Tc phase increases in 64h sample. When we observe microstructure, it is clear that needle shaped phases keep growing as sintering proceeds. From X-ray diffraction results and the fact that the superconducting properties are improved dramatically as sintering time is increased from 32 to 64 hours, these needle shaped phases are considered to be high Tc phases. However, according to EDX analysis chemical compositions of these needle like phases fluctuate from place to place. Typical results are presented in Table 1. Since 110K phase has chemical composition with 2:2:2:3 cation ratio, these needle shaped phases are considered not to be a single 110K phase but the admixture of 80K and 110K phases. This was confirmed by transmission electron microscopic observation. Figure 8 shows the electron diffraction patterns for needle shaped region. Two types of spots corresponding to 80K and 110K phases are superimposed in this figure indicating the coexistence of these two phases. Therefore it is concluded that the growth of 110K phase alone is unlikely to take place in our sample. It is also notable that Pb is detected in every spot, which suggests the superconducting phase contains Pb.

Since 32h sample composed of mainly low Tc phase, high Tc phase is considered to grow from low Tc phase. According to EDX analysis, 32h sample comprised a mixture of low Tc phase, $(Sr,Ca)_3Cu_5O_8$, and CuO. Average chemcial composition of low Tc phase was $Pb_{0.4}Bi_{2.0}Sr_{1.4}Ca_{1.8}Cu_{2.2}$ and compositional mapping indicates that the phases are not homogeneous. Since $(Sr,Ca)_3Cu_5O_8$ phase disappears completely in 64h sample, we may conclude that high Tc phase is formed by the reaction between low Tc phase and $(Sr,Ca)_3Cu_5O_8$. But when we carefully examine compositional mapping and distribution of element in the 32h sample, we can find the phase consisting of Ca and Pb, although X-ray diffraction analysis failed to detect the presence of Ca_2PbO_4. These phases also decreased as sintering proceeded. Hence these phases may also help the growth of high Tc phase.

As for the beneficial effect of Pb addition, a definite answer is not yet given. It has been reported[9] that Pb containing phase with low melting point acts as flux and promote the growth of high Tc phase, which is somewhat different from our observation. And it seems probable that the growing mechanism for high Tc phase is not one fold and depends strongly on the initial condition.

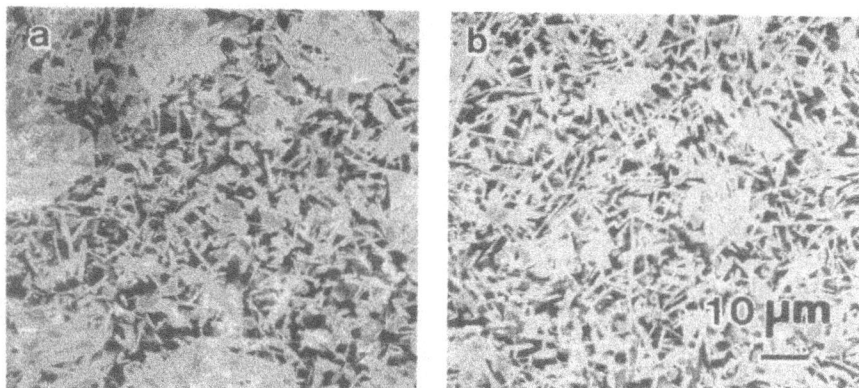

Fig. 6.   Scanning electron micrographs for $Pb_{0.2}BiSrCaCu_2O_y$ sintered at 845°C for (a) 32 hours and (b) 64 hours.

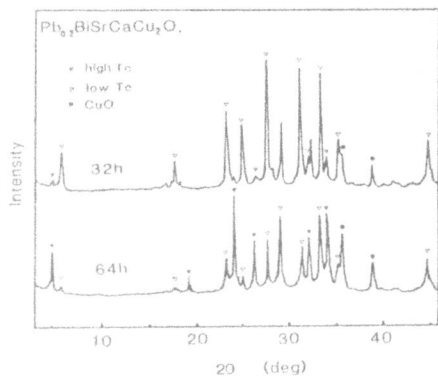

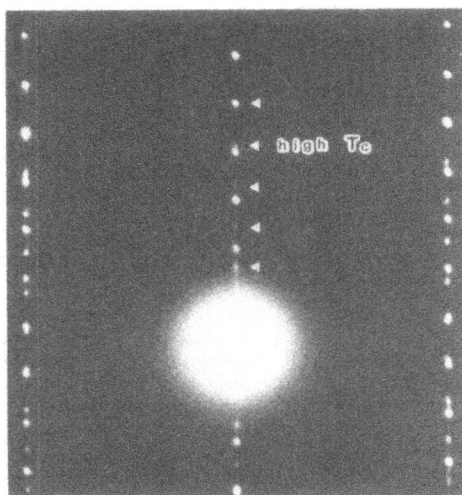

Fig. 7. X-ray diffraction patterns for PbSrCaCuO sintered at 845°C for 32 and 64 hours.

Fig. 8.   Electron diffraction patterns for needle shaped phase. Note low Tc and high Tc phase coexist.

Table 1.   Results of compositional analyses for needle shaped region.

Pb	Bi	Sr	Ca	Cu	Pb	Bi	Sr	Ca	Cu
0.4	2.0	1.3	2.1	3.3	0.4	2.0	1.4	1.8	2.6
0.2	2.1	1.3	2.0	3.4	0.4	2.0	1.3	1.8	2.4
0.4	2.0	1.3	1.9	2.5	0.4	2.0	1.3	1.7	3.1
0.4	2.0	1.2	1.9	4.5	0.4	2.0	1.5	2.0	2.7
0.4	2.0	1.5	2.4	3.0	0.3	2.0	1.3	1.5	2.1
0.4	2.0	1.5	2.1	3.3	0.4	2.0	1.2	1.4	4.0

Here we suppose the possibility that Pb itself helps to lower the free energy of high Tc phase structure. It has been reported by many groups[10] that high Tc phase is not stable and starts to decompose when sintering is continued after the sample reached the optimum state. In fact, high Tc phase decomposed with sintering for 256 hours in the present investigation. It is also found that Pb defuses out of the sample on sintering. Therefore the result can be understood by supposing that a decrease in Pb content caused the reduction of stability and thereby resulted in decomposition of high Tc phase.

CONCLUSION

Pb addition is effective to raise zero resistance temperature and Jc of Bi compound through increasing the volume fraction of high Tc phase. However, magnetic measurement suggests that the deterioration of Jc in magnetic field is worse than that of Y based superconductors, which may be attributable to the absence of dense twin structure in Bi compound.

As for the beneficial effect of Pb, the answer is not clear. But we suppose Pb may stabilize the structure of high Tc phase.

REFERENCES

[1] H. Maeda, Y. Tanaka, M. Fukutomi, and T. Asano, Jpn. J. Appl. Phys., 26, L209, (1988).
[2] M. Murakami, M. Morita, H. Teshima, K. Doi and S. Matsuda, "Microstructure and critical current density of oxide superconductors", Physica C, 153-155, 994-995 (1988).
[3] T. Asano, Y. Tanaka, M. Fukutomi, K. Jikihara, J. Machida and H. Maeda, Jpn. J. Appl. Phys., 27, L1652 (1988).
[4] K. Togano, H. Kuamakura, H. Maeda, K. Takahashi and M. Nakao, Jpn. J. Appl. Phys., 27, L323 (1988).
[5] A. Hayashi, M. Murakami, M. Morita, H. Teshima, K. doi, K. Sawano, M. Sugiyama, H. Hamada and S. Matsuda, "Microstructure and superconducting properties of BiSrCaCuO compound", Proc. 1st ISS held in Nagoya (1988).
[6] M. Takano, J. Takada, K. Oda, H. Kitaguchi, Y. Miura, Y. Ikeda, Y. Tomii and H. Mazaki, Jpn. J. Appl. Phys., 27, L1041 (1988).
[7] H. Mazaki, M. Takano, J. Takada, K. Oda, H. Kitaguchi, Y. Miura, Y. Ikeda, Y. Tomii and T. Kubozonoe, Jpn. J. Appl. Phys., 27, L1639 (1988).
[8] T. Matsushita, B. Ni, Y. Sudo, M. Iwakuma, K. Funaki, M. Takeo and K. Yamafuji, Jpn. J. Appl. Phys., 27, 929-936 (1988).
[9] T. Hatano et. al, Jpn. J. Appl. Phys., 27 (1988) to be published.
[10] e. g. H. Nobumasa, K. Shimizu, Y. Kitano and T. Kawai, Jpn. J. Appl. Phys., 27, L1669 (1988).

SUPERCONDUCTIVITY OF $YBa_2Cu_3O_{7-x}$-METAL COMPOSITES

Nobuhito IMANAKA, Fumihiko SAITO, Hisao IMAI, and Gin-ya ADACHI

Department of Applied Chemistry, Faculty of Engineering, Osaka University
Yamadaoka 2-1, Suita, Osaka, 565 Japan

Abstract - Metal powder such as gold or silver was mixed into the Y-Ba-Cu-O system for the purpose of suppressing non-superconducting phase formation in the grain boundary and for filling pores in the $YBa_2Cu_3O_{7-x}$ sintered bulk. Critical current densities(Jc) appreciably increased by the metal mixing. Especially, 5wt% Au mixing, increased Jc about 4 times larger than that for $YBa_2Cu_3O_{7-x}$ without Au. Tc^{zero} decrease with applying a magnetic field was also suppressed by the metal mixing.

Introduction

Since Bednorz and Müller[1] discovered La-Ba-Cu-O superconductor which shows onset critical temperature(Tc^{on}) at near 30K, many researchers in the world have concentrated on finding other oxides with higher Tc. Chu et al.[2] reported that Tc^{on} near 96K was obtained with the Y-Ba-Cu-O system, which is about 20K higher than the liquid nitrogen temperature. This Tc is greatly enhanced in comparison with the conventional metal alloy superconductor such as Nb-Ti and Nb_3Sn. However, the critical current density (Jc) and critical magnetic field(Hc) are much lower than those for the practical superconductors. These are supposed to result mainly from the fact that the secondary phase which is not a superconductor at all formed in the grain boundary and pores existed between the grains[3].

In this study, several metal powders were mixed into the Y-Ba-Cu-O system so as to prevent the non-superconducting phase from forming and to fill the pores. The superconducting properties, especially, critical current density and magnetic field influence to the Tc^{zero} of $YBa_2Cu_3O_{7-x}$-Metal composites, were compared with those for Y-Ba-Cu-O sample without metal mixing.

Experimental

$YBa_2Cu_3O_{7-x}$-Au Composites

A mixture of Y_2O_3, $BaCO_3$ and CuO(molar ratio 0.5:2:3) was calcined at 973K for 2h in an oxygen atmosphere and then heated at 1213K for 5h in the same atmosphere. The resultant $YBa_2Cu_3O_{7-x}$ powder was mixed with an appropriate amount of Au powder(purity >99.9%, grain size<150um) and then pelletized. The $YBa_2Cu_3O_{7-x}$-Au pellet was sintered at 1213K for 5h in the oxygen flowing atmosphere.

$YBa_2Cu_3O_{7-x}$-Ag Composites

The powder of Y_2O_3, $BaCO_3$, and CuO(molar ratio 0.5:2:3) was mixed with Ag powder (purity>99.9%, grain size<44um). The mixture was calcined at 973K for 2h in the oxygen flowing atmosphere and then heated at 1173K for 5h in the same flowing. The $YBa_2Cu_3O_{7-x}$-Ag preheated sample was pulverized and made into pellets. The pellets were sintered at 1173K for 5h in the oxygen flowing atmosphere.

Measurements

Electrical conductivity was measured by the four probe method. Silver paste was used so as to obtain a good contact between lead wires and the sample. The temperature was measured by Au+0.07at%Fe-Chromel thermocouple.

Results and Discussion

$YBa_2Cu_3O_{7-x}$-Au Composites

Tc^{zero} deviation with the mixed Au wt% is plotted in Fig. 1. The $YBa_2Cu_3O_{7-x}$ sample without Au($YBa_2Cu_3O_{7-x}$ standard sample) showed Tc^{zero} at 89.5K. With increasing the amount of Au mixed, Tc^{zero} slightly decreased. However, the decrease at Tc^{zero} is approximately 2K even for the 40wt% of Au mixing. In the case that Au was added up to 60wt%, the Tc^{zero} greatly decreased down to 82K. The Au mixing up to 40wt% was found to influence little to the superconducting characteristics of Tc^{zero}.

The Jc deviation with Au wt% is presented in Fig. 2. The Jc for the $YBa_2Cu_3O_{7-x}$ standard sample was 88A/cm². By 3wt% Au mixing, Jc increased up to 166A/cm², and showed maximum Jc of 307A/cm² for Au 5wt% addition. The Jc was about 4 times higher than that for the standard. However, the Jc for Au 8wt% decreased abruptly down to 161A/cm². In the case of Au 60wt%, Jc was 7A/cm², which is considerably lower than that of the $YBa_2Cu_3O_{7-x}$ standard. Since most of the $YBa_2Cu_3O_{7-x}$ particle seems to keep no longer in contact with each other, the Jc decreased.

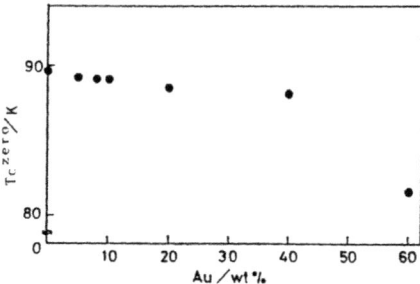

Fig. 1. The variation of Tc^{zero} with the mixed Au wt% for the $YBa_2Cu_3O_{7-x}$-Au composites.

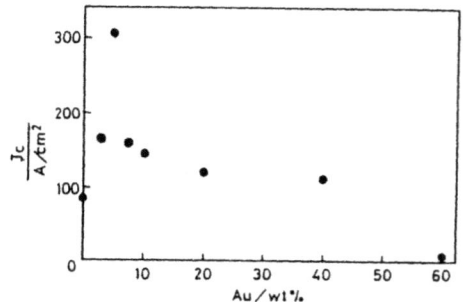

Fig. 2. The variation of Jc with the mixed Au wt% for the YBa_2-Cu_3O_{7-x}-Au composites.

The variation of Jc with the mixed Au wt% for $YBa_2Cu_3O_{7-x}$-Au composites at 77K is shown in Fig. 3. The Jc was obtained from magnetic and resistance measurements, respectively. The Jc from magnetic measurement was approximately twice as large as that from the resistance measurement. Kumakura et al.[4] reported that the Jc from magnetic measurement was more than 10 times higher than that from resistance for the $YBa_2Cu_3O_{7-x}$ sample. Jc from a magnetic measurement includes the eddy current in the inter- and intra-grain, which is not effective current for transport. Therefore, the Au mixed composite was found to decrease the ratio of the eddy current density to the Jc.

Fig. 4 presents the Tc^{zero} deviation with the magnetic flux density for the $YBa_2Cu_3O_{7-x}$ standard and $YBa_2Cu_3O_{7-x}$-Au(5wt%) composite sample. Because this measurement was conducted a few weeks after the sample synthesis, the samples may have been slightly degraded by the moisture in the atmosphere, and Tc^{zero} was about 3K lower than that in Fig. 1. The Tc^{zero} for the standard sample without magnetic field was 86.4K. The Tc^{zero} abruptly decreased by applying a magnetic field. In the case of 500G, the Tc^{zero} decreased down to 81K. The Tc^{zero} decrease was not evident for the magnetic field higher than 500G. The Tc^{zero} became lower than the liquid nitrogen temperature, 77K, when the applied magnetic flux density was greater than 3000G. Tc^{zero} for the $YBa_2Cu_3O_{7-x}$-Au(5wt%) composite also decreased apparently by a magnetic field application. However, the decrease of Tc^{zero} when the magnetic flux density was applied from 0 to 500G was appreciably small in comparison with the case of $YBa_2Cu_3O_{7-x}$ standard sample. Tc^{zero} was still higher than 77K even at the magnetic flux density of 7500G.

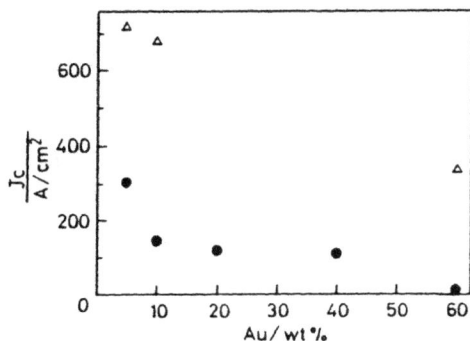

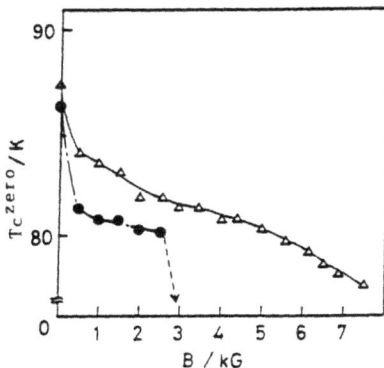

Fig. 3. The variation of Jc with the mixed Au wt% for the YBa_2-Cu_3O_{7-x}-Au composites at 77K.

△Jc from magnetic measurement

●Jc from resistance measurement

Fig. 4. The Tc^{zero} deviation with the magnetic flux density.

●$YBa_2Cu_3O_{7-x}$ standard sample

△$YBa_2Cu_3O_{7-x}$-Au(5wt%) composite

YBa$_2$Cu$_3$O$_{7-x}$-Ag Composites

Fig. 5 presents the Tc^{zero} deviation with the Ag addition. YBa$_2$Cu$_3$O$_{7-x}$ without Ag mixing shows Tc^{zero} at 86K. This Tc^{zero} was about 4K lower than the standard presented in Fig. 1. This is attributed to the fact that the temperature for the heat treatment was 40K lower than that for YBa$_2$Cu$_3$O$_{7-x}$-Au preparation. For the YBa$_2$Cu$_3$O$_{7-x}$-Ag composites, Tc^{zero} maintained almost constant near 90K, which is approximately 4K higher than the standard. Oxygen from the atmosphere easily permeates Ag powder. This oxygen permeation contributed to the YBa$_2$Cu$_3$O$_{7-x}$ superconductor formation.

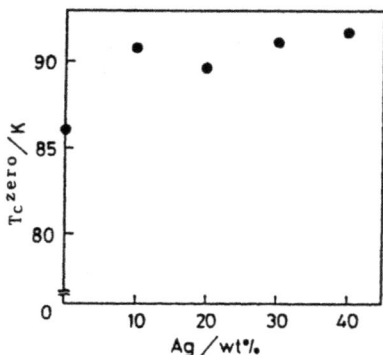

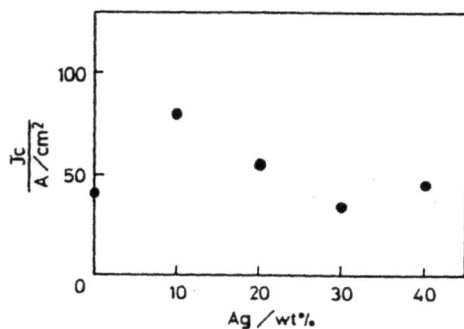

Fig. 5. The variation of Tc^{zero} with the mixed Ag wt% for the YBa$_2$Cu$_3$O$_{7-x}$-Ag composites.

Fig. 6. The variation of Jc with the mixed Ag wt% for the YBa$_2$-Cu$_3$O$_{7-x}$-Ag composites.

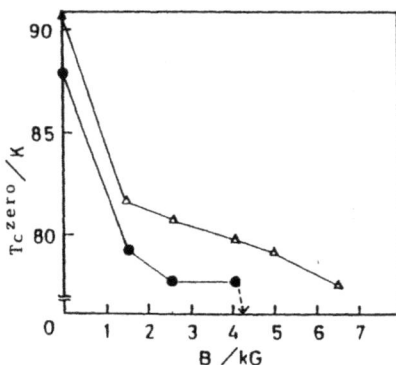

Fig. 7. The Tc^{zero} deviation with the magnetic flux density.

● YBa$_2$Cu$_3$O$_{7-x}$ standard sample

△ YBa$_2$Cu$_3$O$_{7-x}$-Ag(10wt%) composite

The variation of Jc for the $YBa_2Cu_3O_{7-x}$-Ag composites is presented in Fig. 6. The Jc for the $YBa_2Cu_3O_{7-x}$ without Ag was $40A/cm^2$, which was considerably lower than that for the standard in Fig. 2 because of the low heating temperature. An Ag 10wt% added composite shows the highest Jc of $80A/cm^2$ in the $YBa_2Cu_3O_{7-x}$-Ag composites.

Fig. 7 presents the Tc^{zero} variation for the standard sample and $YBa_2Cu_3O_{7-x}$-Ag(10wt%) composite under the magnetic flux density. The $YBa_2Cu_3O_{7-x}$-Ag composite was unstable compared with the $YBa_2Cu_3O_{7-x}$-Au one. The measurements were conducted with freshly prepared rectangular pellets. Tc^{zero} for the standard was 88K, which was about 2K higher than that in Fig. 5. By applying a magnetic field, Tc^{zero} for the $YBa_2Cu_3O_{7-x}$ standard greatly decreased under 80K. Tc^{zero} became lower than 77K when the applied magnetic flux density exceeded 4000G. On the other hand, Tc^{zero} for the $YBa_2Cu_3O_{7-x}$-Ag (10wt%) composite was 91K without a magnetic field. Tc^{zero} decreased the same as the case of the $YBa_2Cu_3O_{7-x}$ standard sample by the magnetic field application. However, the Tc^{zero} still remained >77K even at 6700G.

Conclusion

An addition of metal such as Au or Ag into $YBa_2Cu_3O_{7-x}$ superconductor enhanced the Jc. Especially, Jc for the Au 5wt% mixed composite was approximately 4 times higher in comparison with that for the $YBa_2Cu_3O_{7-x}$ standard sample. The Tc^{zero} decrease under a magnetic field was also considerably suppressed by the metal mixing.

References

1. J. G. Bednorz and K. A. Müller, Z. Phys., B64, 189-193(1986).
2. M. K. Wu, J. R. Ashburn, C. J. Torng, P. H. Hor, R. L. Meng, L. Gao, Z. J. Huang, Y. Q. Wang, and C. W. Chu, Phys. Rev. Lett., 58, 908-910(1987).
3. J. W. Ekin, A. I. Braginski, A. J. Panson, M. A. Janocko, D. W. Capone II, N. J. Zaluzec, B. Flandermeyer, O. F. de Lima, M. Hong, J. Kwo, and S. H. Liou, J. Appl. Phys., 62, 4821-4828(1987).
4. H. Kumakura, M. Uehara, Y. Yoshida, and K. Togano, Phys. Lett. A, 124, 367-369(1987).

PREPARATION OF HIGH-Tc Bi-Sr-Ca-Cu-O SUPERCONDUCTORS

Takahito EMOTO, Junya YAMAMOTO, and Yoshishige MURAKAMI

Laboratory for Applied Superconductivity, Osaka University
2-1, Yamadaoka Suita, Osaka, 565, Japan

Abstract - High-Tc Bi-Sr-Ca-Cu-O superconductors were prepared by the rapid quenching method after the melt-process. The melt-quenched samples of $BiSrCaCu_3O_x$ showed high-Tc transition (Tc(onset)=120K) after annealing at 845-875 °C.

Introduction

The discovery of 90K superconductivity in the Y-Ba-Cu-O system[1] has stimulated the search for higher temperature superconductors. The recent breakthrough discovery of superconductivity above 100K in the Bi-Sr-Ca-Cu-O system[2] and Tl-Ca-Ba-Cu-O system[3,4] generated a great deal of interest similar to that of the Y-Ba-Cu-O ceramics.

It is known that there are at least three distinct phases in this system; $Bi_2(Sr,Ca)_4Cu_3O_x$ with Tc=110K (high-Tc phase), $Bi_2(Sr,Ca)_3Cu_2O_x$ with Tc=80K (low-Tc phase), $Bi_2(Sr,Ca)_2CuO_x$ with Tc=7K (semiconducting phase)[5,6].

The biggest problem concerning the fabrication of this material by the usual solid phase reactions is that the production of the high-Tc phase has always been accompanied by the production of the low-Tc and other additional phases. In order to solve this problem and to increase the high-Tc phase, an addition of excess Ca and Cu[7,8] and prolonged sintering[9] have been attempted. In addition, Pb substitution in this compound and low oxygen pressure treatment during the solid reaction[10] have resulted in the stabilization of the high-Tc phase.

To solve the above problem, we proposed the preparation by melt-process. The increased densification and homogeneity achieved in samples prepared using this technique should increase the transport critical current densities over those attainable in conventionally prepared materials.

As the melting point of the Bi-Sr-Ca-Cu-O system is lower than that of Y-Ba-Cu-O system, preparation by melt-process is profitable for technological applications. For example, various rapid solidification techniques (melt spinning, splat quenching, electron beam and laser annealing, etc.) can be applied for the fabrication of superconducting wires and/or tapes in this system.

In this paper, we report superconducting properties of Bi-Sr-Ca-Cu-O ceramics prepared by the melt-process. For the preliminary studies, we prepared samples by the furnace cooling method and investigated the effects of varying the Cu and Ca content of the starting composition. Furthermore, we prepared the samples by the rapid quenching method and examined the relationships between the sintering temperature and the Tc.

Preparation by Melt-process (furnace cooling method)

Influence of Cu concentration

Samples with the nominal composition of $BiSrCaCu_yO_x$ (y=2.0, 2.5, 3.0,

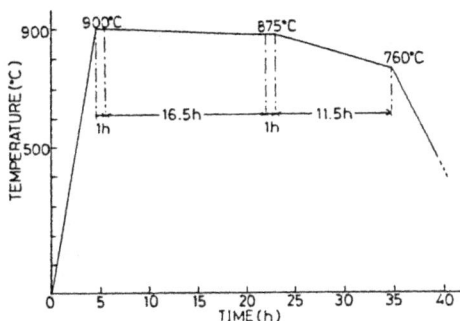

Fig.1. Heat treatment process of preparation by the furnace cooling method

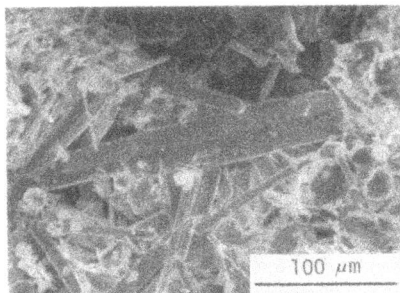

Fig.2. SEM photograph

3.5, 4.0) were prepared from high purity powders of Bi_2O_3, $SrCO_3$, $CaCO_3$ and CuO. These powders were mixed, preheated at 850 °C for 5 hours and subsequently pulverized.

The typical heat treatment was as follows: heating to 900 °C at the rate of 200 °C/h, maintaining at 900 °C for 1 hour, cooling to 875 °C at the rate of 1.5 °C/h, further cooling to 760 °C at the rate of 10 °C/h, and then cooling to room temperature in a furnace in air.(Figure 1)   After melting and resoldification, many small needle crystals with metallic luster were obtained. (Figure 2)

The electrical resistivity was measured by the standard four-probe method.   Voltage and current leads were connected to the samples with silver paste.   The temperature was measured using a Cu-Constantan thermocouple. Figure 3 shows the temperature dependence of the resistivity of these samples. The specimen with y=3.0 shows a slight decrease in resistivity around 110K and showed zero resistivity at 97K.   Irrespective of the difference in Cu content,

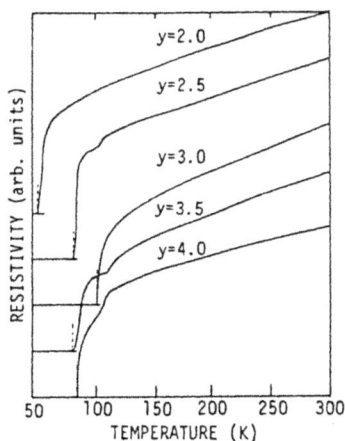

Fig.3. Temperature dependence of electrical resistivity of $BiSrCaCu_yO_x$ prepared by the furnace cooling method

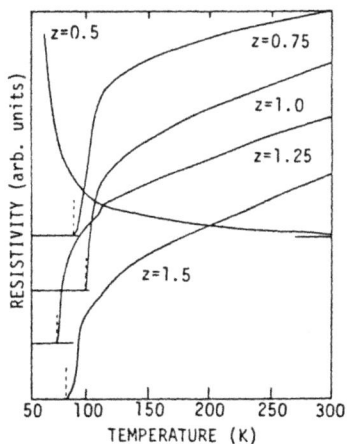

Fig.4. Temperature dependence of electrical resistivity of $BiSrCa_2Cu_3O_x$ prepared by the furnace cooling method

the specimens (y=2.5, 3.5, 4.0) show two distinct superconducting transitions; one centered near 110K and the other around 80K. The specimen with y=2.0 shows no transitions around 110K and 80K, probably due to the growth of a semiconducting or insulating phase.

Influence of Ca concentration

Furthermore, samples with the nominal composition of $BiSrCa_zCu_3O_x$ (z=0.5, 0.75, 1.0, 1.25, 1.5) were prepared under the same heat treatment conditions. Figure 4 shows the temperature dependence of the resistivity of these samples. For the specimen with z=0.5, as the temperature was lowered, the resistance increased, which is characteristic of a semiconductor. As the Ca concentration was increased, the Tc was increased. However excessive Ca content (z>1.25) brings about a decrease of Tc.

From the above results, we concluded that the optimum starting composition to obtain the optimum high-Tc transition is Bi:Sr:Ca:Cu=1:1:1:3.

Preparation by Melt-process (rapid quenching method)

Dependence on the quenching condition before annealing

The compounds were prepared at the composition Bi:Sr:Ca:Cu=1:1:1:3 and preheated under the above conditions. The sintering temperatures and quenching conditions are summarized in Table 1.

At room temperature, all compounds had a resistivity of over 100 times that of the compounds prepared by the furnace cooling method and showed a typical semiconducting resistivity profile.(Figure 5) It is noteworthy that compound A was observed to have a transition about at 60K, but compounds B and C did not have a transition above 40K. This is probably due to some reaction during step 3 in Table 1.

Dependence on the annealing temperature

These compounds were annealed at several temperatures between 845 and 875 °C for 10 hours in air. The heating rate was 200 °C/h. After a 10 hour

Table 1. Sintering temperature and quenching conditions of three compounds

	Step 1	Step 2	Step 3	Step 4
A			───────	
B	heating 900 °C at the rate of 200 °C/h	maintaining 900 °C for 1h	cooling to 885 °C at the rate of 1.5 °C/h	quenching to room temp.
C			cooling to 875 °C at the rate of 1.5 °C/h	

annealing process, the samples were naturally cooled in the furnace. Figures 6, 7 and 8 show the temperature dependence of the resistivity of compounds A, B and C for different annealing temperatures, respectively.

For compound A, the samples annealed at 865 and 875 °C showed two distinct superconducting transitions. Two Tc(onset) were observed near 120K and 95K for the 865°C annealed sample and near 125K and 97K for the 875°C annealed sample, and zero resistivity occurred at 90K for both samples.

Although Tc(onset) near 120K was observed, the ratio of the resistivity-drop near 120K for the 865 °C annealed sample was smaller than that of the 875 °C annealed

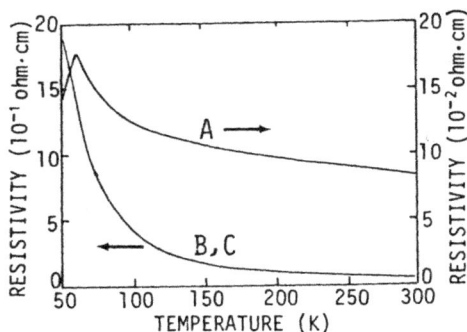

Fig.5. Temperature dependence of electrical resistivity of $BiSrCaCu_3O_x$ prepared by the rapid quenching method before annealing

sample as it is normalized by resistivity at 120K; the ratios of resistivity-drop are about 20% for the 865 °C annealed sample, and about 80% for 875 °C annealed sample.

The samples annealed at 845 and 855 °C showed one superconducting transition about at 100K and 110K and zero resistivity at 95K and 100K, respectively.

For compounds B and C, Tc(onset) near 120K was observed in all samples. The ratio of the resistivity-drop for the 865 °C annealed samples were the largest, which probably indicates that the proportion of high-Tc phase superconductor was highest compared with that of the other samples. These samples showed a clear resistive transition at 120 and 90K, but from the magnetic susceptibility measurement, susceptibility decrease was not observed. This was probably due to the minute amount of the volume fraction of

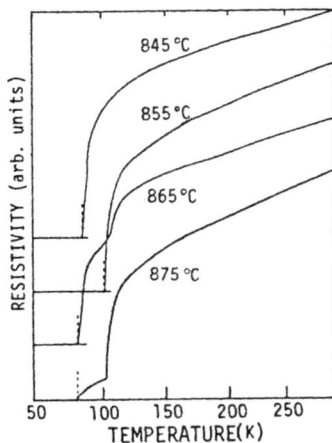

Fig.6. Temperature dependence of electrical resistivity of compound A after annealing

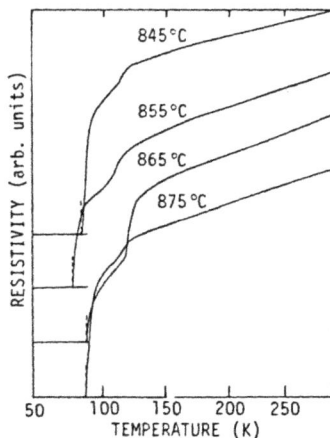

Fig.7. Temperature dependence of electrical resistivity of compound B after annealing

superconducting phases.

From these results, we concluded that the optimum annealing temperatures for the resistivity transition is 855, 865 and 865 °C for compounds A, B and C, respectively.

Conclusion

We have prepared high-Tc Bi-Sr-Ca-Cu-O superconductors by the melt-process. The results obtained in this study are summarized as follows.

1) The optimum starting composition was Bi:Sr:Ca:Cu=1:1:1:3.

2) Before annealing, the samples quenched at high temperatures showed a semiconducting resistivity profile.

3) After annealing, the samples showed a high-Tc transition. However, it seems that the sample also contained the low-Tc phase. Moreover, the amount of the volume fraction of the superconducting phases was minimal.

For future study, we will investigate the applications of this method.

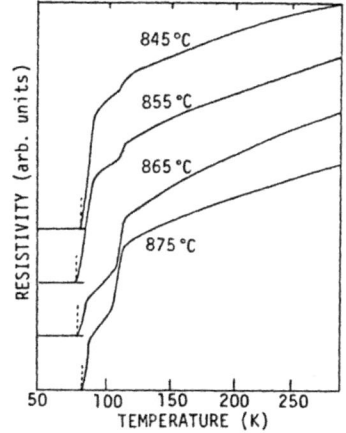

Fig.8. Temperature dependence of electrical resistivity of compound C after annealing

References

1) M. K. Wu, J. R. Ashbun, C. J. Tomg, P. H. Hor, R. L. Meng, L. Gao, Z. J. Huang, Y. Q. Wang and C. W. Chu :Phy. Rev. Lett. 58 (1987) L908

2) H. Maeda, Y. Tanaka, M. Fukutomi and T. Asano :Jpn. J. Appl. Phys. 27 (1988) L209

3) Z. Z. Sheng and A. M. Hermann :Nature 332 (1988) L55

4) Z. Z. Sheng and A. M. Hermann :Nature 332 (1988) L138

5) J. Akimitsu, A. Yamazaki, H. Sawa, and H. Fujiki :Jpn. J. Appl. Phys. 26 (1987) L2080

6) C. Michel, M. Hervieu, M. M. Borel, A. Grandin, F. Deslandes, J. Provost and B. Raveau :Z. Phys. B68 (1987) 421

7) A. Sumiyama, T. yoshitoshi, H. Endo, J. Tsuchiya, N. Kijima, M. Mizuno and Y. Oguri :Jpn. J. Appl. Phys. 27 (1988) L542

8) N. Kijima, H. Endo, J. Tsuchiya, A. Sumiyama, M. Mizuno and Y. Oguri : Jpn. J. Appl. Phys. 27 (1988) L821

9) H. Nobumasa, K. Shimizu, Y. Kitano and T. Kawai :Jpn. J. Appl. Phys. 27 (1988) L846

10) U. Endo, S. Koyama and T. Kawai :Jpn. J. Appl. Phys. 27 (1988) L1476

Y-Ba-Cu-O SUPERCONDUCTOR PREPARED FROM FREEZE-DRIED POWDERS

T.Ito, H.Ohtera, T.Tachiwaki*, A.Sasaki and A.Hiraki

Department of Electrical Engineering, Osaka University,
Suita, Osaka 565,
Department of Chemical Engineering, Doshisha University,
Kyoto, Kyoto 602

Abstract - Superconducting material of the Y-Ba-Cu-O system with T_c of 93.6K has successfully been prepared from freeze-dried fine powders with a mixture of Y, Ba and Cu carbonates. Although a heat treatment of the homogenous freeze-dried powders at high temperatures(≥ 850 ℃) allows a solid phase growth of the superconducting Y-Ba-Cu-O compound, the material obtained is porous. Much better results are attained by means of two successive sinterings : the first one for decomposing process and the second one for forming process followed by press of the powder. The advantage and disadvantage of the present preparation method is discussed.

Introduction

Since Wu et al.[1] found the superconducting material, Y-Ba-Cu-O compound, with T_c of 90K, various attempts have been made to prepare purified and stable materials with high T_c in a form of bulk or thin film.[2] We have applied the freeze-drying method to preparation of the Y-Ba-Cu-O superconductor. The freeze-drying method is known as one of techniques to make fine ceramic powders.[3] This employs a sequence of a rapid freezing process and a drying one at low pressures, keeping composition of the constituent elements except H_2O without selective growth of any particular compounds, when the solution changes to frozen fine powders and finally to freeze-dried powders. Because of usage of a solution, quite good uniformity in composition can be attained for the freeze-dried powders.

Sample Preparation

An aqueous solution of Y, Ba and Cu nitrates with a suitable composition was converted to a product of the carbonates of $Y_2(CO_3)_3$, $BaCO_3$ and $CuCO_3$ through a reaction between the nitrate solution and ammonium carbonate. Then the product was sprayed into a bath filled with n-hexane which was cooled down to -70℃ by using a mixture of acetone and dry ice. The frozen powder thus obtained was transferred into a special vacuum chamber designed for the freeze-drying process, in which the frozen powder was dried through sublimation of H_2O at low temperatures (-20 ℃).[4] At this stage, an assembly of fine (light blue) powder with typical sizes of 300nm was obtained. Then the assembly was decomposed in a temperature-controlled electric furnace at high temperatures (up tp 950 ℃). After this process, fine black powder of Y-Ba-Cu-O compounds was obtained. We tried to prepare the Y-Ba-Cu-O compounds from various solutions, and finally found that the carbonates formed as a sediment during the reaction in the solution were most suitable for the present purpose although the composition of the freeze-dried material changed from that of the starting solution. The composition of Y, Ba and Cu was determined through measurements of X-ray fluorescence. Fluorescence yields of the Y, Ba and Cu Kα -lines emitted from the specimen obtained were calibrated with those of standard samples for which the composition was known. In the case when the sedimental carbonates are selectively used to prepare freeze-dried powders, the amount of Ba should be increased in the starting solution of the nitrates. When all of the product of the carbonates is used, the composition of the starting solution can be retained although the homogeneity becomes less. Details of the preparation will be published elsewhere.[5]

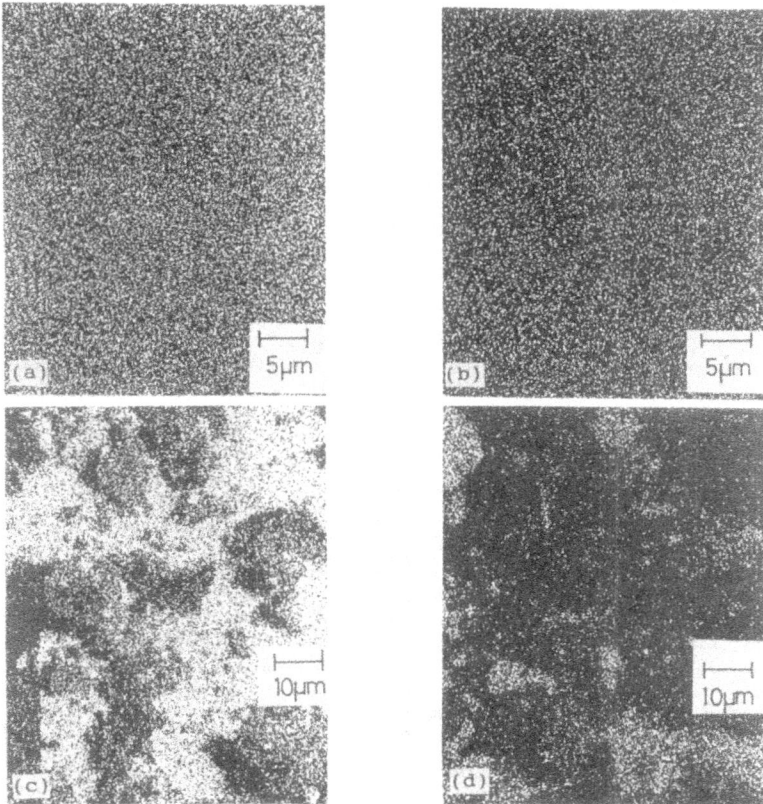

Fig.1. Comparison of EPMA images taken for a specimen prepared from the freeze-drying method and one prepared from the conventional method by mixing three compounds of Y_2O_3, $BaCO_3$ and CuO. (a) / (c) : Cu Lα -line images and (b) / (d): Y Lα -line images. Note the difference in the magnification between (a)/(b) and (c)/(d).

Figure 1 shows typical EPMA (electron probe for microanalysis) images with Y and Cu Lα -lines for the freeze-dried Y-Ba-Cu-O compounds and for a mixture of Y_2O_3, $BaCO_3$ and CuO powders. As expected, homogeneity of the specimen is much better for the former than the latter in the resolution limits (~ 1 μm) of the EPMA apparatus used. The uniformity of the specimen prepared allowed a rapid growth in solid phase. From a thermal analysis, we found that it was neccessary to heat the freeze-dried powders at high temperatures above $\sim$ 850℃ , which can be related to the onset of decomposition of $BaCO_3$. However, the temperature was lower than that for decomposition of pure $BaCO_3$, meaning that pure $BaCO_3$ may not be present in the freeze-dried powder but a mixture of Y-Ba-Cu compounds may be formed at this stage.

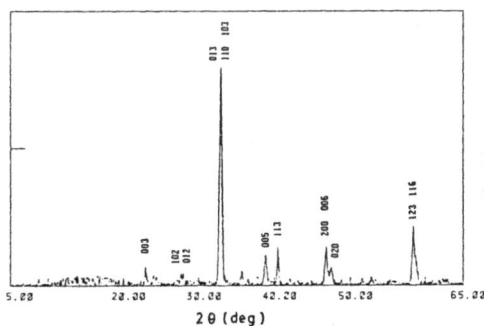

Fig.2. A typical X-ray diffraction pattern taken for a specimen after a sintering treatment following the freeze-drying process. The heating was performed at 900℃ for 2 hours in air.

Fig.3. SEM images taken for the freeze-dried specimens after a sintering (a) at 900℃ for 2 hours and (b) at 850 ℃ for 18 hours.

Figure 2 shows an X-ray diffraction pattern measured for a specimen just after the first heat treatment in air at 900 ℃ for 2 hours which followed the freeze-drying process; At this stage, the superconducting phase of the Y-Ba-Cu-O compound appeared. As is shown in Fig.3(a), a SEM (scanning electron microscope) picture taken for the specimen at this stage indicates substantial growth of the compound, which is assigned to the Y-Ba-Cu-O compound from the diffraction results. When the temperature was decreased to 850℃ in the first sintering process, the growth rate of the Y-Ba-Cu-O compound was decreased. Figure 3(b) shows a SEM image measured for a specimen after the first heat treatment in air at 850℃ for 18 hours. The size of the grown compound was almost same as that shown in Fig.3(a). For the specimen treated only with the first sintering, there was no reproducible evidence in resistance measurements that the specimen became super-conducting. The reason for this is thought due to porous structure of the heat-treated material, which occurred during the decomposition process. Possible presence of unidentified materials at grain boundaries may be another reason.

In order to reduce number and size of pores in the specimen, we employed two successive heat treatments for the freeze-dried powders. The first one is a process for the decomposition of the carbonates, allowing a suitable composition of Y, Ba and Cu without extra elements except oxygen. The second process works just as in the conventional preparation from three powdered compounds of Y_2O_3, $BaCO_3$ and CuO. Figure 4 shows a typical X-ray diffraction pattern taken for a specimen obtained after the two successive sinterings, indicating that the superconducting compound of $YBa_2Cu_3O_{7-x}$ is formed. The reproducibility of the preparation was quite good: we ascribe it to good homogeneity of the freeze-dried specimen.

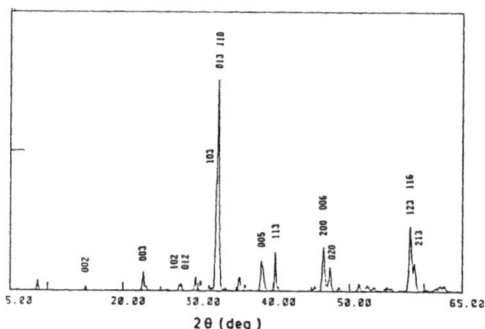

Fig.4. An X-ray diffraction pattern taken for a specimen after the fol-lowing successive heat treatments after the freeze-dried process : at 860℃ for 2 hours and at 900 ℃ for 23 hours.

Temperature Dependences of Electric and Magnetic Property

Electric resistivity (resistance) of the specimens formed in a disk was measured by means of a standard four-point probe method. Figure 5 shows a typical dependence of the resistance on temperature; T_c-zero measured (the temperature below which the resistance was negligibly small) was 93.6K with T_c-onset of 97.0K (the temperature at which a rapid decrease in the resistance started). Magnetic property of the specimen was also measured at the same time when the resistance measurement was carried out. By monitoring change in reactance of a picking coil which was contacted directly to the specimen surface, one can obtain information on AC susceptibility of the specimen. A typical result shown in Fig.5 indicates that the temperature at which the decrease in reactance started was 97K (exactly the same as the above T_c-onset) and the reactance or AC susceptibility changed more gradually than the resistance as a function of temperature.

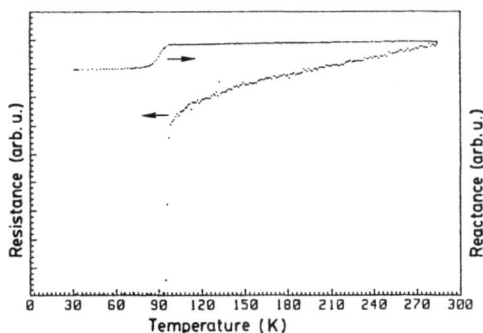

Fig.5. Temperature dependences of resistance and AC susceptibility of a specimen treated with the two-stage sintering. The AC susceptibility can be monitored by measuring the reactance of a probing coil contacted directly to the specimen.

Conclusion

We have successfully prepared the superconducting compound of the Y-Ba-Cu-O system with T_c-zero = 93.6K by means of the freeze-drying method. The present method is charactrized by :
(1) The homogeneity of the freeze-dried specimen before sintering treatments is much better than that of conventionally prepared one. This leads to good reproducibility of the preparation of the superconducting material.
(2) The composition of the freeze-dried powder changes from that in an aqueous nitrate solution when the sedimental carbonates of Y, Ba and Cu are freeze-dried for the best homogeneity. However, the composition is controlled by modification of the composition of the starting solu- tion. On the other hand, one can use all of the reaction product to keep the composition with slight loss in homogeneity.

Acknowledgements

The authors would like to thank Prof. Y.Murakami for his giving them opportunities to measure low-temperature properties at LAS, Osaka University.

References

1) M.K.Wu, J.R.Ashburn, C.J.Torug, P.H.Hor, R.L.Meng, L.Gao,
   Z.J.Huang, Y.Q.Wang and C.W.Chu: Phys. Rev. Lett. 58(1987)1908.
2) Jpn.J.Appl.Phys. 26(4-5) (1987), Part II.
3) F.J.Schnettler, F.R.Monforte and W.W.Rhodes: Sci.Ceram. 4(1968)79.
4) T.Tachiwaki, T.Okada and H.Uyeha: The Science and Engineering
   Review of Doshisha University 28(1987)1.
5) T.Tachiwaki, H.Uyeha, T.Ito and A.Hiraki: to be published.

PREPARATION OF Y-Ba-Cu-O SUPERCONDUCTOR
BY MELT PROCESSING

Shigeoki SAJI[*], Shunichi ABE[**], Toshiya SHIBAYANAGI[*] and Shigenori HORI[*]

* Department of Materials Science and Engineering
Faculty of Engineering, Osaka University
2-1 Yamadaoka, Suita, Osaka 565, Japan
** Graduate Student, Osaka University

Abstract - Y-Ba-Cu-Oxides heated above the solidus temperature were solidified under three cooling conditions, and a 2-stage annealing of the solidified samples was carried out at 950 and then 500 °C in flowing oxygen atmosphere. Microstructure, distribution of various phases and the superconducting transition temperature, T_c were investigated for as-solidified samples and for annealed samples. The Y_2O_3 phase was present in all as-solidified samples and it might decompose into the $YBa_2Cu_3O_x$ and Y_2BaCuO_5 phases during the 2-stage annealing. High T_c (80-98K) superconductors were obtained after annealing of the air-cooled or the furnace-cooled samples. However, the press-quenched and annealed samples showed low T_c ($\sim$30K) , and this is attributed to the lack of oxygen in $YBa_2Cu_3O_x$ phase.

Introduction

Progress toward major applications of the bulk, high T_c superconductors has been hindered by very low J_c at 77K and its severe degradation in weak magnetic field[1]. J_c values at 77K of $YBa_2Cu_3O_x$ superconductors prepared from the sinter method are about 200-1000 A/cm^2 in zero field and about 1-10 A/cm^2 at H=1 T. The poor properties are ascribed to the "weak link problem" in sintered materials. The main sources of the weak links are likely to be at or near the grain boundaries which are closely related to various kinds of defects : non-superconducting phases, impurities, voids, microcracks, crystallographic anisotropy and so on.

On the other hand, melt processing is a favorable method to obtain a high-quality, dense, void-free high T_c ceramic superconductors in a wide variety of shapes. A very dense and preferentially aligned microstructure obtained by the "melt-textured growth" processing[2] for Y-Ba-Cu-O, gave rise to dramatically improved Jc ($\sim$17000 A/cm^2 at 77K in zero magnetic field and $\sim$ 4000 A/cm^2 at H=1 T).

Basic research on melt processing may serve in preparation of large single crystals aiding the understanding of the mechanism of superconductivity. However, there are many unknown factors in the preparation of high Tc superconductors by melt processing. It is author's objective to clarify phase equilibria at higher temperatures and effects of solidification conditions on superconducting properties of Y-Ba-Cu-Oxide.

Experimental

The powder of Y_2O_3(99.99%), $BaCO_3$(99.9%)and CuO(99.5%) was weighed in a ratio of Y:Ba:Cu=1:2:3 and mixed in an alumina mortar. The mixed powder was calcined at 850 °C for 3h. The calcined powders($\sim$10gr) crushed were melted in air in alumina crucibles. Since the solidus temperature is likely to be about

$1000\,°C^{(2)}$, the powders were heated to $1300-1400\,°C$ and held for 20 min prior to cooling. The molten oxide was cooled down to room temperature under three cooling conditions ; press-quenching between two copper plates, PQ, air-cooling in the crucible, AC and furnace-cooling ($\sim2\,°C/min$), FC. A 2-stage annealing (1st at $950\,°C-20h$ and then 2nd at $500\,°C-20\sim100h$) of the solidified oxides was carried out in flowing oxygen atmosphere.

Microstructure of the solidified and annealed samples was observed by optical microscopy(OM), scanning electron microscopy(SEM) and back-scattered electron method(BSE). Crystal structure and composition of various phases were determined by using an X-ray diffraction and an electron probe microanalysis (EPMA). T_c was measured by a four-probe method and by an inductance method using the Meißner effect.

Results and Discussion

(1) As-solidified sample

Various phases detected on X-ray diffraction patterns of as-solidified samples under the three cooling conditions are tabulated in Table 1. In all the as-solidified samples, the Y_2O_3 phase, A in Fig.1 (a)-(c) was present in the matrix composing of $BaCuO_2$, Y_2BaCuO_5, YBa_2Cu_3, CuO or unknown phases. Two types of unknown phases were observed in the secondary electron image of Fig.1(a). The $YBa_2Cu_3O_x$ detected in as-solidified samples might be the tetragonal phase or the second orthorhombic phase (ortho-II), because all the as-solidified samples did not show both the Meißner effect and zero resistance above the liquid nitrogen temperature. It is much interesting that the Y_2O_3 phase is present in all the as-solidified samples. The phase diagram of Y_2O_3-$BaO-CuO$ system [4] is unclear above about $1100\,°C$. A doubt whether Y_2O_3 powders remain unmelted because of short heating time, is eliminated by the following experimental results.

Table 1   Phases detected by X-ray analysis for as-solidified samples under three cooling conditions, press-quenching (PQ), air-cooling (AC) and furnace-cooling(FC).

Sample	Phases					
	Y_2O_3	$Y_1Ba_2Cu_3O_x$	$BaCuO_2$	CuO	Y_2BaCuO_5	unknown
as-PQ	O	−	O	−	O, −	O
as-AC	O	O	−	−	O, −	O
as-FC	O	O	O	O	−	−

O:detected , −:non-detected

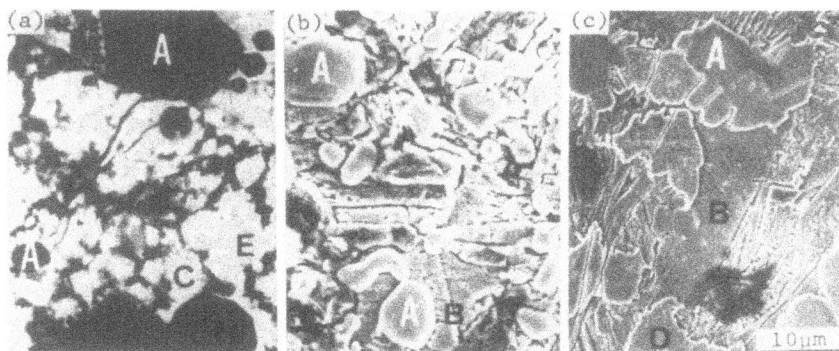

Fig.1 Micrographs of as-solidified samples under three cooling conditions.

(a) press-quenched, secondary electron image (b) air-cooled, SEM

(c) furnace-cooled, SEM

A: Y_2O_3 phase, B: $YBa_2Cu_3O_x$ phase, C: $BaCuO_2$ phase

D: Y_2BaCuO_5 phase, E: unknown phase

Sintered high T_c superconducting Y-Ba-Cu-Oxide in which the Y_2O_3 phase was not detected by both X-ray analysis and EPMA, was heated at 1300 °C for 20min (above the solidus) and then air-cooled to room temperature. In this as-solidified sample, the Y_2O_3 phase was detected by both the methods, as shown in Fig.2. Thus, the Y_2O_3 phase is likely in equilibrium with liquid phase at 1300 °C.

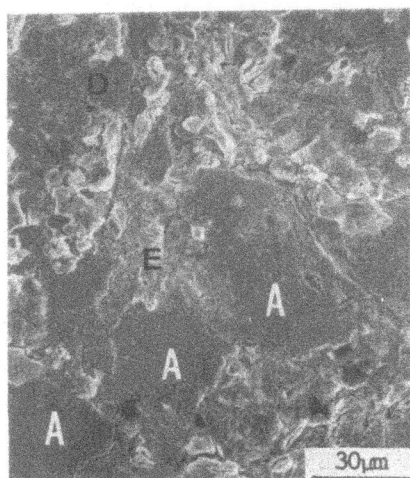

Fig. 2

Micrograph of a solidified sample by air-cooling after heating at 1300 °C for 20 min (above the solidus) of a sintered high T_c superconductor.

A: Y_2O_3 phase, D: Y_2BaCuO_5 phase,

E: unknown phase

(2) Annealed samples

Phases detected on the X-ray diffraction patterns of the variously solidified and annealed (2-stage) samples are listed in Table 2. In all the annealed samples, the Y_2O_3 phase was not detected by X-ray analysis and also rarely visible in SEM observation, as seen in Fig.3 (a)-(c). Annealed samples are mainly composed of $YBa_2Cu_3O_x$, $BaCuO_2$, Y_2BaCuO_5 or CuO phases. Thus, the Y_2O_3 phase and the unknown phases in as-solidified samples seem to decompose into the above mentioned phases during the 2-stage annealing, particularly 1st stage annealing (950°C-20h).

Table 2   Phases detected by X-ray analysis for the variously
cooled and then annealed (2-stage) samples.

Sample	Phases					
	Y_2O_3	$Y_1Ba_2Cu_3O_x$	$BaCuO_2$	CuO	Y_2BaCuO_5	unknown
PQ-ann.	—	O	—	O	O	—
AC-ann.	—	O	O	O	—	—
FC-ann.	—	O	O	—	O	—

O:detected , —:non-detected

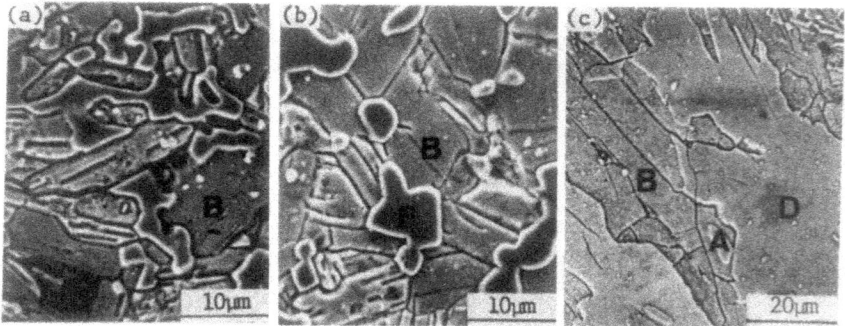

Fig.3   Micrographs of variously solidified and annealed samples. SEM
(a) press-quenched and annealed (b) air-cooled and annealed
(c) furnace-cooled and annealed
A: Y_2O_3 phase, B:$YBa_2Cu_3O_x$ phase, D: Y_2BaCuO_5 phase
F: CuO phase

Figure 4 shows transition curves from normal- to super-conductivity for the annealed samples. AC- and FC-annealed samples show high T_c (80–98K). The two transition temperatures obtained from resistance and inductance curves are fitted together. However, the PQ-annealed sample has a low T_c ($\sim$30K). To clarify the reason, the lattice constant (c-cell dimension) of $YBa_2Cu_3O_x$ phases was measured and the oxygen number , x was estimated by the following equation[5] for the three kinds of annealed samples.

$$x=67.108-5.161xc \text{ -------(1)}$$

where c is the c-cell dimension in angstroms. The results obtained are given in Table 3.

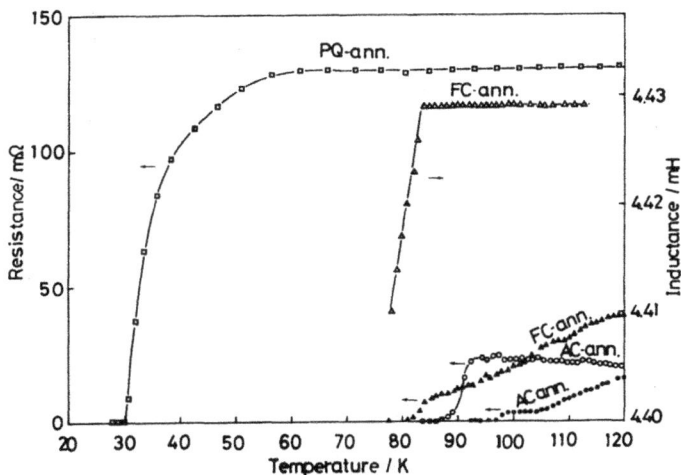

Fig.4   Temperature dependence of resistance and inductance for the variously solidified and annealed samples; PQ-ann., AC-ann., and FC-ann..

Table 3   Lattice constants and oxygen number,x in the variously cooled and annealed samples.

Sample	c-cell demension, Å	oxygen number, x
PQ-ann.	11.743	6.5024
AC-ann.	11.679	6.8327
FC-ann.	11.698	6.7400

The oxygen number of $YBa_2Cu_3O_x$ in the PQ-annealed sample is lower than the others. T_c value (30K) for the PQ-annealed sample nearly agrees with the value (~40K) estimated from the relationships between T_c and oxygen number[6].

Conclusions

The Y_2O_3 phase is in equilibrium with liquid phase at 1300-1400 °C and decomposes into the superconducting $YBa_2Cu_3O_x$ phase and others during 2-stage annealing (at 950 and 500 °C). High T_c (80-98K) superconductors are obtained after the annealing of air- and furnace-cooled samples. However, the press-quenched and annealed sample shows a low T_c (~30K), and this is attributed to the lack of oxygen in the $YBa_2Cu_3O_x$ phase.

Acknowledgments

We wish to thank Prof. Y.Murakami and his students for their help in Tc measurements below liquid nitrogen temperature. We also wish to thank NIPPON YTTRIUM,LTD. for supplying the starting materials.

References

1) H.Kumakura, M.Uehara, Y.Yoshida and K,Togano : "Critical Current Densities in Sintered Ba-Y-Cu-O Compound", Physics Letters A ,124, 367-369 (1987).

2) S.Jin, T.H.Tiefel, R.C.Sherwood, M.E.Davis, R.B.Van Dover, G.W.Kammlott, R.A.Fastnacht and H.D.Keith : "High Critical Currents in Y-Ba-Cu-O Superconductors", Appl.Phys.Lett.,52, 2074-2076 (1988).

3) T.Takabatake, Y.Nakazawa and M.Ishikawa : "Superconductivity in Metamorphic Phase of $Ba_2ErCu_3O_{7-\delta}$", Jpn.J.Appl.Phys.,26, L1231-1232 (1987).

4) K.Oka, K.Nakane, M.Ito, M.Saito and H.Unoki : "Phase-Equilibrium Diagram in the Ternary System Y_2O_3-BaO-CuO", Jpn.J.Appl.Phys.,27, L1065-1067 (1988).

5) A .Ono, S.Sueno, M.Kobayashi and Y.Ishizawa : "Preparation of $Ba_2YCu_3O_{7-\delta}$ with Tc below 65K", Jpn.J.Appl.Phys.,26, L1985-1987 (1987).

6) E.Takayama-Muromachi, Y.Uchida, M.Ishii, T.Tanaka and K.Kato : "High Tc Superconductor $YBa_2Cu_3O_x$ - Oxygen Content vs. Tc Relation",Jpn.J.Appl.Phys. 26, L1156-1158 (1987).

FABRICATION OF YBACUO WITH HIGH TRANSPORT Jc

Masato MURAKAMI, Mitsuru MORITA, and Katsuyoshi MIYAMOTO

R & D Laboratories-I, Nippon Steel Corporation
1618 Ida, Nakahara-ku, Kawasaki 211 Japan

Abstract - Applications of high Tc oxide superconductors are hindered by low critical current density(Jc). Weak-link networks along grain boundaries are responsible for low Jc. This problem was overcome by the quench and melt growth process which enables unidirectional growth of the superconducting phase and minimize the intrusion of second phases. Through this process Jc value of 10^4 A/cm^2 has been achieved at 77K and 1T.

Introduction

One of the most crucial properties for practical applications of oxide superconductors is critical current density(Jc). Jc value exceeding 10^5 A/cm^2 is required in the presence of a significant magnetic field. However, Jc's of bulk sintered Y based oxide superconductors are fairly small and typically 10^2 to 10^3 A/cm^2 at 77K and zero magnetic field[1]. A sharp decrease of Jc also takes place when magnetic field is applied.

These low Jc values in bulk sintered materials are attributed to the weak-links present at grain boundaries. It has been found[2] that there exist various defects along grain boundaries such as second phases and cracks which limit supercurrent. However, a number of experimental results[3-5] also imply that even clean grain boundaries can become weak-links for supercurrent[3]. This is probably due to a strong anisotropy in Jc. Therefore one possible technique to enhance Jc of oxide superconductors will be texture control to achieve preferential grain alignment[6]. In fact, Jin et. al[7] applied the melt textured growth process for the Y based superconductors and obtained Jc value of 4000 A/cm^2 at 77K and 1T.

It is also very important to delineate the effects of grain boundaries on critical current density. For this, a large single crystal must be prepared. Then bicrystals composed of two grains with certain crystallographic orientations may be fabricated, which enables thorough understanding of grain boundary effect. So, we tried to synthesize a large single crystal with good quality[4]. However, it becomes clear from microstructural investigation[4] that superconducting $YBa_2Cu_3O_x$ (123) phase grows by the peritectic reaction between Y_2BaCuO_5(211) and liquid phase(admixture of BaO and CuO), and thereby the size of single crystal is limited depending on the distribution of 211 phases. On the other hand, it is also found from magnetization measurement[4] that weak-links can be reduced in the solidified sample, although transport Jc was only 100 A/cm^2 at 77K in zero magnetic field. Microstructural observation revealed that the connectivity of superconducting phase is reduced by the presence of liquid and 211 phases in such solidified samples, which indicates much higher Jc will be obtainable if we can enhance the connectivity of 123 phases, which can be realized by the modification of the fabrication process.

In this paper we report the fabrication process and some physical properties of YBaCuO with high transport Jc.

Experimental

Precursor powders were prepared by mixing 99.9% pure Y_2O_3, $BaCO_3$, and CuO followed by calcining at 900°C for 8h in air. The powders were then heated to 1300 - 1500 °C in platinum crucibles and either slow cooled or quenched on a copper hearth. Quenched plates were again heated to 1000 - 1300 °C followed by controlled cooling in flowing oxygen.

Electric resistivity was measured by a standard four probe method. Transport critical current density was obtained from V–I characteristic curves using 1 μV/cm criterion. Gold was deposited on the surface of the superconductor in order to reduce contact resistance. Copper wires were attached to the samples by ultrasonic solder. Magnetization measurement was conducted with a vibrating sample magnetometer equipped with a cryostat.

Microstructure was observed with a polarized optical microscope. Compositional analysis was performed using a computer assisted electron probe microanalyzer(CMA).

Results and discussion

Solidification process

Figure 1 shows typical microstructure of YBaCuO compound with Y:Ba:Cu ratio of 1:2:4 slow cooled from 1200 °C in flowing oxygen. According to CMA analysis the sample is composed of four phases: superconducting 123 phase, 211 phase, CuO, and $BaCuO_2$. We also notice that a number of 211 phases are trapped inside the 123 phases as the result of the peritectic reaction between 211 and liquid phases. The 123 phase nucleates at the interface of the two phases and grows according to the following reaction:

$$Y_2BaCuO_5 + L(3BaO + 5CuO) \rightarrow 2YBa_2Cu_3O_{6.5}$$

For this reaction taking place both 211 and liquid phases must be supplied, therefore once 211 phase is surrounded by 123 phase, no further reaction proceeds resulting in the multiphased microstructure as shown in Fig. 1. Thus transport Jc of the solidified sample is inevitably low and typically less than 100 A/cm² at 77K in zero magnetic field due to the weak connectivity of the superconducting phases.

However, we obtained very hopeful data from the solidified samples as to the improvement of Jc. At first, 123 phase was very homogeneous compared with the sintered sample. Secondly superconducting transition was fairly sharp both in resistivity and ac measurements. These results indicate that the quality of the solidified sample is good. Microstructural observation also indicates that preferred orientation of grain growth for the 123 phase is perpendicular to C axis, which is the optimum direction for supercurrent. Therefore it is possible to align grain axis to a preferred orientation by controlled solidification. At last magnetization measurement at 77K suggests that a significant amount of weak-links can be reduced in the solidified sample. In bulk sintered materials, magnetization hysteresis(ΔM) is almost independent of sample size(d), until d is reduced to the level of grain size, which demonstrates that the weak-link net work along grain boundaries inhibit supercurrent resulting in low Jc. This seems to be inevitable in randomly oriented polycrystalline samples. While in the solidified samples as shown in Fig. 2, ΔM becomes small with the decrease of d contrasting with the result of bulk sintered materials. According to the critical state model[8] ΔM and d has a linear relationship if supercurrent flows through the whole sample. Although a completely linear relationship was not obtained in the solidified sample, a strong dependence of ΔM on d shows that a significant amount of

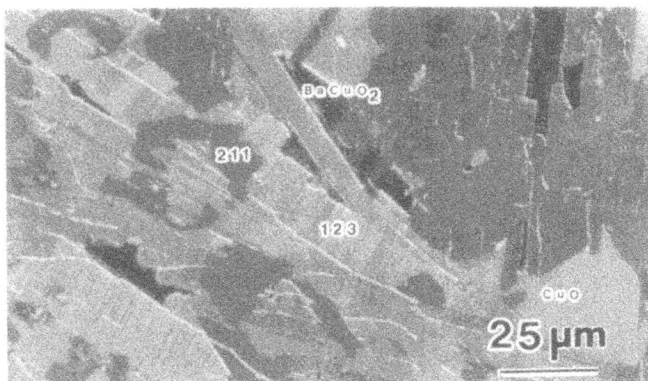

Fig. 1.   Microstructure of YBaCuO with excess CuO slow cooled from 1200°C.

Fig. 2.   Microstructure of YBaCuO with composition slightly off 1:2:3 stoichiometry toward Y rich region slow cooled from 1200 °C.   Note the superconducting phases are well developed and connected to each other.

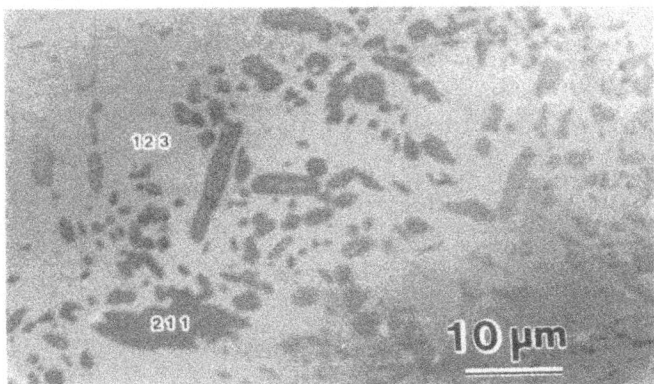

Fig. 3.   Microstructure of YBaCuO quenched and melt processed in flowing oxygen.   Note 211 phases are finely distributed in the 123 phase.

weak-links are suppressed in the solidified sample. If we can enhance the connectivity of the superconducting phase, it will be possible to improve transport Jc.

Modification of the process

In order to enhance the connectivity of the superconducting phases, we employed compositional modification and unidirectional solidification. An addition of excess BaO and CuO promoted liquid phase formation and reduced the connectivity of the superconducting phases. While an excess Y addition promoted the 211 phase formation leading to the improvement in transport Jc. This is ascribed to the improved connectivity owing to the reduced distance of adjacent 211 phases which become nucleation site for the 123 phases. Then the best result was obtained in the sample with chemical composition slightly off 1:2:3 stoichiometry toward Y rich region. In such samples it was difficult to determine transport Jc in zero magnetic field, because a large electric current is needed and heat generated from the electric contacts quenched superconductivity. Therefore we measured Jc in the presence of magnetic field. Jc value exceeding $300A/cm^2$ at 77K in 1T was achieved in this sample. Microstructure of the sample is presented in Fig. 3. It's clear that the connectivity of the superconducting phases is improved and the grains are well aligned which can be confirmed by the twin structure in each grain. (Electric current was applied parallel and magnetic field was applied perpendicular to the longitudinal direction). Although this value is remarkably high compared to the bulk sintered material, it is still two orders of magnitude lower than the value required for practical applications.

From magnetization measurement, it is found that the linearity between ΔM and d was improved, however, a complete linearity was not realized indicating that weak-links are still present in the sample. Microstructural observation reveals that there still exist coarse 211 phases, liquid phase thin layers and cracks which cause weak-link phenomena. This suggests that further improvement in Jc is possible through microstructural control.

Further improvement in Jc

As already shown in the former section, the superconducting phase is formed by the peritectic reaction between 211 and liquid phases. Once these two phases are separated, further reaction does not proceed resulting in poor connectivity of the superconducting phases. Under the condition that the volume of 211 phases is constant, it is desirable to disperse 211 phases as finely as possible in the liquid in order to promote the growth of the 123 phase. We have found that 211 phases become finer as the temperature is raised, but at temperatures above 1200°C they start to decompose into Y_2O_3 and admixture of BaO and CuO. On cooling 211 phases nucleate from Y_2O_3. Therefore the distribution of Y_2O_3 determines the distribution of 211 phases. At higher temperatures, Y_2O_3 becomes finer, then we raised melting temperature up to 1450°C to refine and disperse Y_2O_3 and then quenched the molten sample on a copper hearth and observed its microstructure. The specimen comprised finely dispersed Y_2O_3 and liquid phase(admixture of BaO, CuO and $BaCuO_2$). When we reheat the sample up to 1200°C liquid phase melts again and on the subsequent slow cooling 211 phases nucleate at Y_2O_3 sites and keep growing down to around 1000°C followed by the nucleation and growth of 123 phase. Fig. 4 shows its microstructure. It is notable that the grain size of 211 phases is extremely reduced compared with the sample which was simply slow cooled from 1200°C(Fig. 3). We also found that no grain boundary is observed along growing direction in a fairly large region, although some domain boundaries and second phase thin layers are often observed. When we tried

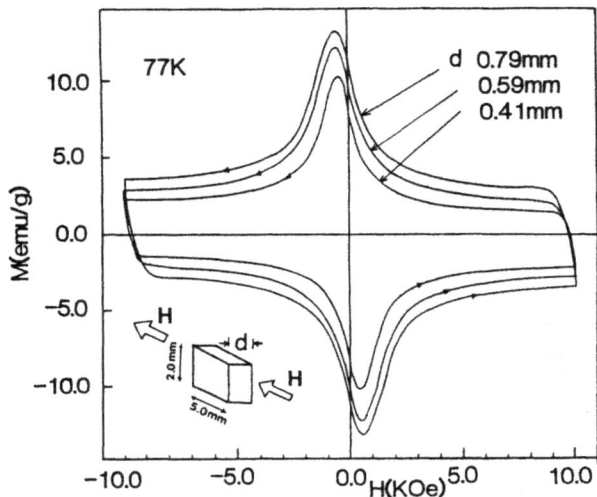

Fig. 4. Magnetization curves for YBaCuO quenched and melt processed. These data were taken as reducing the sample size.

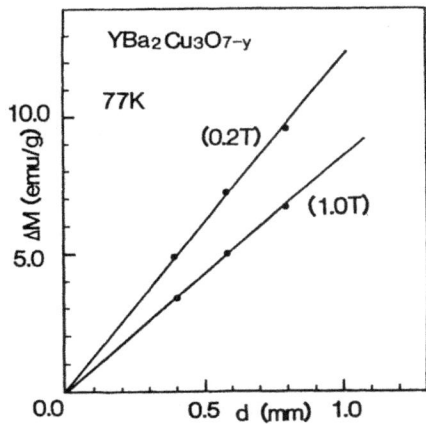

Fig. 5. Relationships between ΔM and d.

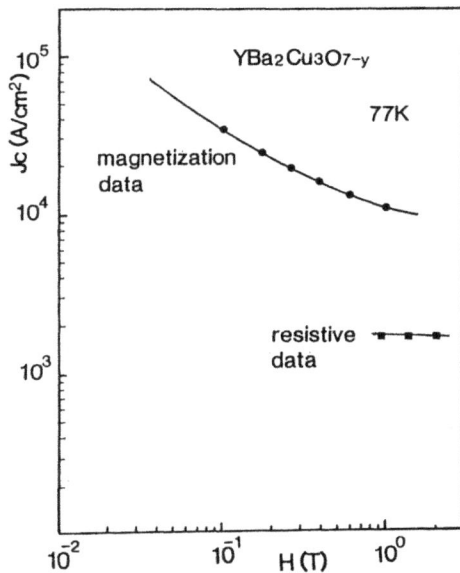

Fig. 6. Magnetic field dependence of critical current density .

to measure transport Jc, we again faced the problem that a significant heat was generated from the electric contacts. The sample quenched into normal state at the current density of 1500 A/cm^2 at 77K and 1T. Only a slight decrease in transport Jc was observed even when we increased magnetic field up to 10T. When we carefully examined the V-I curves, we noticed a small voltage appeared before the sample becomes normal. This voltage may be ascribed to the movement of flux. However, as shown in Fig. 5, the voltage is not reversed even when we reverse the current direction. This suggests that the voltage was due to heat generation from one electric contact. Because of this the sample quenched into normal state constantly at about 1500 A/cm^2 regardless of magnetic field. This result implies that the sample has much higher Jc value. Therefore we performed magnetization measurement.

Fig. 6 presents magnetization curves for this sample taken at 77K as we reduce the sample thickness d. When we plot ΔM versus d, a straight line can be drawn as shown in Fig. 7. This demonstrates that weak-links are almost absent in the sample. And the magnetic behavior of this sample can be understood by the critical state model. Under the condition that linear relationship between ΔM and d is obtained, we can estimate Jc using the following equation:

$$Jc = 20 \ (M^+ - M^-) \ / \ d$$

where M^+ and M^- are the magnetization increasing and decreasing the field process, respectively. Then we obtained Jc value of about 10^4 A/cm^2 at 77K and 1T which is very promising value for practical applications. We believe that high Jc value can be ascribed to the following beneficial effects: (1) no grain boundary along a-b direction can contribute to the elimination of the weak-links and (2) finely dispersed 211 phases work as pinning points.

At present our sample is about 20 mm long in maximum because of the furnace arrangement. Now we are trying to fabricate longer wires by controlled directional solidification.

Conclusion

By a combination of quenching and subsequent melt growth, it is possible to fabricate Y based oxide superconductors with no grain boundary along a-b direction. Magnetization measurements indicate that such samples are almost free from weak-links and yield Jc value exceeding 10^4 A/cm^2 at 77K and 1T.

References

[1] D. C. Larbarestier, S. E. Babcock, X. Cai, M. Daeumling, D. P. Hampshire, T. F. Kelly, L. A. Lavanier, P. J. Lee and J. Seuntjens, Physica C, 1580-1585 (1988).
[2] M. Murakami, M. Morita, K. Sawano, T. Inuzuka, S. Matsuda, and H. Kubo, to appear in Proc. SINTERING '87 ed. by S. Somiya (1987).
[3] K. Funaki, M. Iwakuma, Y. Sudo, B. Ni, T. Kisu, T. Matsushita, M. Takeo and K. Yamafuji, Jpn. J. Appl. Phys., 26, L1445 (1987).
[4] M. Murakami and M. Morita, "Superconducting properties of single crystalline oxide superconductors", Proc. 1988 Tokyo MRS Meeting.
[5] E. Shimizu and D. ito, "Critical current density obtained from particle size dependence of magnetization in YBa$_2$Cu$_3$O$_{7-x}$ powders", Proc. 1st ISS held in Nagoya (1988).
[6] J. W. Ekin, Adv. Ceram. Mat. 2, 586-592 (1987).
[7] S. Jin, T. H. Tiefel, R. C. Sherwood, R. B. van Dover, M. E. Davis, G. W. Kammlott, R. A. Fastnacht, Phys. Rev. B., 37, 7850 (1988).
[8] W. A. Fietz, M. R. Beasley, J. Silcox, and W. W. Webb, Phys. Rev., 136, A335 (1964).

FABRICATION OF HIGH Tc SUPERCONDUCTING TAPE
BY SOLID REACTION METHOD

Hidehito MUKAI, Nobuhiro SHIBUTA, Takeshi HIKATA,
Ken-ichi SATO, Masayuki NAGATA and Hajime HITOTSUYANAGI

Osaka Research Laboratories, Sumitomo Electric Industries, Ltd
Simaya 1-1-3, konohana-ku, Osaka 554, Japan.

Abstract - Ag sheathed high Tc superconducting tapes (wires) of $YBa_2Cu_3O_y$ and $Bi_{1.6}Pb_{0.4}Sr_2Ca_2Cu_3O_y$ were fabricated by the powder-in-tube method. Improvements of the powder core density and the crystal orientation were achieved by the pressing and rolling. The critical current density Jc of 4,140 A/cm^2 in Y-systems and of 6,930 A/cm^2 in Bi-systems were obtained at 77.3 K in a zero magnetic field. Using these samples, we made clear that magnetic dependence of the Jc was largely improved by increasing the Jc at a zero magnetic field. The critical current Ic was uniform within 3 % of distribution in a 15 m long wires. A coil made of the rolled wires could generate 64 gauss of the maximum magnetic fields.

Introduction

In order to apply the high Tc superconducting materiales for practical use, it is required not only the improvement of the superconducting properties but also the development of process technology to produce long, uniform wires.

The powder-in-tube method is one of the most convenient process to fabricate long wires. However, the critical current density Jc of Ag sheathed wires made by the usual solid reaction method is lower than that of the single crystal thin films by 3-4 orders.[1-2] Here, we report the improvement of the superconductivity of Ag sheathed wires made by the pressing and rolling technology.[3-5] The improvement of the magnetic dependence of Jc and the longitudinal uniformity of the superconducting wires are obtained. We also reported the properties of a test coil made of the Ag sheathed wires.

Experimental

Y-system

Samples of nominal composition of $YBa_2Cu_3O_y$ (YBCO) were prepared by thoroughly mixing appropriate amounts of Y_2O_3, $BaCO_3$ and CuO powders, each 4 N purity. The mixed powders were heated up to 880 - 920°C in air in 3 h, maintained at this temperature for 6 - 24 h and cooled in the furnace to room temperature, then ground into powders. After the above procedure was repeated again, the powders pressed into 8 mm diameter pellets and rectangular bars ($3\times3\times20mm^3$). The residual powders, pellets and bars were sintered at 920 - 960°C in air for 6 - 24 h followed by oxygen annealing at 500 - 700°C for 6 - 24 h and cooled in the furnace to room temperature at a rate of 100°C/h. Samples were reground and we checked the superconducting properties for measurering the X-ray diffraction pattern, Tc, Ic and AC susceptibility before putting into silver tube.

Powders were put into silver tube of diameter ranged from 6 mm to 12 mm and the initial packing density was 3.0 - 3.8 g/cm³. The powder and silver tube composite was firstly swaged and then drawn to the wires of 0.23 - 3.0 mm diameter using a drawing machine. Taped wires obtained by pressing and rolling. Short samples of 30 mm length were pressed under the load up to 10,000 kgW by a cold pressing machine, which can make the tapes of 0.05 mm thickness. Rolling process was carried out to fabricate the long wires.

By these methods, we investigated the relation between the Jc value and the reduction ratio of wires of various diameter. In order to clarify the reason for the improvement of Jc with increasing the strain or the reduction ratio, the density change of the powder core was investigated by measuring the cross section of the Ag sheath and the powder core and weights per unit length. X-ray diffraction analysis of the powder core were performed after removing the Ag sheath.

Longitudinal Ic distribution of a 15 m long wire was measured. The short samples of 30 mm length were cut away at every two meters and then sintered together. Using the long cylindrical type furnace, we sintered a 50 cm long wire of different dimensions and measured Ic distribution of the 50 cm long sample. We investigated the bending tolerances using the mandrels of different diameters. After measuring the fundamental properties of the rolled wires described above, we made a coil which was wound by the wires around a ceramic tube of a 50 mm outer diameter.

All the Ag sheathed wires and coils were finaly sintered and annealed in the same manner described before. Ic measurement was performed by the conventional four probe dc method under the criterion of $1 \mu V/cm$.

Bi-system

Appropriate amounts of Bi_2O_3, PbO, $SrCO_3$, $CaCO_3$ and CuO powders with 3 or 4N purity were mixed into composition, Bi:Pb:Sr:Ca:Cu = 1.6:0.4:2:2:3 and were calcined and sintered from 750°C to 870°C for 8 - 200 h in air, then ground into powders. The powders were put into silver tubes. The wires were fabricated by the similar method in the case of YBCO. Finaly the wire samples were sintered from 800°C to 870°C for 8 - 800 h in air. The magnetic field dependence of J_C was measured by the four probe dc method up to 2.5 T. The definition of J_C was $0.1 \mu V/cm$ at this measurement to obtain the accurate value because the value of Ic decreased in the high magnetic field. The magnetic field was applied to perpendicular to the currents, and parallel and perpendicular to the wide plane of Ag sheathed wires.

Results and discussion

Y-system

The Jc of YBCO Ag sheathed wires were increased with increasing the reduction ratio by pressing or rolling processes. Figure 1 relates Jc with the reduction ratio by pressing. The maximum Jc value of 4,140 A/cm² was obtained with the wire of 0.061 mm thickness, which strain by cold drawing is 7.94 and the reduction ratio by pressing is 3.72. Pressed wires revealed the Jc value of 3,000 - 4,000 A/cm² with the good reproducibility. On the other hand, the Jc value of a round wire was only 500 A/cm². Figure 2 shows the variation of the powder core density by the cold drawing process. Densification of the powder core was done with a few passes. In following procedure, the powder core density was saturated to about 80 % of the theoretical density. Further densification of the powder core was made by pressing up to 91 % of the theoretical density. So that Jc increased with increasing the powder core density.

Moreover, the Jc-D/t profiles were highly depended on the strain by the cold drawing. In other words, Jc seems to depend on the powder core size (Fig.1). Figure 4 showes the X-ray diffraction patterns of the taped YBCO core (a) and the powder (b). The intensity of (00ℓ) peaks of the taped core was enhanced. It suggests that the c-axis of grains aligned perpendicular to the wide plain at the boundary between the Ag sheath and the powder core. If the crystal orientation is limited to a 'effective depth', Jc increase with decreasing the absolute size of core. Detailed evaluation of the effective depth of the crystal orientation is now under studied.

Coil by Y-system

Prior to make a coil, we investigated the fundamental properties of the rolled wires, that is, the longitudinal uniformity of the superconductibity and the bending torelance.

The Ic distribution in a 15 m long wire is shown in Fig. 5. The short samples cut away every two meters were sintered at the same time. The deviation from the mean value of 7.6 A was only 3 % or less over the 15 m long wire.

A 50 cm long straight sample was sintered. Figure 6 shows the Ic distribution. Ic change in the right hand side was due to the temperature distribution of furnace; 5°C higher than the middle of the sample. The deviation from the mean value of 1.57 A was only 5 %. Any large deterioration was not observed in the case of sintering of the 50 cm long wire.

The Ic change of the wires when wound to the mandrels by 'wind and react' (W&R) and 'react and wind' (R&W) method is shown in Fig. 7. The Ic of the R&W wires rapidly decrease to 50 % of the straight samples value in exceeding the bending strain to 0.7 %. But the W&R wires keep the initial property up to 0.5 % of the bending strain.

After confirming the fundamental properties of rolled wires, we made a coil which was wound with 7 turns of 12 layers on the ceramic tube of 50 mm outer diameter. Figure 8 shows the coil load line. This superconducting coil could generate 47 gauss at the center of the coil and 64 gauss on the conductor with a transport current of 35 A. The summation of Ic of each wire at a zero magnetic field was 81 A.

Bi-system

The J_C was improved by making the oxide superconductor BPSCCO into homogeneous high T_C phase and aligning the crystal structure. J_C was increased up to $6,930 A/cm^2$ at 77.3 K. Figures 9 and 10 show the properties of the wire which J_C is $5,500\ A/cm^2$.

Figure 9 shows the X-ray diffraction pattern of the wide surface of BPSCCO itself. The series of peaks for the high T_C (110K) phase appear strongly with the characteritic (002) peak and the peaks of the low T_C phases (80K,7K) don't appear. It was clear that the grains of the high T_C phase were oriented with the c-axis perpendicular to the wide plane, because the intensity of the (00ℓ) peaks were strong and that of (200) peak was observed weakly.

Figure 10 shows the magnetic dependence of J_C for various superconductors. Before present work, J_C of the Ag sheathed wires was decreased rapidly at the weak magnetic field of about 100 gauss. The Jc value of $1,000\ A/cm^2$ at a zero magnetic field decreased to $35\ A/cm^2$ at 0.1 T of parallel magnetic field. The wire with J_C of $5,500\ A/cm^2$ at a zero magnetic field shows $1,200\ A/cm^2$ at 0.1 T of parallel magnetic fields to the wide plane of the sample and 550 A/cm^2 at 0.1 T of perpendicular to the wide plane, respectively. This improvement is very important because 0.1 T is a magnetic field strength for the power cable application.

Conclusion

The Jc of the Ag sheathed high Tc superconducting wires was improved by the metal working processes, especialy rolling or pressing process. The tendency of the Jc increase with increasing the strain and the reduction ratio was explained as follows : (a) the powder core densification, (b) the crystal orientation at the boundury between the metal sheath and the powder core.

The longitudinal uniformity of the superconducting properties was confirmed for the long wires made by rolling process. The coil made of that wires could transport the critical current of 35 A and generate maximum magnetic fields of 64 gauss.

The homogeneous high Tc (110 K) phase and the alignment of the grains were achieved in the powder core by the powder-in-tube method. As a result, critical current density was increased to 6,930 A/cm^2 and the Jc in the magnetic field of 0.1 T was improved by about 30 times for BPSCCO Ag sheathed wires.

References

1 T.Nakahara: A paper presented at 'lst international symposium on superconductivity', Nagoya, Japan, August 28-31, 1988.

2 S.Yazu: A paper presented at 'lst international symposium on superconductivity', Nagoya, Japan, August 28-31, 1988.

3 K.Ohmatsu, K.Ohkura, H.Takei, S.Yazu and H.Hitotsuyanagi: Jap. J. of Appl. Phy., 26(1987)1207.

4 T.Hikata and K.Sato: to be published.

5 K.Kawashima, M.Nagata, Y.Hosoda, S.Takano, N.Shibuta, H.Mukai and T.Hikata : '88 Appl. superconductivity conference, Sanfrancisco, CA, August 21-25, 1988. paper MI-6

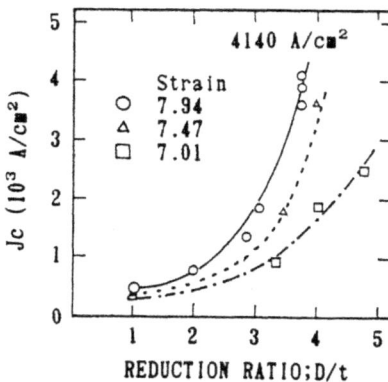

Fig.1. Jc dependence of the reduction ratio.

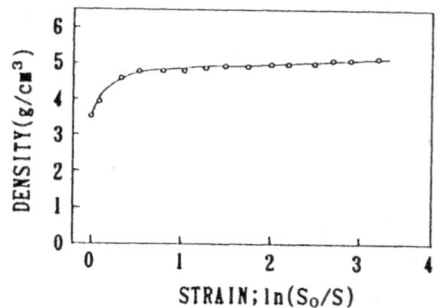

Fig.2. Relation between the powder core density and the strain by cold drawing.

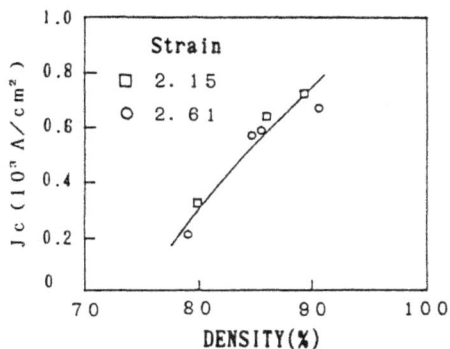

Fig.3. Relation between the critical current density and the powder core density by pressing.

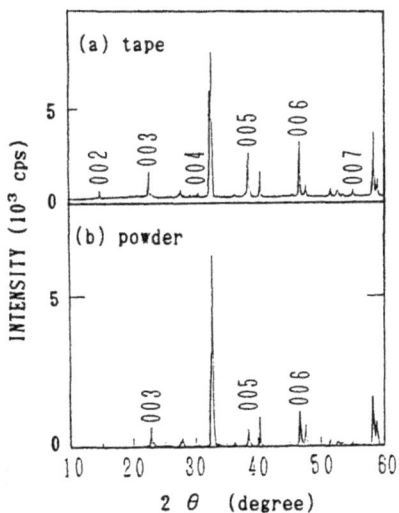

Fig.4. X-ray diffraction measurement of YBCO. Profile (a) is the powder core of pressed tape afte removing the Ag sheath. Profile (b) is the powder which was packed into Ag tube.

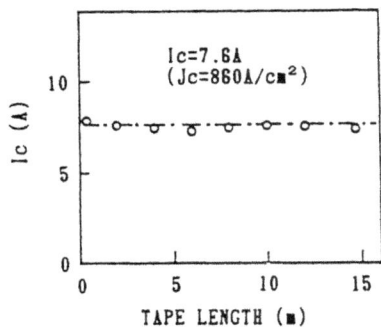

Fig.5. Critical current distribution in a 15 m long wire made by rolling. (0.5 mm thickness × 5 mm width)

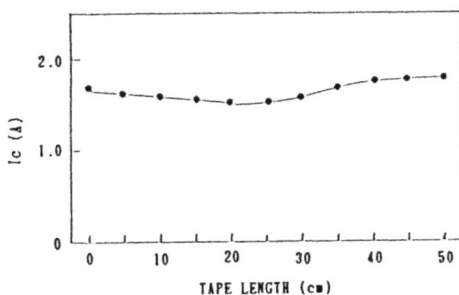

Fig.6. Critical current distribution of a 50 cm long sample. (0.3 mm thickness× 2.3 mm width)

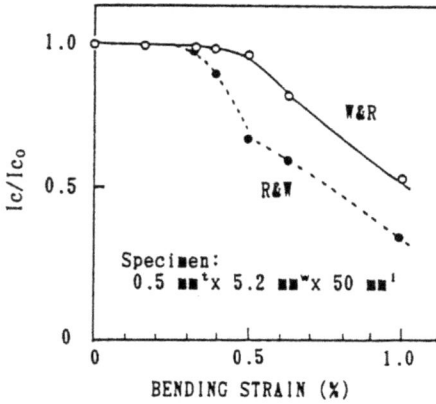

Fig.7. Jc change of the wires wound to mandreles ; open circles represent the wires made by the W&R method, closed circles represent by the R&W method.

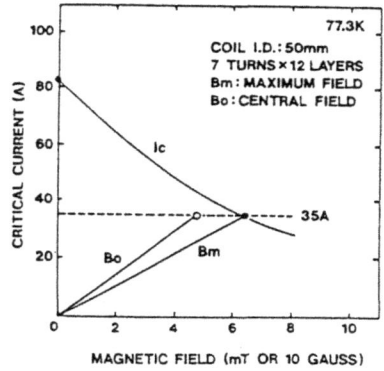

Fig.8. Load line of test coil.

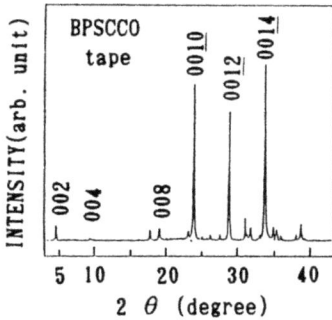

Fig.9. X-ray diffraction pattern of the BPSCCO wire after removing the Ag sheath.

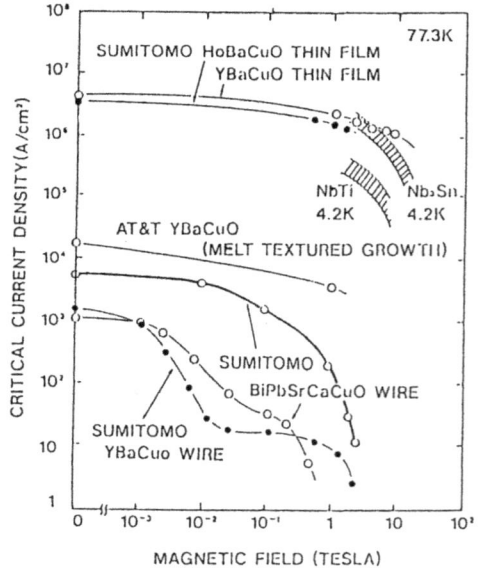

Fig.10. Critical current density dependence on the applied magnetic field of the thin film, the bulk material made by melt textured growth method (AT&T) and the YBCO, BPSCCO Ag sheathed wires.

FORMATION OF THE HIGH-Tc PHASE OF THE Bi-Pb-Sr-Ca-Cu-O
SUPERCONDUCTOR

Utako ENDO and Satoshi KOYAMA

Research and Development Department, Chemical Division
Daikin Industries, Ltd., Nishi-hitotsuya, Settsu-shi, Osaka 566

Tomoji KAWAI

The Institute of Scientific and Industrial Research
Osaka University, Mihogaoka, Ibaraki, Osaka 567

Abstract - A pure 110K phase of the Bi-Pb-Sr-Ca-Cu-O superconductor was obtained by co-decomposition of metal nitrates and
a solid reaction under low oxygen pressure. The best starting
compositions were in the region close to $Bi_{1.84}Pb_{0.34}Sr_2Ca_2Cu_3O_y$
with a little excess of Ca and Cu. It was found that insufficiency of Ca and Cu gives rise to the 80K phase formation, while
their surplus causes formation of Ca_2PbO_4 and a semiconducting
phase. They had better be of a little excess to obtain a pure
110K phase, though their amount delicately affects Tc(0) and
only a small surplus is enough to make it lower. Lead possibly
occupies Bi sites and dopes holes to the system along with Sr
defects.

Introduction

Since superconductivity with onset above 100K in the Bi-Sr-
Ca-Cu-O system was discovered,[1] it was one of the most challenging themes to single out the 110K phase. Among the efforts,
it has been reported that partial substitution of Pb for Bi
increases the volume fraction of the 110K phase.[2,3] Reaction
under low oxygen pressure was also found to be very effective.[4]
We have obtained a pure 110K phase by strictly controlling the
starting composition with additional Pb,[2,3] preparing the powders by a co-decomposition method and reacting them under low
oxygen pressure.[4,5] In the course of study, we found that a
subtle difference in composition, even an order of a per cent,
causes to form other phases or impurities and also affects
Tc(0). Even though samples show the same X-ray diffraction
pattern, some show superconductive transition above 100K, while
others show low-temperature tailing in R-T curves and have Tc(0)
around 90K. This tailing is observed by several groups and it
does not seem to be a unique matter of ours. In this work,
believing it important to control R-T property, we varied the
starting composition and examined the changes in X-ray patterns
and R-T curves. We found that the amount of Ca and Cu considerably affects Tc(0) and formation of other phases and impurities.

Experimental

The Bi-Pb-Sr-Ca-Cu-O samples were prepared by co-decomposition of metal nitrates. The powders of Bi_2O_3, PbO, $Sr(NO_3)_2$, $Ca(NO_3)_2 \cdot 4H_2O$ and CuO were dissolved in nitric acid with the desired cation ratio. The solution was stirred and heated until it became dry to form a solid with a light blue color. The nitrates were then co-decomposed at 800 °C for 30 min. The powder thus prepared was ground and pressed into pellets. They were heated at 835 °C for 84 hours under oxygen pressure of 1/13 atm and then were slowly cooled to room temperature.

The samples with the composition $Bi_x Pb_y Sr_2 Ca_2 Cu_3 O_z$ were prepared varying x and y values and then with x and y fixed at 1.84 and 0.34, Sr, Ca and Cu ratios were varied as $Bi_{1.84} Pb_{0.34} Sr_x Ca_y Cu_z O_w$.

The structure and the superconducting properties of the samples thus prepared were evaluated by powder X-ray diffraction using CuKα radiation, electron probe microanalysis (EPMA), resistivity measurement using a standard four-probe technique and ac susceptibility measurement at 200Hz. The dimension of the samples was 13mm in diameter and 1mm in thickness.

Results and Discussion

The ideal composition of the 110K phase is known to be $Bi_2 Sr_2 Ca_2 Cu_3 O_z$. Consequently, we first fixed the Sr:Ca:Cu ratio as 2:2:3 and varied the ratio of Bi and Pb to (Sr,Ca,Cu). Samples all showed the dominant diffraction peaks of the 110K phase ($2\theta=4.7°$, $23.9°$, $28.8°$, $33.8°$), but those of the 80K phase ($2\theta=5.7°$, $23.2°$, $27.5°$) and/or Ca_2PbO_4 ($2\theta=17.8°$) were also observed. Figure 1 shows the relation between the starting compositions and the X-ray reflection intensities of the 80K phase and Ca_2PbO_4 in $Bi_x Pb_y-(Sr_2 Ca_2 Cu_3 O_z)$. The circles in the figure represent the degree of the 80K phase formation and the curves are its "contour lines". The 80K phase was small inside these curves (point B,D). However, the Ca_2PbO_4 formation became clear as the Pb content was increased (point C,D). The optimum content for Bi and Pb, where no Ca_2PbO_4 and the least 80K phase is formed, is considered to lie in the shaded region in Fig.1. As Pb decreases by more than half after the reaction (84 hours), the sum of Bi and Pb should be a little more than 2, i.e. around 2.2 (point B). Thus we chose the value 1.84 and 0.34 as the Bi and Pb content (point B in the figure) and varied the Sr, Ca and Cu ratio.

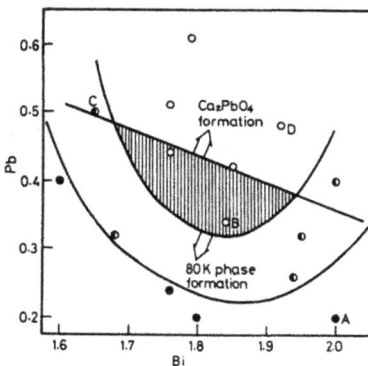

Fig.1. Relation between the amount of Bi and Pb (Sr:Ca:Cu=2:2:3) and the X-ray reflection intensities of the 80K phase in $Bi_x Pb_y Sr_2 Ca_2 Cu_3 O_z$. Circles denote: ● ; the peak intensity >300(cps), ◑ ; >150, ○ ; <150.

Figure 2 shows the relation between compositions and reflection intensities at $2\Theta=27.5°$ (80K phase (a)), $17.8°$ (Ca_2PbO_4 (b)) and $21.9°$ (semiconducting phase (c)). The last peak and one at $2\Theta=7.2°$ seemed to be coupled and were tentatively assigned as that of the semiconducting phase with a single Cu-O layer sandwiched by Bi_2O_2. The 80K phase tends to be formed as Ca and/or Cu is decreased. The composition of the 80K phase being $Bi(Pb)_2Sr_2Ca_1Cu_2O_y$ and relatively rich in Sr, this phase might be left alone if Ca is insufficient to convert it to the 110K phase. Ca_2PbO_4 appears in the region where Ca or Cu is comparatively a large excess, e.g. $Bi_{1.84}Pb_{0.34}Sr_{1.60}Ca_{2.17}Cu_{3.24}O_y$. The 110K phase can be stable and formed preferentially to let remaining Ca form Ca_2PbO_4.

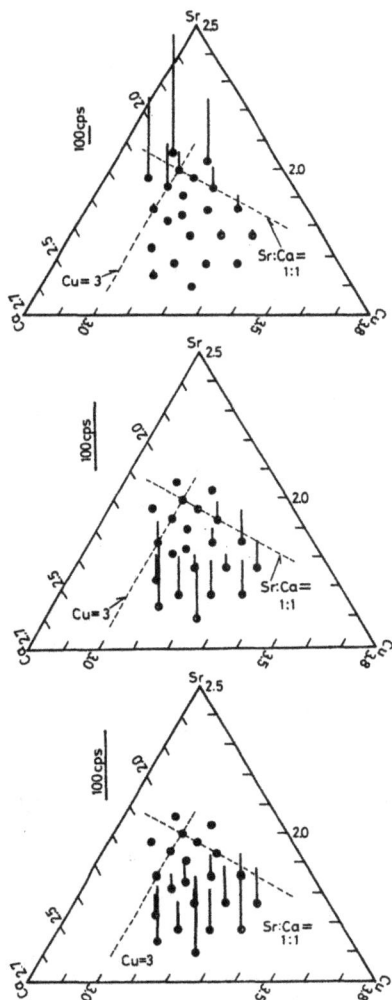

As both Ca and Cu are increased, apparent semiconducting phase tends to appear. In this region, Bi is also remaining because the amount of Sr being small, it is all used up to form the 80K and/or the 110K phase. Thus it is possible that these three remaining elements make up a new phase, which has single Cu-O layered structure.

These results are summarized in Fig.3. The 80K phase was not present in the region below the solid curve (point C,D in Fig.3). The Ca_2PbO_4 formation was not observed above the dotted curve (point A,B,C). A semiconducting phase appeared below the broken curve (point D,E). Compositions in the shaded region gave rise to the pure 110K phase. It is necessary to choose the proper starting compositions within such a narrow region to get the pure 110K phase.

An example of the X-ray pattern for the sample in the shaded region ($Bi_{1.84}Pb_{0.34}Sr_{1.91}Ca_{2.03}Cu_{3.06}O_y$) is shown in Fig.4. There is no indication of the 80K phase or any other impurities. All the peaks are indexed assuming that this 110K phase has a tetragonal unit cell with $a=b=5.396Å$ and $c=37.180Å$.

Fig.2. Relation between ratios of Sr, Ca and Cu (Bi:Pb=1.84:0.34) and intensities of the X-ray reflection peaks at (a) $2\Theta=27.5°$ (80K phase), (b) $17.8°$ (Ca_2PbO_4) and (c) $21.9°$ (semiconducting phase).

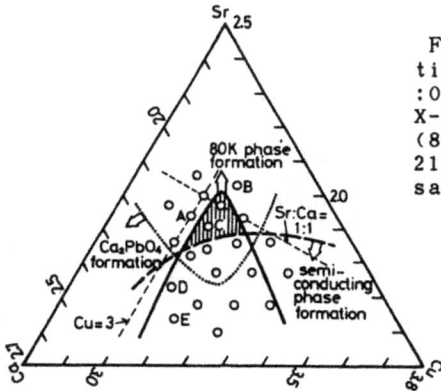

Fig.3. Relation between the ratios of Sr, Ca and Cu (Bi:Pb=1.84 :0.34) and the intensities of the X-ray reflection peak at $2\theta=27.5°$ (80K phase), $17.8°$ (Ca_2PbO_4) and $21.9°$ (semiconducting phase)[4] in the samples $Bi_{1.84}Pb_{0.34}Sr_xCa_yCu_zO_w$.

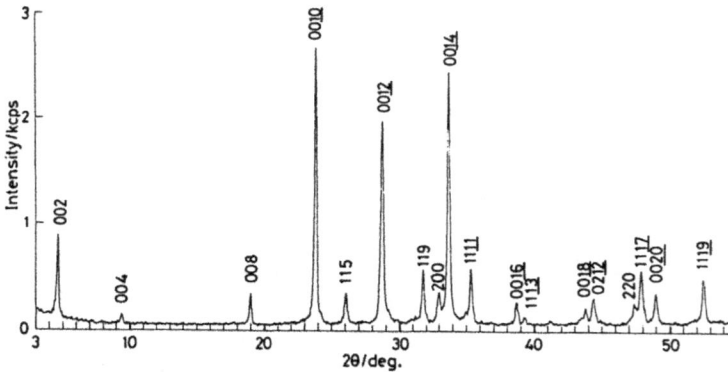

Fig.4. The X-ray diffraction pattern of the sample with the nominal composition $Bi_{1.84}Pb_{0.34}Sr_{1.91}Ca_{2.03}Cu_{3.06}O_y$ prepared at 835 °C for 84 hours ($O_2=1/13$ atm).

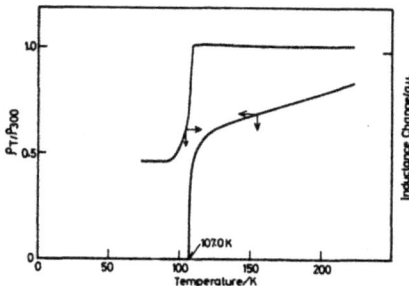

Fig.5. Resistivity-temperature and ac susceptibility-temperature curves of $Bi_{1.84}Pb_{0.34}Sr_{1.97}Ca_{1.97}Cu_{3.06}O_y$.

Figure 5 shows the results of the resistivity and the ac susceptibility measurements on the sample in the shaded region in Fig.3. The resistivity begins to decrease slowly at around 135K and shows a sharp drop at 110K to become zero at 107.0K with ΔTc within 3K, corresponding to the sharp drop in the susceptibility-temperature curve. Susceptibility measurements also showed no presence of the 80K phase. Samples with the composition in the shaded region in Fig.3 all gave similar results.

In order to compare the superconducting property of these samples, we measured temperature dependence of resistivity with the current of 1mA. Zero-resistivity temperatures are written in Fig.6. They exceed 100K in the Sr-rich region, but becomes lower as Sr is decreased. Resistivity-temperature curves accordingly show low-temperature tailing in Ca and Cu rich samples, which became smaller and finally disappeared as they were decreased. Figure 7 shows R-T curves of four samples indicated as A-D in Fig.6. The Ca-rich sample A, which contained Ca_2PbO_4 and the semiconducting phase as impurities, showed two-step like R-T curve and resistivity turned zero only at 86K, while the Sr-rich sample D, which contained about 20% of the 80K phase, showed no tailing and its Tc(0) was 104K. Samples B and C both gave the same X-ray pattern of a pure 110K phase and showed superconductivity below 107K when measured with a small current, e.g. 50µA. When the current was increased to 1mA, however, a tail appeared in the sample B alone to lower Tc(0) to 96K, which is compared with that of the sample C, remaining as high as 102K. It is interesting to note that the critical line that differentiates R-T property lies in the region on the composition diagram where the same X-ray pattern is given. As stated above, Tc(0) exceeding 100K with Sr/Ca ratio of 1.00 (sample C) is suddenly lowered to 96K when Sr/Ca=0.94 (sample B).

Comparing R-T curves of all the samples, the tailing seems to have some correlation with the appearance of the semiconducting phase. It is observed in the samples rich in Ca and Cu, which corresponds to the condition of the semiconducting phase formation. In fact, even if a sample is free from Ca_2PbO_4, its R-T curve has a tail if the semiconducting phase is present.

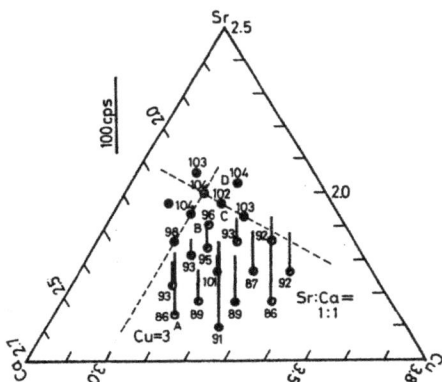

Fig.6. Relation between the ratio of Sr, Ca and Cu (Bi:Pb=1.84:0.34) and Tc(0) measured with the current of 1mA. The bars indicate intensities of the X-ray reflection peak at $2\theta=21.9°$.

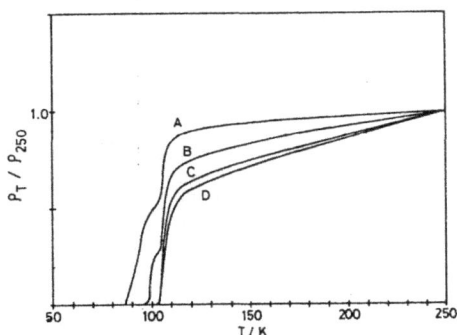

Fig.7. Temperature dependence of the samples noted as A-D in Fig.6:
A; $Bi_{1.84}Pb_{0.34}Sr_{1.64}Ca_{2.25}Cu_{3.11}O_y$,
B; $Bi_{1.84}Pb_{0.34}Sr_{1.91}Ca_{2.03}Cu_{3.06}O_y$,
C; $Bi_{1.84}Pb_{0.34}Sr_{1.97}Ca_{1.97}Cu_{3.06}O_y$,
D; $Bi_{1.84}Pb_{0.34}Sr_{2.03}Ca_{1.90}Cu_{3.06}O_y$.

There are two possibilities as the cause of this tailing. One is the 110K phase being connected with weak links to lower Tc(0). The other is the existence of some new superconducting phase with Tc around 95K. It is suggested from the fact that the onset temperature of the drop of resistivity below 100K is always about 95K. Absence of the 80K phase is shown with ac susceptibility measurement. It must be noted that this new phase, if any, has a triple Cu-O layered structure which gives the same X-ray pattern as the 110K phase. In any case, the composition must be strictly controlled around Sr:Ca:Cu=2:2:3 to let the amount of Ca and Cu be least in order to obtain a pure 110K phase which superconducts above 100K with a large current.

We may now discuss the nature of the 110K superconductor from the above results of composition variation. To begin with, we should state that segregation was not observed in EPMA analysis, suggesting uniformity of samples, hence validity of regarding the Sr:Ca:Cu ratio to be of the superconductors. The ideal structure of the 110K phase gives Cu valence of +2, leading to the necessity of doping holes to make it a superconductor. We propose that Pb^{2+}, substituted for Bi^{3+}, acts as a hole donor. As a result of controlling the Sr:Ca:Cu ratio to three orders, a pure phase was obtained when it was very close to 2:2:3. This fact suggests that additional Pb is substituted for Bi to occupy its sites. Then because the valence of Pb is +2 and that of Bi is +3, one Pb atom can dope one hole to the system. Another candidate for hole donor is Sr defect. Our results also shows that Ca and Cu had better be of a little excess for formation of a pure phase. In other words, a pure phase is obtained with less Sr than the ideal amount, which implies the presence of Sr defects. A small variation in Sr amount can actually change Tc(0) by several degrees, suggesting a possibility of being a hole donor. In short, we believe that the fact that the best Sr:Ca:Cu ratio for obtaining a pure 110K phase is very close to 2:2:3 with a little excess of Ca and Cu leads to regard Pb and Sr defect as hole donors.

References

1. H.Maeda, Y.Tanaka, M.Fukutomi and T.Asano: Jpn.J.Appl.Phys., 27 (1988) L209.
2. S.A.Sunshine, T.Siegrist, L.F.Schneemeyer, D.W.Murphy, R.J.Cava, B.Batlogg, R.B.van Dover, R.M.Fleming, S.H.Glarum, S.Nakahara, R.Farrow, J.J.Krajewski, S.M.Zahurak, J.V.Waszczak, J.H.Marshall, P.Marsh, L.W.Rupp,Jr. and W.F.Peck: Phys.Rev.B38 (1988) 893.
3. M.Takano, J.Takada, K.Oda, H.Kitaguchi, Y.Miura, Y.Ikeda, Y.Tomii and H.Mazaki: Jpn.J.Appl.Phys., 27 (1988) L1041.
4. U.Endo, S.Koyama and T.Kawai: Jpn.J.Appl.Phys., 27 (1988) L1476.
5. S.Koyama, U.Endo and T.Kawai: to be published in Jpn.J.Appl.Phys., 27 (1988) No.10.

PREPARATION AND CHARACTERIZATION OF
HIGH Tc SUPERCONDUCTIVE Tl AND Bi SYSTEMS

Masayoshi TONOUCHI, Yoshito FUKUMOTO, and Takeshi KOBAYASHI

Faculty of Engineering Science, Osaka University
Toyonaka, Osaka 560, Japan

Abstract

Effect of element addition on the superconductivity of rare-earth-element-free high Tc superconductors has been investigated. The element was chosen out of the IIa group. From the measurements of the ac susceptibility and resistance, addition of the appropriate amount of Mg together with Cu was found to raise zero-resistance temperatures of the Bi system.

Introduction

Discovery of rare-earth-element-free high Tc superconductors of BiSrCaCuO compound have led people to another striking progress in high Tc superconductor research. Up to the present, superconductivity above 120K has been realized in the TlBaCaCuO system. In both systems, the number of CuO_2 plane plays an important role in high Tc superconductivity: The more the number increases, the more the Tc does. It suggests that the modulation of the CuO_2 plane would enhance the Tc much higher than 150K. However, the addition of CuO together with Ca element into the 2223 phase of Bi and Tl compounds has not improved the Tc much. Therefore, in order to realize the strong enhancement of the Tc, a role of the Ca element as well as CuO_2 plane should be clarified.

The present paper reports the effect of the additional element on the superconductivity of Bi and Tl compound. The element was chosen out of the IIa group to study the role of Ca element.

Experiment

The compound ceramics were prepared by the conventional solid state reaction method. All of the starting materials were the oxide of the elements. At first, the calcination was done at about 800°C for about 10hr, and then the pellets were sintered at 820-850°C for 48-120hr in air.

The prepared ceramics were characterized by the Tc, Meissner effect, and X-ray diffraction.

Results and Discussions

Figure 1 shows the resistance vs. temperature curve of the $(Bi_{0.7}Pb_{0.3})_2Sr_2Ca_2Cu_3O_x$. The specimen was sintered at 835°C for 120hr. The zero-resistance temperature of 106K was obtained. Figure 2 gives the

114

resistance vs. temperature of the $Tl_2Ba_2Ca_2Cu_3O_x$. The specimen was sintered at 880C for 1hr followed by the annealing at 850C for 5hr. The zero-resistance temperature of 118K was obtained.

Then we tried an addition of new element into the 2223 phase of Tl and Bi compound. In the Tl system, however, we could not characterize the exact properties due to the decrease of the Tl element from the specimen. We have closely studied about the Bi compound. Figure 3 shows the resistance vs . temperature curves of Bi compounds doped with Mg, Ca, and Sr together with Cu. The composition of the prepared ceramics was the $(Bi_{0.7}Pb_{0.3})_2Sr_2Ca_2M_2Cu_5O_x$ (M=Be,Mg,Ca,Sr,Ba). The Mg doped specimen had the zero-resistance temperature of 108K, which was a little bit higher than the one of the non-doped specimen. The enhancement of the Tc was observed in many Mg doped samples. A typical effect was given in Fig.4. The increase of $CaCuO_2$ decreased the zero-resistance temperature. In the Sr doped samples, decrease of both onset and zero-resistance temperatures were observed. This phase was found to have the superconductivity at about 60K which was lower than that of the 2212 phase in Bi compound. The Ba introduction formed the insulating compound.

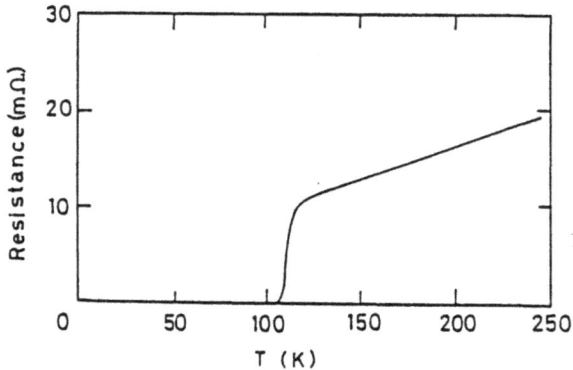

Fig.1 Resistance vs. temperature for Bi(Pb)-Sr-Ca-Cu-O sample.

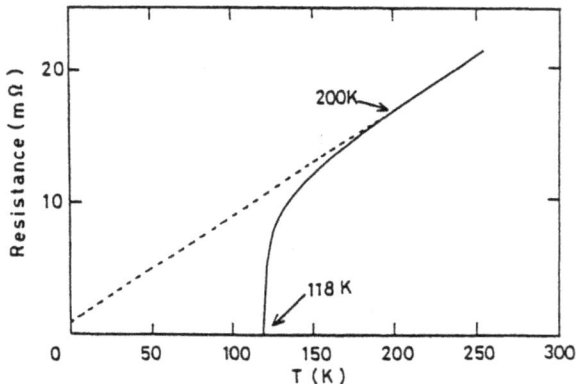

Fig.2 Resistance vs. temperature for Tl-Ba-Ca-Cu-O sample.

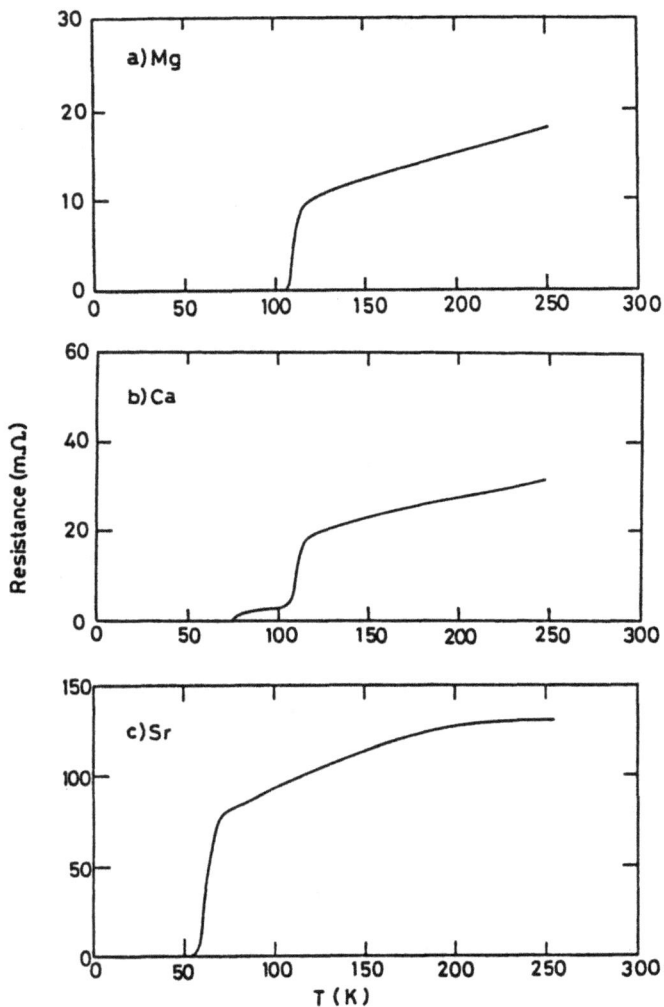

Fig.3 Resistance vs. temperature for the $(Bi,Pb)_2Sr_2Ca_2M_2Cu_5O_X$ (M=Mg(a), Ca(b), and Sr(c)) sample.

116

Conclusion

An introduction of Mg element together with Cu into the 2223 phase of Bi compound were found to enhance the zero-resistance temperature. The Sr doped sample had the superconductivity at about 60K. An introduction of Ba element destroyed the superconductivity of Bi compound.

References

1. H.Maeda,Y.Tanaka,M.Fukutomi, and T.Asano:Jpn.J.Appl.Phys.27(1988)L209.
2. Z.Z.Sheng and A.M.Herman: Nature 332(1988)138.
3. Y.Kubo, Y.Shimakawa, T.Manako, T.Satoh and H.Igarashi:
                              Jpn.J.Appl.Phys.27(1988)L591.

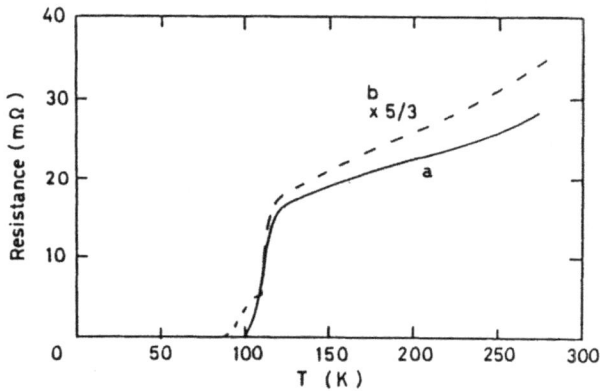

Fig.4 Resistance vs. temperature. Solid and dashed lines are for $(Bi,Pb)_2Sr_2Ca_2Mg_1Cu_4O_X$ and $(Bi,Pb)_2Sr_2Ca_2Cu_3O_X$ samples, respectively.

EFFECT OF HIGH PRESSURE OXYGEN ANNEALING ON SUPERCONDUCTIVITY OF LaxCuO4-y AND La2-xBaxCuO4-y

Rikuo OGAWA, Takayuki MIYATAKE and Kazuyuki SHIBUTANI

Superconducting and Cryogenic Technology Section,
Technology Developement Group, Kobe Steel, Ltd.,
3-18,1 Chome, Wakihamacho, Chuoku, Kobe, 651, Japan

Abstract - The influence of oxygen non-stoichiometry on the superconductivity of LaxCuO4-y ($1.8 \leq x \leq 2.0$) and La2-xBaxCuO4-y($x \leq 0.25$) have been studied by using the high pressure oxygen annealing and the magnetic susceptibility measurement. The undoped samples or less than 0.02 Ba doped samples showed the superconductivity which is closely related to the oxygen deficiency after the high pressure oxygen annealing. In these samples, the enhancement in the susceptibility measurement was found to depend on the cooling rate from the temperature above some critical point. For more than 0.03 Ba doping, high pressure oxygen annealing increased the carrier density and led to the increase in Tc and amount of superconductivity.

Introduction

Since the discovery of high Tc superconductivity in the perovskite related oxides, the important role of oxygen non-stoichiometry on the superconductivity has been pointed out in many papers. We have revealed that the high pressure oxygen annealing(O2-HIP) could decrease the oxygen deficiency and improve the superconductivity in Y-Ba-Cu-O and La-(Ba)-Cu-O systems (1,2). In La2-xBaxCuO4-y system, the superconductivity was observed for $x \geq 0.03$ after conventional annealing. O2-HIP treatment enhanced the superconductivity in undoped La2CuO4-y. At x=0.025, the superconductivity was not observed even if it was treated in O2-HIP. All data in the previous paper were taken by means of VSM (Vibrating Sample Magnetmeter) with cryostat. In order to clarify the phenomena more deeply, the magnetic susceptibility measurement were made by SQUID susceptmeter in the present paper. In this paper, we also report the influence of La deficiency on the superconductivity of LaxCuO4-y system for $1.8 \leq x \leq 2.0$.

Experimental

Samples of LaxCuO4-y($1.8 \leq x \leq 2.0$) and La2-xBaxCuO4-y($0 \leq x \leq 0.25$) were prepared by the solid state reaction of La2O3,BaCO3 and CuO powders. These powders were mixed and calcined at 1050 C for 12hr. After calcined, the powders were throughly ground and carlcined again on the same condition. Reground powders were pressed and annealed at two conditions, 1000° C for 2hr (or 1050° C for 8hr) in the electric furnace using Ar+20%O2 gas mixture of 1 atm. ,or 1000° C for 2hr or 600° C for 2hr in the Hot Isostatic Pressing using the same gas mixture of Ar+20%O2 of 100MPa.

Magnetic susceptibility measurements were carried out by using SQUID susceptmeter (HSM 2000) or vibrating sample magnetmeter with cryostat.

Results and Discussion

Fig.1 and Fig.2 show the critical temperature (Tc) and the amount of diamagnetic signals as a function of Ba doping (x) in La2-xBaxCu4-y system. As shown in Fig.1, the superconductivity of undoped La2CuO4-y is sensitive to the oxygen concentration. After the conventional annealing , the superconductivity is not observed. The special heat-treatment to increase the oxygen concentration, such as a long annealing in pure oxygen(3), a plasma oxidation technique(4) and a high pressure oxygen annealing(O2-HIP) can create the superconductivity. Tc and the amount of diamagnetic signals in undoped La2CuO4-y phase are strongly depend on the oxygen concentration. 600°C O2-HIP treatment increases the Tc and the amount of diamagnetic signals than 1000°C O2-HIP. This means that high oxygen concentration can be achieved by low temperature O2-HIP as described in Y-Ba-Cu-O system(2). At x=0.025, the superconductivity is not observed even if it is treated in O2-HIP. This point,as for Ba doping, is known as the transition point from semiconductor to metal state. The sufficient carrier can not be supplied by decreasing the oxygen deficiency at this transition point. Above this point, substitution of trivalent La with divalent Ba creates the oxygen deficiency in the lattice,and then sufficient carrier density is obtained. The maximun amount of superconductivity was obtained at x=0.075. For 0.03≤x≤0.075, O2-HIP treatment increases the Tc and the amount of diamagnetic signals as shown in Fig.2. For 0.075≤x≤0.25, however,the amount of superconductivity decreases with increasing the Ba doping. Although the increase in Tc is observed in this range of Ba doping after O2-HIP treatment, the amount of superconductivity does not increase at x=0.125 and 0.19 after O2-HIP treatment.

Fig. 3--5 show the results of magnetic susceptibility measurement by using the SQUID susceptmeter. These figures show the magnetic susceptibility of La2CuO4-y(Fig.3), La1.975Ba0.025 CuO4-y(Fig.4), La1.812Ba0.188CuO4-y(Fig.5) as a function of temperature after the conventional annealing and O2-HIP treatment (1000°C x 2hr at 100MPa of Ar+20%O2). The superconductivity of La2CuO4-y is found to be very sensitive to the oxygen deficiency (Fig.3). On the other hand, La1.975Ba0.025CuO4-y sample shows only very small susceptibility change below 30K after O2-HIP treatment (Fig.4). La1.812Ba0.188CuO4-y sample shows the increases in Tc and the diamagnetic signals after O2-HIP treatment(Fig.5). The carriers for superconductivity are considered to be supplied by the different sources between the semiconductive state of La2CuO4-y and the metal state of La2-xBaxCuO4-y. In semiconductive state, x≤0.025, the carriers are supposed to be supplied by La deficiency or oxygen deficiency itself,but in metal state,x≥0.03,they are supplied by Ba doping.

In order to clarify the phenomena of superconductivity in
La2CuO4-y, four samples with La deficiency (LaxCuO4-y
x=1.99,1.95,1.85,1.80) were prepared by the same method. All
samples were heat-treated in O2-HIP (1000°C for 2hr at 100MPa of
Ar+20%O2),and the magnetic susceptibility measurements were
carried out. Yoshizaki et al have reported anomalous susceptibility change in La2CuO4-y depending on the existense of the
magnetic field(5). So paying attention on the conditions of
cooling and heating the samples in the existence of the magnetic
field , the magnetic susceptibility was measured. The results
obtained are shown in Fig.6 and 7. Fig.6 a) shows the susceptibility changes of all samples during heating up from 4.6K after
all samples were once rapidly cooled to4.6K from 300K in the
magnetic field of 0.01T.All samples show the same behavior. No
change was detected among the samples with the different La
deficiency. Fig 6 b) shows the results of x=1.95,1.80, samples
during cooling down slowly(3K/min) from 300K to 4.6K in the
magnetic field of 0.01T compared with Fig.6 a). Samples cooled
down slowly in the magnetic field show the enhancement of
diamagnetic signals. Fig.6 c) shows the results of x=1.99 sample.
The sample was firstly quenched from 300K to 4.6K,then heated up
to 50K. The susceptibility was measured during heating. After that,
this sample was cooled slowly(3K/min). The same susceptibility
change as during heating was obtained during this cooling. After
that, the sample was again heated up to 300K,and then cooled slowly
to 4.6K. The result showed the enhancement of diamagnetic signals.
There exsists some critical temperature around the Neel temperature
as pointed out by Yoshizaki et al. They have reported this
phenomenon was closely related to the existence of the magnetic
field(Zero Field Cooling or Field Cooling). Hoever,we did all
tests in the magnetic fields and could get the same kind results
changing the cooling rates. Our results indicate that slow
cooling from the temperature above the critical temperature
enhances the diamagnetic signals. Fig.7 showed M-H curves of
x=1.99 samples at 4.6K. One was rapidly cooled ,and the other
was slowly cooled, from 300K to 4.6K in no magnetic field.

Conclusion

From the extensive study of the oxygen non-stoichiometry in
LaxCuO4-y and La2-xBaxCuO4-y system by using the high pressure
oxygen annealing(O2-HIP), it is concluded that below 0.02 Ba
doping, superconductivity appears below some oxygen deficiency
, at 0.025 Ba doping , effective carriers are not created by O2-
HIP treatment ,and above 0.03 Ba doping, superconductivity is
again improved by O2-HIP treatment. The enhancement of the
diamagnetic signales in LaxCuO4-y system is found to depend on
the cooling rate from the temperature above some critical point.
Then, this phenomenan might be due to some kind of ordering
of oxygen atoms or defects.

Reference

1) K.Shibutani et al;Proceeding of MRS International Meeting on
Advanced Materials,Tokyo,(1988) to be published

2) T.Miyatake,K.Shibutani and R.Ogawa; Proceeding of MRS Meeting
   Tokyo (1988) to be published
3) P.M.Grant et al ; Phys. Rev. Lett. 58 (1987) p2482
4) J.M.Tarascon et al ; Novel Superconductivity,p705,Plenum Press,NY and
   London (1987)
5) R.Yoshizaki et al; Solid State Commun.,Vol.65,No.12,(1988)p1539

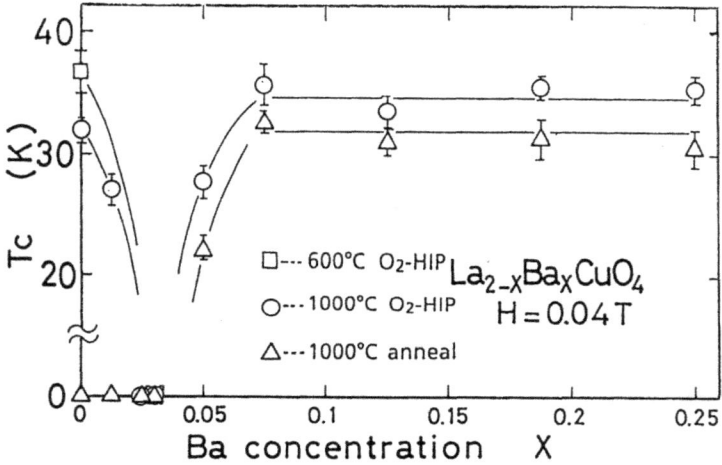

Fig.1 Critical temperature,Tc, as a function of Ba doping,x,
    in La2-xBaxCuO4-y

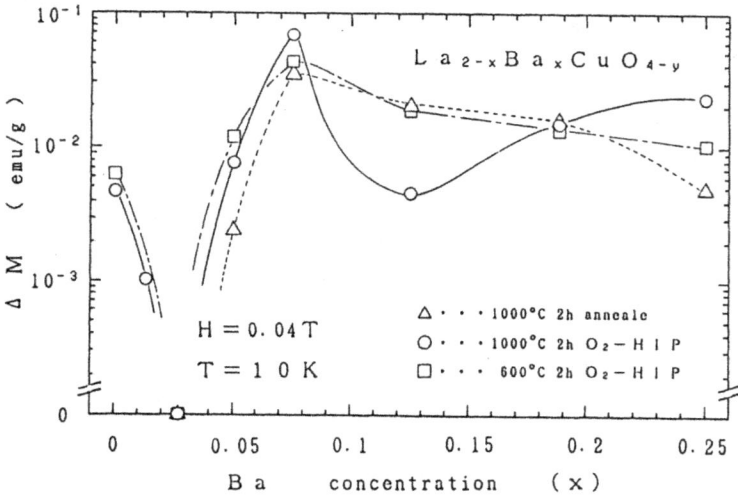

Fig.2 Diamagnetic signals at 10K and 0.04T as a function of Ba
    doping,x, in La2-xBaxCuO4-y

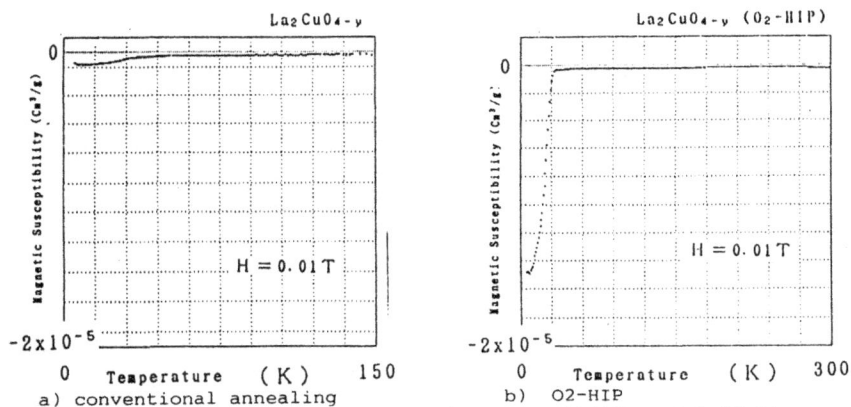

Fig.3 Temperature dependence of magnetic susceptibility of La2CuO4-y

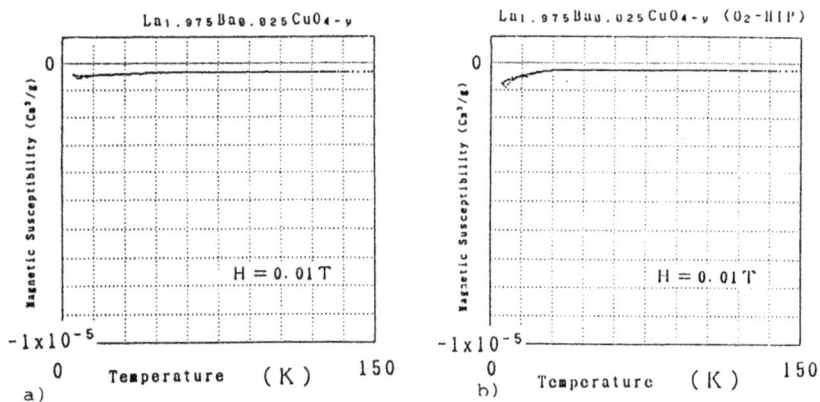

Fig.4 Temperature dependence of magnetic susceptibility of La2-xBaxCuO4-y
x=0.025

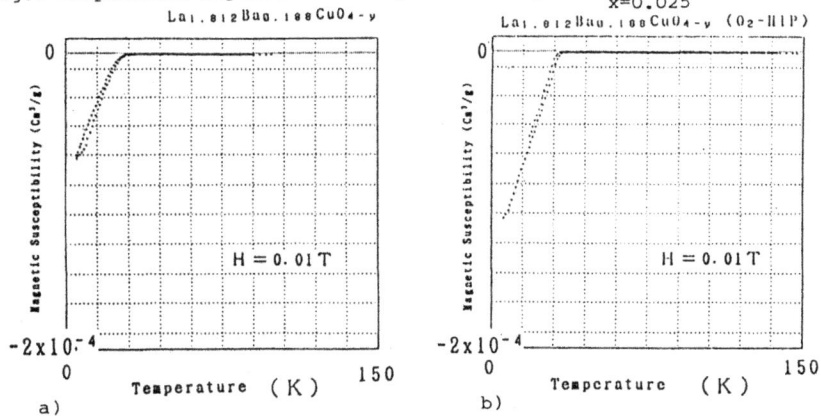

Fig.5 Temperature dependence of magnetic susceptibility of La2-xBaxCuO4-y

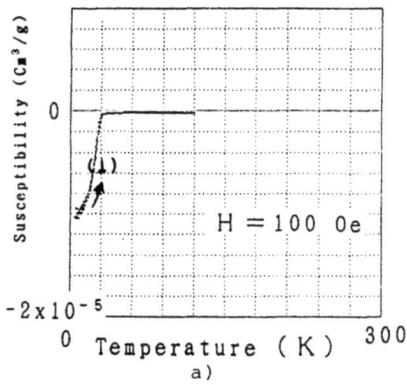

a)

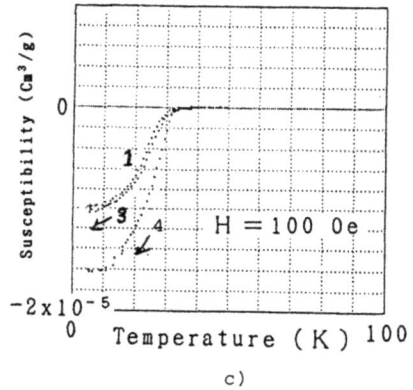

c)

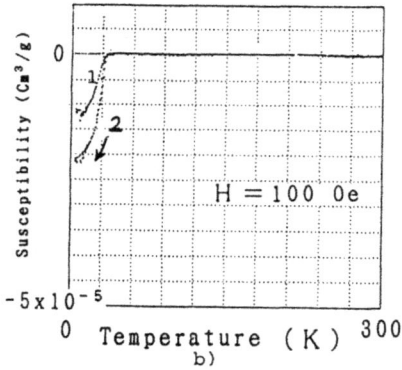

b)

Fig.6 Temperature dependence of
   magnetic susceptibility LaxCuO4-y
a) heating from 4.6K after rapid
   cooling from 300K to 4.6K in
   0.01T, all samples
b) slow cooling (3K/min) from 300K
   to 4.6K in 0.01T (2),x=1.95,1.80
c) slow cooling from 150K to 4.6K(3)
   and slow cooling from 300K to
   4.6K(4) after heating to 300K,
   x=1.99

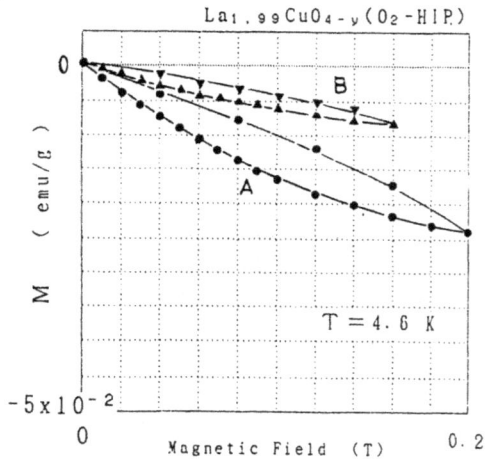

Fig.7 M-H curves of La1.99CuO4-y,A:slow cooling,B:rapid cooling
   in no magnetic field.   (3K/min)   (100K/min)

FABRICATION OF SCREEN-PRINTED HIGH-Tc SUPERCONDUCTING
OXIDE THICK FILMS ON YSZ SUBSTRATES

Toshiaki MAEOKA, Hideya OKADA, Yoshitaka NAGAMORI
Chiyoko HIROSUE, Yoshimi KONDO, Shinobu YANO, Hisako INADA
Susumu OKAMOTO and Kunihiko HAYASHI

Planning Department, Hayashi Chemical Industry Co., Ltd.
Ishihara Kisshoin, Minami-ku, Kyoto, 601, Japan

Abstract - Fine and homogeneous $Y_1Ba_2Cu_3O_{7-x}$ superconducting powder was prepared by oxalic acid-ethanol method and obtained powder was used as a component of screen printing ink. Screen-printed patterns of $Y_1Ba_2Cu_3O_{7-x}$ ink on 8 mol% doped yttria stabilized zirconia (YSZ) substrates were fired in a temperature range between 930°C and 1010°C in an oxygen atmosphere. Superconducting oxide thick film with superconductive transition critical temperature (Tc) arround 90k and critical current density (Jc) of 223A/cm² at liquid nitrogen temperature (77k) was fabricated. Adhesive strength between superconductive oxide thick film and YSZ substrate was also measured.

Introduction

Since the discovery of high-Tc superconductive oxide in the Y-Ba-Cu-O system [1], many studies and reports have been presented aiming at electric application of superconducting wire and electronic device application from theoretical and technological point of view. In general, this material has high-Tc that exceeds boiling point of liquid nitrogen (77K) but its Jc is not so high. As a bulk polycrystal., the highest Jc of the Y-Ba-Cu oxide is reported $10^3A/cm^2$ and usually it shows $10^2A/cm^2$ [2,3]. In a magnetic field the Jc decreases rapidly with the field. It is, therefore, usually told that this superconductive material is very difficult to apply for power electric superconductive wire because of its sensitivity against the external magnetic field.
On the contrary, the sensitivity of the Y-Ba-Cu-O superconductive material is very suitable to detect external magnetic field by measuring flux-flow resistance as output voltage in the mixed state [4]. In this sense, weak external magnetic field can be detected by homogeneous long and narrow line pattern of the Y-Ba-Cu-O superconductive thick film. However, at present the study on thick film fabrication is very few and the maximum Jc of thick film is about $30A/cm^2$ [5]. In thick film fabrication, almost all the researchers have used metal oxides such as yttrium oxide, copper oxide and barium carbonate as starting materials and they have used these metal oxides by mixing and solid states reaction method. Recently in ceramic field, chemically prepared powders have becoming popular for making homogeneous high quality ceramics [6].
In this paper, the $Y_1Ba_2Cu_3O_{7-x}$ superconductive powder was prepared by chemical coprecipitation so called oxalic acid-ethanol method. Homogeneous superconductive thick film oxide with high Jc and high adhesive strength was fabricated on YSZ substrates and evaluated the possibility of the thick film for the actual application for magnetic field sensor [7].

Experimental procedure

Powder preparation - Figure 1 shows a process flow diagram of $Y_1Ba_2Cu_3O_{7-x}$ powder preparation by oxalic acid-ethanol method. The mixture of $Y^{+3}Ba^{+2}Cu^{+2}$ nitrate solution with an accurate molten ratio of Y:Ba:Cu = 1:2:3 was prepared from $Y(NO_3)_3$ $6H_2O$, $Ba(NO_3)_2$ and $Cu(NO_3)_2$ $3H_2O$. The mixture of $Y^{+3}Ba^{+2}Cu^{+2}$ nitrate solution was poured into oxalic acid in ethanol solution and stirred sufficiently. The obtained precipitate was filtered and washed with ethanol and dried in vacume at 80°C for 24 hrs. Thermal reaction of the precipitate was measured by thermogravimetric (TG) and differential thermal analysis (DTA) in air at the heating ratio of 5°C/min. The precipitate was calcined at 810°C for 13 hrs and again calcined at 910°C for 8 hrs and crushed. Reaction of the powder calcined at 910°C was examined by X-ray diffraction (XRD) analysis with CuKα_1 radiation and particle size analyzer and scanning electron microscope (SEM). Molten ratio of the powder was analyzed by inductively coupled plasma emission spectrometry (ICP).

Thick film fabrication - The Y-Ba-Cu-O superconductive powder was mixed with acrylic resin, butadiene rubber and mineral spirit and stirred sufficiently. The viscosity of the superconductive oxide powder containing ink was adjusted to 15000 centi-poise. This superconducting ink was screen printed on the YSZ substrate (size:25x25x1mm thick.) through 250 mesh stainless screen and dried 30 sec by hair dryer. This screen-printing procedure was repeated 4 to 20 times. Then this superconducting ink was dried in an electric oven for 2 hrs at 150°C and fired in an alumina crucible at the firing temperature range between 930°C and 1010°C in an oxygen flow. The heating and cooling rate was 180°C/hr and 138°C/hr respectively. Oxygen flow rate was 1 litter/min.

Evaluation of thick film - Thickness of the fired films was measured by surface texture analyzer and observed by an optical microscope. The SEM photograph of screen-printed thick films sintered at 930°C and 1010°C for 10 min were taken. Temperature dependance of resistivity of the films was measured by dc four probe method from 77K to room temperature. The Jc at 77K was also measured. The Jc was defined by the current density when the generated voltage reached 1 μV/cm from the V-I characteristics of the films. Adhesive strength of the thick films to the YSZ substrate was evaluated by tensile strength test using cyano-acrylate regin. To determine the homogeneity of the film, length dependance of resistance with various thickness was measured at room temperature.

Results and Discussion

Evaluation of powder - Thermal analysis of the coprecipitate is shown in figure 2. In DTA curve endothermic peak around 100°C is due to dehydration of oxalate, exthothermic peak between the temperature range 200°C and 300°C is thought to be thermal decomposition of oxalates and endothermic peak around 900°C is recognized to be decomposition of barium carbonate. In TG curve, in the temperature range between 200°C and 450°C, the weight decreased because of dehydration and decomposition of oxalates. The weight also decreased from 800°C to 1000°C due to the formation of the Y-Ba-Cu-O. The XRD pattern of $Y_1Ba_2Cu_3O_{7-x}$ powder calcined at 910°C for 8 hrs was shown in figure 3 which shows uniform orthorhombic single phase of $Y_1Ba_2Cu_3O_{7-x}$. Figure 4 shows particle size analysis of the powder calcined at 910°C for 8 hrs. The average particle size of this powder was about 0.39μm. The SEM photograph shown in figure 5 also sustains the same result. Molten ratio of Y:Ba:Cu was determined as 1.0:2.0:3.0 by the result of ICP analysis as shown in table 1. The fine Y-Ba-Cu-O powder with accurate molten ratio of Y:Ba:Cu=1.0:2.0:3.0 and uniform single phase of $Y_1Ba_2Cu_3O_{7-x}$ was prepared by oxalic acid-ethanol method.

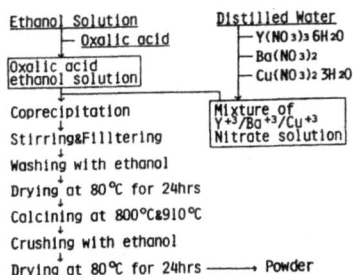

Fig.1. Process flow diagram of $Y_1Ba_2Cu_3O_{7-x}$ powder preparation by oxalic acid-ethanol method.

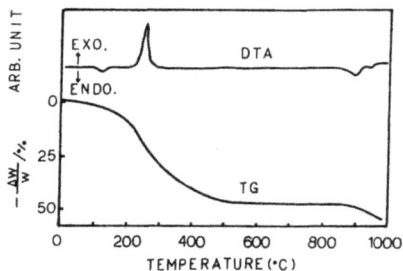

Fig.2. TG and DTA of the coprecipitate at the heating rate of 5°C/min.

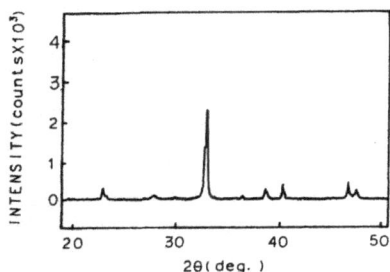

Fig.3. XRD pattern with CuKα1 radiation of $Y_1Ba_2Cu_3O_{7-x}$ powder calcined at 910°C for 8hrs.

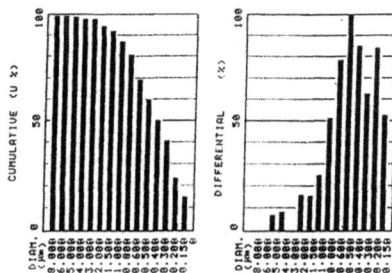

Fig.4. Particle size analysis of $Y_1Ba_2Cu_3O_{7-x}$ powder calcined at 910°C for 8hrs.

Fig.5. SEM photograph of $Y_1Ba_2Cu_3O_{7-x}$ powder calcined at 910°C for 8hrs.

Table 1. ICP analysis of Y-Ba-Cu-O powder calcined 910°C for 8hrs.

	Yttrium	Barium	Copper
Atom(%)	16.5	50.2	49.6
Molten ratio(%)	1.0	2.0	3.01

Evaluation of thick film - Surface of fired thick film was not smooth by the observation of optical microscope (magnification of 30 times). The films have many pores and traces of screen mesh. Figure 6 shows surface condition of the film fired at 1000°C for 10 min measured by surface texture analyzer. Figure 7 shows SEM photographs of the films fired at 930°C and 1010°C for 10 min. These photographs also show the film density is not so high. Temperature dependence of resistivity of the films fired was shown in figure 8. In the temperature range between 930°C and 1010°C, every film shows superconductive behavior. Resistivity has a positive coefficient from Tc to room temperature and shows metalic characteristic. Figure 9 shows V-I characteristics of the films fired. In the firing temperature range between 930°C and 1010°C, the Jc more than $10^2 A/cm^2$ could be obtained. Arrows in the figure depict Jc value, when the generated voltage reaches $1\mu V/cm$. Table 2 shows superconductive characteristics and adhesive strength of screen-printed films fired at the temperature range between 930°C and 1010°C. The Jc increase in accordance with the increase of the firing temperature. This tendency is thought to be caused by the density of the films. The pores of the film decreases when the firing temperature increased as shown in abovementioned figure 7. The highest Jc was 223 A/cm^2 in this study even some traces and pores were existed in the film. Adhesive strength was measured using cyanoacrylate resin. Stainless steel and films were adhered and tensile strength was measured. In spite of the removal of the steel from the film at the tensile strength was 46kgf /cm^2, the film was not peeled from substrate. The adhesive strength, therefore determined more than 46 kgf/mm^2 as shown in table 2. To clarify the homogeneity of the films, length dependance of resistance of screen printed films of a constant width of 0.75mm with various thickness of 10, 40 and 45μm, fired at 980°C for 10 min was measured as shown in figure 10. In the thickness of 10, 40 and 45 m, resistance increases almost linealy with the increment of length up to 25cm. The Jc increases with the increment of firing temperature and this is thought to be caused by the density of the film. The maximum

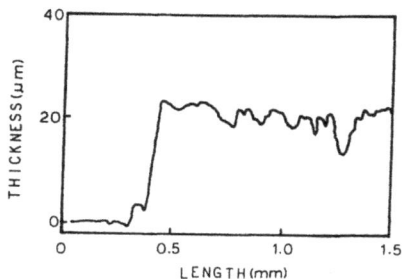

Fig.6. Surface condition of the film fired at 1000°C for 10min.

(A) Fired at 930°C for 10min.

(B) Fired at 1010°C for 10min.

Fig.7. SEM photographs of the films fired at 930°C(A) and 1010°C(B) for 10min.

Jc in this study was 223A/cm^2. Adhesive strength of the film to the substrate was determined more than 46kgf/cm^2. The homogeneity of the film was also certified by length dependance of resistance. As an example, high-Tc super-conducting oxide thick film fabricated on the YSZ substrate is illustrated in figure 11.

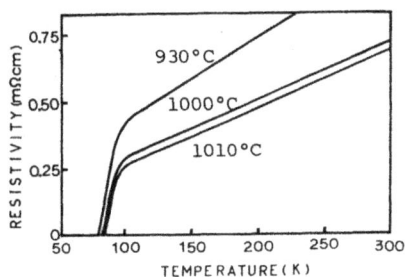

Fig.8. Temperature dependance of re-sistivity of the films fired at 930°C, 1000°C and 1010°C for 10min.

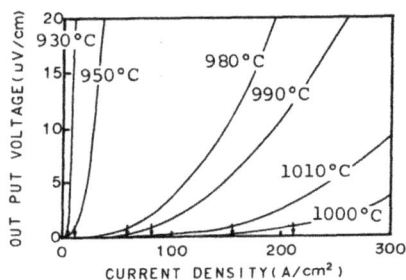

Fig.9. V-I characteristics of the films fired at 930°C, 950°C, 980°C, 990°C, 1000°C and 1010°C for 10min.

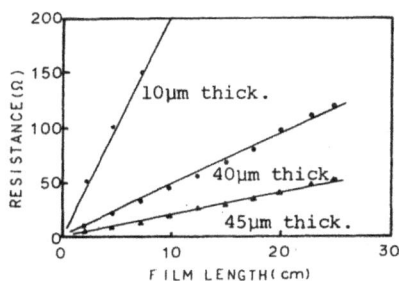

Fig.10. Length dependance of resistance of the films (0.75mm width) with the thick-ness of 10μm, 40μm and 45μm fired at 980°C for 10min.

Fig.11. High-Tc superconducting oxide thick film fabricated on the YSZ substrate.

Table 2. Evaluation of superconductivity and adhesive strength of the films fired at 930°C, 950°C, 980°C, 990°C, 1000°C and 1010°C.

Sample No.	Firing temp. (°C)	Firing time (min.)	Terminal length (mm)	Width (mm)	Thickness (μm)	Tc zero (K)	Jc (A/cm^2)	Adhesive Strength (kgf/cm^2)
1	930	10	250	0.75	20	78	4.0	46.0 >
2	950	10	250	0.75	20	81	12.1	46.0 >
3	980	10	250	0.75	20	83	61.5	46.0 >
4	990	10	250	0.75	20	85	79.5	46.0 >
5	1000	10	250	0.75	20	84	223.0	46.0 >
6	1010	10	250	0.75	20	85	151.8	46.0 >

Conclusions

The results obtained in this study are summarized as follows.

1) Fine (0.39µm) Y-Ba-Cu-O powder with accurate molten ratio of 1.0:2.0: 3.0 and uniform orthorhombic single phase was prepared by oxalic acid-ethanol method.

2) $Y_1Ba_2Cu_3O_{7-x}$ superconductive ink for printing fine pattern was satisfactory developed.

3) The Y-Ba-Cu-O superconductive thick oxide film with Tc of 84K and the Jc of 223A/cm^2 was fabricated.

4) Adhesive strength of the film to the YSZ substrate more than 46.0kgf/cm^2 which has a possibility to actual application could be achieved.

5) Homogeneity of the film was certified by length dependance of resistivity at room temperature.

Acknowlegements

The authers would like to express thier gratitude to Dr. S. Nishijima, Dr. T. Takahata, Professor T. Okada of ISIR Osaka University and Mr. H.Okushiba, Professor T. Hagihara of department of physics, Osaka Kyoiku University, for evaluating films. They also would like to express their thanks to Mr. K. Ueda of Kyoto Prefectual Comprehensive Guidance Center for Small and medium Enterprises for helpful technical suggestions.

References

1) M.K. Wu, J.R. Ashburm, C.J. Torng, P.H. Hor, R.L. Meng, L. Gao, Z.J. Huang, Y.Q. Wang, C.W. Chu, "Superconductivity at 93 K in a New Mixed-Phase Y-Ba-Cu-O Compound System at Ambient Pressure", Phys. Rev. Lett. 58, 9, 908-910 (1987).

2) R.J. Cava, B. Batlogg, R.B. van Dover, D.W. Murphy, S. Sunshine. T. Siegrist J.P. Remeika, E.A. Rietman, S. Zahurak, and G.P. Espinosa, "Bulk Superconductivity at 91 K in Single-Phase Oxygen-Deficient Perovskite $Ba_2YCu_3O_{9-\delta}$", Phys. Rev. Lett. 58, 6, 1676-1679 (1987).

3) J.W. Ekin, A.I. Braginski, A.J. Panson, M.A. Janocko, D.W. Capone 2, N.J. Zaluzec, B. Flandermeyer, O.F. de Lima, M.Hong, J. Kwo, S.H. Liou, "Evidence for weak link and anisotrophy limitations on the transport critical current in bulk polycrystalline $Y_1Ba_2Cu_3O_x$", J. Appl. Phys. 62, 12, 4821-4828 (1987).

4) Y.B. Kim, C.F. Hempstead, A.R. Strnad, "Flux-Flow Resistance in Type-2 Super conductors", Phys. Rev. 139, 4A, A1163-A1168 (1987).

5) J. Tabuchi, A, Ochi, K. Uchiyama, M. Yonezawa, "Fabrication of Screen-Printed High-Tc Superconducting Oxide Thick Films on Various Substrates", Nippon Seramikkusu-Kyokai-Gakujutsu-Ronbun, 96, 4, 450-454 (1988).

6) K. Hayashi, S. Okamoto, Y. Nagamori, T. Maeoka, H. Okada, T. Yamamoto, "Fabrication of Transparent PLZT Ceramics with a High Transmittance and their Application to Optical Light Shutter", Jpn. J. Appl. Phys. 26, Sppl. 26-2 126-128 (1987).

7) S. Nishijima, K. Takahata, T. Okada, H. Okushiba, T. Hagihara, Y. Nagamori, S. Okamoto, K. Hayashi, "Three-Dimensional Flux-Sensor Composed of High-Tc Superconductor" Proceedings of the Applied Superconductivity Conference, ED-13, (1988). (To be published in IEEE Trans. on Magn., MAG-25, 2, March issue (1989).)

PREPARATION OF Y-Ba-Cu-O FILMS BY RF MAGNETRON SPUTTERING

Mitsumasa SUZUKI, Tsutomu OZAKI, and Yutaka SHIMADA*

Department of Electrical Engineering,Tohoku University,
Aoba,Aramaki,Sendai 980,Japan.
*Research Institute for Scientific Measurements,Tohoku University,
2-1-1,Katahira-cho,Sendai 980,Japan.

Abstract - We have prepared thin films of Y-Ba-Cu-O on the cleaved surface
of MgO in an argon-oxygen atmosphere by rf sputtering and rf magnetron
sputtering using targets consisting of Y-Ba-Cu-O powder. Resputtering effects
were shown in rf-sputtered films. The composition of films is very different
from that of targets. Ba and Cu contents are appreciably reduced. However,
such resputtering effects could be considerably reduced for films prepared by
rf magnetron sputtering. Films deposited without heating MgO substrates were
amorphous. After heat treatments of 1 hour at 850-930°C in oxygen gas flow,
these films exhibit superconductivity. The maximum transition temperature is
as high as 80 K. In the case of depositions on substrates heated up to
700°C, as-deposited films show superconductivity with a zero resistance
temperature of above 70 K. Their superconducting critical current density at
4.2 K reaches a level of 1×10^4 A/cm^2 at 1 T. Relations between deposition
conditions and transition temperature for as-deposited superconducting films
are discussed.

Introduction

It is said that the technical applications of new high-T_c oxide superconductors realize at first in the fields of microelectronics and sensors. To
achieve these applications,the development of synthesis techniques for oxide
films of good quality becomes very important. Thin films of oxide superconductors to date have been prepared by different techniques such as electron-beam
evaporation, sputtering, laser evaporation,and very recently chemical vapor
deposition[1-4]. First of all, extensive studies on $YBa_2Cu_3O_{7-x}$ have been
performed using a rf magnetron sputtering system with a single target and the
deposition parameters affecting the formation of high-T_c oxide films have been
investigated. Generally, most of sputtered $YBa_2Cu_3O_{7-x}$ films have been made
from targets compensated on Ba and Cu. Moreover, the target composition
providing high-T_c films changes depending on the deposition temperature. These
facts indicate that the deposition mechanism of the perovskite phase in the
case of sputtering is very complicated. Much information is required for a
better understanding of the growth process of the high-T_c oxide films. In this
paper, we report the results on the preparation of $YBa_2Cu_3O_{7-x}$ films by both
rf and rf-magnetron sputtering techniques which offer different resputtering
effects. Further, preliminarily results on the preparation condition for as-
deposited superconducting films are presented.

Experimental

The preparation of thin films of $YBa_2Cu_3O_{7-x}$ was carried out using the
rf sputtering system(ULVAC,SBR-1104) in the initial stage. However, recently
every film have been made with the modified system capable of producing a

magnetron sputtering mode. For achieving the magnetron sputtering mode a rare-earth cobalt magnet was inserted under the cathode. The powder of Y-Ba-Cu-O which has been made by sintering two times the mixture of Y_2O_3, $BaCO_3$ and CuO in air at $900\,^{\circ}C$ for 10 h have been used as the target material. The target diameter is approximately 90 mm and eight powders with different compositions were prepared to test the effect of target composition on the growth of the high-T_c phase. The sputter deposition was carried out on cleaved-(100) MgO substrates which was kept at a substrate temperature T_d between ~160 and $\sim700\,^{\circ}C$ by a nichrome sheet heater. The rf input power is 100 W. The target-substrate(T-S) spacing D is varied between 25 and 40 mm. Argon and oxygen are used as sputtering gases. Each gas has been adjusted independently with a Needle valve. The oxygen partial pressure P_{O2} investigated in the present experiment ranges from 1.5 to 7.5 mTorr and the total gas pressure P_{Ar+O2} is in the range of 8-25 mTorr. After presputtering for 2 h, films 0.8-1.6 μm thick have been obtained by sputtering for 3-7 h. Films which have been prepared at T_d=160°C show amorphous. These films have been subsequently annealed in flowing oxygen gas at 850-950°C for 1 h to form the superconducting phase. As-deposited superconducting films have been mostly obtained at $T_d\gtreqless650\,^{\circ}C$. The optimum of target composition, T-S spacing and cooling process after deposition have been checked for as-deposited superconducting films. The superconducting transition temperature T_c of films was measured by a standard four-probe resistance technique using a Pt-resistance thermometer. The critical current density J_c of some films was measured at 4.2 K in the magnetic field normal to the film plane up to 15 T. Metal composition and crystal phases were analyzed by EDX(Energy dispersive X-ray analysis) and XRD (X-ray diffract meter), respectively.

Results and discussion

Figure 1 shows the arrangement of a target and substrates in the present

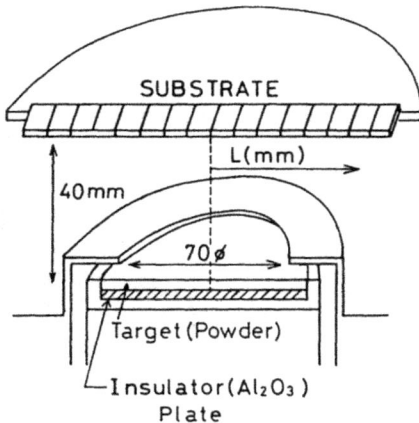

Fig.1 Arrangement of a target and substrates.

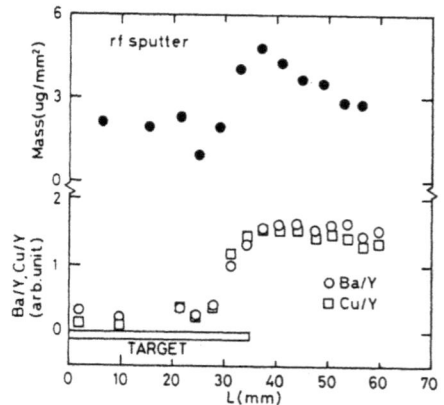

Fig.2 Weight per unit area of deposited film and X-ray intensity ratios of Ba and Cu to Y relating with composition of film as a function of substrate position for rf sputtering.

rf sputtering system. When the rf magnetron sputtering mode is made, the rare-earth cobalt magnet which generate a magnetic field of 300 G near the target surface is set under the target. At first we have attempted to prepare oxide films using a target consisting of Y-Ba-Cu-O powder on the conditions of P_{Ar+O2}=5-50 mTorr and T_d=230 °C by rf sputtering. The deposited film was not uniform as shown in Fig.2. The weight per unit area of the deposit is highly decreased at substrate positions less than 30 mm, where the resputtering effect seems to be very strong. As the substrate position is away from the resputtering region, the weight of deposited films rapidly increases. The maximum deposit is obtained at a substrate position of 35 mm corresponding to the radius of the target shield. The composition of deposited films is also much changed with the substrate position, as shown in the lower part of Fig.2. The X-ray intensity ratio of Cu to Y obtained in the EDX measurement is considerably low at substrate positions less than 30 mm and markedly increases with increasing the substrate position from 30 to 40 mm. Similar dependence on the substrate position is seen for the X-ray intensity of Ba to Cu. When the rf input power was increased from 100 W, the resputtering effects tend to become stronger. These facts suggest that Y is exclusively deposited on the substrate in the resputtering region. Probably,Y oxides which seem to show a strong tolerance against resputtering is formed readily due to the remarkable activity of Y. Other workers have experienced similar resputtering effects in preparing films of Y-Ba-Cu-O by rf sputtering from a sintered disk target and by dc sputtering[5,6]. In the case of rf magnetron sputtering, such resputtering effects are considerably reduced, as shown in Fig.3. The maximum deposit occurs just above the target center, where Ba and Cu are much incorporated in

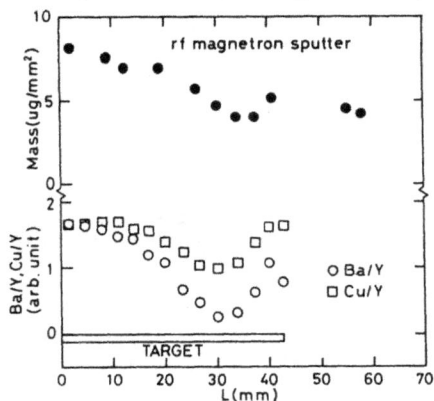

Fig.3 Weight per unit area of deposited film and X-ray intensity ratios of Ba and Cu to Y relating with composition of film as a function of substrate position for rf magnetron sputtering.

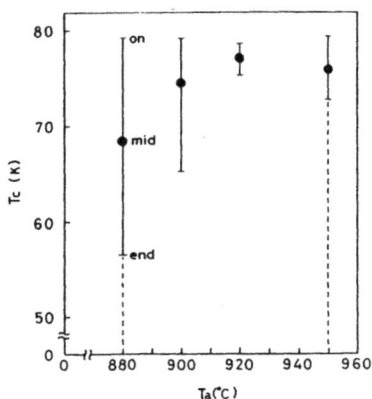

Fig.4 Superconducting transition temperature T_c of post-deposition annealed films as a function of annealing temperature T_a.

films. Hereafter, the preparation of superconducting films of $YBa_2Cu_3O_{7-x}$ has been performed by rf magnetron sputtering.

According to many other reports on thin film formation of $YBa_2Cu_3O_{7-x}$, amorphous films are produced by sputter-deposition at low T_d and a post-deposition annealing is required for transformation to superconducting films. In the present experiment, the optimum annealing temperature has been investigated on Y-Ba-Cu-O films which have been prepared at T_d=160 °C from the target with the 1:3:5 ratio of Y:Ba:Cu. Figure 4 shows the superconducting transition temperature as a function of the annealing temperature T_a for a heat treatment time of 1 h. The optimized annealing temperature is about 920°C. X-ray diffraction analysis on these films reveals that the perovskite phase grows at higher than 850 C and its amount increases with T_a. Resistive transition curves of films which have been deposited from targets with different compositions and subsequently annealed at the optimum temperature of 920°C are given in Fig.5. Values of $T_{c.mid}$ for these films range from 72 to 76 K and are about the same. It is noticeable that differences of $T_{c.mid}$ between obtained films are very small, regardless of the different target compositions.

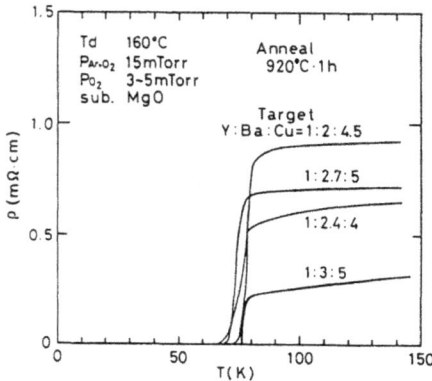

Fig.5 Resistivity vs temperature for films deposited from targets with different composition.

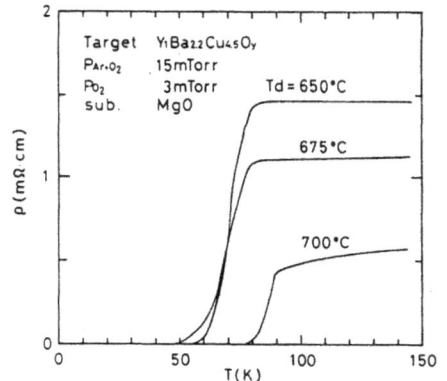

Fig.6 Resistivity vs temperature for as-deposited superconducting films prepared at temperatures between 650 and 700°C.

In order to get as-deposited superconducting films, the deposition temperature must be increased to above 600 C. Our attention have been directed to the effects of the T_d, sputtering gas pressure and target composition on the formation of as-deposited superconducting films. Figure 6 shows resistive transition curves of films deposited at 650, 675 and 700 °C. The target composition is the 1:2.2:4.5 ratio of Y:Ba:Cu and the T-S spacing is 30 mm. The P_{Ar+O2} and P_{O2} are 15 and 3 mTorr, respectively. These may be optimum conditions for preparing high-T_c films. As shown in the figure, all the films indicate as-deposited superconducting. There is a small difference of T_c between two films deposited at 650 and 675 ℃. However, a remarkable increase of T_c and a decrease of normal state resistivity are seen by increasing T_d to 700°C. According to X-ray diffraction analysis, the highest T_c film obtained at T_d=700 ℃ has strong X-ray diffraction line intensities of the perovskite phase and shows a preferential orientation of the c-axis perpendicular to the film plane as shown in Fig.7. The T_d is fixed at 700 C and the effect of the target composition on T_c has been investigated. Figure 8 shows plots of $T_{c,end}$ of films deposited from targets with different compositions against target composition. Clearly, it is seen that the target composition influences on the $T_{c,end}$ of as-deposited superconducting films and its effect is considerably stronger than that for post-deposition annealed films. These are considered to result from a change in the composition of deposited films with T_d. Especially the compensation of Cu to the target seems to give a beneficial effect on the growth of high-T_c films. However, it may be noted that the compensation of Ba is lower than expected from the similar resputtering effects for Ba and Cu in

133

the case of rf sputtering, as shown in Fig.2. The optimum target composition obtained to date is the 1:2.2:4.5 ratio of Y:Ba:Cu close to the target ratio (1:2:4.5) used by K.Wasa et al.[7]. The $T_{c,end}$ and $T_{c,on}$ of the best film are 78.2 and 85.1 K, respectively. On the other hand, an increase in the transition width has been seen in films deposited under the conditions of low P_{Ar+O_2}= 8 mTorr and D=25 or 35 mm, though their onset T_c maintains 85 K. Meanwhile, the transition width of films is affected by the cooling process after deposition, as shown in Fig.9. The film, which have been deposited at 700 °C and exposed in the course of cooling to 450°C for 1 h in a high P_{O_2} of $\sim$20 Torr, show sharper transition than a film quenched. Oxygen may be highly incorporated into the film by following the above mentioned cooling process.

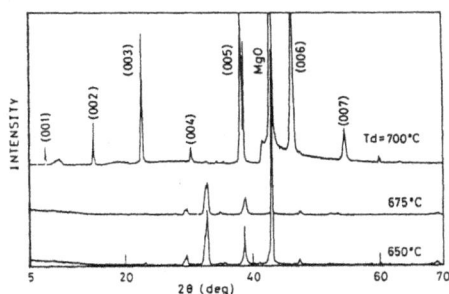

Fig.7 X-ray diffraction patterns of as-deposited superconducting films prepared at T_d=650, 675 and 700°C.

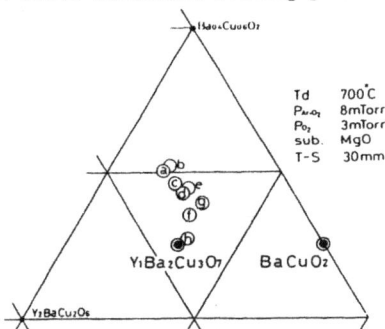

Fig.8 Dependence of T_c in as-deposited superconducting films on target composition.

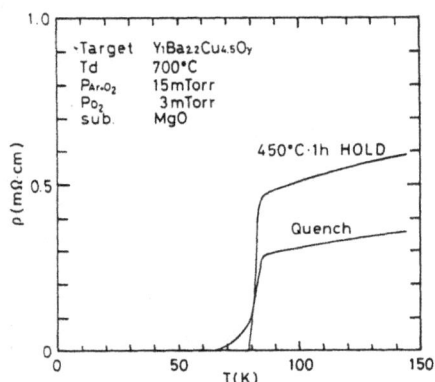

Target composition			Tc(K)		ρ_x(mΩ cm)
	Y	Ba	Cu	end - on	
a	1	2	4.5	25	80	4.82
b	1	2.2	5	50	90	2.66
c	1	2.2	4.5	64	89	1.88
d	1	2.4	4.5	46	85	0.86
e	1	2.7	5	35	87	5.39
f	1	2.4	4	18	72	4.71
g	1	3	5	35	84	3.69
h	1	2.2	3.3	15	75	8.29

Fig.9 Resistivity vs temperatures for a film quenched and for a film exposed to 450°C for 1 h during the course of cooling.

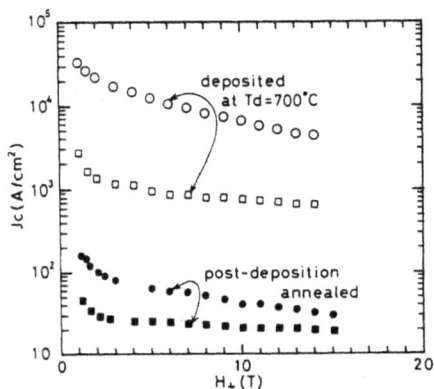

Fig.10 Critical current density J_c at 4.2 K of as-deposited superconducting films and post-deposition annealed films.

134

Figure 10 shows the field dependent critical current density, J_c, at 4.2 K of as-deposited superconducting films and post-deposition annealed ones. No patterning treatments on films for J_c measurements have been done. The applied field was oriented perpendicular to the film plane. Two films indicative of as-deposited superconducting show appreciably high J_c. The J_c of the best film is 3.3×10^4 A/cm^2 at 1 T and 4.3×10^3 A/cm^2 at 14 T. Most of the post-deposition annealed films exhibit a marked reduction of J_c. Their J_c values are in the range of 10-10^2 A/cm^2 at 14 T and are close to a level for bulk samples obtained by sintering. Such a reduction in J_c for post-deposition annealed films may be due to smaller contact areas between grains, since X-ray diffraction line intensities of the perovskite phase in both of as-deposited superconducting films and post-deposition annealed ones are about the same.

Summary

We have prepared post-deposition annealed films indicative of superconductivity, and further produced as-deposited superconducting ones on the cleaved surface of MgO by rf magnetron sputtering using targets consisting of Y-Ba-Cu-O powder. Although the T_c of post-deposition annealed films less depends on the target composition, that of as-deposited superconducting films is highly influenced by the target composition, sputtering gas pressure, and T-S spacing. As-deposited superconducting films with $T_{c,on}$=91 K and $T_{c,end}$=74-78 K have been obtained using a target with the 1:2.2:4.5 ratio of Y:Ba:Cu on the conditions of T_d=700 °C, T-S spacing=30 mm and argon-oxygen gas pressure=15 mTorr. The J_c at 4.2 K of an as-deposited superconducting film with $T_{c,end}$=74 K, is 3.3×10^4 A/cm^2 at 1 T and 4.3×10^3 A/cm^2 at 14 T. However, the J_c of post-deposition annealed films is considerably lower and close to a level of 10 A/cm^2 which is seen for bulk samples prepared by sintering.

The measurement of J_c was performed at the High Field Laboratory of Superconducting Materials, Tohoku University.

References

1) B.Oh, M.Naito, S.Aranson, P.Rosenthal, R.Barton, M.R.Beasley, T.H.Geballe, R.H.Hammond, and A.Kapitulnik, "Critical current densities and transport in superconducting YBa$_2$Cu$_3$O$_{7-x}$ films made by electron beam coevaporation", Appl.Phys.Lett.51, 852-854(1987).

2)Y.Enomoto, T.Murakami, M.Suzuki, and K.Moriwaki, "Largely Anisotropic Superconducting Critical Current in Epitaxially Grown Ba$_2$YCu$_3$O$_{7-x}$ Thin Film", Jpn.J.Appl.Phys.26, L1248-L1250(1987).

3) X.D.Wu, D.Dijkkamp, S.B.Ogale, A.Inam, E.W.Chase, P.F.Miceli, C.C.Chang, J.M.Tarascon, and T.Venkatesan, "Epitaxial ordering of oxide superconductor thin films on (100) SrTiO$_3$ prepared by pulsed laser evaporation", Appl.Phys.Lett.51, 861-863(1987).

4) A.D.Berry, D.K.Gaskill, R.T.Holm, E.J.Cukauskas, R.Kaplan, and R.L.Henry, "Formation of high T_c superconducting films by organometallic chemical vapor deposition", Appl.Phys.Lett.52, 1743-1745(1988).

5) S.I.Shah and P.F.Carcia, "Superconductivity and resputtering effects in rf sputtered YBa$_2$Cu$_3$O$_{7-x}$ thin films", Appl.Phys.Lett.51, 2146-2148(1987).

6) Y.Saito, K.Sakabe, K.Ishihara, K.Manaka, and S.Suganomata, "Preparation of Y-Ba-Cu-O Films by dc sputtering", Jpn.J.Appl.Phys.27, 1103-1104(1988).

7) K.Hirochi, H.Adachi, K.Setsune, O.Yamazaki, and K.Wasa, "Thickness Dependence of Superconductivity in As-sputtered Er-Ba-Cu-O Thin Films", Jpn.J.Appl.Phys.26,L1837-L1838(1987).

SYNTHESIS OF HIGH-TEMPERATURE SUPERCONDUCTING
Y-Ba-Cu-O THIN FILM BY ICB

Kenichiro YAMANISHI*, Yasuyuki KAWAGOE*, Seiji YASUNAGA*,
Katuhiro IMADA** and Ken SATO**

*   Product Development Laboratory
** Materials and Electronic Devices Laboratory
    Mitsubishi Electric Corporation
    Amagasaki, Hyogo, 661, Japan

Abstract - As-Grown Y-Ba-Cu-O thin films with high transition temperature have been synthesized by utilizing Ionized Cluster Beam (ICB) codeposition of Y, Ba and Cu in the activated oxygen atmosphere. The activated
oxygen was generated by silent discharge and was ejected to the substrate.
The pressure in the chamber was kept at 1.3×10^{-2} Pa, and the substrate was
heated from 600 °C to 650 °C. The transition temperature to zero-resistance
state (Tc, zero) of the as-grown film on the MgO(100) substrate was above
the boiling point of liquid nitrogen.

Introduction

Various techniques have been proposed to produce high-temperature
superconducting thin films [1-3]. Techniques which synthesize the superconducting thin films by using multiple sources should allow to control the
composition ratio of the film materials exactly. Therefore the techniques
could be applied to forming high quality thin films. The kinetic energy and
the chemical reactivity of ionized clusters assist the formation of films
with high density and high adhesive strength and the control of crystal
phases of films at low substrate temperature [4]. These facts have been
demonstrated by the formations of an Au thin film with smooth surface, a
single cristal Al thin film and a TiO$_2$ thin film with single crystalline
structure [5-7]. Some of these films are already applied to practical use,
and also the Y-Ba-Cu-O superconducting thin film has been already obtained
by using the combined method of ICB codeposition and post-annealing [8].
The method to eject evaporants through nozzles is also effective in
obtaining uniform distribution of film thickness and in stabilizing the
deposition rate in oxygen gas, because the crucible structure prevents the
evaporant from oxidization.
    The present paper describes the development of the reactive ICB
deposition apparatus with three sources and the application to synthesis of
the high-temperature superconducting Y-Ba-Cu-O thin film without annealing.
The films were formed by the codeposition of Y, Ba and Cu in oxygen gas.
The as-grown films have shown the transition to zero-resistance state (Tc,
zero) above the boiling point of liquid nitrogen.

Experimental Arrangement

The apparatus equipped with three ICB sources is shown in Fig. 1. Each
ICB sorce was set in the chamber with the axis at an angle of 20° to the
line normal to the substrate. The deposition rate of the each source can be

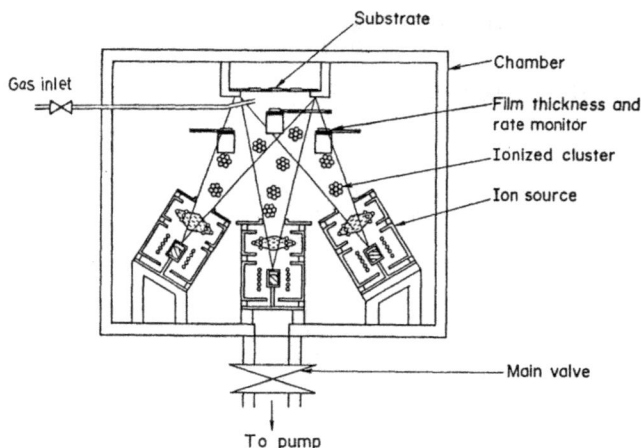

Fig. 1   The schematic of the reactive ICB apparatus equipped
with three ICB sources.

measured independentry by using the deposition rate monitor. Ionization
ratios od clusters and the acceleration voltages of the sources can be also
controlled independently, and the maximum acceleration voltage is 8 kV. The
substrate can be rotated and heated up to 900 °C. The distribution of film
thickness is within 4% over the substrate diameter of 100 mm. A cryogenic
pump was used and ultimate pressure in the chamber was below 1×10^{-4} Pa.
     As evaporants, Y, Ba and Cu metals were used. The crucible made of
refractory metal was used. Three elements were codeposited at the rate of
about 10 nm/min on MgO(100) substrate. Here, the composition of the film
under growth was controlled by adjusting the evaporation rate of each metal
evaporant. The oxygen partial pressure in the chamber was between from
0.65×10^{-2} Pa and 2.6×10^{-2} Pa, the substrate temperature was varied from 500
°C to 800 °C and the acceleration voltage was set below 2 kV. Oxygen gas
was ejected forward the substrate through the nozzle from the distance of
20 mm. Growth conditions in each method are listed in Table.

Table 1   Growth conditions of Y-Ba-Cu-O thin films.

Evaporant material	Metal Y, Ba , Cu (3N)
Crucible material	Refractory metal
Substrate	MgO (100)
Deposition rate	10 (nm /min)
Acceleration voltage	0~2 (kV)
Pressure	$0.65{\sim}2.6 \times 10^{2}$ (Pa)
Substrate temperature	500 ~ 800 (°C)

The measurement of electrical resistance was carried out in vacuum by a conventional four-probe technique using Au or In contacts. The crystal phase was analysed by X-ray diffraction. The composition ratio and depth profil of each element were analysed by ICP method and auger electron spectroscopy respectively.

Results and Discussions

Oxygen partial pressure and substrate temperature are very important in achieving the crystallization in an as-grown film From the viewpoint of assisting crystallization, the role of additional energy was studied by application of the acceleration voltage. Oxygen partial pressure was kept in the range under 2.6×10^{-2} Pa to prevent the ICB sources from oxidizing. The upper limit of substrate temperature was set at 800 °C to decrease the diffusion of substrate elements.

The crystallinity of the superconducting phase is observed in the film deposited at high oxygen partial pressure and adequate substrate temperature. The X-ray diffraction studies revealed the superconducting phase most clearly in the film synthesized under the following conditions.

The acceleration voltage was 0.7 kV, the substrate temperature was from 600 °C to 650 °C, the film thickness was 0.5 μm and the pressure was 1.3×10^{-2} Pa. Figure 2 shows the diffraction pattern. The film shows the transition to superconducting state and the starting temperature (Tc, onset) was about 20K as shown in Fig. 3.

Though the amount of oxygen supplied to the substrate was sufficient to synthesize the superconductor, the amount consumed by the reaction seemes to by low. To solve this problem, the addition of activated oxygen to the oxygen gas was tried. Practically the oxygen gas containing 5% of O_3 which was generated by silent discharge was used. The crystallinity of the super-conducting phase is observed clearly in the film as shown in Fig. 2. SEM analysis revealed that the size of crystal in the film was about 1000 Å and was 1/10 compared with that of the crystal in the film formed in the gas without O_3. X-ray diffraction pattern shows the crystallinity is also different in both cases. Activated oxygen is effective to obtain the film with high transition temperature and Tc,zero of the film was 70K as shown in Fig. 3. Figure 4 shows depth profiles of elements of the film.

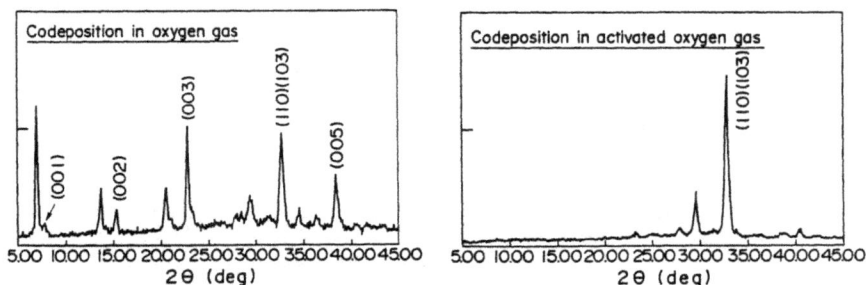

Fig. 2 The X-ray diffraction patterns of the as-grown Y-Ba-Cu-O films

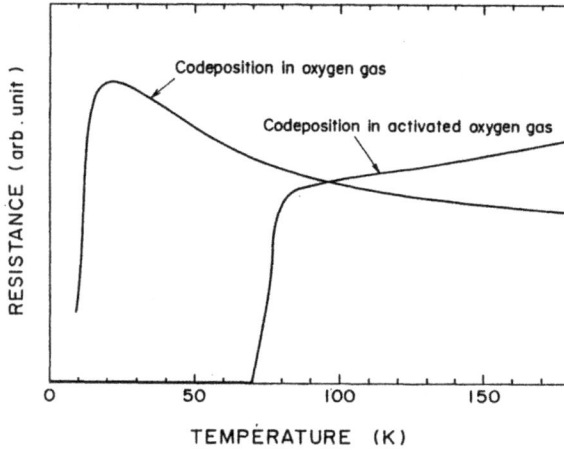

Fig. 3 The electrical resistance-temperature relationships
of the as-grown Y-Ba-Cu-O films.

This film shows very little interdiffusion of the substrate material
compared with the film post-annealed at the high temperature of 930 °C [8].
Furthermore the films of 0.1 μm thickness were formed by controlling evaporation rate of each evaporant more precisely. Then the film with Tc,zero
above the boiling point of liquied nitrogen was obtained as shown in Fig. 5.

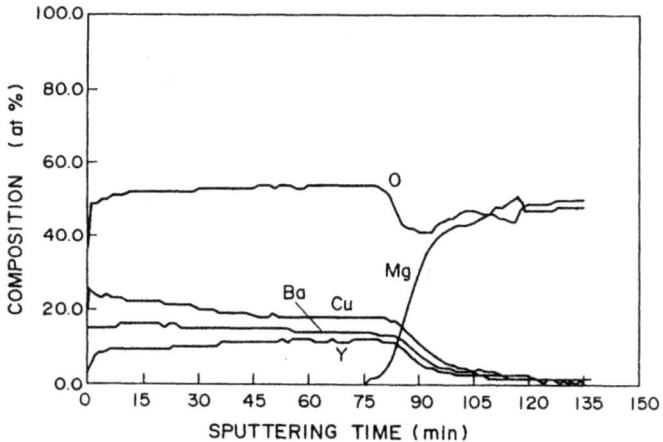

Fig. 4 Depth profiles of constituent elements of the as-grown film.

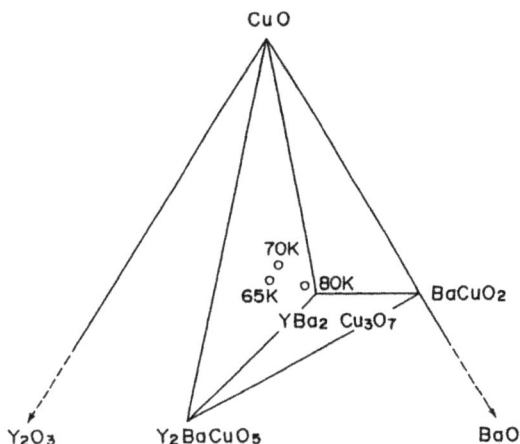

Fig. 5 Dependence of Tc,zeros on composition ratios
of the as-grown films.

    These results suggest that the as-grown method using multiple ICB
allows to synthesize the superconducting thin film at low temperature, so it
has the possibility to obtain the high quality thin film.

Conclusions

    The results obtained in this study are summarized follows.
    1) The Y-Ba-Cu-O thin film with Tc,zero above the boiling point of
liquid nitrogen was synthesized by using the codeposition technique in
oxygen gas without post-annealing.
    2) The ICB method using multiple sources can be applied to obtaining
films with high Tc,zero by adjusting the composition ratio of film materials.
    3) The energy of ionized clusters and activated oxygen play a very
important role in forming high quality thin films at low substrate temperature.

References

1) M.Hong, S.H.Liou, J.Kwo and B.A.Davidson, Appl. Phys. Lett. $\underline{51}$ (9) 694
    (1987).
2) J.Kwo, T.C.Hsieh, R.M.Fleming, M.Hong, S.h.Liou, B.A.Davidson and
    L.C.Feldman, Appl. Phys. Lett, $\underline{51}$ (14) 1112 (1987).
3) M.Naito, R.H.Hammond, Bo.Oh, M.R.Hahn, J.W.P.Hsu, P.Rosenthal,
    A.F.Marshall, M.R.Beasley, T.H.Geballe and A.Kapitulnik, J.Mater. Res. $\underline{2}$
    (6) 713 (1987).
4) T.Takagi, I.Yamada and K.Matsubara, Thin Solid Films, $\underline{58}$ 9 (1979).
5) K.Yamanishi, H.Tsukazaki and S.Yasunaga, Proc, Int'l Workshop on ICBT,
    139 (1986).

6) I.Yamada, C.J.Palmstrøm, E.Kennedy, J.W.Mayer, H.Inokawa and T.Takagi, Mater. Res. Soc. Symp. Proc. <u>37</u> 401 (1985).

7) K.Fukushima, I.Yamada and T.Takagi, Proc. of 9th Symp. on ISIAT'85, 363 (1985).

8) K.Yamanishi, S.Yasunaga, K.Imada, K.Sato and Y.Hashimoto, Mater. Res. Soc. Symp. Proc. <u>99</u>, 343 (1988).

FORMATION OF OXIDE SUPERCONDUCTOR THIN FILMS WITH EXCIMER LASER

Masaki KANAI, Tomoji KAWAI and Shichio KAWAI

The Institute of Scientific and Industrial Research,
Osaka University, 8-1, Mihogaoka, Ibaraki, Osaka 567

Abstract - Thin films of Bi-Sr-Ca-Cu-O and Bi-Pb-Sr-Ca-Cu-O superconductor were synthesized by a laser ablation method and also by a multi-target successive deposition technique. The Pb doping decreased the melting temperature, making the formation of high Tc phase much easier. This effect of Pb doping was investigated in detail.

Introduction

Since the discovery of Bi-Sr-Ca-Cu-O superconductor with Tc higher than liquid nitrogen temperature by Maeda and his coworkers,[1] many studies have been done on this system. This system contains two superconducting phases with Tc of 80K and 110K. The crystal structure of the 80K phase has two $Cu-O_2$ layers between the adjacent Bi_2O_2 layers[2)3)], and 110K phase has three $Cu-O_2$ layers[4)], respectively.

Recently, it has been reported that the Pb doping for the Bi-Sr-Ca-Cu-O system leads to the easier formation of 110K phase in bulk ceramics.[5] We prepared Bi-Sr-Ca-Cu-O (BSCCO) and Bi-Pb-Sr-Ca-Cu-O (BPSCCO) thin films by a laser ablation method.[6] This method has a feature in which the composition of the film does not change significantly in comparison with that of the target, when the substrate is not heated. In the present study, these thin films were prepared in three different ways as shown below.

No.1: Samples were made by sintering of amorphous films. This is a similar way in preparing the bulk ceramics.

No.2: As-grown films were prepared so as to stack the crystalline layers on the substrate permitting the deposition and crystallization to proceed simultaneously.

No.3: Samples were obtained by successive deposition from multi-targets. This allows the film composition to be controlled by change of deposition time through each target.

These films were characterized by X-ray diffraction and Tc measurement. The properties of Pb doped films were compared with those of the Pb undoped ones prepared through the No.1 technique .

Experimental

The films were prepared by pulses of an ArF excimer laser focused on the target placed in vacuum chamber.[7] Emitted atoms or molecules from the target were accumulated on a substrate placed at the opposite side to form a film. Laser intensity was $1 - 10J/cm^2$ pulse on the target after

focusing, and the repetition rate was 10 - 20 Hz.   The substrate used was
a MgO(100) single crystal.
    In the 1st technique as noted in the above, the films were deposited on
the substrate at room temperature, followed by post annealing at high
temperature in air to crystallize the films. The film thickness was $2\mu m$ in
most cases.   Target of Pb undoped system was synthesized by sintering
mixtures of Bi_2O_3, $CaCO_3$, $SrCO_3$ and CuO with appropriate ratios, while
in the target for Pb doped system Bi_2O_3, PbO, $CaCO_3$, $SrCO_3$ and CuO were
used.
    In the 2nd technique, as-grown films were prepared on the heated
substrates under oxygen gas flow.   The substrate temperature (Ts) was from
$400°C$ to $800°C$ and the film thickness was about $2000Å$. The target used was
the same as mentioned above.
    In the 3rd technique, the films were prepared by successive deposition
from the multi-targets of sintered Bi_2O_3, $SrCuO_y$ and $CaCuO_y$ pellets.   One
cyclic process was made by successive deposition from Bi_2O_3 target for 20
sec, $SrCuO_y$ target for 10 sec, $CaCuO_y$ for 60 or 80 sec and $SrCuO_y$ for 10
sec. Films were prepared with 20 cycles on heated substrate under oxygen
gas flow.(Ts= $600°C$)   Post-annealing in air was also carried out.   The
thickness of deposit through 1 cyclic deposition was about $100Å$.
    Resistance-temperature curves of the films were obtained with a standard
four probe technique using a calibrated germanium sensor.   Gold wire leads
were connected through an indium electrode.   An X-ray diffraction was
measured with a Rigaku RAD-C system using $CuK\alpha$, and the film composition
was investigated by EPMA type EMX-2A from Shimazu Ltd.

Result and Discussion

    The X-ray diffraction pattern and Tc measurement of the films prepared
by the No.1 method, namely post-annealing of accumulating on substrate at
room temperature, show the following results. Pb undoped BSCCO film heated
at $890°C$ for 1 minute indicated semiconductive R-T behavior and had (002)
peak at 2θ= 7.2 in X-ray diffraction pattern.[7]   It is shown that the
structure of this film has a single $Cu-O_2$ layer between Bi_2O_2 layers. This
phase also appeared when the film was sintered at $800°C$. [2] The BSCCO film
heated at a temperature between $890°C$ and $800°C$ showed 80K class
superconductive transition having (002) peak at 2θ= 5.7 in diffraction
pattern.[7]   The structure of this film appears to contain double $Cu-O_2$
layers between Bi_2O_2 layers.   For the high Tc phase of BSCCO film, a
decrease of resistivity around 110K was observed with heating at 885 C for
1 minute.   The diffraction peak of the high Tc phase, however, did not
appear in diffraction pattern of this film. High Tc phase appeared when
sintered just below the melting temperature(about $890°C$).   But the
temperature range is narrow and the heating at the experimental
temperature   for longer time may lead to the decomposition of the film.
Therefore, it is not very easy to obtain the film of high Tc phase in Pb
undoped BSCCO system. In contrast, Pb doped BPSCCO film sintered at lower
temperature, $850°C$, for 15 min showed a slight decrease in resistivity
around 110K.   There was no evidence of high Tc phase, however, in the
diffraction pattern in a similar way to the Pb undoped film (Fig.1 a).
Sintering for longer time increased the content of high Tc phase.   The X-

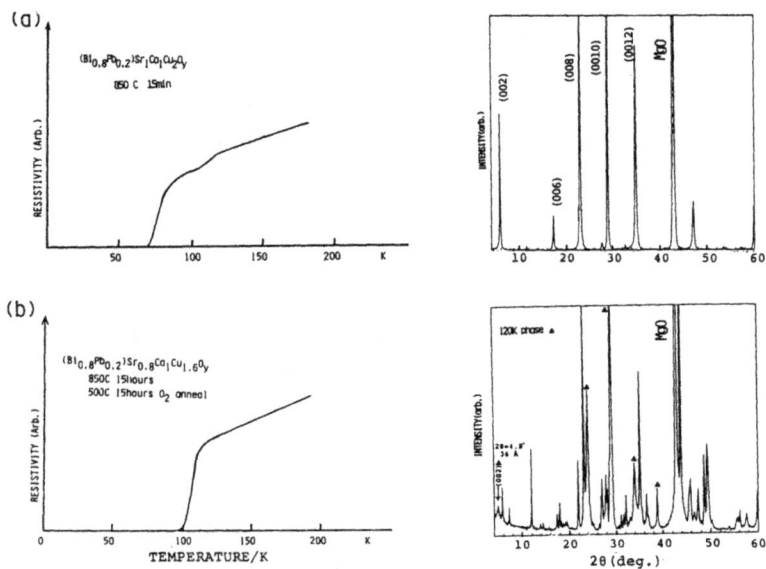

Fig.1 The X-ray diffraction patterns and R-T curves of Bi-Pb-Sr-Ca-Cu-O
thin films. The film thickness was 2μm. (a) The sample was sintered
at 850°C for 15 minutes, (b) at 850°C for 15 hours

ray diffraction pattern showed high Tc phase with BPSCCO sintered at 850°C
for 15 hours. This films revealed Tc^{zero} at 95K. Heating of Pb doped
films below 850°C formed oriented film of 80K phase with the c-axis
perpendicular to the surface of substrate. Heating above 850°C led to
the formation of semiconductive phase having a single $Cu-O_2$ layer and to
decomposition of the films. These BPSCCO films had the melting temperature
around 850°C, so the heating at the temperature just below the melting
point was essential to obtain high Tc phase for both BSCCO and BPSCCO
films. These behavior of BSCCO and BPSCCO films on sintering temperature
was shown in Fig.2.

A BPSCCO film was obtained, which contained 30 – 40% of high Tc phase
on X-ray diffraction pattern by post-annealing of amorphous film. (Fig.3)
Sintering temperature was 830°C and the thickness was about 4000Å. This
film decomposed above 830°C, though the films shown in Fig.1 having 2μm
thickness decomposed above 850°C. Thinner films decomposed at further
lower temperature regardless of the target composition, so the
decomposition temperature may depend on the film thickness rather than the
composition of the target.

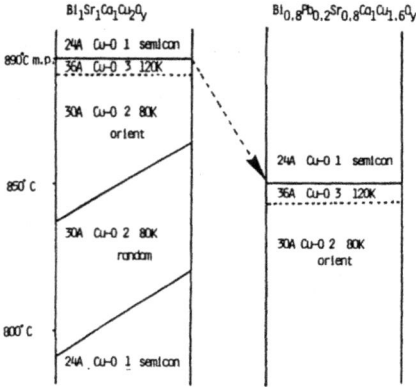

Fig.2 The diagram of the produce in BSCCO and BPSCCO films on sintering temperature

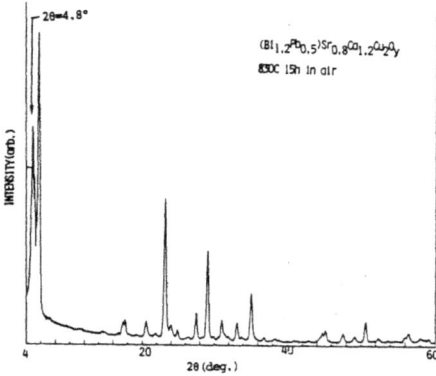

Fig.3 The X-ray diffraction pattern of BPSCCO thin film including high Tc phase. The film thickness was 4000Å. The target is $Bi_{1.2}Pb_{0.5}Sr_{0.8}Ca_{1.2}$-$Cu_2O_y$ and the sample was heated at 830°C for 15 hours.

400℃
~600℃
low

substrate
temperature

high
730°C

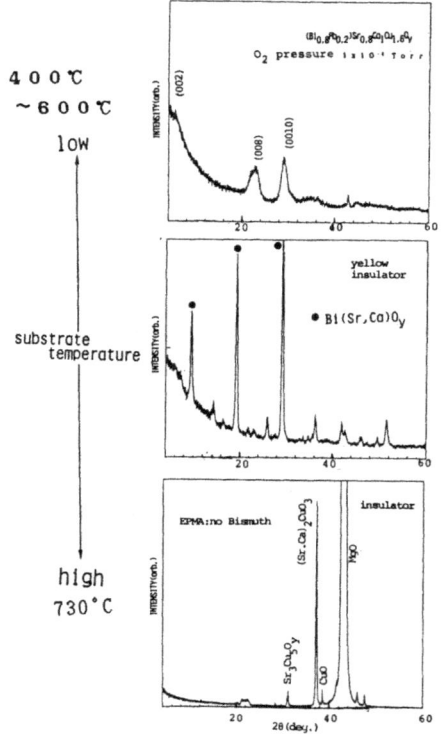

Fig.4 Variation of X-ray diffraction patterns of BPSCCO as-grown films with substrate temperature. Target used was $Bi_{0.8}Pb_{0.2}Sr_{0.8}Ca_1Cu_{1.6}O_y$, and oxygen pressure was 1×10^{-2} Torr

The sample prepared by the No.2 method, a crystallized film in as-grown state, had the composition being changed from that of the target. Variation of X-ray diffraction pattern of the films with the substrate temperature is shown in Fig.4. The target composition and oxygen pressure were fixed(Bi:Pb:Sr:Ca:Cu= 0.8:0.2:0.8:1:1.6, O_2 pressure= 1×10^{-2} Torr). X-ray diffraction pattern of the film synthesized with Ts through 400 to 600 C consisted of the peaks representing a precursor of 80K phase. Higher substrate temperature formed a yellow film having diffraction peaks of other phase. This film was insulator and decrease of Ca and Cu was observed from EPMA analysis. The product is presumed to be $Bi(Sr,Ca)O_y$ considering from the peak position in the diffraction pattern. The film prepared at higher substrate temperature of about $750°C$, was colorless insulator. In this film, Bi was not detected from EPMA analysis, indicating sudden decrease of Bi at higher region of substrate temperature about $750°C$. Oxygen pressure was an important factor to obtain a crystallized as-grown film. As seen from the diffraction pattern, an as-grown film mainly consisting of 80K phase could be obtained by increasing the oxygen pressure during the deposition.(Fig.5)

Fig.6 shows X-ray diffraction patterns of the films prepared by multi-target method and post-annealed at 800°C for 3 hours (No.3 method). The film deposited from Bi_2O_3 target for 20 sec, $SrCuO_y$ for 10 sec, $CaCuO_y$ for 60sec and and $SrCuO_y$ for 10 sec in a single cyclic procedure, shows the diffraction pattern of 80K phase having double $Cu-O_2$ layers and of semiconductive phase having a single $Cu-O_2$ layer. On the other hand, the sample deposited from $CaCuO_y$ for 80 sec consisted of 80K phase, and the semiconductive phase disappeared in this film. These results indicate that it is possible to control the composition of the film; that is, to control the number of the $Cu-O_2$ layer of these Bi based compound, by the change of the deposition time for each target.

Conclusion

1 By Pb doping into Bi-Sr-Ca-Cu-O films, the melting temperature of the

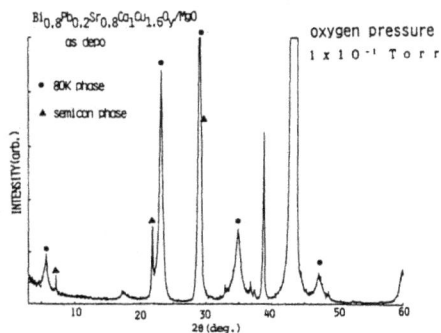

Fig.5 The X-ray diffraction pattern of as-grown BPSCCO film. Oxygen gas pressure was 1×10^{-1} Torr.

(a)

(b)

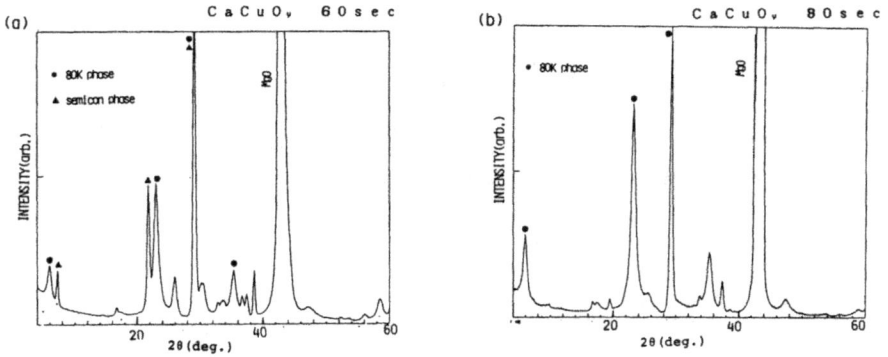

ig.6   The X-ray diffraction patterns of Bi-Sr-Ca-Cu-O thin films prepared
by successive deposition from multi-targets. These films were sintered
at 800°C for 3 hours in air. (a) The sample was deposited from Bi_2O_3
target for 20 sec, $SrCuO_y$ for 10 sec, $CaCuO_y$ for 60sec and and $SrCuO_y$
for 10 sec in a single cycle. (b) The deposition time from $CaCuO_y$
target became longer, 80 sec in 1 cycle.

films can be reduced by 30 - 40 C, making the formation of high Tc phase
easier.
2   With increase in substrate temperature, the contents of Ca and Cu in
the films decreased, and sudden decrease of Bi occurred sequentially.
3   As-grown films having diffraction peaks of 80K phase could be
obtained. Relatively high oxygen pressure was required to get enough
crystallized as-grown films.
4   It is possible to control the number of $Cu-O_2$ layers of Bi-Sr-Ca-Cu-O
compound by a successive deposition using multi targets.

Reference

1) H.Maeda, Y. Tanaka, M.Fukutomi, and T.Asano: Jpn.J.Appl.Phys. 27 (1988)
   L209
2) E.Takayama-Muromachi, Y. Uchida, A.Ono, F.Izumi, M.Onoda, Y.Matsui,
   K.Kosuda, S.Takekawa and K.Kato: Jpn.J.Appl.Phys. 27 (1988) L365
3) M.A.Subramanian, C.C.Torardi,   J.C.Calabrese,   J.Gopalakrishnan,
   K.J.Morrssey, T.R.Askew, R.B.Flippen, V.Chowdahri, and A.W.Sleight:
   Science 239 (1988) 1015
4) H.Nobumasa, K.Shimizu, Y.Kitano and T.Kawai: Jpn.J.Appl.Phys. 27 (1988)
   L846
5) S.A.Sunshine, T.Siegrist, L.F.Schneemeyer, D.W.Murphy,   R.J.Cava,
   B.Batlogg, R.B.van Dover, R.M.Fleming, S.H.Glarum, S.Nakahara,
   R.Farrow, J.J.Krajewski, S.M.Zahurak, J.V.Waszczak, J.H.Marshall,
   P.Marsh, L.W.Rupp.Jr. and   W.F.Peck: Phys.Rev.B 38 (1988) 893
6) T.Kawai, M.Kanai and M.Kawai: Mater.Res.Soc.Symp.Proc. 99 (1988) 327
7) M.Kanai, T.Kawai, M.Kawai and Shichio Kawai: Jpn.J.Appl.Phys 27 (1988)
   L1293

PREPARATION AND CHARACTERIZATION OF HIGH Tc
SUPERCONDUCTING BiSrCaCuO FILMS

Yoshito FUKUMOTO, Masayoshi TONOUCHI and Takeshi KOBAYASHI

Faculty of Engineering Science, Osaka University
1-1 Machikaneyama, Toyonaka, Osaka, 560, Japan

ABSTRACT - We have systematically investigated thin film formation and its characterization of the new high-Tc superconducting oxide Bi-(Pb)-Sr-Ca-Cu-O (B(P)SCCO) system. As-grown sputtered films with Tc,zero of 50K were formed without post-annealing. After controlled annealing process, Tc,zero rose up to 90K and most of the resistance dropped at even higher temperature. Post annealing effects were studied using amorphous films deposited at lower temperature. ESCA study revealed that the Bi 4f binding energy closely correlated with the film superconductivity. In addition, a strong dependence of the substrate materials (MgO single crystal and YSZ polycrystal) on the superconductivity was observed.

Introduction

Since the discovery of a high Tc superconductor Bi-Sr-Ca-Cu-O (BSCCO) system without any rare earth element by Maeda et. al [1], great efforts have been made to study this material [2-3]. This system has at least two superconducting phases with different Tc; the higher one reaches about 115K, which is much higher than that of Y-Ba-Cu-O (YBCO) system. Partial substitution of Pb for Bi (BPSCCO) has been reported to be effective in increasing fraction of the high-Tc phase [4]. Thin film formation of this BSCCO system has been done by many researchers [5-6]. However, because of the co-existence of other phases and lack of the knowledge of this material nature, a huge gap still exists in the thin film formation between YBCO and BSCCO systems.

In this paper, thin film formation and its characterization of the B(P)SCCO system has been investigated. An as-grown superconducting thin film formation and synthesis of high Tc phase by post-annealing are described in the first section. In the second section the chemical condition of the BSCCO film surface and the substrate material dependence on the film formation are discussed.

Experimental

The BPSCCO thin film formation was carried out by reactive rf-diode sputtering at the elevated temperature. The sputtering conditions are summarized in Table 1. The film thickness was about 1μm. The sputtered films were characterized systematically by electrical measurements, X-ray diffraction, scanning electron microscope, EPMA and ESCA.

Table 1.   Sputtering Conditions

Substrate	(100) MgO and YSZ
Target	$Bi_{3.0}Pb_{1.0}Sr_{1.0}Ca_{1.0}Cu_{2.4}O_y$ [A] $Bi_{2.4}Sr_{1.0}Ca_{1.0}Cu_{2.4}O_y$ [B]
Gas	$Ar + O_2$ (25-40%) 6Pa
Power	80W
Substrate temperature	300-700°C

Results and Discussion

As-grown BPSCCO Film

To obtain as-sputtered BPSCCO films, the substrate temperature was set to be above 600°C and the target [A] was used.   As a substrate, we chose (100) MgO single crystal from the result of next section.   Just after complete deposition, the oxygen gas was fully fed to the chamber and the substrate was slowly cooled down (20°C/minute).

Typical temperature dependence of the film resistance is shown in Fig.1 as a parameter on the substrate temperature during deposition.   The sputtering gas was $Ar+25\%O_2$ of 6Pa.   The films deposited below 640°C never showed superconductivity, and the superconducting properties were improved accompanying with rising the substrate temperature.   However, the films deposited at above 700°C were melted and missing even the conductivity.

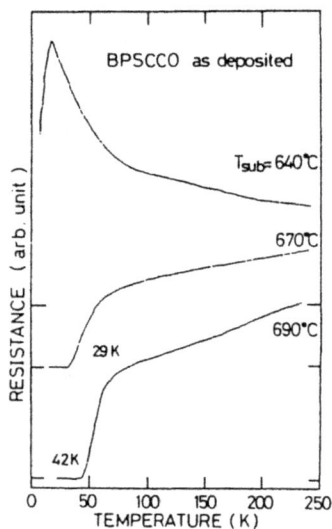

Fig.1. Temperature dependences of the resistance of the as-sputtered films on substrate temperature during deposition.

The above mentioned resistance properties are very broad, indicating that its crystallinity is not so good. Therefore we optimized the O_2 composition of the sputter gas up to 40%. As a result, a film with Tc,zero of 50K was obtained (Fig.2). Figure 3 shows the X-ray diffraction pattern of this film. We can see that the film has c-axis normal orientation with the lattice constant c of 3.07nm.

Figure 4 is the scanning electron micrograph of the as-grown film. Although small precipitations are dispersed, the underlying film surface is almost smooth. From EPMA measurement, these small precipitations mainly contains of Cu. As indicated by the R-T curve in Fig.2, the as-grown film mainly crystallized in the low Tc 2212 phase. Therefore these precipitations may be the crystallizations of excess element from 2212 phase.

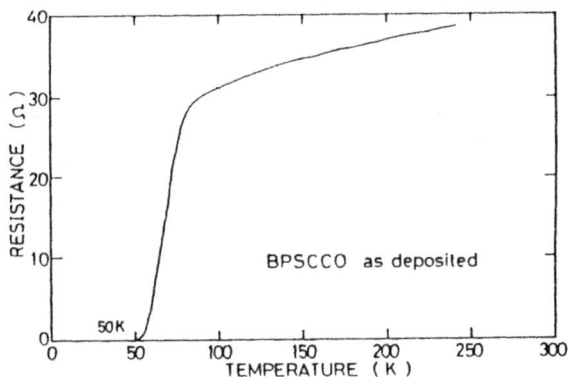

Fig.2. Temperature dependence of the resistance of the as-sputtered film.

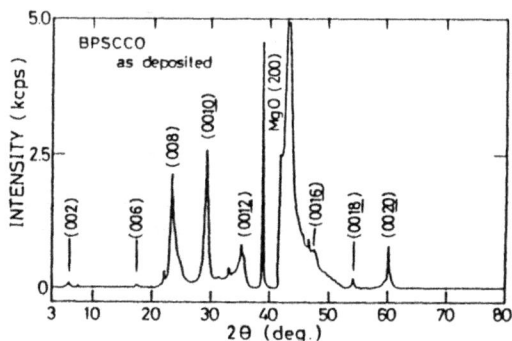

Fig.3. X-ray diffraction pattern of the as-sputtered film.

150

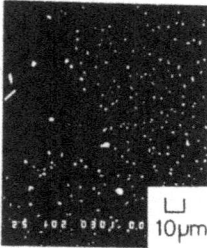

Fig.4. SEM photograph of the as-sputtered film.

The as-deposited films were then annealed at 850°C for 50 hours in air. Figure 5 show the temperature dependences of the annealed films which were deposited at 640 and at 690°C respectively. The film deposited at 640°C includes less volume of high Tc phase than the one deposited at 690°C. This result indicates that the substrate temperature during deposition has great influence on the superconductivity after post-annealing especially on synthesis of high Tc phase.

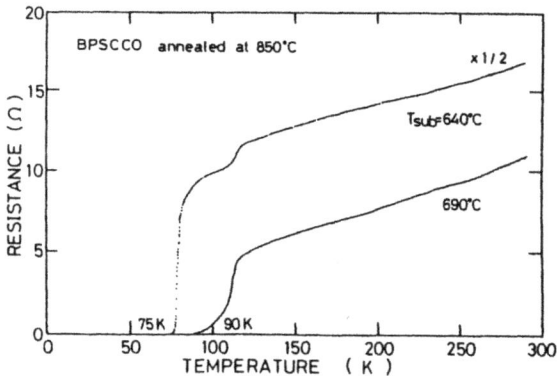

Fig.5. Temperature dependences of the resistance of the film annealed at 850 C for 50 hours in air.

Post-annealing Effects

The post-annealing effects on the chemi-physical state of the BSCCO film was studied using amorphous films deposited at lower temperature. The target used here was [B] and the substrate temperature was about 300°C.

The chemical condition of the BSCCO film surface was measured by ESCA. Figure 6 shows the spectra at the Bi 4f region for two samples of superconductive (a) and non-superconductive (b). As can be seen in this figure, the binding energy clearly shifted from 157.9eV to 158.6eV. The relation between the Bi 4f $_{7/2}$ binding energy and the zero resistance temperature of the film is given in Fig.7. The peak energy shifted continuously from 157.7eV to 159.6eV with the degradation of superconductivity from 70K class superconductor to non-superconductor. It shows that the superconducting properties has close correlation with the

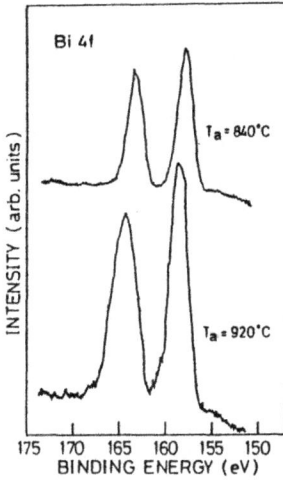

Fig.6. ESCA spectra of Bi 4f level of post-annealed film surface.

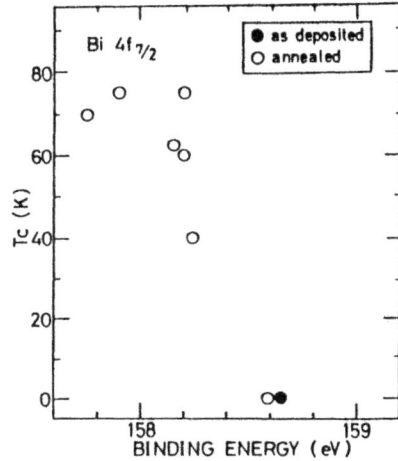

Fig.7. Tc,zero vs Bi 4f binding energy

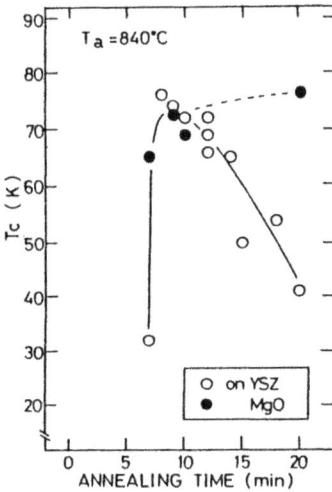

Fig.8. Annealingtime dependencesof zero-resistance temperature for different substrate materials.

chemical state of the Bi ion.   On the other hand, no meaningful change was observed for both Cu and O spectra.   In the case of YBCO system, the Y 3d binding energy was the same regardless of film superconductivity. Therefore this peak shift in Bi 4f region is peculiar to BSCCO system.

The substrate materials influence on superconductivity was also examined.   For two different substrate materials, polycrystalline YSZ and (100) MgO single crystal, the observed zero resistance temperatures are plotted in Fig.8 as a function of annealing time.   Films on both substrates exhibited superconductivity at around 70K annealed for 5-7 minutes. The substrate influence on the superconducting properties appeared in annealing for 15-20 minutes.   The Tc of the films on MgO was slightly improved, but the one of the films on YSZ degraded significantly.   Not being shown the data here, the film on YSZ substrate revealed X-ray diffraction intensity by 1/20 weaker than that on MgO substrate.   This result indicates that a large difference also exists in crystallinity as well as superconductivity.   From these results the difference in substrate materials, MgO and YSZ, can be attributed to the interaction between the films and substrate materials.

Conclusion

The results obtained in this study are summarized as follows.
1) As-grown sputtered BPSCCO thin films with Tc,zero of 50K are formed without post-annealing.
2) After   controlled   annealing process Tc,zero rose up to 90K and most of the resistance dropped at even higher temperature.   The superconducting properties after annealing were also affected by the substrate temperature during deposition.
3) Bi   4f binding   energy   has   close   correlation with the film superconductivity.
4) A   strong dependence of the substrate materials (MgO single crystal and YSZ polycrystal) on the superconductivity was observed.

References

1) H.Maeda, Y.Tanaka, M.Fukutomi and T.Asano, "A   New   High-Tc Oxide Superconductor without a Rare Earth Element", Jpn.J.Appl.Phys., 27, 2, L209-L210 (1988).
2) E.Takayama-Muromachi, Y.Uchida,   A.Ono,   F.Izumi, M.Onoda, Y.Matsui, K.Kosuda, S.Takekawa and K.Kato, "Identification of the Superconducting Phase in the Bi-Ca-Sr-Cu-O System", Jpn.J.Appl.Phys., 27, 3, L365-L368, (1988).
3) T.Kijima, J.Tanaka, Y.Bando, M.Onoda and F.Izumi, "Identification of a High-Tc Superconducting Phase in the Bi-Ca-Sr-Cu-O System", Jpn.J.Appl.Phys., 27, 3, L369-L371 (1988).
4) M.Takano, J.Takada, K.Oda, H.Kitaguchi, Y.Miura, Y.Ikeda, Y.Tomii and H.Mazaki, "High-Tc Phase Promoted and Stabilized in the Bi,Pb-Sr-Ca-Cu-O System", Jpn.J.Appl.Phys., 27, 6, L1041-L1043 (1988).
5) M.Fukutomi, J.Machida, Y.Tanaka, T.Asano, T.Yamamoto and H.Maeda, "New Technique for Preparation of BiSrCaCuO Thin Films with Tc of 100K and Above", Jpn.J.Appl.Phys, 27, 8, L1484-L1486 (1988).
6) H.Asano, M.Asahi, Y.Katoh and M.Michikami, "Low Temperature Growth of High-Tc Bi-Sr-Ca-Cu-O Films by Magnetron Sputtering", Jpn.J.Appl.Phys, 27, 8, L1487-L1488 (1988).

Ln-Ba-Cu-O THIN FILMS ON ITO COATED GLASS

Kozo FUJINO

Central Research Laboratory, Nippon Sheet Glass Co.,Ltd
Konoike, Itami, Hyogo,664,Japan

Abstract -Thin films of the high Tc Ln-Ba-Cu-O(Ln=Y,Er) superconductors have been prepared by RF magnetron sputtering onto transparent electro-conductive indium tin oxide (ITO) film coated glass substrate. ITO film has been found to become a diffusion barrier and to prevent the strain between the superconducting film and substrate. The superconducting film deposited on ITO/GLASS below 700 C has shown onset transition temperatures up to 88K.

Introduction

In recent years, thin films of the newly discovered high-temperature superconductor oxide Ln-Ba-Cu-O were attempted to deposit on a variety of substrates by several techniques. However most of successful thin film works to date were limited to the use only of expensive single crystal substrates e.g.,(100)MgO and (100)SrTiO$_3$[1-3]. There have not been many studies reported on the growth of thin films onto glass substrate with and without buffer layers[4-5]. Moreover, few works have exhibited good superconducting properties with high zero resistance transition temperature. Problem is cracks owing to stress and chemical interaction between the superconducting film and the substrate.

There are some advantages for that ITO coated glass has been chosen. First ITO has almost same thermal expansion coefficient of the superconductor Y-Ba-Cu-O. For this reason it seems that the formation of cracks in the film due to different expansion coefficient decreases largely because of the use of ITO as a buffer layer. Secondly, besides buffer layer ITO is capable of becoming the transparent and electro-conductive materials available for optoelectronic and microelectronic devices.

Experimental

Silicate glasses were used as a substrate. Prior to the deposition of Ln-Ba-Cu-O films, they were coated with ITO film of the thickness of 200nm by electron beam evaporation method. ITO film deposited above 300 C has a sharp peak corresponding to (222) in X-ray diffraction pattern.

All the Ln-Ba-Cu-O films were grown in a RF magnetron sputtering system. The preparation of the 3 inch single Ln-Ba$_2$-Cu$_{4.5}$-O sputtering target was made by mixing the starting materials of metal oxides and sintering them at 900 C. Figure 1 is a scanning electron microscope photograph of the typical surface of the Y-Ba-Cu-O sputtering target. A difference between before and after sputtering with long time can be seen clearly in the surface morphology. Accordingly, the virgin target was arranged each sputtering time in the system to get the reproducible experimental results. The sputtering conditions of Ln-Ba-Cu-O films are listed in Table 1. The film deposition was carried out at relatively low temperatures below 700 C without post anneal. Since the desired atomic ratios of the superconducting phase are Ln:Ba:Cu=1:2:3, the film composition was adjusted in the vicinity of the value by controlling sputtering conditions. Stoichiometry was determined by inductively coupled plasma emission spectroscopy.

Table 1. Sputtering conditions.

Substrate	ITO/Glass
Substrate temperature	550-700 C
Sputtering gas	$Ar+O_2$
Gas pressure	1-5Pa
RF power	50-100W
Growth rate	4-12nm/min.
Thickness	500nm

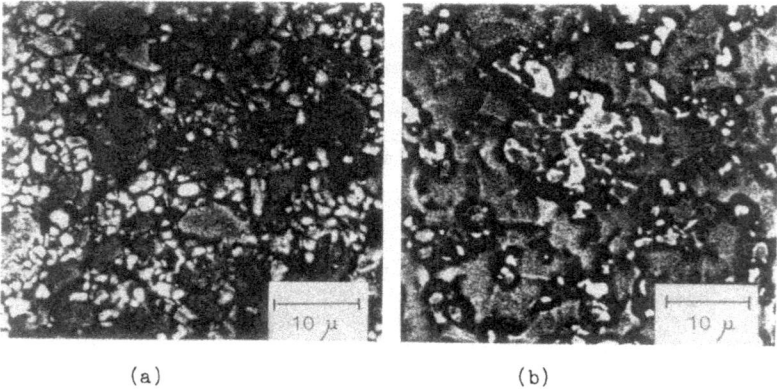

(a)   (b)

Fig.1. SEM photographs of the surface of the Y-Ba-Cu-O target for (a)as-prepared and (b)many times sputtered.

The temperature dependence of the resistivity was performed by four-point prove technique. Indium current and voltage contacts were bonded to the Ln-Ba-Cu-O film.

Results and Discussion

The photograph of the Y-Ba-Cu-O film on ITO/glass is shown in Fig. 2. The color of the film on ITO/glass is black while the film directly on glass is transparent. This indicates that the Y-Ba-Cu-O film and glass react chemically and that ITO film can serve as a diffusion barrier. In Fig.3 is shown SEM photograph of this sample which has two parts of ITO/glass and glass substrate. The surface morphology in Fig.3(b) which is quite rough implies that the mutual diffusion between the superconducting film and glass occurs completely. So, auger depth profile on the sample of Er-Ba-Cu-O/ITO/glass was measured as shown in Fig.4. This data prove that ITO film plays a role as a barrier layer.

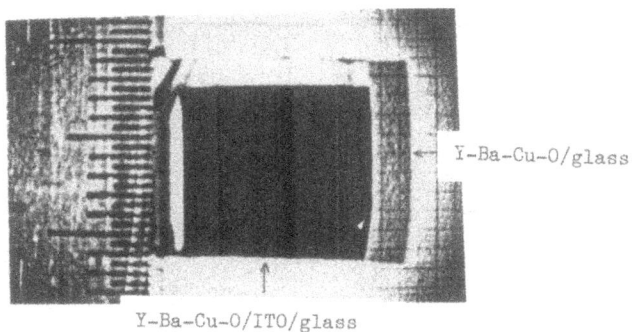

Fig.2. Photograph of a sample which consists of ITO/glass and glass substrate.

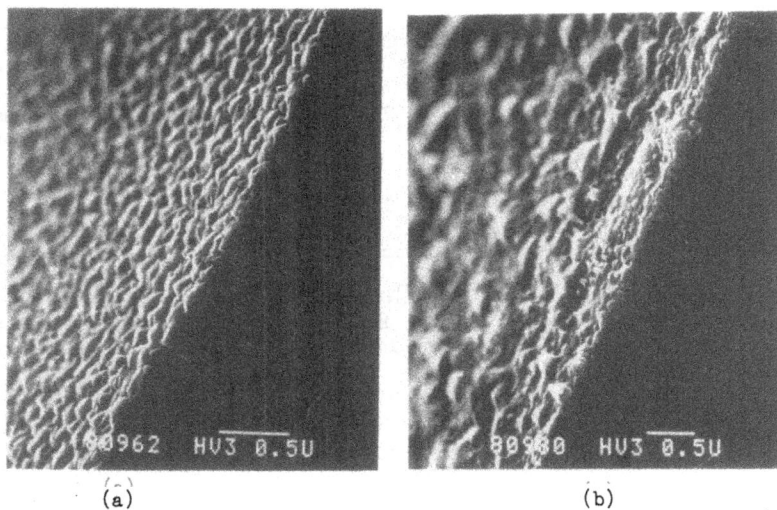

(a)   (b)

Fig.3. SEM photograph of the cross section of the Y-Ba-Cu-O film (a)for on ITO/glass (b)for on glass.

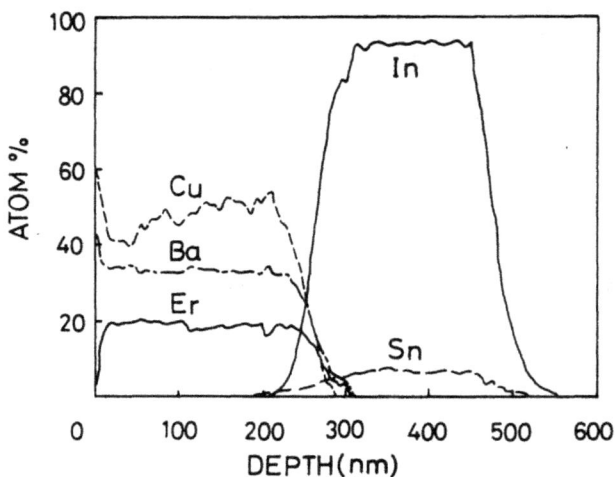

Fig.4. Auger depth profile for the Er-Ba-Cu-O film deposited on ITO/glass.

When the specimens were fabricated using the substrate of ITO coated silica glass, microcracks in the superconducting film occur occasionally as can be seen in Fig.5. This crack in superconducting film goes through ITO film into the substrate. The reason for generating cracks can be that the thermal stress would remain at the superconducting film due to the difference of thermal expansion coefficients between the film and substrate[6]. With decrease in the substrate temperature, it is considerable to produce the strain despite the existence of ITO buffer layer. Table 2 lists thermal expansion coefficient of various materials[7-8]. The thermal expansion coefficient of silica glass substrate seems to be extremely small compared with that of the superconducting film.

In order to suppress the strain, I attempted to use the silicate glass with relatively high coefficient, 8×10^{-6}, in stead of silica glass. The attempt succeeded in that any microcrack could not be observed in the Y-Ba-Cu-O film on ITO/silicate glass.

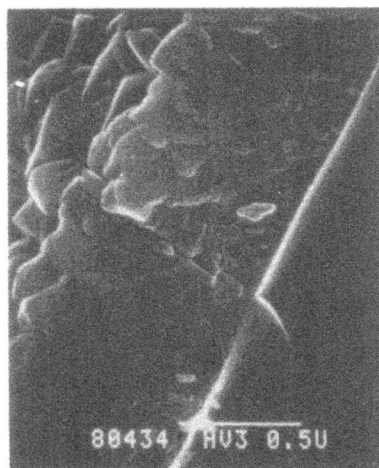

Fig.5. SEM photographs of Y-Ba-Cu-O film on ITO/silica glass

Table 2. Thermal expansion coefficients.

Materials	Thermal expansion coefficient
$YBa_2Cu_3O_y$	$1.3\text{--}1.7 \times 10^{-5}$
ITO	1.0×10^{-5}
Silica glass	5×10^{-7}
Silicate glass (except for Silica)	$3\text{--}10 \times 10^{-6}$

The temperature dependence of the resistivity was measured as shown in Fig.6. The Y-Ba-Cu-O film in Fig.6 has the critical onset temperature of 88K and zero resistance at 27K. This Y-Ba-Cu-O film was deposited on ITO coated silicate glass at 630 C. From the observation of SEM of cross section, the film looks like columnar texture and fairly dense.

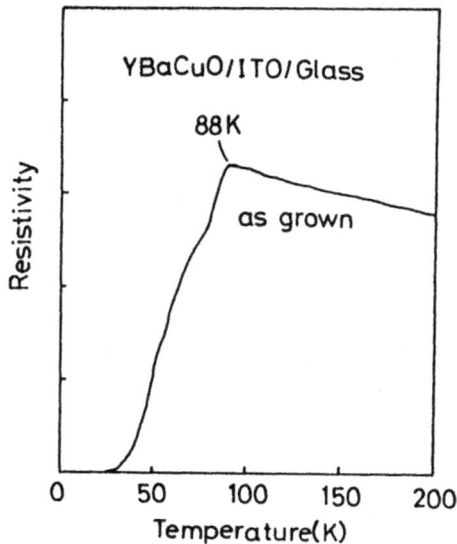

Fig.6. Temperature dependence of the resistance of Y-Ba-Cu-O film deposited on ITO/silicate glass.

Conclusion

It was impossible to deposit the superconducting film directly on the glass substrate without ITO film as a buffer layer, because glass and superconducting film react severely. My preliminary results show that Y-Ba-Cu-O films deposited on ITO/glass substrate below 700K exhibit onset temperature of around 90K without post anneal. Considerably smooth surface of Y-Ba-Cu-O could be obtained in the as-deposited state. ITO film is expected to be a effective buffer layer and unique candidate because of the electro-conductive and transparent property.

References

1)S.IShah and P.F.Carcia Appl.Phys.Lett.51,25,2146(1988)
2)Y.M. Chiang,S.L.Furcone, J.A.Ikeda,and D.A.Rudman MRS Symp.Proc.,99,307 (1987)
3)J.Kwo,T.C.Hsieh,M.Hong,R.M.Fleming,S.H.Liou and B.A.Davidson,MRS Symp.Proc., 99,339(1987)
4)M.Aslam, R.E.Soltis, E.M.Logothetis, R.Ager, M.Mikkor, W.Win, J.T.Chen, and L.E.Wenger Appl.Phys.Lett.53,2,153(1988)
5)A.Mogro-Campero and L.G.Tuner Appl.Phys.Lett.52,14,1185(1988)
6)W.Y.Lee, J.Salem, V.Lee, T.Huang, R.Savoy, V.Deline, and J.Duran Appl. Phys. Lett. 52,26,2263(1988)
7)T.Hashimoto,K.Fueki,A.Kishi,T.Azumi,and H.Koinuma Jpn.J.Appl.Phys. 27,2,L214 (1988)
8)H.M.O'Bryan and P.K.Gallagher,Adv.Ceram.Master.2,640(1987)

LOW-TEMPERATURE GROWTH OF NdBaCuO EPITAXIAL FILMS

Masahiro Iyori, Yuji Yoshizako, Masayoshi Tonouchi and Takeshi Kobayashi

Faculty of Engineering Science, Osaka University, 1-1 Machikaneyama
Toyonaka, Osaka 560, Japan

Abstract - The oxide superconducting NdBaCuO and YBaCuO films were
epitaxially grown on SrTiO$_3$ substrate at the temperature down to 520 °C by
the reactive rf magnetron sputtering. However, the regulated trilayered-
perovskite structure, peculiar to 1-2-3 phase, was missing for the
temperature lower than 560 °C. The films grown at below and above 500 °C had
an amorphous and epitaxial phases,respectively. Though we do not have
systematic data yet, a reduced growth rate seemed to be of favor for the
epitaxial growth.

Introduction

Since the discovery of the high Tc superconducting oxides [1], an
energetic effort has been paid on the study of the systems [2]. Because of
their potential characteristics applicable to the electronic devices such as
transistors, a number of film formation techniques has been studied and
established [3,4]. We have reported the preliminary experiments on the
multiple heteroepitaxial growth and superlattice formation of the high Tc
superconductors [5], and also reported the double heterostructure formation
of YBaCuO/MgO/YBaCuO systems. A real high Tc superconductor device requires
the combined layered-structure with an attractive material such as a
semiconductor. However, there still existence a lot of problem in film
formation processes. Among them, a development of the low temperature growth
technique of the epitaxial films is of prime importance.

In the present work, we report epitaxial growth of the YBaCuO and
NdBaCuO films at as low as 520 °C by the rf magnetron sputtering. Nd system
was chosen from the multi-valent characteristic of Nd ion. Owing to this
unique characteristic, Ba vacancy can be easily occupied by the excess Nd
atom in some case, resulting in the near-stoichiometric crystal synthesis
[6]. Whereas, the atom replacement has a serious problem that the
superconducting properties are likely to degrade due to a large difference
of the atomic radius between Ba atom and Nd. Our aim to study Nd system,
however, is outside this problem: the NdBaCuO may serve as an suitable one
side layer of the hetero-junction, and may be epitaxially grown at the
temperatures much lower range. Our present experiment answers this question.

Experimental Procedure

YBaCuO and NdBaCuO films were prepared in the conventional rf magnetron
sputtering system. (110) oriented SrTiO3 wafer was used as the substrates.
The deposition was done at the substrate temperature Ts from 480 to 670 °C
under the condition listed in Table 1. The powder target was prepared by
sintering Y2O3(or Nd2O3), CuO, and BaCO3 at 900 °C 48 hr., and grinding it
into powder. The substrate temperature was monitored by measuring the
substrate holder temperature.

Characterization of the prepared films was done by RHEED measurement and X-ray diffraction measurement.

Table 1.   Deposition conditions

Sputtering Gas	: Ar + O_2 (50%)
Total Gas Pressure	: 2 - 4 Pa
Substrate	: $SrTiO_3$ (110)
Substrate Temperature	: 480 - 670 °C
RF Power	: 50 W
Target	: $NdBa_2Cu_{4.5}O_x$
	$YBa_2Cu_{4.5}O_x$
Deposition Rate	: 5 - 12 nm/min
Film Thickness	: 120 nm

Result and Discussion

The deposition at above 630 °C easily provided us the NdBaCuO epitaxial films. Figure 1 (a) gives the RHEED patterns from the epitaxial NdBaCuO films.   The film prepared at total gas pressure P_t of 2Pa gave the clear pattern corresponding to the trilayered perovskite structure. Whereas the one prepared at P_t of 4Pa did the pattern corresponding to nothing more than perovskite structure, which explained by the disappearance of the 1/3- fold periodic pattern between the intense signals (Fig.1 (b)). This might be attributed to the existence of a number of staking fault and/or the developed non-stoichiometry. The gas pressure less than 1 Pa, unfortunately, led to the unstable plasma in our system. Thus we carried out the depositions under the P_t of 2Pa.

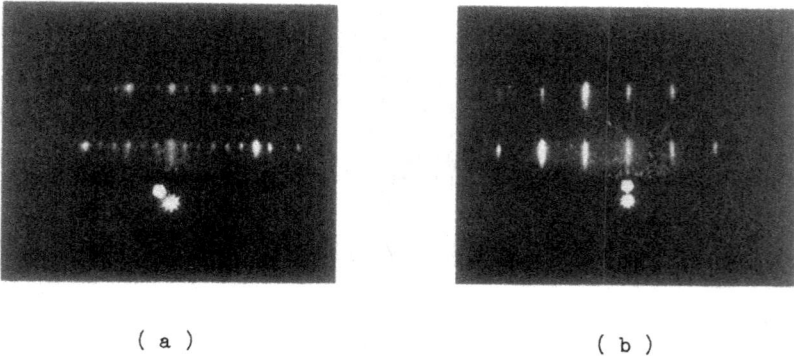

(a)   (b)

Fig. 1.   RHEED patterns of NdBaCuO films on (110)$SrTiO_3$.
(a) deposited with total gas pressure of 2 Pa.
(b) deposited with total gas pressure of 4 Pa.

Crystalline quality dependence on the substrate temperature was then examined by RHEED measurements. Figures 2 (a), (b), and (c) give the RHEED patterns taken from the as-grown NdBaCuO films prepared at 580°C, 540°C, and 500°C, respectively. First two gave the spotty patterns, which revealed that the epitaxial growth of the high Tc films were realized at as low as 500°C. Below the value, amorphous phase of NdBaCuO film was formed, which explained by the halo pattern. The films deposited above 560°C had the clear pattern corresponding to the trilayered perovskite structure. Between 530°C and 560°C, the films with poor crystalline quality were formed. The results are summarized in Table 2. Within our experiments, we obtained epitaxial films and amorphous films and observed no polycrystalline region, which differs from the result reported so far.

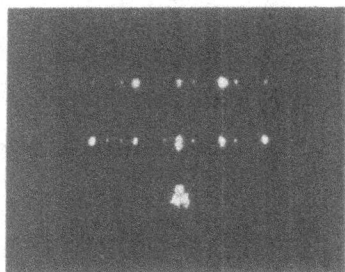

(a)

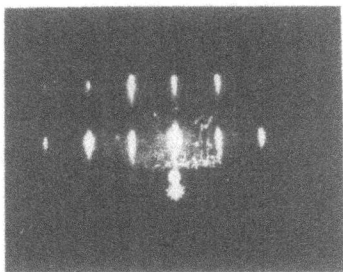

(b)

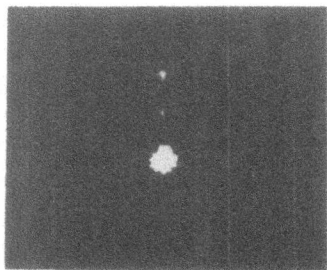

(c)

Fig. 2.   RHEED patterns of NdBaCuO films on (110)SrTiO$_3$.

(a) deposited at 580°C.
(b) deposited at 540°C.
(c) deposited at 500°C.

Table 2. Substrate temperature dependence of the RHEED patterns.

		P_t (Pa)	
		2.0	4.0
Ts (°C)	670	◎	◎
	630	◎ ←——	○
	600	◎	
	580	◎	
	560	○	
	540	○	
	520	○	
	500	△	
	480	✕	

◎ : similar to Fig. 2 (a)

○ : similar to Fig. 2 (b)

△ : similar to Fig. 2 (c)

✕ : no diffraction pattern

Ts : substrate temperature

Pt : total pressure

YBaCuO epitaxy was examined in the same way. Figure 3 gave the RHEED patterns from the YBCO films prepared at 600°C and 550°C. The both patterns corresponding to trilayered perovskite structure. However, the film deposited at 550°C showed such a pretty weak 1/3 fold periodic pattern that it had the poor crystalline quality. Dependence of the YBaCuO epitaxial growth on the substrate temperature was found to be almost same as the NdBaCuO epitaxy.

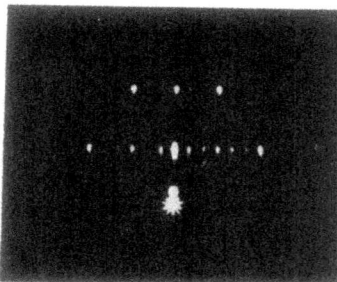

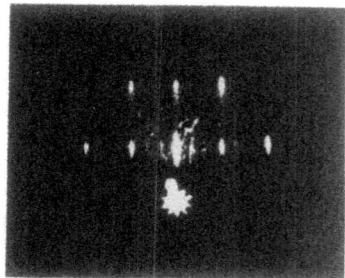

(a)   (b)

Fig. 3. RHEED patterns of YBaCuO films on (110)SrTiO$_3$. (a) deposited at 600°C. (b) deposited at 550°C.

Conclusion

Epitaxial high Tc superconducting NdBaCuO films were prepared at as low as 520 C. Above and below the value, the film crystalline phase changed from epitaxial one to amorphous one. Substrate temperature dependence of the YBaCuO epitaxy was found to be almost same as the NdBaCuO.

References

1) J. G. Bednorz and K. A. Müller, "Possible High Tc Superconductivity in the Ba-La-Cu-O System", Z. Phys. B-Condensed Matter 64, 189-193 (1986).

2) S. Uchida, H. Takagi, K. Kitazawa and S. Tanaka, "High Tc superconductivity of La-Ba-Cu Oxides", Jpn. J. Appl. Phys. 27 1-2 (1987).

3) Y. Enomoto, T. Murakami, M. Suzuki and K. Moriwaki, "Largely Anisotropic Superconducting Critical Current in Epitaxially Grown $Ba_2YCu_3O_{7-x}$ Thin Film", Jpn. J. Appl. Phys. 26 1248-1250 (1987).

4) S. Witanachchi, H.S. Kwok, X.W. Wang, and D.T. Shaw, "Deposition of superconducting Y-Ba-Cu-O films at 400 C without post-annealing", Appl. Phys. Lett. 53, 234-236 (1988).

5) T. Kobayashi, M. Tonouchi, Y. Yoshizako, M. Iyori and K. Fujino, "Multiple-Heterostructure Growth of High Tc Oxide Superconductors", JOURNAL OF THE JAPAN SOCIETY OF Powder and Powder Metallurgy 35, 392-396 (1988).

6) H. Nozaki, S. Takekawa, and Y. Ishizawa, "Superconductivity of the System $Nd_{1+2x}Ba_{2(1-x)}Cu_3O_y$ (0<x<0.4):Effect of the Magnetic Nd^{3+} Ion Occupying the Ba-Site", Jpn. J. Appl. Phys. 27 31-33 (1988).

PREPARATION OF HIGH TEMPERATURE SUPERCONDUCTING FILMS WITH ION BEAM SPUTTERING APPARATUS

H. KUWAHARA, N. NAKAMURA and E. KAMIJO,

Surface Modification Res. Dept., Nissin Electric Co., Ltd.
47 Umezu-Takase-cho, Ukyo-ku, Kyoto 615, Japan

K. YOSHIMURA and S. TANEMURA

Ceramic Science Dept., Govt. Ind. Res. Ins., Nagoya
1 Hirate-cho, Kita-ku, Nagoya 462, Japan

Abstract - The Bi-Pb-Sr-Ca-Cu-O thin film has been prepared by the method of ion beam sputtering. The $Bi_2SrCaCu_2$ oxide and the Pb metal were co-sputtered adjusting the composition ratio of Pb/Bi+Pb between 20% to 47%. Although Pb was in the film after deposition, it was not detected after annealing. The film annealed at 840 C for 2 hours showed the superconducting transition at 78 K. The film was also prepared with the oxide target only. However, it did not show a superconducting transition after the same annealing conditions. The Pb addition was shown to be effective to produce the 80 K phase of the Bi-Sr-Ca-Cu-O thin film.

Introduction

Since the discovery of the Bi-Sr-Ca-Cu-O(BSCCO) superconducting material by Maeda et al.[1], extensive investigations to prepare the BSCCO thin film have been carried out[2-6]. Although a single phase of this system seems to be very hard to produce, some features such as very high Tc and stable crystal structure against water are very attractive.

In synthesis of the Bi compounds, the Pb assist is known to be useful. Sunshine et al.[7] and Takano et al.[8] reported that the Pb addition caused preferential growth of the 110 K phase of the BSCCO system(high Tc phase) in bulk.

The method of Ion Beam Sputtering(IBS) has been shown to be effective for growing thin films of the Y-Ba-Cu-O high Tc superconducting oxide[9-10]. However, by using an usual IBS apparatus with single ion source and single target, the composition (Bi, Sr, Ca and Cu) of the film strongly depends on the composition of the target under operating conditions such as substrate temperature, oxygen partial pressure around the substrate, the ion species and accelerating energy. Since this Bi system has basic 4 metal components, it is expected to be very difficult to adjust the composition in the film to the precise stoichiometry.

We have started preparations of the BSCCO thin film using the IBS method. In this paper, the characterization of thin films containing the so-called low BSCCO phase is presented.

Experiments

The BSCCO thin films were prepared by the IBS method. The

acceleration voltage of ion beam for sputtering was 500 V and ion current 30 mA. The operating pressure of Argon gas in the vacuum chamber was 2.2 - 5.5 x 10^{-5} torr and partial oxygen pressure was fixed at 1.1 x 10^{-5} torr. The substrate temperature was kept 300 °C during deposition. The deposition rate and thickness of as-grown film were about 50 Å/min and 0.8 micron meter respectively. The substrate material is MgO(110).

The composition of the oxide target was Bi:Sr:Ca:Cu = 2:1:1:2. The deposition was made with and without Pb metal. The composition of the as-grown film was measured by the method of inductively coupled plasma atomic emission spectroscopy. In case of deposition without Pb metal, the composition of as-grown film was Bi:Sr:Ca:Cu = 2.00:1.82:2.06:3.00. The composition with Pb was Bi:Sr:Ca:Cu:Pb = 2.00:1.70:1.86:2.90:0.5. It is noted that the composition ratio of Pb/(Bi+Pb) was varied from 20% to 47%. However there seemed to be no effects on the annealed sample within the range of Pb contents. After deposition, samples were annealed in oxygen atmosphere at various furnace temperature between 800 - 860 °C for 2 hours. The increment and decrement of the furnace temperatures were 100 °C/hr. The film annealed at 860 °C became virtually transparent due to the evaporation of large amount of the film material during annealing.

Results and Discussions

The measurements of electrical resistivity was carried out using a standard four probe method. The temperature dependence of resistivity strongly depended on the annealing temperature. Preliminary measurements by X-ray diffractmeter with high temperature attachment(at 850 °C) indicated that the intensity of the (002) line of the 80 K phase grew and showed a saturation after 2 hours. The Increment of temperature rise to 850 °C from room temperature was 15 °C/min. The annealing time of 2 hours was not varied because of this measurement. Films without lead did not become a single phased superconducting material. Although some samples showed resistivity drops around 70 K, resistivity did not become zero.

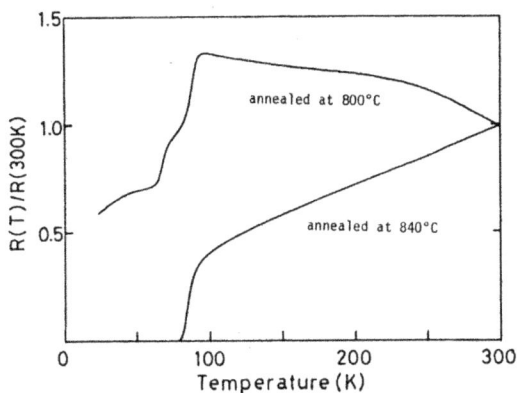

Fig. 1   Normalized resistance vs. temperature for samples with Pb annealed at 800 and 840 °C.

The samples with lead annealed at 800 °C and 820 °C showed resistivity drops at around 90 K and 70 K. The resistivity of the sample annealed at 840 °C became zero at 78 K. Fig. 1 shows the resistivity vs. temperature of the samples with lead annealed at 800 °C and 840 °C.

The diffraction pattern of of the as-deposited films showed no sharp peaks. The amorphous-like broad patterns were detected for all films. The diffraction pattern of the sample annealed at 840 °C which showed the superconducting transition at 78 K is shown in Fig. 2(a). To assign the index of the peaks, the results of the crystal structure analysis for the 80 K phase of BSCCO system, which is also called as the (2212) phase, were referred[11-15]. As assigned in Fig. 2(a), strong (00L) peaks were observed. The c-axis of the film must be oriented perpendicular to the substrate. The averaged lattice constant was c = 30.86 Å and agreed with the value obtained by Sunshine et al.[7]. More detailed analysis for this pattern was given in elsewhere[16]. The diffraction pattern annealed at 800 °C is also given in Fig. 2(b). As expected from the results of the resistivity behavior, the similar pattern of the (2212) phase with same broadness and unidentified lines was observed.

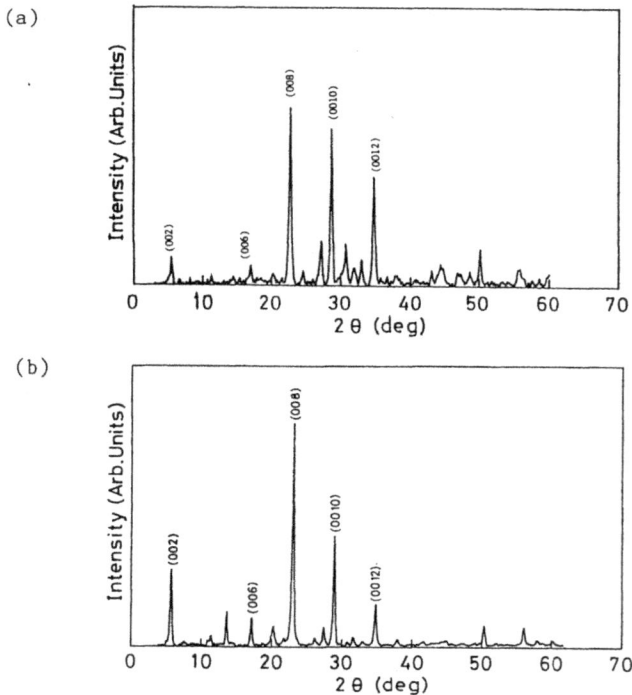

(a)

(b)

Fig. 2 X-ray diffraction patterns; (a) the sample annealed at 800 °C and (b) at 840 °C.

The SEM observation was made on the surface of the samples with lead after annealing. The surface of all the samples annealed at 800, 820 and 840 °C were rather rough. Fig. 3 shows the SEM image of the samples annealed at 800 °C. In this image, there exist two kinds of structures. One has a needle like shape and another granular like shape. Differences between these two structures were clearly observed by EPMA(Electron Prove Micro Analyzer). The composition of the needle shaped part was Bi:Sr:Ca:Cu = 2.0:2.1:1.0:2.5, and that of the granular part 2.0:2.4:1.8:3.6. It is noted that Pb element was not observed. It must be evaporated from the sample during the annealing process. The needle shaped part must correspond to the (2212) phase, since the resistivity behavior around 80 K and the x-ray diffraction pattern confirmed the existence of the (2212) phase and its composition was very close to the composition of this part.

Fig. 3   The SEM observation of the film annealed at 800 °C.

The SEM image of the sample annealed at 840 °C is shown in Fig. 4. The overall surface of this film was also rough, however some part of the surface was quite smooth. The EPMA measurements showed that the composition of this sample was remarkably uniform and its value was Bi:Sr:Ca:Cu = 2.0:1.9:1.0:1.9. The element of Pb was not detected on this film. All the measurements done for this sample suggested that there was only one single phase of (2212) in this film.

Conclusion

The BSCCO thin films were prepared on the MgO(110) substrate with and without the Pb element by the IBS method. The thin films annealed between 800 – 860 °C were examined. The content of the Pb element was varied between 20 to 47%, however Pb was completely evaporated from the film during annealing process. We did not find any clear effects on the annealed film originated from the amount of Pb within this range before

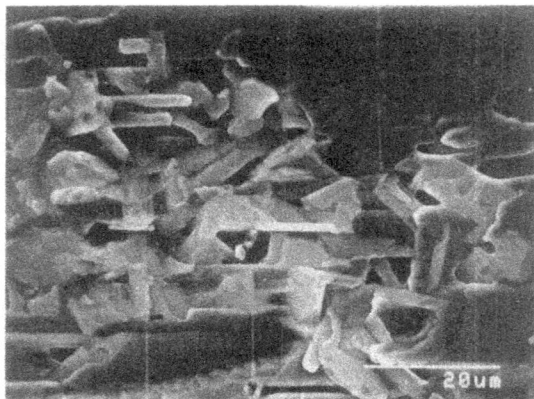

Fig. 4   The SEM observation of the film annealed at 840 °C.

annealing. However, the film without Pb did not become superconductive by annealing.

The film prepared with Pb annealed at 840 °C made a superconductive transition at 78 K. The x-ray analysis and the EPMA measurements showed that the film consisted of the single (2212) phase and had strong c-axis orientation perpendicular to the substrate. It has been shown that the Pb assist is also very helpful to produce the single phased (2212) thin film.

We would like to thank Prof. U. Mizutani(Nagoya Univ.) for the resistivity measurements, Dr. N Ishizuka(GIRI, Nagoya) for the ICP analysis and Dr. Y Murase(GIRI, Nagoya) for the EPMA measurements respectively. The x-ray measurements and SEM observation by Mr. K. Ogata and his group(Nissin) are greatly acknowledged.

References

1) H. Maeda, Y. Tanaka, F. Fukutomi and T.Asano, Jpn. J. Appl. Phys. Lett. 27, L209(1988)
2) H. Koinuma et al., Jpn. J. Appl. Phys. 27, L376(19888)
3) M. Nakao et al., Jpn. J. Appl. Phys. 27, L378(1988)
4) M. Fukutomi et al., Jpn. J. Appl. Phys. 27, L632(1988)
5) K. Kuroda, M. Mukaida, M. Yamamoto and S. Miyazawa, Jpn. J. Appl. Phys. 27, L625(1988)
6) Y. Ichikawa et al., Phys. Rev. B38, L765(1988)
7) S. A. Sunshine et al., Phys. Rev. B38, 893(1988)
8) M. Takano et al., Jpn. J. Appl. Phys. 27, L1041(1988)
9) K. Yoshimura, S. Nogawa and S. Tanemura, Proc. SPIE 948, 99(1988) (High-Tc Superconductivity: Thin Films and Devices, Newport Beach)
10) T. Yotsuya et al., to be published in Proc. ISS88(Nagoya).
11) Y. Syona et al., Jpn. J. Appl. Phys. 27, L569(1988)
12) T. Kajitani et al., Jpn. J. Appl. Phys. 27, L589(1988)

13) M. Onoda, Y. Yamamoto, E. Takayama-Muromachi and S. Takenaka, J. Jpn. Appl. Phys. $\underline{27}$, L833(1988)
14) J. M. Tarascon et al., Phys. Rev. B$\underline{27}$, 9382(1988)
15) S. Sueno et al., Jap. J. Appl. Phys. $\underline{27}$ L1463(1988)
16) K. Yoshimura, H. Kuwahara and S. Tanemura, to be published in Proc. ISS88(Nagoya)

Ln-Ba-Cu-O Thin Film Deposited by Ion Beam Sputtering

Tsutom YOTSUYA, Yoshihiko SUZUKI, and Soichi OGAWA
Osaka Prefectural Industrial Technology Research Institute, 2-1-53
Enokojima, Nishi-ku, Osaka 550

Hajime KUWAHARA
Nissin Electric Co., Ltd., 47 Umezu-takase-cho, Ukyo-ku, Kyoto 615

Tetsuro TAJIMA
Daikin Industries Ltd., 1304 Kanaoka-cho, Sakai, Osaka 591

Kohei OTANI
Hitachi Zosen Technical Research Laboratory, 1-3-22 Sakurajima, Konohana-
ku, Osaka 554

Junya YAMAMOTO
Osaka University, Laboratory for Applied Superconductivity, 2-1 Yamadaoka,
Suita, Osaka 565

ABSTRACT-The oxide superconducting thin films of LnBaCuO(Ln=Y and Yb)
by using the ion beam sputtering were successfully fabricated. The as-grown
films had already oxygen deficient perovskeit structure with Tc_0=54 K. The
film fabricated on the MgO(001) substrate showed c-axis orientation. After
deposition, the films were annealed around 900°C in the oxygen atmosphere.
After heat treatment, the films showed superconducting transition at 88 K
for YBaCuO and 77 K for YbBaCuO system.

INTRODUCTION

Recent discovery of high-Tc oxide superconductors[1,2,3], has caused
great impact not only on basic physics but also on the field of
superconductor's applications. The large efforts were made for fabricating
high quality thin films for the electronic applications. Because the thin
film of the oxide superconductor achieved relatively high critical current
density compared with that of bulk compound. The high Tc oxide thin films
were made by various method such as rf-magnetoron sputtering[4], laser
ablation[5], electron beam deposition[6], and spray-pyrolysis[7]. We have
studied high quality of thin film of oxide superconductors by using the
ion beam sputtering method(IBS). Because the operating pressure was two
order of magnitude lower than that of the rf-magnetoron sputtering system,
so good adhesion with substrate and high quality without gaseous impurity
were expected. Furthermore, the amount of oxygen deficiency is well known
to play an important roll for the superconducting behavior in the oxide
superconductor. Since operating pressure was as low as 10^{-4} Torr, low
energy oxygen ion beam could be used to realize oxygen treatment during
deposition. Considering these feature, we adopted the IBS method.

EXPERIMENTS

The $LnBa_2Cu_3O_x$ thin films(Ln=Y and Yb) were deposited by the IBS method. The sputtering chamber was evacuated by a conventional rotary pump and a cryo-pump down to 1×10^{-6} Torr. After evacuation, oxygen of 1×10^{-4} Torr for reactive gas and Ar of 5×10^{-4} Torr for ion beam gas were introduced into the chamber. As shown in Fig. 1, the target was mounted at 45 degree to the Ar ion beam. The acceleration voltage was applied between the target and the anode. The energy of the ion beam was 400 eV and the ion beam current was 400 mA. The deposition rate was about 0.7 Å/sec for oxide target. The typical film thickness was about 1∿1.6μm. The substrate temperature was elevated up to 600°C by using lamp heaters. After evaporation, an atmospheric oxygen gas was filled in the chamber and then substrate temperature was lowered to room temperature. The MgO(100) and $SrTiO_3$(100) single crystals were used for substrate.

After deposition all the films were heat treated in a tubular furnace under dry oxygen flowing. As the annealing condition strongly depended on substrate material and film compositions, optimized conditions were carefully selected.

The characterization of the films were carried out by electrical resistance measurements, X-ray diffract meter, inductively coupled plasma atomic emission spectroscopy (ICPAES), and scanning electron microscope (SEM) observation.

The measurements of the electrical resistance were carried out by ordinal four probe method. The electrodes were connected on the film surface with silver paste. The typical distance between potential leads were about 2 mm, and supplied current was 0.5 mA.

Chemical etching using dilute HNO_3 acid was made in order to measure critical current density (J_c). In this case the electrical contacts were made by indium solder. The J_c was defined as the value when the voltage of 1μV was detected.

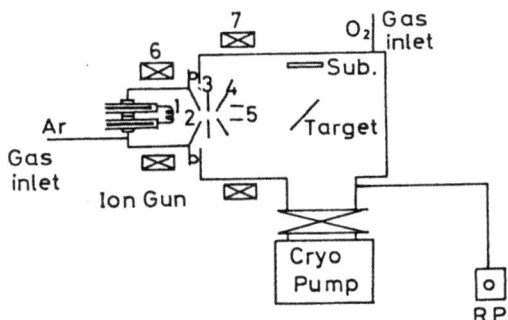

Fig. 1. A schematic diagram of the Ion Beam Sputtering Apparatus.

RESULTS and DISCUSSIONS

The as-grown YBCO film deposited on the MgO(100) substrate showed already oxygen-deficient perovskeit structure. The c-axis orientation was strongly observed for this case. The best superconducting properties for YBCO film deposited on MgO(100) substrate was obtained from the film of which composition was closed to Y:Ba:Cu=1:2:3. The target used had the composition of Y:Ba:Cu = 1:2:3.6. This film composition was almost constant over the substrate temperature range between 500 to 600°C. As shown in Fig. 2, the resistivity of the typical as-grown film was about 9mΩcm at room temperature and it decreased as temperature lowered from room temperature to 120 K. However resistivity hump was observed around 110 K, and then it became zero at 54 K. It was difficult to improve the superconductive transition temperature by optimized sputtering conditions.

To determine the crystalline structure of these films, the X-ray diffraction analysis was carried out by using Cu Kα radiation. Diffraction lines due to 00ℓ(ℓ=1,2,3,4,5,6) were strongly detected. This means that the c-axis of the film was oriented to the normal to the substrate surface. The lattice constant c of the as-grown film was as large as 12 Å. This value was too large compared with bulk YBCO material, since it is 11.85 Å even for the tetragonal phase[9].

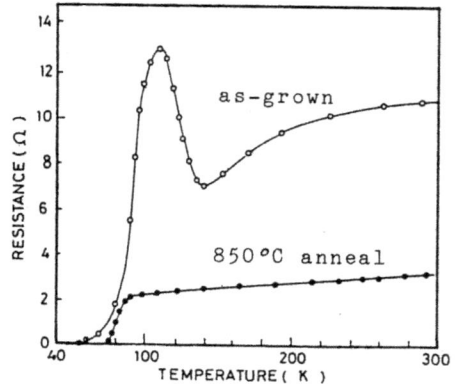

Fig. 2. Temperature dependence of the electrical resistance for the YBCO film.

It must be impossible to explain this large lattice constant in terms of oxygen deficient only. Before annealing, Cu, Ba and Y were distributed uniformly. However, non-uniform copper rich area was observed by the EPMA measurements. This means that excess Cu atoms were expelled from the grown crystal after annealing. And before annealing, the excess atoms of Cu were inserted into the perovskeit crystalline structure, so that the c-axis was enlarged by the excess atoms. On a early stage of annealing(500∿600°C) the film surface was very smooth, but it was deteriorated by the high temperature(800∿900°C) annealing process[8].

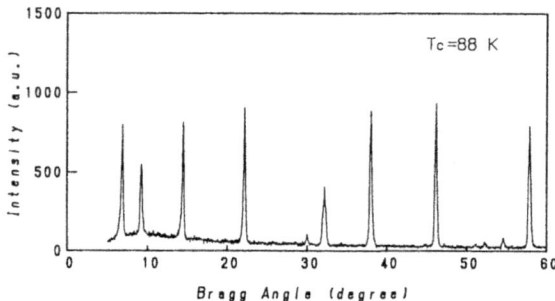

Fig. 3. The X-ray diffraction pattern of the YBa2Cu3Ox thin film after 900°C 30min heat treatment. The MgO(100) was used for the substrate.

The heat treatment effects on the superconducting properties and crystalline structure were carefully examined for the YBCO thin films. The X-ray diffract pattern is shown in Fig. 3. The lattice parameter c was shortened as small as 11.67 Å by heat treatment. In addition to expelling the excess atoms, oxygen was taken into the perovskeit structure during annealing.

After annealing at 900°C for 30 minutes, the resistivity at room temperature was decreased to about 200μΩcm and largest resistance ratio(R_{300K}/R_{100K}) was 3.3. As shown in Fig. 4, the superconducting transition was observed at 88 K. The extrapolated line of the resistance goes through nearly equal zero at absolute zero temperature. The critical current density achieved was as large as 10^5 A/cm^2 at 70 K.

The YBCO film deposited on the SrTiO$_3$(100) substrate was also studied. The best results in terms of the transition temperature(Tc=88 K) were obtained for the target composition of Y:Ba:Cu=1:2.5:5. It is noted that the composition of the films was Cu rich compared with the composition of Y:Ba:Cu=1:2:3. The resistance measured at room temperature was decreased as

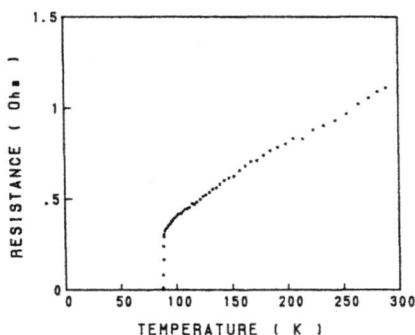

Fig. 4. Electrical resistance of the YBa$_2$ Cu$_3$ O$_x$ film as a function of temperature.

annealing temperature was increased. On the contrary to the MgO substrate, the resistivity of the YBCO film rapidly increased as the annealing temperature exceeded more than 850°C. We considered that YBCO thin film strongly reacted with the STO substrate at and above 850°C. X-ray diffraction measurements showed that there were a few unknown peaks and the structures included more than a single phase. However the ab-plane was preferentially normal to the substrate surface. And these X-ray results seemed to be consistent with the findings by the SEM observation. The needlelike crystal growth was observed by SEM. These crystal growth was not observed for the MgO substrate. The typical length of the needlelike crystal was about 10 μm and the diameter was about 500 nm. The orientation of ab-plane normal to the substrate surface was observed only for the YBCO films deposited on the SrTiO$_3$ substrate. The longitudinal direction of needlelike crystal observed by SEM must be the c-axis direction considering the X-ray results.

The valence of ytterbium varies between 2^+ and 3^+. This means that in the 1-2-3 perovskeit structure, the ytterbium atom not only occupies Y site but could also be substituted with barium atom. It is noted that by this possible substitution the requirement of the composition ratio for the stoichiometry may become easy. The ytterbium barium copper oxide film (YbBaCuO) was sputtered with alloy target of which composition was Yb:Ba:Cu=1:2.5:3.7.

The deposition rate of the YbBaCuO films was continually monitored. The deposition rate was as large as 110 Å/min for the virgin target. However it was gradually decreased as sputtering was carried out. After 100 hours the rate was down to 40 Å/min, this value was similar to the value obtained from the YBCO oxide targets. It is supposed that the sur-

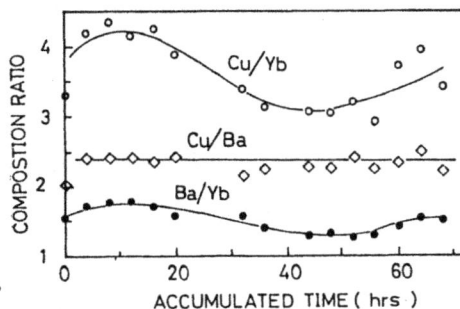

Fig. 5. The composition ratios for the Yb-Ba-Cu-O film as a function of accumulated time.

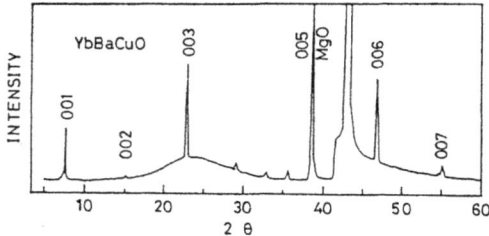

Fig. 6. X-ray diffract pattern of the c-axis oriented YbBaCuO film. The composition of the film was Yb:Ba:Cu=1:1.7:4.2.

face of the alloy target was oxidized during sputtering so the surface condition became similar to the oxide targets. The composition of the films were independent of sputtering conditions such as substrate temperature, oxygen partial pressure, deposition rate under present experimental conditions. Although the composition rate of Cu/Yb and Ba/Yb was oscillatory changed as the sputtering was carried out. But the ratio of Ba/Cu was remained constant. Figure 5 shows the composition ratio as a function of accumulated time. This was caused by the segregation existed in the alloy target. As shown in Fig. 6, the best c-axis oriented YbBaCuO film was also obtained for the film composition of Yb:Ba: Cu=1:1.7:4.2. The Tc_0 was 54 K for as-grown film and it was improved to 77 K after heat treatment.

CONCLUSION

We have investigated LnBaCuO(Ln=Y, and Yb) thin film by the method of ion beam sputtering. The YBCO film was prepared by the oxide targets and YbBCO was by the alloy target. The as-grown films deposited on MgO(100) showed c-axis orientation for both compounds and had superconducting transition temperature of 54 K. The characteristics of the superconductivity was improved after heat treatment. The obtained results are summarized as follows.

(1) The c-axis orientation films on MgO were obtained for YBCO and YbBaCuO system. The as-grown films showed superconductive transition up to 54 K. The lattice parameter was as large as 12 A for both film.

(2) The superconducting properties were improved by heat treatment in the oxygen flowing furnace, however the best heat treatment condition depended on substrate material and film composition.

(3) The YBCO film deposited on the MgO substrate and the $SrTiO_3$ substrate showed superconducting transition up to 88 K.

(4) Even though the film composition of YbBCO was not still close to 123 composition, the superconducting transition of 77 K was obtained after annealing.

(5) The alloy target has many advantages such as handling, packing density, deposition rate, however the segregation in the target caused the variation of the film compositions.

We would like to thank Dr. T. Hamada of Osaka Pref. Ind. Tech. Res. Institute(OPITRI) for X-ray analysis and useful discussions, and Mr. F. Uratani of OPITRI and Mr. M. Yoshikawa of Hosokawa-Micron Co. Ltd. for ICPAES measurements. The YbBaCu alloy target was prepared by Dr. T. Degawa and Mr. T. Kawae of Mitsui Engineering & Ship Building Co. Ltd.. This work was supported by the Osaka Cooperative Research Project for High-Tc Supercondoctors.

REFERENCES

1. J.G.Bednotz and K.A.Muller; Z. Phys. **B 64** (1987) 189 .
2. M.K.Wu, J.R.Ashburn, C.J.Torng, P.H.Hor, R.L.Meng, L.Gao, Z.J.Huang and C.W.Chu; Phys. Rev. Letts. **58** (1987) 908.
3. H.Maeda, Y.Tanaka, M.Fukutomi and T.Asano; Jpn. J. Appl. Phys. **27** (1988) L209.
4. Y.Enomoto, T.Murakami, M.Suzuki, and K. Moriwaki; Jpn. J. Appl. Phys., Letts., **51** (1987) L1845.
5. J.Narayan, N.Biunno, R.Singh, O.W.Holland, and O.Auciello; Appl. Phys. Letts, **51** (1987) 1745.
6. P.Chaudrai, R.H.Koch, R.B.Laibowitz, T.R.McGuire, and R.J.Gambino; Phys. Rev. Letts., **58** (1987) 2684.
7. M.Kawai, T.Kawai, H.Masuhira, and M.Takahasi; Jpn. J. Appl. Phys. Letts., **26** (1988) L1740.
8. T.Yotsuya,Y.Suzuki, S.Ogawa, H.Kuwahara, K.Otani, T.Emoto and J.Yamamoto; Procceeding of ICMC-88(Shenyang, China) to be published.
9. M.Kogachi, S.Nakanishi, K.Nakahigashi, S.Minamigawa, H.Sasakura, N.Fukuoka, and A. Yanase; Jpn. J. Appl. Phys. 27 (1988) L1228

MgO GROWTH ON (110) NdBaCuO EPITAXIAL FILMS

Yuji Yoshizako, Masahiro Iyori, Masayoshi Tonouchi and Takeshi Kobayashi

Faculty of Engineering Science, Osaka University, 1-1 Machikaneyama
Toyonaka, Osaka 560, Japan

Abstract - Formation of an MgO/YBaCuO single-heterostructure and an YBaCuO/MgO/YBaCuO double-heterostructure have been attempted by the conventional rf magretron sputtering with multi-targets. The RHEED, XRD, SEM observations revealed that the paramount double-heterostructure with (110) crystal orientation has been epitaxally grown, at least in part of the wafer. The remaning problem is the structural improvement of the MgO grown layer which, in turn, will promise the hopeful success of the epitaxy in growing this kind of new heterostructures.

Introduction

Epitaxal growth technipue have produced phenominal success in semiconductor device and physics, owing to the ideal property of epitaxally grown films. Introduction of the epitaxial technology to the high Tc superconductintg oxide film growth is quite attractive because one can get not only the improved superconductivity but the highly anisotropic conduction due to the regulated two-dimensional electron path.

From this point of view, formation of epitaxally grown insulator/superconductor junctions was examined. This time, MgO was adopted as an insulator, and YBaCuO as a superconductor. MgO is a substrate most commonly used for YBaCuO depositon and have been proved good insulator in former NbN SIS junctions. As a first trial, formation of (110)YBaCuO/MgO/(110)YBaCuO double-heterostructure was also done. Only this structure may allow the two-dimensional electron tunneling, which could open a new electron physics in both the scientific and engineering field.

Growth of (110)MgO

Previous to the formation of double-heterostructure, hetero epitaxial growth of (110)MgO was investigated. A grown MgO film has (100) natural orientation. To form (110)YBaCuO/MgO/YBaCuO double-heterostructure, (110)MgO middle layer ought to be required. MgO films were deposited by rf-magnetron sputtering whth sintered MgO disk target (3 inch diamater). Single-crystal (110)SrTiO$_3$ substrate was adopted in this preliminary experiment. SrTiO$_3$ is nearly equivalent to YBaCuO in crystal structure and lattice constant. Figure 1 shows RHEED patterns of MgO film deposited at sputtering condition listed in Tab.I. These reveal the feasibility of the hetero epitaxial growth of (110)MgO on (110)SrTiO$_3$ substrate.

Tab.I   MgO sputtering conditions

Sputtering gas	Ar
Applied rf voltage	0.8 kV
Anode current	60 mA
Gas pressure	1.0 Pa
Substrate temperature	500 C
Substrate	(110)SrTiO$_3$
Deposition rate	1.5 nm/min
Thickness	50 nm

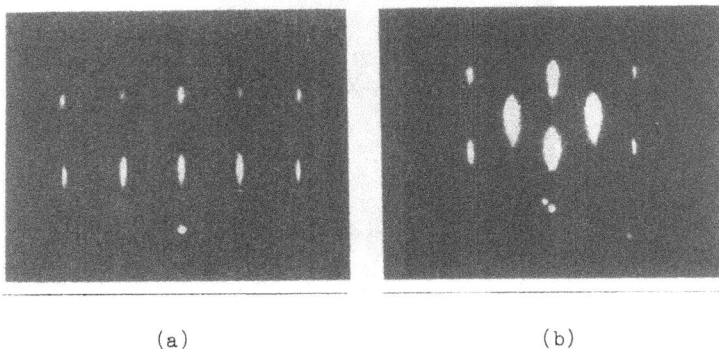

(a)   (b)

Fig.1 RHEED patterns observed on (a)[110] and (b) [001]
      azimuth from MgO film. The film was deposited on
      (110)SrTiO$_3$ substrate at 500 °C.

Preparation of (110)YBaCuO

(110)YBaCuO film was deposited on (110)SrTiO$_3$ by rf-magnetron
sputtering [1,2].   The target used was Y$_1$Ba$_2$Cu$_{4.5}$O powder which was made by
1sintering the mixture of BaCO$_3$, Y$_2$O$_3$ and CuO at 900 C for 10 hr in O$_2$
atmosphare.   The sputteing conditions are summarized in Tab.II.
    Figure 2 shows RHEED pattern obserced on (110) azimuth from (110)YBaCuO
prepared on (110)SrTiO$_3$.   This indicates epitaxial growth of YBaCuO
trilayered perovskite structure with long range order.   But the pattern is
not streak, so improvement of surface smoothness is a remaining problem.

Tab.II   (110)YBaCuO sputtering conditions

Sputtering gas	Ar + O_2 (50%)
Applied rf voltage	1.5 kV
Input power	50 W
Gas pressure	2.0 Pa
Substrate temperature	600 °C
Substrate	(110)SrTiO$_3$
Deposition rate	5 nm/min
Thickness	150 nm

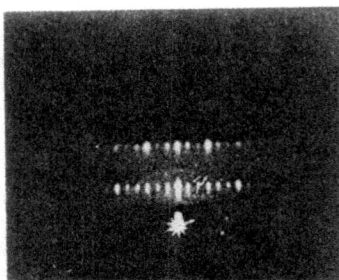

Fig.2 RHEED pattern from (110)YBaCuO film on (110)SrTiO$_3$.

MgO growth on (110)YBaCuO

Superconducting YBaCuO exhibits orthorhombic-tetraganal phase transition at about 500 °C in vaccum [3], above which the superconductivity is missing. MgO deposition on YBaCuO must be carried out at lower temperature than 500 °C. Substrate temperature dependance of MgO crystallization at sputtering conditions of Tab.I was investigated by RHEED observation.

Under substrate temperature 200 °C, ring pattern was observed (Fig.3). At 300-500 °C spot pattern shown in Fig.4 appeared, indicating that (100)MgO grain started to grow epitaxally on (110)YBaCuO parallel to [001] azimuth. At temperature 550 °C, new pattern was observed, as shown in Fig.5. These spots could be assigned as illustrated in Fig.5, suggesting appearance of epitaxally grown (110)MgO grains. This result proves the possibility of (110) and (100)MgO epitaxal formation on (110)YBaCuO. But improvement of sputtering conditions and sputtering gas atmosphare is necessary to obtain a fine single-phase crystal at lower temperature.

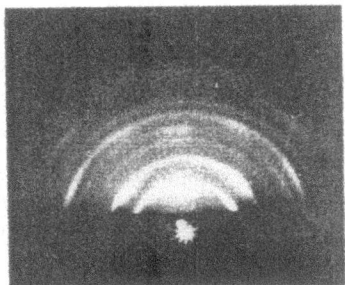

fig.3 RHEED pattern from MgO film deposited
on (110)YBaCuO at 200 C.

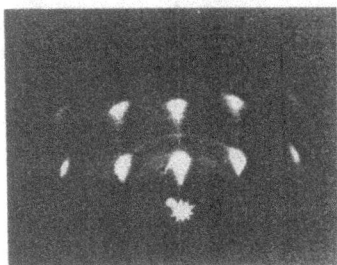

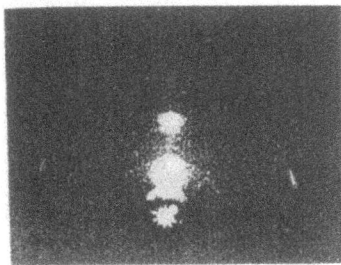

(a)   (b)

Fig.4 RHEED patterns observed on (a)[010] and (b)[120]
azimuth from MgO film. The film was deposited
on (110)YBaCuO on 300 C.

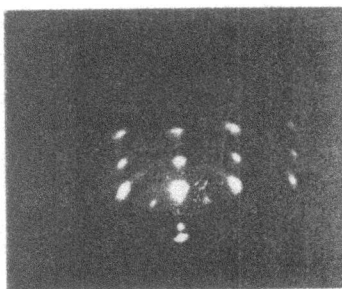

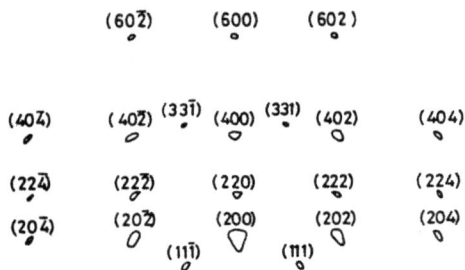

Fig.5 RHEED pattern observed on [110] azimuth
from MgO film. The film was deposited
on (110)SrTiO$_3$ at 550 C.

Formation of YBaCuO/MgO/YBaCuO Double-Heterostructure

YBaCuO/MgO/YBaCuO double-heterostructure was grown as a final stage.
The MgO middle layer was deposited under the sputtering condition of Tab.I
and substrate temperature 550 C.   Top layer of YBaCuO was formed at the
condition of Tab.II except the substrate temperature of 670 C.
Figure 6 shows RHEED patterns observed from YBaCuO/MgO/YBaCuO double
heterostructure.   These patterns indicate the epitaxal growth of
(110)YBaCuO/MgO/YBaCuO structured grains, assuring the successful formation
of this new heterostructrues immediately after the improvement of MgO
crystal structure.

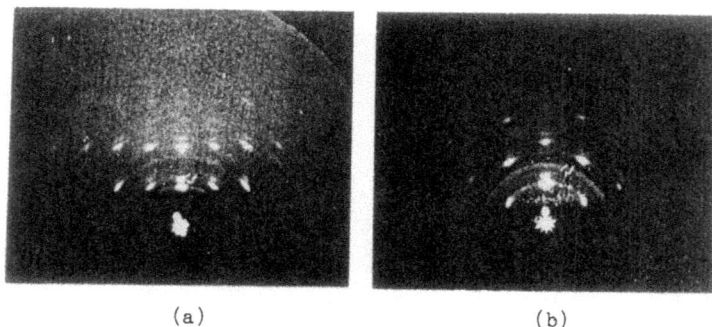

(a)   (b)

Fig.6 RHEED patterns observed on (a)[110] and (b)[001] azimuth
     from   YBaCuO/MgO/YBaCuO   double-hererostructure.

Conclusion

Epitaxal growth of (110)MgO on (110)SrTiO$_3$ substrate was carried out by
rf-magnetron sputtering.   Formation of an YBaCuO/MgO/YBaCuO double-
heterostructure have been attempted.   The RHEED observation revealed the
existence of cryatal grains having epitaxally grown heterostructute.
Improvement of the crystallinity of the MgO layer is a key to establish the
double-heterostructure epitaxy.

References

1) M.Tonouchi, Y.Yoshizako, M.Iyori and T.Kobayashi; Appl. Supercon. Conf.
   (San Fransisco, 1988).
2) T.Kobayashi, M.Tonouchi, Y.Yoshizako, M.Iyori and K.Fujino; JOURNAL OF
   THE JAPAN SOCIETY OF Powder and Powder Metallurgy 35, 392-396 (1988).
3) M.Kogachi, S.Nakanishi, K.Nakahigashi, S.Minamigawa, H.Sasakura,
   N.Fukuoka and A.Yanase; Jpn. J. Appl. Phys. 27(1988) L1228.

SUPERCONDUCTIVITY RELATING TO Tl LOCALIZED
STATE IN NARROW-GAP SEMICONDUCTOR PbTe

Hironaru MURAKAMI, Yoichi MIZOMATA, and Ryozo AOKI

Dept. of Physics, Kyushu University 33 Fukuoka , 812 Japan

Abstract - Tl doped $PbTe_0$ shows superconducting Tc up to 1.5K at extremely low carrier density of $p \leq 10^{20}/cm^3$. Another dopant Na is introduced to this Pb(Tl)Te system. In the course of Na concentration increase, the superconducting Tc, hole carrier dencity p, and hole mobility μ showed two peaks distinctly in correspondence with each other. From these facts, it is considered that the superconductivity in this polarizable semiconductor is much related with the resonant scattering at the semi-localized Tl impurity state.

Introduction

A typical IV-VI semiconductor PbTe has been paid attention by its characteristic properties of high mobility carriers and infra-red optical activity due to narrow energy gap of $\sim 0.2eV$.
   Doping by Tl element on this PbTe gives rise to a semilocalized impurity state at energy levels locating out of the gap and closely below top of the valence band. By introducing another acceptor dopant Na to this Pb(Tl)Te system, the Fermi level in the valence band is shifted down with the Na concentration. As it passes over the Tl impurity levels, a resonant scattering of the hole carrier takes place on the virtual bound Tl impurity state.
   It is interesting that at the same sample condition for this resonant scattering, superconducting transition temperature Tc was found rising up near to 2K even at extremely small carrier concentration of $p \leq 10^{20}$ by Chernik et al [1], and physical properties of this Pb(Tl)Te system have been discussed with great interest regarding a new superconducting non-phonon mechanism [2]. Many investigations have been carried out on bulk sintered samples prepared by the ceramic method [1].

Experiments

   We could prepare fine film samples by a novel technique including (PbTe+Tl_2Te) calcing and co-deposition with Na_xTe in a hot wall-cell [3].
   The resonant scattering state is reflected on the hole carrier mobility μ via the scattering relaxation time, and on the hole carrier density p from the density of virtual bound states. Those characteristic quantities μ and p were evaluated from the measurement of electrical resistivity ρ and the Hall voltage at 77K. The superconducting Tc was determined at the midpoint of the resistive transition curve.
   Those obtained results are summarized in fig.1 against the Na doping concentration which is relatively scaled in abscissa by the Na evaporation source temperature T(NaTe) in the hot wall-cell.

182

Here we find clear correspondence of superconducting Tc with the carrier parameters μ and p at the two levels of virtual bound Tl impurity state, which enhances Tc in about twice. We cannot understand this enhancement only as due to the increase of the density of states at the impurity levels, since the ratio of the Tc height at the two levels is found inconsistent with that of the carrier density p. Consequently, it turns out that the resonant carrier scattering condition with the virtual bound impurity state is evidently correlated with the superconductivity and enhances its Tc in the covalent-ionic bonding semiconductor PbTe.

References

1) I.A. Chernik, S.N. Lykov, and N.I. Grechko; Soviet Phys. Solid State 24 (1982) 1661.
2) V.I. Kaidanov and Yu.I. Ravich: Sov. Phys. Usp. 28 (1985) 31.
3) H. Murakami, T. Migita, Y. Mizomata, Y. Inoue, and R. Aoki; Jpn. J. Appl. Phys., Ser 1 S.C. Mat. (1988) 135.

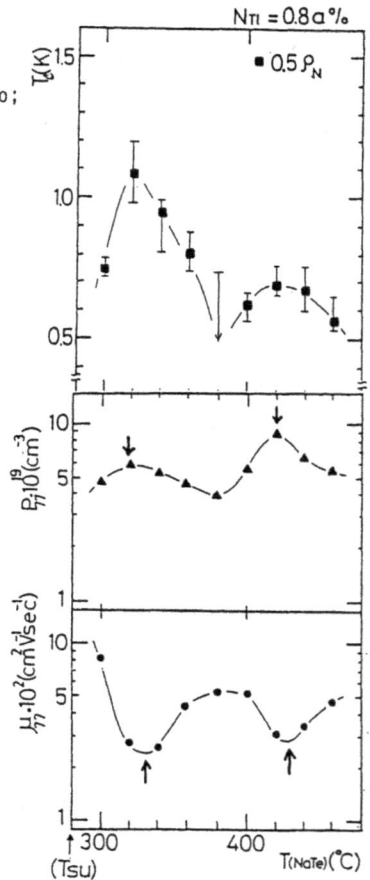

Fig.1 Dependence of the Superconducting Tc, hole carrier density p, and the hole mobility μ on Na dopant relative concentration T(NaTe) in Pb(Tl0.8 at.%)Te.

NEUTRON IRRADIATION EFFECTS IN ADVANCED SUPERCONDUCTORS

Hiroyuki YOSHIDA, Hisao KODAKA, Kiyomi MIYATA, Yoshihiko HAYASHI,
and Kozo ATOBE*

Research Reactor Institute, Kyoto University,
Kumatori-cho, Sennan-gun, Osaka 590-04, Japan
*Naruto University of Education,
Naruto-cho, Tokushima 772, Japan

Abstract - Effects of neutron irradiation on superconducting transitions were studied by susceptibility and resistivity measurements for A15 type compounds, Laves-phase compounds and oxide superconductors. For A15 superconductors, the transition temperature (T_c) decreased with increasing neutron fluence and showed large drop started at about 5×10^{18} n/cm^2 (E>0.1 MeV). Post-irradiation annealing gave recovery of T_c, but the behaviors were different for the materials with different composition and microstructure. The Laves-phase compounds showed less degradation than the A15 superconductors. For oxide superconductors very sensitive transition change was observed, including the radiation-induced superconductivity.

Introduction

Much works has been done for neutron irradiation effects on the superconducting properties of the A15 compounds and the following phenomena have been reported. (i) critical current (I_c) increases in the low fluence range and decreases due to T_c depression at the higher fluence. (ii) T_c decreases in the high neutron fluence range. The former has been discussed in the connection with irradiation induced pinning centre for magnetic flux and the change in the upper critical magnetic field H_{c2} [1-3]. The large T_c depression has been explained by loss integrity of Nb chains related to the degree of long range order [4,5], production of atomic displacements leading to strong electron scattering resulting a depression in density of state [6], and irradiation induced low T_c regions acting as normal state regions in the superconducting matrix [7,8].

Disordering is known as a general phenomenon caused by irradiation in the materials with superlattice structure. The A15 superconductors, such as Nb_3Sn, Nb_3Al, Nb_3Ge and V_3Ga, change their T_c depending on the degree of order [5,10,11]. So Laves-phase compounds, such as V_2Zr and $V_2(Zr,Hf)$, are expected to show less sensitivity to radiation, but little investigation has been done previously on neutron irradiation effect. Recently oxide superconductors have been extensively studied and the following phenomena were reported. (i) very sensitive depression of T_c [11-14], (ii) relatively large increase in I_c [14], and (iii) radiation induced superconductivity [11,15,16].

Experimentals

The samples used were several Nb_3Sn composites prepared by different techniques and arc-melted Nb_3Al, $Nb_3(Al,Ge)$, $Nb_3(Al,Si)$ and $Nb_3(Al,Ge,Si)$ buttons [9]. The V-Zr, V-Zr-Hf and V-Zr-Hf-Ti Laves-phase compounds with different compositions were also prepared by arc-melting. The La_2CuO_{4-y}, $LnBa_2Cu_3O_x$ (Ln=Y, Er and La) oxides were prepared by sintering [12]. Their superconducting transitions were measured by resistivity and susceptibility methods before and after the reactor irradiations. The irradiations were performed using the long period irradiation facility (360 K) and the loop irradiation facility (20 and 350 K) at the Kyoto University Research Reactor (KUR).

Results and Discussions

Figure 1 shows the compositional changes of T_c(onset) for the as-melted bottoms of Nb-Al-Ge, Nb-Al-Si and V-Zr-Hf systems. These specimens showed sharp transition in their inductance-temperature curves measured by the susceptibility method, except the multiple component compounds. The highest T_c(=19.0 K) appeared around the composition of $Nb_2(Al_{0.8}Ge_{0.2})$ in the Nb-Al-Ge(Si) system, and T_c =11.5 K around the $V_2(Zr_{0.8}Hf_{0.2})$ in the V-Zr-Hf system. In the figure the composition dependence of T_c(onset) is also shown for Nb-Al-Ge and Nb-Al-Si systems after the annealing for 5 hours at $1500^{\circ}C$ and then for 2 hours at $900^{\circ}C$. Among these samples the several small specimens with or without the annealings were used for the irradiation experiments.

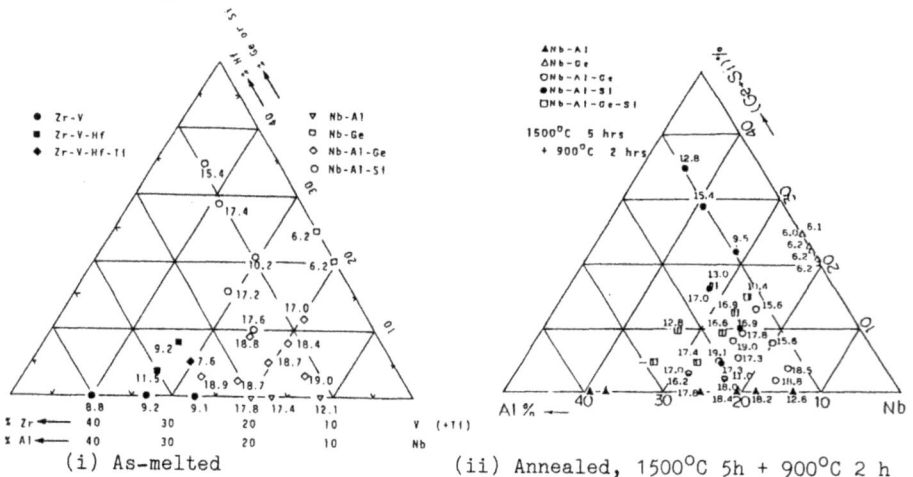

(i) As-melted

(ii) Annealed, $1500^{\circ}C$ 5h + $900^{\circ}C$ 2 h

Fig. 1. Composition dependence of T_c(onset) in Nb-Al-Ge(Si) and V(-Ti)-Zr-Hf systems

Figure 2 shows the typical changes of superconducting properties for Nb, A15, Laves-phase and oxide superconductors, where the changes in J_c and T_c are platted as the relative values $\Delta J_c/J_{c0}$ and $\Delta T_c/T_{c0}$ against the values of J_{c0} and T_{c0} before irradiation, respectively. The $\Delta J_c/J_{c0}$ was obtained by Okada et al.[18], the T_c/T_{c0} curves for A15 and $YBa_2Cu_3O_x$ were referred from the data obtained by Sweedler et al.[5] and Müller et al. [13], respectively. For Nb no detectable change was observed in the present neutron fluence. Almost all the data for A15 superconductors appeared to scatter around the curve proposed by Sweedler, which suggest that the T_c degradation occurred by the same mechanism of radiation induced disordering. For the Laves-phase V_2Zr compound change relatively smaller than that of A15 superconductors was observed after the neutron irradiation.

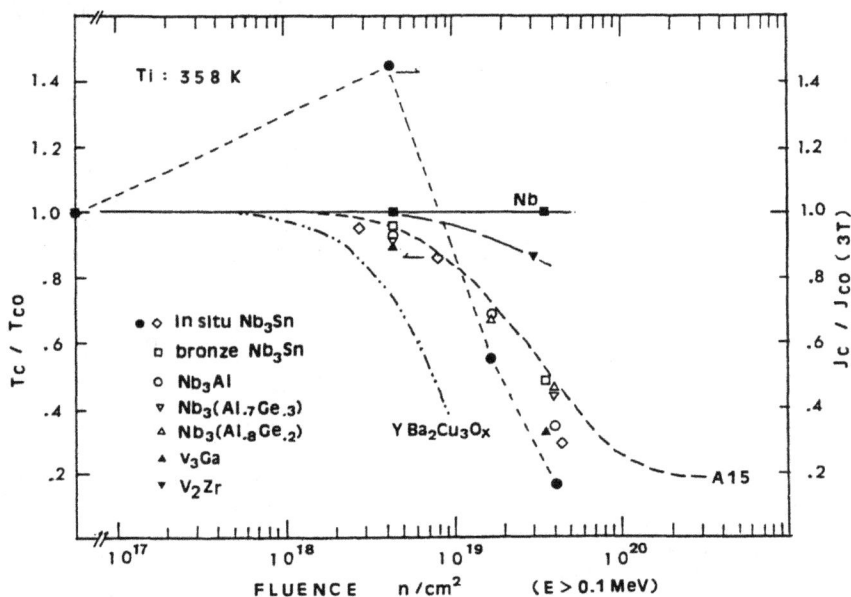

Fig. 2. Changes of superconducting properties caused by neutron irradiation for A15 and Laves-phase compounds. The data are compared with those of other superconductors.

During the post-irradiation annealings the superconducting properties generally show recovery, which might correspond to anihilation of radiation-induced defects and reordering. Figure 3 shows examples of the transition curves obtained during the isochronal annealing for the Nb_3Al and $Nb_3(Al_{0.77}Ge_{0.23})$ irradiated heavily with reactor neutrons [10]. For the Nb_3Al the recovery occurred at two stages about 725 K and 973 K. The recovery behavior is very similar to that observed the in-situ Nb_3Sn wire [9]. It was estimated that the 725 K stage corresponds to the recovery due to anihilation of radiation-induced point defects, and the 973 K stage to reordering from disordered state. After the isochronal annealings up to 1023 K the transition curve recovered to the almost the same curve before the irradiation [10]. In the $Nb_3(Al_{0.77}Ge_{0.23})$ a different behavior was observed as seen in Fig. 3 (ii). The transition curves recovered up to about 773 K in the same manner, however at above the temperatures the curves showed very broad transition suggesting the presence of two phases. The higher T_c component showed an increase of transition temperature with increasing the annealing temperature, and the T_c(onset) finally achieved to 25 K after the 1023 K annealing which is much higher than the original T_c = 19.0 K before the irradiation [10]. The gradual transition curves show multiple components and one of them having the highest T_c = 25 K in this system forms during the post-irradiation annealings.

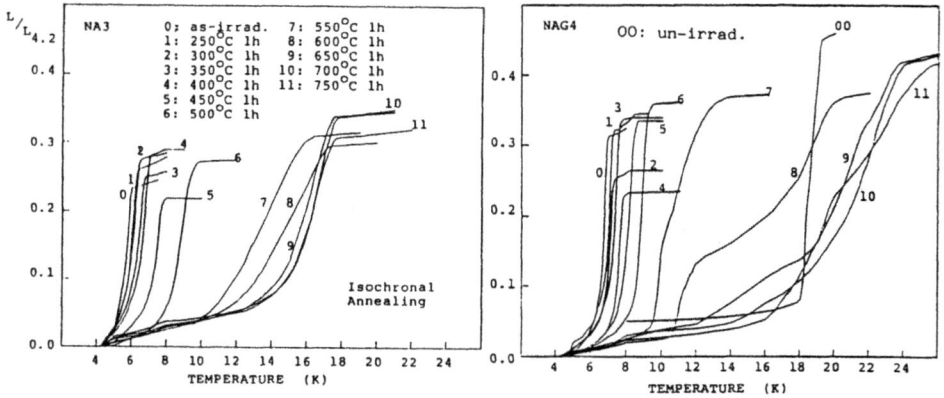

(i) Nb_3Al (ii) $Nb_3(Al_{0.77}Ge_{0.23})$

Fig. 3. Recovery behaviors of superconducting transition for Nb_3Al and $Nb_3(Al,Ge)$ during isochronal annealings after neutron irradiation up to 3.6×10^{19} n/cm^2.

Is is known that the stoichiometric La_2CuO_4 shows semiconductor type behavior in resistance-temperature curve and La_2CuO_{4-y} containing small amounts of oxygen defects and substitutional divalent ions shows a sharp drop just below the resistivity peak like filamentary superconductor. The facts lead to an expectation that the superconductivity of La_2CuO_4 should be sensitively changeable by neutron irradiation. Figure 4 shows the example obtained by the in pile measurements using the loop irradiation facility. The resistivity drops appeared at very low fluence of neutron irradiation, where the numbers at the curves shows fast neutron fluence of $\times 10^{14}$ n/cm^2 (E>0.1 MeV). The temperatures of peak (T_p) and drop to zero (T_0) increased with increasing of irradiation fluence and they achieve to 41.7 and 48.9 K, respectively, at the highest neutron fluence in this series of experiment [16]. The increase was not yet saturated because of very low fluence, however the residual resistivity above T_p also increased suggesting the bulk matrix property. The origin of irradiation induced superconductivity is not yet clear but should be strongly related with oxygen vacancies or interstitials induced by neutron irradiation [17].

Bi-Sr-Ca-Cu oxide is also of interest to study of irradiation effects because it shows two stage superconducting transitions at 85 and 110 K corresponding to the two structures of $Bi_2Sr_2CaCu_2O_x$ and $Bi_2Sr_2Ca_2Cu_2O_x$, respectively. The resistance-temperature curves obtained for Bi-Sr-Ca-Cu oxide superconductor by a different series of in-pile experiment are also shown in Figure 4. After neutron irradiation the residual resistivity above T_0 increased corresponding to irradiation induced defects and a new tail at lower temperature side clearly appeared. The 110 K transition shows little temperature shift, however the 85 K transition remarkably shifts to lower temperature side. During holding for 168 hours at room temperature without neutron irradiation, the resistivity changed due to internal irradiation of α particles from Po^{210} nuclei (half live=138.4 days) produced by thermal neutron transmutation reaction of $Bi^{209}(n,\gamma)Po^{210}$. The internal irradiation seems to give less effect on transition rather than residual

resistivity. The shift of lower temperature transition and the new tail
can be understood if there is a third structure containing higher content
of elements with high neutron cross-section, i.e. $Bi_2Sr_2Ca_2Cu_3O_x$ for their
transition at about 60 K. This leads to a consideration that the 110 K
transition corresponds to $BiSr_2Ca_2Cu_3O_x$ phase with the same structure as
the 125 K phase of $TlBa_2Ca_2Cu_3O_9$ [19].

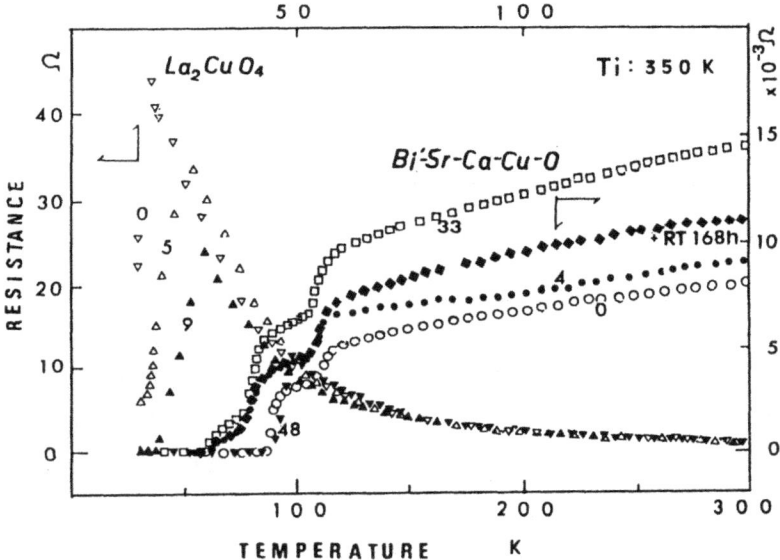

Fig. 4. Changes of resistance-temperature curves caused by reactor
irradiation at 350 K for La_2CuO_{4-y} and Bi-Sr-Ca-Cu-O.
The irradiation conditions are as follows:

		irrad. time	fast neutrons $0\times10^{15}n/cm^2$	thermal neutrons $0\times10^{16}n/cm^2$	gamma ray 0×10^7
La oxide	▽	0 h	0	0	0
	△	6.5	4.9	3.9	1.3
	▲	36.5	9.2	7.2	2.1
	▼	189.5	47.8	37.4	10. 8
Bi oxide	○	0	0	0	0
	●	3	4.6	2.5	3.6
	□	21	33.0	17.3	25.0

Conclusion

Effects of neutron irradiation on superconducting transition in the
advanced superconductors studied can be summarized as follows:

1) For A15 compounds a large depression of T_c occurs due to radiation
disordering.

2) The depression of T_c in A15 compounds recovered by post-irradiation
annealing. The behavior depends on composition and microstructure.

3) Less depression of T_c was observed in Laves-phase compound.

4) Sensitive changes, i.e. T_c decrease, and radiation induced superconductivity, were observed in high T_c oxides depending on composition and
structure.

188

Acknowledgments

The author would like to express many thanks to Prof. T.Okada of Osaka University for collaboration in large part, and to Mr. M.Okada and other members of KUR for kind support. Thanks are also due to Dr. H. Hitotsuyanagi of Sumitomo Electric Company for the Bi-Sr-Ca-Cu-O samples.

References

1) B.S.Brown, T.H.Blewitt, and D.J.Woznick, J. Appli. Phys. 46, 5163.
2) S.T.Sekla,"Effect of irradiation on the critical current of alloy and compound superconductors", J. Nucl. Mater. 72, 91-113(1978)
3) Y.Hirano, S.Nishijima, M.Fukumoto, T.Okada, H.Kodaka, and H,Yoshida, "Neutron irradiation effects on in situ Nb_3Sn superconducting wires", IEEE Trans. Magn. 21, 779-782 (1985).
4) M.Fahnle,"Irradiation effects in Nb_3Sn"Phys.Sta.Sol(b)92,K127-129(1979).
5) A.R.Sweedler, D.E.Cox, and S.Molehlecke,"Neutron irradiation effect of superconducting compounds", J. Nucl. Mater. 72, 50-69 (1978).
6) P.M.Poate, R.C.Dynes, L.R.Testardi and R.H.Hammond,"Comments on defect production and stoichiometry in A-15 superconductors", Phys. Rev. Lett. 37, 1308-1311 (1976).
7) C.S.Pande,"Effect of nuclear irradiation on superconducting transition temperatures of A15 materials", Solid State Commu.24,241-245(2977).
8) H.Yoshida,M.Takeda and H.Hashimoto,"Lattice defects and strain domains in high field A15 superconductors", Proc. Inter. Sympo. Flux Pinning Electromagn. Prop. Supercon., 160-162 (1985).
9) H.Yoshida, H.Kodaka, M.Takeda, and T.Okada, "Radiation induced changes of microstructure and properties in A15 superconductors", Mater. Sci. Forum Vol.15-18, 1141-1146 (1987).
10) H.Yoshida,H.Kodaka and Y.Hayashi,"Superconducting transition caused by disordering and reordering in Nb_3AlGe crystals",Scri.Metll.22,1-4(1988).
11) B.A.Aleksashin,I.F.Berger,S.V.Verkhovskii,V.L.Voronin,B.N.Goshchitskii, S.A.Davydov, A.E.Karkin,V.L.Kozhevnikov,A.V.Mirmelshtein,K.N.Mikhalyov, V.D.Parkhomenko, and S.M.Cheshnitaskii, "Effect of disordering on the properties of high-temperature superconductors", USSR Academy Sci. Rep.'Problems of high-temperature superconductivity' No.1, 1-21 (1988).
12) H.Yoshida, K.Atobe, K.Miyata, and H.Kodaka, "Superconductivity and irradiation effects in Ln-Ba-Cu oxide superconductors", Proc. Sintering '87,(in press); H.Yoshida and K.Atobe, "Radiation induced changes of superconductivity in oxides", Proc. MRS (Tokyo),(in press).
13) P.Muller,H.Gerstenberg,M.Fisher,W.Schindler,J.Strobel,G.Sanmnn-Ischenko, and H.Kammermeier,"Low temperature neutron irradiation of high-T_c-super-conductors $YBa_2Cu_3O_7$ and $Y_{1.2}Ba_{0.8}CuO_4$",Solid State Commu.65,223(1988).
14) S.T.Sekula,D.K.Christen,H.R.Kerchner,J.R.Thompson,L.A.Boatner,B.C.Sales, "Fast neutron damage studies of $La_{1.85}Sr_{0.15}CuO_4$",Jpn.J.Appl.Phys. 26, Suppl.26-3, 1185-1186(1987).
15) H.Yoshida and K.Atobe,"Changes of conductivity and superconductivity in Ln-Ba-Cu oxides by neutron irradiation",Physica C 153-155,337-338(1988).
16) H.Yoshida and K.Atobe, "Superconductivity induced by irradiation in La_2CuO_4", Physica C 156, 225-229 (1988).
17) A.Slupice, B.Giordanengo, R.Tournier, M.Hervieu, A.Maignan, C.Martin, C,Michel, and J.Provost, "Bulk superconductivity in $Tl_2Ba_2CaCu_2O_8$ and $TlBa_2Ca_2Cu_3O_9$ phases", Physica C 156, 243-248 (1988).
18) T.Okada, M.Fukumoto, K.Katagiri,K.Saito,H.Kodaka and H.Yoshida,"Effects of irradiation and strain in a bronze processed multifilamentary Nb_3Sn superconducting composite", IEEE Trans.Magn. 23, 972-975 (1987).

ELECTRONIC STRUCTURES OF TRANSITION METAL OXIDES BY THE TWO-BAND HUBBARD MODEL. VALENCE-BOND FULL CI STUDY

Kizashi YAMAGUCHI, Masayoshi NAKANO, Hideo NAMIMOTO and Takayuki FUENO

Faculty of Engineering Science, Osaka University, Toyonaka,
Osaka, 560, Japan

Abstract - Two-band Hubbard Hamiltonians of the model clusters for doped transition metal oxides were exactly diagonalized by the full valence-bond (VB) configuration interaction (CI) method. The populations of hole and spin densities were calculated for one-electron oxidized clusters by the full VB CI method. The positive holes are mainly populated over the oxygen sites in the case of copper and nickel oxides, whereas they are essentially localized on the metal sites in the case of iron and manganese oxides. The spin-flip (SF) type excitation energy calculated for copper spin systems is lower than the spin-unflip (SU) excitation energies such as the charge-transfer (CT) energies between copper and oxygen sites. However, the CT energies decrease with the increase of the concentration of holes. This indicates that both spin and charge fluctuations are contributable to the formation of Cooper pairs in the high-Tc oxide superconductors.

Introduction

Ab initio unrestricted Hartree-Fock (UHF) and UHF Møller-Plesset (MP) perturbation calculations [1-4] of the clusters of transition metal oxides and halides have revealed that copper oxygen bonds in the high-Tc oxide superconductors belong to the intermediate correlation regime: for example, the CuOCu bond is covalent, but it also exhibits the diradical character as in the case of oxygenated dipoles [5,6]. Therefore, the electronic properties of the copper oxygen bonds should be sensitive to the environmemtal effects such as substitutions of copper ions by other transition metal ions and concentration of holes introduced. This labile chemical valency enables us to propose spin and charge fluctuation models [1-4] for the high-Tc superconductivity, giving a conventional equation for the transition temperature (Tc) [7]

$$k_B \, Tc = \Delta E(ex) \, \exp \, (-1/\lambda) \qquad (1)$$

where $\Delta E(ex)$ is the charge-transfer (CT) excitation energy Δ between copper and oxygen sites in the charge-fluctuation model [1], while it is the magnetic excitation energy $nJab$ (Jab: effective exchange integral between copper spins; n= 2-6) in the spin-fluctuation [1] or J [7] model. However, if the magnetic excitation is considered to be the bitriplet-type excitation (bitron) [8], the charge and spin fluctuations could be cooperative to the high-Tc superconductivity, leading to a mix mechanism, in which $\Delta E(ex) = \Delta(\Delta/nJab)^\lambda$ [9]. In a previous paper [10], the Δ and Jab values for undoped transition metal oxides were examined on the basis of the two-band Hubbard model. In order to obtain deeper understanding of doped transition metal oxides, we here perform the exact diagonalizations of two-band Hubbard Hamiltonians for one-electron oxidized clusters of transition metal oxides.

Model clusters and computational methods

Figure 1 illustrates the model clusters examined here. Ia and Ib are the CuO2 clusters within the CuO2 plane of several oxide superconductors. IIa (T'=T) denotes the CuO4 cluster in the plane, where IIb(T'≠T) is considered as the CuO4 cluster involved in the one-dimensional chain of the R-Ba-Cu-O system. III is the CuO5 cluster involved in the La-Ba-Cu-O (40K), R-Ba-Cu-O (90K), Bi-Ca-Sr-Cu-O and Tl-Ba-Ca-Cu-O (120K) systems. VI is the Cu2O3 cluster within the CuO2 plane. V and VI are the model clusters for examination of the interplane interaction in the R-Ba-Cu-O system.

As previously [10], the two-band Hubbard Hamiltonians involving the Cu and O sites are considered for the copper oxide clusters I-VI as well as all other clusters in which the copper ions are replaced by other transition metal ions (M=Ni, Co, Fe, Mn). The $d\gamma$-orbitals are taken as the frontier orbitals for transition metal ions M(+m), while the 2p (x,y,z)-orbitals are considered for the oxygen anions (O(-1), O(-2)). The one electron orbital ($-\varepsilon$ aa) and on-site Coulombic repulsion (Udd) energies for the M- and O-sites as well as the transfer integrals (T) between M(+m) and O(-n) are given in the ref. 10. The intersite Coulombic repulsion energy V was neglected since the spin density wave (SDW)-type instability is examined here [11]. The full valence-bond (VB) configuration interaction (CI) calculations were carried out for the exact diagonalizations of the model Hamiltonians. The parameters used in the model Hamiltonians are summarized in Table 1.

Populations of holes in the transition metal oxides

Recent experiments indicated that the carriers for the high-Tc super-conductivity are   Cooper pairs of holes introduced in the oxides. In order to elucidate the populations of holes in the transition metal oxides, the full VB CI calculations were carried out for the clusters Ia(b), IIa and IIIa (T'=T). Since the formal charges are assumed to be +2 and -2 for the transition metal ion and oxygen anion, respectively, the hole densities for the metal(M)- and oxygen (O) sites are defined by

$$\Delta\rho\,(M) = \rho\,(M) - 2.0, \quad \Delta\rho\,(O) = \rho\,(O) + 2.0 \tag{2}$$

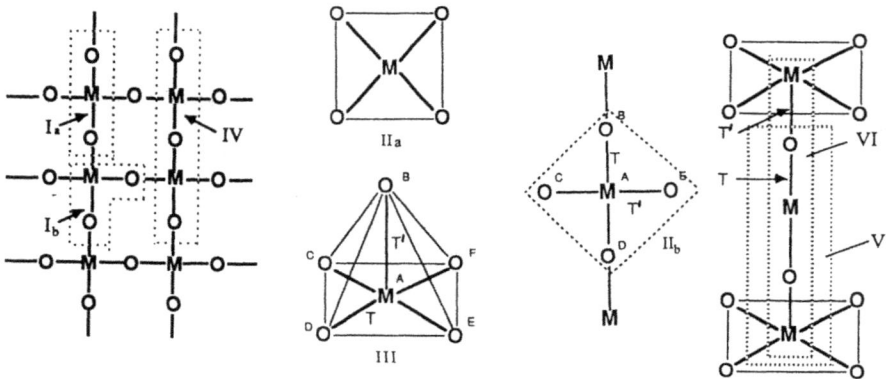

Fig. 1 Model clusters for the high-Tc oxide superconductors

where $\rho(X)$ denotes the charge density at the site X by the full VB CI method. Table 2 summarizes the hole densities in the ground and first excited states of I-III, together with the charge-transfer (CT) excitation energy from the oxygen anion to the transition metal ion. From Table 2, the hole is mainly populated over the oxygen sites in the ground state of the copper oxide clusters, in compatible with many experimental results [3]. The situation is rather similar in the case of the nickel oxide clusters. On the other hand, the hole is essentially localized on the metal site in other transition metal oxides (M=Co, Fe, Mn).

The hole is mainly populated over the oxygen sites in all the first excited states examined in Table 2. Therefore the first excited state is the CT state from the oxygen anions to the transition metal ion in all the clusters. The CT excitation energy is about 1.0 - 1.5 eV for copper oxides, being smaller than 2.0 eV for the copper oxides without holes [10]. On the other hand, it increases with the replacement of copper ions by other transition metal ions (M=Ni, Co, Fe, Mn). Thus, the CT excitation energy $\Delta(\delta)$ in the copper oxides is sensitive to concentrations (δ) of holes and substitutions of transition metal ions, leading to the following equation

$$\Delta(\delta) = \Delta(0)(1 - \delta) \qquad (3)$$

where $\Delta(0)$ is the CT excitation energy for the undoped copper oxide and δ expresses the environmental effects such as the hole concentration.

The full VB CI calculations were also carried out for the binuclear clusters IV within the CuO2 plane. Table 3 summarizes the populations of the hole and spin densities in the ground, first- and second-excited states, together with the corresponding excitation energies. Table 3 shows that the hole is mainly polulated over the oxygen sites in copper and nickel clusters, while the spins are essentially localized on the metal sites. On

Table 1 Parameters in the two-band model

$$d\gamma1 = (dx^2 - dy^2)/\sqrt{2}$$

$$d\gamma2 = [2dz^2-(dx^2+dy^2)]/\sqrt{6}$$

$$d(Cu) = \alpha(d\gamma1)^1(d\gamma2)^2 +$$
$$\beta(d\gamma1)^2(d\gamma2)^1$$

$$d(M) = (d\gamma1)^1(d\gamma2)^1$$
$$(M=Ni, Co, Fe, Mn)$$

$$<px|H|d(Cu)> = <py|H|d(Cu)>$$
$$= T$$
$$<pz|H|d(Cu)> = T'$$
$$<px|H|d(M)> = <py|H|d(M)>$$
$$= T$$
$$<pz|H|d(M)> = T'$$

Table 2 Populations of holes for transition metal oxides

Model System	Ground state		Excited State		Δ
	M	O	M	O	
Ia (CuO2)$^{-1}$	0.129	0.435	-0.104	0.552	1.08
(Ib)(NiO2)$^{-1}$	0.348	0.326	-0.041	0.520	1.18
(CoO2)$^{-1}$	0.637	0.182	-0.026	0.513	1.96
(FeO2)$^{-1}$	0.840	0.080	-0.014	0.507	2.82
(MnO2)$^{-1}$	0.914	0.043	-0.011	0.505	3.98
IIa (CuO4)$^{-5}$	0.111	0.222	-0.200	0.300	1.40
(NiO4)$^{-5}$	0.321	0.170	-0.099	0.275	1.54
(CoO4)$^{-5}$	0.553	0.112	-0.068	0.267	2.31
(FeO4)$^{-5}$	0.756	0.061	-0.039	0.260	3.04
(MnO4)$^{-5}$	0.853	0.037	-0.030	0.258	4.13
IIIa (CuO5)$^{-7}$	0.104	0.179	-0.228	0.246	1.53
(NiO5)$^{-7}$	0.310	0.138	-0.121	0.224	1.67
(CoO5)$^{-7}$	0.527	0.095	-0.085	0.217	2.44
(FeO5)$^{-7}$	0.726	0.055	-0.050	0.210	3.14
(MnO5)$^{-7}$	0.829	0.034	-0.039	0.208	4.20

the other hand, both holes and spins are substantially localized on the
iron and manganese ions in their clusters. The cobalt cluster corresponds
to the intermediate case. The spin density on the central oxygen O(2) is
negative in sign in the ground state of all the clusters, indicating that
the VB configuration $[2(\uparrow\downarrow\uparrow)-(\uparrow\uparrow\downarrow)-(\downarrow\uparrow\uparrow)]/\sqrt{6}$ is largely contributable to the
state: the spin coupling between spins on the metal ions is formally triplet-
type because of the antiferromagnetic exchange interaction (J) between the
M and O(2) ions [3]. This indicates that the magnitude of the antiferromagnetic exchange interaction between copper spins in the CuO2 plane decreases
with the increase of the hole concentration

$$Jab(\delta) = Jab(0)(1 - \delta)\qquad(4)$$

where Jab(0) means the J-value for the undoped copper oxide. The equation
is also applicable to the pπ-hole on oxygen as examined previously [12].
    In the first excited state of each cluster, the populations of the hole
and spin densities are close to those of the ground state except for that the
sign of the spin density on the central oxygen (O2) becomes positive. This
indicates that the first excited state is responsible for the spin-flip (SF)
excitation of spins on the metal sites: $[(\uparrow\uparrow\downarrow)-(\downarrow\uparrow\uparrow)]/\sqrt{2}$. Namely the spins
on the metal ions are singlet-coupled. The SF excitation energies decrease
in the order: Cu→Ni→Co→Fe→Mn. The same tendency was also recognized for
undoped transition metal oxides [3,10]. From Table 3, the second excited
state can be regarded as the CT excited state from the oxygen anions to the
copper site. The CT excitation energy is 1.26 eV for doped copper oxide,
being smaller than 2.0 eV for the undoped one [10]. This supports the
equation (3). The CT energies increase in the order: Cu→Ni→Co→Fe→Mn.

Jahn-Teller Distortions of Transition Metal Oxides

    The X-ray analysis of the parent copper oxide Ln2CuO4 revealed the
Jahn-Teller distorted CuO6 structure with the elongated axial Cu-O bonds.
Similarly, the axial Cu-O bond is significantly long for the CuO5 unit in

Table 3 Populations of holes (HD) and spin (SD) densities for the
cluster IV (O(1)-M-O(2)-M-O(1))

M		Ground State			First Exc. State			Second Ex. State		Δ1	Δ2
	O(1)	M	O(2)	O(1)	M	O(2)	O(1)	M	O(2)		
Cu HD	0.230	0.027	0.486	0.349	0.015	0.270	0.426	-0.102	0.352	0.50	1.26
SD	0.068	0.491	-0.117	-0.034	0.405	0.258	0.383	-0.007	0.248		
Ni HD	0.150	0.161	0.378	0.241	0.156	0.205	0.354	-0.040	0.372	0.40	1.34
SD	0.023	0.529	-0.105	-0.013	0.413	0.200	0.118	0.320	0.124		
Co HD	0.086	0.298	0.232	0.119	0.324	0.112	0.346	-0.026	0.358	0.36	2.12
SD	0.011	0.521	-0.062	0.001	0.444	0.111	0.115	0.325	0.119		
Fe HD	0.041	0.403	0.114	0.048	0.428	0.048	0.340	-0.014	0.347	0.23	2.92
SD	0.004	0.510	-0.029	0.002	0.474	0.048	0.113	0.329	0.116		
Mn HD	0.023	0.444	0.065	0.025	0.461	0.027	0.339	-0.011	0.344	0.17	4.06
SD	0.003	0.505	-0.015	0.002	0.484	0.027	0.113	0.330	0.115		

a) eV, b) Δ1: first (SF) excitation energy, Δ2: second (CT) excitation
energy

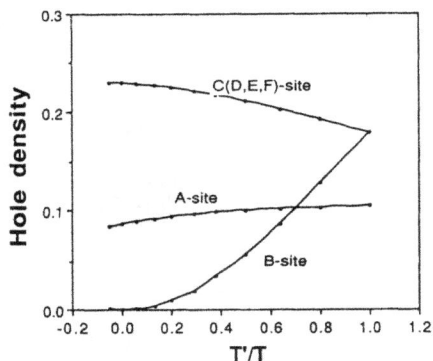

Fig. 2 Deformation of CuO5 cluster

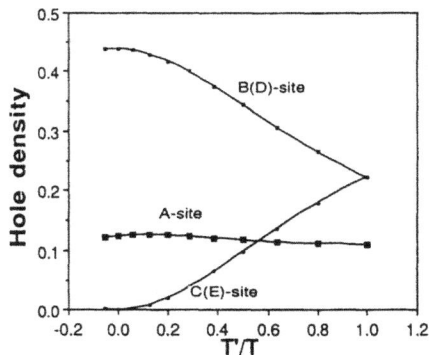

Fig. 3 Deformation of CuO4 cluster

YBa2Cu3Ox. In order to examine changes of hole populations with these Jahn-Teller distortions, the full VB CI studies were performed for the deformation process from CuO5 (T'=T) to CuO4 (IIa)+O(T'=0) through the intermediate structure with T'≠T. Figure 2 illustrates variations of hole densities on the copper and oxygen sites with t=T'/T. Fig.2 shows that the hole density on the axial oxygen (B-site) decreases smoothly with the elogation of the Cu-O (B) bond, while the hole densities on other sites reduce to those of the square planer form IIa. Therefore, the positive hole is mainly populated over the oxygen sites linked with the copper ion.

Figure 3 illustrates variations of hole densities with the parameter t, which is responsible for deformation of the square planer form IIa to the orthorhombic form IIb within the one-dimensional chain in YBa2Cu3Ox. The hole density on the copper atom remains almost constant with this deformation, while the hole density on the B(D) site (see model II in Fig. 1) increases with the deformation. This implies that the positive holes are introduced into the 2pz orbitals of oxygen anions in the Ba-O plane if the

Table 4 Populations of holes for oxide cluster V O(1)-M(1)-O(2)-M(2)

M	t	O(1)	M(1)	O(2)	M(2)
Cu	1.0	0.388	0.105	0.504	0.002
	0.5	0.434	0.130	0.444	-0.009
	0.0	0.438	0.124	0.438	0.000
Ni	1.0	0.264	0.293	0.391	0.052
	0.4	0.329	0.329	0.340	0.002
	0.0	0.347	0.305	0.347	0.000
Co	1.0	0.143	0.516	0.234	0.107
	0.4	0.217	0.548	0.228	0.007
	0.0	0.257	0.485	0.257	0.000
Fe	1.0	0.064	0.673	0.111	0.152
	0.5	0.110	0.748	0.121	0.021
	0.0	0.167	0.666	0.167	0.000
Mn	1.0	0.035	0.725	0.063	0.177
	0.5	0.065	0.836	0.073	0.027
	0.0	0.114	0.772	0.114	0.000

Table 5 Populations of holes for the oxide cluster VI M(1)-O-M(2)-O-M(1)

M	t	M(1)	O	M(2)	Δ
Cu	1.0	-0.003	0.473	0.061	0.683
	0.5	-0.009	0.447	0.124	1.289
	0.0	0.000	0.438	0.124	2.454
Ni	1.0	0.039	0.335	0.251	0.549
	0.4	0.002	0.336	0.324	1.389
	0.0	0.000	0.347	0.305	2.384
Co	1.0	0.077	0.197	0.451	0.478
	0.4	0.007	0.224	0.538	1.181
	0.0	0.000	0.257	0.485	2.019
Fe	1.0	0.106	0.094	0.600	0.282
	0.5	0.020	0.117	0.725	0.593
	0.0	0.000	0.167	0.666	1.233
Mn	1.0	0.121	0.054	0.651	0.204
	0.5	0.025	0.071	0.809	0.424
	0.0	0.000	0.114	0.772	0.923

holes are actually populated over this plane, as concluded from the photo-emission experiments for YBa2Cu3Ox [13].

Interplane interactions

The interactions between CuO2 planes through the one-dimensional chain in YBa2Cu3Ox were examined, assuming the model clusters V and VI. The interplane interaction is expressed with the parameter t (see Fig. 1). Tables 4 and 5 summarizes the populations of positive holes over the oxygen and metal sites, as well as the SF excitation energies $\Delta(\delta)$ for model VI. From Table 4, the holes are populated over the oxygen sites in the copper and nickel oxide clusters, indicating the introduction of holes in the Ba-O plane. On the other hand, the holes are mainly localized on the iron and manganese ions at the second site M(2) in the model cluster V. This implies the trapping of holes by these ions introduced in the one-dimensional chain in YBa2Cu3Ox. The situations are quite similar to the case of model VI.

The lowest spin-conserving excitation is the SF-type excitation in the cluster VI. Therefore the excitation energy is the largest for the copper cluster [1-4]. There are spin-nonconserving magnetic excitations with lower excitation energies for model VI.

Conclusions

The present full VB CI calculations on the basis of the two-band Hubbard Hamiltonians and previous ab initio UHF and UHF MP calculations [1-4] have indicated that the holes are mainly populated over oxygen sites in the case of La-Ba-Cu-O, R-Ba-Cu-O, Bi-Sr-Ca-Cu-O and related species. The attractive interaction between holes could be attributable to (1) the spin fluctuation (bitriplet excitation) [1,7], (2) the charge fluctuation (CT excitation) [1,14] or (3) the fluctuation of both spin and charge associated with local vibrations (phonons) [10]. Full VB CI calculations of more larger clusters are necessary for further discussions of these possibilities.

References

1) K. Yamaguchi et. al., Jpn. J. Appl. Phys. 26, L1362-1364 (1987).
2) K. Yamaguchi et. al., Jpn. J. Appl. Phys. 26, L2037-2040 (1987).
3) K. Yamaguchi et. al., Jpn. J. Appl. Phys. 27, L509-512 (1988).
4) K. Yamaguchi et. al., Physica C153-155, 1213-1214 (1988).
5) K. Yamaguchi, J. Mol. Structure (Theochem) 103, 101-130 (1983).
6) K. Yamaguchi et. al., " Appl. Quantum Chem. " (V. H. Smith et. al. Eds, D. Reidel, Tokyo, 1986) 155-184.
7) K. Yamaguchi and T. Fueno, Jpn. J. Appl. Phys. 27, L393-396 (1988).
8) K. Yamaguchi et. al., Proc. Sympo. Superconductivity (The Chemical Society of Japan, Fukuoka, 1987)
9) I. Tutto and J. Ruvalds, Phys. Rev. B19, 5641-5645 (1979).
10) K. Yamaguchi et. al., Jpn. J. Appl. Phys. 27, L1835-1838 (1988).
11) K. Yamaguchi, Chem. Phys. Lett. 33, 330-335 (1975).
12) K. Yamaguchi et. al., Proc. MRS Sympo. Tokyo, 1988, in press.
13) A. Bianconi et. al., Solid State Commun. 63, 1009-1013. (1987).
14) K. Yamaguchi et. al., Kagaku 42, 583-589 (1987).

RAMAN SPECTRA OF Bi-Sr-Ca-Cu-O SYSTEM

Masanori HANGYO, Shin-ichi NAKASHIMA, Makoto NISHIUCHI,
Akiyoshi MITSUISHI and Tomoji KAWAI[*]

Department of Applied Physics, Osaka University, Suita, Osaka 565, Japan
[*]The Institute of Scientific and Industrial Research, Osaka University,
Ibaraki, Osaka 567, Japan

Abstract - Raman spectra of Bi-Sr-Ca-Cu-O system have been measured for various phases. The spectra of three phases, i. e. 2201, 2212 and 2223(Pb-doped) phases are found to be similar to each other, which reflects the similarity of the structures of these phases. Raman microprobe is used to characterize the ceramic and thin film materials. It is shown that a small amount of impurity phases are included in the samples, which are evaluated to be almost single phase by the X-ray diffraction. Raman microprobe is found to be useful to check the uniformity of the samples.

Introduction

A new high-T_C oxide superconductor, Bi-Sr-Ca-Cu-O (BSCCO), attracts much attention from the view point of the superconducting mechanism and technological applications. There are at least three phases in this system; 2201($T_C \approx 7K$), 2212(80K) and 2223(120K) phases. The structures of these phases are shown in Fig. 1. The composition is represented by a general formula $Bi_2Sr_2Ca_{n-1}Cu_nO_{4+2n}$. These phases can be obtained by inserting Ca and CuO_2 planes successively into the 2201 phase. The 2212 phase is modulated incommensurately along the b-axis. A steep increase of T_C with n suggests that the insertion of Ca and CuO_2 may change the hole concentration in the Cu-O planes and/or the interaction between the holes. It is worth while to study the relation between the structure and the phonon spectra of this system to obtain information on the mechanism of the super-conductivity in this system.

There are several reports on the Raman spectra of BSCCO[1-5]. However, they are limited to only the 80K phase except for a paper by Cardona et al.[3], who measured the 2201 phase also. Especially, spectra of the 2223 phase have not been reported until now probably because of the difficulty in the preparation of the single-phase samples.

Here, we measured the Raman spectra of the 2201, 2212 and 2223 phases and compared the spectra of three phases. We also measured the Raman spectra of ceramic and thin film samples by a Raman microprobe to evaluate the homogeneity of the sample microscopically.

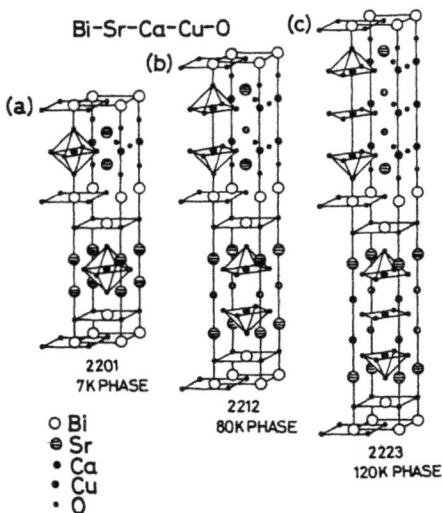

Fig.1. Crystal structures of Bi-Sr-Ca-Cu-O.

Experimentals

Bulk ceramic samples of the 2212 phase were made by sintering a mixture of powder reagents of Bi_2O_3, $SrCO_3$, $CaCO_3$ and CuO as reported before[6]. Ceramic samples of the 2223 phase were made by sintering at low oxygen pressure the powder obtained by codecomposition of metal nitrates, the cation ratio of which was Bi:Pb:Sr:Ca:Cu=0.8:0.2:0.8:1.0:1.4[7]. The 2201 sample was made by sintering a mixture of Bi_2O_3, $SrCO_3$ and CuO, the cation ratio of which was Bi:Sr:Cu=1:1:1 or 2:2:1. Oriented films of the 2212 phase were prepared by the laser sputtering method[8]. Films containing Pb were also made. The samples were characterized by the X-ray diffraction method.

Macroscopic Raman scattering from the area of about 5×0.1 mm^2, was measured with a quasi-backscattering geometry using the 4880 A line of an Ar ion laser of c.a. 100 mW. The scattered light was dispersed by a Spex 1403 double monochromator equipped with photon counting electronics. Raman microprobe measurements were conducted by a dilor XY spectrometer with an optical microscope, which is a combination of a filtering double monochromator and a spectrometer with a 50 cm focal length equipped with an optical multichannel detector. To prevent the samples from damage by the strongly focused laser light, power levels of the laser light were limited to less than 5 mW and the laser spot was somewhat defocused on the sample (several microns in diameter).

Raman Spectra of Bi-Sr-Ca-Cu-O System

Raman spectra of Bi-Sr-Ca-Cu-O samples prepared by various methods, which are almost the single 80K phase except for the sample (e), are shown in Fig. 2. The sample (a) was sintered at 867°C for 17 hours and quenched in air and the sample (b) was sintered at 867°C for 20 hours and furnace-cooled. The sample (c), which was sintered at 875°C for 40 hours, was kindly supplied by Yotsuya of Osaka Prefectural Industrial Technology Research Institute. The cation ratio of the starting mixture of powders for this sample was Bi:Sr:Ca:Cu= 2:2:2:3. Samples (d) and (e) are laser sputtered films. The sample (e) contains Pb and is found to be a mixture of three phases by the X-ray diffraction. In all spectra, Raman bands are seen at 467, 632 and 665 cm^{-1}. A broad band is also seen at around 300 cm^{-1}. A broad band peaked at around 550 cm^{-1} appears for all samples except for the sample (c). The relative intensity of the 550 cm^{-1} band varies considerably with the preparation method.

The Raman spectra of the 80K phase have been reported by several groups for ceramic and single crystal samples[1-5]. The structure above 200 cm^{-1} in the spectra (a)-(e) essentially coincides with that of the single crystal. However, the 550 cm^{-1} band does not appear in the spectra of the single crystal. According to Nishitani et al.[9], Raman spectra of ceramic BSCCO samples with $T_c = 65K$ show a strong broad band peaked at $\sim$550 cm^{-1}. This band is identified as a spectrum of a impurity phase by a Raman microprobe as will be mentioned in the next section.

Figure 3 shows the Raman spectra of the 2201, 2221 and 2223 phases from the top to the bottom, respectively. The whole structure is quite similar for three phases, although the scattering intensity of the 120 K phase is considerably weaker than the other two phases and the S/N ratio is poor. The bands above 300 cm^{-1} must be assigned to the vibrations of oxygen (The 550 cm^{-1} band is due to the impurity phase). The facter group analysis of the vibrational modes for three phases are listed in Table 1 assuming body-centered tetragonal structure (D_{4h}). The Raman active modes are A_{1g}, B_{1g} and E_{2g}. Even if we limit ourselves to the vibrational modes of oxygen atoms, the number of Raman active mode increases with going from the 2201 phase to the 2212 and 2223 phases. However, no additional lines appear above 300 cm^{-1} with increasing the number of CuO$_2$ planes in the unit cell. Therefore, the 467 and 620 - 670 cm^{-1} bands may be assigned to the vibrations of the oxygen atoms in the SrO and BiO$_2$ planes if we take into account that the local structures around the SrO and BiO$_2$ planes are preserved with inserting the Ca and CuO$_2$ planes. Further assignment of the observed modes requires the polarization measurement of single crystal of three phases.

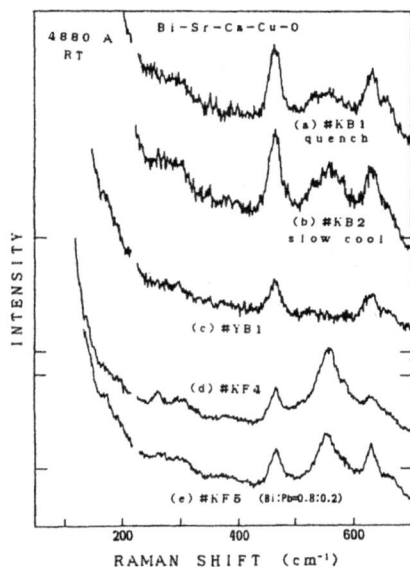

Fig.2. Raman spectra of BSCCO (80K phase) prepared by various methods.

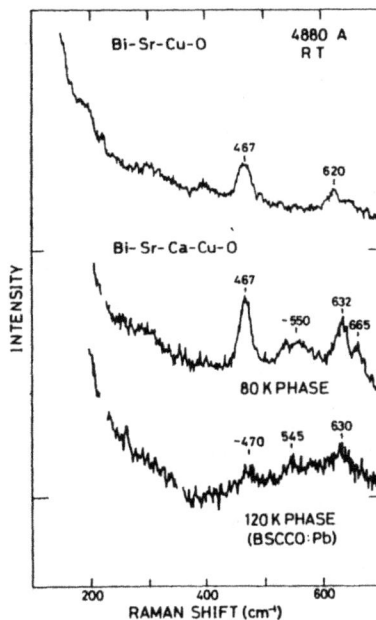

Fig.3. Raman spectra of BSCCO of the 7K, 80K and 120K phases.

Table 1. Factor group analysis of the phonon modes of BSCCO.

STRUCTURE	SUBUNIT	REPRESENTATION
2201	Bi$_2$Sr$_2$CuO$_6$	$4A_{1g}+4E_g+6A_{2u}+B_{2u}+7E_u$
2212	Bi$_2$Sr$_2$CaCu$_2$O$_8$	$6A_{1g}+B_{1g}+7E_g+7A_{2u}+B_{2u}+8E_u$
2223	Bi$_2$Sr$_2$Ca$_2$Cu$_2$O$_{10}$	$7A_{1g}+B_{1g}+8E_g+9A_{2u}+2B_{2u}+11E_u$

Raman Microprobe Measurements

The Raman microprobe measurements were carried out to evaluate the homogeneity of the ceramic and thin film samples. Figures 4 and 5 show the Raman spectra obtained from different positions on the ceramic sample (b) in

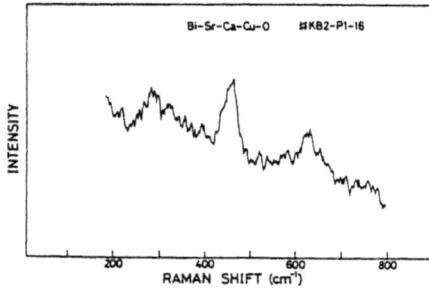

Fig.4. Microprobe Raman spectrum
of the sample (b) in Fig. 2.

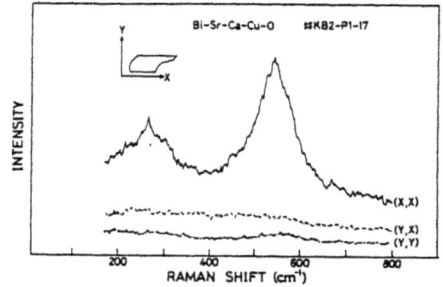

Fig.5. The same as Fig.4, but of the
different position.

Fig. 2. This sample consists of micro-crystals with dimensions of several microns. Figure 4 is similar to the macroscopic Raman spectrum of this sample (the spectrum (b) in Fig. 2) except for the absence of the 550 cm^{-1} band. Judging from the shape of the micro-crystal, incident laser light may be parallel to the c-axis in the case of Fig. 4. We measured the spectra at many points chosen randomly and found that most points show the spectra essentially the same as Fig. 4. However, some long-shaped micro-crystals give the spectra as shown in Fig. 5, the scattering intensity of which is considerably stronger than that of Fig. 4. These spectra show a remarkable polarization dependence as seen in Fig. 5. The above microprobe measurements clearly show that this ceramic sample contains a small amount of impurity phase, which gives rise to the spectra in Fig. 5, in addition to the 80K phase, which gives rise to Fig. 4. The Raman microprobe measurements on samples (a), (d) and (e) in Fig. 2 show that this impurity phase is included also in these samples. Further, Fig. 2 indicates that the film samples contain larger amount of this impurity phase than the ceramic samples. This impurity phase precipitated between the grains of the superconducting phase may decrease T_c.

In order to compare the spectrum of micro-crystal of the 2212 phase with those of other phases, we also measured Raman spectra of the micro-crystals

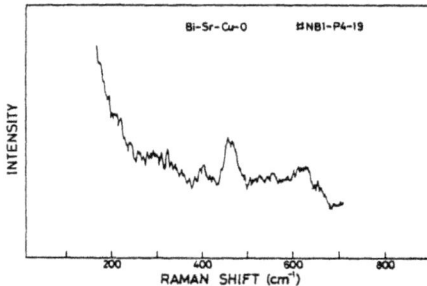

Fig.6. Microprobe Raman spectrum
of the 7K phase sample.

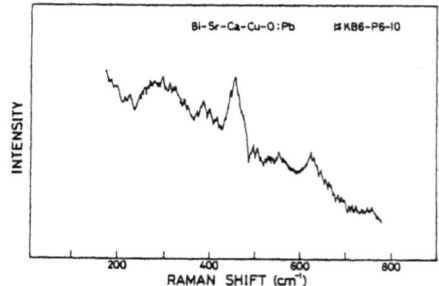

Fig.7. Microprobe Raman spectrum
of the 120K phase sample.

contained in the ceramic samples of the 2201 and 2223 phases. Figures 6 and
7 are the spectra of the micro-crystals contained in the 2201 and 2223 phase
ceramics, respectively. Both spectra are essentially the same as Fig. 4.
This again indicates that the Raman spectra of three phases are quite
similar.

Finally, we show strongly polarized spectra observed for the micro-
crystal in the ceramic 2201 phase sample in Fig. 8. The crystal has a
rectangular shape as shown in the inset. If the polarization of the incident
light is parallel to the long edge of the sample, the spectrum is similar to
Fig. 6. On the other hand, the polarization of the incident light is
perpendicular to the long edge, only the 630 cm^{-1} band appears strongly.
Cardona et al.[3] pointed out that the 630 cm^{-1} band may be strong in the
(c,c) polarization although they could not confirm it by single crystals. If
the Y coordinate of the micro-crystal in Fig. 8 corresponds to the c-axis of
the 2201 phase, the polarization property of the 630 cm^{-1} band is consistent
with that suggested by Cardona et al. The polarization measurements on the
single crystal with well defined crystal axes are needed to clarify this
point.

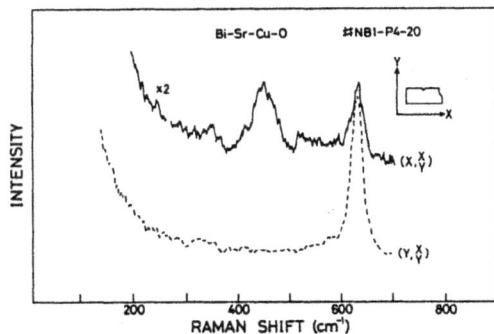

Fig.8. Microprobe Raman spectra of the
micro-crystal with a rectangular shape
in the 7K sample.

Conclusion

The Raman spectra of Bi-Sr-Ca-Cu-O system are measured. The spectra of
the 2201, 2212 and 2223 phases are similar to each other, which reflects the
structural similarity of three phases. The difference of the layer stacking
of three phases are not manifested clearly in the spectra. From this result,
the bands at around 470 and 620-670 cm^{-1} are assigned to the vibrations of
oxygen in the SrO and Bi_2O_3 planes.

The Raman microprobe measurements reveal the existence of some impurity
phase in the ceramic and film samples. The Raman microprobe is found to be
effective to detect a small amount of impurity phases and may be successfully
applied to the characterization of high-T_c superconducting materials.

Acknowledgment - The authors would like to thank Mr. T. Yotsuya of Osaka
Prefectural Industrial Technology Research Institute for supplying a part of
the BSCCO samples used in this study.

References

1) L.A.Farrow, L.H.Greene, J.M.Tarascon, P.A.Morris, W.A.Bonner, and G.W.Hull, "Raman scattering from the $Bi_2Sr_2CaCu_2O_{8+y}$ superconductor", Phys. Rev., B38, 1, 752-754 (1988).

2) M.Stavola, D.M.Krol, L.F.Schneemeyer, S.A.Sunshine, R.M.Fleming, J.V.Waszczak, and S.G.Kosinski, "Raman scattering from single crystal of the 84-K superconductor $Bi_{2.2}Ca_{0.8}Sr_2Cu_2O_{8+\delta}$", Phys. Rev., B38, 7, 5110-5113 (1988).

3) M.Cardona, C.Thomsen, R.Liu, H.G. von Schnering, M.Hartweg, Y.F.Yan, and Z.X.Zhao, "Raman scattering on superconducting crystals of $Bi_2(Sr_{1-x}Ca_x)_{n+1}O_{(6+2n)+\delta}$ (n=0, 1)", Solid State Commun., 66, 12, 1225-1230 (1988).

4) G.Burns, G.V.Chandrashekhar, F.H.Dacol, M.W.Shafer, and P.Strobel, "Phonons in the high temperature $Bi_2Ca_{n-1}Sr_2Cu_nO_{4+2n}$ superconductors", Solid State Commun., 67, 6, 603-607 (1988).

5) S.Sugai, H.Takagi, S.Uchida, and S.Tanaka, "Raman scattering in Bi-Sr-Ca-Cu-O single crystals", Jpn. J. Appl. Phys., 27, 7, L1290-L1292 (1988).

6) H.Nobumase, K.Shimizu, Y.Kitano, and T.Kawai, "High T_C phase of Bi-Sr-Ca-Cu-O superconductor", Jpn. J. Appl. Phys., 27, 5, L846-L848 (1988).

7) U.Endo, S.Koyama, and T.Kawai, "Preparation of the high-T_C phase of Bi-Sr-Ca-Cu-O superconductor", Jpn. J. Appl. Phys., 27, 8, L1476-L1479 (1988).

8) M.Kanai, T.Kawai, M.Kawai, and S.Kawai, "Formation of Bi-Sr-Ca-Cu-O thin films by a laser sputtering method", Jpn. J. Appl. Phys., 27, 7, L1293-L1296 (1988).

9) R.Nishitani, N.Yoshida, Y.Sasaki, N.Kuroda, Y.Nishina, H.Yoshida, Y.Okabe, T.Takahashi, S.Saito, and Y.Koike, Preprints of the Autumn Meeting of the Physical Society of Japan, Hiroshima, October 1988.

BASIC STRUCTURES OF HIGH Tc OXIDE SUPERCONDUCTORS
-- FROM THE VIEW POINT OF SOLID STATE CHEMISTRY--

Tomoji KAWAI, Shichio KAWAI, Shigeyuki TANAKA[1], Sadao TAKAGI[1],

Takeshi HORIUCHI[1] and Kiyoshi OGURA [1]

The Institute of Scientific and Industrial Research, Osaka University, Mihogaoka, Ibaraki, Osaka 567 Japan.

1) Faculty of Science and Engineering, Kinki University, Kowakae, Higashiosaka, Osaka 577 Japan.

Abstract - The conditions for obtaining oxide superconducting materials are summarized by surveying the compounds so far discovered. The crystal structures and the mechanism so far revealed were used to make a concept for the formation of the copper containing high Tc superconductor.

Introduction

The existence of oxide superconductor whose Tc is over 30K had not been confirmed before the discovery of La-Ba-Cu-O superconductor by Bednorz and Muller. Since their discovery, variety of oxide superconductors whose Tc ranges from 30K to 125K have been established. The purpose of this paper is to obtain characteristics of the structure of the high Tc oxide superconductors; what crystal structure is needed for the high Tc material?

Oxide Superconductors over 30K

Recently discovered oxide superconductors whose Tc are over 30K are summarized in Tab. 1. This table shows that there are variety of combinations of elements to form the high Tc superconductors, and the high Tc oxide superconductors are not rare compounds. However, the basic structures are classified into only four types; cubic perovskites, K_2NiF_4 type, oxygen deficient triple layered perovskites and Auribillius type layer compounds (expressed as $(MO)_nA_2Ca_{n-1}Cu_nO_{2+2n}$).

Table 1

La system	Tc	note
1. $(La_{1-x}M_x)_2CuO_4$ $x=\leqq0.2$ M=Ba,Sr,Ca	25–40K	K_2NiF_4 structure
2. $La_1Ba_2Cu_3O_7$	90K	
3. $La_{2-x}Na_xCuO_4$	40K	K does not work
4. $(Bi_{1-x}La_x)SrCuO$	40K	
5. $La_{2-x}CuO_{4-y}$	40K,90K	
6. $La_2Ba_3LuCu_6O_y$	50K	

Y system		
1. $Y_1Ba_2Cu_3O_{7-y}$	90K	Y can be replaced by Ln ion Ln=La to Yb except Ce,Pr,Tb
2. $Y_2Ba_4Cu_8O_{20-x}$	80K	

Bi system		
1. $Bi_2Sr_2Cu_1O_y$	semiconductor, 7K,20K	single Cu–O_2
2. $Bi_2Sr_2Ca_1Cu_2O_y$	85K	double Cu–O_2
3. $Bi_2Sr_2Ca_2Cu_3O_y$	110K	triple Cu–O_2 layers
4. $Bi_2Sr_2Ca_3Cu_4O_y$	90K	
5. $Ba(Pb_{1-x}Bi_x)O_3$	13K	cubic
6. $(Ba_{1-x}K_x)BiO_3$	30K	cubic
7. $(Bi_{0.1}La_{1.8})Sr_{0.1}Cu O_y$	42K	
8. $(Bi_{1-x}Pb_x)_2Sr_2Ca_2Cu_3O_y$	110K	
9. $Bi(Sr_{0.75}Nd_{0.25})CuO$	20K	Nd,Pr,La

Tl system		
1. $Tl_2Ba_2Cu_1O_y$	20K,80K	single Cu–O_2
2. $Tl_2Ba_2Ca_1Cu_2O_y$	105K	double Cu–O_2
3. $Tl_2Ba_2Ca_2Cu_3O_y$	125K	triple Cu–O_2

4. $Tl_1Ba_2CuO_y$ 0K   single Tl-O

5. $Tl_1Ba_2Ca_1Cu_2O_y$ 80K   "

6. $Tl_1Ba_2Ca_2Cu_3O_y$ 110K   "

7. $Tl_1Ba_2Ca_3Cu_4O_y$ 120K   "

8. $Tl_1Ba_2Ca_4Cu_5O_y$ 110K   "

9. $Tl_1Ba_3Ca_2Cu_4O_y$ 160K(?)   "

10. $Tl_1Sr_2(Ca)Cu_3O_y$ 105K   "

11. $(Tl_{0.5}Pb_{0.5})Sr_2Ca_2Cu_3O_y$ 120K   "

12. $(Tl_{1-x}K_x)_2Ba_2Ca_2Cu_3O_y$ 120K   "
     $x=0.7-0.9$

Other systems

1. $Nd_{1.6}Sr_{0.2}Ce_{0.2}CuO_4$ 27K Nd_2CuO_4(?) or not?

2. $(Tl_{0.75}Bi_{0.25})_{1.33}(Sr_{0.5}Ca_{0.5})_{2.7}Cu_2O$ 75K

Common Structural Features for the High Tc Superconductors

The common features of the compounds listed in Tab.1 can be summarized as follows;

1. Mostly copper containing oxides.

2. Formal charge of copper is adjusted by oxygen and metal cations to be between 2+ and 3+. Not only alkaline earth metal(2+) or rare earth metal(3+), but also alkaline metal ion(+1) can be used for this adjustment.

3. Perovskite structure or perovskite related structure. Octahedral $Cu-O_6$ unit or $Cu-O_5$ unit.

4. The structures have two dimensionality. Oxygen deficient layer, rock salt layer, Bi_2O_2, Tl_2O_2 or Tl_1O_1 layers play a role in forming layer structures.

5. Both large ions and small ions are contained to stabilize crystal structures and enhance the formation of layer structure.

6. Antiferromagnetic property and superconductivity of these compounds are closely related.

ABO_3

EX)

1) $Ba(Bi_xPb_{1-x})O_3$

2) $(Ba_{1-x}K_x)BiO_3$

A_2CuO_4 (K_2NiF_4 type)

EX)

1) $(La_{1-x}Sr_x)_2CuO_4$

2) $(La_{1-x}Na_x)_2CuO_4$

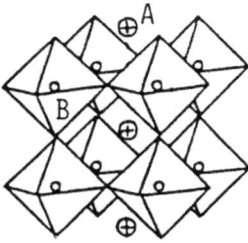

A = Ba,K   B = Bi,Pb
Tc = 13 - 30 K

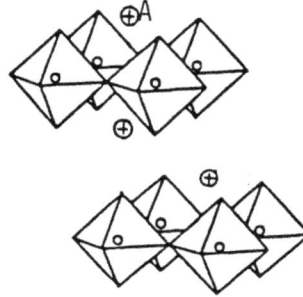

A = La,(Ba,Sr,Ca)
Tc = 20 - 40 K

$(MO)_nA_2CuO_y$

EX)

1) $Bi_2Sr_2Cu_1O_6$

2) $Tl_2Ba_2Cu_1O_6$

Fig.1 (a)
Schematic representation of the basic structure for ABO_3, A_2CuO_4 and $(MO)_nA_2CuO_y$ compounds.   Octahedral CuO_6 unit are arrayed to form planes.

 MO
A_2CuO_y
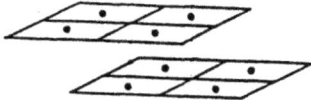

M = Bi,Tl   n = 1,2   A = Sr,Ba

Tc = 7 - 20 K
  = 20 - 80 K

$ABa_2Cu_3O_y$

EX) $Y_1Ba_2Cu_3O_y$

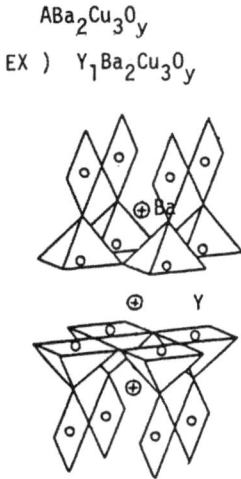

A Y,Lanthanide
Tc= 90 K

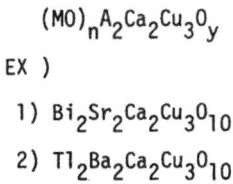

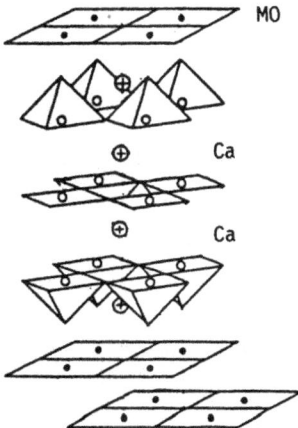

$(MO)_nA_2Ca_2Cu_3O_y$

EX)

1) $Bi_2Sr_2Ca_2Cu_3O_{10}$

2) $Tl_2Ba_2Ca_2Cu_3O_{10}$

M= Bi,Tl   n = 1,2   A= Sr,Ba
Tc = 110 K (M = Bi) , 125 K (M = Tl)

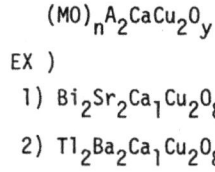

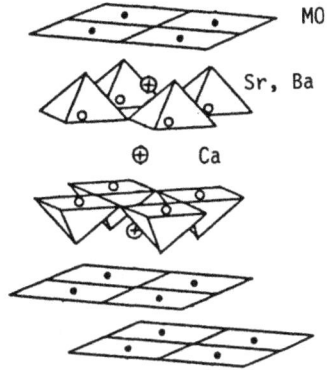

$(MO)_nA_2CaCu_2O_y$

EX)

1) $Bi_2Sr_2Ca_1Cu_2O_8$

2) $Tl_2Ba_2Ca_1Cu_2O_8$

M = Bi,Tl   n = 1,2   A = Sr,Ba
Tc = 85 K   (Bi =M)
    = 105 K   (Tl = M)

Fig.1(b)

Schematic representation of the basic structure for $ABa_2Cu_3O_7$, $(MO)_nA_2CaCuO_{10}$ and $(MO)_nA_2Ca_2Cu_3O_y$. Pyramid type CuO_5 units are arrayed to form planes.

7. The value of Tc may have relations with number of layers, copper valence, Cu-O bond length and/or angle of O-Cu-O planes.

Copper seems to be essential for these kinds of high Tc superconductors. Whether the mechanism of the superconductivity of Ba-Bi-Pb-O and Ba-K-Bi-O compounds is the same as those of copper containing compound is not clear.

The essential structures for these compounds are schematically shown in Fig. 1. Group one has octahedral $Cu-O_6$ unit connected each other forming $Cu-O_2$ planes, and another group has pyramid type $Cu-O_5$ unit connected each other forming $Cu-O_2$ planes. The oxygen out of the $Cu-O_2$ plane seems necessary. Actually, $NdCuO_4$ structure does not show superconductivity. On the other hand, the connection of the layers perpendicular to the $Cu-O_2$ plane through oxygen does not yield superconductivity. The Tc seems to be higher for the compounds having pyramid structure than for the octahedral structure.

Mechanism of the Copper Containing High Tc Superconductor

The mechanism of high Tc superconductivity so far revealed seems to be consistent with the above structures. It has been clarified that the oxidation state of copper remains 2+ having single spin on the Cu^{2+} site, and the holes created by doping are located at the oxygen 2p state. The spin fluctuation at the Cu^{2+} state or the charge fluctuation may lead to the formation of the hole pair to give rise to the superconductivity. The two dimensional nature enhances these kinds of quantum effect. Thus the structures listed above is consistent with this mechanism.

Desirable Structure for the Copper Containing Superconductor

Based on the above discussions, the representative structure for the high Tc oxide superconductor is considered as follows.(Fig.2) It has Cu-O_2 plane whose unit structure is six coordination octahedral perovskite structure or oxygen deficient five coordination pyramid type structure. This structure have two dimensional nature with oxygen deficient layer , rock salt layer or the layers other than perovskite structure. The formal charge of copper should be adjusted by the cations in the A site of the perovskite or the charges of the layers to get the value between 2.1 to 2.3. To get high Tc value, the distance of Cu-O should be short and Cu-O plane should not be buckled.

Methods for the Synthesis of a Desired Structure

The principleof forming layered structures is considered for two

cases. One is to form the perovskite having a layered nature. The first
principle is to satisfy the tolerance factor for the A site cations of
the perovskite, in which the size of the cations are restricted. Then the
combination of large ion with smaller ion leads to the super-lattice
layer formation yielding the two dimensional nature. Furthermore, the use
of Y or Ca causes the oxygen deficient layers because they have the
coordination number of 8 to oxygen ligands due to the small ionic radii.
These combination often produces the compounds which have two dimensional
nature. Another one is to insert the layers between perovskite structure
taking advantage of the nature of VB or IIIB groups due to the bonding
through p orbital.   These combination is really necessary for making the
layered structure represented in Fig. 2.

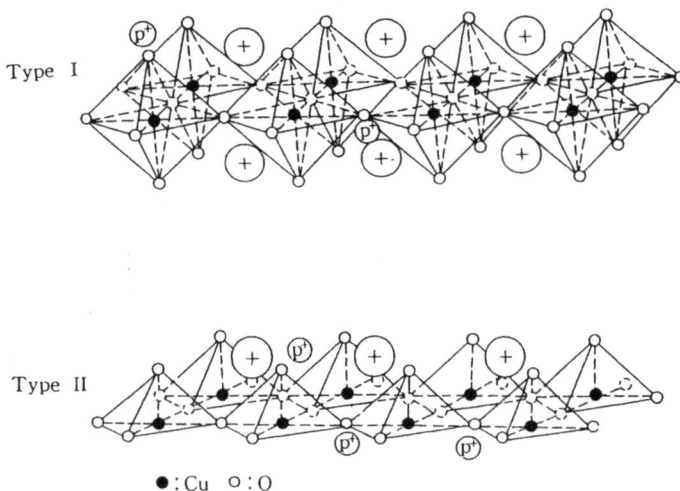

●:Cu ○:O

Fig.2   Schematic representation of the structure for the high Tc
superconductor.   The CuO_5 pyramid or CuO_6 units are arrayed to form Cu-O_2
plane.   The cations adjust the formal charge of Cu to be around 2.2.
Holes(p) are located at the oxygen site, and these holes make Cooper
pair to give rise to the high temperature superconductivity.

ELECTRICAL CONDUCTIVITY AND SEEBECK COEFFICIENT OF
Y-Ba-Cu-O AND Bi-Sr-Ca-Cu-O SYSTEMS AT HIGH TEMPERATURE.

Hiroshi Nagai, Masaru Yokota, Kazuhiko Majima and Ken Ohobayashi*

Department of Materials Science and Engineering, Osaka University,
2-1 Yamadaoka, Suita, Osaka 565, Japan.
* Graduate Student of Osaka University, Suita, Osaka 565, Japan.

Abstract - The high temperature behaviors of Y-Ba-Cu-O and Bi-Sr-Ca-Cu -O systems were investigated solely by electrical conductivity and thermo-electric power measurements and X-ray diffraction. An obvious slope change was detected in the temperature dependence of the electrical resistivity and thermoelectric power for both the $YBa_2(Cu_{1-x}Fe_x)_3O_y$ and $BiSrCaCu_2O_z$ systems.   Arrhenius plots of the electrical conductivities of $YBa_2(Cu_{1-x}Fe_x)_3O_y$ consisted of five straight lines.   The temperatures at the intersects of the straight lines corresponded to the phase transition temperatures between the orthorhombic(I), orthorhombic(II) and tetragonal phases.

Introduction

Recent studies have revealed that the superconducting properties of Y-Ba-Cu-O [1]-[7] and Bi-Sr-Ca-Cu-O [8][9] systems are strongly affected by the preparation process including the sintering temperature and time, annealing temperature, cooling rate, quenching temperature and/or partial pressure of oxygen during synthesis.   These facts strongly suggest that it is important to clarify the high temperature properties of these materials.   However, many unresolved questions still remain concerning the high temperature behaviors of these materials.   The purpose of this study is to systematically investigate the high temperature behaviors of these materials in relation to superconducting properties.

Experimental procedures

Samples with the nominal compositions of $YBa_2(Cu_{1-x}Fe_x)_3O_y$ (x= 0-0.03) and $BiSrCaCu_2O_z$ were prepared by a solid state reaction in air.   Starting materials were high-purity powders of Y_2O_3 , $BaCO_3$, CuO, Fe_2O_3 , Bi_2O_3 , $SrCO_3$, $CaCO_3$.   The powders were mixed, calcined, ground, pressed into pellets and sintered in air for 24 hr.   They were slowly cooled in the furnace to room temperature.   The sintering temperatures were 900 ℃ for the Y-Ba-Cu-O system and 868℃ for the Bi-Sr-Ca-Cu-O system.   The samples thus obtained were examined by a powder X-ray method using Cu kα radiation. The superconducting properties were investigated by resistivity and magnetization measurements.   Resistivity and thermoelectric power measurements were chosen to detect the phase transition and thermal behavior of these materials at high temperatures. The ambient gases were O_2, air and O_2/Ar mixtures and the oxygen partial pressure was continuously monitored by an oxygen sensor of stabilized zirconia during the measurements.

Results and Discussion

$YBa_2(Cu_{1-x}Fe_x)_3O_y$

As typical examples, Figs.1(a) and (b) show the resistivity curves versus temperature for $YBa_2Cu_3O_y$ and $YBa_2(Cu_{0.99}Fe_{0.01})_3O_y$ from room temperature to 900℃ under various oxygen partial pressures, respectively. The resistivity curves were very reversible during several heating and cooling cycles and no hysteresis was observed. A slope change in the temperature dependence of the resistivity was clearly observed at the temperatures marked with arrows in all cases. X-ray diffraction analyses of the samples quenched from above and below the marked temperatures into liquid nitrogen revealed that an orthorhombic-tetragonal transition occurred at the marked temperatures. Freitas and Plaskett [10] also reported a slope change in the temperature dependence of the resistivity at which the orthorhombic-tetragonal transition occurred. The transition temperature decreased with decreasing partial pressure of oxygen and is in good agreement with the transition temperatures reported by Jorgensen et al [11]. In order to obtain more information about the thermal behavior of this material, the electrical conductivities of $YBa_2(Cu_{1-x}Fe_x)_3O_y$ are plotted against 1/T for various oxygen partial pressures in Fig.2. It can be seen from these figures that the Arrhenius plots of the electrical conductivities consisted of five straight lines in all cases. It should be pointed out that each of the five straight lines has almost the same slope under different Po_2. Furthermore, the samples quenched from the region with the same slope under different Po_2 showed almost the same X-ray diffraction patterns. It has been reported by several authors [12]-[15] that the oxygen concentration of this material decreases with increasing temperature. It has been generally agreed that the orthorhombic structure with $6.7 \leqq y \leqq 7.0$ exhibits a 90 K class superconductor and another orthorhombic structure with the random absence of oxygen along a one-dimensional chain ($6.4 \leqq y \leqq 6.7$) exhibits a 60 K superconductor and the tetragonal structure with $y \leqq 6.4$ shows a nonsuperconducting semiconductive behavior. It can be considered from these results that the different slopes in the Arrhenius plots of the electrical conductivity may result from the different activation energy due to the difference in the enthalpy, ΔH_f, for forming the oxygen vacancies. The temperatures at the intersects are plotted against the oxygen partial pressures at the respective intersects in Fig.3. X-ray diffraction analyses and Tc measurements of the samples quenched from various temperature regions,[A]-[E] in Fig.3, revealed that the samples quenched from region [A] were the orthorhombic phase and their Tc's were around 90 K. The samples quenched from regions [B] and [C] were also orthorhombic, but Tc's of these materials were 60~ 80 K. Their X-ray diffraction patterns were similar to that of OrthoⅡ reported by Nakazawa et al. [16] The samples quenched from

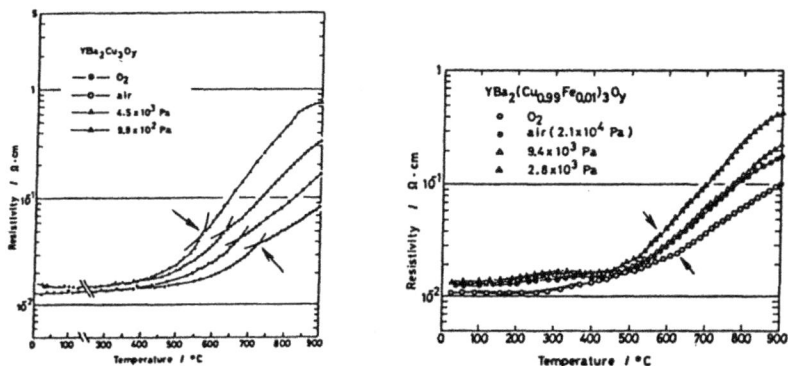

Fig.1   Electrical resistivity of $YBa_2(Cu_{1-x}Fe_x)_3O_y$ as a function of temperature under various P_{O_2} . (a) $x=0$ and (b) $x=0.01$

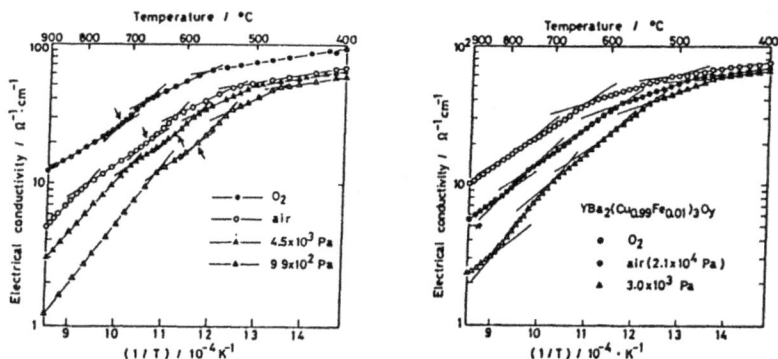

Fig.2   Electrical conductivity of $YBa_2(Cu_{1-x}Fe_x)_3O_y$ as a function of $1/T$ under various P_{O_2} .   (a) $x=0$ and (b) $x=0.01$.

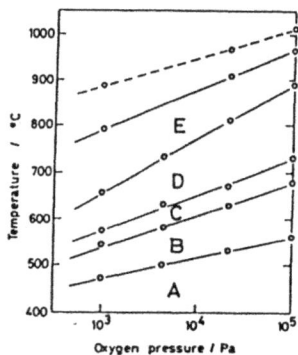

Fig.3   Phase equilibria for $YBa_2Cu_3O_y$ as a function of P_{O_2} .

regions [D] and [E] were tetragonal. Although the slopes in the Arrhenius plots of the electrical conductivity for the regions [D] and [E] were apparently different from each other, no significant difference in the property was detected except for the marked increase in the C cell parameter of the tetragonal phase in the [E] region.

As examples, Figs.4(a) and (b) show the temperature dependence of the thermoelectric power for $YBa_2Cu_3O_y$ and $YBa_2(Cu_{0.99}Fe_{0.01})_3O_y$ under various oxygen partial pressures, respectively. An apparent slope change was also detected in the temperature dependence of the thermoelectric power. The Arrhenius plots of the thermoelectric power also consisted of five straight lines in all cases similar to that for the electrical conductivity. The temperatures at the intersects are in good agreement with those for the electrical conductivity. The temperatures at the intersects were plotted versus Fe concentration(x) in Fig.5. This figure clearly shows that the transition temperatures decrease with increasing x. The Tc(onset) values for $YBa_2(Cu_{1-x}Fe_x)_3O_y$ are plotted versus the quenching temperatures in Fig.6. It was found that the Tc values were around 90 K for the samples quenched from region [A] in Fig.5 and around 60 K for those quenched from region [C]. The Tc values of samples quenched from region [B] steeply decreased with increasing quenching temperature. It can be considered from these results that the intersects of the straight lines in the Arrhenius plots of the electrical conductivity and thermoelectric power correspond to the phase transition temperatures.

BiSrCaCu₂Oₓ

As a typical example, Fig.7 shows the resistivity curves versus temperature for $BiSrCaCu_2O_z$ in air. The resistivity increased with increasing temperature up to 650°C and became constant or slightly decreased and then increased again. However, during the cooling process the change was not detected. Since it has been reported that the decrease in the oxygen content of this material at high temperature is not large, the cause of the slope change in the temperature dependence of the electrical resistivity is considered to be different from that for the Y-Ba-Cu-O system, but is not clear yet. Further detailed research is in progress.

Conclusions

The high temperature behaviors of $YBa_2(Cu_{1-x}Fe_x)_3O_y$ and $BiSrCaCu_2O_z$ were investigated solely by resistivity and thermoelectric power measurements and X-ray diffraction. The results obtained were as follows :
1) The resistivity of $YBa_2(Cu_{1-x}Fe_x)_3O_y$ increased with increasing temperature and decreasing partial pressure of oxygen. A slope change in the temperature dependence of the resistivity was clearly observed. The temperature at the slope change corresponds to the orthorhombic-tetragonal transition temperature. The temperature decreased with decreasing partial pressure of oxygen and increasing Fe content(x).
2) The Arrhenius plots of the electrical conductivity consisted of five

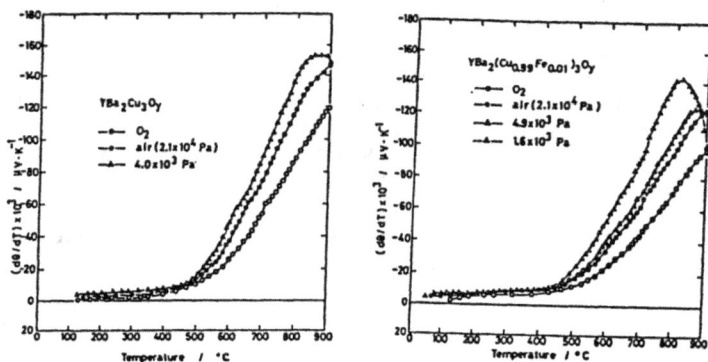

Fig.4 Thermoelectric power of $YBa_2(Cu_{1-x}Fe_x)_3O_y$ as a function of
temperature under various Po_2 . (a) x=0 and (b) x=0.01

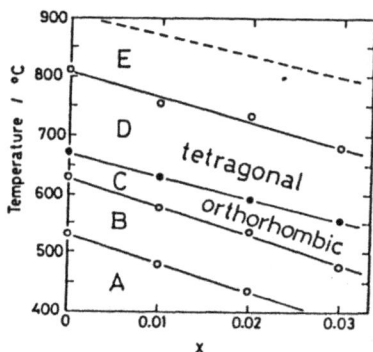

Fig.5 Phase equilibria of
$YBa_2(Cu_{1-x}Fe_x)_3O_y$ as
a function of x.

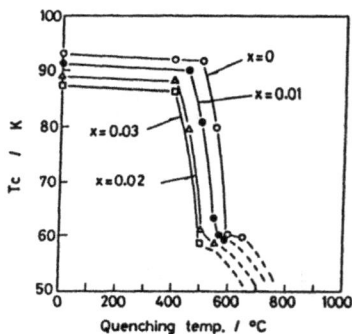

Fig.6 Tc(onset) of $YBa_2(Cu_{1-x}Fe_x)_3O_y$ as a function of
quenching temperature.

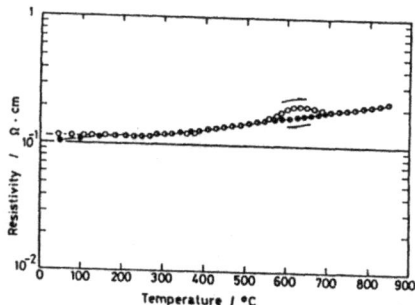

Fig.7 Electrical resistivity of $BiSrCaCu_2O_x$ as a function of
temperature in air.

straight lines.   It can be considered from the results of X-ray analyses
and Tc measurements that the temperatures at the intersects of the
straight lines correspond to the ortho I (Tc ~90 K)-ortho II (Tc ~60 K)-
tetragonal phase(nonsuperconductor) transition temperatures.
3) The temperature dependence of the thermoelectric power also shows
almost the same tendency as that of the electrical conductivity.
4) The resistivity of $BiSrCaCu_2O_x$ also increased with increasing temperature up to 650 ℃ and became constant or slightly decreased and then
increased again.   However, no change in the temperature dependence of the
resistivity was detected during the cooling process.

References

(1) T.Hatano, A.Matsushita, K.Nakamura, Y.Sakka, T.Matsumoto and
     K.Ogawa : Jpn.J.Appl.Phys.26(1987),L721.
(2) Y.Kubo, T.Yoshitake, J.Tabuchi, Y.Nakabayashi, A.Ochi, K.Utsumi,
     I.Igarashi and M.Yonezawa : Jpn.J.Appl.Phys.26(1987),L768.
(3) M.Oda, T.Murakami, T.Enomoto and M.Suzuki : Jpn.J.Appl.Phys.
     26(1987),L804.
(4) S.Leng, N.Narita, K.Higashida and H.Mazaki : Jpn.J.Appl.Phys.
     26(1987),L1394.
(5) O.Fukunaga, Y.Ishizawa and T.Tanaka : Yogyo-Kyokai-Shi,
     95(1987),663.
(6) E.Takayama-Muromachi, Y.Uchida, K.Yukino, T.Tanaka and K.Kato :
     Jpn.J.Appl.Phys.26(1987),L665.
(7) H.Sawada, T.Iwazumi, Y.Saito, Y.Abe, H.Ikeda and R.Yoshizaki :
     Jpn.J.Appl.Phys.26(1987),L1054.
(8) H.Maeda, Y.Tanaka, M.Fukutomi and T.Asano, Jpn.J.Appl.Phys.
     27(1988)L209.
(9) Y.Tanaka, M.Fukutomi, T.Asano and H.Maeda, Jpn.J.Appl.Phys.
     27(1988)L548.
(10) P.P.Freitas and T.S.Plaskett : Phys.Rev.B. 36(1987),5723.
(11) J.D.Jorgensen, M.A.Beno, D.G.Hinks, L.Soderholm, K.J.Volin, R.L.
     Hitterman, J.D.Grace, I.V.Schuller, C.U.Segre, K.Zhang and M.S.
     Kleefisch : Phys.Rev.B. 36(1987),3608.
(12) K.Kishio, J.Shimoyama, T.Hasegawa, K.Kitazawa and K.Fueki :
     Jpn.J.Appl.Phys.26(1987),L1228.
(13) H.Oyanagi, H.Ihara, T.Matsubara, M.Tokumoto, T.Matsushita, M.
     Hirabayashi, K.Murata, N.Terada, T.Yao, H.Iwasaki and Y.Kimura :
     Jpn.J.Appl.Phys.26(1987),L1561.
(14) Y.Kubo, Y.Nakabayashi, J.Tabuchi, T.Yoshitake, A.Ochi, K.Utsumi,
     H.Igarashi and M.Yonezawa : Jpn.J.Appl.Phys. 26(1987)L1888.
(15) S.Nakanishi, M.Kogachi, H.Sakakura, N.Fukuoka, S.Minamigawa,
     K.Nakahigashi and A.Yanase : Jpn.J.Appl.Phys.27(1988)L329.
(16) Y.Nakazawa, M.Ishikawa, T.Takabatake, K.Koga and K.Terakura :
     Jpn.J.Appl.Phys.26(1987,L796.

^{57}Fe MOSSBAUER STUDY OF HIGH-T_c Y-Ba-Cu OXIDE SUPERCONDUCTOR

Saburo NASU, Yasukage ODA, Takao KOHARA*, Koichi UEDA*,
Teruya SHINJO**, Kunisuke ASAYAMA and Francisco E. FUJITA

Department of Material Physics, Faculty of Engineering Science,
Osaka University, Toyonaka, Osaka 560, Japan
*Himeji Institute of Technology, Himeji, Hyogo 671-22, Japan
**Institute for Chemical Research, Kyoto University,
Uji, Kyoto-fu 611, Japan

Abstract - ^{57}Fe Mossbauer measurements have been performed for Fe-doped $YBa_2(Cu_{1-x}Fe_x)_3O_{7-y}$ (x=0.005 - 0.10) at various temperatures between 0.1 K and 473 K. Fe atoms mainly substitute at Cu1 chain sites, while the small portion of Fe atoms occupy the Cu2 plane sites indicating a antiferromagnetic long range order in oxygen deficient compounds. Fe atoms at Cu1 chain sites shows magnetically broadened Mossbauer spectra suggesting a spin-glass behavior at low temperatures. The line shapes due to Fe at Cu1 chain sites are quite similar while the freezing temperatures are quite different for super- and non-superconducting specimens. Conversion electron Mossbauer spectrum has been measured using disk shaped specimens.

Introduction

Since the discovery of high-T_c superconducting oxides, much efforts have been devoted to investigate the chemical and physical properties of these oxides. Hyperfine interaction studies utilizing the ^{57}Fe Mossbauer effects have been performed by several research groups including us [1-2] and a microscopic nature of the Fe-doped 1-2-3 compounds has been reported.

In this investigation, we performed the ^{57}Fe Mossbauer measurements for the Fe-doped 1-2-3 compounds in order to clarify the magnetic behavior of ^{57}Fe atoms and the site-assignments of the Mossbauer spectra. For the ^{57}Fe Mossbauer measurements, we used ^{57}Fe enriched $YBa_2(Cu_{1-x}Fe_x)_3O_{7-y}$ powder specimens which were prepared by solid state reaction method described previously [3]. Specimens for x<0.02 are orthorhombic and those of x>0.03 and quenched from 1173 K are tetragonal in their structure. Mossbauer studies under external high field [2] clearly showed that the Fe atoms in this 1-2-3 compound have localized magnetic moments even at the superconducting states of the specimen and indicates the magnetic hyperfine structure in Mossbauer spectra depending on the specimen condition, Fe concentration and temperature. Fe atoms which were doped into 1-2-3 compounds surely occupy the two Cu sites that are Cu2 plane sites in the CuO_2 plane and Cu1 chain sites sandwiched by the two Ba layers. The Fe atoms do not distribute randomly in these sites and occupy mainly the Cu1 chain sites. Fe at the Cu2 plane site shows a long-range antiferromagnetic order in an oxygen deficient compound which was prepared by quenching from 1173 K and the Néel temperature determined was 423 K, being in agreement with the neutron diffraction experiment [4].

Site-assignments of the spectra

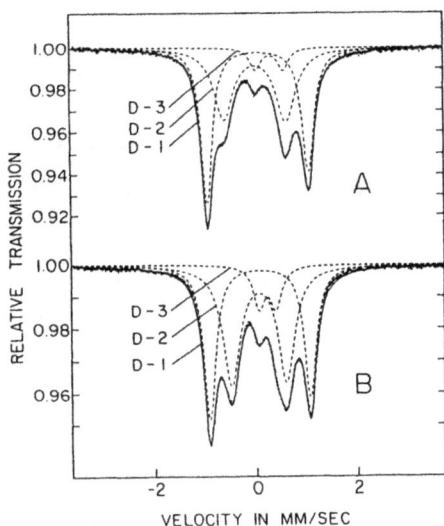

Fig.1 ^{57}Fe Mössbauer spectra of Fe-doped 1-2-3 compounds at 300 K. A:x=0.02 and B:x=0.08. Velocity scale is relative to bcc Fe at 300 K.

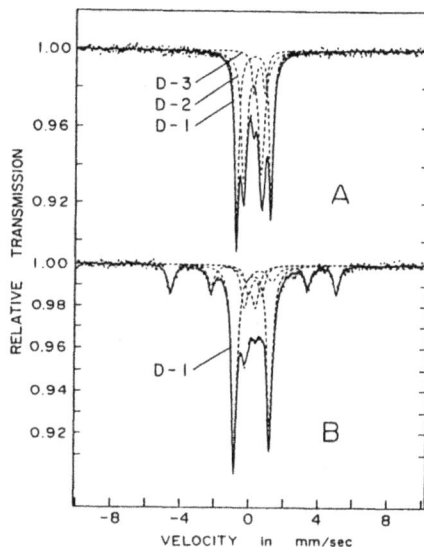

Fig.2 ^{57}Fe Mössbauer spectra of x=0.08 specimens. A: Slowly cooled superconductor and B:Quenched semiconductor.

Figure 1 shows two typical ^{57}Fe Mössbauer spectra at 300 K obtained from x=0.02 and x=0.08 specimens which were prepared by slow cooling from 1173 k in air. Mössbauer spectra were well analysed using three different quadrupole-split-doublets. The specimen of x=0.02 is orthorhombic and T_c (R=0) is 76 K and x=0.08 specimen is tetragonal and T_c(R=0) is 34 K. Magnitude of the quadrupole splitting and isomer shift values of each components obtained from both specimens are quite similar to each other, although the crystal structures are different: one is orthorhombic and the other tetragonal. Above result suggests that the microscopic environments around Fe in these compounds are quite similar and the hyperfine interaction parameters depends mainly on the number of oxygen coordination around Fe. As shown in Fig. 1 we denote each of the quadrupole-split-doublets as D-1, D-2 and D-3 components from the largest to smallest quadrupole-splitting.

Figure 2 shows two spectra obtained from x=0.08 specimens. Spectrum A was obtained from the superconducting specimen which was slowly cooled from 1173 K. Spectrum B was obtained from the non-superconducting specimen quenched from 1173 K which was high-temperature tetragonal phase. Spectrum B in Fig. 2 clearly shows the existence of a magnetically-split-sextet and a D-1 component, while the D-2 component disappeared completely, which was seen clearly in spectrum A. Above results suggest that the D-2 component is attributed to the Fe in Cu1 chain site, since the structural difference between the slowly cooled specimen and the quenched one is the oxygen arrangement around Cu1 chain sites. In order to clarify the origin of the magnetically-split-sextet component in spectrum B, the specimen was annealed for 48 hours at 723 K under oxygen flowing condition. Complete disappearance of the magnetic component and appearance of a paramagnetic doublet corresponding to D-3 were observed. Neutron diffraction experiments

[4] evidenced the existence of the long-range antiferromagnetic order at Cu2 plane sites in oxygen deficient 1-2-3 compounds and suggested that the magnetically-split-sextet in spectrum B is due to Fe in the magnetically ordered Cu2 plane corresponding to the D-3 component in slowly cooled specimen. After annealing at 723 K in oxygen after quenching from 1173 K for x=0.018 specimen, the intensity of D-1 component decreased while that of D-2 component increased and D-4 component having a negative isomer shift value was appeared, suggesting that the D-1 component is also due to the Fe at Cu1 chain sites. Contribution of the D-4 component in each spectrum as shown in Fig. 1 and 2 is rather small and has not been used for the fitting-procedure. Intensities of D-1 and D-2 components are much larger than that of D-3 component and suggests that Fe in the 1-2-3 compound does not distribute randomly at Cu sites and occupied mainly Cu1 chain sites instead of Cu2 plane sites.

Magnetic behavior of Fe at Cu sites

In order to understand the magnetic behavior of Fe at the Cu2 plane and also Fe at Cu1, we measured carefully the spectra at various temperatures from 0.1 K to 473 K. Figure 3 and 4 show the typical Mossbauer spectra at several temperatures obtained from superconducting x=0.08 and non-superconducting x=0.08 specimen, respectively. In these spectra the sextet components having well resolved large hyperfine field were observed at low temperatures and at high temperatures these components were converted to the D-3 component in Fig. 1. After the subtraction of these sextets the magnetically broadened complex components were found at lower temperatures. The shape of the inner part of the spectra as shown in Fig. 3 and 4 is rather similar to each other, which suggested that the magnetic ground states and the spin structure of Fe are same for D-1 and D-2 components. Temperature dependence of well resolved magnetic sextet is shown as open squares in Fig. 5. The Néel temperature of this component determined experimentally was 423 K and agreed well with the value determined by neutron diffraction technique. At lower temperatures below about 200 K for the quenched specimen with x=0.08, the magnetically broadened complex components were observed as shown in Fig. 4. Open triangles and circles showed the extra-broadening of the width in spectrum compared with that of the paramagnetic D-1 component at high temperature. At 0.1 K a spectrum essentially identical to that obtained at 4.2 K as shown in Fig. 3 was observed. This result suggests that the magnetic broadening is not due to the dynamical effect by slow relaxation time of the magnetic interaction but the random orientation of the magnetic moments as in spin-glass. The experimental points shown by black squares, triangles and circles are obtained from the superconducting x=0.08 specimen of which spectra were shown in Fig. 3. From the comparison of these two temperature dependence of the spectral shapes, it was found that the temperatures at which the magnetic splitting and broadening started to occur in spectrum are quite different depending on the existence of the superconductivity. Spectral shape was not affected by the appearance of the superconductivity but the strength of magnetic interaction is affected largely by the existence of the superconductivity. The magnetic splitting of D-3 component for superconducting specimens is not seen in spite of its large magnetic moment, until the D-1 and D-2 components become to be magnetic and show the broad spectrum. The temperature dependence of the magnitude of hyperfine field is unusual, deviating remarkably from the usual Brillouin-like curve. This behavior is accounted for if the exchange field at the chain sites is much smaller than that at the plane sites. This hypothesis

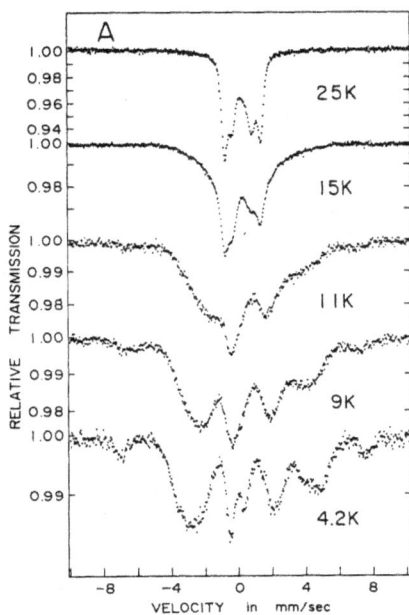

Fig.3 ^{57}Fe Mössbauer spectra as function of temperature for x=0.08 superconductor. Velocity is relative to bcc Fe at 300 K.

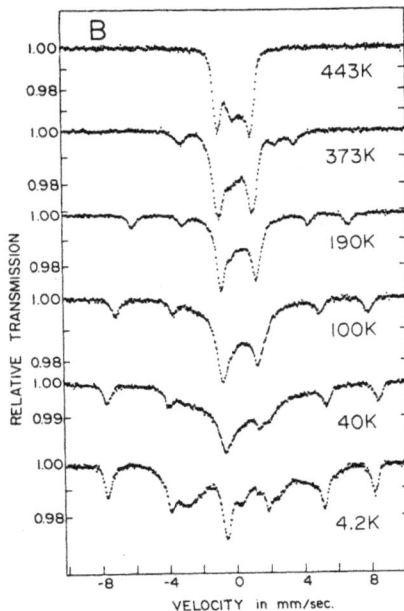

Fig.4 ^{57}Fe Mössbauer spectra as function of temperature x=0.08 non-superconductor. Velocity is relative to bcc Fe at 300 K.

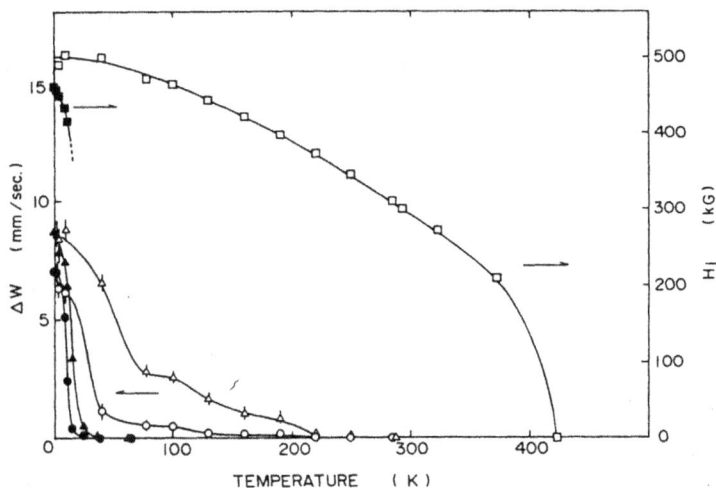

Fig.5   Temperature dependence of the hyperfine fields obtained from x=0.08 specimens. Open circles are obtained from quenched specimens. Full circles are obtained from superconductors.

is supported by the result from neutron diffraction study that the Cu atoms in the chain sites are almost non-magnetic. Temperature dependence of the hyperfine fields which is shown in Fig. 5 can be summarized as follows. In the oxygen deficient non-superconducting state, Fe at Cu2 plane sites shows a long-range magnetic order coupled strongly to the Cu2 planer antiferromagnet. The Fe atom at Cu1 chain sites behaves as a paramagnet separatated magnetically from Cu2 planer antiferromagnet at high temperature and freezed magnetically showing a broad complex spectrum at low temperature. In superconducting state, the Fe atom at Cu2 chain sites is completely paramagnetic until the magnetic freezing of Fe at Cu1 chain sites occurred. The magnetic coupling of the Fe at Cu2 plane sites depends largely on the existence of the superconducting Cu2 and it is most probable that the Cu2 plane plays an essential role in the superconductivity.

Conversion electron Mossbauer spectra

Mossbauer measurements for 1-2-3 compounds have been performed using powder specimens in transmission geometry. It is worthwhile to measure the conversion electron Mossbauer spectrum using disk shaped specimen and compare with the results for powder specimens, since the 1-2-3 compound is not always stable and specimen preparation from disk to powder may cause the oxygen desorption or absorption. Figure 6 shows the conversion electron Mossbauer spectrum (upper figure) by disk shaped specimen and the absorption Mossbauer spectrum (lower figure) by powder specimen, x=0.04. Mossbauer parameters determined from both of the spectra in Fig. 6 for D-1, D-2 and D-3 components

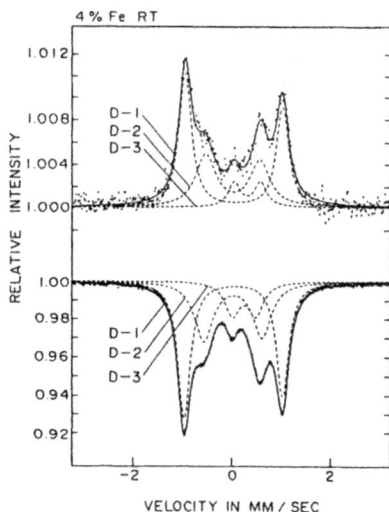

Fig.6 ^{57}Fe conversion electron Mössbauer spectrum (upper figure) and transmission gamma-ray Mössbauer spectrum (lower figure) obtained from x=0.04 specimen at 300 K. Velocity is relative to bcc Fe at 300 K.

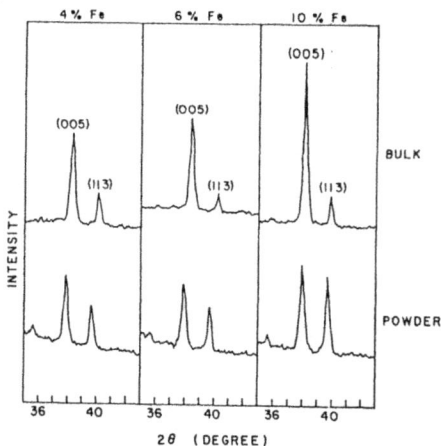

Fig.7 X-ray diffraction pattern for Cu radiation obtained from Fe-doped 1-2-3 compounds. Upper patterns are for disk shaped specimens (Bulk) and lower patterns are for powder specimens.

did not show any large difference except for the intensity ratio in D-1 component. The intensity ratio in the quadrupole-split-doublet depends on the angle between the gamma-ray propagation direction and the principle axis of the electric field gradient. Figure 7 shows the X-ray diffraction pattern which clearly shows the preferred orientation of c-axis to the perpendicular direction to the surface for the disk shaped specimens, since the intensity of (005) is enhanced for disk shaped specimens compared with that of the powder specimens. Since the X-ray diffraction pattern showed the preferred orientation of the c-axis of a perpendicular direction to the surface, the principle axis is the most probably parallel or perpendicular to c-axis depending on the sign of the electric field gradient for the D-1 component.

Summary

The results obtained are summarized as follows.
1) ^{57}Fe Mossbauer spectra in Fe-doped 1-2-3 compounds consist of mainly 3 different quadrupole-split-doublets, those are denoted D-1, D-2 and D-3 components.
2) Fe atoms occupy Cu sites and do not distribute randomly. Fe atoms occupy mainly Cu1 chain sites.
3) The D-3 component is due to Fe at the Cu2 plane sites and shows well resolved sextet in oxygen deficient compounds due to the long-range antiferromagnetic order of the Cu2 plane. The Néel temperature determined from temperature dependence of the hyperfine field was 423 K.
4) Fe at Cu2 plane sites in superconducting states is paramagnetic until the magnetic freezing at Cu1 chain sites occurs.
5) D-1 and D-2 components attributed to the Fe at the Cu1 chain sites show the magnetically broadened complex spectra at low temperatures and suggest that the magnetic freezing occurred as in spin-glass even when the specimen is superconductor.
6) Temperature dependence of the hyperfine field is unusual, deviating from the usual Brillouin-like curve.
7) Conversion electron Mossbauer spectra were measured using disk shaped specimens and the direction of the principle axis of the electric field gradient was discussed.

References

1) H.Tang et.al.,Phys.Rev.,B36, 4018 (1987), J.M.Coey and K.Donnelly, Z.Phys. B67, 513 (1987), X.Z.Zhou et.al.,Phys. Rev., B36, 7230 (1987), G.Gomez et.al., Phys. Rev. B36, 7226 (1987), M.Takano and Y.Takeda, Jpn. J. Appl. Phys., 26, L1862 (1987), C.W.Kimball et.al., Physica, 148B, 309 (1987), T.Tamaki et.al., Solid State Comm., 65, 43 (1988), E.R.Bauminger et.al., Solid State Comm., 65, 123 (1988), Q.A.Pankhurst et.al., J. Phys. C, 21,L7 (1987) and E.Baggioo-Saitovitch et.al., Phys. Rev., B37, 7967 (1988).
2) S.Nasu et.al., Physica, 148B, 484 (1987), Proc. of JIMS-5, Kyoto (1988) and Proc. of MRS Symposium-D, Tokyo, (1988) to be published.
3) Y.Oda et.al., Jpn. J. Appl. Phys., 26, L1660 (1987).
4) J.M.Tranquada et.al., Phys. Rev. Lett., 60, 156 (1987).

EFFECT OF OXYGEN DEFICIENCY AND METAL SUBSTITUTION ON RAMAN SPECTRA OF $YBa_2Cu_3O_{7-\delta}$

Masanori HANGYO, Shin-ichi NAKASHIMA, Makoto NISHIUCHI, and Akiyoshi MITSUISHI

Department of Applied Physics, Osaka University
Yamada-Oka 2-1, Suita, Osaka 565, Japan

Abstract - Raman spectra have been measured on $YBa_2Cu_3O_{7-\delta}$ (YBCO) samples which contain various amount of oxygen deficiency and various substitutional metal atoms. It is found that the 502 cm^{-1} mode, which is assigned to the stretching vibration of the oxygen atoms in the BaO layers in the z-direction, shifts downwards monotonically with decreasing the oxygen content 7-δ. This mode also shifts downwards when a part of the Cu atoms are replaced by Fe, Co, Ga and Al. These results are discussed in relation to the structural change around the Cu ions and around the substitutional sites of dopant metals in YBCO.

Introduction

The superconducting properties of $YBa_2Cu_3O_{7-\delta}$ is sensitive to the oxygen content: it changes from a superconductor to a semiconductor when 7-δ is changed from 7 to 6. Figure 1 shows the structure of YBCO. The amount of the oxygen in the Cu_1-O_4 plane strongly depends on the process of the sample preparation. When the site O_4 denoted by the broken circle in the figure is fully occupied by the oxygen atoms (7-δ = 7), the sample is orthorhombic and superconducting with T_c = 92 K. On the other hand, when the oxygen atom is completely absent at the site O_4 (7-δ = 6), the sample is tetragonal and semiconducting. The structural change from orthorhombic to tetragonal phase occurs at around 7-δ = 6.5. This structural change from orthorhombic to tetragonal means the break of the Cu_1-O_4 chains along the b-axis and will modify the vibrational spectra of YBCO.

The orthorhombic-to-tetragonal transition is also induced by the introduction of some species of dopant metal ions in the Cu sites (Fe, Co, Ga). However, the orthorhombic-to-tetragonal transition is not always induced by the substitution of Cu atoms. In fact, the introduction of Ni and Zn ions does not induce the orthorhombic-to-tetragonal transition. The vibrational spectra will give valuable information about the difference of the effect of these substitutional metal ions.

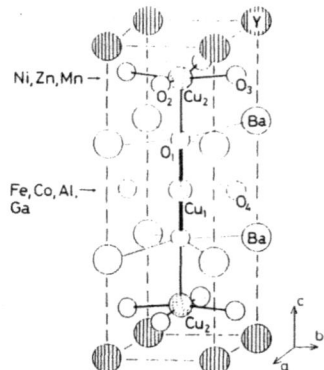

Fig.1. Crystal structure of YBCO.

In this paper, we have measured the Raman spectra of YBCO with various oxygen deficiency and dopant metals in order to obtain the structural and vibrational information on YBCO system. A part of the present work was reported in refs. [1,2].

Sample Preparation

YBCO samples with various oxygen content were prepared by annealing 92 K ceramic YBCO samples at T_q (350 $\sim$ 1000°C) in air and then quenching into liquid nitrogen. Samples containing various substitutional metal atoms were prepared also by the sintering in the same way as in preparing the 92 K sample. Doping of the metal atoms was made by mixing Y_2O_3, $BaCO_3$, CuO and appropriate metal oxides (MnO_2, Fe_2O_3, CoO, NiO, ZnO, Ga_2O_3 and Al_2O_3). The samples were denoted by the nominal dopant composition x as $YBa_2(Cu_{1-x}M_x)_3O_{7-\delta}$, hereafter.

The YBCO samples with various oxygen content and $YBa_2(Cu_{1-x}M_x)_3O_{7-\delta}$ (M=Fe, Co, Ni) were characterized by the powder X-ray diffraction. As a rough estimation of the Meissner effect at liquid nitrogen temperature, the samples were immersed in liquid nitrogen and the repulsion between the samples and a permanent magnet was observed. For quenched YBCO, the repulsion disappeared for T_q higher than about 600°C. The repulsion disappeared at x $\simeq$ 0.02 for Fe- and Co-substituted samples and at x $\simeq$ 0.05 for Ni-substituted samples, respectively. For Mn, Zn, Ga and Al, the repulsion disappeared at the compositions in the range 0.01< x <0.05.

Raman Measurements

Raman spectra were measured with a quasi-backscattering geometry using the 4880 A line of an Ar ion laser. The incident radiation of powers less than 100 mW was line-focused on the sample surface with a cylindrical lens to avoid heating or damage of the sample. The scattered light was detected by a Spex 1403 double monochromator equipped with photon counting electronics.

Effect of Oxygen Content

Figure 2 shows the Raman spectra of YBCO quenched from various temperatures. Several lines are seen in the range of 100 $\sim$ 700 cm^{-1}. The spectrum quenched from 350°C is essentially the same with that of YBCO cooled slowly after the sintering at 930°C. From the polarized Raman spectra of single crystals, the lines at 148, 340, 444 and 502 cm^{-1} have been assigned to the translation of Cu_2 along the z-direction, the out-of-phase bending of the Cu_2-O_2 and Cu_2-O_3 bonds, the in-phase bending of the Cu_2-O_2 and Cu_2-O_3 bonds, and the stretching of the Cu_1-O_1 bond along the z-direction, respectively[3]. These are all the A_g modes in the D_{2h} symmetry. The Raman bands shift continuously with changing T_q.

Kishio et al.[4] measured the oxygen content as a function of temperature and partial pressure of oxygen. Using their result, T_q can be related to 7-δ assuming that the oxygen content at T_q is preserved upon the rapid quenching into liquid nitrogen. The frequencies of the Raman bands are plotted as a function of the oxygen content in Fig. 3. The frequencies of the 502 and 148 cm^{-1} bands increase almost linearly with the oxygen content, whereas that of the 444 cm^{-1} band decreases.

The effect of the oxygen content on the Raman spectra of YBCO has been investigated by several groups[1,5-7]. However, there are some disagreements among the reported data. For example, Kirillov et al.[6] reported that the 500 and 438 cm^{-1} lines in the orthorhombic phase merge into one line at 472 cm^{-1} in the tetragonal phase abruptly at 7-δ = 6.4, although other groups reported that both 502 and 444 cm^{-1} lines in the orthorhombic phase change their frequencies continuously with the oxygen content and both lines survive in the tetragonal phase. The reason for this disagreement is not clear at present, but in the spectra of Kirillov et al. the disorder-activated infrared active 632 cm^{-1} band appears strongly, which indicates poor quality of their samples. Our result almost coincides with the result of Macfarlane et al.[7] except for the behavior of the 444 cm^{-1} band.

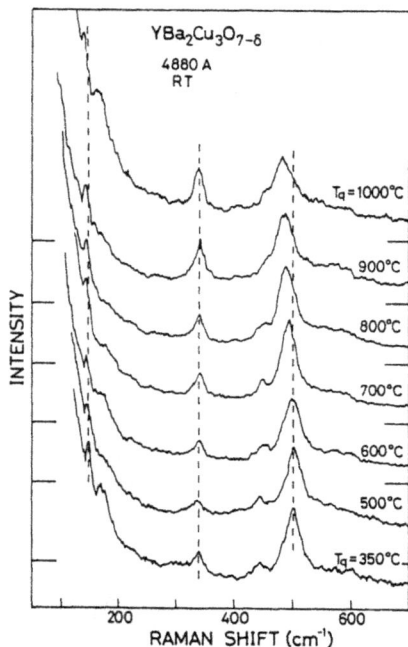

Fig.2. Raman spectra of YBCO quenched from various temperatures T_q.

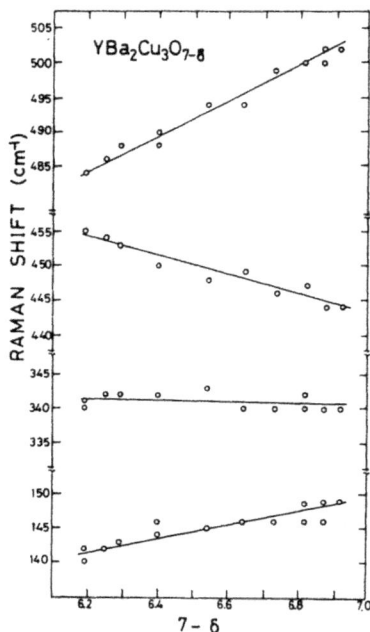

Fig.3. Raman frequencies plotted as a function of $7-\delta$.

Effect of Metal Substitution

Figure 4 shows the dependence of the Raman shift of $YBa_2(Cu_{1-x}M_x)_3O_{7-\delta}$ on the dopant content for Fe, Co and Ni substitution. It is seen that the 502 cm^{-1} band shifts downwards with increasing x for Fe and Co, whereas the frequency of this band is almost independent of x for Ni. On the other hand, the 444 cm^{-1} band shifts downwards for Ni, whereas it is almost independent of x or slightly upshifts for Fe and Co. We also examined the dependence of the Raman shift on the dopant content for Mn, Zn, Ga and Al. Here, we concentrate our attention to the 502 cm^{-1} band, which is most sensitive to the dopant species and their content. The dependence of the frequency of the 502 cm^{-1} band on the dopant can be classified into two groups: one shows the considerable downshift with increasing dopant content and the other is almost independent of the dopant content. Table 1 summarizes the behavior of the 502 cm^{-1} band for various dopants.

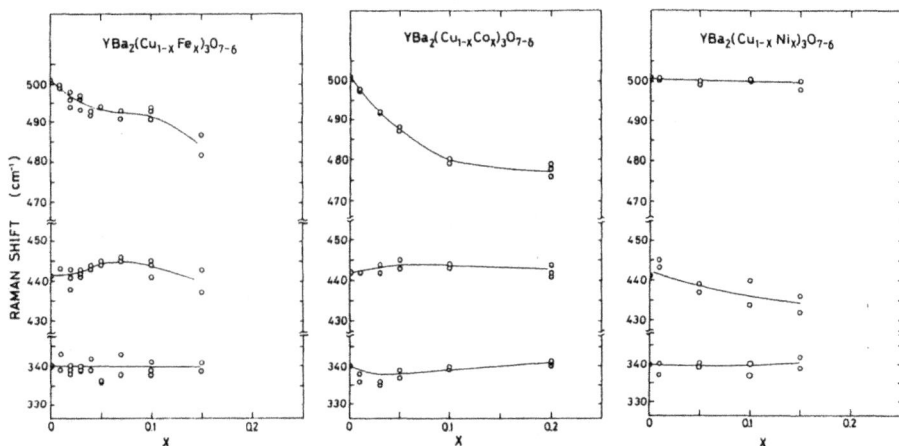

Fig. 4. Raman frequencies as a function of the concentration of the dopant for Fe, Co and Ni.

Table 1. Relation among ortho-tetra transition, shift of the 502 cm^{-1} band and substitutional site determined by the neutron scattering for various dopant metals.

DOPANT METAL	ORTHO-TETRA TRANSITION	SHIFT OF 502 cm^{-1} BAND	SUBSTITUTIONAL SITE
Mn	?	NO	?
Fe	YES	YES	Cu_1
Co	YES	YES	$Cu_1 \gg Cu_2$
Ni	NO	NO	Cu_2
Zn	NO	NO	Cu_2[10], $Cu_1 \gg Cu_2$[11]
Ga	YES	YES	?
Al	YES	YES	Cu_1

Discussion

At first, we discuss the shift of the Raman band with the oxygen content. The fact that the 502 cm^{-1} band is most strongly affected by the oxygen content is expected because this vibration corresponds to the Cu_1-O_1 stretching and the occupancy of the O_4 site may affect the circumstance around Cu_1-O_1. However, the distance between Cu_1 and O_1 decreases with decreasing the oxygen content. If the frequency of the 502 cm^{-1} mode is predominantly governed by the restoring force of the Cu_1-O_1 bond, this band is expected to show the upward shift with decreasing the oxygen content from the simple consideration of the bond length-force constant relation. Removal of O_4 means the disappearance of the repulsion between O_1 and O_4, which shortens the distance between Cu_1 and O_1. The formal valency of Cu_1 in $YBa_2Cu_3O_7$ is +3 (the valency of the Cu ion in YBCO is not still settled). If the valency of Cu_1 does not change by the removal of O_4, the ionic bonding force between Cu_1 and O_1 is enhanced by the contraction of the Cu_1-O_1 bond. Therefore, the valency of the Cu_1 may be reduced by the removal of O_4, which weaken the ionic bonding force leading to the softening of the 502 cm^{-1} band. However, there is no evidence of the change of the Cu_1 valency.

The large upward shift of the 444 cm^{-1} band with decreasing the oxygen content also seems to contradict to a simple consideration: since the O_1-O_2 (O_1-O_3) distance increases with decreasing the oxygen content, the repulsion between O_1 and O_2 (O_1 and O_3) is expected to be reduced and the 444 cm^{-1} band may shift downwards. This contradicts to the observed result. The softening of the 502 cm^{-1} mode and the hardening of the 444 cm^{-1} mode are still puzzling problems.

Next we proceed to the effect of substitutional metals. As stated above, the dopant metals are classified into two groups according to the behavior of the 502 cm^{-1} band. These metals occupy the Cu sites. In YBCO, there are two inequivalent Cu sites, i. e. Cu_1 and Cu_2. In Table 1, we listed the substitutional sites for various metals obtained by the neutron and X-ray analysis together with whether the orthorhombic-to-tetragonal transition occurs by the substitution or not. It is seen that there is a close correlation among the shift of the 502 cm^{-1} band, occurrence of the orthorhombic-to-tetragonal transition and the substitutional sites. The shift of the 502 cm^{-1} band occurs when the dopant metal occupies the Cu_1 site, whereas the shift does not occur when the dopant occupies only the Cu_2 site. This seems quite reasonable because the substitution of Cu_1 will affect the Cu_1-O_1 vibration considerably. One may think that the oxygen deficiency occurs by the introduction of the dopant and it brings about the downward shift of the 502 cm^{-1} band. However, the oxygen content changes only slightly with the Cu substitution and the shift cannot be explained by the oxygen content. Therefore, the shift is brought about really by the replacement of Cu by dopant metals.

Branching or domain formation of the Cu_1-O_4 chains induced by the dopant occupying the Cu_1 site brings about the orthorhombic-to-tetragonal transition as pointed out by Takayama-Muromachi et al.[8] and Shimakawa et al.[9]. From the above discussion, the replacement of the Cu_1 site by dopant atoms induces both the shift of the 502 cm^{-1} band and the orthorhombic-to-tetragonal transition.

As to the substitutional site of Zn, two contradictory neutron experiments have been reported. Xiao et al.[10] reported that Zn occupies the Cu_2 site only and Kajitani et al.[11] reported that it occupies preferentially the Cu_1 site. The present result supports the former result. The substitutional site seems to be closely related to the valency of dopant metals. Ni and Zn, which have the definite valency of +2, occupy the Cu_2 site, whereas Ga and Al, which have the definite valency of +3, occupy the Cu_1 site. Substitutional sites for various dopant metals obtained in this study are shown in Fig. 1: Ni, Zn, Mn occupy only the Cu_2 site and Fe, Co, Al and Ga occupy only the Cu_1 site or both Cu_1 and Cu_2 sites.

Conclusion

The Raman lines are found to shift systematically with the oxygen content for YBCO. This shift is attributed to the change of the valency of Cu ions. The 502 cm^{-1} band shifts with increasing the content of Fe, Co, Ga and Al, whereas it is almost independent of the content of Mn, Ni and Zn. By comparing this result with the neutron and X-ray diffraction data, it is concluded that the 502 cm^{-1} mode shifts when the Cu_1 site is replaced by the dopant metals.

References

1) M.Hangyo, S.Nakashima, K.Mizoguchi, A.Fujii, and A.Mitsuishi, "Effect of oxygen content on phonon Raman spectra of $YBa_2Cu_3O_{7-\delta}$", Solid State Commun., 65, 8, 835–839 (1988).

2) M.Hangyo, S.Nakashima, M.Nishiuchi, K.Nii, and A.Mitsuishi, "Effect of metal substitution on Raman spectra of $YBa_2(Cu_{1-x}M_x)_3O_{7-\delta}$ (M=Fe, Co, Ni)", Solid State Commun., in press.

3) See for example, R.Liu, C.Thomsen, W.Kress, M.Cardona, B.Gegenheimer, F.W. de Witte, J.Prade, A.D.Kulkarni, and U.Schroder, "Frequencies, eigenvectors, and single-crystal selection rules of k=0 phonon in $YBa_2Cu_3O_{7-\delta}$: theory and experiment", Phys. Rev., B37, 13,7971-7974 (1988).

4) K.Kishio, J.Shimoyama, T.Hasegawa, K.Kitazawa, and K.Fueki, "Determination of oxygen nonstoichiometry in a high-T_c superconductor $Ba_2YCu_3O_{7-\delta}$", Jpn. J. Appl. Phys., 26, 7, L1228-L1230 (1987).

5) C.Thomsen, R.Liu, M.Bauer, A.Wittlin, L.Genzel, M.Cardona, E.Schönherr, W.Bauhofer, and W.König, "Systematic Raman and infrared studies of the superconductor $YBa_2Cu_3O_{7-x}$ as a function of oxygen concentration ($0 \leq x \leq 1$)", Solid State Commun., 65, 1, 55-58 (1988).

6) D.Kirillov, J.P.Collman, J.T.McDevitt, G.T.Yee, M.J.Holcomb, and I.Bozovic, "Raman spectra of $YBa_2Cu_3O_x$ superconductors with different oxygen content", Phys. Rev., B37, 7, 3360-3363 (1988).

7) R.M.Macfarlane, H.J.Rosen, E.M.Engler, R.D.Jacowitz, and V.Y.Lee, "Raman study of the effect of oxygen stoichiometry on the phonon spectrum of the high-T_c superconducting $YBa_2Cu_3O_x$", Phys. Rev., B38, 1, 284-289 (1988).

8) E.Takayama-Muromachi, Y.Uchida, and K.Kato, "Superconductivity of $YBa_2Cu_{3-x}M_xO_y$ (M=Co, Fe, Ni, Zn)", Jpn. J. Appl. Phys., 26, 12, L2087-L2090 (1987).

9) Y.Shimakawa, Y.Kubo, K.Utsumi, Y.Takeda, and M.Takano, "The effect of annealing in high-pressure oxygen on $YBa_2(Cu_{1-x}M_x)_3O_y$ (M=Co, Ni, Zn)", Jpn. J. Appl. Phys., 27, 6, L1071-L1073 (1988).

10) G.Xiao, M.Z.Cieplak, D.Musser, A.Gavrin, F.H.Streitz, C.L.Chien, J.J.Rhyne, and J.A.Gotaas, "Significance of plane versus chain sites in high-temperature oxide superconductors", Nature, 332, 17, 238-240 (1988).

11) T.Kajitani, K.Kusaba, M.Kikuchi, Y.Syono, and M.Hirabayashi, "Crystal structures of $YBa_2Cu_{3-\delta}A_\delta O_{9-\gamma}$ (A=Ni, Zn and Co)", Jpn. J. Appl. Phys., 27, 3, L354-357 (1988).

226

EFFECT OF NON-STOICHIOMETRY IN METALLIC COMPOSITION AND HEAT TREATMENT TEMPERATURE ON THE SUPERCONDUCTING PROPERTIES OF SPUTTERED Y-Ba-Cu-O FILMS

H. Kajikawa, T. Hase, M. Okuda, K. Nishimura and Y. Kawate

Electronics Technology Center, Kobe Steel, Ltd., 1-5-5,
Takatsukadai, Nishi-ku, Kobe, 673-02, Japan

Abstract - We studied the effect of non-stoichiometry in metallic composition and heat treatment temperature on the structural and superconducting properties of sputtered Y-Ba-Cu-O films on $SrTiO_3$ (100) substrates. Better superconducting properties (Tc and Jc) were obtained in the films with non-stoichiometric composition of $Y_1Ba_{2-2.3}Cu_{2.3-2.7}O_x$ rather than in the films with stoichiometric composition of $Y_1Ba_2Cu_3O_x$. By changing the metallic composition and the heat treatment temperature, the crystal orientations and the lattice constant a_0 were controlled to obtain Tc of 87.4 K and Jc at 77 K in excess of 10^4 A/cm^2.

Introduction

In our preceding work [1], we investigated the effect of heat treatment conditions on the crystal orientations and morphology of sputtered Y-Ba-Cu-O superconducting films using a two step heat treatment method. We found that in the films of $Y_1Ba_{2.0-2.4}Cu_{2.0-2.8}$, the heat treatments in the first step at 840, 880 and 920 °C resulted in c-type, a-type and ac-type films, respectively, and the fine grained films with smooth surfaces of Ra <1000 A were obtained under the heat treatment below 880 °C.

In this work, we studied the effect of both non-stoichiometry in metallic composition and the heat treatment temperature on the crystal orientations, the lattice constant a_0, and Tc and Jc.

Experimental

Film fabrication

Initially, amorphous and insulating Y-Ba-Cu-O films were deposited on $SrTiO_3$ (100) substrates by means of RF diode sputtering. The sputtering conditions are shown in Table 1. The deposition time was adjusted so that the film thickness became about 1 micron. In order to form the crystalline superconducting phase of $YBa_2Cu_3O_{7-y}$, a two step heat treatment was made in an oxygen atmosphere as shown in Fig.1. In the first step, the films were kept for 25 minutes at a temperature between 800°C to 950°C to form the tetragonal phase, and in the second step, the temperature was kept at 550 °C for 2 hours to form the orthorhombic phase. In this work, the rate of the temperature change between the heat treatment processes were rather higher than the rates used in another works to avoid unnecessary influences on the crystallization during the interval of the temperature changes.

Film characterization

The metallic composition of the films was measured before the heat treatment by electron probe micro-analysis (EPMA) calibrated using the powder of bulk $YBa_2Cu_3O_{7-y}$ as a standard. The diameter of the electron beam used was about 100 microns.

The film structure was analyzed by X-ray diffraction and the a_0 of the shorter lattice constant in the basal plane was estimated from the difference in 2θ between the peak of I(200) for $YBa_2Cu_3O_{7-y}$ and the peak of I(300) for $SrTiO_3$ in the X-ray diffraction pattern. The surface morphology was investigated by scanning electron microscopy (SEM) when necessary.

The superconducting properties of Tc and Jc at 77 K were measured by the conventional d.c. four-point-probe method without external magnetic field. The strips used for the measurement were about 2 mm in width and 10 mm in length. The electrodes at intervals of about 1 mm were formed by an In-Cd solder with an ultrasonic soldering tool. Because the contact resistance was rather large and 6000 micro-ohm cm², it will be improved.

Results and Discussion

Crystal orientations

Figure 2 shows the ratios of I(002) to I(200) for $YBa_2Cu_3O_{7-y}$ vs. heat treatment temperature (HTT) obtained from the films having the same metallic composition of $Y_1Ba_{2.4}Cu_{2.7}$. From this result, it was expected that the crystal orientaions were controlable by HTT. But as pointed out in our previous work [1], they seemed to be affected by the metallic composition of the films besides HTT. Figures 3(a)-3(c) show the crystal orientations of the films having the metallic composition of $Y_{10-30\%}Ba_{30-50\%}Cu_{40-50\%}$ under the heat treatment (HT) at 840 °C, 880 °C and 920 °C, respectively. The definition of "r", "a", "ac" and "c" types of the films is shown in Table 2, in reference to [1-3]. It is noticeable that the crystal orientations were ruled mainly by the yttrium content and the predominant types of the films change from "r" or "c" to "a" through "ac" as the yttrium content increased for each HTT. The origin of those crystal orientations is not clearly explained at this stage. From the viewpoint of lattice matching between $YBa_2Cu_3O_{7-y}$ (orthorhombic, a_0=3.82 Å, b_0=3.89 Å, c_0=3x3.89 Å) and $SrTiO_3$ (cubic, a_0=3.90 Å), the a-axis oriented grains are the most favourable. So it is natural to think that there exists something which stabilizes c-axis oriented grains during the crystallization and their quantity may be controlled by both HTT and the yttrium content. One of them is probably $BaCuO_2$ which is produced at a lower temperature in comparison with $YBa_2Cu_3O_{7-y}$ during the HT and it may be reduced when the yttrium content in the films is more than stoichiometric amount [4]. Figure 4 shows the ratios of I(002) for $YBa_2Cu_3O_{7-y}$, to I(29.3°) for $BaCuO_2$ in highly oriented films. From this result, c-axis oriented grains are found to be dependent on the amount of $BaCuO_2$ and our presumption is confirmed.

Lattice constant a_0

Figure 5 shows the dependence of Tc on the lattice constant a_0. Tc became higher as the a_0 became smaller, as is usual in bulk $YBa_2Cu_3O_{7-y}$ samples. Figures 6(a)-6(c) show the distribution of a_0's of the films having the metallic composition of $Y_{10-30\%}Ba_{30-50\%}Cu_{40-50\%}$ under the HT at 840 °C, 880 °C and 920 °C, respectively. In each diagram, roughly estimated contours are drawn. The dependence of a_0 on HTT and the film composition is very complicated in contrast with the crystal orientations. It is, however, likely that second phases such as $BaCuO_2$ or Y_2BaCuO_5 share more oxygen and suffocate $YBa_2Cu_3O_{7-y}$ during the heat treatment, because the a_0 is dependent on oxygen defficiency "y" and became smaller in the a- or ac-type films with less $BaCuO_2$ than in the c-type films with a lot of $BaCuO_2$, under the HT at 880 °C and 920 °C.

Superconducting properties

The highest Tc's obtained in the present work were 84.9 K, 87.4 K and 80 K for the a-type, ac-type and c-type films, respectively. The highest Jc's for each type were 210 A/cm^2, 1.25x10^4 A/cm^2 and 330 A/cm^2, respectively. The highest Tc and Jc were obtained in the ac-type films with the composition of $Y_1Ba_{2.3}Cu_{2.7}$ and $Y_1Ba_{2.0}Cu_{2.3}$ under the HT at 880 °C. This is due to the following facts; (1) the c-type films had large a_0, (2) the region including the ac- or a-type films and the region including smaller a_0 were well overlapped between Fig.3(b) and Fig.6(b), and Fig.3(c) and Fig.6(c), where metallic composition of the films was non-stoichiometric, (3) the films under the HT at 920 °C consisted of coarse grains with a diameter >1 micron [1] and seemed to contain a lot of weak links, and (4) in our process ac-type films were crack-free but there were a lot of cracks in the a-type films, under the HT at 880 °C [1].

Since the films with Tc > 80 K and Jc of 5-6x10^3 A/cm^2, indeed, had low copper content of $Y_1Ba_{2-2.3}Cu_{2.3-2.7}$, we may expect to obtain reproducibility of films with good quality if we try to work in such a non-stoichiometric region.

Conclusions

We studied the effect of both non-stoichiometry in metallic composition and the heat treatment temperature on the crystal orientations, the lattice constant a_0, and Tc and Jc of sputtered Y-Ba-Cu-O films on $SrTiO_3$ (100) substrates. The results are summarized as follows.

1) The crystal orientations of the films were controlled mainly by the yttrium content. The c-axis, mixed c- and a-axis, and a-axis oriented grains became predominant in turn as the yttrium content increased.

2)   The lattice constant a_0 depended (in a very complicated manner) both on the metallic composition and the heat treatment temperature.

3) Better superconducting properties were obtained not in the films with stoichiometric composition, but in the films with non-stoichiometric composition of $Y_1Ba_{2-2.3}Cu_{2.3-2.7}O_x$. In those films, both the crystal orientations and the lattice constant a_0 were able to be well-controlled, and thus the highest Tc of 87.4 K and the highest Jc at 77 K in excess of   10^4 A/cm^2 were obtained.

References

1. H.Kajikawa, T.Hase, M.Okuda,   K.Nishimura and   Y.Kawate,   Extended Abstract of 5th Workshop on FED "TOPICAL MEETING ON HIGH-TEMPERATURE SUPERCONDUCTING ELECTRON DEVICES", 95 (1988).
2. K.Char, A.D.Kent, A.Kapitulnik, M.R.Beasley, and T.H.Geballe, Appl. Phys. Lett., 51(17),1370 (1987).
3. M.Naito, R.H.Hammond, B.Oh, M.R.Hahn, J.W.P.Hus, P.Rosentahl, A.F.Marshall, M.R.Beasley and A.Kapitulnic, J.Mater.Res.,2(6), 713 (1987).
4. K.Doi, M.Murakami, S.Matsuda and A.Takayama, Proceedings of ISS'88 (1988).

Table 1. Sputtering conditions.

Target	$Y_1Ba_{5.7}Cu_6O_x$ $\phi 100 \times {}^t5mm$
Substrate	$SrTiO_3(100)$
Sputtering gas	$Ar + O_2$ $0.5 \geq O_2/(Ar+O_2) \geq 0.2$
Gas pressure	$\doteq 1Pa$
RF input power	$200 \sim 300$ W
Substrate temperature	300 ℃
Growth rate	$\geq 15\text{Å}/min$

Table 2. Definition of the film types.

(r)ː $Max[I(103),(110)]/Max[I(200),I(002)]>3.3$, (random-type)

(a)ː $I(002)/I(200)<0.01$, (a-type)

(c)ː $I(002)/I(200)>0.17$, (c-type)

(ac)ː $0.01<I(002)/I(200)<0.17$, (ac-type)

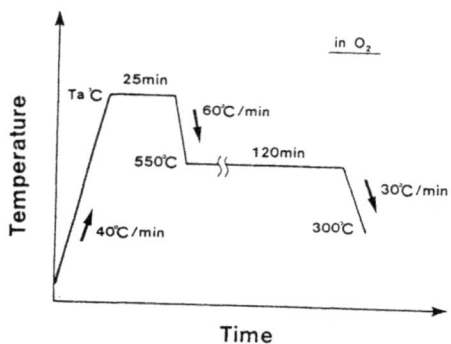

Fig. 1. Heat treatment process.

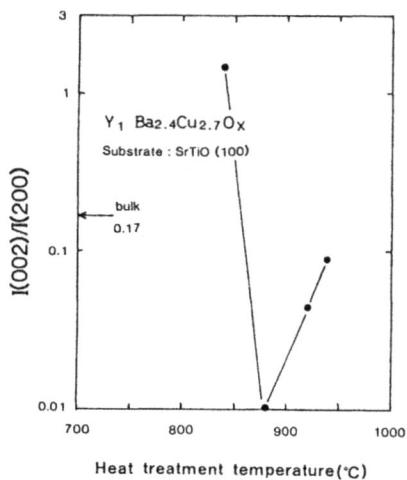

Fig. 2. Effect of heat treatment
temperature on the crystal
orientations.

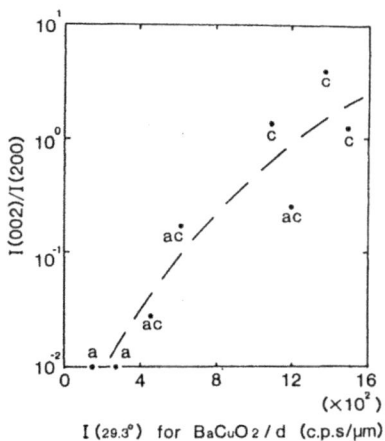

Fig. 4. Crystal orientation
vs. I(29.3°) for BaCuO$_2$.

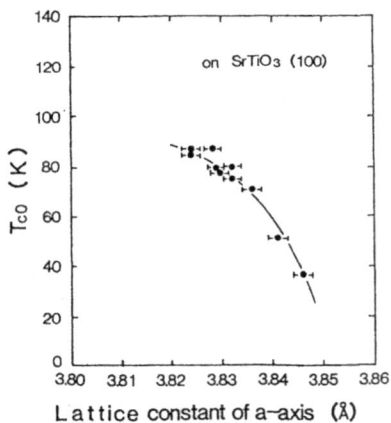

Fig. 5. Tc vs. lattice constant.

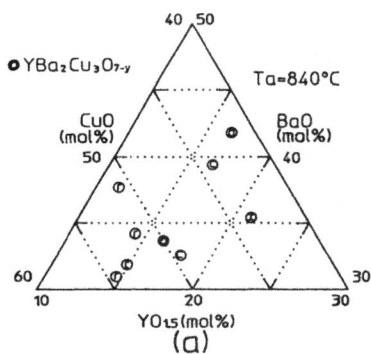

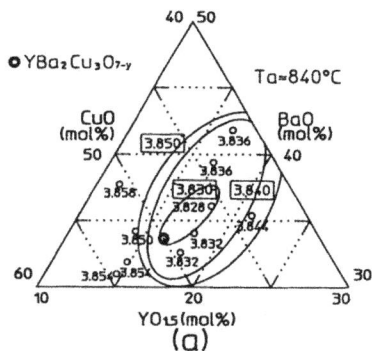

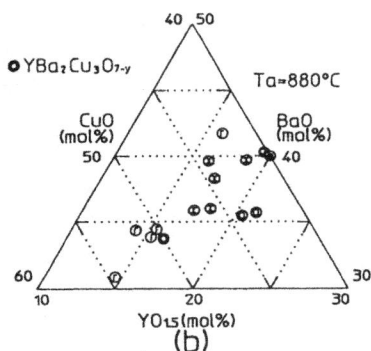

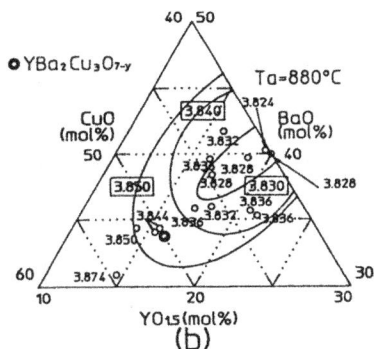

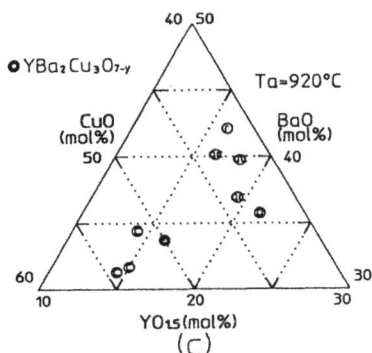

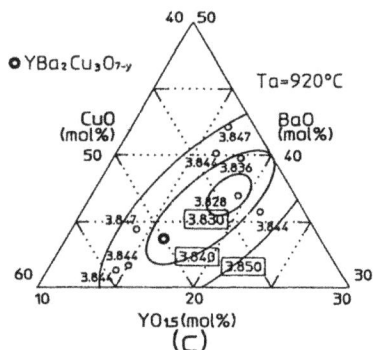

Fig. 3. Film types vs. composition under heat treatment at 840°C (a), 880°C (b) and 920°C.

Fig. 6. Distribution of lattice constant a_0 of the films under heat treatment at 840°C (a), 880°C (b) and 920°C (c).

ELECTRICAL PROPERTIES OF SINGLE-CRYSTAL LnBa$_2$Cu$_3$O$_y$
(Ln=Eu,Dy,Ho) UNDER A PULSED MAGNETIC FIELD UP TO 50T

Makoto HIKITA, Yukimichi TAJIMA, Hiroyuki FUKE[*],
Kiyohiro SUGIYAMA[*], Muneyuki DATE[*], and Akio YAMAGISHI[**]

NTT Opto-Electronics Laboratories, Tokai, Ibaraki 319-11, Japan
[*]Department of Physics, Faculty of Science, Osaka University,
Toyonaka, Osaka 560, Japan
[**]The Research Center for Extreme Materials, Osaka University,
Toyonaka Osaka 560, Japan

Abstract - A wide-temperature-range profile of the anisotropic magnetoresistance has been studied using high quality single crystals of 123-compounds. The experiments were carried out for high fields under a pulsed magnetic field up to 50T and for low fields under a static superconducting magnet. The magnetoresistance and normal resistivity in the a-b plane basically behave like conventional metals. The anisotropic parameters of H$_{c2}$(0) and ξ(0) are discussed based on the results estimated from different methods of dH$_{c2}$/dT and fluctuation conductivity.

Introduction

90K-class oxide superconductors are expected to have higher upper critical field H$_{c2}$ and anisotropic electrical properties between the a-b plane and the c-axis. In earlier works[1-5], anisotropic H$_{c2}$(0) was only estimated from lower field data obtained near T$_c$. This paper presents the experimental magnetoresistance data of high-quality single crystal LnBa$_2$Cu$_3$O$_y$ (Ln=Eu, Dy, Ho) under a pulsed magnetic field up to 50T[6-8]. Wide temperature range magnetoresistance of crystals were directly obtained and discussed the characteristics. On the other hand, coherent lengths, ξ_{ab}(0) and ξ_c(0), are estimated from dH$_{c2}$/dT data and fluctuation conductivity which is evaluated from excess conductivity above T$_c$. We also discuss these estimated values.

Experimental

Single crystals of LnBa$_2$Cu$_3$O$_y$ (Ln=Eu, Dy, Ho) were grown from CuO rich Ln-Ba-Cu-O molten compounds[9]. YBa$_2$Cu$_3$O$_y$ single crystals were grown by a new procedure using only multi-layered pellets[10]. All as-grown crystals were annealed in an oxygen at 900 °C for 5 hours, cooled 450-500 °C, and held at 450-500 °C for 20-200 hours. Crystals annealed at 450-500 °C for longer times exhibit lower resistivity. The polarized microscope photograph of annealed crystal

Fig.1. Polarized microscope photograph of single-crystal DyBa$_2$Cu$_3$O$_y$

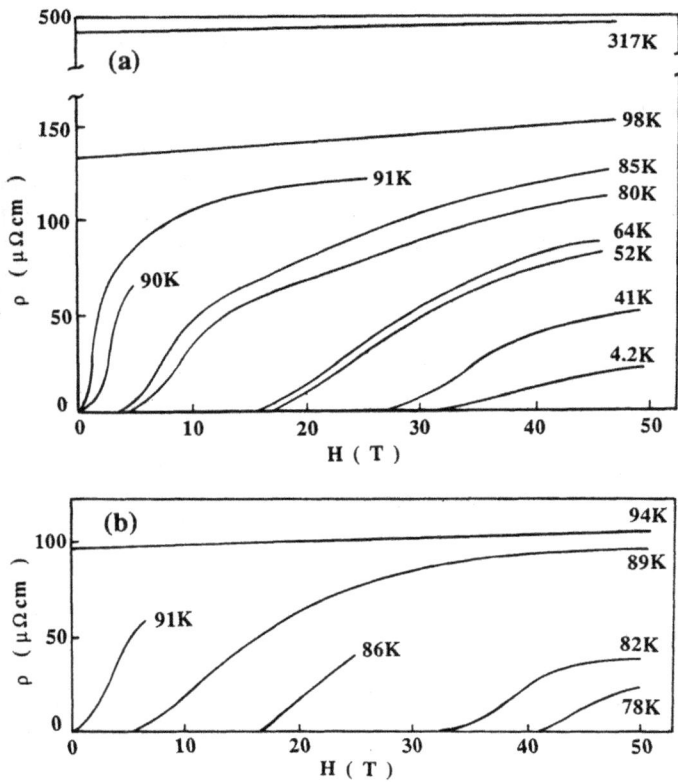

Fig.2. Magnetoresistance (MR) in the a-b plane of DyBa$_2$Cu$_3$O$_y$ applied field (a) perpendicular and (b) parallel to the a-b plane

$DyBa_2Cu_3O_y$ is shown in Fig.1. Resistance was measured by four-terminal method. Electrical contact was made with 25-μm-diameter gold wires with conductive silver paste on the gold films evaporated on the largest facet of the a-b plane. High field magnetoresistance was measured under a pulsed magnetic field up to 50 T at *the Research Center for Extreme Materials* in *Osaka University* where the sample temperature was measured with a calibrated Au-Fe/Ag thermocouple[11]. The pulsed width was 0.4 msec. Low field magnetoresistance was measured under a static field up to 12T, where a calibrated carbon-glass thermometer was used for the temperature measurement.

Results and Discussion

High Field Study

Single crystals of $EuBa_2Cu_3O_y$, $DyBa_2Cu_3O_y$, and $HoBa_2Cu_3O_y$ were used for high field studies. Figure 2 shows magnetoresistance (MR) in the a-b plane of $DyBa_2Cu_3O_y$ applied field perpendicular ($MR_{\perp ab}$) and parallel ($MR_{\parallel ab}$) to the a-b plane under a pulsed magnetic field. Resistance onset (T_{co}) of $MR_{\perp ab}$ is observed at the temperature down to 4.2K. On the contrary, $MR_{\parallel ab}$ is only observed above 78 K. Negative MR is not found up to room temperature in the experiment range. These basic results are similar to those for $EuBa_2Cu_3O_y$ and $HoBa_2Cu_3O_y$. Figure 3 shows $H_{c2}(T_{co})$ of crystals obtained from $MR_{\perp ab}$ and $MR_{\parallel ab}$. As the $H_{c2}\perp(T_{co})$ curves exhibit upward curvature near T_c, anisotropic ratio between $H_{c2}(T_{co})$ of perpendicular and parallel to a-b plane is larger near T_c, The lower temperature profile of $H_{c2\perp}(T_{co})$ less than 77K resembles conventional high-H_{c2} materials such as A15 compounds.

Table 1   Parameters of single crystals estimated from high-field data

	T_c	ρ	$\dfrac{dH_{c2}^{\perp ab}}{dT}$	$\dfrac{dH_{c2}^{\parallel ab}}{dT}$	$H_{c2}^{\perp ab}(0)$	$H_{c2}^{\parallel ab}(0)$	$\xi_{ab}(0)$	$\xi_c(0)$
	(K)	(μΩcm)	(T/K)	(T/K)	(T)	(T)	(Å)	(Å)
Eu (T_{co})	94	170	0.41	3.8	27 **	250	35	3.8
Ho (T_{co})	93	47	0.95	2.7	40 **	-	29	-
(T_{co})			0.54	3.1	33 **	200	32	5.3
Dy ($T_{c0.5}$)	92	140	1.0	-	63	-	23	-
($T_{c0.9}$)			3.1	-	200	-	13	-

*T_{co}, $T_{c0.5}$, and $T_{c0.9}$ are defined as temperatures at $\rho=0$, $\rho=0.5\rho_n$, and $\rho=0.9\rho_n$, respectively.

** directly obtained value from high field measurements

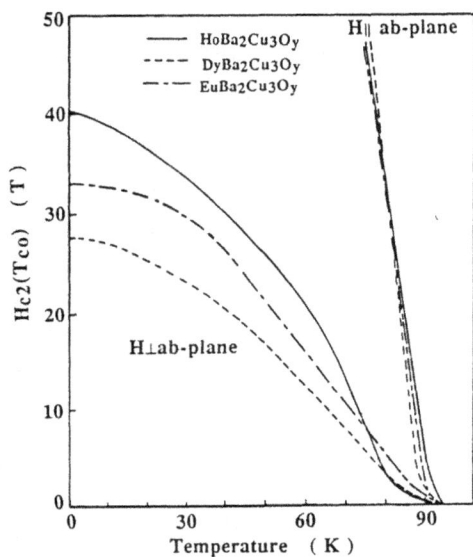

Fig.3 $H_{c2}(T_{co})$ curves of single-crystal obtained from $MR_{\perp ab}$ and $MR_{\parallel ab}$

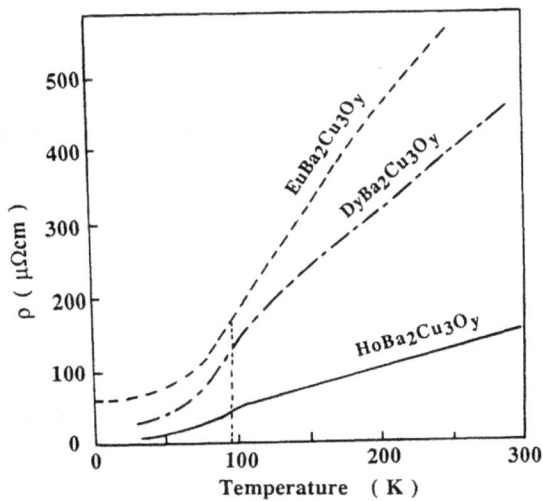

Fig.4 Temperature dependence of normal resistivity in the a-b plane of single crystals. curves are combined with measured values above T_c and estimated values from high-field data below T_c.

Estimated parameters from the data are described in Tab. 1, where $H_{c2\perp}(0)=\Phi_0/2\pi\xi_{ab}^2(0)$, $H_{c2\parallel}(0)=\Phi_0/2\pi\xi_{ab}(0)\xi_c(0)$, and $H_{c2}(0)=0.69T_c(dH_{c2}/dT)$ (WHH relation [12]) are used. In the case for $DyBa_2Cu_3O_y$, $dH_{c2\perp}/dT$ and $H_{c2\perp}(0)$ are obtained for T_{co}, $T_{c0.5}$ and $T_{c0.9}$. The WHH relation is well explained the experimental data of $H_{c2\perp}(0)$ of $EuBa_2Cu_3O_y$ and $DyBa_2Cu_3O_y$. However, the experimental $H_{c2}(0)$ of $HoBa_2Cu_3O_y$ is substantially lower than those estimated from $dH_{c2\perp}/dT$. It suggests that zero resistivity is broken at lower temperature than that of thermodynamically defined H_{c2}. It is very important problem for a practical superconducting magnet and measurements for a basic physics. Future precise investigations for the different results between crystals will be necessary.

The normal resistivities of crystals in the a-b plane below T_c is roughly estimated from the $MR\perp$ data as shown in Fig.4. Figure 4 also shows the normal resistivity above T_c measured without magnetic field. The temperature dependent resistivity behaves like a conventional metal in all crystals.

Low Field Study

The magnetic field dependence of the excess conductivity above T_c is obtained from the $MR_{\perp ab\text{-plane}}$ data of single-crystal $YBa_2Cu_3O_y$ above T_c. Fluctuation conductivity is calculated based on the recent theory proposed for weekly-coupled 2-dimensional superconducting layers by Hikami and Larkin[13]. In this calculation, only Aslamazov-Larkin contribution is considered, and Maki-Thompson term is not because electron phase braking time is small (<10-13 sec)[14,15]. Through this procedure, the $\xi_{ab}(0)$ of 13 Å and $\xi_c(0)$ of 2 Å are obtained from the best fit [16]. These values are half those obtained from the high field data of dH_{c2}/dT.

One of the reasons for the difference of $\xi(0)$ between estimation methods is interpreted by a giant flux creep model proposed by Yeshurun and Malozemoff[17]. They claimed that giant flux creep make a magnetoresistance curve near $\rho(H)=0$ change. However, the $H_{c2\perp}(T_{co})$ curve below 30K is not satisfied with the relation of 1- $T/Tc \propto H_c(T_{co})^{2/3}$ which is found for the flux creep model by Yeshurun and Malozemoff. In the present stage, we believe that the $\xi_{ab}(0)$ and $\xi_c(0)$ possess the limited values between 13-30 Å and 2-6 Å , respectively, estimated by the different two methods for 123-compounds.

Conclusions

A wide-temperature-range profile of the anisotropic magnetoresistance for 123-compound single crystals has been studied. The essential results are as follows.
From the high magnetic field study.
(1) The temperature dependence of the H_{c2} curve except that of $HoBa_2Cu_3O_y$ is well described by the WHH relation.
(2) The normal resistivity in the a-b plane shows characteristics like a conventional metal.

(3) The magnetoresistance at temperatures below and above T_c is positive.

From the low magnetic field study.

(4) The estimated $\xi_{ab}(0)$ and $\xi_c(0)$ from excess conductivity considering Aslamazov-Larkin contribution of thermodynamically fluctuation are half those obtained from $dHc2/dT$.

The extended study of magnetoresistance below and above Tc is need to understand intrinsic physical parameters of 123-compounds.

Acknowledgements - The authors would like to thank Y.Katayama, Y.Yamada, A.Yamaji and T.Murakami for their continuous support and encouragement throughout the course of the study. They also would like to thank A.Katsui and T.Konaka for the crystals, M.Suzuki for helpful discussions and T.Ishii for the experiments.

References

1)T.K.Worthington, W.J.Gallagher, and T.R.Dinger, Phys. Rev. Lett. 59, 1160(1987)

2)Y.Iye, T.Tamegai, H.Takeya, and H.Takei, Jpn. J. Appl. Phys. 26 , L1057(1987)

3)M.Hikita, Y.Tajima, A.Katsui, Y.Hidaka, T.Iwata, and S.Tsurumi, Phys. Rev. B36, 7199(1987)

4)J.S.Moodera, R.Meservey, J.E.Tkaczyk, C.X.Hao, G.A.Gibson, and P.M.Tedrow, Phys. Rev. B 37, 619(1988)

5)Y.Hidaka,Y.Enomoto, M.Suzuki, M.Oda, A.Katsui and T.Murakami, Jpn. J. Appl. Phys., 26, 1726(1987)

6)Y.Tajima, M.Hikita, T.Ishii, H.Fuke, K.Sugiyama, M. Date, A. Yamagishi, A.Katsui, Y. Hidaka, T. Iwata and S.Tsurumi, Phys. Rev. B 37, 7956(1988)

7)A.Yamagishi, H.Fuke, K.Sugiyama, M.Date, Y.Tajima, M.Hikita, T.Ishii, A.Katsui, Y.Hidaka, T.Iwata and S.Tsurumi, Physica C, 153-155, 1459(1988)

8)A.Yamagishi, H.Fuke, K.Sugiyama, M.Date, Y.Tajima, and M.Hikita; to be published in Proceeding of The 2nd Internatinal Symposium on High Field Magnetism (1988, Leuven)

9)A.Katsui, Y.Hidaka, andH.Ohtsuka; Jpn. J. Appl. Phys., 26, L1521(1987)

10)T.Konaka, I.Sankawa, M.Sato, and M.Hikita, J. Cryst. Growth, 91, 278(1988)

11)K.Okuda, S.Noguchi, A.Yamagishi, K.Sugiyama, and M.Date, Jpn. J. Appl. Phys.,26, L822(1987)

12)N.R.Werthamer, E.Helfand, and P.C.Hohenberg, Phys. Rev. 147, 295(1966)

13) S.Hikami and A.I.Larkin, Mod. Phys. Lett. B 2, 693(1988)

14)Z.Schlesinger, R.T.Collins, D.L.Kaiser, and F.Holtzberg, Phys. Rev. Lett. 59, 1958(1987)

15)M.Gurvich and A.T.Fiory, Phys. Rev. Lett. 59, 1337(1987)

16)M.Hikita and M.Suzuki, preprint

17)Y.Yeshurun and A.P.Molozemoff, Phys. Rev. Lett. 60, 2202(1988)

MEASUREMENTS OF CRITICAL CURRENTS IN OXIDE SUPERCONDUCTOR
BY PULSE CURRENT TECHNIQUE

Kenji SHIMOHATA, Syoichi YOKOYAMA, Masao MORITA,
Tadatoshi YAMADA and †Mitsunobu WAKATA

Mitsubishi Electric Corp. Central Research Lab.
8-1-1 Tsukaguchi-Honmachi Amagasaki, Hyogo, 661, Japan
†Mitsubishi Electric Corp. Materials and Electric Devices Lab.
1-1-57 Miyashimo Sagamihara, Kanagawa, 229, Japan

Abstract - We have measured critical currents of bulk oxide superconductor by pulse current technique and high field magnetization by pulse magnet. We have found that the intrinsic critical current of $YBa_2Cu_3O_{7-x}$ is higher than 30,000A/cm^2 at B=20Tesla.

Introduction

Since Bednorz and Müller reported the possible high T_c superconductivity[1], many high T_c oxide superconducters have been discovered. [2] [3] [4]

In the high T_c oxide family, it is reported the superconducting transition temperature of $YBa_2Cu_3O_{7-x}$(YBCO) is higher than 90K, and critical current density (J_C) and upper critical field(H_{C2}) of YBCO single crystal are more than 10^6A/cm^2 ($J_c\perp$, B=0) and 30T($H_{c2}\perp$) for a-b plane, or 10^4A/cm^2($J_{c//}$, B=0) and 7T($H_{c2//}$) for c-axis at 77K. [5] But bulk YBCO has very low transport critical current density, tipicaly 10^3A/cm^2 at B=0. We had reported intrinsic transport critical current of bulk YBCO by pulse current technique was more than 10^4A/cm^2 at 77k(Fig.1). [6]

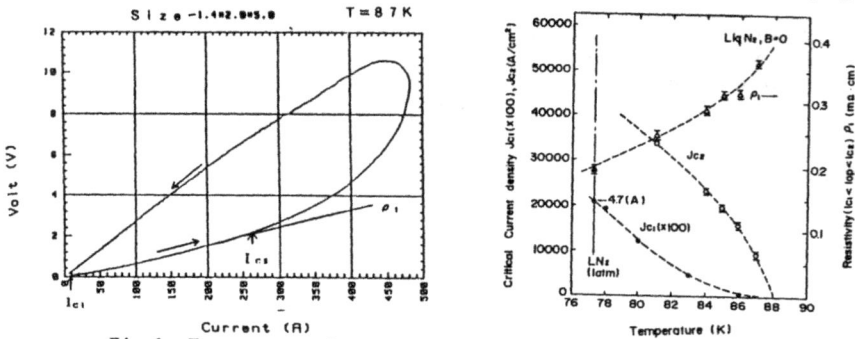

Fig.1 Temperature dependence of the critical current density
of $YBa_2Cu_3O_{7-x}$ sample by pulse I-V measurement.

Experimental Procedure

Figure 2 shows the block diagram of pulse I-V measurement system. Figure 3 shows the time dependence of pulse magnetic field and pulse current. Magnet is copper solenoid type and needs liquid-N_2 cooling. Parameters of pulse magnet are shown in table 1. Frequency of pulse magnetic field is 20Hz and that of pulse current is 2.5kHz, so the YBCO sample feels quasi-static magnetic field during current flow.

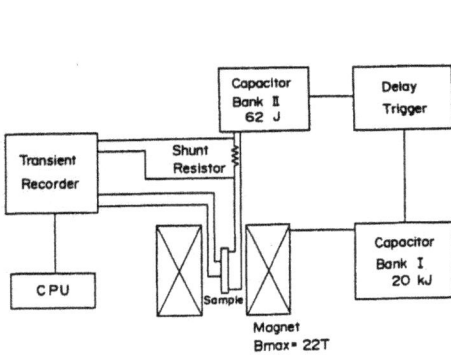

Fig.2   Pulse I-V measurement system.

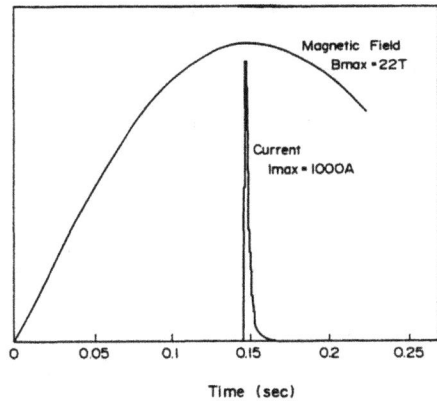

Fig.3   Time dependence of the magnetic field and the current.

Table 1.   Parameters of pulse magnet

Inner diameter	18mm
Outer diameter	60mm
Coil length	80mm
Turn number	370Turn
Conductor(Cu)	2mm ϕ
Inductance	1.6mH

The pulse I-V characteristic of YBCO at B=5T and T_i(initial temperature)=77K is shown in Fig.4. When the maximum pulse current (I_{max}) is 380A, there is no hysteresis. But when I_{max} is 740A, hysterisis appears during current increase and decrease by joule heating. So we measured $YBa_2Cu_3Ag_{0.5}O_{7-x}$(YBCAO) to reduce the contact resistivity and resistivity of normal state, and we adopted the data of current increase region only. Then the rise of temperature by joule heating effect is estimated less than 1K.

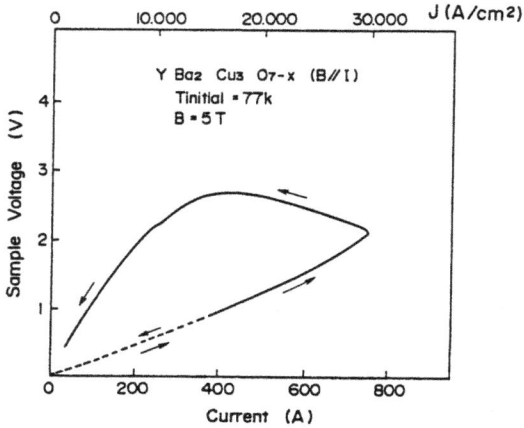

Fig.4   Pulse I-V characteristic of YBa$_2$Cu$_3$O$_{7-x}$ at B=5Tesla.

Magnetization was measured by pick up coil. Magnetoresistance was measured by four probe method with constant current. These methods have been developed by M. Date. [7]

Experimental result

Magnetization measurement

Figure 5 shows the magnetization data of YBCO at 77K. When we assume that the typical grain size of YBCO is 10 μm, we can estimate critical current density (J$_{CM}$) by using Been model from magnetization data. Figure 6 shows the J$_{CM}$ as a function of external magnetic field. In low magnetic field, J$_{CM}$ is same order of J$_{c//}$ and looks vanish toward B=7T(=B$_{c2//}$). This result is reasonable compared with single crystal data. But even in high magnetic field, J$_{CM}$ remains. This means that the path of shielding current becomes narrow at B=7T. Because shielding currents cannot flow through the region that the magnetic field ingradient for c-axis is over 7T.

Magnetoresistance measurement

Figure 7 shows the magnetoresistance measurement data of bulk YBCO. The sample dimension is 2.5x1.2x15.0mm^3 (voltage tap distance is 4mm), the resistivity at 100K is 0.75mΩ·cm and critical current at 77K is 4.5A(J$_c$=150A/cm^2). Up to B=7T, resistivity is very low($\rho_\perp$=0.05mΩ·cm ~1/10 ρ_{normal}) and there is no difference between $\rho_{B//I}$ and $\rho_{B\perp I}$. Over B=7T, $\rho_{B\perp I}$ is larger than $\rho_{B//I}$. It seems that the flux flow starts at B=7T.

Pulse I-V measurement

Figure 8 shows the magnetic field dependence of pulse I-V measurement. In low current region(I<400A), the higher the magnetic field is, the higher the resistivity($\rho_\perp$) is. Magnetic field dependence of $\rho_\perp$ agrees with magnetoresistance data. But drastic change like Fig.1 can not be seen up to B=20T and J=3×10⁴A/cm². The reason of this difference comes from the lack of orientation of bulk sample. Namely the angle between magnetic field and c-axis differs from each grain.

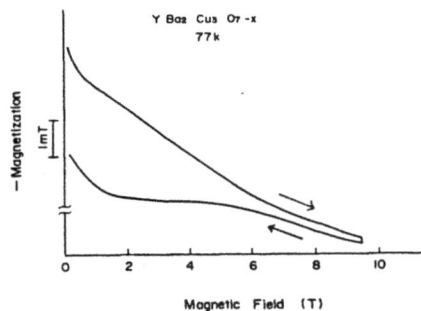

Fig.5 Magnetization data of YBa₂Cu₃O₇₋ₓ.

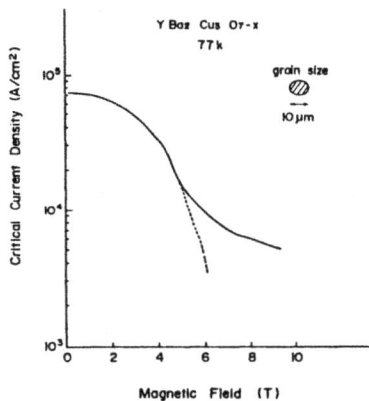

Fig.6 Estimated critical current density by using Been model. The dot line is guided for the eye.

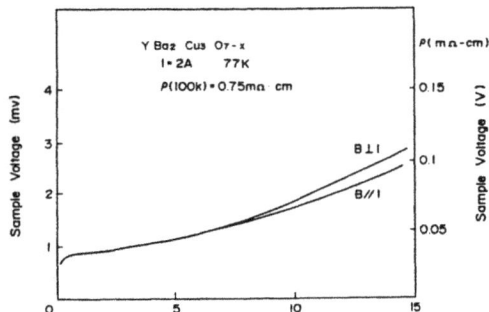

Fig.7 Magnetoresistance data of YBa₂Cu₃O₇₋ₓ by pulse magnet.

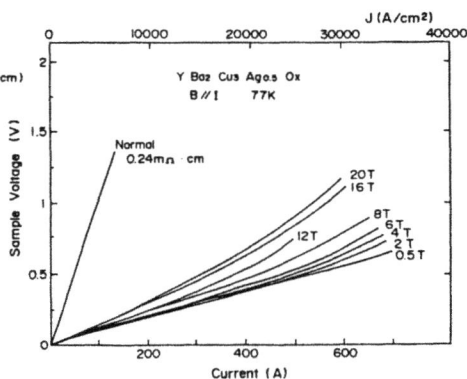

Fig.8 Magnetic field dependence of pulse I-V data in bulk YBa₂Cu₃Ag₀.₅O₇₋ₓ.

242

Conclusion

Intrinsic critical current density(J_c) of bulk YBCO is comparable of single crystal one. J_c of bulk YBCO is over $3 \times 10^4 A/cm^2$ at B=20T, 77K.
Magnetoresistance for B//I differs from one for B⊥I over B=7T.

References

(1) J.G.Bednorz and K.A.Müller,"Possible High T$_c$ Superconductivity in the Ba-La-Cu-O System", Z.Phys.B 64,189(1986).

(2) M.K.Wu, J.R.Ashburn, C.J.Torng, P.H.Hor, R.L.Meng, L.Gao, Z.J.Huang, Y.Q.Wang, and C.W.Chu,"Superconductivity at 93K in a New Mixed-Phase Y-Ba-Cu-O Compound System at Ambient Pressure", Phys.Rev.Lett.58,908(1987).

(3) H.Maeda, Y,Tanaka, M.Fukutomi, T,Asano,"A New High Tc Oxide Superconductor Without Rare Earth Element", Jpn.J.Appl.Phys.Lett, Vol27, 209(1988).

(4) Z.Z.Sheng and A.M.Harmann, Nature(London)332,55(1988).

(5) For Example,Proceedings of the 18th International Conference on Low Temperature Physics Part 2.

(6) S.Yokoyama, T.Yamada, M.Wakata,"Extended Abstract(The 49th Autumn Meeting,1988)The Japan Society of Applied Physics.

(7) T.Sakakibara, H.Morimoto, M.motokawa and M.Date, High Field Magnetism(North Holland Publishing Company), 299,167(1983)

TEMPERATURE DEPENDENCE OF CRITICAL CURRENT DENSITIES FOR
SINGLE-CRYSTAL BiSrCacuO AND POLYCRYSTAL BiPbSrCacuO

Misao KOIZUMI, Daisuke ITO, Yutaka YAMADA,
Satoru MURASE, and Shunji NOMURA

Toshiba R & D Center, 4-1, Ukishima-cho, Kawasaki-ku,
Kawasaki, 210, Japan

Abstract - The magnetization in a single-crystal $Bi_{2.2}(Sr,Ca)_{2.6}Cu_2O_y$
and in polycrystalline $(Bi_{0.7}Pb_{0.3})_2Sr_2Ca_2Cu_3O_y$ in magnetic fields both
parallel and perpendicular to the basal plane in the temperature range from
5K to near Tc have been measured. A similar temperature dependence of critical current densities for both the single-crystal and polycrystalline samples
was found. The magnetization critical current densities Jc at zero magnetic
field rapidly decreased with temperature increase below $\sim$20K and gradually
decreased above $\sim$20K. In the higher temperature region above $\sim$20K, the temperature dependence of Jc showed the relation of $Jc \propto [1-T/Tc]^\gamma$. Here, γ was 2
under magnetic field applied parallel to the basal plane of the single-crys-
tal sample, and γ was obtained as 4 under a perpendicular field to the basal
plane. On the other hand, γ was obtained as 2 for the polycrystalline
sample. The Jc of both the single-crystal and polycrystalline samples in
temperature 5K were determined as $\sim 2 \times 10^4$ A/cm^2 and $\sim 5 \times 10^5$ A/cm^2, respectively.

Introduction

In 1988, Maeda, et al. (1), discovered a new oxide superconductor
BiSrCacuO with critical temperature Tc above 105K, higher than that of
$Y_1Ba_2Cu_3O_7$ by more than 10K. This new material possesses a layered structure
analogous to that of YBaCuO superconductor. However, there are important
differences between the two materials, such as the absence of twinning and
invarient oxygen content of BiSrCacuO system. In particular, the comparison
between these single-crystals is important for the understanding the critical
current characteristics of the oxide superconductor.
Recently, studies on the critical current densities for the single-
crystal BiSrCaO system were reported by R. B. van Dover, et al. (2), and
J. J. Lin, et al. (3). However, a systematic measurement of the temperature
dependence of critical current densities for the single-crystal BiSrCacuO
system has not been reported yet. This work reports on the temperature
dependence of critical current densities in single-crystal
$Bi_{2.2}(Sr,Ca)_{2.6}Cu_2O_y$. The authors measured magnetization as a function of
the applied field with the field applied either parallel or perpendicular to
the basal plane in the temperature range from 5K to near Tc. The authors
also measured magnetization for polycrystalline $(Bi_{0.7}Pb_{0.3})_2Sr_2Ca_2Cu_3O_y$.
The critical current densities were also determined from magnetization curves
by applying the Been's model.

Experimental

Single-crystals of $Bi_{2.2}(Sr,Ca)_{2.6}Cu_2O_y$ were grown by a self-flux method
as described previously (4). A plate-like crystal with average dimensions

0.3 x 1.25 x 1.5 mm was used in the present study. The crystal was mechanically isolated. On the other hand, sintered polycrystalline samples $(Bi_{0.7}Pb_{0.3})_2Sr_2Ca_2Cu_3O_y$ in the form of 2 cm diameter pellets were prepared by a conventional solid state reaction process as described previously (5). A pellet was cut into a rectangular bar sample with dimensions 2.15 x 2.3 x 3.5 mm and used in the present study. The samples were remeasured after grinding into a powder with an average diameter of $\sim$50 μm. The magnetization measurements were performed using a SQUID magnetometer (Quantum Design, model MPMS) in the temperature range from 5K to near Tc, in a magnetic field up to 5T with the field direction parallel or perpendicular to the basal plane (corresponded to the a-b plane) of the single-crystal sample.

Results and Discussion

Figure 1 shows the Meissner (flux expulsion) effect for single-crystal $Bi_{2.2}(Sr,Ca)_{2.6}Cu_2O_y$ in low perpendicular magnetic field. The data were taken by cooling the samples through the superconducting transition in an external field as shown in the figure. A superconducting transition was observed at 80K in Fig. 1. After correction for the demagnetization factor, the Meissner signal was $\sim$15 % of an ideal superconductor. This value was similar to the single-crystal YBaCuO system determined by Dinger, et al. (6). On the other hand, this value was one fifth of the single-crystal BiSrCaCuO obtained by Lin, et al.(3). The same transition temperature was obtained under the parallel magnetic field.

Figure 2(a) and 2(b) show typical magnetization curves for the single-crystal sample at temperature 5K in parallel and perpendicular fields, respectively. In comparison of the two curves, strong anisotropy in the magnetization curves to field direction was not observed here. A similar result was also obtained by Lin, et al. (3). However, in a higher temperature region above 20K, the anisotropy appeared in the critical current densities as follows.

The critical current densities Jc were determined from the magnetization curves using a simple relation between Jc and magnetization, $Jc = 30\Delta M/r$ (Jc in A/cm^2, ΔM in emu/cm^3, r in cm), where ΔM is the width of a magnetization curve, and r is the radius of a disk sample.

Figure 3 shows magnetic field H dependence of the Jc for the single-crystal sample for various temperature and the two field directions. In Fig. 3, the anisotropy of the Jc with respect to field direction is small in the temperature range below $\sim$20K. However, in the temperature range above $\sim$20K, it is shown that the anisotropy clearly appears in the Jc. Also, the anisotropy shows that the Jc obtained under a field parallel to the a-b plane of the single-crystal is somewhat higher than that for the perpendicular field, as shown in Fig.3. This result differs from the anisotropic behavior of YBaCuO oxides (6). The physical meaning of the result in Fig. 3 is unclear at present. Further work to make this result clear is under way.

Figure 4 shows the temperature dependence of the zero field Jc obtained from magnetization curves in parallel and perpendicular fields for the single-crystal sample. The results show that the Jc rapidly decreases with temperature increase below $\sim$20K and gradually decreases above $\sim$20K. In higher temperatures above $\sim$20K, the temperature dependence of the Jc exhibits $Jc \propto [1-T/Tc]^\gamma$, where γ is 2 for the parallel field to the a-b plane and γ is 4 for perpendicular field. We defined the Jc at zero magnetic field for temperature 5K as $Jc \sim 2 \times 10^4$ A/cm^2. This value is two orders smaller than that observed in the single-crystal of YBaCuO (6). According to van Dover, et al. (2), this low Jc is attributed to the weakly coupled regions which are supposed to exist in the crystal; induced shielding currents would circulate only within each region, because the effective sample size would be smaller than the measured bulk size.

We compared the Jc of polycrystalline $(Bi_{0.7}Pb_{0.3})_2Sr_2Ca_2C_3O_y$ system with that for a single-crystal $Bi_{2.2}(Sr,Ca)_{2.6}Cu_2O_y$. Figure 5 shows the Meissner effect for polycrystalline bulked sample. Similar results were obtained for the after-grinding powder sample. It showed the two step superconducting transitions. The transitions appeared at 108K and around 70K, respectively. The transition temperature 108K very well agreed with zero resistance temperature by a independent resistivity measurement. This sample contained low-Tc phase (i.e. 70K) as well as a high-Tc phase, as shown in Fig. 5. In this case, the Meissner signal was 16 % of an ideal superconductor.

Figure 6(a) and 6(b) show typical magnetization curves for the powdered polycrystalline samples at temperature 5K and 70K, respectively. Figure 6 shows that the magnetization at approximately zero field drastically decreases at higher temperature. For the bulk sample, almost the same magnetization characteristics were observed.

Figure 7 shows the magnetic field H dependence of the critical current density Jc for the powdered polycrystalline samples for various temperatures. In Fig. 7, Jc peaks were observed in Jc v.s H curves, the peaks become remarkable with increasing temperature.

Figure 8 shows the temperature dependence of the Jc at zero magnetic field for the powdered polycrystalline samples. The temperature dependence of the Jc showns a result similar to that for the single-crystal sample under a parallel magnetic field, as shown in Fig. 3. In the higher temperature region above $\sim$20K, the slope for the Jc versus temperature curve γ was 2. The zero field Jc at temperature 5K was obtained as Jc $\sim 5 \times 10^5$ A/cm^2.

As described above, a similar temperature dependence of the critical current density was observed for single-crystal and polycrystalline samples. These results showed that the Jc rapidly decreased with increasing temperature below $\sim$20K and gradually decreased above $\sim$20K. It is considered that this temperature dependence of Jc is intrinsic property for these BiSrCaCuO superconductor systems.

Conclusion

Magnetization was measured for single-crystal $Bi_{2.2}(Sr,Ca)_{2.6}Cu_2O_y$ and for polycrystalline $(Bi_{0.7}Pb_{0.3})_2Sr_2Ca_2Cu_3O_y$ in magnetic fields up to 5T in the temperature range from 5K to near Tc. The temperature dependence of critical current densities observed for the single-crystal was similar to that of polycrystalline samples. The zero field Jc rapidly decreased with increasing temperature below $\sim$20K and gradually decreased above $\sim$20K. In the higher temperature region above $\sim$20K, the Jc related to the temperature as Jc $\propto [1-T/Tc]^\gamma$. Here, γ was 2 under a magnetic field applied parallel to the basal plane of the single-crystal sample, and γ was 4 under a perpendicular field to the basal plane. On the other hand, γ was 2 for the polycrystalline sample. The Jc of both the single-crystal and polycrystalline samples in temperature 5K were determined as $\sim 2 \times 10^4$ A/cm^2 and $\sim 5 \times 10^5$ A/cm^2, respectively.

The authors would like to thank Mr. F. Umibe for his review of the English manuscript.

246

References

(1) H. Maeda, Y. Tanaka,M.Fukutomi, and T. Sano: Jpn. J. Appl. Phys. 27, L209 (1988).
(2) R. B. van Dover, L. F. Schneemeyer, E. M. Gyorgy, and J. V. Waszczak, "Critical current densities in single-crystal $Bi_{2.2}Sr_2Ca_{0.8}CuO_{8+}$" Appl. Phys. Lett., 52 (22), 1919-1912 (1988).
(3) J. J. Lin, E. L. Benitez, S. J. Poon, M. A. Subramanian, J. Gopalakrishnan, and A. W. Sleight "Superconducting properties of single-crystal $Bi_2Sr_2CaCu_2O_{8+y}$"Phys. Rev.,B38 (7) 5095-5097 (1988).
(4) S. Nomura, T.Yamashita, H. Yoshino, and K. Ando "Single crystal growth of BiSrCaCu Oxides and its superconductivity" Proc. Int. Supercon. Soci., Nagoya, (1988).
(5) Y. Yamada, S. Murase, M. Koizumi, M. Tanaka, D. Itou, S. Takeno, I. Suzuki, and S. Nakamura "Pb substitute Bi-Sr-Ca-Cu-O superconductor" Proc. Int. Supercon. Soci., Nagoya, (1988).
(6) T. R. Dinger,T. K.Worthington, W. J. Gallagher, and R. L. Sandstom "Direct observation of electronic anisotropy in single-crystal $Y_1Ba_2Cu_3O_{7-x}$" Phys. Rev. Lett., 58 (25), 2686-2690 (1987).

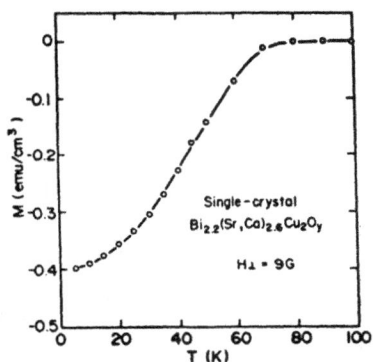

Fig.1 Meissner effect for single-crystal $Bi_{2.2}(Sr,Ca)_{2.6}Cu_2O_y$ in low perpendicular magnetic field.

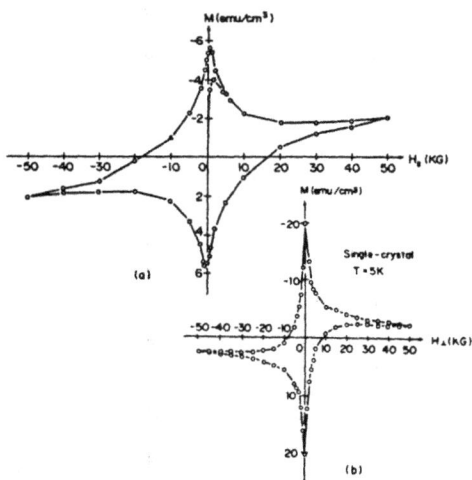

Fig.2 Typical magnetization curves for single-crystal $Bi_{2.2}(Sr,Ca)_{2.6}Cu_2O_y$ at temperature 5K in (a) parallel and (b) perpendicular field.

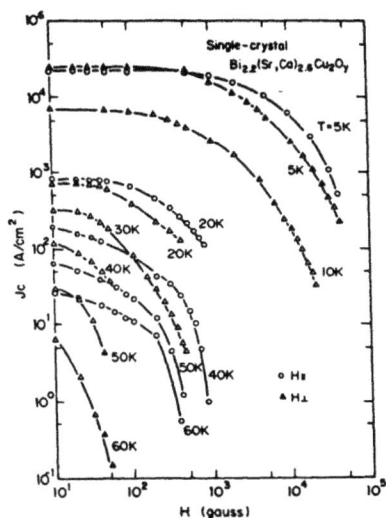

Fig.3 Magnetic field dependence of critical current for single-crystal $Bi_{2.2}(Sr,Ca)_{2.6}Cu_2O_y$ as a function temperature for two field direction.

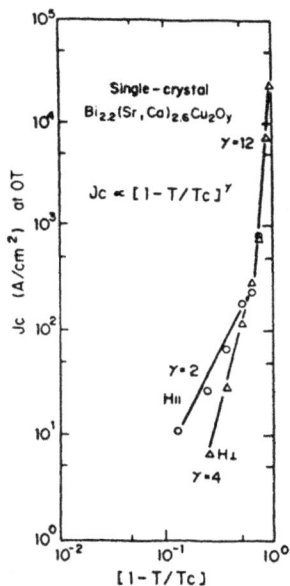

Fig.4 Temperature dependence of zero field current densities for the single-crystal $Bi_{2.2}(Sr,Ca)_{2.6}Cu_2O_y$ in parallel and perpendicular fields.

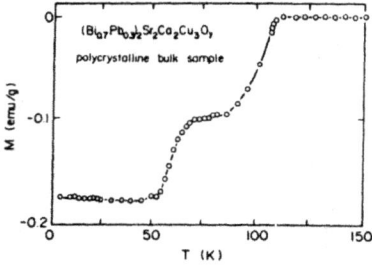

Fig.5 Meissner effect for polycrystalline $(Bi_{0.7}Pb_{0.3})_2Sr_2Ca_2Cu_3O_y$
bulk sample.

(a) T = 5K

(b) T = 70K

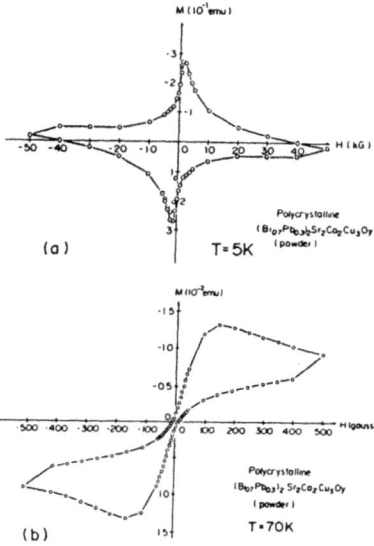

Fig.6 Typical magnetization curves
for powdered polycrystalline
$(Bi_{0.7}Pb_{0.3})_2Sr_2Ca_2Cu_3O_y$ at temperature of (a) 5K and (b) 70K

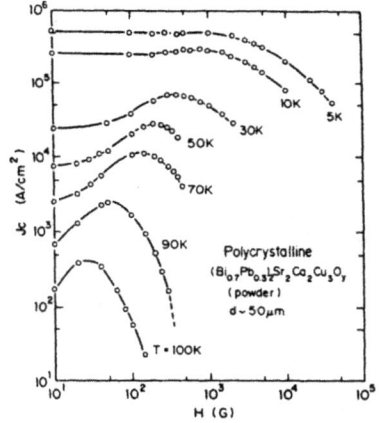

Fig.7 Magnetic field dependence of
critical current densities for
the powdered polycrystalline
$(Bi_{0.7}Pb_{0.3})_2Sr_2Ca_2Cu_3O_y$ for function of temperature.

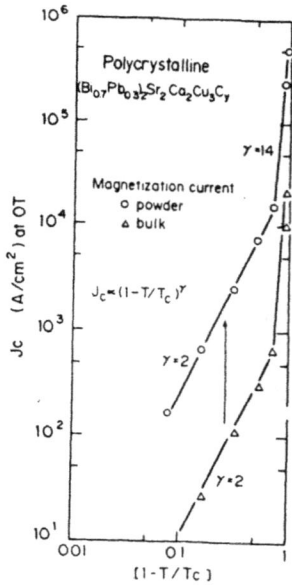

Fig.8 Temperature dependence of
zero field critical current
densities for polycrystalline
(a) powder and (b) bulk sample.

CHARACTERIZATION OF MICROSTRUCTURE IN AG-SHEATHED TAPES OF $Ba_2LnCu_3O_{6+X}$ OXIDES (Ln=Y, Gd and Ho)

Tomoo TAKAYAMA, Kozo OSAMURA, Shojiro OCHIAI, and Hitoshi TABATA

Department of Metallurgy, Kyoto University, Sakyo-ku, Kyoto 606 Japan

Abstract - The present aim of investigation was to make clear the influence of each step in the powder-in-tube technique for preparing the silver sheathed composite superconductors with high critical current density. Substitution of BaO_2 for $BaCO_3$, repetition of calcining, cold-working and hot-pressing of the composite, utilization of the hydrogen gas reduced powder have been attempted. The most effective process to improve the critical current density was the combination of the utilization of reduced powder and the cold-rolling. Especially the increase of critical current density was prominent at high magnetic field up to 15 T at 4.2 K.

Introduction

A key property to materialize superconducting oxides for the applications in superconducting magnets is their critical current density. However, a low critical current density has only been achieved for the composites of $Ba_2YCu_3O_{6+x}$ sintered powder sheathed with silver. For preparing the silver sheathed composites, the process with so many steps has to be treated. Therefore it is important to characterize the influence from a specific step on the microstructure and make clear the interrelation with the whole process. Especially in the case of silver sheathed composites, crack generates during a heat treatment because of the thermal stress and results in a serious degradation of the critical current density(1). In order to eliminate the generation of cracks, the hydrogen reduction process has been studied(2). In the present study, each process has been discussed to optimize the condition of the preparation and the effect of the hydrogen reduction has been investigated to obtain high critical current density.

Experimental Procedure

The process employed here is schematically expressed in Fig.1. A given amount of $BaCO_3$ (or BaO_2), Ln_2O_3 (Ln=Y,Gd,Ho) and CuO powders was weighed and mixed together. The mixture was calcined at 1173 K for 173 ks and then ground. This step was repeated up to three times. In a part of experiments, the calcined powder was reduced by hydrogen gas and its partially reduced powder was used for the further process, by which the powder was filled in a silver tube. Then the composite consisted of oxide powder sheathed with silver tube was cold-rolled to form a tape. Some of specimens were hot-pressed uniaxially with 67 MPa for 3.6 ks at various temperatures. They were heat-treated in oxygen atmosphere at 1203 K and cooled gradually. In order to avoid any confusion of description, a notation is used in the text as Y-tape (or Ho-tape) specimen, when the oxide is $Ba_2LnCu_3O_{6+x}$ (Ln = Y or Ho). The superconducting critical current was measured by means of four probe method at 77 K at zero magnetic field. Also the critical current was measured in the magnetic field from zero to 15 T at 4.2 K at High Field Laboratory for Superconducting Materials of Tohoku University. The microstructural invesitigation was performed using optical and scanning electron microscopes.

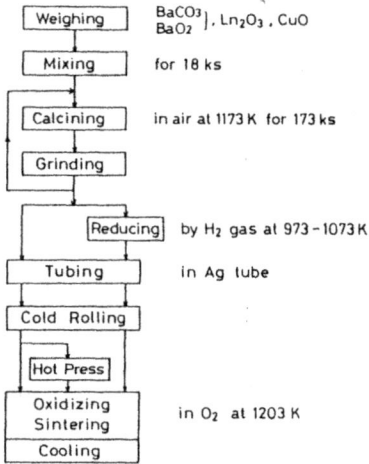

Fig. 1 Process for preparing the Ag sheathed specimens employed here.

Table 1 Comparison of critical current density at 77 K of the Y-tape specimens made from $BaCO_3$ powder and from BaO_2.

thickness	from $BaCO_3$	from BaO_2
2.0 mm	2.1 MA/m^2	1.9 MA/m^2
1.0 mm	3.2 MA/m^2	3.1 MA/m^2

Table 2 Effect of repeating time of calcining on critical current density at 77 K and critical temperature.

repeating time	Jc (MA/m^2)	Tc (K)
one time	1.3	85.6
three times	2.5	87.1

Experimental Results and Discussion

The effect of substitution from $BaCO_3$ to BaO_2 powder was examined using Y-tape specimens. As shown in Table 1, it was found that the substitution did not bring any change for the critical current density. Table 2 shows the effect of repeating time of calcining. In this case, the bulk specimens without silver sheath were used. It is clear that the repetition by three times improved the critical current density.

Some of Y-tape specimens were hot-pressed at temperatures of 773, 973 and 1173 K. As shown in Table 3, the critical current density decreased with increasing temperature, consisting with the observation by which micro-cracks exist more frequently for the specimens pressed at the higher temperature.

By the cold-rolling, a texture of crystal orientation tended to generate towards the rolling direction. Table 4 shows the result of X-ray diffraction analysis. The "powder" specimen corresponds to the powder ground from the former tape specimen. The diffraction intensities of (0 0 1) planes became relatively large for the tape specimen. As mentioned later, the critical current increased with increasing the degree of cold-working.

Table 3   Influence of hot-press on critical current
          density at 77 K.

press temperature (K)	Jc (MA/m^2)
as rolled	4.1
773	1.5
973	1.3
1173	0

Table 4   Comparison of relative X-ray diffraction
          intensity of the powder with that of the
          tape.

2θ (deg.)	h k l	I/Io (%)	
		powder	tape
17.64	0 0 2	–	10
26.60	0 0 3 0 1 0	8	17
37.99	0 1 3	94	98
38.35	1 1 0 1 0 3	100	100
45.10	0 0 5 0 1 4	18	43
47.27	1 1 3	16	17
54.84	0 0 6	29	40
55.84	2 0 0	12	13
60.64	1 1 5	6	5

Table 5   Influence of removal of the sheath on
          critical current density at 77 K.

	with sheath	without sheath
wire	4.9 MA/m^2	7.0 MA/m^2
tape	2.9 MA/m^2	3.4 MA/m^2

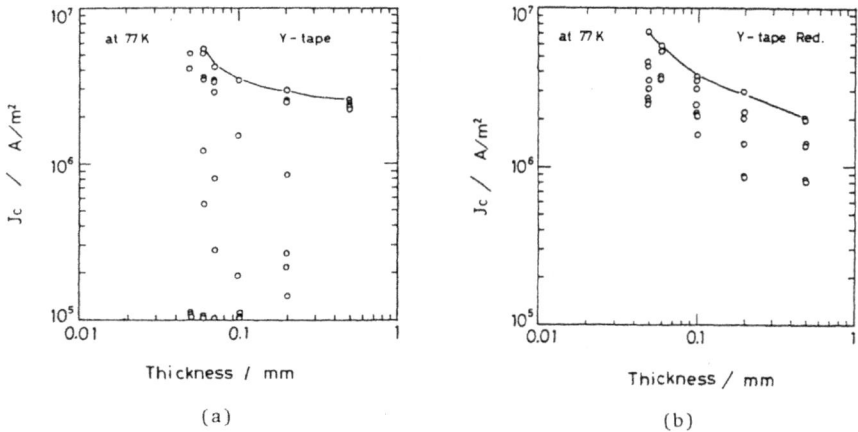

Fig. 2   Change of critical current density at 77 K as a function of
         tape thickness for the specimens perepared by using the
         original oxide powder (a) and by using the hydrogen gas
         reduced powder (b).

The influence of sheathing was mentioned as follows. After cold
rolling, silver sheath was removed from one of the specimens. And both
specimens with and without sheath were heat-treated with together. From Table
5, it was found that the specimens without sheath give the higher critical
current density. As already reported(1), it is attributed to micro-cracks
generated during heating due to the thermal tensile stress acting on the
oxide. As shown in Fig. 2(a), the observed values of critical current were
scattered much more for the thinner specimens less than 0.2 mm. Micro-cracks
seems to influence more seriously for these thinner specimens. So we
attempted to eliminate the generation of micro-cracks, utilizing the partially
reduced powder.

When the oxide was reduced by hydrogen gas, it decomposed into BaO, Ln_2O_3
and CuO and by further reduction, CuO decomposed up to metallic copper. The
partially reduced powder was filled in the silver tube, cold-rolled and
oxidized at high temperatures. Through such a process, an elimination of
cracks might be expected, because of heating and re-oxidization of the
partially metallic mixture. As shown in Fig. 2(b), the scattering of the
observed values has been reduced, especially the number of degraded data was
largely eliminated , when the hydrogen gas reduced powder was used.

The annealing time dependence of the critical current density is shown in
Fig. 3 for the Y-tape specimens. It was found that the critical current
density became highest at the time between 86 and 173 ks and larger for the
thinner specimen. The largest value for each thickness was plotted as a
function of thickness as shown in Fig. 4. The critical current density tended
to increase with decreasing thickness. The corresponding microstructures are
indicated in Fig. 5.

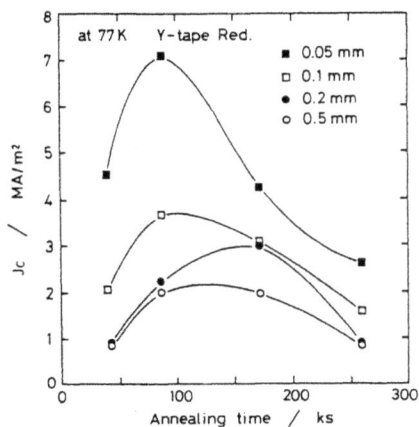

Fig. 3 Annealing time dependence of critical current density at 77 K for the Y-tape specimens.

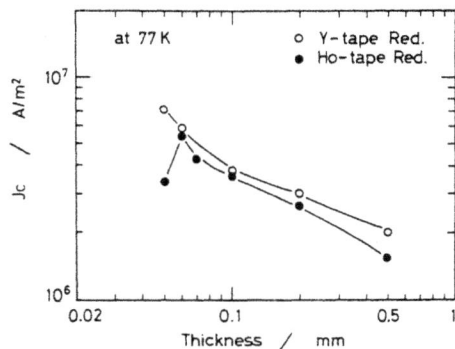

Fig. 4 The largest value of critical current density at 77 K for the specimens with respecitive thickness.

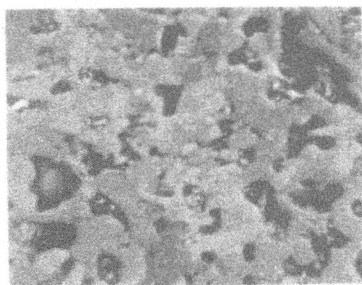

0.5 mm   Jc = 2.0 MA/m²

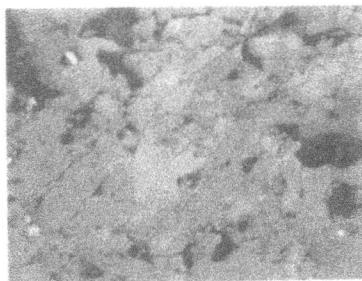

0.2 mm   Jc = 2.2 MA/m²

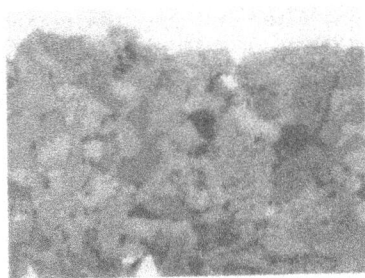

0.1 mm   Jc = 3.7 MA/m²

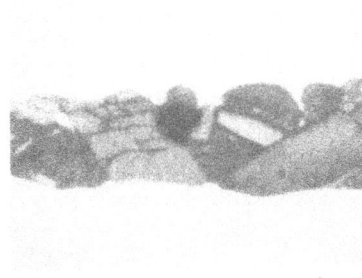

0.05 mm   Jc = 7.1 MA/m²

Fig. 5 Optical micrographs of the Y-tape specimens. Thickness and the critical current density are indicated below each photo.

254

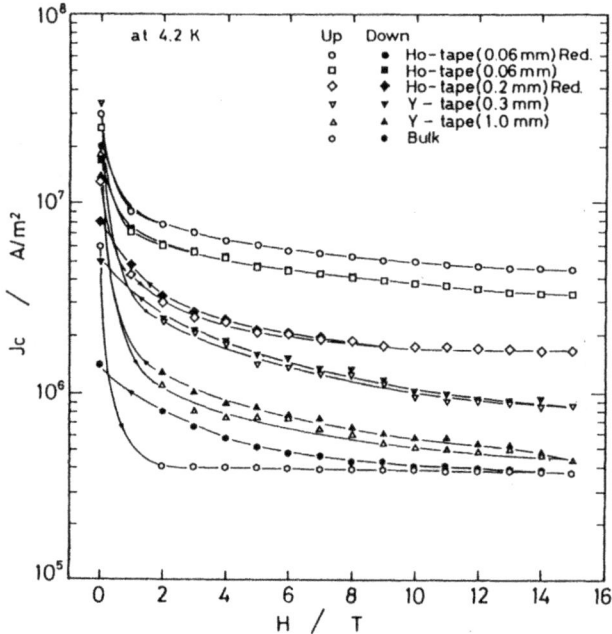

Fig. 6 Magnetic field dependence of the critical current density at
4.2 K for Y- and Ho-tape specimens. Bulk indicates the
specimen prepared without Ag sheath, and Red. is attached
when the hydrogen gas reduced powder was used.

The magnetic field dependence of critical current density at 4.2 K is
shown in Fig. 6 for various specimens. Comparing with the data for the bulk
specimen, the critical current density increased over the whole region of
magnetic field for the thinner tape specimens. For the specimens with the
same thickness, the hydrogen gas reduced powder improved the critical current
density.

Acknowledgement: The authers wish to express their gratitude to Prof.Y Muto,
Prof.S Noto, Dr.K Watanabe and Messer K Kudo, K Sai and Y Ishikawa of HFLSM at
Tohoku University for their help in the critical current measurements. This
study was made possible by a Scientific Research Grant in-Aid (Project No.
62124055) from the Ministry of Education

References

1) O. Kohno et al, Jpn. J. Appl. Phys., 26, 1653 (1987).
2) K. Osamura et al, Proc. of ICEC12 (July, 1988, Southampton)

High Resolution Observation of High Tc Superconductor based on Bi–Ca–Sr–Cu–O System

Y. Tsubokawa, H. Endoh, R. Shimizu, J. Ogiwara[*],
K. Harada and Y. Taniguchi

Faculty. of Engineering, Dept. of Applied Physics,
Osaka University, Suita, Osaka, JAPAN 565
[*]Central Research Laboratory, Sumitomo Cement Co. Ltd.
Funabashi, Chiba, JAPAN 274

Abstract --- Basic experiments to characterize high Tc superconductor based on Bi–Ca–Sr–Cu–O system were carried out by using electron microscope equipped with an electron energy loss spectrometer and cold stage cooled by liquid nitrogen. Radiation damage by electron irradiation with different primary energies is also discussed.

Introduction

The development and characterization of high Tc superconductors based on Bi–Ca–Sr–Cu–O system have just begun. High resolution electron microscopy has been playing an important role in studying atomic structure of these materials. The present work aims to obtain basic information on how to characterize the materials by high resolution analytical electron microscopy.

Basic property

Figure 1 shows temperature dependence of resistivities of five kinds of superconductors based on Si–Ca–Sr–Cu–O system produced by a project team of Sumitomo Cement Co. Ltd.. The resistivities of these materials show double transitions. The higher transition leads to the onset of superconductive property and the lower transition leads to zero-resistance. The sample named 'R5' which has the highest temperatures of the onset and zero-resistance, 120K and 95K respectively, is cleaved in an agate mortar filled with ethyl alcohol. Thin fine flakes of this sample were prepared to the specimens for observation under electron microscope.

The electron diffraction patterns corresponding to b^*–c^* reciprocal lattice plane and a^*–b^* or a^*–c^* reciprocal lattice plane are shown in Figs. 2(a) and 2(b), respectively. These electron diffraction patterns suggest that the lattice constants of the sample 'R5' are estimated as follow ; a = 0.54nm , b = c = 2.7nm , $\alpha = \beta = \gamma = 90°$ and that there are five-times modulations along b^* and c^* axes. It should be comfirmed that these data on the unit cell is not affected by structure anomaly.

High resolution observation

High resolution observations by JEM–4000EX electron microscope with the accelerating voltage of 400kV are carried out. Figures 3(a) and 3(b) are high resolution images observed from <001> direction and its enlargement, respectively. Bright lines along vertical direction in Figs. 3(a) and 3(b) correspond to (100) planes. Dark lines along horizontal direction in Fig. 3(a) correspond to the five-times modulation along b^* axis.

Figure 4 shows high resolution image observed from <011> direction. Figure 4(b) is the upper part of Fig. 4(a) in higher magnification. The top areas in both figures show the images of edge of the sample. The modulation structure keeps the regularity in thicker area but grain-like structures appear near the edge. The vertical fringes show (200) planes and the horizontal fringes with longer period in Fig. 4(a) correspond to the five-times modulation along <010> and <001>.

Observation by a cold stage

Behavior of superconductors at low temperature can be observed by electron microscope installed cold stage in the specimen chamber. Figure 5 shows the outer view and specifications of a cold stage, which can be cooled by liquid nitrogen, for JEM-200CX electron microscope. Figure 6 shows high resolution observation by JEM-200CX attached the cold stage under the accelerating voltage of 200kV. Figure 6(a) shows the observation without supplying the liquid nitrogen. The horizontal lattice fringes resolve 0.27nm. Figure 6(b) shows the observation with the cold stage cooled by liquid nitrogen. Both resolution and contrast in this image are inferior to the image of Fig. 6(a) but (011) lattice fringe with the spacing of 0.38nm can be resolved in whole area and there are some areas where (100) lattice fringe with the spacing of 0.27nm is observable in faint contrast. Comparing these two images and from corresponding electron diffraction patterns, it is suggested that there is no structure difference between the two thermal conditions.

EELS analysis

Core loss spectra of oxygen and bismuth are measured by electron energy loss spectrometer. The core loss intensity of oxygen K-edge(540eV) is high enough to be detected as shown in Fig. 7. But the S/N ratio is rather small to form an image by the electrons interacted with oxygen atoms. It is strongly required to develop effective methods to get the information on atomic location of oxygen atoms using the EELS signal. Core loss intensity of bismuth was not detected in the present experiment.

Radiation damages

Radiation damage of specimen by electron irradiation is an essential problem in high resolution observations. Energy dependence of the radiation damage is studied to get low damage condition. Figure 8 shows the structure differences after electron irradiation for 10 minutes at the accelerating voltages of 100kV-400kV. These pictures indicate that radiation damage decreases for higher accelerating voltage. These experimental results suggest that the radiation damage is caused by inelastic scattering process of penetrating electrons.

Conclusion

The result obtained in the present work is summarized as follows ;
(1) The lattice constants of the sample were determined by electron diffraction patterns.
(2) The high resolution images in the direction of <010> and <001> were observed under JEM-4000EX electron microscope at the accelerating voltage of 400kV.
(3) The high resolution images with the resolution of 0.38nm was observed

using the cold stage cooled by liquid nitrogen installed in JEM-200CX microscope at the accelerating voltage of 200kV.

(4) The core loss spectrum of oxygen K-edge has detected with sufficiently high signal to noise ratio.

(5) Energy dependence of damages induced by electron irradiation was observed. By the electron irradiation for more than ten minutes, the structure of this sample suffered being damaged at 100kV and 200kV, but hardly damaged at 300kV and 400kV.

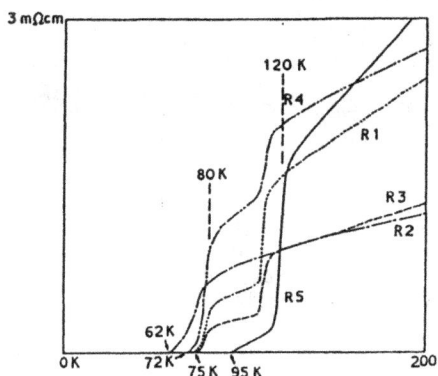

Fig.1. Temperature dependence of the resistivities for different five Bi-Ca-Sr-Cu-O samples. The electron microscopic observation was made for the R5 sample, whose on-set and off-set temperatures are 120K and 95K, respectively.

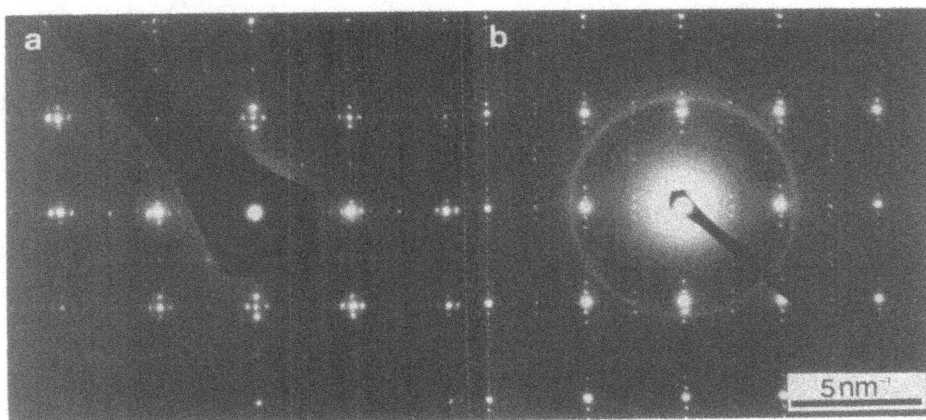

Fig.2. Electron diffraction patterns of the R5 sample. (a) b-c plane, (b) a-b plane. Unit cell dimensions are identified to be a=0.54nm and b=c=2.7nm. with ring pattern of Au-reference sample.

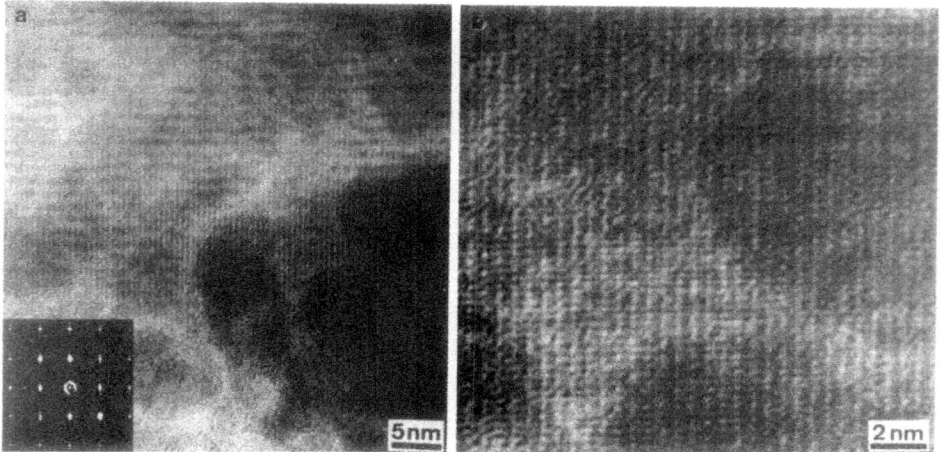

Fig.3. High resolution electron micrographs of the R5 sample obtained with JEM-4000EX with an incident electron beam along [010] direction at different magnifications. The vertical white lines in (b) correspond to the structure modulated by the double of the basic periodicity whereas the horizontal black lines in (a) correspond to the structure with periodicity five times longer than the basic periodicity.

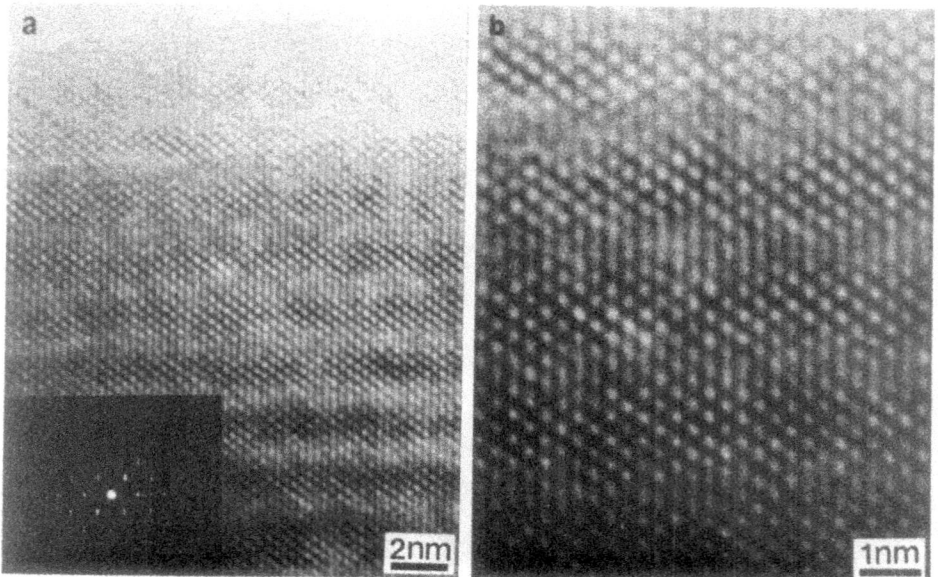

Fig.4. High resolution electron micrographs of the R5 sample with an incident electron beam along [011] direction at different magnifications. (b) is a magnified area of upper part of (a) (near to the sample-edge). Note that the regular structure becomes disturbed.

cooling mode   : conduction with copper rod

cooling limit : below -160°C

temperature sensor   :
            thermocouple (copper-constantan)

tilting angle : single tilt (±45°)

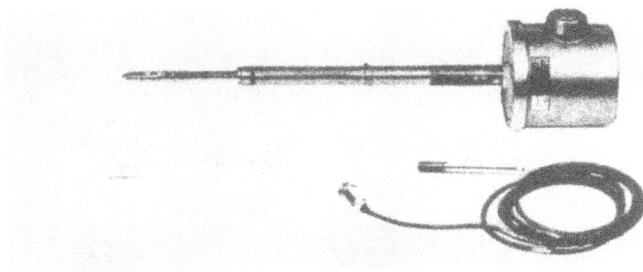

Fig.5. Parts of the cold-stage holder of JEM-200CX used for observation under electron microscope at nearly liquid nitrogen temperature.

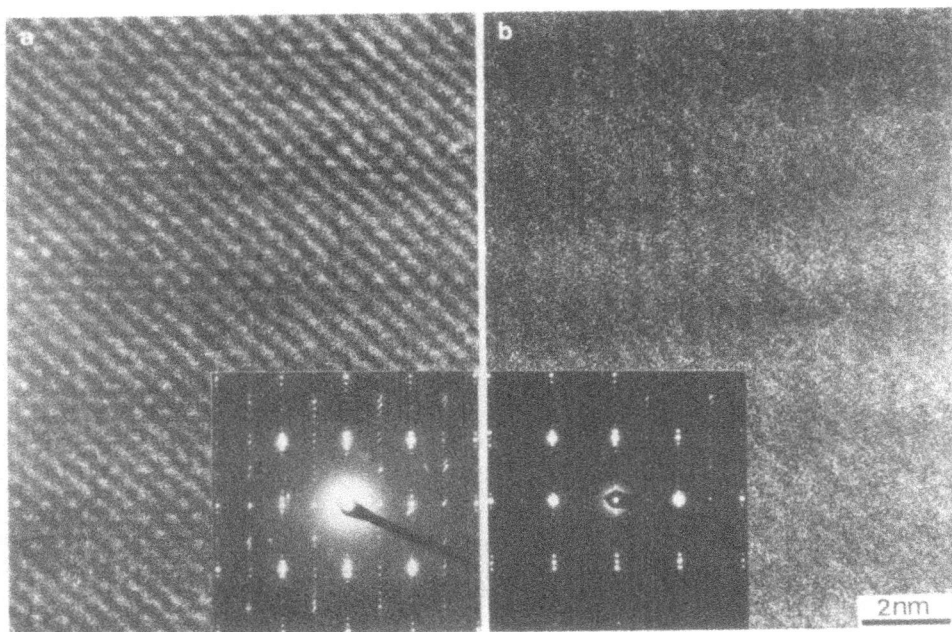

Fig.6. High resolution electron micrographs obtained with a liquid nitrogen cold-stage ; (a) before cooling and (b) nearly at liquid nitrogen temperature. Any marked differences are not observed in (a) and (b), both the lattice image and diffraction pattern.

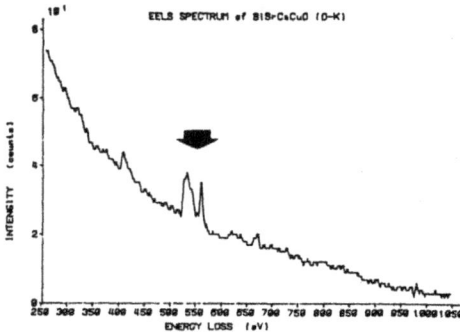

Fig.7. Energy loss spectrum of the Bi–Ca–Sr–Cu–O with relatively sharp peak of the oxygen K-edge. The peaks for Bi were not clearly detected, embedded in the background noise in this case.

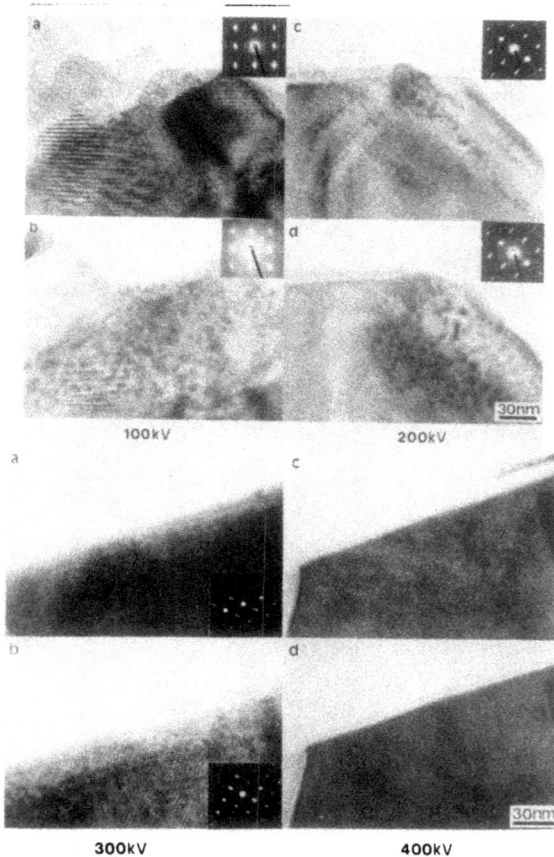

Fig.8. Radiation damages in the Bi–Ca–Sr–Cu–O sample caused by irradiation of 100keV, 200keV, 300keV and 400keV electrons for observation under electron microscope. Note that 400keV electrons hardly incured any damages during observation whereas 100keV electrons did cause marked damaging.

HIGH RESOLUTION ELECTRON MICROSCOPE IMAGES OF
OXIDE SUPERCONDUCTORS TAKEN AT 4.2 K

Hiroyuki YOSHIDA*, Yasuhiro YOKOTA**, Hatsujiro HASHIMOTO**,
Masashi IWATSUKI*** and Yoshiyasu HARADA***

*Research Reactor Institute, Kyoto University, Kumatori-cho, Osaka, Japan
**Science University of Okayama, Ridai-cho, Okayama 700, Japan
***JEOL, Nakagami-cho, Akishima, Tokyo 196, Japan

Abstract - Multi-beam imaging of high resolution electron microscopy
has been applied to observation of the microstructures in high T_c oxide
superconductors using a superconducting cryo electron microscope at 4.2 K.
Atomic resolution images showing twin structure were clearly imaged for the
(a,b) planes of $ErBa_2Cu_3O_x$. A bright contrast often appeared at the twin
boundary. The layered structures of $Bi_2Sr_2CaCu_2O_x$ and $Bi_2Sr_2Ca_2Cu_3O_x$ were
also observed.

Introduction

Since the discovery of high T_c superconductors above 90 K in $YBa_2Cu_3O_x$
many efforts have been devoted to investigate the relationship between
microstructure and superconducting properties [1-5]. The high resolution
electron microscopy has been expected for finding a relation between the
structure and superconductivity. The oxide superconductors with T_c = 90 K,
such as $YBa_2Cu_3O_x$ and $ErBa_2Cu_3O_x$, the crystal structure of oxygen deficient
perovskite is known to shows the transition from tetragonal to orthorhombic
structure introducing many twins depending on oxygen content and annealing
condition. The layered structure has been reported as the origin of the
strong anisotropy of superconducting properties, and the twin structure has
also been discussed in a relation with their superconductivity [2-5].

After the finding of Bi-Sr-Ca-Cu oxide superconductors with higher T_c
[6], the multiple phases have been discussed for the two-stages of the
superconducting transitions at 85 K and 110 K. The crystal models have been
presented based on the studies by electron and neutron diffraction [7,8] as
well as high resolution electron microscope observation [7,9-11] for the
lower and higher T_c phases. The structural moduration was also found by
these studies.

For investigating on the relation between the superconducting property
and the microstructure, high resolution electron microscope observation at
the superconducting state has been perticularly expected.

Experimantals

Two kinds of oxide superconductors were used for the present work.
The specimen having nominal composition of $ErBa_2Cu_3O_x$ was prepared by the
usual sintering method [12]. A sharp superconducting transition appeared at
90 K in the previous measurement[12]. The specimen of Bi-Sr-Ca-Cu oxide was
measured to show two-step transition at near 110 and 80 K by a previous

electrical resistance measurement [13].

Electron microscope observations were performed by the multi-beam imaging method using a superconducting cryo electron microscope JEM-2000SCM [14], which was operated at 160 KV. The shielding type superconducting lens and the specimen stage are located in a liquid helium cryostat. As a top of the specimen holder was set in the stage during observations, a heat link from room temperature could be avoided. The equipment enabled the specimen to be kept at 4.2 K without thermal drift and vibration, and resulted a high resolution images resolving 0.26 nm lattice of Nb_3Sn in the previous work [15,16]. The tilting goniometer for adjusting the crystal orientation is not yet developed, the thin foils with suitable orientation were selected for the observations. The magnetic field at the specimen position was 1.4 T when the electron microscope was operated at 160 kV.

Results and Discussions

Figure 1 shows an example of high resolution images of $ErBa_2Cu_3O_x$, where two thin crystals with different orientations occasionally coexist. The upper one was photographed under the condition at which the incident electron beam is normal to the c-axis. The layer structure image with c=1.17 nm which consists of bright and dark lines of ErO/BaO/ErO can be seen corresponding to the structure model as shown in Figure 2. A strong contrast layer with wider distance than the regular layers is seen in Figure 1 as indicated by a big arrow suggesting an Er or Cu rich plane [3]. The bottom crystal is oriented at the (a,b) plane being parallel to the electron beam, where both the lattices of 0.39 nm are resolved. There is twin structure, whose boundaries are indicated by small arrows. Along the boundaries, broad bright contrast appeared sometimes followed by broad dark lines. The twin structure in $YBa_2Cu_3O_x$ has been reported by many authors.

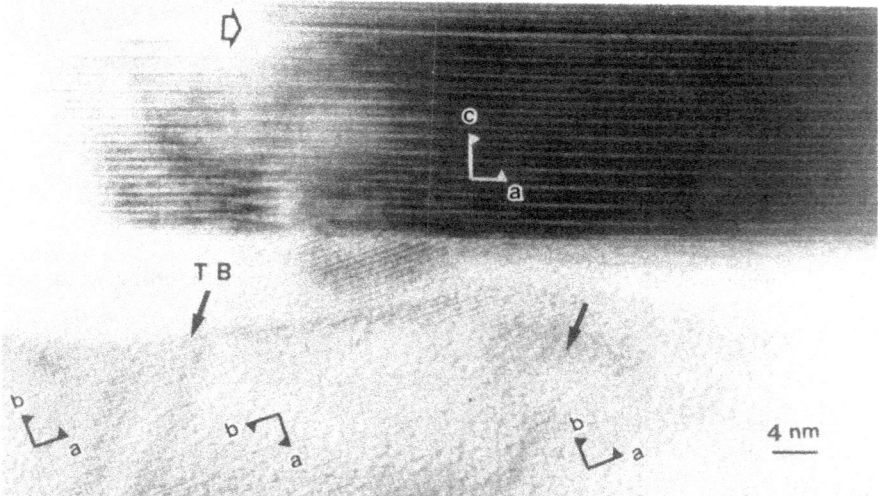

Fig. 1. Cryo electron microscope image of two thin $ErBa_2Cu_3O_x$ crystals with different orientations.

The twin structure is also seen in the multi-beam image of ErBa$_2$Cu$_3$O$_x$ as shown in Figure 3, where the contrast change is seen in each twin. The structure images corresponding to the atomic lattice spacing are clearly resolved. As the atomic arrangement is usually regular even at the twin boundaries, there is scarcely lattice imperfection and distortion. However the broad bright and dark lines often appeared along the twin boundaries. The origin of the contrast is not yet clear but there may be some relation with the magnetic flux distribution in the superconducting state.

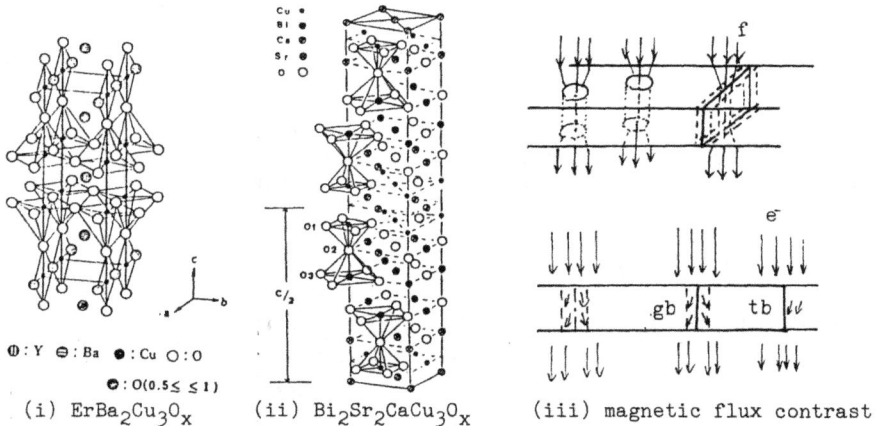

(i) ErBa$_2$Cu$_3$O$_x$ (ii) Bi$_2$Sr$_2$CaCu$_3$O$_x$ (iii) magnetic flux contrast

Fig. 2. Structure models of ErBa$_2$Cu$_3$O$_x$ and Bi$_2$Sr$_2$CaCu$_2$O$_x$, and schematic illustration of contrast due to magnetic flux

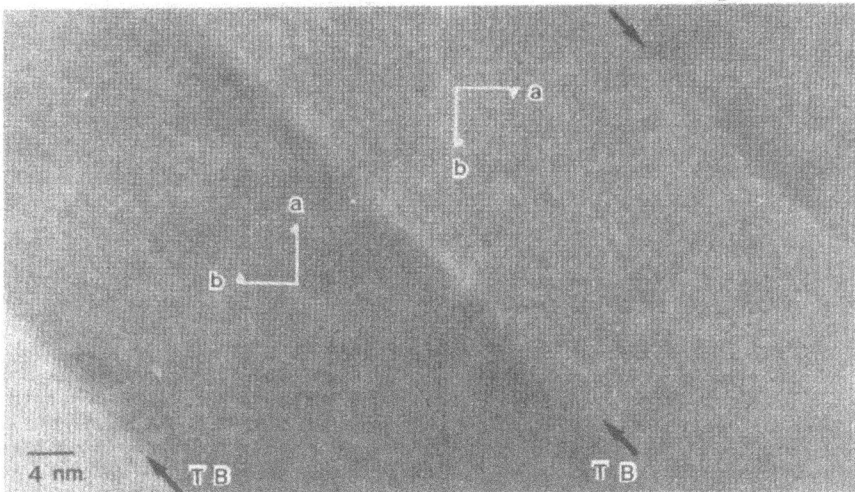

Fig. 3. Twin structure in ErBa$_2$CuO$_x$ at 4.2 K

As the crystal structure of superconducting LnBa$_2$Cu$_3$O$_x$ (Ln=Y and rear earth elements) is an oxygen deficient perovskite with the orthorhombic structure (Fig. 2), the twin structure is believed to form during the transition from tetragonal structure to orthorhombic one in the annealing process of the sample preparation. The contribution of the twin structure to superconducting properties has been discussed for the boundaries acting as the paths for superconducting electrons and as the pinning centres for preventing the movement of magnetic flux. Among the mixed twin structures a part containing fine microtwins was reported to show higher T_c [5]. For the A15 superconductors the grain boundaries are believed to act as pinning centres, where many lattice imperfections at the grain boundaries are effective for the contribution. Although the twin boundary in principle contains less lattice imperfections, the contribution to the pinning force seems to be good enough for the oxide superconductors because of their short coherent length along the a and b directions which is very closed to the lattice parameter. This may correspond to the broad bright and/or dark contrast appearing at the twin boundaries in $ErBa_2Cu_3O_x$ at low temperatures. In Figure 2 (iii) a schematic illustration of the origin of flux line contrast distributed in the matrix of superconducting state, as well as contrast due to flux lines localized at grain boundary or twin boundary, is shown.

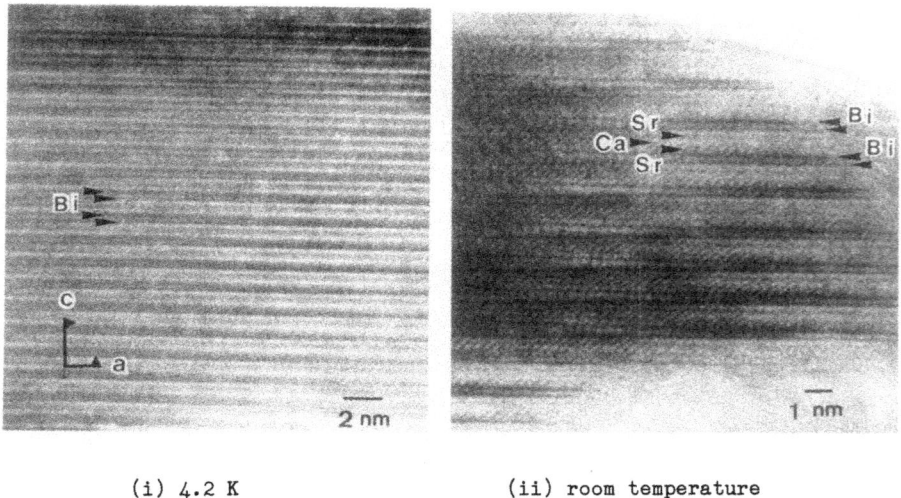

(i) 4.2 K   (ii) room temperature

Fig. 4. High resolution electron microscope images of Bi-Sr-Ca-Cu oxide

In Figure 4 examples of the multi-beam images for the Bi-Sr-Ca-Cu oxide specimen are shown, which were photographed when the electron beam is normal to the c-axis. The layered structure with c/2 = 1.54 nm seems to consist of the periodicity of BiO/SrO/Ca/SrO/BiO planes and the darkest lines correspond to the BiO planes and the CuO_2 shows too weak in contrast. The composition of $Bi_2Sr_2CaCu_2O_x$ is the same as the structure reported as the lower T_c (85 K) phase [7-9]. In the multi-beam image taken at room temperature there is another layer structure with a wider periodicity than the above structure. The layer structure corresponds to $Bi_2Sr_2Ca_2Cu_3O_x$, which is reported as the higher T_c (110 K) phase by other authors [10]. The both structures are similar to the two structures of 105 K and 125 K phases in Tl-Ba-Ca-Cu-O superconductors [10], while $TlBa_2Ca_2Cu_3O_9$ is reported as the 125 K phase [11].

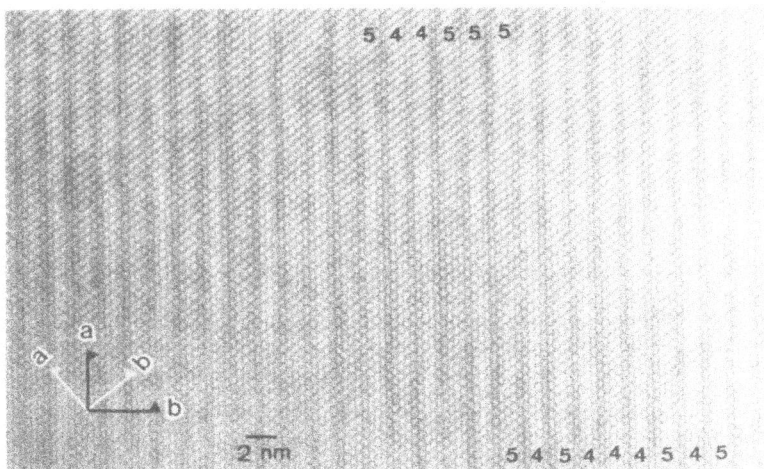

Fig. 5. Modulated structure observed in Bi-Sr-Ca-Cu oxide
at room temperature

    Figure 5 shows another example of the multi-beam images taken at the
electron beam parallel to the c-axis. The atomic lattice spacings with
a=b=0.54 nm are clearly resolved, however broad and wavy contrast strongly
appeared along the a-direction indicated by black arrow suggesting some
structure modulation with a periodicity of 5 atomic distance. The a- and
b-directions indicated by white arrows correspond to the ones as reported
by Tarascon et al. [7]. They also reported a modulation in the electron
diffraction patterns. If the structure model presented by them is rotated
by 90 degree we obtain the model illustrated in Figure 2, whose a- and b-
directions are indicated by black arrows in Figure 5. According to the
modulation on neutron diffraction patterns, Kajatani et al. discussed the
superlattice structure (a=0.54 nm, b=2.70 nm) with a periodic arrangement
of oxygen vacancies. The sub-unit cell is very close to the model shown
in Fig. 2, however the periodicity due to oxygen deficiency might not give
strong modulated contrast appeared in the images. A superlattice of five
atomic distance can be considered if the position change between Sr and Ca
atoms occurs periodically for cancellation of the composition difference.
The periodicity of modulation was observed to vary sometimes from five to
four as shown in the image at room temperature. The buckling of the layer
of $SrO/Ca/CuO_2$ planes between the rigid BiO/BiO planes is considered as an
origin of the strain contrast forming the modulated image. The contrast
of modulation appeared relatively weak at 4.2 K, which may correspond to
the strain reduced at superconducting state.

                              Conclusion

    The results obtained in this study are summarized as follows.
    1) $ErBa_2Cu_3O_x$ shows layered structure along the c-axis corresponding
to the crystal structure. This microstructure contains twin structure
which may be concerned with superconducting properties.
    2) $Bi_2Sr_2CaCu_2O_x$ and $Bi_2Sr_2Ca_2Cu_3O_x$ with layer structure are recog-
nized as the 85 K phase and the other phase. The modulation in the high

resolution images also observed.

Acknowledgements

The authors would like to express their thanks to Dr.K.Atobe of Naruto University of Education and Dr.H.Hitotsuyanagi of Sumitomo Electric Industry for their kind supply of the Er-Ba-Cu-O and Bi-Sr-Ca-Cu-O superconductors, respectively.

References

1) M.K.Wu,J.R.Ashuburn,C.J.Trong,P.H.Hor,R.L.Meng,L.Gao,Z.J.Huang,Y.A.Wang and C.W.Chu,"Superconducting at 93K in a new mixed-phase Y-Ba-Cu-O compound system at ambient pressure",Phys.Rev.Lett. 58, 908-912 (1987).

2) K.Hiraga, D.Shindo, M.Hirabayashi, M.Kikuchi, and Y.Shono, "High resolution electron microscopy of high T_c superconductor Y-Ba-Cu-O", J. Electron Microsc. 36, 261-269 (1987).

3) D.J.Li, H.Shibahara, J.P.Zhang, L.D.Marks, H.O.Marcy, and S.Song, "Synthesis and structure of copper- and yttrium-rich $YBa_2Cu_3O_{7-x}$ super-conductors", Physica C 156, 201-207 (1988).

4) Z.Hiroi, M.Takano, Y.Takeda, R.Kanno, and Y.Bando, "Microdomain structure in $YBa_2(Cu_{1-x}Fe_x)O_{7-y}$ observed by electron microscopy", Jpn J. Appl. Phys. 27, L580-L583 (1988).

5) H.Hayashi, Y.Yokota, S.Morita, N.Hanada, and S.Hayashi, "Separation and characterization of the superconducting $YBa_2Cu_3O_{7-x}$ powder", Jpn J. Appli. Phys. 27, (in press)(1988).

6) H.Maeda, Y.Tanaka, M.Fukutomi and T.Asano,"New high-T_c superconductors without rare earth element", Jpn J. Appli. Phys. 27, L209-L210 (1988).

7) J.M.Tarascon,Y.Le Page,P.Barboux,B.G.Bagley,L.H.Green,W.R.McKinnon,G.W. Hull,M.Giroud,D.M.Hwang,"Crystal substructure and physical properties of superconducting $Bi_4(Sr,Ca)_6Cu_4O_{16+x}$",Phys.Rev.B37,9382-9385(1988).

8) T.Kajitani, K.Kusaba, M.Kikuchi, N.Kobayashi, Y.Shono, T.B.Williams, and M.Hirabayashi, "Structural study of high T_c superconductor $Bi_{2-x}(Ca,Sr)_3Cu_{2+x}O_{9-y}$", Jpn J. Aplli. Phys. 27, L587-L590 (1988).

9) H.Maeda, Y.Tanaka, M.Fukutomi, T.Asano, K.Togano, H.Kumakura, M.Uehara, S.Ikeda, K.Ogawa, S.Horiuchi, andY.Matsui,"New high-T_c superconductors without rare earth element", Physica C 153-155, 602-607 (1988).

10) E.A.Hewat, P.Bordet, J.J.Capponi, C.Chaillout, J.L.Hodeau and M.Marezio, "Superstructure of the superconductor $Bi_2Sr_2CaCu_2O_8$ by high resolution electron microscopy", Physica C 153-155, 619-620 (1988).

11) A.Sulpice, B.Giordanengo, R.Tournier, M.Hervieu, A.maignan, C.Martin, C.Michel, and j.Provost,"Bulk superconductivity in $Tl_2Ba_2CaCu_2O_8$ and $TlBa_2Ca_2Cu_3O_9$ phases", Physica C 156, 243-248 (1988).

12) H.Yoshida,K.Atobe,K.Miyata and H.Kodaka,"Superconductivity and irradiation effects in Ln-Ba-Cu oxide superconductors",Proc.Sintering'87(1988).

13) H.Yoshida, and K.Atobe, "Irradiation induced changes of superconductivity in oxides", Proc. MRS Sympo. Tokyo, (in press) (1988).

14) M.Iwatsuki, H.Kihara, K.Nakanishi, and Y.Harada, "Cryo electron microscope with superconducting lens", Proc. 11th Int. Cong. on Electron Microscopy, Kyoto, p. 251-254 (1986).

15) H.Yoshida, H.Hashimoto, Y.Yokota, and M.Iwatsuki, "High resolution electron microscope observations of super- and normal-states of A15 superconductors", Jpn J. Aplli. Phys. 26 Suppl. 26-3, 943-944.

16) H.Yoshida, Y.Yokota, H.Hashimoto, M.Iwatsuki, and Y.Harada, "High resolution electron microscope images of superconducting A15 compounds obsereved at 4.2 K", J. Electron Microsc. 36, 228-234 (1987).

THE CHARACTERIZATION OF THE Y-Ba-Cu-O SYSTEM COMPOUND PREPARED BY SOLID STATE REACTION, METAL ALKOXIDE SOLUTIONS AND SOL-GEL METHOD

Masa-aki MUROYA[1], Yasuyuki TAKEMOTO[2], Tadashi HATAYAMA[1], Katsuhiko YAMAMOTO[1], Toshihiro YOSHIDA[1] and Eiichi TADA[3]

[1]Osaka Electro-Communication University, 18-8, Hatsumachi, Neyagawa, Osaka, 572 Japan, [2]Kobe University of Marcantile Marine, Fukae-minami, Higashinada, Kobe, 658 Japan, [3]Faculty of Engineering Science, Osaka University, 1-1, Machikaneyama, Toyonaka, 560 Japan.

Abstract- A number of specimens in Y-Ba-Cu-O system compounds capable forming films and disks were prepared by three different procedures, such as solid state reacion, mixing of Y-, Ba- and Cu-alkoxides and hydrolysis of their mixed alkoxides. The characterizations of these specimens were carried out by means of X-ray diffraction(XRD), XPS, resistivity and FT-IR. Orthorhombic $YBa_2Cu_3O_{7-\delta}$ was produced in the specimens prepared from each starting material, but the conditions for the production of such phase depend on the atmosphers and temperatures of heat-treatment. It is revealed that copper concentration on the surface layer increases with elevating the heat-treatment temperature, while yttrium and barium diffused from the surface layer to the bulk phase. High Tc superconducting phenomena were observed in those specimens.

Introduction

It is well known that Y-Ba-Cu-O(YBCO) system compound is one of high Tc superconducting materials. It has been indicated by several workers(1-6) that thin and thick YBCO films were formed by using several preparation methods such as screen printing, sputtering, plasma spray decomposition, decomposition of acid salt and sol-gel. However, there are still some ambiguities on the process of formation in the materials.

In this report, disk and thick film of YBCO system compound were prepared by different starting materials and procedures, and their structural features were investigated as a function of the heat-treatment temperature.

Experimental

Film or disk specimens of YBCO compound were prepared in the following three different procedures. 1) Commercial powders of Y_2O_3, $BaCO_3$ and CuO

were weighed in molar ratio, Y:Ba:Cu=1:2:3, and then mixed for 2 h (so called SR1). This mixed powder was calcined at the temperature of 902°C for 5 h in air and compressed by 500 kg/cm² for 20 min after slowly cooled down. Discoidal specimen(SR2), with 20 mm diameter and about 1 mm thickness, thus obtained from the procedures as mentioned above. These SR1 and SR2 samples were used for further heat-treatment at temperatures ranging from 408°C to 952°C in air or oxygen for 1 h.   2) Yttrium-, barium- and copper-ethoxide, which were provided from Hakusui Chemical Industry Ltd, were mixed at about 70°C, exposing to nitrogen for 3.5 h. Specimen with dark blue color was thus obtained as a highly viscous solution(LR) which was suitable for dip coating. 3) Yttrium-, barium- and copper-ethoxide, mentioned above, were mixed in molar ratio of Y:Ba:Cu=1:2 :3. This mixed solution was added into distilled water, pH of the solution was adjusted to 3~6 by using a hydrochloric acid and an ammonia disolved water, and then transparent sol(HR) thus obtained from the procedures.
    The transparent sol(HR) and the high viscous solution(LR) were divided into two portion. One of these was used in dip coating and the characterizations of these materials were carried out by other portion.
    The X-ray powder diffraction was measured by the same method as reported ealier(8). The XPS experiments were performed using a Shimazu Model ESCA 750 X-ray photoelectron spectrometer utilizing a MgKα (hν =1251 eV) radiation. Spectra were taken at 1.15 eV resolution and reproduced at about ±0.1 eV on repeated runs. The infrared spectra were measured by JASCO Model 5MP FT-IR spectrophotometer. IR spectra of a coated film and a powdered sample were taken with DR and Polarizing Refrection technique. Kubelka Munk transform was carried out in the case of powdered samples. Electrical resistance was measured by using a four-prove method.

Results and Discussion

Characterization of SR

    Fig. 1 illustrates the XRD patterns of the SR1(lines a) and the SR2(b) against the heat-treatment temperature. The diffraction line at about 32.5° , which is corresponded to orthorhombic YBCO compound, is not appeared on the SR1(a4) heated at 408°C. Its diffraction line is barely found by heat-treated at the temperature of 804°C(a8), clearly observed at 902°C(a9), but not coincided with that of YBCO because the diffraction line at about 32.5° is not splitted. On the other hand, the diffraction line at about 32.50° on the SR2 sample(b) heated at 853 °C is splitted to those of 32.40° and 32.68° correspond with (103), (101) and (013) of orthorhombic phase, as is well known. The SR2 is already calcined at 902 °C before the heat-treatment. Therefore, it shows that the performance of preheat-treatment is effective procedure for its increasing the ortho-rhombic phase. By heating up to 952°C, the transformation to tetragonal is already started. It is suggested that the transition of chemical state is expected on the SR specimens.

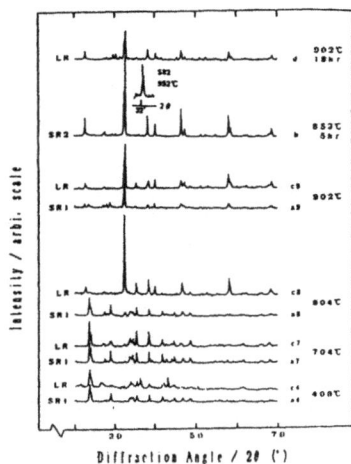

Fig. 1. The XRD patterns of
         SR1, SR2 and LR.

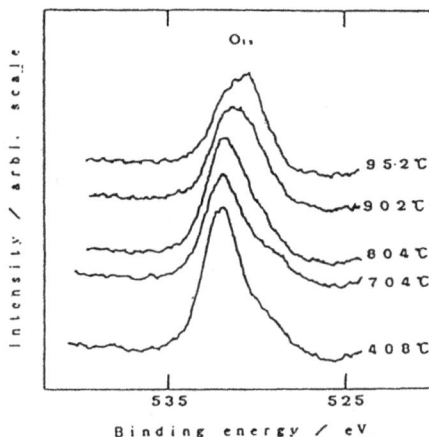

Fig. 2. The XPS spectra of O_{1s}
         core level of SR1.

Fig. 2 shows the XPS spectra of O_{1s} core level on the SR1 versus the
heat-treatment temperature. The O_{1s} core level spectrum of the sample
heated at 408°C is asymmetry, as has been suggests that the presence of
more than one discrete chemical species on the surface. The binding energies of main and shoulder signals are 532.6 and about 530 eV, respectively. The intensity of such shoulder signal increases with rising the temperature, and then its signal becomes a main peak at 952°C. It may be true
that the transformation of its O_{1s} spectrum is related to the phase transition. This point of view may be supported from the results that the
progress of phase transition obtained from XRD experiments(Fig. 1) is
similar to that of the behavior in the temperature dependence of O_{1s}
signal.

The superconducting phenomena are expected to the SR2 sample. The
temperature dependence of resistivity on the SR2 was illustrated in a of
Fig. 3. The resistivity decreases with the temperature decrease of the
material and then its value was not detectable at the temperature of 92° K
(T_c-end).

Characterizations of LR and HR

Fig. 4 shows the infrared spectra on the samples(LR and HR). As seen in
spectrum a, the strong absorption bands ascribed probably to N-H, C-H and
C=O vibrations, and the weak absorption bands due to O-H and C-O-C

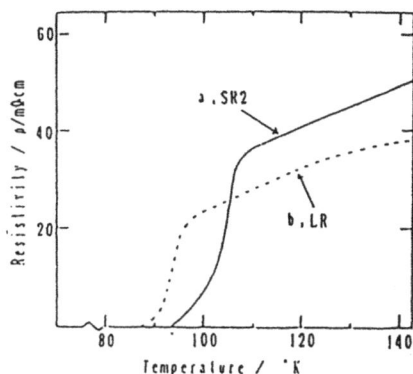

Fig. 3. The temperature
dependency of resistivity
of SR2 and LR.

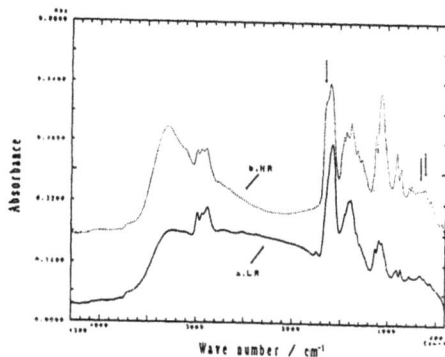

Fig. 4. Infrared spectra of
LR and HR dried at room
temperature.

vibrations were observed on the LR dried at room temperature. The bands of
N-H and C=O are originated in containing materials in starting alkoxides.
The vibration spectra having related to Y, Ba and Cu are not detectable.
On the other hand, the spectrum of HR sample(spectrum b) is almost
analogous to that of LR, but the shoulder band at 1630 cm⁻¹ and the weak
bands at 654 and 608 cm⁻¹ are different from that of the LR.

By heating the LR at 408°C, C=O band clearly remained in its situation,
but N-H bands(N-H str. and bend.) are almost disappeared, as is seen in
a of Fig. 5. The spectrum changes observed in LR were also found in the
HR. When the temperature was elevated, the intensity of C=O band becomes
drastically small at 754°C(d), and then its band disappears at about 804
°C. On the other hand, the C=O band is also observed in the HR, and the
temperature dependence of the C=O band is similar as that of the LR.

The absorption band assigned to metal-oxygen-metal vibration is not
clear in this experiment but the absorption band at below about 700 cm⁻¹,
as is seen in Fig. 5, is possibly correspond to its vibration. The presence of C=O specie found in these samples is sufficiently expect from XPS
spectra.

The XPS spectra were measured, and their spectra on the LR sample
heat-treated at various temperatures were shown in Fig. 6. In the spectrum
at 408°C, a signal at 290.0 eV draw near to higher side of signal at 285.0
eV assigned to C₁ₛ is found, the intensity of such signal decreases with
elevating the temperature, and then the signal can not be detected. The
thermal behavior of the signal is similar to that of C=O band observed in
IR spectra. Thus the signal is considered that the origin of its signal
can be found in C₁ₛ of C=O specie. Also one of these signals of the O₁ₛ
at 532.6 and 530.2 eV is possibly corresponded to its oxygen with C=O
specie.

In other binding energy region, the signals of Ba3d₅/₂(780.7 eV),
Ba3d₃/₂(796.2), Y3d₅/₂(156.7) and Y3d₃/₂(158.0) were clearly

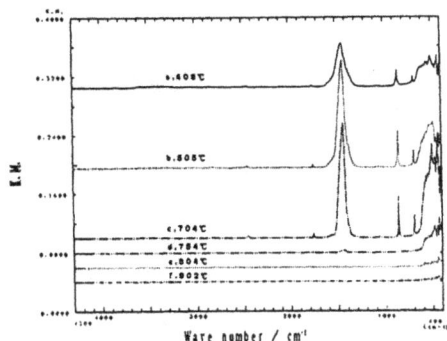

Fig. 5. IR spectra of LR
heat-treated at various
temperatures.

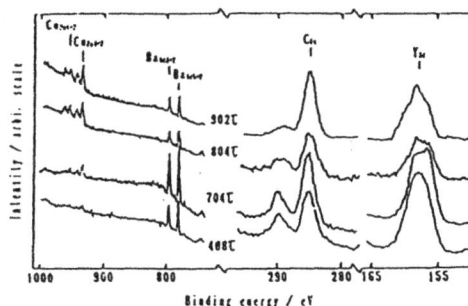

Fig. 6. XPS spectra of LR
heat-treated at various
temperatures.

observed, and the intensities of these signals decrease with elevating
the temperature. Hereupon the signals assigned to $Cu2p_{1/2}$(953.7 eV),
$Cu2p_{3/2}$(934.5) and Cu_{Auger}(335.3) were obscurely observed on the sample
heated at 408°C, but the intensities of these signals increase with
rising the temperature. The behavior of these XPS signals suggests that
copper diffuses from bulk phase to surface layer, but ytterium and barium
diffuses into bulk structure, in course of structural change by the
heat-treatment. The tendency of these phenomena was also found for the SR.

In order to see morphology of the LR, XRD pattern was measured as a
function of heat-treatment temperature, and the patterns were shown in
lines c of Fig. 1. No diffraction line at about 32.5° is detected on the
sample at 408°C, and it is relatively analogous almost to that of the SR1
heated at same temperature(a4), The diffraction line at about 32.5° is
slightly observed on the LR(c7) at 704°C, drastically appears by heated at
804°C(c8), and splitted to 32.40° and 32.68° by heated at 902°C(c9). Such
thermal morphologies of this sample are not the same as that of SR1, that
is, the orthorhombic phase obtained by LR preparation was appeared at
lower temperature region in comparison with that of SR1.

After heating this LR at 902°C for 18 h, the temperature dependence of
resistivity was measured, and its data was shown in the curve b of Fig. 3.
The T_c-end point was at 88° K. The production of superconductor was
confirmed in the LR specimen.

On the other hand, the morphology of HR is also almost analogous to
that of the LR, but the experimental conditions, such as pH, amount of
water for hydrolysis and preparation temperature, are dominated in its
morphology of this specimen.

Conclusion

The results of the characterization of Y-Ba-Cu-O system compound prepared by different starting materials and procedures can be summarized as follows:

(1) The orthorhombic phase of $YBa_2Cu_3O_{7-\delta}$(SR2) was produced by the solid state reaction of powdered starting materials. Superconducting phenomena were detected in the temperature of 92° K(T_c-end) on the orthorhombic phase of this material.

(2) Also orthorhombic phase was produced by metal-alkoxide solutions using the procedure of mixing-low temperature heating and of hydrolysis.

(3) Contrastive surface diffusions of each component in copper, barium and yttrium were observed in the temperature range from about 400°C to about 900°C.

Acknowledgement

The authers wish to express his deep gratitude to Mr. F. Uchida, Hakusui Chemical Industry Co., Ltd, for supplying Metal-alkoxides for this research. The authers would like to thank Mr. N. Ohoshima for FT-IR experimental help to this reseach program.

References

1) J. Tabuchi, A. Ochi, K. Utsumi and M. Yonezawa,
Nippon Seramikkusu-Kyokai-Gakujutsu-Ronbunshi, 96, 450(1988).
2) R. C. Budhani, Sing-Mo H. Hzeng. H. J. Doerr and R. F. Bunshah, Appl.
Phys. Lett., 51, 1277(1987).
3) B. Y. Jin, S. J. Lee, S. N. Song, S. J. Hwn, J. Thiel and K. R.
Ketterson, Adv. Ceram. Mater., 2, 436(1987).
4) M. Awano, M. Tanigawa, H. Takagi, Y. Torii, A. Tsuzuki, N. Murayama
and E. Ishii, Nippon Seramikkusu-Kyokai-Gakujutsu-Ronbunshi, 96,
426(1988).
5) T. Kumagai, H. Ykota, K. Kawaguchi, W. Kondo and S. Mizuta, Chem.
Lett., 1987, 1645.
6) T. Monde, H. Kozuka and S. Sakka, Chem. Lett., 1988, 287.
7) M. Tatsumisago, H. Sato and T. Minami, Chem. Express, 3, 311(1988).
8) M. Muroya and S. Kondo, Bull. Chem. Soc. Japan, 43, 3453(1970).

CHARACTERIZATION OF HIGH T_c SUPERCONDUCTORS BY LUMINESCENCE METHOD

Yasufumi FUJIWARA and Takeshi KOBAYASHI

Faculty of Engineering Science, Osaka University
1-1, Machikaneyama, Toyonaka, Osaka, 560, Japan

Abstract - We have systematically investigated high T_c superconductors by photo- (PL) and x-ray excited thermally stimulated luminescence (TSL) methods. In PL measurements, characteristic PL signals due to electronic transitions between weakly crystal-field split spin-orbit levels of trivalent rare-earth (Re) ions have been observed in the infrared region in the Nd-Ba-Cu-O and Er-Ba-Cu-O systems. The energy position and spectral shape strongly change, depending on a microscopic field around Re ions. No luminescence has been observed in the Eu-Ba-Cu-O and Yb-Ba-Cu-O systems at present. In TSL measurements, characteristic visible TSL signals have been obtained in some kinds of the Re-Ba-Cu-O (Re = Nd, Eu, Gd, Ho, Er), Tl-Ba-Ca-Cu-O and Bi-Sr-Ca-Cu-O systems, suggesting that some traps exist in these systems. The glow curves significantly depend on composition, preparation conditions and cumulative x-ray irradiation time.

Introduction

Since the discovery of superconductivity in YBa_2Cu_3O [1], the transition temperature T_c being above liquid-nitrogen temperature, much effort has been paid to the field of physics and applications. Main subjects to be challenged in this field are as follows; search of other new oxide superconductors with an even higher T_c, elucidation of superconducting mechanism, fabrication of high-quality wire and thin film, proposal of new-concept devices and so on.

Recently, we have reported the first observation of radiant PL from the Nd- and Er-systems [2], which was never obtained in conventional metal superconductors. Tissue and Wright have also succeeded to detect visible PL of some Re ions, Pr^{3+} and Eu^{3+}, incorporated substitutionally into the $La_{1.85}Sr_{0.15}CuO_4$ ceramic compound. [3] Cooke et al have observed several TSL glow peaks in the 80 - 275 K temperature range in single-phase $GdBa_2Cu_3O$ and two-phase $Re_{1.5}Ba_{1.5}Cu_2O$ (Re = Ho, Eu) samples. [4] Successive luminescence studies [5-8] give us the following two feasibilities. One is to use the luminescence signal as a characterization probe for these high T_c superconductors. The other is to develop new devices with both superconducting and light-emitting functions. Pawar et al have reported the phenomenon of electroluminescence (EL) under the application of electric field in high T_c Y-Ba-Cu-Zr-O superconductors. [9]

In the present paper, preliminary results of PL and TSL measurements in several kinds of high T_c superconductors will be described and discussed.

Experimental

The samples studied in this work were prepared by conventional solid-state reaction technique using appropriate mixture of standard reagents.

In PL measurements, a cw Ar^+ laser operating at 514.5 nm was utilized as a photoexcitation source. The luminescence from a sample was analyzed by a grating monochromator and detected by a Ge p-i-n photodiode or a photomultiplier according to the wavelength region. The output was fed to a lock-in amplifier and processed by a computer-controlled signal-averaging system.

Prior to TSL measurements, a sample was mounted on a Cu cold finger of a cryostat and cooled conductively down to liquid-nitrogen temperature. In the dark, it was irradiated by an x-ray from a Cu target equipped in a commercially available x-ray diffraction machine. Then the sample temperature was raised in a controllable manner by a neighboring heater and monitored by a thermocouple. The typical heating rate was about 0.1 K/sec. The luminescence signal was detected by a photomultiplier with sensitivity for visible light and processed by a conventional lock-in system.

Results and Discussion

PL Results

In the Er-Ba-Cu-O system with all kinds of composition including well-known 1-2-3 composition, characteristic luminescence has been observed in the 0.8 eV region, consisting of a series of sharp emission lines. [2] They were tentatively assigned to electronic transitions between crystal-field split spin-orbit levels of Er^{3+} ions. Typical spectral line-width was from 1.5 meV to 2.0 meV. Similar luminescences concerning Nd^{3+} ions have been also observed in the Nd-Ba-Cu-O system in three different energy regions; 0.91 eV, 1.15 eV and 1.35 eV. Details of the related spin-orbit levels were described in another paper [2].

The PL spectrum is strongly sensitive to the local environment of Re ions. As described previously [2], for example, the intensity of the higher-energy PL line increases with the decrease of Er composition, x, in $Er_xBa_{1-x}CuO$ compounds. It can be explained by the population change of 4f electrons belonging to the ground and excited levels. Furthermore, the energy position and the spectral shape of the Er- and Nd-related luminescence in the Er- and Nd-Ba-Cu-O systems are significantly different from those in the starting sources, Er_2O_3 and Nd_2O_3. [5,6] It is quite acceptable that these observations are due to a difference of a microscopic crystal arrangement surrounding Re ions.

Such a luminescence has never been observed in the Eu- and Yb-Ba-Cu-O systems, though Eu_2O_3 and Yb_2O_3 exhibit radiant luminescence under the same measurement conditions, which might reflect a difference of their band structures and be due to a resonance between the upper band and the excited level. Similar results were obtained in YBa_2Cu_3O with 10 % and 100 % of Eu. [10]

The relationship between the superconducting and light-emitting properties is not clear at present. Recently, Andreev et al have observed a cathodeluminescence (CL) initiated by a superconducting transition in $Y_{1.2}Ba_{0.8}CuO$. [11] In addition, Luff et al have reported that the temperature dependence of the CL intensity reveals a number of dips which are related to anomalies in resistivity and to specific heat data. [12]

TSL Results

TSL measurements have been performed in some kinds of the Re-Ba-Cu-O (Re = Nd, Eu, Gd, Ho, Er), Tl-Ba-Ca-Cu-O and Bi-Sr-Ca-Cu-O systems. Characteristic TSL signals have been successfully observed in all the systems studied in the present work, suggesting that some traps are included in these systems. Figure 1 shows TSL glow curves obtained in $Gd_xBa_{1-x}CuO$ with Gd composition of 0.3, 0.33 and 0.4 prepared under the same condition. It should be noticed that the obtained glow curves strongly depend on Gd composition. Although Cooke et al reported TSL results in YBa_2Cu_3O [4], their glow curve is not coincide with ours. The reason will be discussed later.

The TSL glow curve obtained in TlBaCaCuO with initial ratio of Tl:Ba:Ca:Cu = 2:2:2:3 (sample A), exhibiting superconducting transition at 97 K, is shown in Fig. 2. The glow curve consists of four obvious peaks from peak A to peak D in order of the lowest temperature. A similar glow curve was also obtained in TlBaCaCuO with initial ratio of Tl:Ba:Ca:Cu = 3:2:2:3 (sample B), though the intensity was about 35 times lower than that of the former sample. [7,8] Taking it into consideration that a volume fraction of the superconducting phase in sample B is about 10 times smaller than that in sample A, the TSL signal might originate from the superconducting phase.

These glow curves were analyzed by a well-known general-order kinetic equation [5] describing the TSL intensity as a function of temperature. The equation includes three variables which characterize a related trap; thermally activated energy E, frequency factor s and kinetic order l. These variables were estimated by a least squares fit of the above-mentioned equation onto the experimentally obtained glow curves. The features of the trap related to each glow peak depicted in Figs. 1 and 2 are listed in Tab. 1. A parenthesis indicates a value estimated by the fit onto an initial-

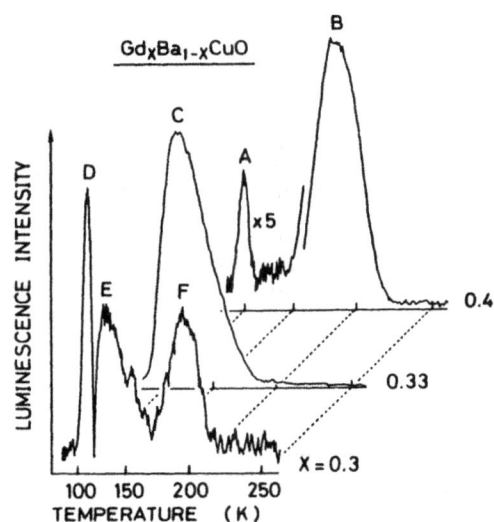

Fig. 1   TSL glow curves of $Gd_xBa_{1-x}CuO$ with x of 0.3, 0.33 and 0.4.

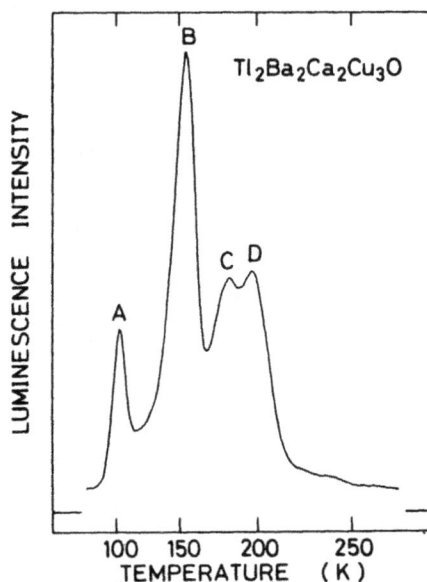

Fig. 2   TSL glow curve of $Tl_2Ba_2Ca_2Cu_3O$.

Table 1   Characteristics of traps related to glow peaks shown
in Figs. 1 and 2.

material	peak	E (eV)	s (sec^{-1})	1
$Gd_xBa_{1-x}CuO$ (x = 0.4)	A	0.11	1.42 x 10^3	1.21
	B	0.19	5.13 x 10^2	1.21
(x = 0.33)	C	(0.08)	(1.17 x 10)	(1.91)
(x = 0.3)	D	0.24	1.25 x 10^9	1.01
	E	(0.19)	(1.42 x 10^5)	(1.91)
	F	0.30	3.08 x 10^5	1.21
$Tl_2Ba_2Ca_2Cu_3O$	A	0.12	1.12 x 10^4	1.21
	B	0.24	6.75 x 10^5	1.11
	C	–	–	–
	D	0.27	5.91 x 10^4	1.11

rise part of the glow curve, because the glow peak could not be completely
fitted with an assumption of a single peak. On the other hand, since only
peak Ccould never be picked up using optical filters in the Tl sample, the
analysis for peak C was not successfully performed. As can be seen in Tab.
1, the activation energy of the related traps ranges from 0.1 eV to 0.3 eV
regardless of the system, suggesting that chemical structures of the traps
are alike.

The wavelength region of TSL signals gives us some information about
the luminescence mechanism, e.g. position of the recombination center and
recombination path. Figure 3 shows preliminary results concerning the
dependence of the wavelength on the intensity of peak B normalized by that
of peak A observed in $Gd_{0.4}Ba_{0.6}CuO$, which was measured by using band-pass
and high-pass filters.   An error bar displays the wavelength at the
transmittance of 0 % and 50 % against the peak transmittance of the used
filter. Since the intensity of peak B measured by a high-pass filter with a
cut-off wavelength of 0.65 μm was lower than that with a cut-off wavelength
of 0.58 μm, it can be roughly estimated that the peak wavelength of peaks A
and B exists in the visible region from 0.4 μm to 0.6 μm. The normalized
intensity is dependent on the wavelength, indicating that the spectral
region differs from each other. This observation might reflect a difference
in recombination processes and/or multiphase properties. A similar
investigation was performed for the Tl sample. [7,8] As a result, we have
found that the peak wavelength of the glow peaks ranges from 0.3 μm to 0.6
μm, and that their wavelength regions differ from one another, and those of
peaks B, C, A and D are arranged in order of the shortest wavelength.

The obtained TSL glow curves significantly depend on sample preparation
conditions [5] and cumulative x-ray irradiation time [5-8]. As for the
cumulative x-ray irradiation time dependence of the glow curve, two kinds of
dependences are observed. One is that the irradiation time influences only
the intensities of glow peaks. The other is that a new peak appears and

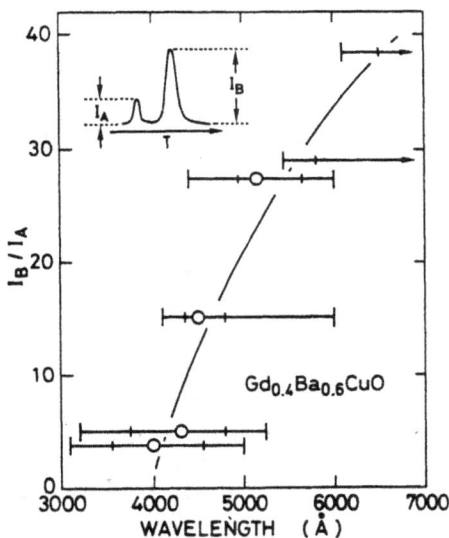

Fig. 3 Wavelength dependence of the peak intensity of glow peak B obtained in $Gd_{0.4}Ba_{0.6}CuO$, which is normalized by that of glow peak A.

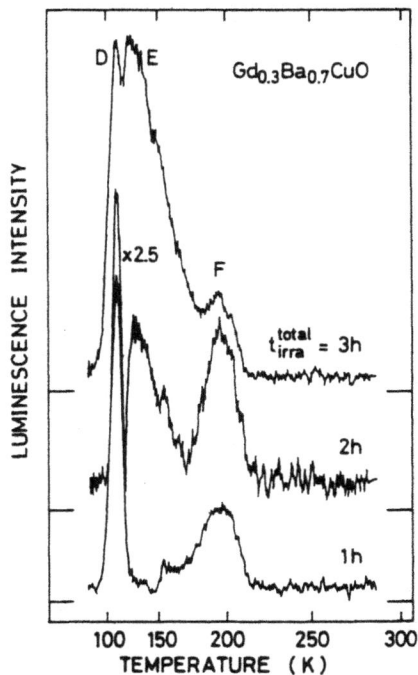

Fig. 4 Cumulative x-ray irradiation time dependence of TSL glow curves obtained in $Gd_{0.3}Ba_{0.7}CuO$.

grows with the increase in irradiation time. The typical example of the latter case is shown in Fig. 4. The new peak is, however, very unstable. If we preserve the long irradiated sample for about a half day at room temperature, the new glow peak almost completely disappears and displays abehavior similar to the above-mentioned unique one for successive x-ray irradiation. This suggests that though the trap related to the new glow peak is created by x-ray irradiation, it is metastable and is relaxed by the room temperature annealing.

Conclusion

High T_c superconductors have been systematically investigated by PL and TSL techniques. In PL measurements, characteristic luminescences have been observed in the infrared region in the Nd- and Er-systems. Their energy positions and spectral shapes are significantly sensitive to the local environment around Re ions. Such a luminescence has never been observed in the Eu- and Yb-systems at present. In TSL measurements, characteristic

278

visible luminescences have been successfully obtained in some kinds of Re-
Ba-Cu-O (Re = Nd, Eu, Gd, Ho, Er), Tl-Ba-Ca-Cu-O and Bi-Sr-Ca-Cu-O systems,
indicating that some traps exists in these systems. The thermally activated
energies of related traps range from 0.1 eV to 0.3 eV regardless of the
system. TSL glow curves strongly depend on composition, preparation
conditions and cumulative x-ray irradiation time.

References

1) M. K. Wu, J. R. Ashburn, C. J. Torng, P. H. Hor, R. L. Meng, L. Gao, Z.
J. Huang, Y. Q. Wang, and C. W. Chu, "Superconductivity at 93 K in a
mixed-phase Y-Ba-Cu-O compound system", Phys. Rev. Lett., 58, 9, 908-910
(1987).
2) Y. Fujiwara, T. Takahashi, Y. Fukumoto, M. Tonouchi, and T. Kobayashi,
"Radiant Er-related luminescence of high T_c superconducting Er-Ba-Cu-O
system", Proc. 18th Int. Conf. on Low Temperature Physics (Kyoto, Japan),
2123-2124 (1987).
3) B. M. Tissue and J. C. Wright, "Laser spectroscopy as a probe of
structural changes in Eu doped $La_{2-x}Sr_xCuO_4$", J. Luminescence, 40&41,313
-314 (1988).
4) D. W. Cooke, H. Rempp, Z. Fisk, and J. L. Smith, "Thermally
stimulated luminescence from rare-earth-doped barium copper oxides",
Phys. Rev. Lett., 36, 4, 2287-2289 (1987).
5) Y. Fujiwara, T. Nishino, and T. Kobayashi, "Characterization of LnBaCuO
(Ln = Er, Eu, Gd...) ceramics by photo- and thermally stimulated
luminescences", Proc. 1988 MRS Int. Meet. on Advanced Materials (Tokyo,
Japan), (1988). (in press)
6) Y. Fujiwara and T. Kobayashi, "Characterization of high T_c superconductor by luminescence methods", Proc. 1988 Applied Superconductivity
Conf. (San Francisco, U.S.A.), (1988). (in press)
7) Y. Fujiwara, M. Tonouchi, and T. Kobayashi, "Luminescence study of new
high T_c oxide superconductors", Proc. 1st Int. Symp. Superconductivity
(Nagoya, Japan), (1988). (in press)
8) Y. Fujiwara, M. Tonouchi, and T. Kobayashi, "Thermally stimulated luminescence from high T_c superconducting Tl-Ba-Ca-Cu-O system", Jpn. J.
Appl. Phys., 27, 9, L1706-L1708 (1988).
9) S. H. Pawar, H. T. Lokhande, C. D. Lokhande, R. N. Patil, B. Jayaram, S.
K. Agarwal, A. Gupta, and A. V. Narlikar, Solid State Commun., 67, 1,
47-49 (1988).
10) B. M. Tissue and J. C. Wright, "Observation of sharp-line lanthanide
fluorescence in high temperature superconductors", J. Luminescence, 37,
117-121 (1987).
11) V. N. Andreev, B. P. Zakharchenya, S. E. Nikitin, F. A. Chudnovskii, E.
B. Shadrin, and E. M. Sher, "Cathodeluminescence in a high-temperature
superconductor Y-Ba-Cu-O", JEPT Lett., 46, 10, 492-495 (1987).
12) B. J. Luff, P. D. Townsend, and J. Osborne, "Cathodeluminescence
measurements of high T_c superconductors", J. Phys. D: Appl. Phys., 21,
663-665 (1988).

OBSERVATION OF THE HIGH-Tc PHASE AND DETERMINATION OF THE Pb POSITION
In Bi-Pb-Sr-Ca-Cu OXIDE SUPERCONDUCTOR.

Hitoshi NOBUMASA,Takahisa ARIMA,Kazuharu SHIMIZU,
Yuji OTSUKA[+],Yukio MURATA[+]
and Tomoji KAWAI[++]

Composite Material Lab.,Toray Ind.Inc.
2-1,Sonoyama 3-chome,Otsu,Shiga,520,Japan
[+]Toray Research Center Inc.
1-1,Sonoyama 1-chome,Otsu,Shiga,520,Japan
[++]The Institute of Scientific and Industrial Research,
Osaka University,Ibaraki,Osaka,567,Japan

Abstract - The high-Tc phase of Bi-Pb-Sr-Ca-Cu oxide superconductor was observed directly by high-resolution transmission electron microscopy (HREM), and the atomic positions of Pb in the crystal were determined by high-resolution analytical electron microscopy(HRAEM) with high spatial resolution 17 Å using 5 to 10 Å probe diameter. The HREM observation revealed that the crystal consists only of the triple Cu-O layered structure with $c/2=18$ Å without any intergrowth and that the crystal structure is modulated along the b-axis. The HRAEM indicated that the Pb atoms were located in the Bi-O layers with an atomic ratio of $Pb/Bi \approx 0.1$.

Introduction

Since the Bi-Sr-Ca-Cu oxide superconductor was found by Maeda et al.[1], many investigations on this material by HREM have been reported[2-12]. This material always has intergrowths of triple Cu-O layers and single Cu-O layers among double Cu-O layers.
Recently, high-Tc single phase was obtained by partial Pb substitution[13, 14], and the following two points are the most important matters of concern at the present stage. The first one is whether the crystal structure of the Pb substituted sample is same as that of the Pb free one, and the second one is where the Pb atoms are located in the crystal structure.
We have made a Bi-Pb-Sr-Ca-Cu oxide superconducting film which mainly consisted of the high-Tc phase[15], and observed the crystal structure by HREM and analyzed the positions of Pb atoms by HRAEM. The crystal consists only of the triple Cu-O layers with $c/2=18$ Å without intergrowth, and the Pb atoms are definitely located in the Bi-O layers with an atomic ratio of Pb to Bi to be 0.1.

Experimental

The Bi-Pb-Sr-Ca-Cu oxide superconducting film was prepared on a MgO single crystal (100) by a spray pyrolysis method[15]. The X-ray diffraction pattern showed that this film mainly consisted of the high-Tc phase with a c-axis length of 18 Å, and orientated with the c-axis perpendicular to the MgO(100) surface. Tc_{zero} of this film was 103.5 K.
For both HREM observation and HRAEM , ultrathin sections were made by using an ultramicrotome (LKB Ultrotome V, LKB Produkter, Bromma, Sweden) equipped with a diamond knife. The thin sections were placed on a holey carbon, and electron microscopic observation was carried out by Hitachi

H-800UHR at 200kV. The HRAEM was also carried out using the dedicated STEM (VG HB501) with a UTW energy dispersive detector (Kevex3400-0383). This STEM was equipped with a field emission gun(FEG) for high current at specimen plane more than 1nA. Therefore large amount of X-ray for elements was excited from very small area less than 10 Å diameter.

The analysis was carried out at two positions in the layered structure, i.e. at the Bi-O layer and at the triple Cu-O layers by fitting the probe to the center of each position. Hereafter we express the former as "position-A", and the latter as "position-B". Fig.1 (a) and (b) show schematicaly the analytical positions by using the structure model of the high-Tc phase proposed by J.M.Tarascon et al.[16]. The distances between atoms refered to the structure image which was reported by Matsui et al.[12].

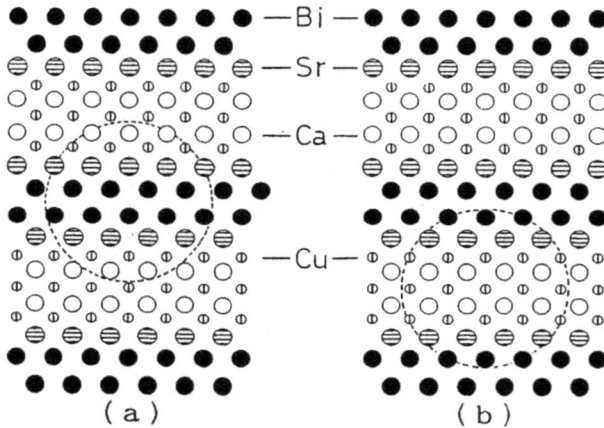

Fig.1 Schematic representations of the probe positions of HRAEM (dashed circle line) on the structure model of the high-Tc phase in which oxygen atoms are omitted.
(a) "position-A"   (b) "position-B"

The minimum probe size of HB501 is in the range of 5 to 10 Å as above mentioned, but the spatial resolution on this particular sample was somewhat larger in the present measurement, about 17 Å as judged from the comparison between the peak intensities[Ca(K_α),Bi(L_β)] for "position-A" and "position-B". Two reasons for the expansion of the spatial resolution may be given. One is that the material is not so stable under the intense electron beam subjected to present analysis that some diffusion occured presumably during electron irradiation. The other may be that ultrathin section placed on the holey carbon were not rigid enough to prevent the specimen drift at high magnification more than 10^7 times. Under these circumstances, high spatial analysis was carried out by following procedures taking account of specimen drift and damage, (1) acquisition time for a given layer - for example "position-A" - was 5 seconds at each time, (2) after that, electron probe was moved electrically to other point at the given layer, and the X-ray signals for 40 times at the given layer were integrated. Therefore total acquisition time was 200 seconds at both positions, A and B. And also the electron dose to specimen was limited as little as possible in order to minimize the specimen damage. By monitering STEM image, it was confirmed that the serious specimen damage was prevented to the minimum. Using these techniques, we could successfully determine the position of Pb in the Bi-Pb-Sr-Ca-Cu-O, as discussed later.

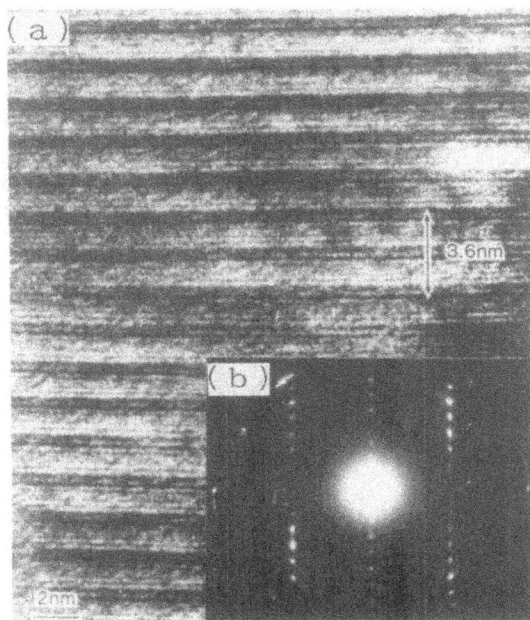

Fig.2
High-resolution structure image for the a-c planes (a) and an electron diffraction pattern (b) of the high-Tc phase in the Bi-Pb-Sr-Ca-Cu oxide superconductor.

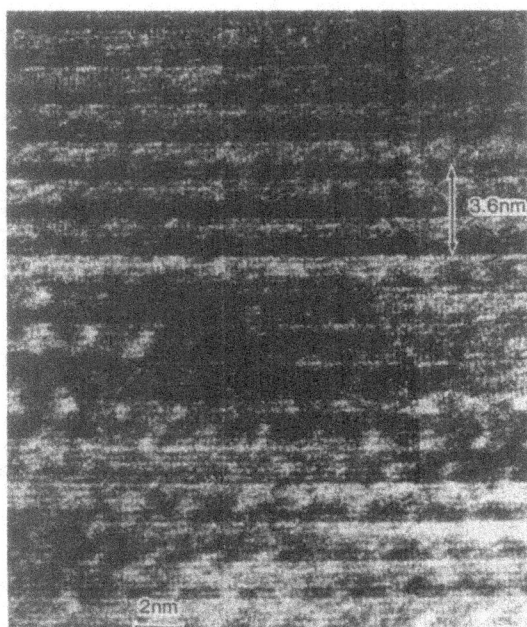

Fig.3
High-resolution structure image for the b-c planes of the high-Tc phase in the Bi-Pb-Sr-Ca-Cu oxide superconductor.

Results and discussions

Fig.2(a) and (b) show the electron micrograph and an electron diffraction pattern for the a-c plane, and also Fig.3 shows the electron micrograph for the b-c plane. It is very obvious from these micrograph that the crystal consists only of the triple Cu-O layered structure between Bi_2O_2 layers with $c/2=18$ Å. There are intergrowth regions in the Pb free Bi-Sr-Ca-Cu oxide superconductor, as indicated by the broadening $2\theta=4.8°$ peak in the X-ray diffraction pattern[16]. For our sample, the $2\theta=4.8°$ peak was very sharp in agreement with TEM results[15]. Therefore it is found not only from X-ray diffraction pattern but also from TEM observation that Pb stabilizes the high-Tc phase.

A modulated structure along the b-axis was also observed with a periodicity of about 24 Å, and these results are the same as in the case of the Pb free sample. However, in some areas, the contrast change based on the modulation structure was not so clear. We think that the modulation structure depends on the Pb content in the material which decreases with longer sintering time. The atomic ratio of Pb to Bi in our sample was 1:9 as mentioned later.

Fig4 (a) and (b) show the results of the HRAEM on the two positions, A and B in Fig.1. The vertical axis is normalized by the intensity of $Ca(K_\alpha)$ signal. With comparison between (a)and (b), it is clear that the Bi signal intensity at "position-A" was much higher than that at "position-B". It was therefore confirmed that the two positions were discriminated and analyzed separately. The Pb to Bi atomic ratios at "position-A" and "position-B" were determined to be 1:9 and 1:8 respectively. Even though there was some cross talk which was due to the expanded spatial resolution, we could make the following discussion. If Pb atoms would be located in a Sr site, a Ca site or a Cu site, the atomic ratio of Pb to Bi at "position-B" would be much larger than that at "position-A" as shown in Fig.4(b) by a dashed line which was estimated assuming uniform Pb distribution, because the contribution of Pb in Cu-O layers are superimposed. The experiment showed that although the Bi and Pb signal intensities at the "position-A" are much higher than at "position-B" ,the Pb to Bi atomic ratio is about the same at both positions. Therefore, we can conclude that Pb atoms are located in the Bi-O layers. Fig.5 shows the result of the HRAEM obtained by using STEM mode when the analytical area was 120 Å × 90 Å. The peak intensity of $Bi(L_\beta)$ normalized by Ca takes the middle position between those of "position-A" and "position-B", and the ratio of Pb to Bi is 1:9. The chemical composition of this film was analyzed totally by an inductively coupled plasma atomic emission spectrometry (ICP). The atomic ratio of Pb to Bi was found to be 1:9, which is consistent with HRAEM which Bi and Pb behaved in the same manner.

Lastly, we would like to discuss the effect of the Pb substitution in Bi-O layers. As for the copper-oxide superconductor, there may be a consensus that the carriers are holes, for instance, in the case of $(La_{1-x}Sr_x)_2CuO_4$ or $YBa_2Cu_3O_{6.5+y}$. In the former case, Sr is a donor, and in the latter case, the excess oxygen is a donor. In the structure models of Bi compound superconductor proposed by Tarascon et al., however, valencies are completely balanced. Although in the case of Bi-Pb-Sr-Ca-Cu-O compound Pb may be a hole donor, because Pb is divalent and Bi is trivalent, in the case of Pb free sample the high-Tc phase also appears. Then we should study other possibility for the hole donor.

On further examination using the HRAEM, it seems that Bi atoms are also replaced by Sr atoms. In the case of that the spatial resolution of the HRAEM was about 17 A, the EDX signal of Sr at the "position-A" should be like the profile shown by dashed line in Fig.6, calculated from the EDX signal of Bi and Sr at the "position-B" (Fig.7), assuming the structure model suggested by

Tarascon et al. is correct. But in the experimental data, the Sr signal at the "position-A" is larger, so we think some of the Sr atoms are located in the Bi-O layers. The Sr atoms must be divalent, so Sr might be a hole donor. In that case, the model of Tarascon is a parent material, and the Sr substituted structure is a real superconductor.

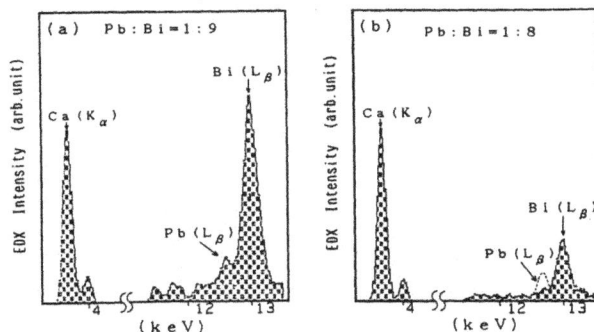

Fig.4
The spectra of the HRAEM on the two positions.
(a) "position-A"
(b) "position-B"

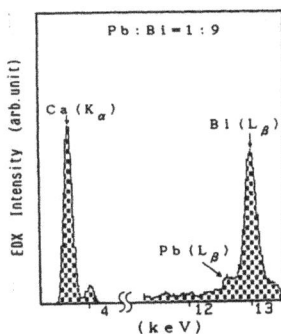

Fig.5
The spectra of the HRAEM when the analytical area was 120 Å X 90 Å.

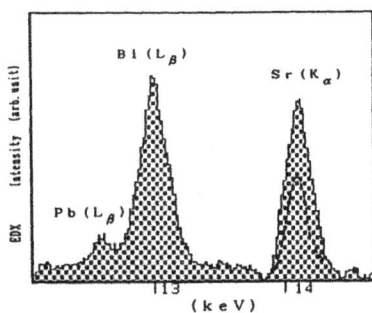

Fig.6
The spectra of the HRAEM on the "position-A".

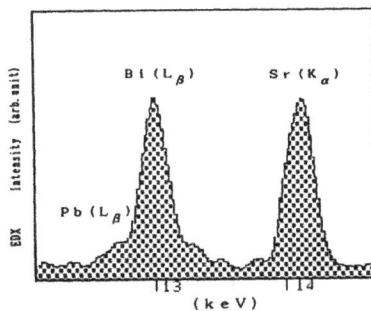

Fig.7
The spectra of the HRAEM on the "position-B".

Conclusions

The HREM observation of Bi-Pb-Sr-Ca-Cu oxidesuperconductor revealed that the crystal consists only of the triple Cu-O layered structure. Any intergrowth was not observed for this compound. The sample (Pb:Bi=1:9) showed a modulated structure along the b-axis with a periodicity of 24 A . Furthermore, the HRAEM carried out for "position-A" and "position-B" clearly indicated that Pb and Sr atoms are located in the Bi-O layers. This result indicates that these atoms might be a hole donor in this compound superconductor.

References

1) H.Maeda,Y.Tanaka,M.Fukutomi and T.Asano : Jpn. J. Appl. Phys. 27 (1988) L209.
2) Y.Bando,T.Kijima,Y.Kitami,J.Tanaka,F.Izumi and M.Yokoyama : Jpn. J. Appl. Phys. 27 (1988) L358.
3) Y.Matsui,H.Maeda,Y.Tanaka and S.Horiuchi : Jpn. J. Appl. Phys. 27 (1988) L361.
4) Y.Matsui,H.Maeda,Y.Tanaka and S.Horiuchi : Jpn. J. Appl. Phys. 27 (1988) L372.
5) Y.Syono, K.Hiraga, N.Kobayashi, M.Kikuchi, K.Kusaba, T.Kajitani, D.Shindo, S.Hosoya, A.Tokiwa, S.Terada and Y.Muto : Jpn. J. Appl. Phys. 27 (1988) L569.
6) K.Hiraga,M.Hirabayashi,M.Kikuchi and Y.Syono : Jpn. J. Appl. Phys. 27 (1988) L573.
7) N.Kijima,H.Endo,J.Tsuchiya,A.Sumiyama,M.Mizuno and Y.Oguri : Jpn. J. Appl. Phys. 27 (1988) L821.
8) Y.Matsui,H.Maeda,Y.Tanaka,E.Muromachi,S.Takekawa and S.Horiuchi : Jpn. J. Appl. Phys. 27 (1988) L827.
9) S. Ikeda, H.Ichinose,T. Kimura,T. Matsumoto, H.Maeda,Y.Ishida and K.Ogawa : Jpn. J. Appl. Phys. 27 (1988) L999.
10) D.Shindo,H.Hiraga,M.Hirabayashi,M.Kikuchi and Y.Syono : Jpn. J. Appl. Phys. 27 (1988) L1018.
11) S.Horiuchi,H.Maeda,Y.Tanaka and Y.Matsui : Jpn. J. Appl. Phys. 27 (1988) L1172.
12) Y.Matsui,S.Takekawa,H.Nozaki,A.Umezono,E.Muromachi and S.Horiuchi : Jpn. J. Appl. Phys. 27 (1988) L1242.
13) U.Endo,S.Koyama and T.Kawai : Jpn. J. Appl. Phys. 27 (1988) L1476.
14) S.Koyama,U.Endo and T.Kawai : Jpn. J. Appl. Phys. 27, No.10 (1988).
15) H.Nobumasa,K.Shimizu,Y.Kitano and T.Kawai : Jpn. J. Appl. Phys. 27 (1988) L1669.
16) J.M.Tarascon, W.R.McKinnon, P.Barboux, D.M.Hwaang, B.G.Bagley, L.H.Green, G.Hull,Y.LePage,N.Stoffel and M.Giroud : to be published in Phys. Rev. B.

THE STRESS-STRAIN RELATIONSHIP FOR MULTILAYERS
OF THE HIGH Tc SUPERCONDUCTING OXIDES

Hiroaki Hidaka & Hiroshi Yamamura
TOSOH CORPORATION, TOKYO RESERCH CENTER
2743-1, HAYAKAWA AYASE KANAGAWA, JAPAN 252

ABSTRACT

The calculation of the stress-strain relationship for multilayers of the high Tc superconducting oxides was performed with regard to a wide variety. The elucidation of this relationship is expected quite helpful for the preparation of high-quality multilayers of these materials. This calculation is possible to do in the same way of Timoshenko's bi-metal treatment. We did for the first time computation of the residual stress and strain, and the state of stress and strain for these multilayers has been acquired in detail by this calculation.

INTRODUCTION

The investigational work for practical application of high critical temperature superconducting materials has been active recently. Paticularly, device application of thin films of these materials has been tried by many workers[1,2,3]. The application to SQUID, three-terminal device or optical detector has been reported. The preparation of high-quality multilayers using high Tc superconducting oxides is very important for these applications. However, it is well known that these multilayers are highly stressed mostly due to the large discrepancy of the thermal expanstion coefficients between the substrate material and grown layers[4]. These stress and strain impair physical properties such as critical temperature and, occasionally, induce cracking inside the film. Therefore, the elucidation of stress-strain relationship is expected quite helpful for the preparation of high-quality multilayers using superconducting oxides.
In this work, we did for the first time the calculation of the residual stress and strain with regard to a wide variety of multilayers. This calculation is possible to do in the same way of Timoshenko's bi-metal treatment[6]. The stress-strain relationship for these maltilayrs were revealed and occasinally, that in a case of using silicon substrate with buffer layer, which is very significant from a practical standpoint such as three terminal device, were done in detail.

COMPUTAITIONAL METHOD

The calculation of the stress-strain relationship for epitaxial multilayers is able to be .performed by the same way of Timoshenko's bi-metal treatment[5]. That way was briefly introduced before in previous paper[5]. First of all, it is supposed that multilayers overspread infinitely and the influence of gravity is ignored. In this case, the state of stress and strain is exactly alike at each point around the plane of the film. The thickness of the layer k is here indicated as a_k, and the radious of curvature of that is done as ρ . Equations which shall be solved are set up following.

$$\sum_{k=1}^{n} P_k = 0 \tag{1}$$

$$(1/\rho)\cdot\sum_{k=1}^{n}(I_k/S_k) = \sum_{k=1}^{n}(z_{k-1} + a_k/2)\cdot P_k \tag{2}$$

$$(S_k/a_k)\cdot P_k + \alpha_k\Delta T_k + a_k/(2\rho) = (S_{k-1}/a_{k-1})\cdot P_{k-1} + \alpha_{k-1}\Delta T_{k-1} - a_{k-1}/(2\rho) \tag{3}$$

$$(2 \leqq k \leqq n)$$

In these equations, P_k is the force which acts in the layer k, α_k is the coefficient of thermal expantion of the layer k. ΔT_k is the deviation in temperature for the film preparation. I_k is the moment of inertia;$a_k^3/12$. S_k is the elastic compliance ;$(1-\nu(k))/E(k)$ ($\nu(k)$:Poisson's ratio, $E(k)$:Young's modulus). That is to say, a set of these equations is the eigenvalue problem in (n+1) dimensions. Solving this problem, forces; P1,P2,...,Pn and the curvature; ρ are aquired with regard to a wide variety of multilayers. According to these values, stresses such as that in undersurface of the layer 1 and strains are calculated. These calculations are performed in practice using a personal computer by the way of numerical program such as Gaussian method.

RESULTS AND DISCUSSION

$Y_1Ba_2Cu_3O_{7-x}$ thin film on various substrates

For preparation of superconducting oxide thin film, various materials have been proposed as substrates by many workers[4]. For example, yttria stabilized zilconia (YSZ), strontium titanate, magunesium oxide or glass has been done up to the present. The physical properties of these materials are shown in table 1[4,7,8,9]. The coefficient of thermal expantion of substrate material is differrent from that of $Y_1Ba_2Cu_3O_{7-x}$ (YBCO) as shown in this table. This difference is a problem for high temperature preparation of superconducting oxide thin film because of occurrence of residual stress and strain in that film. Figure 1 shows stress profile which is obtained from the calculation explaned above. In this treatment, the thickness of substrate is 1 mm, and that of thin film YBCO is 1 μm, and the preparation temperature (T_s) is 700°C. Values of stress and strain are calculated for room and liquid nitrogen temperatures(TOs) respectively. In the case of a

Table 1. Physical Properties of Materials

Material	α $(\times 10^{-6})$	E $(\times 10^4 MPa)$	ν
YBCO	16.9	10.4	0.35
MgO	13.8	24.9	0.18
$SrTiO_3$	11.1	30.3	0.23
YSZ	10.3	20.0	0.25
CaF_2	24	7.6	0.28
Si	2.4	13.1	0.22
Pyrex Glass	3	1.17	0.20

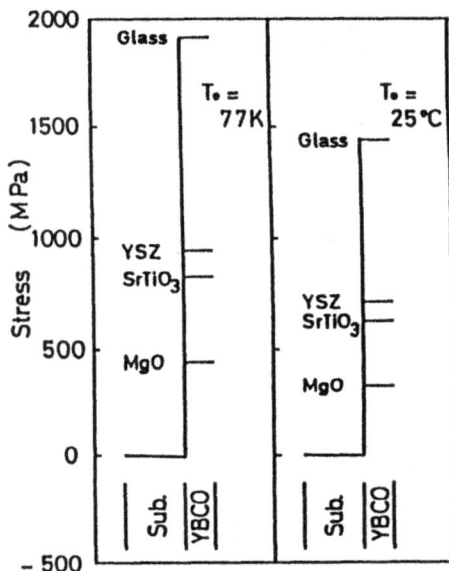

Figure 1. Calculated stresses of YBCO/Substrate structures.

pyrex glass for substrate, large tensile stress, 1910 MPa occurs in YBCO thin film at 77 K, supposing that the deformation is completely elastic between T_s and T_0. Originating in that large stress, this thin film spreads as 1.2% around the plane. This tensile stress gives rise to crack formation or deterioration of physical properties of the film such as a critical temperature(T_c). The rupture strength, in reference, for sapphire single crystal is approximately 500 or 1500 MPa. On the other hand, in cases of YSZ, SrTiO3 and MgO for substrate, values of the residual stress are 944 MPa, 830 MPa and 443 MPa respectively, and that of strains are 0.6%, 0.5% and 0.3% respectively. In cases mentioned here, tensile stresses occur in thin films, however, those values are smaller than that in a case of pyrex glass. Therefore, the possibility of crack formation or property deterioration is supposed to be comparatively small. According to these results and the report of Koinuma et al.[4], the upper critical stress for practical use of YBCO thin film is estimated to be 1000 or 1500 MPa. That value is reasonable in comparison with the rupture strength of sapphire.

Multilayers on silicon substrate

We have done a trial of usage of silicon wafer for substrate material. We reported in previous paper[5] that the superconducting film is quite fine on silicon substrate with MgO buffer layer from the mesurement of depth profiles of Auger electron spectroscopy(AES). Figure 2 shows the dependence of preparation temperature at the calculation on residual stress of YBCO thin film at 77K. The solid line and the dashed and dotted line in this figure are cases of YBCO thin film on Si substrate with and without MgO buffer layer respectively. Si substrate is 500 um thick and, YBCO and MgO films are 1μm thick. The decrease in residual stress when forming a buffer layer is very small. And, according to the critical stress 1000 or 1500 MPa for cracking of thin film, the preparation temperature shoud be appropriate to be 400°C or less for forming these thin films well. Figure 3 shows the dependence of buffer layer thickness on the residual stress of YBCO thin film prepared on Si substrate at 77K. The decrease in that stress when increasing the thickness of that layer from 0.1 μm to 1 μm is small as shown in this figure.

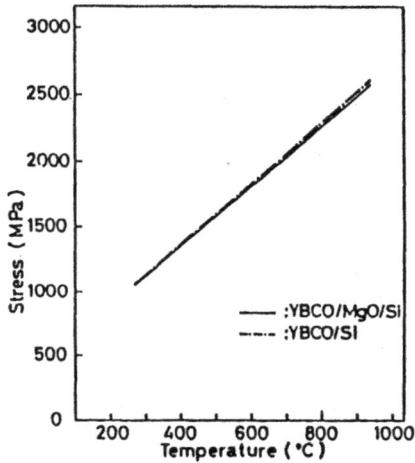

Figure 2. Dependence of YBCO deposition temperature on that stress.

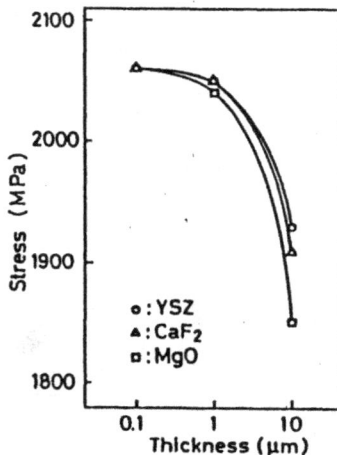

Figure 3. Dependence of buffer layer thickness on the residual stress.

In order to release that stress by the way of forming buffer layer, it shall be necessary to thicken the layer for 10 μm or more. And, the dependence of coefficient of thermal expantion and Young's modulus of buffer layer on the residual stress of YBCO layer at 77K were studied by the calculation. Si substrate is 500 um thick and YBCO and buffer layers are 1 μm thick. The stress of YBCO layer decreases as the coefficient of thermal expantion and the Young's modulus decrease. However, for the perpose of releasing the residual stress, following that result, that values are necessary to be considerably large.

Figure 4 shows the profile of residual stress of multilayer which is constructed with YBCO/MgO/SrTiO$_3$/Si at 77K. Thicknesses of Si substrate and SrTiO$_3$, MgO and YBCO thin films are 500 μm, 1μm, 1μm and 1μm respectively. The residual stress of YBCO thin film is the value of 2018 MPa and this is similar to that of the film without buffer layer. Moreover, buffer layers between YBCO film and Si substrate are very stressed as shown in this figure. Table 2 shows the stress in YBCO thin film in a wide variety of multilayers. That stress scarcely decreases for each combination as shown in this table and the thickness of grown film is too small to release residual stress. That reason may be due to the assumption that the deformation caused by thermal shurinkage is only elastic. Therefore, in practical use, the plastic deformation of buffer layer in the preparation is very important mechanism in order to release residual stresses. This may be, for example, achieved with the adoption of annealing prosess at suitable temperatures. Needless to say, it is very helpful for practical use to lower the preparation temperature or, if possible, to heighten the critical temperature of superconducting thin film.

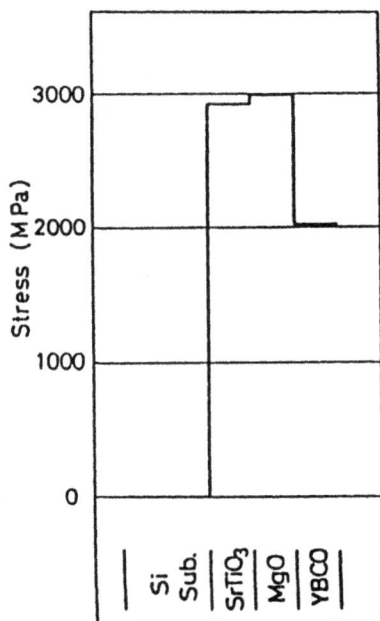

Figure 4. Profile of residual stress of
a multilayer.

Table 2. The stress in YBCO film in a wide
variety of multilayers

Structure	Stress (MPa)
YBCO/MgO/SrTiO3/Si	2018
YBCO/MgO/CaF2/Si	2025
YBCO/CaF2/MgO/Si	2025
YBCO/CaF2/YSZ/Si	2034
YBCO/SrTiO3/Glass/Si	2040
MgO/YBCO/SrTiO3/Si	2018
CaF2/YBCO/YSZ/Si	2034
YSZ/YBCO/MgO/Si	2026

CONCLUSION

We did the computation of the residual stress and strain for oxide superconducting multilayers, and the state of stress and strain for that layers has been acquired in detail. Dependence of substrate material, preparing temperature or film thickness on residual stress was made clear. And, it is suggested that a multilayer prepared on Si substrate is highly stressed and that the intensity of the stress is too large and crack formation or property deterioration are induced.

ACKNOWLEDGEMENTS

The authors would like to thank Professor T.Kobayashi of Faculty of engineering Science, Osaka University for his constant encouragement and usefull suggestion.

REFERENCES

1) H.Tanabe,S.Kita,Y.Yoshizako and T.Kobayashi,Jpn.Jppl.Phys.26,1961(1987).
2) K.Hashimoto,U.Kabasawa,M.Tonouchi and T.Kobayashi, Shingaku-gihou SCE88-25(1988).
3) Y.Enomoto & T.Murakami,J.Appl.Phys.59,3807(1986).
4) T.Hashimoto,K.Fueki,A.Kishi,T.Azumi and H.Koinuma,Jpn.J.Appl.Phys.27 L.214(1988).
5) H.Hidaka & H.Yamamura, Proceeding of ISS'88 in presss.
6) S.Timoshenko,J.O.S.A.&R.S.I.,11 233(1925).
7) S.Kudou,'Bunkougakuteki-seishituwo-shutoshita-kisobusseizuhyou',Kyouritu-Shuppan(1972).
8) J.L.Tallon,A.H.Shuitema & N.E.Tapp,Appl.Phys.Lett.52 507(1988)
9) R.O.Bell & G.Rupprecht,Phys.Rev.129 90(1963).

STRUCTURE STUDY ON Bi-Sr-Ca-Cu-O THIN FILMS PREPARED BY LOW TEMPERATURE PROCESS

Tomoaki MATSUSHIMA*, Kumiko HIROCHI, Hideaki ADACHI,
Kentaro SETSUNE and Kiyotaka WASA

Central Research Laboratories, Matsushita Electric Industrial Co., Ltd.,
Moriguchi, Osaka 570, Japan
*Research and Development Laboratory, Matsushita Electric Works, Ltd.,
Kadoma, Osaka 571, Japan

Abstract - Thin films of Bi-Sr-Ca-Cu-O system were prepared by low temperature process using rf magnetron sputtering. A crystal structure of the as-deposited films was affected quite sensitively by the substrate temperatures between $550°C$ and $600°C$. The film with the high-Tc phase structure was successfully deposited at $570°C$. This film showed a low superconducting transition temperature. After annealing of this film at $570°C$ under an oxygen pressure of 0.1 atm, the high-Tc phase separated partially into Ca-Cu-O phase. The low temperature preparation of Bi-Sr-Ca-Cu-O film with the high-Tc phase structure seems to be difficult.

Introduction

A new superconducting Bi-Sr-Ca-Cu-O system has been discovered by Maeda et al.[1]. This compound was found to have two superconducting phases with different superconducting transition temperatures Tc of 80K (the low-Tc phase) and 110K (the high-Tc phase)[2]. A crystal growth process of the high-Tc phase has not been understood well yet. Recently, thin films of Bi-Sr-Ca-Cu-O system were prepared by several reseachers using sputtering[3,4] and electron beam evaporation[5]. These studies reported that high temperature ($>850°C$) post annealing process was necessary to prepare the films with the high-Tc phase ($c \sim 37Å$). From the view point of the device process, the elimination of high temperature post annealing process is desirable. The preparation of the films by low temperature process is important for fundamental physical studies as well as for device applications.

We have reported earlier about the low temperature process for the prepararion of Ln-Ba-Cu-O films[6]. In this paper, we tried to prepare Bi-Sr-Ca-Cu-O films by low temperature process in the same way as Ln-Ba-Cu-O films. The crystal structures and superconducting properties of these films were studied. The film with the high-Tc phase was prepared at the substrate temperature as low as $570°C$. A change in the crystallinity of the film deposited at $570°C$ was investigated in relation to the post-annealing process at the same temperature of $570°C$ to prevent a crystal growth induced by higher temperature treatment. Moreover, a role of oxygen on the annealing was investigated.

Experimental Procedure

Thin films of Bi-Sr-Ca-Cu-O system were deposited on (100) MgO single crystal substrates by rf magnetron sputtering. The detailed sputtering conditions are listed in Table 1.

Sputtering target was obtained by reacting the mixture of Bi2O3(99.99%), SrCO3(99.99%), CaCO3(99.99%) and CuO(99.9%)at 860 C for 8h in air. The target composition was Bi1.9SrCa1.5-Cu1.5Oy. The substrate temperature was monitored by measuring the surface temperature of a reference

Table 1. Sputtering conditions.

Target	Bi1.9SrCa1.5Cu1.5Oy
Substrate	MgO (100)
Sputtering Gas	Ar/O2=3/2
Gas Pressure	0.4 Pa
RF Input Power	110 W
Growth Rate	100 Å/min

film set on the reverse side of a sample holder using infrared radiation thermometer. After deposition for 30 min., an oxygen gas was introduced into the sputtering chamber and the substrate was quickly cooled down to room temperature for about 30 min.. Thin films of about 0.3 μm thickness were obtained by our deposition. The crystal structure analysis was carried out by X-ray diffractometer using Cu target. The resistivity of the films was measured by a standard dc four-probe method with evaporated Au electrodes. The diamagnetization measurement was done by a rf-SQUID susceptometer in a magnetic field of a few Oersted as residual field. An annealing under higher oxygen pressure was carried out using a conventional electric furnace with a shield stainless steel tube in which the films were enclosed.

Results and Discussion

It is well known that there are three types of crystal on Bi-Sr-Ca-Cu-O system[7];Bi2Sr2Ca2Cu3Oy (the high-Tc phase), Bi2Sr2CaCu2Oy (the low-Tc phase) and Bi2Sr2CuOy. In X-ray diffraction patterns, these crystal structures are characterized by 2θ angle of (002) line; Bragg peaks located at 2θ=4.9 , 5.9 and 7.2 are corresponded to the high-Tc phase, the low-Tc phase and Bi2Sr2CuOy phase, respectively. Figure 1 and Figure 2 show X-ray diffraction patterns of the as-deposited films at different substrate temperatures of 550°C, 570°C and 600°C. Broad and periodic peaks were observed in X-ray diffraction patterns of these films. In the film deposited at 550 °C, as shown in Fig.1 (a), a broad peak at around 2θ=4 was observed. Since this peak is not coincided with the peak of above three phases and the other compounds, this peak is probably due to a new Bi-Sr-Ca-Cu-O phase. A detailed study on this new phase is undertaken now. When the substrate temperature was raised to 570 °C, a broad peak at around 2θ=5 was observed as shown in Fig.1 (b). As mentioned above, the existence of this

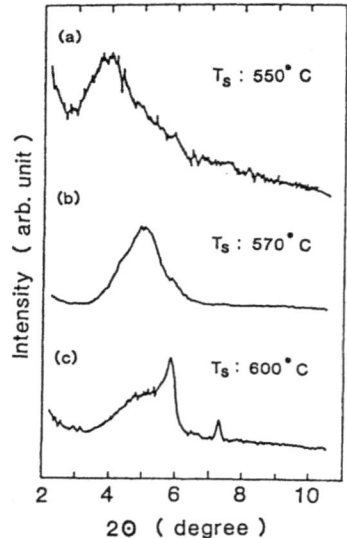

Fig.1. X-ray diffraction patterns (2θ between 2° and 10°) of the as-deposited films.

peak confirms the presence of the high-Tc phase. However this peak suggests an occurrence of stacking faults caused by Bi2O2 layers, because the broad peaks of X-ray diffraction patterns are generally resulted by stacking faults in crystal structure. Although the other peaks located at higher angle were also broad as shown in Fig.2(b), Bragg angles 2θ of these peaks were coincided with those of the high-Tc phase which was obtained by bulk sample [7]. Thus, these peaks were assigned as (00n) lines of the high-Tc phase and the diffraction line of the low-Tc phase or Bi2Sr2CuOy phase was not appreciably observed as shown in Fig.2(b). A peak width of (0012) line was broader than that observed in the film annealed at high temperature (>850 °C) [3]. This broadening suggests that the crystallinity of this film is lower than that of the film annealed at high temperature. A lattice constant c_2 calculated from these lines is 35.3 Å. This value is smaller than that obtained by bulk sample [2].

When the substrate temperature was raised to 600 °C, a sharp peak at 2θ=5.9 and a weak peak at 2θ=7.2 appeared as shown in Fig.1 (c). These peaks were characteristic (002) line of the low-Tc phase and (002) line of the Bi2Sr2CuOy phase (c~24Å), respectively. From these results, it was found that the crystal structure of the as-deposited films was affected quite sensitively by the substrate temperatures between 550 °C and 600 °C. The film with the high-Tc phase was successfully deposited at 570 °C, but the crystallinity of this film is low.

Figure 3 shows the temperature dependence of the resistivities of the films. The film deposited at 550 °C had a higher resistivity than those of other two films at room temperature and showed a semiconductor like resistivity behavior. The film deposited at 570 °C showed the superconducting transition with an onset temperature (Tc on) of around 80K and zero resistance temperature (Tc zero) of 23K. The film deposited

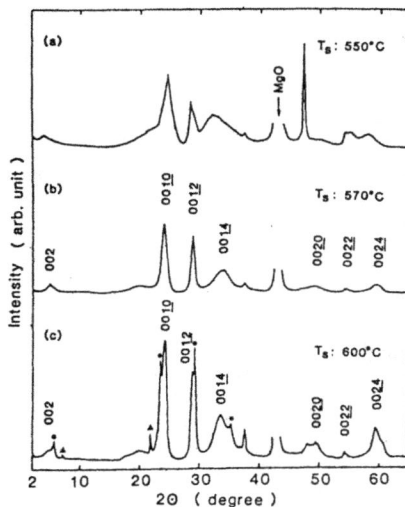

Fig.2. X-ray diffraction patterns (2θ between 2° and 65°) of the as-deposited films, closed circles:the low Tc phase, triangle:Bi2Sr2CuOy phase.

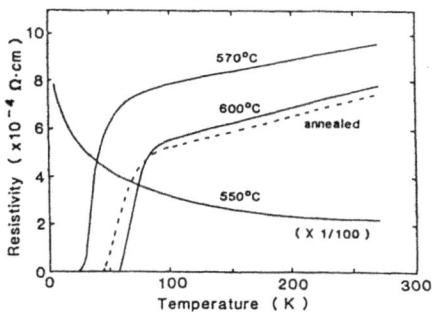

Fig.3. The temperature dependence of the resistivity for the as-deposited films.

at 600 °C showed Tc on of 80K and
higher Tc zero of 62K than that of
film deposited at 570 °C. The
superconductivity of this film is
probably caused by the low-Tc phase.
In X-ray diffraction pattern, as shown
in Fig.2(c), the peak width of the
low-Tc phase is sharper than that of
the high-Tc phase. Thus the
crystallinity of the low-Tc phase is
higher than that of the high-Tc phase
when the films are deposited at the
substrate temperature lower than
600 °C.

The diamagnetization measurement
was done for the film deposited at
570 °C, as shown in Fig.4, a weak
diamagnetization was observed below
around 42K. In the vicinity of 80K,
the diamagnetization could not be
observed. The obtained value of the
diamagnetization at 4.2K was
-29(emu/cc) and this value is two
order of magnitude less than that of
the film annealed at high temperature
[8]. This result indicated that a
volume fraction of the
superconductivity of this film is very
low.

In order to investigate a crystal
growth of the high-Tc phase grown at
570 °C, the films were annealed at the
same temperature of 570 °C for 30h to
prevent a crystal growth induced by
higher heat treatment. Annealing was
carried out under the oxygen pressure
of 0.1 atm, 1 atm and 5.5 atm. X-ray
diffraction patterns of the annealed
films were shown in Fig.5. When the
film was annealed under the oxygen
pressure 1 atm, the Tc on increased
from 23K to 45K as shown in
Fig.3(indicated by broken line), but
the Tc on was not changed. No
significant change was found in X-ray
diffraction patterns between before
and after annealing as shown in Fig.5
(b). The annealing under the oxygen
pressure of 5.5 atm showed same
results of the superconductivity and
X-ray analysis(Fig.5(c)) as obtained
by the film annealed under the oxygen
pressure 1 atm. When the film was
annealed under the oxygen pressure of
0.1 atm, the film was transformed
into an insulator. From the analysis

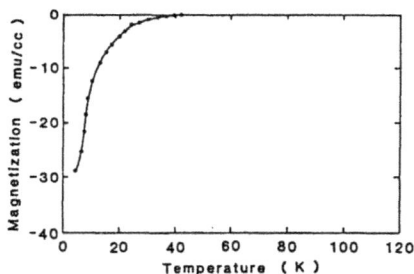

Fig.4. The temperature dependence
of the diamagnetization for the
film deposited at 570 °C.

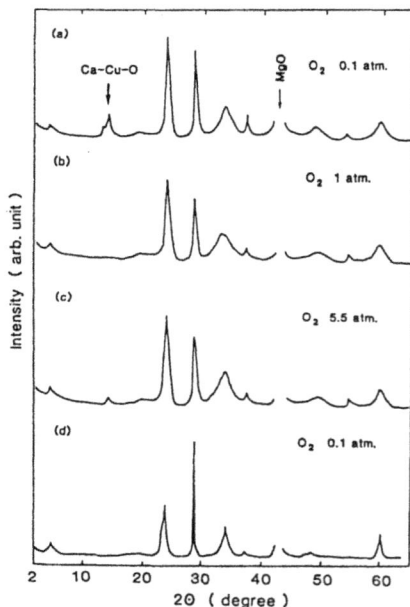

Fig.5. X-ray diffraction patterns
of the films annealed under
the various oxygen pressure.

of X-ray diffraction pattern, as shown in Fig.5 (a), it was found that the high-Tc phase separated partially into Ca-Cu-O phase. This phase separation is probably due to the unstable crystallinity of the high-Tc phase of this film. On the contrary, as shown in Fig.5 (d), the film annealed at 885°C did not show the phase separation after annealing at 570° C under the oxygen pressure of 0.1 atm. The temperature dependence of the resistivities of this film before and after annealing at 570° C under oxygen pressure of 0.1 atm were shown in Fig.6. After annealing Tc on was not changed, but the Tc zero decreased to 65K. This decreasing is probably due to the outdiffusion of the oxygen from the film. The results of X-ray analysis and the superconducting transition temperatures of the films annealed at 570°C for 30 h under the various oxygen pressure were summarized in Table 2. From these results, it was found that the crystal structure of the high-Tc phase grown at 570°C is unstable and that the oxygen affects the superconductivity of Bi-Sr-Ca-Cu-O films.

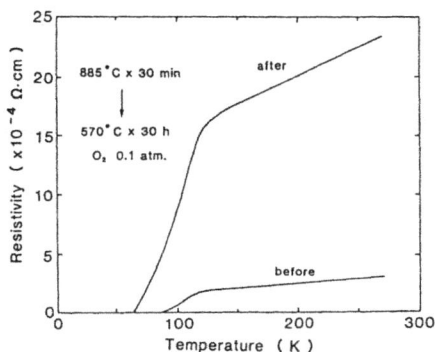

Fig.6. The temperature dependence of the resistivities for the films before and after annealing at 570°C under the oxygen pressure of 0.1 atm.

Table 2. Structures and superconducting transition temperatures of the films annealed at 570°C under the various oxygen pressures.

Oxygen pressure (atm)	Structure	Tc on (K)	Tc zero (K)
as-deposited	high-Tc phase	23	80
0.1	high-Tc + Ca-Cu-O phases	--- insulator ---	
1.0	high-Tc phase	80	45
5.5	high-Tc phase	80	43
0.1*	high-Tc phase	110	65

*The film was pre-annealed at 885°C under the oxygen pressure of 1atm.

Conclusion

We have tried to prepare Bi-Sr-Ca-Cu-O thin films by low temperature process using rf magnetron sputtering. The film with the high-Tc phase was prepared at the substrate temperature as low as 570 ℃, but the crystallinity of this film is lower than that of the film annealed at high temperature. This film showed zero resistivity temperature of 23K and subsequence annealing at 570 ℃ under the oxygen pressure of 1 atm increased Tc zero to 45K. On the contrary, after annealing under the oxygen pressure of 0.1 atm, the film was transformed into the insulator. Oxygen affects the superconductivity of Bi-Sr-Ca-cu-O film at the temperature as low as 570 ℃. The low temperature preparation of Bi-Sr-Ca-Cu-O film with the high-Tc phase structure seems to be difficult. Further study on the crystal growth of the high-Tc phase of Bi-Sr-Ca-Cu-O compound is desired.

Acknowledgments

We would like to thank Dr. T.Nitta for his support of this work and Dr. Y. Ichikawa for his useful discussion, and also thank Dr. S.Michida and Mr. R.Tahara of Matsushita Electric works, for their encourage throughout this work.

References

1) H. Maeda, Y.Tanaka, M. Fukutomi and T. Asano: Jpn. J. Appl. Phys. 27 (1988) L209.
2) E. Takayama-Muromachi, Y. Uchida, Y.Matsui, M. Onoda and K. Kato: Jpn. J. Appl. Phys. 27 (1988) L556.
3) Y. Ichikawa, H. Adachi, K. Hirochi, K. Setsune, S. Hatta and K. Wasa: Phys. Rev. B38 (1988) 765.
4) M. Fukutomi, J. Machida, Y. Tanaka, T. Asano, H. Maeda and K. Hoshino: Jpn. J. Appl. Phys. 27 (1988) L632.
5) T. Yoshitake, T. Satoh, Y. Kubo and H. Igarashi: Jpn. J. Appl. Phys. 27 (1988) L126.
6) T. Kamada, K. Setsune, T. Hirao and K. Wasa: Appl. Phys. Lett. 52 (1988) 1728.
7) J. M. Tarascon, W. R. McKinnon, P. Barboux, D. M. Hwang, B. G. Bagley, L. H. Greene, G. W. Hull, Y. LePage, N. Stoffel and M. Giroud: (submitted to Phys. Rev. B)
8) S. Hatta, Y. Ichikawa, K. Hirochi, k. Setsune, H. Adachi and k. Wasa: Jpn. J. Appl. Phys. 27 (1988) L855.

STRUCTURE AND SUPERCONDUCTING PROPERTIES OF SPUTTERED Gd-Ba-Cu-O
THIN FILMS BY LOW TEMPERATURE PROCESS

Shigenori HAYASHI, Takeshi KAMADA, Kentaro SETSUNE,
Takashi HIRAO, Kiyotaka WASA, and Akihisa MATSUDA[*]

Central Research Laboratories, Matsushita Electric Industrial
Co.,Ltd., Moriguchi, Osaka 570, Japan
[*]Amorphous Materials Section, Electrotechnical Laboratory,
Tsukuba, Ibaraki 305, Japan

Abstract - Structure and superconducting properties of sputtered Gd-Ba-
Cu-O thin films were studied from the viewpoint of deposition conditions and
oxygen content, in the context of low temperature processing. Deposition
process by rf magnetron sputtering at the substrate temperature of 600 °C was
found to be highly dependent on plasma conditions. Annealing and evolution
studies   revealed   the   dynamics   of   oxygen   and   its   influences   on
superconducting properties and on a crystal structure.

Introduction

Since the discovery of high-Tc superconducting oxides[1,2], thin films
of these materials have attracted much attention due to their application
potential to electronic devices[3,4]. Low temperature preparation of
superconducting thin films and accurate understanding of their structural
and physical properties are of great importance.

Superconducting thin films have been successfully prepared at low
temperature by rf magnetron sputtering[5]. Their superconducting properties
and crystal structure are dependent both on deposition conditions and on
oxygen content[6-9]. In the present work, deposition process of Gd-Ba-Cu-O
thin film has been studied utilizing a plasma controlled sputtering
apparatus, and the effects of oxygen on their structural and superconducting
properties have been also studied by annealing and evolution experiments.
The preparation of the films was found to be highly affected by plasma
conditions as well as substrate temperature and target-substrate spacing. On
the other hand, by means of annealing in a vacuum and in an oxygen
atmosphere at low temperature, significant change in structural and superconducting properties originated in the orthorhombic-to-tetragonal phase
transition has been observed reversibly[10]. Evolution study has revealed
different behavior of oxygen between orthorhombic and tetragonal phases.

Experimental

The Gd-Ba-Cu-O thin film was deposited on a MgO (100) single-crystal
substrate by rf magnetron sputtering[5]. The standard sputtering conditions
are listed in Table 1. The sputtering composite oxide target was prepared by
sintering a   reacted mixture of Gd_2O_3, $BaCO_3$ and CuO at 900 °C in   air.
Metal compositions of the` films were determined to be $Gd_1Ba_2Cu_3O_x$ by
inductively coupled   plasma (ICP) emission spectrometry. The substrate
temperature was monitored by measuring the surface temperature of a
reference Ga-Ba-Cu-O film on MgO using a two-wavelength infrared thermometer.

After the deposition, in-situ two-step annealing was performed. In the
first step, oxygen gas was introduced into a sputtering chamber to the
pressure of 1 atm without a change in substrate temperature. After a

holding time of 30 minutes, in the second step, the substrate temperature was held at about 350 °C for 30 minutes and finally cooled down to room temperature. By means of this low temperature process without post-annealing of higher than 850 °C, high quality Gd-Ba-Cu-O thin films with high Tc and sharp transition (Tc onset = 90 K and Tc end = 75-86 K) were obtained.

Figure 1 shows a plasma controlled sputtering apparatus used in this study. Plasma conditions can be modulated by external magnetic field. The mirror field perpendicular to the target was applied by a coil current flowing in the same direction. When the coil current is zero, the system operates as a conventional rf magnetron sputtering apparatus. Change in excited species from plasma was analyzed by optical emission spectrum (OES) measured just below a substrate. Figure 2 shows the OES from plasma under the standard sputtering conditions listed in Table 1. Emission lines at 778 nm from O atom and at 812 nm from Ar atom were especially focused on in this study.

Table 1.   Sputtering conditions.

Target	$Gd_1Ba_2Cu_{4.5}O_x$
Substrate	MgO (100)
Substrate Temperature	600 °C
Sputtering Gas	Ar/O_2 = 3/2
Gas Pressure	0.4 Pa
Rf Input Power	160 W
Growth Rate	80 Å/min
Target-Substrate Spacing	25 mm
Deposition Time	60 min

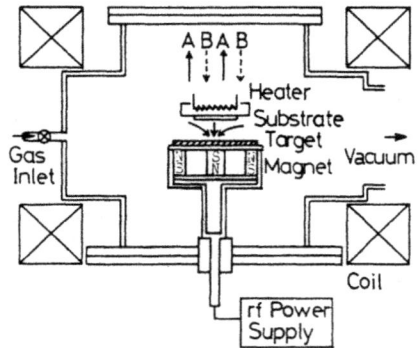

Fig.1. Plasma controlled sputtering apparatus.

In order to investigate the role of oxygen, prepared thin films were annealed in a quartz tube furnace flood with oxygen gas. The annealed films were quickly removed from the furnace to prevent the absorption of oxygen during the cooling time as much as possible. Resistivity of the films was measured by a dc four probe method with evaporated Au electrodes. Crystal orientations and phases were analyzed by X-ray diffraction (XRD) method using CuKα radiation. An evolution study on oxygen was carried out in an apparatus equipped with a mass spectrometer and an infrared imagefurnace.

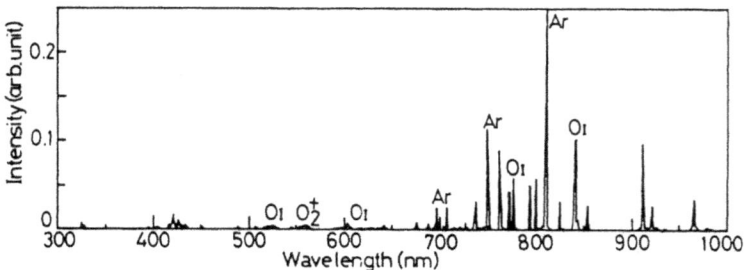

Fig.2. Optical Emission Spectrum (OES) from plasma measured just below substrate under the standard sputtering conditions listed in Table 1.

Results and Discussion

The preparation of Gd-Ba-Cu-O thin film was highly dependent on substrate temperature and on target-substrate spacing. Figure 3 shows the dependence of Tc end on substrate temperature under a target-substrate spacing of 35 mm and 25 mm. The optimum substrate temperature is considered to be lowered by reducing the target-substrate spacing to be 25 mm. The OES analysis showed an increase of emission intensities of plasma with a decrease of the target-substrate spacing. This fact suggests that it is important to control the plasma conditions.

In order to investigate the effects of plasma conditions, the mirror field was applied under the standard sputtering conditions listed in Table 1. Figure 4 shows the effects of applied mirror field on Tc end and on OES of plasma. Permanent magnets beneath the target cause a local magnetic field with B = +500 G in B(positive) direction at the center of the target, and the applied mirror field is characterized by its resulting additional field at the same position. The prepared 0.5 μm film when external magnetic field was zero exhibited a Tc end = 74 K. When the mirror field with B = +110 G was applied, a plasma was radially extended resulting in a decrease of OES intensity, and a deposited film showed a Tc end = 76 K. When the mirror field was applied in A (negative) direction, however, a plasma was radically compressed resulting in an increase of OES intensity. No film was deposited with B = -150 G, and a damaged film with Tc end = 20 K was deposited with B = -100 G.

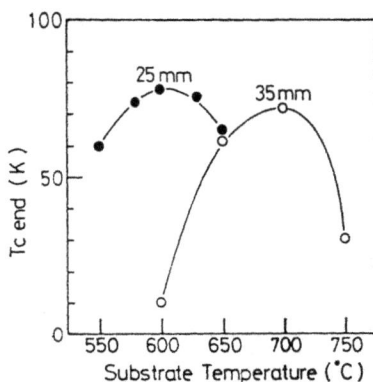

Fig.3. Dependence of Tc end on substrate temperature under a target-substrate spacing of 35 mm and 25 mm.

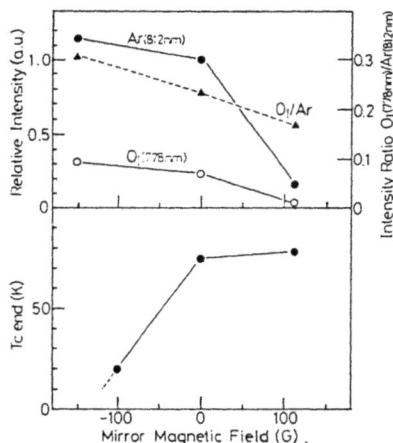

Fig.4. Effects of applied mirror field on Tc end and on OES of plasma.

When reducing the target-substrate spacing, excited species from plasma might be effectively utilized to lower the optimum substrate temperature for deposition. When a plasma was modulated by the mirror field, however, some change in excited species from plasma led to advancing or damaging effects on the deposition, depending on the direction and strength of applied field. The results suggest that it is important to control the plasma conditions for the fabrication of superconducting thin films.

Structural and superconducting properties of the prepared film change dependent on oxygen content as well as on deposition conditions. When annealed in a vacuum at 600 °C for 2 hours, a prepared superconducting film under standard sputtering conditions was transformed into an insulator. The XRD pattern also changed dramatically as shown in Fig.5. Based on the studies about crystal structures of bulk Y-Ba-Cu-O compounds[6,8], patterns of Figs. 5(a) and 5(b) could be assigned to orthorhombic (a=3.90 Å, b=3.85 Å, c=11.70 Å) and tetragonal (a=b= 3.88 Å, c=11.86 Å) phases, respectively[11]. The high relative intensities of the (00n) peaks in both patterns indicate a preferential orienta-tion of the c-axis perpendicular to a substrate surface. The observed phase transition has been considered to be caused by the outdiffusion of the oxygen atoms

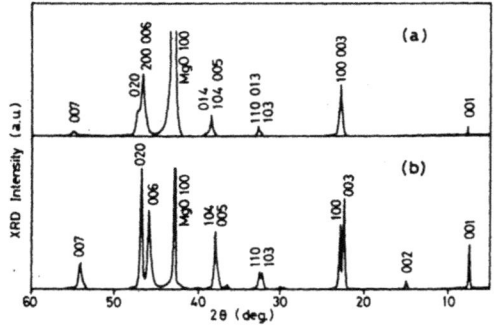

Fig.5. X-ray diffraction patterns of the Gd-Ba-Cu-O film. (a)high super-conducting state (b)annealed state at 600 °C for 2 hours in a vacuum.

on the a-axis inducing the elongation of the c-axis[6]. Structural deforma-tion derived from oxygen vacancies could be evaluated by the degree of the elongation of the c-axis. By an adequate annealing procedure in an oxygen atmosphere, the superconducting properties and XRD pattern of the film were returned to the initial state.

Employing the films converted into a nonsuperconducting tetragonal phase (c=11.83-11.84 Å) by annealing in a vacuum, the dependences of superconducting properties and crystal structure on oxygen annealing temperature were investigated. Figure 6 shows the changes of the c-axis lattice parameter upon annealing in an oxygen atmosphere at various temperatures. When annealed at 250 - 300 °C, the crystal phase still remained tetragonal after 15 hours. In contrast, at the annealing temperature of 350 °C, a superconducting orthorhombic phase appeared in several hours. With in-creasing the annealing time, the orthorhombic phase became predominant accompanied by a decrease in the c-axis value. The annealing time as long as 24 hours was required to attain equiliblium or to restore the superconducting properties of the initial state. As for the temperatures of

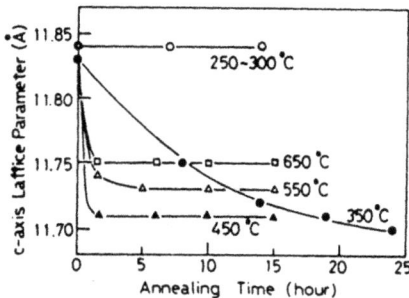

Fig.6. Changes of the c-axis lattice parameter of the Gd-Ba-Cu-O film annealed at various temperatures in an oxygen atmosphere.

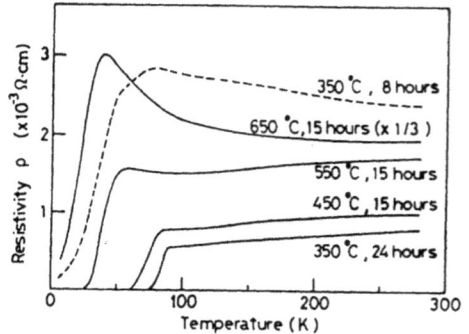

Fig.7. Temperature dependences of the resistivity of the Gd-Ba-Cu-O film annealed at various temperatures in an oxygen atmosphere.

450 °C, 550 °C and 650 °C, only in a few hours, the c-axis values reached to constant values of the orthorhombic phase. The temperature dependence of the resistivity of these films is shown in Fig.7. The dependence of the film annealed at 350°C for 8 hours bending around at 50 K and 80 K (indicated by a broken curve in Fig.7) suggests a mixed state of two phases. Analogous with the Y-Ba-Cu-O bulk ceramics, the second ortorhombic phase (ortho-II)[12,13] with a lower Tc of about 50 K must be taken into consideration in distinction from the 90 K orthorhombic phase(ortho-I). These two phases seem to be stable at around 350°C and 550°C in an oxygen atmosphere respectivly.

The relationships between Tc, the resistivity ρ (290K) and the c-axis lattice parameter revealed here are shown in Fig.8. The nonsuperconducting tetragonal phase was prepared in the range of c > 11.75 Å by annealing in a vacuum. It is deduced that the orthorhombic-to-tetragonal transition induced by oxygen annealing takes place around c=11.75 Å. It should be noted that, in the order of oxygen annealing temperature, the superconducting transition temperatures at both Tc onset and Tc end decrease in correspondence with the elongation of the c-axis. This fact suggests that the structural deformation dependent on oxygen vacancies takes place within the orthorhombic phase including the ortho-I-to-ortho-II phase transition. The ortho-I and ortho-II phases seem to be assigned to the regions of c=11.70-11.72 Å and c=11.73-11.75 Å, respectively.

Figure 9 shows the results of the evolution study on oxygen. The evolved oxygen from the films was detected with raising the ambient temperature at the rate of 10°C/min in a high vacuum. As mentioned above, a film in the tetragonal phase was transformed from a prepared superconducting film in the ortho-I phase by annealing in a vacuum. A film in the ortho-II phase was prepared from a film in the tetragonal phase by annealing in an oxygen atmosphere at 550 °C.

Three evolution peaks were observed for the films in the orthorhombic phase while only the third peak at the highest temperature side was observed for the tetragonal phase. From the results obtained above, the profile below 550°C is considered to be caused by the evolution of oxygen from the ortho-I phase, and evolution peaks around at 600°C and 750°C are

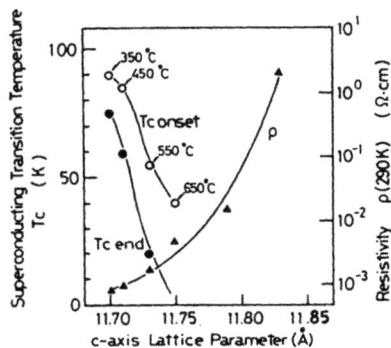

Fig.8. Relationships between the superconducting transition temeprature Tc, the resistivity ρ at 290 K and the c-axis lattice parameter in the Gd-Ba-Cu-O film.

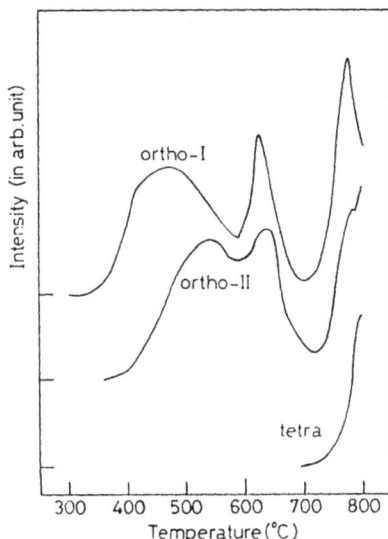

Fig.9. Results of evolution study on oxygen in the Gd-Ba-Cu-O film.

considered to be derived from the ortho-II and tetragonal phases, respectively. The oxygen related to two peaks at the lower temperature side below 600 ℃ in the orthorhombic phase should play an important role on the superconducting mechanism. The dynamics of oxygen has been investigated by thermogravimetric analysis in bulk ceramics[14], however, such a structural profile has not been reported, the evolution analysis will provide useful information to clarify the role of oxygen in high-Tc superconducting systems.

Conclusion

It has been demonstrated that superconducting properties and a crystal structure of sputtered Gd-Ba-Cu-O thin film are significantly affected by deposition conditions and oxygen content. Especially, effects of plasma conditions on deposition process were investigated by means of external magnetic field applicaton. We have found that there exists the orthorhombic-to-tetragonal phase transition, closely related to the behavior of oxygen, in the Gd-Ba-Cu-O film prepared at low temperature similar to the case of Y-Ba-Cu-O bulk ceramics. From a technological view point, it is necessary to prepare the films in a single phase at lower temperarure by controlling plasma conditions and to establish a technique for controlling the oxygen content. High quality thin films without grain boundaries should realize the high superconducting critical current density, and these films should be prepared by low temperature process.

Acknowledgments

The authors would like to thank Dr.T.Nitta for his continuous encouragement. We also thank Mr.M.Kitagawa and Mr.A.Yoshida for their useful advice.

References

1) J.Bednorz and A.K.Muller: Z. Phys. B64 (1986) 189.
2) M.K.Wu, J.R.Ashburn, C.J.Torng, P.H.Hor, R.L.Meng, L.Gao, Z.J.Huang, Y.Q. Wang and C.W.Chu: Phys.Rev.Lett 58 (1987) 908.
3) R.H.Koch, C.P.Umbach, G.J.Clark, P.Chaudhari and R.B.Laibowiz: Appl.Phys. Lett. 51 (1987) 200.
4) H.Adachi, K.Setsune and K.Wasa: Phys.Rev. B35 (1987) 8824.
5) T.Kamada, K.Setsune, T.Hirao and K.Wasa: Appl.Phys.Lett. 52 (1988) 1726.
6) F.Izumi, H.Asano, T.Ishigaki, A.Ono and F.P.Okamura: Jpn.J.Appl.Phys. 26 (1987) L665.
7) M.A.Beno, L.Soderholm, D.W.Capone II, D.G.Hinks, J.D.Jorgensen, Ivan K. Schuller, C.U.Serge, K.Zhang and J.D.Grace: Appl.Phys.Lett 51 (1987) 57.
8) E.Takayama-Muromachi, Y.Uchida, K.Yukino, T.Tanaka and K.Kato: Jpn.J. Appl.Phys. 26 (1987) L665.
9) K.Kishio, J.Shimoyama, T.Hasegawa, K.Kitazawa and K.Fueki: Jpn.J.Appl. Phys. 26 (1987) L1228.
10) S.Hayashi, T.Kamada, K.Setsune, T.Hirao, K.Wasa and A.Matsuda: Jpn.J. Appl.Phys. 27 (1987) L1257.
11) H.Asano, K.Takita, H.Katoh, H.Akinaga, T.Ishigaki, M.Nishino, M.Imai and K.Masuda: Jpn.J.Appl.Phys. 26 (1987) L1410.
12) T.Takabatake,Y.Nakazawa and M.Ishikawa: Jpn.J.Appl.Phys. 26 (1987) L1231.
13) M.Inoue, T.Takemori and T.Sakuda: Jpn.J.Appl.Phys. 26 (1987) L2015.
14) M.Kikuchi, Y.Syono, A.Tokiwa, K.Oh-ishi, H.Arai, K.Hiraga, N.Kobayashi, T.Sasaoka and Y.Muto: Jpn.J.Appl.Phys. 26 (1987) L1066.

SUPERCONDUCTING PROPERTY MEASURING SYSTEM
BY MAGNETIZATION METHOD

K. Ikisawa, T. Mori, N. Takasu

Tsuruimi Research Labs. NKK Corp.

2-1, Suehiro-cho, Tsurumi-ku Yokohama, 230 Japan

Abstract

Superconducting property measuring system (CMS-370B) for high
temperature oxide superconductor has been developed. This system adopts
magnetization measurement. The superconducting properties are able to be
measured automatically and continuously changing the temperature and
external magnetic field. The critical current density as a function of
temperature and magnetic field of high temperature superconductor
$YBa_2Cu_3O_{7-y}$ (YBCO) has been measured. It was confirmed that this system
having the high performance and the accuracy gave the significant
contribution to the superconducting material development.

Introduction

Since the discovery of high critical temperature oxide superconductor
as well as YBCO gave the potentiality of new application to the liquid
nitrogen temperature superconductor in addition to the conventional liquid
helium one. Recently, the research and development in the high temperature
superconductor is expected to make steady progress in the area of not only
the development and theoretical analyses but also the practical application
of the processing technology for making wire and the thin films. This fact
assure that the property of superconductor must be defined and it is
important to estimate based on the appropriate measuring results.
    The properties of superconductor, critical current density (Jc),
critical magnetic field (Hc) and critical temperature (Tc), was used to be
measured individually by using the transport current measurement of four
terminal method. There, however, is the problem that the contact resistance
is caused to the local excess heat generation at the current terminal as to
four terminal method.
    Our new series of measuring system, CMS-370, are synthetic
superconducting property measurement apparatus using DC magnetization
method. The temperature of the specimens can be changed extensively from
77K to 300K in the liquid nitrogen. This paper presents the details of the
systems and the measuring results obtained from this system.

Superconducting property measuring system (CMS-370B)

Main feature

The main feature of new system is described below.

1)Jc measurement by magnetization method
   Critical current density is calculated from magnetization of the samples. Each measurement time is 20 seconds.
2)Employment of hybrid cryostat
   The hybrid cryostat is combined with two type of dewar. The durable outer dewar is made of SUS 304 which is superior to mechanical protection and inner one is GFRP which is superior to high adiabatic performance and weight reduction.
3)High accuracy temperature control system
   The temperature control system consists of the sample holder including the electric heater and high performance programmable controller. The uncertainty of the temperature can be controlled within 0.2°C.
4)Quick measurement by automatic data acquisition system
   Microcomputer (HP9000,Model310) is able to supervise the system and analysis the measuring data quickly.

Construction and specification

The system configuration is outlined in Fig.1 and the construction of the measuring part is shown in Fig.2. This system consists of cryostat, bias magnet, sample holder with temperature controller and microcomputer system. The system CMS-370B which was used in this experiment is the high rank one of exclusive use in liquid nitrogen temperature, CMS-370A not having the temperature controller. The temperature can be changed extensively from 77K to 300K and external magnetic field up to 1.2T. It is specially designed for high temperature superconducting materials.
The principal specification of the system is shown in table 1.

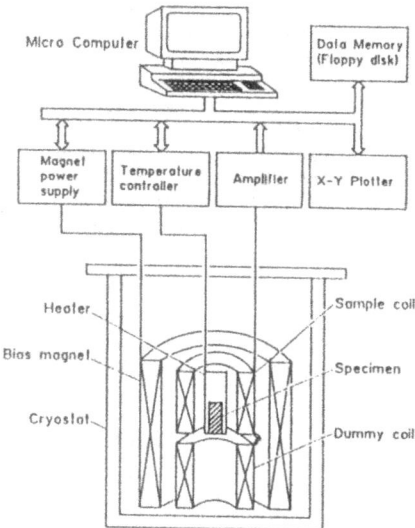

Fig. 1 Overviews of the synthetic superconducting property measurement apparatus.

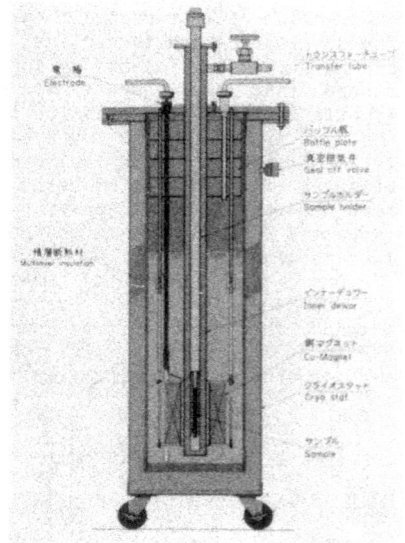

Fig 2 The construction of
the measuring part

Table 1 The Principal Specification
of The System

STANDARD SPECIFICATION AND COMPONENT			
TYPE		CMS-370A	CMS-370B
Measuring Method		Magnetization Method	
Magnet	Type	Copper Wier Magnet	
	Size	$\phi 200 \times 140$	
	Magnetic Field	MAX 2.0T	MAX 1.2T
Cryostat	Type	SUS304 + GFRP	
	Insulation	Super Insulation	
	Coolant	L N₂	
	Insert Dewar	—	GFRP Dewar
	Dewar Size	—	$\phi 60$
Sample Holder	Type	A-I	B-I B-II
	Sample Size	$\phi Dmax = 8mm$ $tmax = 3mm$	
	Heater	—	Cartridge Heater
Power Supply Unit	Type	DC GP060—100R	
	Volts/Current	60V/100A	
	Controller	Function Generator YHP 8116A	
Temperature Controll Unit	Type	—	B-I B-II
	Temperature range	77K	10K~300K
		—	PID + Program
Measuring Unit	Data Sampling Unit	YHP—3852A	
	Analyzer	HP 9000Series Model 310	
	Recorder	YHP 7550A	

(1) Bias magnet

The principals of the bias magnet is shown in table 2. This is the simplest kind of solenoid, a circular cylindrical winding of copper wire to stainless steel bobbin. The central magnetic field is 1.2T at the maximum allowable current of 100A. The homogeneity of the sample space is higher than 1%.

Table 2 The principals of the bias magnet

Wire material	Copper wire (ϕ 2.0mm)
Inner Dia.	63.50 mm
Outer Dia.	160.53 mm
Height	158 mm
Number of turn	2041
Resistance of coil	0.508 Ohm at 77K
Rated current	100 A
Rated magnetic field	1.32 T

(2)Cryostat

The principals of the cryostat is shown in table 3. The inner dewar made of GFRP is constructed by filament winding method using E-grass as grass fiber and epoxy as resin. The bottom multilayer plate which is formed by vacuum pressure immersion is screwed and bonded to the dewar.

(3)Sample holder

The arrangement of the sample holder is shown in Fig.3. The sample is heated by the cartridge heater installed in the copper block. The temperature is monitored at the adjacent point of the heater block from the sample by platinum thermistor and controlled by means of the heater output.

Table 3 NKK hybrid cryostat

O.D.	Material	SUS 304
	Size	$\phi 340 \times h1000$
I.D.	Material	GFRP
	Size	$\phi 250 \times h940$
	Construction	Filament winding method
Insulation		Super insulation
Coolant		LN$_2$

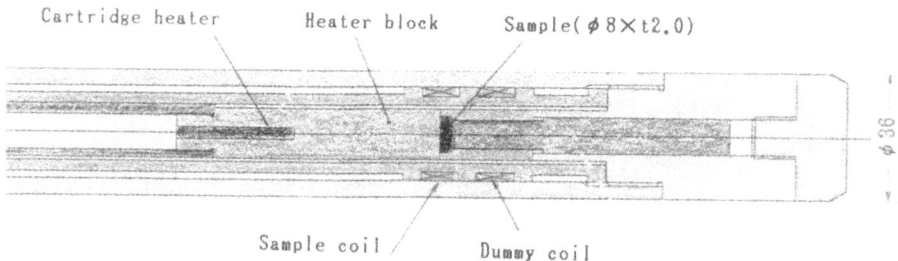

Fig 3   Sample holder

Measuring result

The properties of high Tc oxide superconductor were actually measured by this system.

Specimens

YBCO samples were prepared for the measurement. The specimens were molded from the coprecipitate composite powder which was pulverized into 1μm after presintering. They were sintered in the air at 1223K for 10hr, 743K for 5hr and furnace cooled for 9hr. The average grain size and packing factor of sintered products were 1.7μm and 72.4% respectively.

Estimation of Jc

Figure 4 shows the external field dependence of Jc. And also the results estimated by 4 terminal measurement are shown in order to compare with the magnetization measurement. The Jc derived from the magnetization was higher by several orders of magnitude compare with that from 4 terminal method. This means that the paths of transport and magnetization currents are different. It is reported that the screening currents flow only inside of the grain. The grain boundary is considered to prevent the transport current. Because of this in the Jc estimation with magnetization method the definition of diameter of superconductor, de, is important as shown in Fig.4. The de should be the mean diameter of the grain in this case.

Figure 5 and 6 show the temperature dependency of the magnetization and the critical current density respectively. The Jc decreased almost linearly with increasing the temperature. It is confirmed that the dependency is similar to that of conventional superconductors.

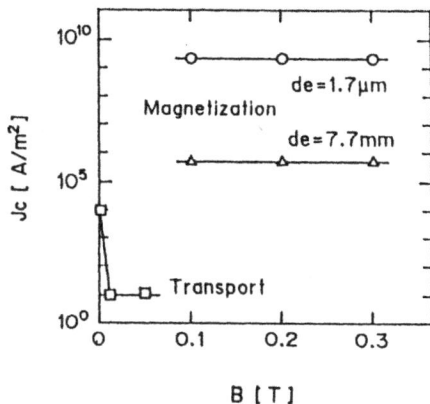

Fig. 4 Jc which was estimated by the transport and magetization measurement

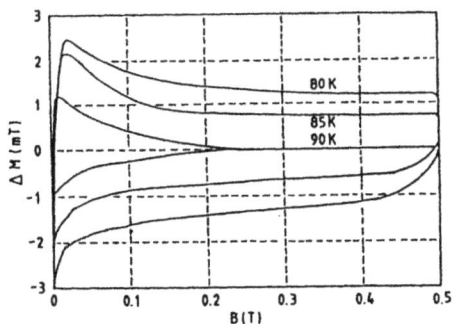

Fig. 5 Temperature dependence of magnetization

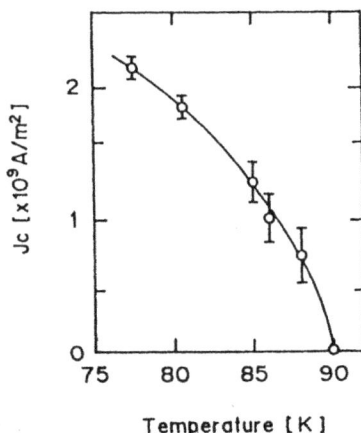

Fig. 6 Temperature dependence of Jc obtained by magnetization measurement

Conclusions

The new equipment which is able to measure the superconducting properties automatically and synthetically has been developed. The superconducting property of YBCO was measured changing the temperature and external magnetic field and following conclusions were drawn.

(1)New apparatus which has the high performance and the accuracy give the significant contribution to the development of superconductors.
(2)The critical current density, Jc, estimated from magnetization is higher by several order of magnitude compared with those from transport current measurement.
(3)The Jc from magnetization decrease almost linealy with the temperature. This dependency is similar to that of conventional superconductors.

Acknowledgments

We should like to express our great appreciation for the various instructions of Dr. T. Okada and Dr. S. Nishijima, ISIR Osaka University, in the development of this functional apparatus.

Reference

[1] T. Matsushita, M. Iwakura, Y.Sudo, B. Ni, T. Kisu, K. Funaki, M. Takeo and K. Yamafuji, "Estimate of attainable Critical Current Density in Superconducting $YBa_2Cu_3O_{7-y}$," Jpn. Appl. Phys., Vol. 26, pp. L1524-1526, SEPTEMBER 1987.

[2] J. W. Ekin, A. J. Panson and B. A. Blankenship, "Method for Making Low-Resistivity Contacts to High Tc Superconductors," Appl Phys Lett., Vol.52, No. 4, 25, JANUARY 1988.

Ohmic Contact to $YBa_2Cu_3O_{7-\delta}$ Ceramics

Tadaoki KUSAKA, Yoshihiko SUZUKI, Akira AOKI,Takahiro AOYAMA[*],
Tsutom YOTSUYA and Soichi OGAWA
Osaka Prefectural Industrial Technology Research Institute,
Nishiku, Osaka 550 JAPAN
[*] Daihen Co. Tagawa, Yodogawaku, Osaka 532 JAPAN

ABSTRACT

Three types of contact formation techniques on $Y_1Ba_2Cu_3O_{7-\delta}$(YBCO) ceramics
have been studied. These techniques are spot welding, ultrasonic soldering
and thin film deposition. The electrical properties of the contacts are
analyzed to find a low resistive contact. The contacts are grouped into two
classes according to the temperature dependence of their resistances. The
interface between thin film contact and YBCO are analyzed by ESCA. Causes
of resistance for both classes of contacts are proposed. Minimum resistivity
(about 1.9×10^{-7} Ωcm^2)is achieved for the contact of silver wire directly
bonded by spot welding and thermally annealed in oxygen.

INTRODUCTION

It is difficult to make an ohmic contact with low resistivity for oxide
superconductor because it has low carrier density. Forming a low resistive
electrode on high Tc superconductor is one of major problems for practical
electric applications. Several different types of contacts on oxide
superconductors including silver epoxy, gold-paste, pressed indium,
ultrasonic soldering and thin film have been previously reported[1,2,3], but
these contact's resistances were slightly too high for practical
applications. J. van der Maas et.al. have reported on a low resistive
contact prepared by silver epoxy, but the processing temperature for the
contact formation was very high (900°C). J.W.Ekin recently have reported
the lowest resistive contact which was formed by gold thin film but this
process need very careful treatment[4]. Low resistance, low temperature
conventional process are needed for contact formation. Accordingly we have
studied three contact formation techniques: ultrasonic soldering, vacuum
deposition and electric discharge spot welding.

EXPERIMENTAL

$Y_1Ba_2Cu_3O_{7-\delta}$(YBCO) ceramic pellets were prepared from the powders produced
by the gelation of citric salts[5].
Three contact formation techniques were examined;
(1) Copper wire (0.1 ϕ) was soldered on a YBCO pellet. Five types of solders
were used with an ultrasonic soldering iron. They are commercially
available from the Asahi-glass Co., Ltd with the exception of indium. A
YBCO ceramic pellet was placed on a 140°C hot plate and the ultrasonic
soldering iron temperature was adjusted according to soldering materials.

(2) Seven kinds of metal thin films were prepared by vacuum deposition.
Platinum film was deposited by the electron beam technique and the others
were deposited by conventional vacuum evaporation using tungsten boat.

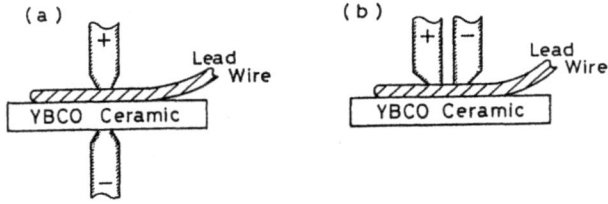

Fig.1 Configurations of spot welding method
(a) series welding type (b) parallel gap type

Table 1 Experimental conditions of spot welding

	Voltage (V)	Pulse Width (msec)	Energy (W·sec)	Weight (Kg/cm^2)
Series type	----	----	50	60
Parallel type	1	30	----	90

Table 2 Contact resistivities

Method	Electrode Material	R_c (Ωcm^2) R.T.	80K
Ultrasonic soldering	#186	9.0×10^{-2}	
	#297	6.9×10^{-2}	
	In	6.0×10^{-3}	1.6×10^{-2}
	#123	1.6×10^{-3}	
	#143	5.0×10^{-4}	1.1×10^{-3}
Vacuum deposition	Sn	5.3×10^{-3}	2.0×10^{-2}
	Zn	5.0×10^{-4}	2.6×10^{-3}
	In	5.5×10^{-4}	1.2×10^{-3}
	Cu	2.9×10^{-4}	3.5×10^{-4}
	Pt	4.6×10^{-4}	2.7×10^{-4}
	Ag	5.3×10^{-4}	1.3×10^{-4}
	Au	1.3×10^{-4}	5.8×10^{-5}
	Ag*	2.1×10^{-4}	2.4×10^{-5}
Sputtering deposition	Pt	1.1×10^{-4}	1.6×10^{-5}
	Ag	1.0×10^{-4}	5.8×10^{-5}
	Ag*	1.7×10^{-4}	8.1×10^{-5}
Spot welding Series type	Cu	Non Ohmic	
Parallel type	Ag	1.0×10^{-3}	1.7×10^{-3}
	Ag*	1.8×10^{-5}	1.9×10^{-7}

* annealed in O_2 500°C 1hr

Platinum and Silver were also deposited by DC 4 electrodes sputtering system. Pressure during sputtering was 5×10^{-5} Torr and target voltage was 500 V. Copper wire was then soldered to the films using ultrasonic vibration. Ultrasonic soldering on thin films lower the contact resistance without ultrasonic vibration. The detail discussion on the effect of ultrasonic soldering was already discussed[5].

(3) Copper and silver wires were welded directly on a YBCO pellet by electric discharge. Two types of spot welding methods, which are parallel gap type (Fig.1-b) and series welding type (Fig.1-a), were examined. The apparatus used here are NW-30C(series type) and MCW-550(parallel type) by Nippon Avionics. The experimental conditions are shown in Table 1.

RESULTS AND DISCUSSION

Table 2 shows the contact's resistivities.
A) Ultrasonic soldering
    Ultrasonic solderings #123 and #143 have low melting point, while #297 has a rather high melting point. Therefore, Table 2 shows that the contacts formed by the solders with low melting points have ohmic characteristics, and those with high melting points have non-ohmic characteristics. Solders with high melting points required heating to high temperatures to form contacts. Soldering at high temperatures may decrease the oxygen content near the YBCO's surface and consequently form a semiconducting or insulating surface layer. It is not yet understood what role each atom in the solder plays in contact formation on YBCO ceramics.

B) Thin film contacts
    The electrical properties of the thin film contacts can be grouped into two classes according to their resistivities' temperature dependences. The first class, which includes copper,indium, zinc, and tin film contacts, has high resistivities that increase at low temperatures. The temperature dependence of the tin contacts resistance is illustrated in Fig. 2 as a typical example of this class. This dependence is similar to those of

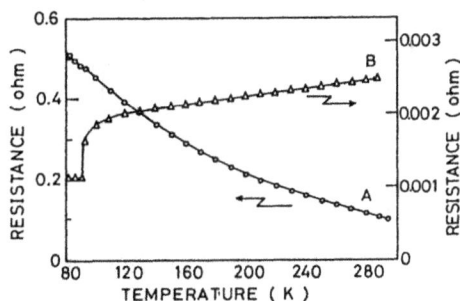

Fig. 2 The temperature dependence of the thin film contact's resistances (A) Sn/YBCO (B) Au/YBCO

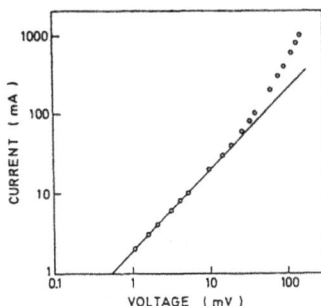

Fig. 3 V-I characteristics of the tin film contact

insulators and semiconductors.   Typical V-I characteristics of the tin
contacts are shown in Fig. 3.   The same characteristics were observed when
reverse voltage was applied to the contact.   While an ohmic property is
observed at low voltages, the V-I characteristics deviate from the straight
line and contact resistance decreases at high voltages.   The tin contact
without ultrasonic soldering had a non-ohmic property that was more clearly
observed.

Possible explanations of the reduction of contact resistance at high bias
voltage include:
(1)   The temperature at the contact increases by joule heating of high
current because of high contact resistance. The increase of temperature made
the contact resistance low because this dependency is semiconductive.
(2)   When an insulative or semiconductive layer exists between
superconductor YBCO and metal contact, possible high field conduction
processes, such as space-charged-limited current, Schottky emission and
Poole-Frenkel emission, shoule be taken into consideration.   If we assume
the thickness of the layer 200Å, its resistivity is estimated to be 2-20 $k\Omega cm$
which is very large. The linear dependence of Log(I/V)- SQRT(V)
characteristics seems to suggest the Poole-Frenkel mechanism.   However the
carrier density is estimated about 10^{17}-10^{19} /cm3. We wonder if this carrier
density is too large to explain the conduction mechanism as Poole-Frenkel
model.   Further studies are needed. We can imagine in both explanation that
there is a thin semiconductive or insulative layer at the interface.

The second class, which includes gold, silver, and platinum films, has
low resistivities, with temperature dependences opposite to those of the
first class.   Figure 2 shows the temperature dependence of the gold

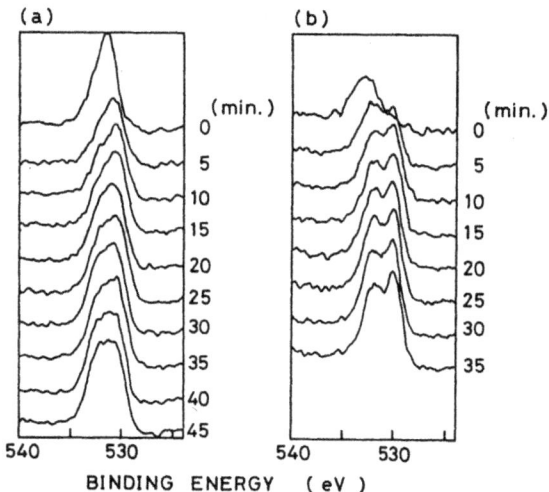

Fig.4 ESCA spectra of O1s state for
    (a) Sn/YBCO and (b) Au/YBCO
  Thesurface of contact/YBCO are etched
  and the change of ESCA spectra is shown.

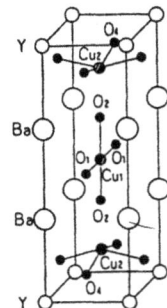

Fig.5   Oxygen configuration
    in the orthorombic
    crystal of YBCO

contact's resistivity. This dependence is rather metallic compared to that of tin contact. Ohmic properties are observed even in the high current region. This suggests that the previously described insulating layer does not exist, or is very thin. The contact's resistance may be due to a large amount of the defect states at the interface or in thin layer.

Figure 4 shows ESCA spectra of O1s state for Au/YBCO and Sn/YBCO. The change of ESCA spectra by etching the surface is shown in the figures. Adsorbed oxygen or oxygen on tin oxide is detected at 0 min. After 5 or 10 min. etching, YBCO surface are exposed. Two peaks at 529.6 eV and 531.6 eV are detected for Au/YBCO, while the peak at 532 eV can hardly be observed in Sn/YBSO. After long time etching, two peaks can be observed for both samples. It is said that the peak at 529.6 eV is originated from O_2 and/or O_4 site and that at 531.6 eV from O_1 site of orthorombic YBCO (Fig.5)[7,8,9] It has also been reported that the peak at 529.6 eV was observed and that of higher energy didn't appear for tetragonal phase of YBCO.[8,9] Therefore, Figure 5 indicates that the interface of Sn-YBCO includes tetragonal phase of YBCO or lack of O1 site oxygen in YBCO. The satelite peak of $Cu2p_{1/2}$ can hardly been observed at the interface of both samples. This means that orthorombic phase of YBCO surface is degraded by the deposition of metal film. Thermal annealing in oxygen at 500 °C improves the contact resistance to be 20 $\mu\Omega cm^2$.

C) Spot welding

The contact by spot welding without annealing shows nonohmic characteristics, whose temperature dependence is semiconductive. The region where the current flowed was found to be partly melted and its color turned green in the case of series spot welding. Figure 6 shows X-ray diffraction pattern of the YBCO after current flow treatment. The split between two peaks at 2θ=32.6 and 32.8 is obscure while it is clear for orthorombic YBCO pelet. Anomalous peaks, which may be originated from Y_2O_3, are observed and the length of c-axis is large (c=11.74 A) compared with that of orthorombic YBCO pellet (c=11.69). These facts suggest that the region's crystal pahse, oxygen content and atomic composition are changed by a large

(a)   (b)

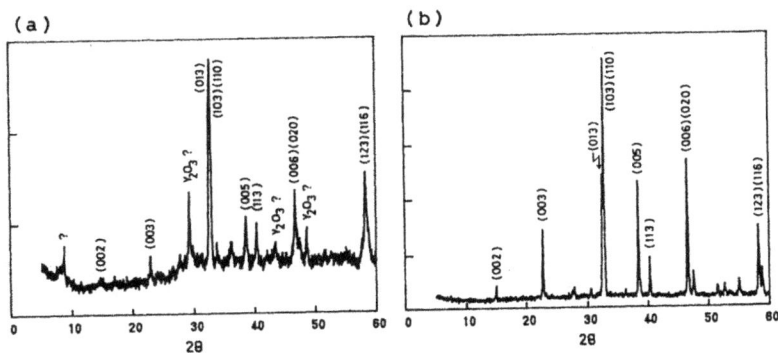

Fig.6   X-ray diffraction patterns of YBCO pellet.
(a) pellet after current flow treatment
(b) orthorombic YBCO pellet

amount of current flow.   In the case of parallel gap, the current flows in the wire or wire-YBCO interface and the degradation of YBCO is supposed to be small.   Thermal annealing for parallel gap samples improves the contact resistance to 1.9×10^{-7} Ωcm^2 which is one of the best result obtained ever. It is thought that the thermal annealing in oxygen gas introduces oxygen into the interface of Ag/YBCO and recovers the degradation.

CONCLUSION

Three types of contact formations on YBCO ceramics were examined to achieve low contact resistance with a low processing temperature.   The contacts   prepared by ultrasonic solderings and Cu, Sn, Zn and In thin films showed high contact resistivity.   The contact resistances increased with decreasing temperature. The Ag, Au and Pt thin film contacts showed low resistivities.   These contact's temperature dependences were opposite to those of the former group and excellent ohmic properties were obtained. Their contact resistances may be determined by a large amount of defect states at the interface or those in the thin layer. Two types of thin film contacts are analyzed by ESCA which reveals that high resistance contacts includes tetragonal phase of YBCO or oxygen deficient layer.   The semiconductive properties of high resistance contact are thought to be caused by the insulating or semiconducting layer at the YBCO-thin film interface. Ag wire was directly spot welded and was annealed in Oxygen gas to achieve the lowest contact resistance (1.9×10^{-7} Ωcm^2) which is one of the lowest resistance obtained ever.

ACKNOWLEDGMENT

We would like to thank Professor J.Shirafuji of Osaka University for his kind discussion and encouragement. They also thank   Mr. K.Ohtani of Hitachi Zosen Co., Ltd. for his support of this experiment.
This study was supported by Osaka cooperative research project for high Tc superconductors.

REFERENCES

1. Y. Maeno, M. Kato, and T. Fujita, Jpn. J. Appl. Phys. vol.26 L329 (1987)
2. J. van der Maas, V. A. Gasparov, and D. Pavuna, Nature vol.328, 603 (1987)
3. K. Takeuchi, Y. Okabe, M. Kawasaki and H. Koinuma, Jpn. J. Appl. Phys. 26, L1017 (1987)
4. J.W.Ekin, T.M.Larson, N.F.Bergren, A.J.Nelson, A.B.Swartzlander, L.L.Kazmerski, A. J. Panson and B. A. Blankenship; Appl. Phys. Lett. 52 (1988) 1819
5. M.Hirabayashi: Bull.Jpn. Inst. of Metals 26 (1987) p943
6. T.Kusaka, Y.Suzuki,T.Yotsuya, S.Ogawa, T.Aoyama and H.Imokawa : Proc. 5th Int. Work Shop on Future Electron Devices, Miyagi, Japan (1988) p205
7. H.Ihara, M.Hirabayashi, N.Terada, Y.Kimura, K.Senzaki, M. Akimoto, K.Bushida, F.Kawashima and R.Uzuka : Jpn. J. Appl. Phys. 26 (1987) L460
8. T.A.Sasaki, Y.Baba, N.Masaki and I.Takano : Jpn. J. Appl. Phys. 26 (1987) L1569
9. H.Watanabe, K.Ikeda, H.Miki and K.Ishida : Jpn. J. Appl. Phys. 27 (1987) L783

HIGH Tc SUPERCONDUCTING THREE-TERMINAL DEVICE UNDER QUASI-PARTICLE INJECTION

Koichi HASHIMOTO, Uki KABASAWA, Masayoshi TONOUCHI AND Takeshi KOBAYASHI
Faculty of Engineering Science, Osaka University, 1-1
Machikaneyama, Toyonaka-shi, Osaka, Japan.

Abstract

A new type of the current injection type three terminal device was fabricated using the high Tc YBaCuO thin epitaxial films, wherein the hot quasi-particle injection effect on the superconducting current was closely examined. The zero bias drain current was efficiently suppressed by the injection of the hot quasi-particles through the gate electrode. Though it is quite speculative, a comparison of the experimental results and analyses based on the familiar BCS theory intimates that the main mechanism of the current modulation is the non-equilibrium superconductivity due to accumulation of the excess quasi-particles.

Introduction

Since a discovery of a new oxide superconductor by Muller and Bednorz [1], the superconducting transition temperature Tc has jumped in excess of the liquid nitrogen boiling point (77K). Most recently, synthesis of BiSrCaCuO and TlBaCaCuO system recorded higher than 100K and drew the superconducting electronics up for the practical use.

One of the advantages inherent in the high Tc superconducting materials is a simple formation of the Josephson junction by only making a waist portion in the poly-crystalline superconducting film. The junction made thus is named a grain boundary Josephson junction (GBJJ for short). Till now, many papers have been published dealing with the fabrication of the GBJJ and applications to the millimeter wave detection and/or the magnetro sensor SQUID operative at 77K [2]. However, the Josephson junction is, in principle, a two terminal diode, and therefore, it can not work as an active device. Further more, quite limited application is allotted to the GBJJ.

To solve these problems, new superconducting devices other than the Josephson diode have been explored. A three terminal device is a promising candidate, being the main subject addressed in this text. We have five kinds of the three terminal devices, all of which comprised the metal superconducting electrode: The Josephson field effect transistor (JOFET) [3], the quasi-particle injection transistor (QUITERON) [4], the superconductor-base semiconductor insulating transistor (SUBSIT) [5], the superconductor base hot-electron transistor (Super-HET) [6],[7], and the micro bridge with the current injection gate [8]. In the present work, we did the first attempt in fabricating the high Tc superconducting transistor. The device comprises the hot quasi-particle injector, and the current modulation is brought about by the injection. Up to now, the current gain higher than 2 has been obtained.

316

Device Structure

Figure 1 shows the schematic drawing of the fabricated device. The device has silver ohmic contacts for the source and drain electrodes, and aluminum gate to form the three terminal structure. As a superconductor film, we prepared epitaxial high Tc thin films. The active region with 10μm wide was formed to obtain higher modulation effect.

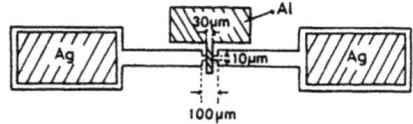

Fig. 1   A top view of the device.

Film Preparation

As high Tc superconductor films, we used most popular oxide-superconductor, YBaCuO films. The film was grown on (100) oriented MgO in argon and oxygen mixed atmosphere by using r.f. magnetron sputtering under the following conditions: gas pressure 4Pa (Ar 2.5Pa, O_2 1.5Pa), substrate temperature at 670°C, r.f.power 50W, deposition rate 6nm per minute, and a target with the composition of $Y_1Ba_2Cu_{4.5}O_y$. We, then, obtained (001)-oriented epitaxial films with 160nm and/or 90nm thick.

Device Preparation

We fabricated the device by procedures as mentioned below (Fig. 2). The film was patterned by chemical etching employing phosphoric acid (H_3PO_4) [9]. This wet etching is a new technique we proposed suitable for the high Tc film. There is no sign of degradation in the superconductivity after the etching process. As the etching mask, we used the photoresist AZ1350J. It is not only

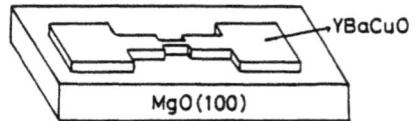

1 The Patterning with employment of a chemical etching in phosphoric acid.

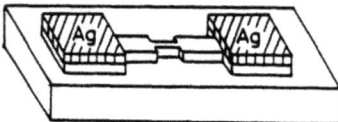

2 Silver is evaporated on the film as a source or drain terminal. Then the silver evaporated film is annealed in oxygen atmosphere to obtain a better contact.

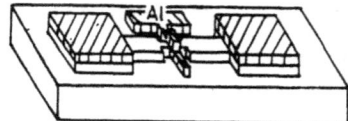

3 Aluminum is evaporated as a gate terminal.

Fig. 2   The procedure of the device fabrication.

quite resistant to the H_3PO_4 solution, but also very safe to the superconductivity of the high $T\overset{.}{c}$ film. Then, silver was evaporated on the film to compose the source and drain contacts, followed by the annealing at 500°C for 4hrs in the oxygen atmosphere [10]. This resulted in a very low specific contact resistance lower than $10^{-}\overset{.}{h}cm^2$. During the device process, the film surface is likely to miss not only its superconductivity but also the normal conductivity. However, it rather facilitated the tunneling injector formation by a simple deposition of aluminum electrode onto the junction portion. In this way, the hot quasi-particle injection with the average energy of few tens mV was automatically obtained.

Measurement

All measurements were made at liquid helium temperature where the non-equilibrium superconductivity is more likely to proceed under the quasi-particle injection. For simplicity, the drain current-voltage curves were measured without employing the four terminal method. The test specimen was mounted on DIL IC header, and wire-bonded with the gold wire (25umϕ). A typical example of the drain current-voltage curve (no-injection) is given in Fig. 3, where residual resistance is enough small to measure. It shows that contact is excellent.

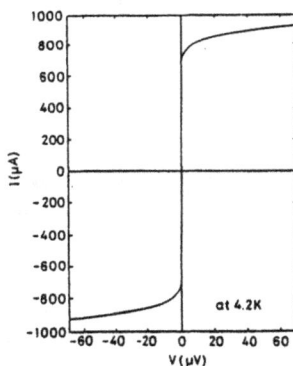

Fig. 3   I-V   characteristic   of the device under no-injection.

Result

Figure 4 shows an example of current-voltage characteristic of an injector. This characteristic proves that hot quasi-particles can be injected, and it might be arise from Metal-Insulator-Superconductor(MIS) structure whose insulator is formed at aluminum gate-YBaCuO film interface when aluminum is evaporated.

We have two sort of the device. One is using 160nm thick epitaxial YBaCuO film, and the other is using 90nm thick epitaxial YBaCuO film. First, we then report the former named #1.

Figure 5 reveals quasi-particle injection effect on the drain current-voltage characteristics.

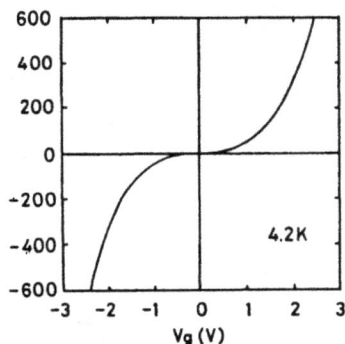

Fig. 4   I-V characteristic of the injector.

With increase in gate current Ig, superconducting critical current Ic decreases, which shows that Ic is modulated by Ig. Modulation characteristic of Ic is given in Fig. 6. The current gain is higher than 2 in the range of Ig from -40 to 40µA, especially, below 10µA of Ig, the gain reaches above 5.

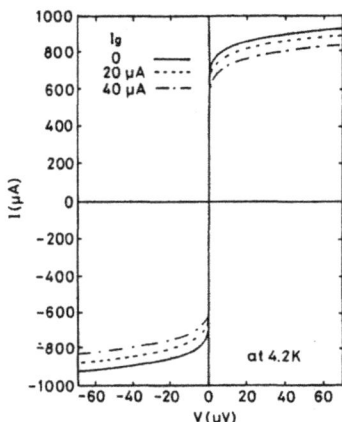

Fig. 5   I-V   characteristics   of the device under quasi-particle injection of #1.

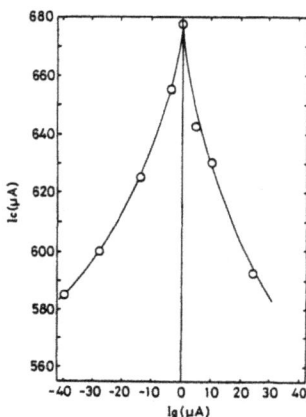

Fig. 6   Modulation characteristic of #1.

By the way, the observed modulation characteristics with a shape like a steep hump, we call this A-type, was reapparent in the other device(using 160nm thick epitaxial Y BaCuO film). The present experimental data were well curve-fitted as shown in Fig. 7 by the analysis of non-equilibrium superconductivity based on BCS theory [11-12]. Whereas, the device #2 from the epitaxial wafer with 90nm thick provided the modulation characteristic little bit different from A-type. Although the drain current-voltage curves look quite like those of #1(see Fig. 8), their modulation characteristics differed from each other(Fig. 9). However, the shape of modulation of device #2 is called B-type. Since there exhibits no essential difference in the quality of the epitaxial film,

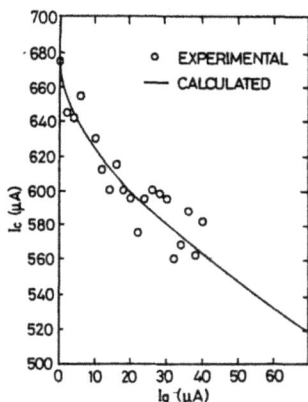

Fig. 7   A   comparison   between experimental and theorical results.

there must be some other cause which gives rise to a modulation characteristic A or B. We, then, investigate these two types by using non-equilibrium theory [11-12]. As a result, we have found that when transition temperature Tc is much higher than measurement temperature Tm(=4.2K), modulation characteristic exhibits A type and that when Tc is little higher than Tm, it exhibits B type(Fig. 10). This result intimates that devices is performed by non-equilibrium superconductivity.

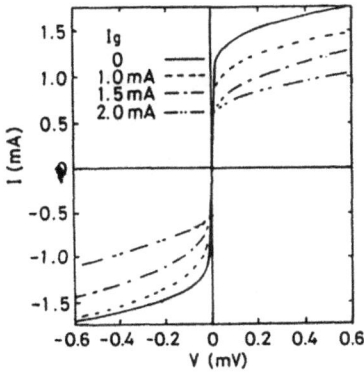

Fig. 8   I-V characteristics of the device under quasi-particle injection of #2.

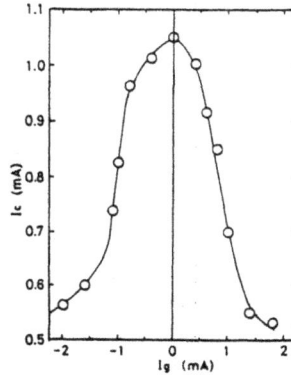

Fig. 9   Modulation characteristic of #2.

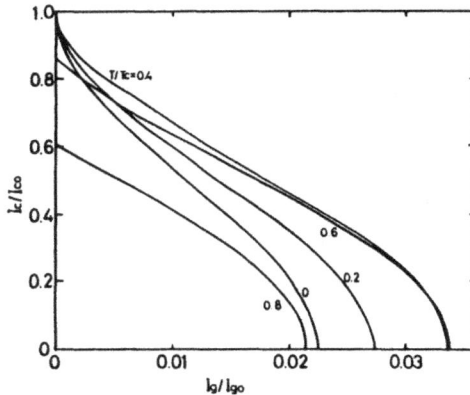

Fig. 10   Modulation by non-equilibrium superconductivity

Conclusion

(1) Three terminal devices using high Tc superconductor, YBaCuO thin films(160nm or 90nm thick) were fabricated.
(2) The prepared device exhibited the drain current modulation effect through the gate current injection. The current gain higher than 2

was obtained.
(3) The modulation characteristic was not depend on quality of superconducting films. From theorical point of view, the modulation might be promoted through development of the non-equilibrium superconductivity.

References

1) J.G.Bedonorz and K.A.Muller; Z.Phys. B-Condensed Matter$\underline{64}$ 189(1986)
2) H.Tanabe, S.Kita, Y.Yoshizako, M.Tonouchi and T.Kobayashi; Jpn.J.Appl.Phys.$\underline{26}$ 1961(1987)
3) T.D.Clark, R.J.Prace and A.D.C.Grassie; J.Appl.Phys.$\underline{51}$ 2736(1980)
4) S.M.Faris, S.I.Raider, W.J.Gallagher and R.E.Drake; IEEE Trans.Magn.MAG-$\underline{18}$ 1293(1983)
5) D.J.Frank, M.J.Bradt and A.D.Davidson; IEEE Trans.Magn.MAG-$\underline{21}$ 721(1985)
6) M.Tonouchi, H.Sakai and T.Kobayashi; Jpn.J.Appl.Phys.$\underline{25}$ 705(1986)
7) H.Sakai, Y.Kurita, M.Tonouchi and T.Kobayashi; Jpn.J.Appl.Phys.$\underline{25}$ 835(1986)
8) S.Sakai and H.Tateno; Solid.St.Devices Conf.Proc.$\underline{21-1}$ 331(1981)
9) Y.Yoshizako, M.Tonouchi and T.Kobayashi; Jpn.J.Appl.Phys.$\underline{26}$ 1533(1987)
10) Y.Tzeng, A.Holt and R.Ely; Appl.Phys.Lett.$\underline{52}$(2) 155(1988)
11) C.S.Owen and D.J.Scalapino; Phys.Rev.Lett.$\underline{28}$ 1599(1972)
12) A.Rothwarf and B.N.Taylor; Phys.Rev.Lett.$\underline{19}$ 27(1967)

MEMORY CHARACTERISTICS OF A CERAMIC SUPERCONDUCTING RING

Masahiko HASUNUMA, Akio TAKEOKA, Shoji SAKAIYA, Toshimasa HIRANO,
Minoru TAKAI, Yasuo KISHI and Yukinori KUWANO

Functional Materials Development Center, SANYO Electric Co., Ltd.

100 Dainichi Higashimachi, Moriguchi, Osaka 570, Japan

Abstract   We investigated the residual magnetic field characteristics of ceramic superconducting rings with asymmetric current paths. The residual magnetic field of the rings to which external currents were applied appeared when one of the branch currents exceeded the critical current of the ring. The field saturated when both branch currents exceeded the critical current and it showed hysteresislike characteristics. The experimental data agreed well with theoretical estimation. A new persistentcurrent-type memory device, which has no Josephson junction in its loop, was developed and operated. We called it the Superconducting Asymmetric Inductance type Memory (SAIM).

Introduction

Since the discovery of a ceramic superconductor in the Y-Ba-Cu-O system, extensive research has been carried out to develop new superconducting devices. Memory devices using persistent current have been investigated in Nb or intermetallic compounds for years.[2,3] We investigated the residual magnetic field characteristics of ceramic superconducting rings with asymmetric current paths to develop a new persistentcurrent-type memory device for the first time.[4,5] The residual magnetic field of a ring, supplied with external currents, appeared at large external currents and showed hysteresislike characteristics. A new persistentcurrent-type memory device was developed on the principle of the hysteresislike characteristics. This paper reports and discusses the hysteresislike characteristics of the residual magnetic fields of superconducting rings and the fundamental operation of the SAIM device.

Hysteresislike Characteristics

Experiments

Ceramic superconducting rings were prepared through a solid state reaction of Y_2O_3, $BaCO_3$ and CuO powders. The mixed powders were calcined at 960 °C in air for 5 hours, ground well and heated again at 960 °C in air for additional 5 hours. They were ground again and pressed into a ring shape. The ring-shaped samples were then sintered at 960 °C for 6 hours under oxygen.[6,7] The rings had outer diameters of 14.0 mm, inner diameters of 10.0 mm, and were

2.0 mm thick. The superconducting characteristics of the rings were measured with the standard four-probe technique. The zero resistivity temperature was 90 K and the current density was 33 A/cm^2.

Figure 1 shows the experimental scheme to measure the magnetic field of a superconducting ring supplied with an external current. Copper wires were wound around the ring and soldered to apply the external current. Before winding the wires, both sides of the ring were mechanically polished and cleaned, then thin copper layers were deposited, in a vacuum, only onto the areas to be soldered. Two samples having different current-path ratios were used for this experiment. The positions of the two soldered points of the samples were determined so as to divide a ring into two paths at a ratio of 1·2 or 1·3. A superconducting ring, cooled in liquid nitrogen, was supplied with external currents. When the current reached a desirable value, the power supply was turned off and the residual magnetic field through the ring was measured with a Hall-effect probe.

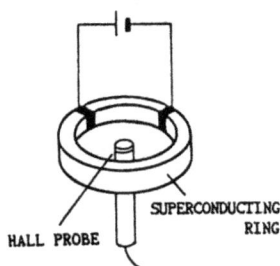

Fig.1 Schematic view of measurements.

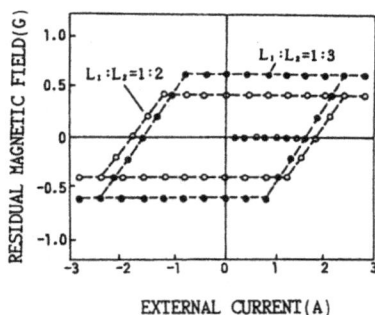

Fig.2 External current dependence of the residual magnetic field of rings.

Results

Figure 2 shows the residual magnetic fields of superconducting rings when external currents were supplied and removed. The residual magnetic field of the sample with paths at a ratio of 1·2 was zero at external currents below 1.8 A and that of the sample with paths at a ratio of 1·3 was zero at external currents below 1.6 A. When large external currents were applied, the residual magnetic field appeared and was proportional to the external current up to 2.4 A. When the external current reached 2.4 A, the residual field saturated and maintained its saturation state at any external currents less than 2.4 A or small inverse external currents. When large inverse external currents were applied, the residual field decreased and saturated again. Now the residual magnetic field showed hysteresislike characteristics.

Theoretical Analysis

A circuit for a superconducting loop with asymmetric current paths is shown in Fig. 3. The inductance of the shorter path is expressed as L_1 and that of the other path as L_2. Superconducting current with parallel circuits splits into two branch currents which are inversely proportional to the inductances. Thus the ratio I_1/I_2 is expressed as

$$\frac{I_1}{I_2} = \frac{L_2 - M_{12}}{L_1 - M_{12}}, \qquad (1)$$

where M_{12} is the mutual inductance and I_1 and I_2 are the branch currents. The total magnetic flux is expressed as

$$\Phi = (L_2 - M_{12})I_2 - (L_1 - M_{12})I_1. \qquad (2)$$

If the equation (1) is satisfied, the flux is zero and residual flux does not appear because the current splits to preserve the magnetic flux throughout the loop at zero.

Fig.3 Superconducting loop with asymmetric current paths.

$L_1 : L_2 = 1 : 2$

As the external current I_A equals $I_1 + I_2$, I_1 and I_2 are given by

$$I_1 = \frac{L_2 - M_{12}}{L_1 + L_2 - 2M_{12}}I_A, \qquad I_2 = \frac{L_1 - M_{12}}{L_1 + L_2 - 2M_{12}}I_A. \qquad (3)$$

The mutual inductance M_{12} is negligible because magnetic coupling between the two paths is very small. Increasing the external current, the branch current I_1 of the shorter path reaches the critical current Ic. Then the external current is divided into the branch currents to keep the shorter path in the superconducting state, which results in the appearance of the magnetic flux. A persistent current is induced to preserve the prior induced magnetic flux when the external current is reduced to zero. The total magnetic flux is expressed as

$$\Phi = L_2 I_A - (L_1 + L_2)I_c. \qquad (4)$$

This equation shows that the flux is proportional to the external current.

If external current increases further, I_2 increases up to Ic and the flux saturates. The saturated residual magnetic flux is given by

$$\Phi = (L_2 - L_1)I_c. \qquad (5)$$

The residual magnetic field maintains its saturation state while the external currents are less than twice the critical current or the inverse external currents are small. In this saturated state, the loop is always in the superconducting state to preserve the saturated residual magnetic flux, and the external current will not cause any change in the magnetic field through the loop. When a large inverse external current is applied and the current of the

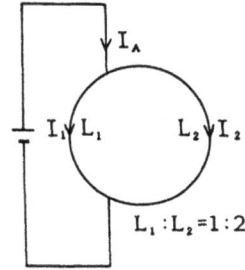

shorter path reaches Ic, the external current splits to keep the shorter path in the superconducting state and increases the branch current in the other path to decrease the magnetic flux. As the residual flux is in the saturated state and the current of the shorter path is Ic, the external current is given by the following equation.

$$I_A = -\frac{2L_1}{L_2}Ic. \qquad (6)$$

Increasing the inverse external current decreases the residual magnetic field. When inverse external current reaches 2 times Ic, the residual magnetic field saturates again and now shows hysteresislike characteristics. The relation between external current and residual flux is shown in Fig.4. The inductance L of the superconducting ring is calculated as 1.2×10^{-8} H. The experimental data agree well with theoretical hysteresis properties discussed above.

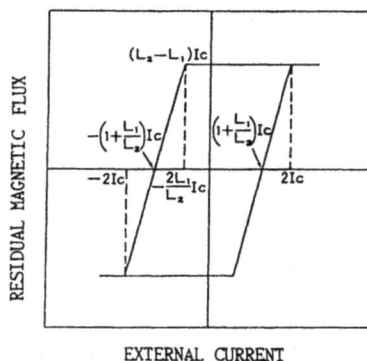

Fig.4 Schematic diagram of the residual magnetic field.

Fundamental Memory Operation

Figure 5 shows the structure of a new memory cell utilizing the difference between the inductances of the current paths in a superconducting loop as discussed above. The memory cell consists of a superconducting loop and a read line. The superconducting loop has asymmetric current paths divided at a ratio of 1:2. The superconducting loop includes an address line A, and a bias line B. The read line R consists of a bar-shaped ceramic superconductor which has a narrow part as a read junction.

A " 1" is stored as a persistent current in the loop, and a "0" as no current. Writing a " 1" is achieved by applying twice the critical current of the loop to line A and B. The currents I_A in line A and I_B in line B are given by the following equations.

$$I_A = (1+\frac{L_1}{L_2})Ic, \qquad I_B = (1-\frac{L_1}{L_2})Ic. \qquad (7)$$

After removing the A and B currents, a persistent loop current is stored. Writing a "0" is achieved by applying an inverse current $-I_A$ to line A. If a " 1" was previously stored, the loop current becomes zero resulting in a "0" state. If a "0" was previously stored, nothing will happen.

For reading, a read current is applied to line R. With a " 1" stored, the read junction has a resistance resulting in a voltage in the read line R. With a "0" stored, the loop current is zero and, thus, the read line R remains unaffected. Figure 6 shows a fundamental memory operation. The currents I_A and I_B were calculated as 1.8 A and 0.6 A, respectively, when the current loop was divided at a ratio of 1:2. Writing a "0" and a "1" was verified successfully. We call this memory the Superconducting Asymmetric Inductance-type Memory (SAIM).

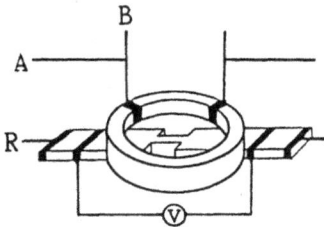

Fig.5 Schematic diagram of a fundamental memory cell.

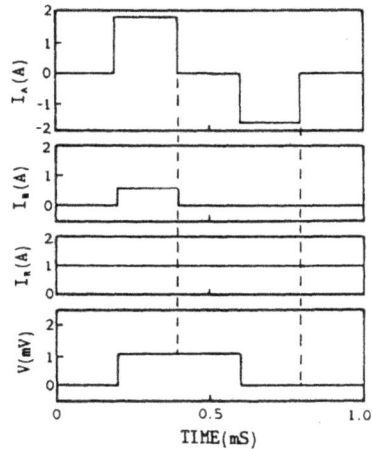

Fig.6 Fundamental memory operation.

Conclusions

The residual magnetic field of a superconducting ring with asymmetric paths showed hysteresislike characteristics. The experimental data of the residual magnetic field agreed well with theoretical hysteresis properties. A new persistentcurrent-type memory device was developed on the principle of the hysteresislike characteristics and a fundamental memory operation was successfully verified. A superconducting loop with asymmetric current paths such as the SAIM suggests a quite new persistentcurrent-type memory device.

References

1) M.K.Wu, J.Ashburn, C.J.Torng, P.H.Hor, R.L.Meng, L.Gao, Z.J.Hang, Y.Q.Wang and C.W.Chu, "Superconductivity at 93 K in a New Mixed-Phase Y-Ba-Cu-O Compound System at Ambient Pressure", Phys. Rev. Lett. 58, 908 (1987).

2) W.H.Henkels, "Fundamental criteria for the design of high-performance Josephson nondestructive readout random access memory cells and experimental configuration", J. Appl. Phys. 50, 8143 (1979).

3) H.H. Zappe, "A Subnanosecond Josephson Tunneling Memory Cell with Nondestructive Readout", IEEE J. Solid-State Circuits SC-10, 12 (1975).

4) A.Takeoka, M.Hasunuma, S.Sakaiya and Y.Kuwano, "Memory Characteristics of Ring-shaped Ceramic Superconductors", FED HiTcSc-ED Workshop, June 2-4,

1988, Japan, pp.283-286
5) A.Takeoka, M.Hasunuma, S.Sakaiya, T.Hirano and Y.Kuwano, "Memory Character-istics of Ring-shaped Ceramic Superconductors" , to be published in IEEE Trans. on Mag. Mag-25.
6) R.J.Cava, B.Batlogg, R.B.van Dover, D.W.Murphy, S.Sunshine, T.Siegrist, J.P.Remeika, E.A.Rietman, S.Zahurak and G.P.Espinosa, "Bulk Superconducti-vity at 91 K in Single-Phase Oxygen-Deficient Perovskite $Ba_2YCu_3O_{9-x}$" , Phys. Rev. Lett. 58, 1676 (1987).
7) S.Uchida, H.Takagi, K.Kitazawa and S.Tanaka, "High Tc Superconductivity of La-Ba-Cu Oxides" , Jpn. J. Appl. Phys. 26, No.1, L1 (1987).

THREE-DIMENSIONAL FLUX-SENSOR COMPOSED OF HIGH-Tc SUPERCONDUCTOR

H.Okushiba, S.Nishijima+, K.Takahata+, T.Okada+ and T.Hagihara

Department of Physics, Osaka Kyouiku University, Tennoji, Osaka 543,
+ISIR Osaka University, Ibaraki, Osaka 567

Abstract - Three-dimensional flux-sensor has been developed aiming at the practical application of high-Tc oxide superconductor Y-Ba-Cu-O. Three-dimensional flux-sensor was designed and the resolution was estimated.

Introduction

The flux flow resistivity of the high-Tc oxide superconductor Y-Ba-Cu-O is influenced on the magnetic field considerably [1]-[4]. The flux flow resistivity also depends on the angle between the current and magnetic field [5]. It means that the magnetic field can be determined three-dimensionally with flux flow resistivity. The SQUID shows the sensitivity from 10^{-12} to 10^{-10} [T] and the Hall elements from 10^{-6} to 10^{1} [T]. The flux-sensor of which sensitivity from 10^{-9} to 10^{-7} [T] could have various advantages for practical application. In this work three-dimensional flux-sensor with sensitivity from 10^{-9} to 10^{-7} [T] have been developed using high-Tc superconductor.

Samples and Experimentals

The nitrates of yttrium, barium and copper were coprecipitated by oxalic acid in ethanol. This coprecipitate was calcined in an oxygen atmosphere (O_2) and the powder regrounded was identified as single phase $YBa_2Cu_3O_{7-x}$ by X-ray diffraction analysis. The superconducting paste prepared by mixing this powder and organic vehicles was printed on yttria stabilized zirconia (YSZ). Various patterns were sintered in O_2 and critical temperature (Tc) was evaluated as 90 [K] by four-terminal measurement. The specifications of developed sensors were presented in Table 1 and some of them were shown in Fig. 1.

Table 1   Specifications of the sensor

Sample No.	Length [mm]	Width [mm]	Thickness [μm]	Terminal distance [mm]	Note
#1	50	1	50	30	
#2	20	(diameter 4mm)		10	Column-like bulk sample
#3	20	1	10	10	#3, #4 and #5 were fabricated
#4	20	1	30	10	by identical condition
#5	20	1	60	10	
#6	300	0.75	44	50,100,150,200,250	

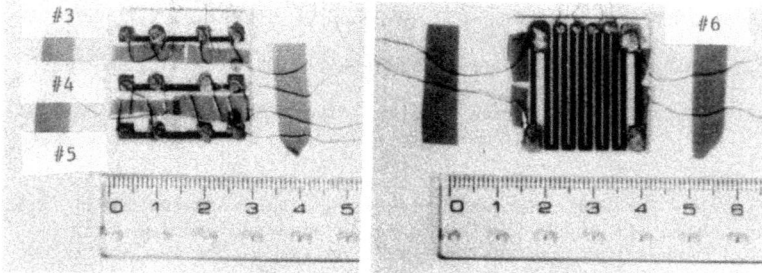

Fig.1 Photographs of the sample #3, #4, #5 and #6.

In the various external magnetic field, voltage versus current (V-I) characteristics were studied by four-terminal method at liquid nitrogen temperature. The electrodes were soldered with the silver paint. Induced voltage was measured by the integrating digital voltmeter (HP-44701A) controlled by micro computer system.

Results and Discussions

Stability of Output Voltage

First of all, the stability of the output voltage was investigated because the heating at the current electrode should be the most fundamental problems in the continuous operation of the flux-sensor. At the several currents, the output voltage of the sample #1 had been measured. There was no change in the voltage of the sample for 10 hours shown as Fig. 2. It means that the effects of the heating at the electrodes are negligible and we do not need to pay attention to the measuring period.

The stability of the voltage in the several external field was also measured. The voltage of the sample #2 is constant at each field for more than 6 hours at 100 [mA] of transport current as shown Fig. 3. The flux flow was confirmed to be stable in the wide range of magnetic field from 10 [Gauss] to 1 [T]. In this test, it was also confirmed that heating at the electrodes is small enough to measure the field.

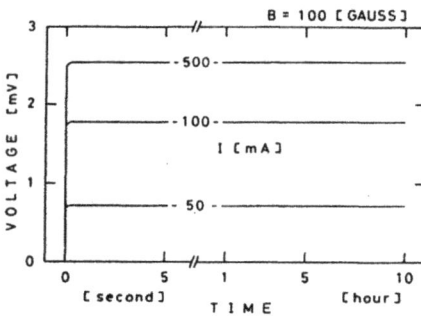

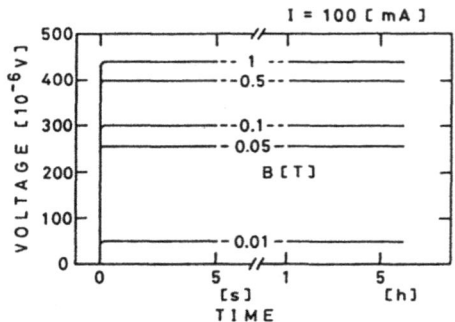

Fig.2 Stability of the output voltage at several currents for 10 hours.

Fig.3 Stability of the output voltage in several field for 6 hours.

Hysteresis of Output Voltage

The V-I characteristics were measured for the sample #1 with changing the external field. The output voltage at each field with the transport current of 20 [mA] are plotted in Fig. 4. Though the direction of the flux flow is changed corresponding to the direction of the field, the voltage measured in the four-terminal method is always positive, and hence the voltage is symmetric with respect to the V-axis. When the negative marking was given to the output voltage induced by the opposite direction of the flux flow, it can be seen that the output voltage shows the usual hysteresis against external field as shown in Fig.4. Once the specimen was warmed up to room temperature (RT), the virgin state was found to be realized again.

The hysteresis of critical current density (Jc) against external field in the Y-Ba-Cu-O has been reported. This could be understood as the effect of the trapped flux in the grain. The critical current in the grain is high compared with that in grain boundaries. When the transport current exceeds the critical current of the grain boundary under the external field, the grain boundary comes to be normal whereas in the grain superconducting state remains. In the grain, therefore, the flux which existed in the grain just before the normal transition of grain boundary should be trapped. The hysteresis of output voltage is thought to be equivalent to that of Jc. In order to eliminate the trapped flux, it is necessary to raise the temperature of the sample higher than Tc. In this work, every measurement was performed after the warming up of the specimen to RT. In practical application the heater should be arranged to warm the sensor higher than Tc just before the measurements.

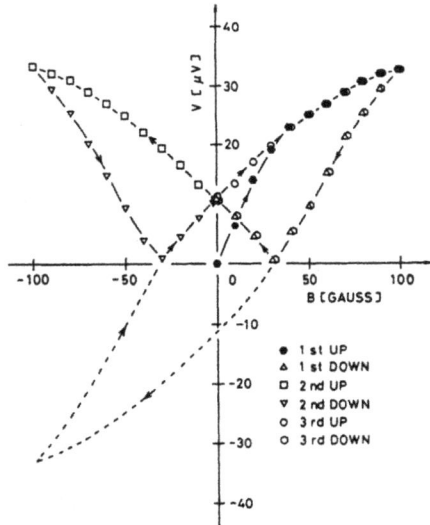

Fig.4 Output voltage for sample #1 at 20 [mA] in each field. The output voltage shows the hysteresis against the field.

Basic Data for Design

The design of the sensor to get high resolution are to be considered. Figure 5 shows the output voltage versus thickness for the samples #3, #4 and #5 in several external field at identical current density of 100 [A/m²]. The effect of the thickness could be neglected. The same experiments were performed on the sensors with 25 [μm] in thickness changing the width of the pattern as 0.5, 1, 2 and 3 [mm]. The obtained data coincide with those in Fig. 5. It was also seen that the effect of the width could be ignored. In order to avoid the heating at the electrodes, the transport current should be low enough. Consequently, it is more effective to design thin and narrow pattern to get high resolution.

The relationship between the output voltage and the pattern length were investigated. The voltage, for the sample #6 at the transport current of 100 [mA] in each field as 5, 10 and 50 [Gauss], was proportional to the pattern length shown as Fig.6. For this reason, the resolution for the sensor could be increased by making the pattern longer.

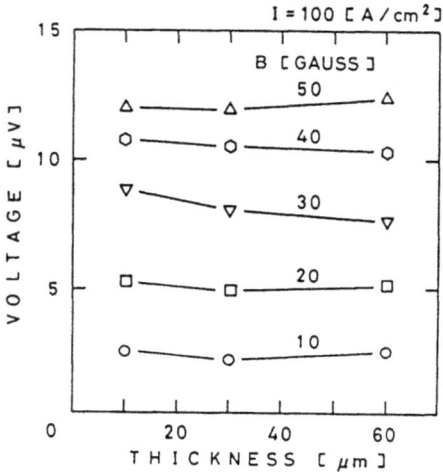

Fig.5 Voltage dependence of thickness for the samples #3, #4 and #5.

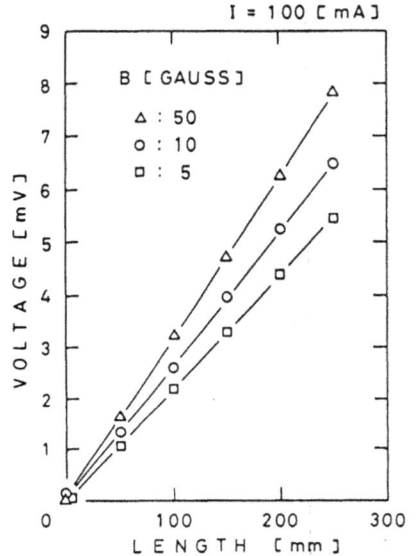

Fig.6 Voltage vs. pattern length for sample #6 at 100 [mA] in each field.

Three-dimensional Flux-Sensor

The voltages plotted in Fig.7 were obtained from the V-I characteristics of the sample #6 where the magnetic field applied perpendicular to the sample surface. As the field intensity increased, the output voltage increased monotonously. That is to say, the magnetic field could be decided by measuring the output voltage. When the transport current density is set lower than the Jc, the sensitivity is zero.

The angular dependence of output voltage was examined. Figure 8 shows the relationship between the angle and the output voltage. The directional angle θ, ϕ are defined as inserts in the figure. The voltage was measured at the transport current of 100 [mA] and field intensity of 10 [Gauss]. The output voltages were affected equally by both directional angle θ, ϕ. These data suggest that the field direction can be sensed.

The output voltage V is given as a function of the intensity of field, directional angle θ and ϕ;

$$V = F (|B| , \theta , \phi).$$

If three sensors are arranged perpendicularly each other, the measured voltages can determine the three unknown values in the equation. When it cannot be solved analytically, it can be calculated numerically. From a different point of view, it is possible to measure the flux three-dimensionally by rotating the sensor mechanically.

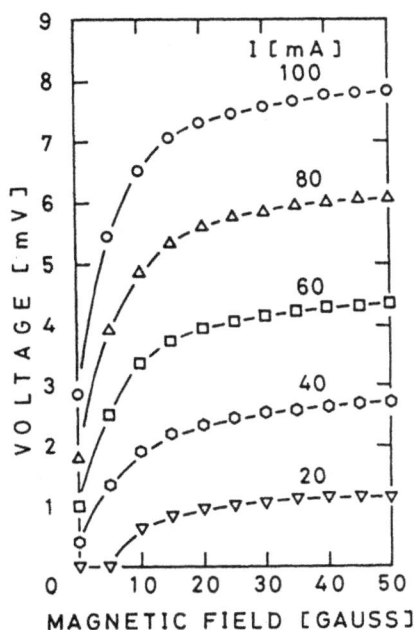

Fig.7 Field dependence of the output voltage for sample #6 at several currents in the terminal distance of 250 [mm].

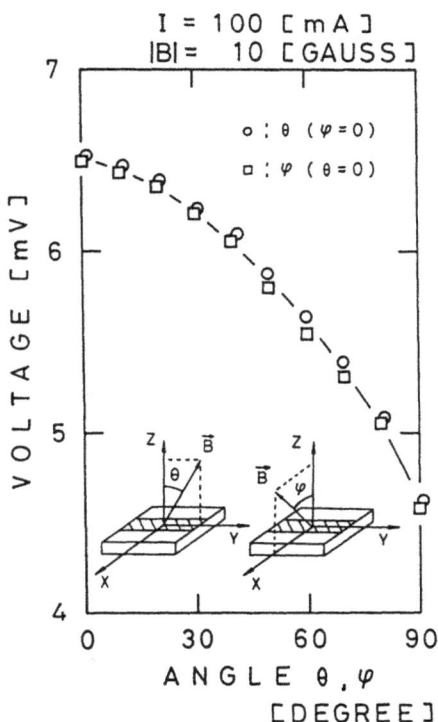

Fig.8 Angle dependence of the output voltage for sample #6 at 100 [mA] in the field intensity of 10 [Gauss].

Resolution

The output voltages plotted in Fig.6 were the average of ten measurements. Table 2 listed the mean value, the mean square error and the resolution at the transport current of 100 [mA] in each field. In the view point of the practical application, the resolution were estimated based on the following assumptions.

(1) The output voltage changes linearly between the measured points.
(2) The mean square error were constant of ±0.5 [μV].
(3) The field always applied the sample perpendicularly.
(4) The two measured values can be distinguished when the difference of the two measured values is larger than the mean square error.

Since the mean square error is ±0.5 [μV], more than 1 [μV] difference should be induced to determine the field intensity. For examples in the region of 0~5 [Gauss], we can divide the 5 [Gauss] into 2599 points using the sensor. Then, the resolution is 2×10^{-3} [Gauss] (2×10^{-7} [T]). As the field increased, the resolution decreased.

332

Table 2   Concrete Lists of the value

Field [Gauss]	Mean value [μV]	Mean square error [μV]	Resolution [mGauss]
0	2846.9	0.5	2
5	5444.6	0.5	5
10	6515.3	0.6	5
15	7051.5	0.3	10
20	7307.9	0.5	20
25	7459.1	0.4	33

Conclusions

The three-dimensional flux-sensor with high resolution has been successfully developed and following conclusions were drown.

(1) The problems of electrodes can be neglected for practical application of flux sensor. The flux flow state is confirmed to be stable and hence the long term operation of flux sensor is possible.

(2) The voltage induced by the flux flow shows the hysteresis against the external field. The phenomenon was understood as the effect of the trapped flux in the grain. Consequently the heat up to higher temperature than Tc is necessary for the flux sensor just before the measurements.

(3) Three-dimensional flux-sensor composed of the thick film were designed and fabricated. The factors which affect the sensitivity were clarified and the design of the flux-sensor comes to be possible.

(4) The resolution was estimated in the range of 0 ~ 5 [Gauss] as 2 [mGauss] (2×10^{-7} [T]).

Acknowledgments

This work is in part supported by the project of cooperative work between Osaka University and Hayashi Chemical Industry in 1988.

References

[1] M.Oda, Y.Hidaka, M.Suzuki and T.Murakami, " Anisotropic superconducting properties of $Ba_2YCu_3O_{7-x}$", Phys. Rev., B38, 252-255 (1988).
[2] J.W.Ekin et al., "Evidence of weak link and anisotropy limitations on the transport critical current in the bulk polycrystalline $YBa_2Cu_3O_{7-x}$", J. Appl. Phys., 62, 4821-4828 (1987).
[3] A.D.Wieck, "Superconducting contacts on $YBa_2Cu_3O_{7-x}$ in magnetic fields", Appl. Phys. Lett., 53, 1216-1218 (1988).
[4] L.H.Allen et al., "Temperature and field dependence of the critical current densities of Y-Ba-Cu-O films", Appl.Phys.Lett., 53, 1338-1340 (1988).
[5] Y.B.Kim, C.F.Hempstead and A.R.Strnad, "Flux-Flow Resistance in Type-II superconductors", Phys. Rev., 139, A1163-A1172 (1965).

EVALUATION OF PERFORMANCE OF MAGNETIC SHIELD
WITH OXIDE SUPERCONDUCTORS

Shigeo YOSHIDA, Masatomo KAINO, Takaharu NISHIHARA, and Atsushi IEUJI

Central Research Laboratory, Shimadzu Corporation
1, Nishinokyo-kuwabaracho, Nakagyo-ku, Kyoto 604, Japan

Abstract - Magnetic shielding capability has been investigated on the YBCO oxide superconductors aiming at a practical application of oxide superconductor. Experiments were carried out using small cylindrical samples at 77K and 4.2K for axial and radial magnetic field. Shielding capabilities for two different temperatures are compared. The transport critical current of sample and the estimated current necessary for the field cancellation are also compared.

Introduction

In recent years the strong magnetic sources have been widely used in many fields in science and industry, and the shielding techniques for the magnetic field have become very important. Among them the superconducting shielding is very interesting technique, because it is substantially passive and generates only the inverse field necessary for the cancellation of the original field naturally. There have been several works on the ordinary superconductors [1-2]. The oxide superconductor is very attractive material in the view point of cooling temperature, so that the shielding capability of it has been investigated. In this paper, the experimental results for YBCO cylindrical samples will be descrived and the relation between shielding capability and transport critical current density will be discussed.

Experimentals

Sample Preparation

The oxide or carbonate powders of Y_2O_3, $BaCO_3$ and CuO were mixed in a ball mill with ethylalcohol for 20 h. Dryed mixture was calcined in an air atmosphere at 1193K for 12h. The calcined powder was ground by hand, pressed by means of a cold isostatic press and sintered in an oxygen atmosphere at 1203K for 24h. Sintered YBCO block was machined to form a cylinder. Two samples (sample #1 and #2) were prepared. The preparation conditions described above are for sample #2. The preparation process for sample #1 was similar to that for #2 on the whole. The dimensions of the sample were $\phi25$-$\phi20$-46 (sample #1, outer diameter-inner diameter-length) and $\phi25$-$\phi18$-30 (sample #2) respectively.

The critical temperature Tc and the transport critical current density Jc were measured for the fraction of each sample. Tc (end point) was around 90 K for two samples. Jc at 77K were 33A/cm² for sample #1 and 110A/cm² for sample #2 at zero field. These values of Jc were obtained with four probe method at the potential difference across the voltage taps of 1μV (it means 3μV/cm criterion).

Experimental Set Up

Figure 1 and 2 show the experimental set up. A hall probe was fixed in the center of the hole of the sample. The magnetic fulx Bex, which had been calibrated beforehand to the magnet current, was applied to the sample.

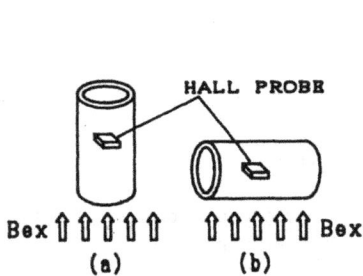

Fig.1. Experimental set up showing axial measurement(a) and radial measurement(b).

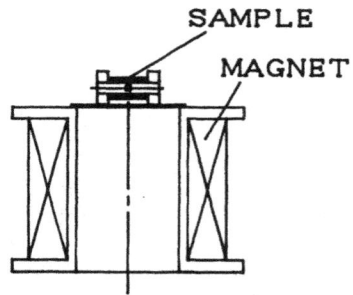

Fig.2. Experimental set up showing magnet and sample.

The shielding capabilities for sample #1 were measured both for the axial field as shown in Fig.1 (a) and for the radial field as shown in Fig.1(b) at liquid nitrogen temperature (77K). The experiment at liquid helium temperature (4.2K) was also carried out for the axial field for sample #1. For sample #2, the axial field capability at 77K was measured.

Results and Discussions

Figure 3 and 4 show the internal flux density Bin measured with the hall probe as a function of the applied flux density Bex for axial and radial fields at 77K. As shown in these figures, Bin was kept at zero as Bex increased to 4 gauss (axial field) or 2.5 gauss (radial field). Then Bin gradually increased and closed to Bex. This tendency is similar for both direction of the applied field and it is shown that the radial field is able to be shielded with the sample as well as the axial shield.

When the field decreased to zero, about 5 gauss of the field was trapped in the sample. Bin started from this value at the second run.

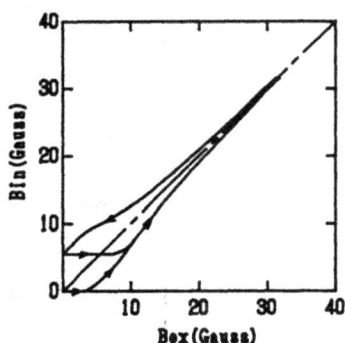

Fig.3. Hall probe output Bin as a
function of external field
Bex at 77K for axial field
(sample #1).

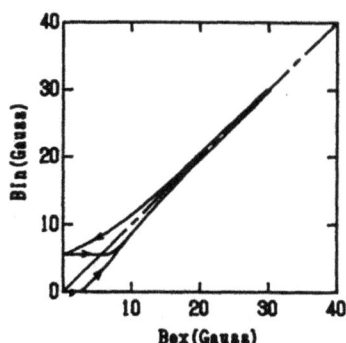

Fig.4. Hall probe output Bin as a
function of external field
Bex at 77K for radial field
(sample #1).

Figure 5 shows Bin-Bex curve at 4.2K for the axial field. The limit value of Bex for zero Bin increased by 6 times (22 gauss) compared with Fig.3. The trapping field also incerased by about 4 times.

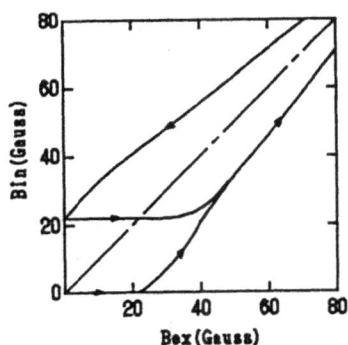

Fig.5. Hall probe output Bin as a
function of external field
Bex at 4.2K for axial field
(sample #1).

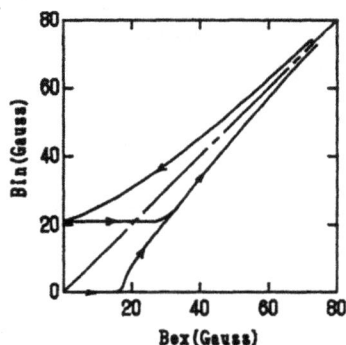

Fig.6. Hall probe output Bin as a
function of external field
Bex at 77K for axial field
(sample #2).

The axial shielding capability for sample #2 at 77k is presented in Fig.6. The increase in the shielding capability of sample #2 (18 gauss) compared with #1 (Fig.3) is attributed to the difference in Jc. As previously described, Jc was 33A/cm^2 for sample #1 and 110A/cm^2 for sample #2. The difference in the cross section area for two samples was within 15 percent and, therefore, the difference in the shielding capability is mainly due to the transport critical current density. The current density which generates 20 gauss at the center of the sample was estimated assuming the uniform current distribution for the dimension of sample #2 and the obtained value was 50A/cm^2. This value will be nearly equivalent to the density necessary for the cancellation in Bex of 20 gauss and the order of it is consist with the measured transport Jc(110A/cm^2) considering the decrease in Jc by the applied field Bex. This consideration is also consistent with the conclusion of other work showing the accordance between the shielding capability and the transport Jc, not the magnetization Jc [3].

By recent works it has been reported that the second critical fields Hc$_2$ of oxide superconductors are much higher than the ordinary superconductors and, that is, oxide superconductors have more posibilities for the large field shielding than the usual superconductors. The magnetic stability [4], the fabrication technique for large superconducting network and the improvement in transport Jc in large field will be the key technologies for the large field shielding from now on.

Conclusion

Magnetic shielding capability has been investigated on the YBCO oxide superconductor for the cylindrical samples. Following conclusions were drawn.
(1) The shielding capability substantially depends upon the transport critical current.
(2) The radial field can be shielded with the cylindrical YBCO shield as well as the axial field.
(3) Oxide superconductor is useful for the fine field shielding at present. But it has the possibility for large field shielding in future.

Acknowledgment

The authors would like to thank Prof. T.Okada and Dr S.Nishijima of ISIR Osaka University for their helpful discussions and experimental aid.

References

[1] S.Nishijima, K.Takahata, I.Miyamoto, T.Okada, S.Nakagawa and M.Yoshiwa, "MAGNETIC SHIELDING NETWORK WITH SUPERCONDUCTING WIRES" , IEEE Trans. on Magn., Vol.Mag-23, No2, pp. 611-614, MARCH 1987.
[2] T.Okada, K.Takahata. S.Nishijima, S.Nakagawa and M.Yoshiwa, "MAGNETIC SHIELDING WITH SUPERCONDUCTING WIRES" , IEEE Trans. on Magn., Vol.24, No.2, pp. 895-898, MARCH 1988.
[3] T.Okada, K.Takahata, S.Nisijima, S.Yoshida and T.Hanasaka, "APPLICABILITY OF OXIDE SUPERCONDUCTOR TO MAGNETIC SHIELDING" submitted to IEEE Trans. on Magn.
[4] T.Ogasawara, "Conductor Design Issues for Oxide Superconductors" , Teionkougaku (Cryogenic Engineering), 23, No.4, (1988) pp.45-52, (in Japanese).

A NEW PROPOSAL OF APPLICATION OF LIQUID NITROGEN COOLED SUPERCONDUCTING POWER CABLES TO ELECTRIC POWER SYSTEM

* Hiromasa FUKAGAWA, * Hiroshi SUZUKI, * Ataru ICHINOSE and ** Shirabe AKITA
Central Research Institute of Electric Power Industry
* 2-6-1 Nagasaka, Yokosuka, Kanagawa, Japan
** 2-11-1 Iwatokita, Komae, Tokyo, Japan

Abstract-Superconducting wires which are available at the liquid nitrogen temperature have not been developed yet. Therefore, assuming that new types of high Tc superconducting tapes or filaments for ac current can be developed near future, We performed a conceptual design and the cost evaluation of liquid nitrogen cooled superconducting cables and proposed their concrete application to power systems.

Introduction

Since a lot of ceramic type high Tc superconducting materials have been discovered, superconducting techniques have been expected to be applied to electric power apparatuses. The most practical application of superconductivity to electric power apparatuses is considered electric power transmission cables, because high Hc is not required for superconductors of power cables, compared with those of magnets or transformers. According to the cost evaluation, if liquid nitrogen cooled superconducting transmission cables can be developed, it will be a competitor of conventional power cables even in the case of lower transmission capacity. In this case, the transmission voltage of liquid nitrogen cooled cables can be 66kV class and the cost of the electric power system will also be expected to be decreased drastically.

The Conceptual Design Condition

The conceptual design structure of liquid nitrogen cooled ac superconducting cables is shown in Figure 1. It is a semiflexible structure with three phase cable cores in a pipe. Superconductors are used to both the current conductor and the shielding conductor. We studied 2 cases of cooling system as shown in Figure 1(a) and 1(b). The former cooling method is named TYPE-1 (the inner and outer cooling system) and the latter is TYPE-2 (the outer cooling system). In addition a tape type superconductor is used for TYPE-1 and a multifilament superconducting wire is used for as a cable current conductor of TYPE-2. In the structure designed, the outer diameter should be limited 420mm so that it can be installed instead of 275kV conventional underground transmission lines. Cases studied in this evaluation and design conditions assumed are shown in Table 1.

(1) AC Loss of TYPE-1 Superconducting cables

It is well known that the surface loss density of Nb_3Sn tape only is typically 10-20 $\mu w/cm^2$ at 500 A/cm and 8K. Considering some factors which may increase the ac loss of a superconducting cable, we selected 3 cases of the conductor surface loss (20, 100 and 1000 $\mu w/cm^2$). The equation used for the conceptual conductor design is as follows:

$$Wo = A (Jx/500)^3$$

$$Wc = 2\pi dWo$$

where, Wo:ac loss per unit superconducting area(W/cm^2), A : surface loss density$(\mu W/cm^2$ at 500A/cm), d: outer diameter of the current conductor or inner diameter of the shield conductor, Jx:current density(A/cm)

(2) AC Loss of TYPE-2 Superconducting cables

In this case, ac loss is hysterisis loss by conductor current itself. The equations used are as follows:

a . Cylindrical Compact Conductor

$$wc = \int_0^{r_0} pF(r) dr = 2d^3 \ \mu_0 Jc \ f \ I \ N/D \ 9\pi$$

b. Hollow Conductor

$$Ws = \int_{r_1}^{r_2} pF(r) dr = (1/9\pi) d^3 \ \mu_0 JcfIN (2\gamma_2{}^3 - 3rl\gamma_2{}^2 + \gamma_1{}^3)/(\gamma_2{}^2 - \gamma_1{}^2)^2$$

where, F(r):density function, I : conductor current, P : loss per a superconducting filament= γ_0 $(8/3\pi)d \ \mu_0$ Jc H f, d:diameter of superconducting filament, μ_0 : permeability, H : magnetic field, Jc : current density, f : frequency, N : number of superconducting filament, D:outer diameter of cylindrical compact conductor, γ_1, γ_2 : inner or outer diameter of hollow conductor.

Considering about 230K temperature difference, the multilayer structure by the insulating tapes is recommendable. Ac breakdown voltages of the insulation paper filled with liquid nitrogen and insulating oil are almost same. From our experienced in the R&D of cryogenic resistive cables, PPLP is the most recommendable insulating material. Ac electric insulation stress adopted in the insulating design is 15 kV/mm.

Multilayer superinsulation, vacuum powder insulation and form insulation are considered. In form insulation the outer diameter of the cable is over 420mm that is the upper limit for the conceptual design. Vacuum powder insulation has about 10 times thermal conductivity as large as superinsulation, better construction and lower vacuum (below 10^{-2} mmHg) than superinsulation. We studied two cases (superinsulation method and vacuum powder insulation method).

Results of Conceptual Design

We developed a calculation code of a conceptual design on superconducting, cables. The optimum condition is the minimum electric power for cooling a longer cable length.

(1) TYPE-1 Superconducting Cables

All 275 kV cases of vacuum powder insulation are impossible for being designed. In the case of 66 kV cables, if the surface loss density is below 100 μW/cm^2, 5GVA class superconducting cables can be designed. The cooling power in 66 kV cables tends to increase with the transmission capacity due to dominant conductor loss but that in 275 kV cables does not increase due to less conductor loss and dominant dielectric loss. The optimum voltage is 66kV in 2.5GVA and 154kV in 5 GVA. In these cases, current density of a superconductor can be below 700 A/cm.

(2) TYPE-2 Superconducting Cables

The smaller is the diameter of a filament, the smaller the conductor loss. The effect of a smaller filament is decreased in vacuum powder insulation in comparison with superinsulation. Even if the current density of a superconductor increases more than 10^4 A/cm^2, the cable length to be cooled does not almost increase. In order to obtain the same characteristics as 66 kV TYPE-1 superconductong cables, the diameter of a superconducting filament needs less than about 50μm. When a superconducting power cable is applied to electric power system as a source line or a connecting line between power substations, the phase angle difference between the voltage of the transmitting end and that of the receiving end should be within 15°-20°because of the stability in power system. The relation between voltage phase angle difference θ and effective power P is given as follow:

$P \approx (V^2/X)\sin \theta$

where, X : reactance of a cable line, V : system voltage

The maximum voltage phase angle is selected 15 degree from the system stability. The longest cable distance to be transmitted stably in the most optimum desigh in the case of TYPE-1 superconducting cables with 100 μW/cm^2 current loss and vacuum powder insulation is shown in Table 2.

Cost Evaluation Results

We used a conventional cost evaluation method as follow :

Transmission cost = total cost for a life time/
(transmission capacity × transmission length)

Total cost for a life time = construction cost for a life time (A) + loss cost for a life time (B)

    A = construction cost × the yearly rate

    B = loss cost (C) + cooling cost (D)

    C = (load loss × loss factor + no load loss)
       × loss cost × 8760

    D = (load loss × loss factor + no load loss
      + fluid loss + penetration heat)
      × refrigilator efficiency × loss cost × 8760

construction cost=cable cost including installation cost + culvert cost + cost of cooling system

In order to compare LN2 cooled superconducting power cable system with other cable systems, we also studied 275 kV or 500 kVconventional OF cables, original 275 kV cryogenic resistive cable and liquid helium cooled superconducting cable.

When the transmission cost of external water cooled 275 kV OF cable is 100%, the cost in the other cables is shown in Figure 2. The cost of liquid nitrogen cooled superconducting cables with 2GVA/route is the same that of the conventional cables. And there is no cost difference between the superinsulation cable system and the vacuum powder insulation one. It can be concluded that vacuum powder insulation is available from the point of view of operation and reliability.

Application of Superconducting Cables to Transmission System

According to the cost evaluation, 66kV superconducting cables can be applied instead of 275 kV conventional OF cables, as shown in Figure 3. One is an application example of 66kV superconducting cables instead of 275 kV underground cables in the conventional system where electric power can be transmitted in the city by 66kV underground transmission lines through 275/66 kV underground power substation, 275 kV underground transmission line (maximum length is 30 km) and 500/275 kV suburban suburban substation. The other is that 66kV superconducting transmission system is connected directly with 66 kV superconducting generation power plants constructed on the coast side of the city. In this case, 22/275 kV and 275/66 kV substations in the conventional system can be omitted. In addition, the transmission capacity can be increased greatly by replacing the conventional cables into superconducting cables in the present culvert.

Conclusion

We performed a conceptual design and the cost evaluation of liquid nitrogen cooled superconducting cables and clarified their concrete application to electric power systems. The results are as follows:

1. The optimum transmission voltage is 66kV for 2.5GVA/cct or 154kV for 5.0GVA/cct. This means an excellent economical merit of neglecting underground power substations in the metropolitan cities.

2. Instead of 6 ccts of 275kV conventional OF cables installed in a culvert (the total transmission capacity is 2,500MVA), only one circuit of 66kV liquid nitrogen superconducting cable can transmit the same capacity. Therefore if all conventional cables in one culvert can be replaced by new cables, the transmission capacity can be increased 5-6 times.

3. In addition if a superconducting generating machine with 66kV armature winding can be developed, we need not power transformers but we can connect the generating machine and transmission cables directly.

4. Even if the system voltage of power cables is 66 kV, superconducting transmission cables can be available without compensation equipments for a longer distance than 50 km.

5. AC current loss of the new superconducting material should be lower than $100uW/cm^2$ at 500 A/cm.

Table 1 The studied cases and the design conditions

Transmission Capacity	1GVA	2.5GVA	5GVA
Voltage (Impulse Voltage)	66kV (350kV)	154kV (750kV)	275kV (1050kV)
Fault Current & Duration Time	30kA-0.4sec	40kA-0.4sec	50kA-0.4sec
Outer Diameter of Whole Cable	420mmΦ		
Material & Thickness of Inner/Outer Pipe	SUS/Steel 5mm respectively		
Cooling Condition Temperature Pressure Drop	inlet 77k outlet 85k 15 atm or less		
Electric Insulation	PPLP-LN² ε×tanδ 2.3×0.001 Gmax Imp 95kV/mm AC 15kV/mm		
Thermal Insulation	Vacuum Powder Insulation Thermal Conductivity 2×10⁻³ W/mk Multi Layer Thermal Insulation 2×10⁻⁴ W/mk (Superinsulation)		
Cable Structure	TYPE-1		TYPE-2
Cooling Method	Inner/Outer Cooling		Outer Cooling
Conductor Shape	Tape Cylindrical Conductor		Multi-Filament Conducor
Conductor Structure	Superconductor —Cu		Superconducting Filament —Cu 2mmΦ
Current Density	1000A/cm or less		10³,10⁴,10⁶A/cm²
AC Loss	20,10²,10³ μW/cm² at 500A/cm		Caluclating Histerisis Loss by Self-Current
Shape of Stablization Material	Cu, Side of Current Conductor (Inner Cooling Tube) Side of Shielding Conductor (Tape)		Cu, Both Side 2mmΦ Mluti-Filament

Table 2 The transmission line constants and characterristic of high Tc superconducting power cable

Cable structure		TYPE-1 (vacuum powder insulation) (AC loss 10² μW/cm²)								
Basic Electric Characterisic	Transmission Capacity (GVA)	1			2.5			5		
	Voltage(kV)	66	154	275	66	154	275	66	154	275
	Impulse Voltage(kV)	350	750	1050	350	750	1050	350	750	1050
Items	Unit	(8750A)	(3750A)	(2100A)	(21870A)	(9375A)	(5250A)	(43740A)	(18750A)	(10500A)
Resistance R	μΩ/km/phase	1.36	0.41	0.32	2.92	1.54	0.73	4.95	2.47	0.86
Capacitance C	μF/km/phase p.u.	1.65 0.000226	0.72 0.00536	0.57 0.01354	1.78 0.00244	0.72 0.00536	0.57 0.01354	1.92 0.00263	0.80 0.00596	0.57 0.01354
Inductance L	μH/km/phase p.u.	12.91 0.000931	23.10 0.000306	27.07 0.000112	11.99 0.000865	23.10 0.000306	27.07 0.000112	11.09 0.000800	20.74 0.000275	27.07 0.00012
Serge Impedance Loading SIL	GW	1.56	4.19	11.0	1.68	4.19	11.0	1.81	4.66	11.0
Limited Line Length Phase Angle=15 degrees	km	280	850	2310	120	340	920	65	190	460
Voltage of Receiving End in Line Length 30km	p.u.	0.998	1.00	1.00	0.992	0.998	1.00	0.981	0.996	0.999
Phase Angle between Voltage of Transmitting End and That of Receiving End in Line Length 30km	degree	1.6	0.5	0.2	3.8	1.3	0.5	7.0	2.4	1.0
Reactive Power by Capacitance in Line Length 30km	GVar	0.12	0.28	0.70	0.13	0.28	0.70	0.14	0.31	0.07

p.u. is 1GVA fundamental

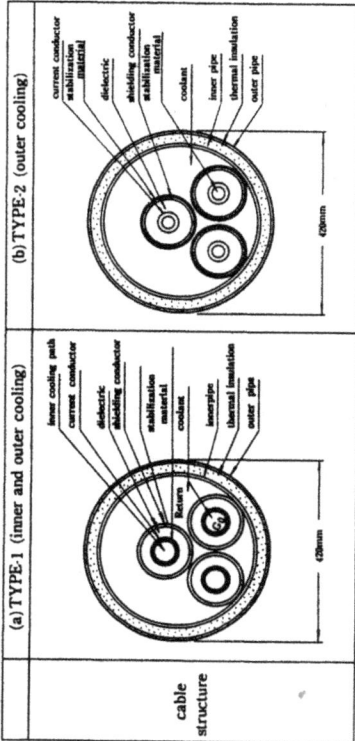

Fig.1 The conceptual designed structures of high Tc superconducting power cable

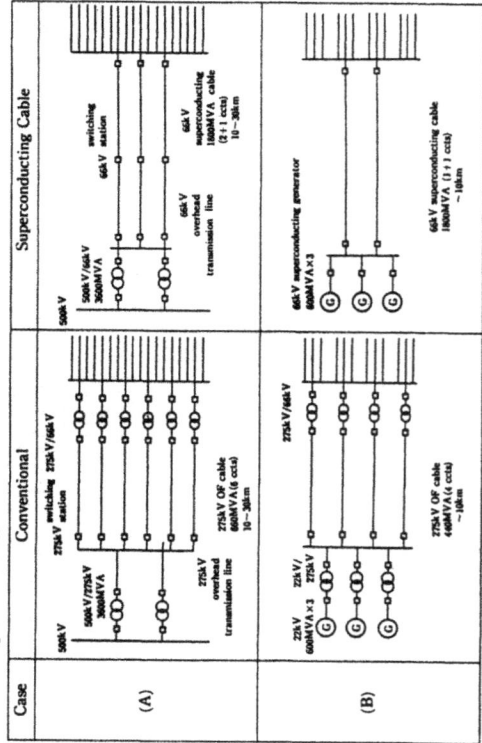

Fig.2 The comparison of transmission cost in all kinds of cables

Fig.3 The comparison of application High Tc superconducting Power cable to usual transmission method

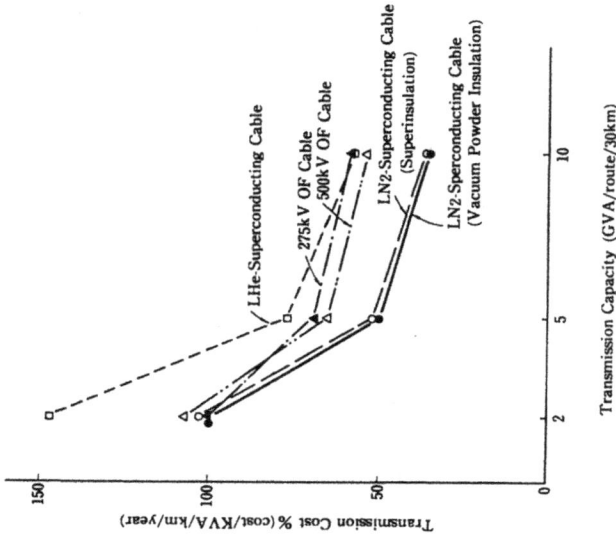

THE DESIGN OF SUPERCONDUCTING POWER CABLE USING HIGH Tc SUPERCONDUCTORS

*Naotaka ICHIYANAGI, *Fumiaki ENOKUBO, **Yoshio FURUTO

The Furukawa Electric Company Ltd.
*   6 Yawatakaigandori Ichihara Chiba Japan
** 2-6-1 Marunouchi Chiyodaku Tokyo Japan

Abstract - The feasibility of 66kV High Tc superconducting cable was studied. Since the Hc_1 of high Tc superconductor is small, it has to be used between Hc_1 and Hc_2 in order to transmit a large current. Therefore AC loss due to the penetration of magnetic field is a very important problem for such a cable. The loss depends on the critical current of material and transmission current.
Our feasibility study reveals that $Jc > 10^4$ A/cm^2 is required for 5000A class and $Jc > 10^5$ A/cm^2 for 10000A class.

Introduction

Since the discovery of high Tc superconducting materials, the feasibility of their applications has been studied in many fields. Their application to power cables is considered most feasible among the field of power supply equipment, because they are used in a relatively low magnetic field. The fundamental design of superconducting wire and the feasibility of replacing 275kV 660MVA/2cct conventional cable in tunnel by 66kV 660MVA/cct(5800A) high Tc superconducting power cable are discussed in this paper.

The Design of Superconducting Wire

A superconducting wire is designed according to the process shown in Fig.1. There are "tape type wire" and "string type wire" which are composed of a superconducting part and a stabilizer part as shown in Fig.1. Only the tape type wire is considered here as a general one. The thickness of the superconducting layer is determined by the transmission current and critical current of material supposing that current density is constant and equal to critical current (Bean Model).

$$ts = Iop/\pi DJc \qquad (1)$$
ts : thickness of SC thin layer,   Iop: design current
D : diameter of thin conductor
Jc : critical current of SC material

When the superconducting layer is quenched by a fault current, it streams in the stabilized layer. The stabilized layer is heated up by this current. Therefore the thickness of stabilized layer should be so designed that the temperature of conductor may not rise too high. The temperature rise is determined by the value of fault current, duration time, the conductivity of the stabilizer and heat capacity. The temperature rise by fault current against total cross section of copper stabilizer is shown in Fig.2. The required cross sectional area is around 100mm^2 for a fault current of less than 50kA*0.4sec to keep the conductor temperature below room temperature. Therefore the total thickness of the composite wire is determined by the diameter of the conductor considering the above conditions.

The strain appears in the superconducting wires when they are wound in conductor shape which drastically reduce the critical current of superconducting material. The critical current against strain measured in our laboratory so far is shown in Fig.3 for instance. This Figure indicates the superconducting wire should be used under a strain smaller than 0.1%. The strain of the wire in terms of the thickness of it is shown in Fig.4. It is obvious that the smaller the strain is the bigger the critical current is required.

| Thickness of SC layer |
| Thickness of stabilizer |
| Stability of SC layer |
| Flexibility of wire |
| AC loss |

SC layer / stabilizer

tape type   string type

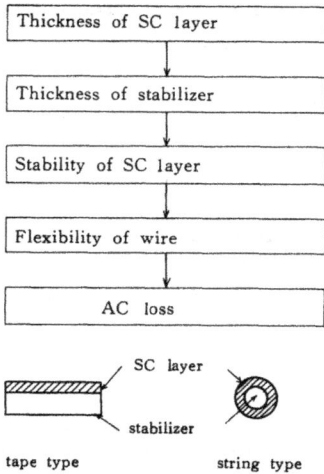

Fig. 1 Design of SC wire

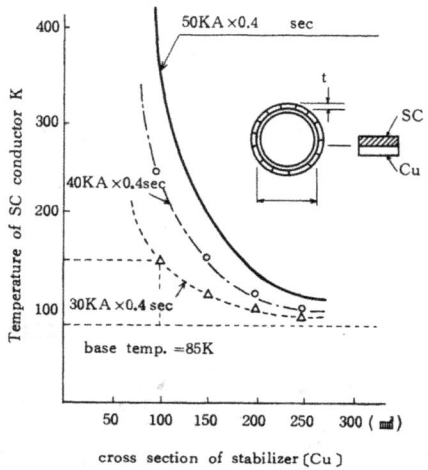

Fig. 2 Temperature rise SC conductor
by fault current

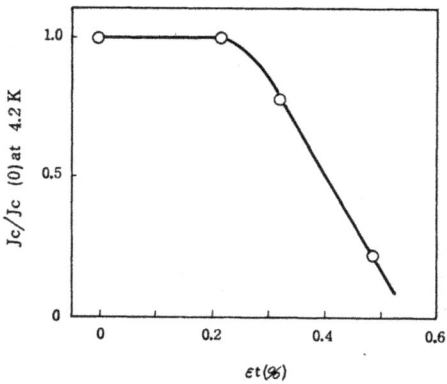

50KA×0.4 sec

40KA×0.4sec

30KA×0.4 sec

base temp. =85K

Temperature of SC conductor K

cross section of stabilizer〔Cu〕

Fig. 3 Jc variation relative to tensile strain

(Y−B−C−O oxide superconductor)

Jc/Jc (0) at 4.2 K

εt (%)

Strain (%)

$D=\emptyset 40$

$\emptyset 60$

$\emptyset 80$

$Jc=10^5$ A/cm

$Jc=10^4$ A/cm

Superconductor wire thickness (t)

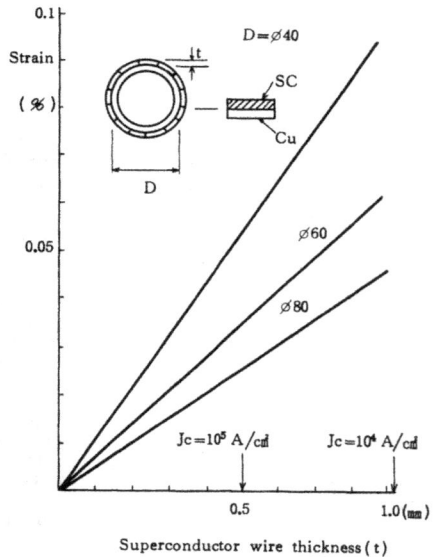

Fig. 4 Wire thickness and strain
(5800A conductor)

Stability of Superconducting Layer

In order to get rid of growing a bud of normal conducting region, flux jump which is the main cause must be suppressed. The dynamic stability condition due to dumping of magnetic flux and adiabatic stability condition due to selfenthalpy should be satisfied for it. Although physical constants of high Tc superconducting material are not sure, the estimated stable condition is that the thickness of superconducting layer is thinner than 6.5mm for $Jc=10^5$ A/cm^2. The condition is easily achieved, because the superconducting layer is supposed to be thinner than 1mm.

AC Loss of Superconducting Cable

The conventional superconducting cable using Nb or Nb_3Sn conductor is so designed that magnetic field at conductor surfaces is smaller than Hc_1 to make conductor loss minimum. In this case the current flows only in the penetration depth layer which is called surface current. The surface current of the conventional material is around 500~700 A/cm by the following equation.

$Is=Iop/\pi D<Hc_1$ (A/cm)　　　(2)
　　Is: surface current　Iop: design current　D: diameter of conductor

However the Hc_1 of high Tc superconducting material (50~100 Oe) is much smaller than conventional material and then surface current is estimated only 28~56 A/cm. Therefore high Tc superconducting conductor has to be used under the surface magnetic field between Hc_1 and Hc_2 in order to transmit a large current with a conductor size limited. In this case the current streams inside the superconducting layer and magnetic field penetrate inside as well which generates AC loss mainly due to magnetic hysteresis. An example of magnetic hysteresis of high Tc superconducting material is shown in Fig.5.

When the current distribution in superconducting layer is assumed to obey the Bean model, the AC loss (Ph) due to hysteresis is generally expressed by the following equation (3) for the surface magnetic field being low enough.

$Ph=2\sqrt{2}/3 \cdot \mu_0 (Iop/Pe)^3 Pef/Jc$　　　(3)
　　Pe : perimeter of conductor　　　f : frequency (50Hz)

The AC loss calculated by Equation (3) against transmission current and critical current is shown in Fig.6 and compared with that of normal conductor. As the energy, ten times more, is required to eliminate the loss generated at liquid Nitrogen temperature, the total loss including cooling energy should be small enough. The various properties of superconducting wires at different critical currents are compared in Table 1. In order to keep the strain less than 0.1% and AC loss of the conductor less than several w/m (limitation to make cooling station span long enough), the critical current should be more than 10^4 A/cm^2 for 5000A transmission and 10^5 A/cm^2 for 10000A transmission.

Example of High Tc Superconducting Cable Design

The 66kV 660MVA/cct high Tc superconducting cable is designed to replace the conventional main transmission line, that is, 275KV 660MVA/2cct in tunnel. The cable structure and each loss are shown in Fig.7. Assuming $Jc=10^4$ A/cm^2, inlet temperature of $LN_2=65k$, inlet pressure of $LN_2=20$ atm, the cooling station span is around 20~30km. The cable design and properties are changed according to practical conditions, for instance, limitation of cable size.

However the critical current of high Tc superconducting material must be higher than 10^4 A/m^2 at least for the first stage.

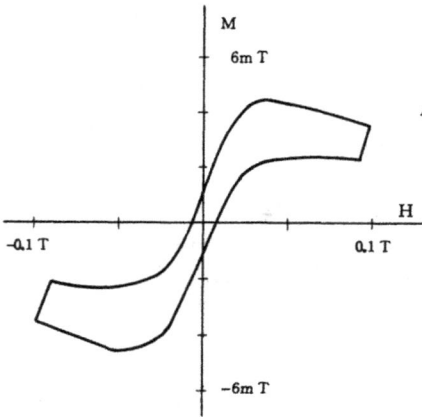

Fig. 5 Magnetization curve (77K)

of Y -B -C -O oxide superconductor

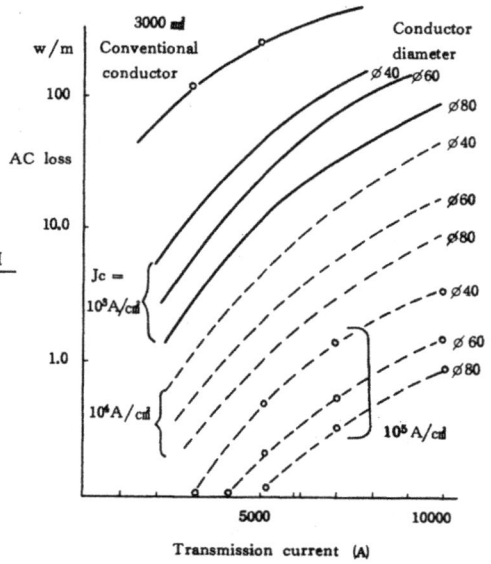

Fig. 6 AC loss

Table 1　Properties of high Tc superconducting conductor (D=φ 60mm)

item	unit	calculated value (tape type)					
		5000A(265A/cm)			10000A(530A/cm)		
rated current	A						
critical current	A/cd	10^4	10^5	10^6	10^4	10^5	10^6
thickness of SC layer (without margin)	mm	0. 27	0. 027	0. 0027	0. 53	0. 053	0. 0053
thickness of stabilizer	mm	0. 5	0. 5	0. 5	0. 5	0. 5	0. 5
strain	%	0. 096	0. 066	0. 063	0. 129	0. 069	0. 063
AC loss	W/m	2. 08	0. 208	0. 0208	16. 7	1. 67	0. 167
adiabatic stability	mm	65. 5	6. 55	0. 655	65. 5	6. 55	0. 655
dynamic stability	mm	7. 4	2. 44	0. 74	5. 3	1. 67	0. 53
minimum propagation zone	mm	56. 5	56. 5	56. 5	28. 3	28. 3	28. 3

$Ks = 5. 5 W/m \cdot k$　　$Cs = 1. 0 \times 10^6 J/m^3 \cdot k$　　$\rho s = 1 \times 10^5 \Omega \cdot m$　(at 77k)

$K = 500 W/m \cdot k$　　$C = 1. 8 \times 10^6 J/m^3 \cdot k$　　$\rho = 20 \times 10^{15} \Omega \cdot m$

Heat inleak
2 w /m

Dielectric loss
0.3w / m

AC loss (external conductor)
2.1w / m

AC loss (internal conductor)

3.3w/m

Semi-synthetic paper insulation

Superconductor tape

Al pipe

Stainless steel pipe (SUS)

285 ϕ

215 ϕ

76 ϕ

52 ϕ

Thermal insulation layer

Superconductor

Stabilizing Cu

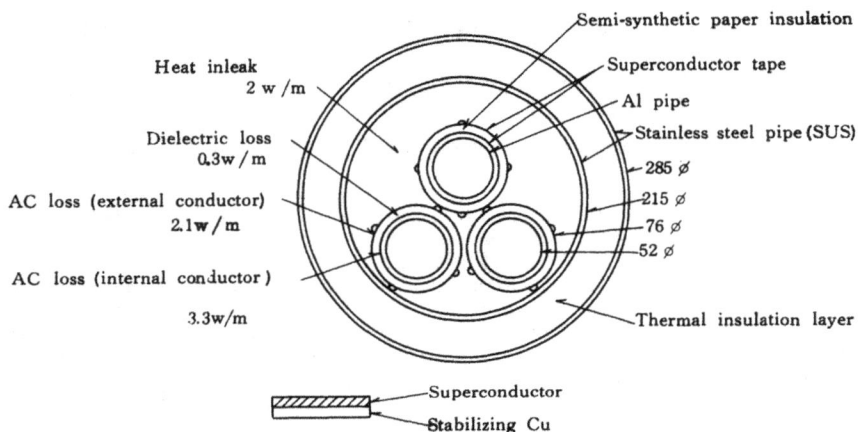

Fig. 7 An example of superconducting cable structure [2]

The cooling stations are assumed to be built at spacing of 20 to 30km along a 66kV, 5,800A(Jc =10^{4}A/cm) line.

Conclusion

The degradation of critical current by strain and large AC loss should be considered, when high Tc superconducting material is applied to a power cable.

These properties depend much on the critical current, and the critical currents are required more than 10^4 A/cm^2 for 5000A transmission and 10^5 A/cm^2 for 10000A transmission.

References

1) Naotaka ICHIYANAGI, Fumiaki ENOKUBO, Yoshio FURUTO
   JIEE National Conference 1304 (1987)

HIGH-Tc SUPERCONDUTOR ANODES FOR RELATIVISTIC ELECTRON BEAM DIODES

Hidenori MATSUZAWA, Yoshiharu ISHIBASHI, Kazunori IRIKURA, Kenji OKAMOTO, Tomoaki OSADA, Akihide MOCHIZUKI, Haruhisa WADA, and Tetsuya AKITSU

Faculty of Engineering, Yamanashi University, Kofu 400, Japan

Abstract - A cylindrical, high-Tc Bi-compound anode (axial length of 30 mm and inner diameter of 10 mm) was used for beam diodes to focus relativistic electron beams (REBs, 270 keV, 1.5 kA, pulse width of less than 5 ns) at pressures of the order of 0.1 -Torr Ne. Even such a narrow anodes focused the REBs, and the optimum pressure was 0.15 Torr, being the same as that for the 20-mm diameter anode. The REBs emerged from the exit of the anode were focused more than for the previous anode with 20-mm diameter. Narrower inner-diameter cylinders seem to be used as lenses and guides for REBs.

Introduction

We proposed superconducting lenses for charged particle beams, especially for relativistic electron beams (REBs), and showed successfully their utility in previous papers [1,2]. In the present paper, further experimental results are reported for the diode configuration where the anode had a 10-mm inner diameter, being narrower than the previously used ones. The possibility is shown that narrower, cylindrical, high-Tc tubes are employable as REB lenses.

The principle of the lenses is as follows (Fig. 1): When intense electron beams are injected into the apertures of superconducting tubes, self-magnetic field of the electron beams are compressed by the tubes because of the Meissner effect or the skin effect. The electron beams are focused with those compressed magnetic field, like a self pinch. These lenses are expected to have high focusing-ability, because superconductor tubes consume no energy. On the contrary, the normal conductor tubes dissipate energy while preventing the magnetic field from penetrating into the normal conductors. The energy consumed is supplied from the electron beams.

When relativistic electron beams travel through a gas-filled, grounded metallic tube, space-charge limited currents I_L were given for normal conductor tubes [3,4]. In deriving the equation for the currents, stationary cases were supposed where the work necessary to produce the self magnetic field was ignored. Therefore, experimental investigations with superconducting tubes may provide data for reexamining the equation of I_L.

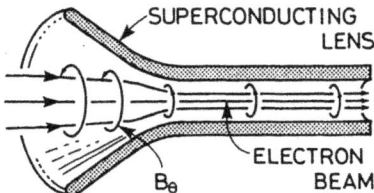

Fig. 1. Principle of superconducting lenses for REBs.

Experimental Apparatus

Figure 2 shows the REB diode used. The superconductor anode had a 30-mm length and a 10-mm inner diameter. Pressures of the diode chamber were controlled by flowing Ne gas of the lowest thermal conductivity among the gases (He, H_2, and Ne) which do not liquefy at the boiling point of nitrogen. These high-pressure operation of diodes provides higher REBs and charge-neutralization [5], and makes it easy to transport REBs through metallic guide tubes [6, 7]. The Faraday cup was separated from the diode chamber with a 20-μm-thick titanium foil. The Faraday cup which was kept at pressures of the order of 10^{-4} Torr collected electron currents of kinetic energies higher than 60 keV [8]. High-voltage pulses (450 kV or more and pulse duration time of 40 ns) were supplied from a coaxial Marx-type generator (ten stages of 8100-pF modules) [9].

Bi-compounds were used for anodes instead of Y-compounds, because the Bi-compounds have a critical temperature of about 100 K which is 10-K higher than that of Y-compounds. Figure 3 shows the temperature characteristics of resistance of Bi-compounds used. We employed commercially available, pre-sintered powders.

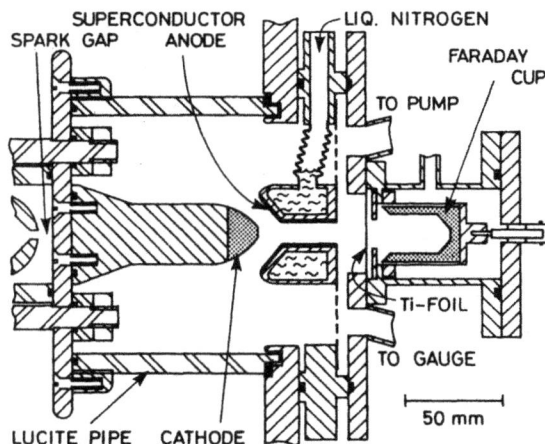

SUPERCONDUCTOR   LIQ. NITROGEN
SPARK GAP   ANODE

FARADAY CUP

TO PUMP

Ti-FOIL

TO GAUGE

50 mm

LUCITE PIPE   CATHODE

Fig. 2. REB diode used. Diode pressures were controlled by flowing Ne gas. Faraday cup was kept at 10^{-4} Torr.

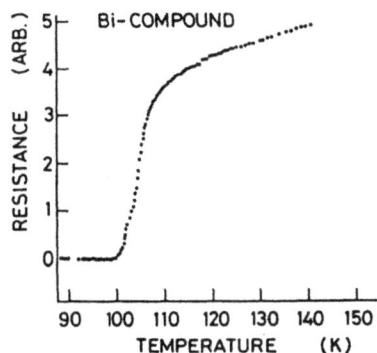

Fig. 3. Temperature characteristics of resistance of Bi-compounds used for REB anodes.

Experimental Results

Figure 4 shows REBs as a function of the distance between the cathode and the top of the anode. When the cathode was spaced at 8 mm from the anode, best impedance matching was realized among the diode and the high-voltage pulse generator. As the cathode-anode distance was kept 0 mm in the previous papers, experimental results in the present paper are for the 0-mm separation unless otherwise stated.

Figure 5 shows the pressure dependences of REBs generated. Open and solid circles indicate the peak values of REBs and their full widths at half maximum (FWHM), respectively. The error bars mean the highest and the lowest REBs at the respective time. When Fig. 5 is compared with the previous data [1, 2] for the 20-mm inner diameter anode, the optimum pressures were the same for both cases, but the pressure ranges over which effective generation and transport of REBs were achieved were much narrower for the 10-mm inner diameter anode. These narrowing in the pressure range is ascribed to the higher degree of charge neutralization: Higher currents are confined in narrower region inside the anode. Similar characteristics were observed even for a 145-mm-long anode [2] because of the same requirement.

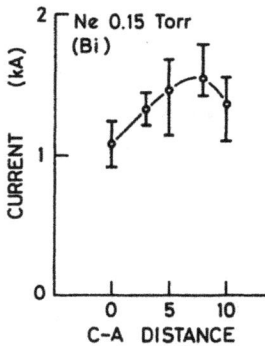

Fig. 4. REBs generated as a function of cathode-anode distances. Best impedance matching is for about 10-mm separation.

Fig. 5. Pressure dependences of REBs and their full widths at half maximum. Optimum pressure is 0.15 Torr of Ne. Aperture diameter was 26 mm.

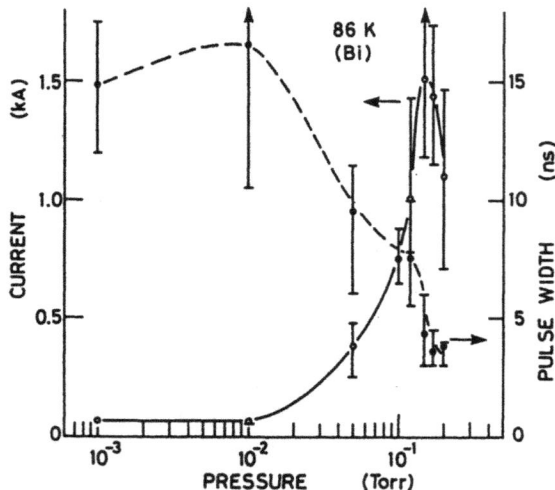

Figures 6a and 6b show the radial distribution of REBs as a function of apertures for copper and Bi-compound anodes with a 10-mm-inner diameter, respectively. In Fig. 6a, cryogenically-cooled copper anode had higher REBs than for room temperature operation, but the REBs did not increase linearly with the apertures: The REBs expanded as a hollow-like distribution. The REBs in Fig. 6b, on the other hand, increased proportionally with the aperture and expanded more for the cathode-anode distance of 8 mm than for 0 mm, probably because of higher space-charge effect. From comparison between Figs. 6a and 6b, superconducting anodes are certainly superior to normal conducting anodes which are cryogenically cooled.

Figure 7 shows the axial-distance dependences of REBs for the superconducting anode. The distance was measured from the exit of the anode. There existed a region (30 to 50 mm from the exit) where REBs were transported almost with no expansion and balance of forces was obtained between the expansion force due to space-charge and compressing magnetic force. The diameter of the aperture was 10 mm.

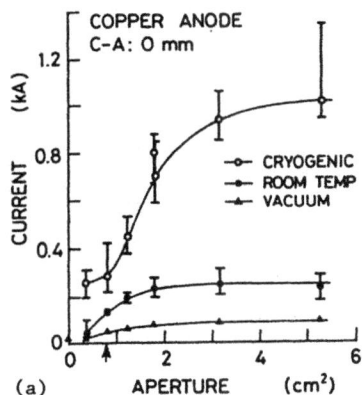

(a)

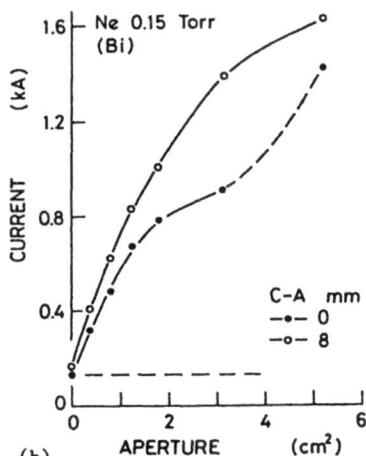

(b)

Fig. 6. Aperture dependences of REBs. (a) copper anode and (b) superconducting anode. Cathode-anode distances are (a) 0 mm and (b) 0 and 8 mm. Ne pressure was 0.15 Torr.

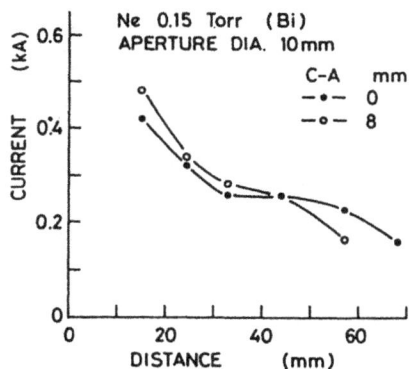

Fig. 7. Axial distance dependences of REBs. Distance was measured from exit of anode. Data are for cathode-anode distances of 0 and 8 mm.

Discussions

The experimental results shown above indicate that superconducting anodes are certainly better in focusing REBs than normal conducting anodes. The differences between them are ascribed to only their electrical resistivities. When resistivity is zero, no difference is apparently observed between the Meissner effect and the skin effect for pulsed electron beams whose duration time is much less than the diffusion time of magnetic field: In either case, pulsed magnetic field is rejected when it is applied to conductors from outside. The Meissner effect plays excellent roles when continuous magnetic field is applied.

Qualitative explanations are given on the mechanisms of focusing of REBs in the following (Fig. 7): When REBs with a radius a are traveling in a free space, their self-magnetic field is distributed as in Fig. 7a. We suppose here that the REBs are injected into superconducting tubes with a radius b. The self field is confined between the REBs and the tubes as in Fig. 7b. Total magnetic flux corresponding to the shaded area is equal in amplitude to that in Fig. 7a, and no energy is consumed to induce current on the wall. The REBs are compressed untill the magnetic pressure balances with expanding pressure of the REBs due to their thermal motion in radial direction and radial gradient of plasma density. When normal conducting tubes are used (Fig. 7c), induced currents on the wall are less than REBs, and a small amount of energies of REBs is consumed continuously via the self-magnetic field as Ohmic loss to induce the currents. The total magnetic flux corresponding to the shaded area is less than that in Fig. 7b. The highest field H_a'' is therefore less than H_a' for the superconducting tube, and the achievable REB-radius is larger than that in Fig. 7b.

Space-charge limiting currents I_L, which can flow through a conducting tube filled with low pressure gases, were given [3,4] by

$$I_L = \frac{17 \ (\gamma^{2/3} - 1)^{3/2}}{\{1 + 2\ln(\frac{b}{a})\}(1 - f_e)} \qquad [kA]$$

where $\gamma = (1 - \beta^2)^{-1/2}$, $\beta = v/c$, v is the velocity of electrons, c is the velocity of light, and f_e is the degree of charge neutralization given by

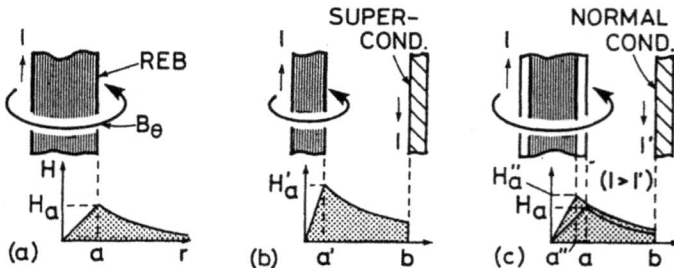

Fig. 7. Self-magnetic field distribution of REBs for free space (a), for superconducting tubes (b), and for normal conducting tubes (c). Magnetic field is completely confined in case of (b). For case of (c), peak value H_a'' of magnetic field is less than H_a' for superconducting tubes in (b) because of Ohmic loss.

n_i/n_e where n_i and n_e are the flowing number density of ions and electrons, respectively. In deriving the equation [3], 'stationary case' was assumed where no consideration was taken into the work necessary to produce the self-magnetic field. Conductors were also assumed to be 'perfect conductors'. Thus the expression for I_L may be modified for superconducting tubes.

For superconducting anodes with a 20-mm-inner diameter [1, 2], the focused REBs had a 2-mm diameter. Kinetic energy of the REBs is 270 keV [10], then $\gamma = 1.4$ and $I_L = 0.4 / (1 - f_e)$ kA. The total current observed for that anode was 0.8 kA, so the total current can be regarded to be within the limiting current I_L and no violation of the equation is found yet. Beam spot sizes for the 20-mm-diameter anode were estimated from damage patterns of plastic films, caused by irradiation of REBs. For the anode with a diameter of 10 mm, no beam spots were observed sucessfully. To discuss the validity of the equation for I_L for superconducting tubes, we are trying another method to measure the beam diameter.

References

1) H. Matsuzawa, O. Ohmori, H. Yamazaki, J. Ueno, A. Furumizu, A. Saito, T. Takahashi, and T. Akitsu, "Enhanced generation of relativistic electron beams with high-pressure operated cryogenic diodes", *7th Inter. Conf. on High-Power Particle Beams*, No. RP4 (Karlsruhe, July 4-8, 1988).

2) H. Matsuzawa, O. Ohmori, H. Yamazaki, J. Ueno, A. Furumizu, A. Saito, T. Takahashi, and T. Akitsu, "High-Tc superconductor lenses and guides for intense charged particle beams", *1st Inter. Symp. on Superconductivity*, No. PVD-5 (Nagoya, August 28-31, 1988).

3) L. S. Bogdankevich and A. A. Rukhadze, "Stability of relativistic electron beams in a plasma and the problem of critical currents", Soviet Physics, 14, 163-179 (1971).

4) W. W. Destler, P. G. O'Shea, and Z. Segalov, "Experimental study of propagation of intense relativistic electron beams in nonconducting vacuum drift tubes after passage through a localized plasma source", J. Appl. Phys., 61, 2458-2462 (1987).

5) R. M. Miller, *An Introduction to the Physics of Intense Charged Particle Beams* (Plenum, New York, 1982).

6) S. E. Graybill, "Dynamics of pulsed high current relativistic electron beams", IEEE Trans. Nucl. Sci., NS-18, 438-446 (1971).

7) P. A. Miller, J. B. Gerardo, and J. W. Poukey, "Relativistic electron beam propagation in low-pressure gases", J. Appl. Phys., 43, 3001-3007 (1972).

8) L. Pages, E. Bertel, H. Joffre, and L. Sklavenittis, "Energy loss, range, and bremsstrahlung yield for 10-keV to 100 MeV electrons in various element and chemical compounds", Atomic Data, 4, 1-127 (1972).

9) H. Matsuzawa and T. Akitsu, "Output voltage waveform improvement of the coaxial Marx-type high-voltage generator", Rev. Sci. Instrum., 56, 2287-2289 (1985).

10) H. Matsuzawa and T. Akitsu, "High-pressure operation of a beam diode for relativistic electron beams", J. Appl. Phys., 63, 4388-4391 (1988).

Section 2. Superconducting Magnets and the Related Materials

PROGRESS IN THE UNDERSTANDING AND MANIPULATION OF
MICROSTRUCTURE IN HIGH J_C NB-TI ALLOY COMPOSITES

P. J. LEE[*], J. C. MCKINNELL[*], and D. C. LARBALESTIER[*+]

The Applied Superconductivity Center, University of Wisconsin-Madison
1500 Johnson Drive
Madison, WI 53706 USA

[*]Also Materials Science Program
[+]Also Dept. of Materials Science and Engineering

Abstract - The development of high J_c microstructures in Nb46.5wt.%Ti under standard processing techniques is reviewed. Microstructures of alloys of higher Ti content given the same processing are shown to be significantly different. The effect of composition on precipitate morphology has recently been found to be quantifiable, predictable and controllable. The implications of these results in terms of low field applications is discussed.

Introduction

NbTi alloys in the composition range of Nb46.5wt.%Ti to 50wt.%Ti have become the dominant commercial superconductors. The development of high J_c microstructures in Nb46.5wt.%Ti has only recently been fully characterized and quantified.[1,2,3] The driving force behind these extensive studies of the Nb46.5wt.%Ti alloy has been the desire for increased J_c wire for high energy physics applications where the flexibility and high J_c of this alloy has made this the standard material for magnet designs in the 5T to 8T range. The excellent high field performance of Nb46.5wt.%Ti is related to the high H_{c2} of this composition. The principal commercial application of superconducting wire, however, is MRI where the magnetic fields are relatively low (0.35-4T). There is also a very large possible future market for NbTi for superconducting magnetic energy storage, SMES, where low to mid magnetic fields will probably be required (3-5T) with cooling by liquid He II (1.8 K). Under these conditions where high H_{c2} is less important, an early study by McInturff and Chase[4] indicated that superior J_cs could be obtained using higher Ti content alloys. Since that early study, improved processing techniques have resulted in a two-fold increase in J_c for the Nb46.5Ti alloy throughout the field range 2-8T.[5] Using these new techniques in a recent study, these high Ti content alloys have again been shown to exhibit superior low field current carrying capability.[6] The microstructures generated in these alloys, however, are inferior in homogeneity to those developed in Nb46.5wt.%Ti and have increased filament hardness.[6,7] If these microstructural problems can be solved, then the low field J_c capabilities of NbTi alloy should be further enhanced.

The paper reviews the essential elements of high J_c NbTi microstructures and examines the latest attempts to control the microstructure of high Ti alloys.

Nb46.5wt.%Ti Final Wire Microstructures

The microstructural development of high J_c Nb46.5wt.%Ti wires has been extensively studied by TEM.[1,2,3] The highly anisotropic nature of these heavily cold worked wires necessitates examination in both transverse and longitudinal cross-section. The results of these studies are summarized in the schematic diagram in Figure 1, which is a three dimensional representation of the α-Ti ribbon morphology in a high J_c wire. Quantification of the microstructure indicates that the mean ribbon thickness is approximately 1 nm with an average of 4 nm of β-NbTi between each ribbon.[3,7] The ribbons rarely exceed 3 nm in thickness in an optimized wire. The α-Ti precipitates extend for more than 2 μm parallel to the drawing axis of the wire. The microstructures of such wires are very uniform across the wire cross-section. α-Ti precipitates make up 20-26% of the filament. Only the α-Ti and β-NbTi phases are observed within the filaments. If diffusion barriers are not used the remnants of brittle reaction products between the Cu stabilizer and Nb-Ti filaments can also be found at the Cu-NbTi interface.

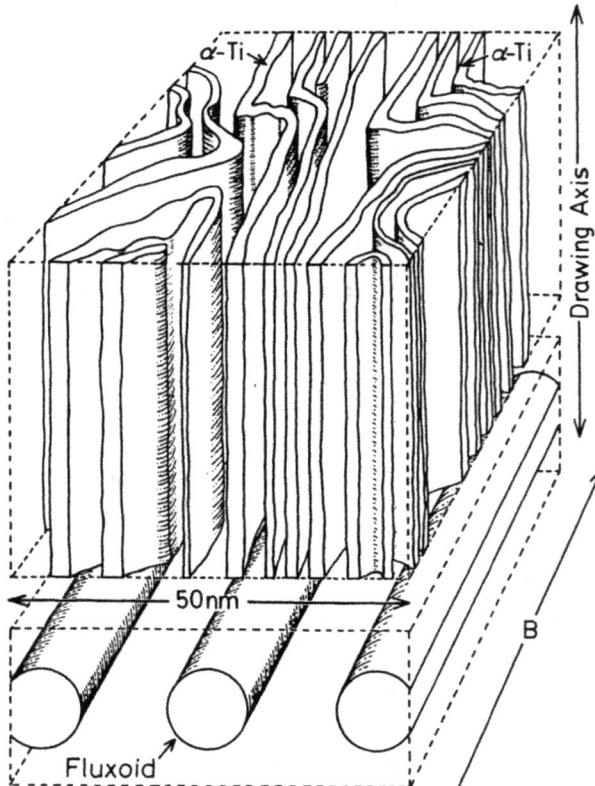

Fig.1. Schematic illustration of α-Ti ribbon morphology in a high J_c Nb46.5wt.%Ti filament. The equilibrium fluxoid spacing at 5T, 4.2 K is compared below.

Development of High J$_c$ Microstructure in Nb46.5wt.%Ti

In order to obtain the uniform microstructure found in the optimized wires the initial NbTi rod must be chemically homogeneous. A schematic illustration of the processing sequence is shown in Figure 2. The initial precipitation is given when the NbTi has received a cold work drawing strain of at least 5. The microstructure after this heat treatment is illustrated in Figure 2b, and consists of α-Ti precipitates at β-NbTi grain boundary triple points and a thin (less than 4 nm) grain boundary film of α-Ti. The heat treatment is typically for 10-80 hours in duration and at between 370°C and 420°C. The effect of temperature and duration of heat treatment is small at this stage. Lower temperatures favor the film precipitation and at 300°C only grain-boundary film is produced.[8] The distribution and quantity of triple-point α-Ti precipitates is controlled by the initial β-NbTi grain

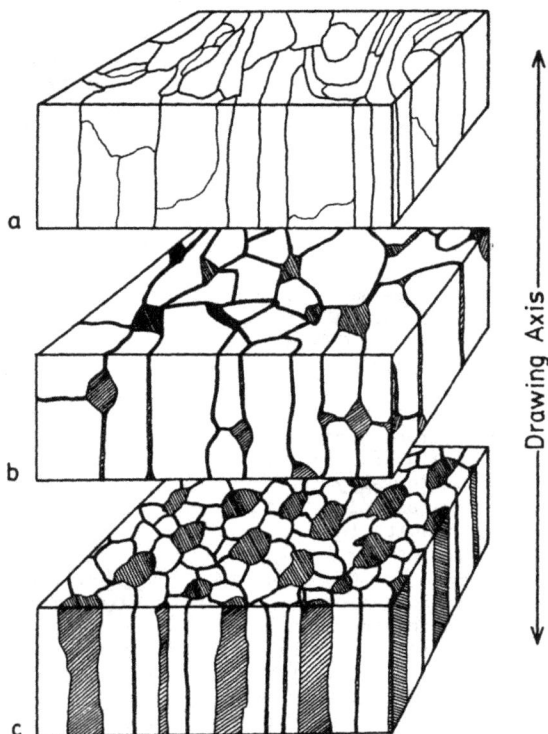

Fig.2. A schematic illustration of the development of microstructure in a conventionally processed Nb46.5wt.%Ti high J$_c$ superconductor, a) β-NbTi before initial heat treatment, b) precipitation of α-Ti (shaded areas) at grain boundaries and grain boundary triple points produced by initial heat treatment, and c) increase in volume of precipitate and refinement of microstructure by subsequent drawing and heat treatment cycles prior to final drawing strain.

size. The precipitate size after initial heat treatment can vary
significantly after initial heat treatment, depending on grain size but the
total volume of precipitate does not exceed 11% of the filament. Further
heat treatments are required to produce the 20-26 volume % of precipitate
found in high J_c material.[3,8] Additional heat treatments are applied after
additional cold work, typically a cold work strain of 1.15. In general the
longer, the more frequent, and the higher the temperature of heat
treatments, the higher is the ultimate J_c of the final wire.[9] These
conditions correspond to an increased volume of precipitate and increased
transverse cross-sectional area.[8] The microstructure after final heat
treatment is illustrated in Figure 2c. The additional heat treatments have
resulted in a uniform distribution of α-Ti precipitates which are similar in
dimension to the β-NbTi grain size. In order to obtain the high J_c
microstructure illustrated in Figure 1 from the final heat treatment
microstructure in Figure 2(c) an additional large cold work strain is
required. The folded ribbon microstructure is a result of the incompatible
deformation characteristics of the hexagonal α-Ti and the BCC β-NbTi. The
folding of the ribbons results in a large increase in the density of
interfacial surface. Final drawing strains required to produce optimized
wire range from 4-6 (98.2-99.8% area reduction), the larger the size of
precipitate at final heat treatment, the larger is the strain required for
optimization. Despite this considerable cold work strain, both phases
exhibit remarkable ductility and we have not so far observed any evidence of
precipitate fracture during normal drawing.

High Titanium Alloys

Applying the heat treatments described above to higher titanium content
alloys results in precipitation that is neither homogeneous nor conducive to
good ductility.[6] For the high Ti alloys, triple point α-Ti competes with
"string of pearls" α-Ti along grain boundaries, Widmanstätten α-Ti needles
in the β-NbTi grains and the metastable ω phase. All these forms of
precipitation can be found in the electron micrograph shown in Figure 3,
which is a transverse cross-section of a Nb53wt.5Ti alloy after initial heat
treatment (at a prestrain of 5) of 40 hours at 420°C. The ω and
Widmanstätten α-Ti considerably increase the hardness of the NbTi and reduce
the drawability of the composite.[6] The precipitation of α-Ti in the string
of pearls morphology results in a very uneven distribution of
precipitation. Increasing the temperature of heat treatment suppresses
ω-precipitation. The higher the Ti content the greater the density of
Widmanstätten α-Ti precipitation.[7] Despite these drawbacks good J_cs can be
obtained from higher Ti content alloys provided that they can be drawn.
Additional heat treatments result in a ripening process that favors the
larger triple-point α-Ti precipitates eliminating most of the needle-like α-
Ti and blunting those that remain resulting in reduced hardness. By using
three heat treatments (80hr/420°C, prestrain of 5) on a Nb58wt.% alloy, a J_c
in excess of 7400 A/mm^2 at 2T has been obtained which is well in excess of
that obtainable with Nb46.5wt.%Ti.[6] Unlike high J_c Nb46.5wt.%Ti wire which
has a strong optimization peak in the J_c versus final drawing strain curve,
the high Ti alloys often exhibit a behavior almost independent of strain.
This result would be expected of a microstructure that contains a large
variation in precipitate size. It suggests that the performance of these
alloys can be further enhanced by creating a more uniform microstructure.

If Nb46.5wt.%Ti is given its first heat treatment at a prestrain of 2

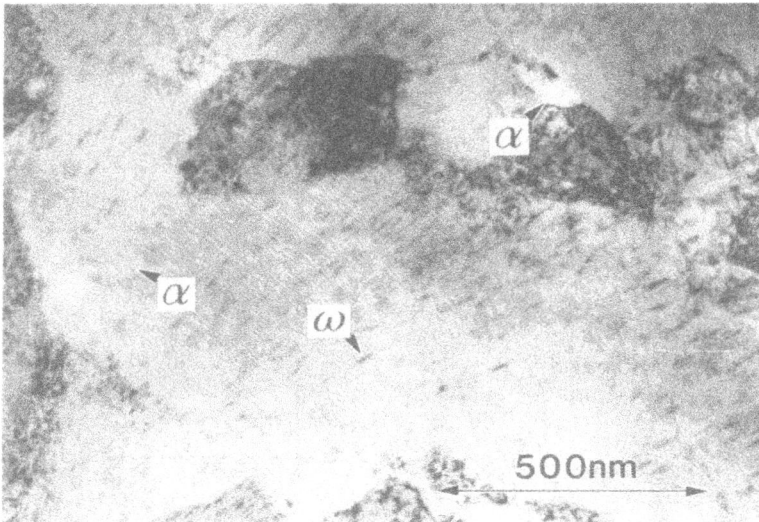

Fig.3. TEM micrograph of transverse cross-section of Nb53wt.%Ti filament
after 1 HT (40hrs/420°C) at a prestrain of 5.

rather than 5, both acicular α-Ti and ω will also be formed.[9] Thus
increasing pre-strain favors triple-point α-Ti precipitation. By applying
this observation to high Ti content alloys, triple-point α-Ti only
microstructures have recently been obtained in alloys of up to 58wt.%Ti.[11]
The relationship between pre-strain and precipitate morphology is
illustrated in Figure 4. With the microstructure now under greater control,
the increased precipitation rate found in higher Ti content alloys can be
quantified. The results of such a quantification[11] are illustrated in
Figure 5 in which the volume of precipitate produced by one and two heat
treatments are compared for three alloys, Nb46.5wt.%Ti, Nb49wt.%Ti, and
Nb53wt.5Ti. By increasing the Ti content by 6.5wt.% the volume of
precipitate was increased by more than 50%.[11] Furthermore, the volume of
precipitate produced in only two heat treatments in the 53wt.% alloy (26
volume %) would require six similar heat treatments in Nb46.5wt.%Ti.[8]

The control now obtainable over the microstructure of high Ti content
NbTi alloys promises further advances in the low field performance of binary
NbTi alloys. Such control requires an increased prestrain that reduces the
available strain space for heat treatment, the increased precipitation rate,
however, results in a reduction in the number of heat treatments required.

Acknowledgments

This work was supported by the US Department of Energy, Division of
High Energy Physics and the Electric Power Research Institute.

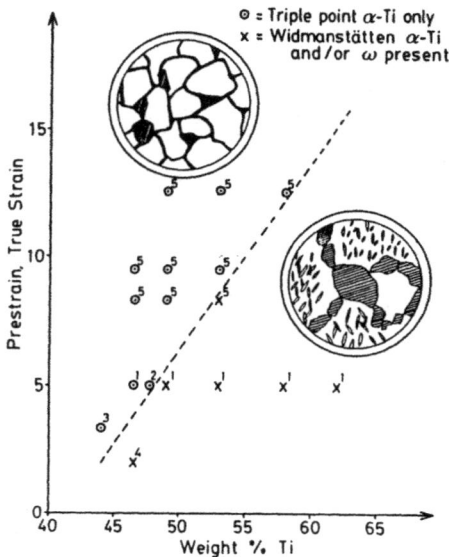

Fig.4. Plot of prestrain versus composition showing the precipitate morphology after initial heat treatment. 1 = ref. 7, 2 = ref. 3, 3 = ref. 12, 4 = ref. 9, 5 = ref. 11

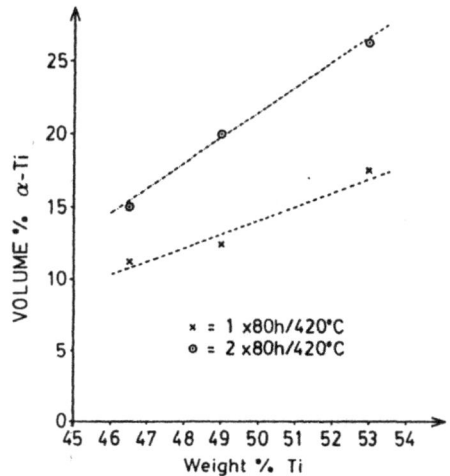

Fig.5. A plot of composition versus volume of α-Ti precipitation after 1st and 2nd heat treatments, data from Lee et al.[11]

References

1) A.W. West and D.C. Larbalestier, Met. Trans. A, 15, 843 (1984).

2) D.C. Larbalestier and A.W. West, Acta Metall., 32, 1871 (1984).

3) P.J. Lee and D.C. Larbalestier, Acta Metall, 35, 2523 (1987).

4) A.D. McInturff and G. Chase, J. Appl. Phys., 44, 2378 (1973).

5) D.C. Larbalestier, A.W. West, W. Starch, W. Warnes, P.J. Lee, W.K. McDonald, P. O'Larey, K. Hemachalem, B. Zeitlin, R. Scanlan, and C. Taylor, IEEE Trans., MAG-21, 269 (1985).

6) J. McKinnell, P.J. Lee, R. Remsbottom, D.C. Larbalestier, P.M. O'Larey, and W.K. McDonald, Adv. Cryogenic Eng. Mat., 34, 1001 (1988).

7) P.J. Lee, D.C. Larbalestier, and J. McKinnell, Adv. Cryogenic Eng. Mat., 34, 967 (1988).

8) P.J. Lee and D.C. Larbalestier, J. Mat. Sci., 23 (1988).

9) M.I. Buckett and D.C. Larbalestier, IEEE Trans. Mag., MAG-23, 1638 (1987).

10) Li Chengren and D.C. Larbalestier, Cryogenics, 27, 171 (1987).

11) P.J. Lee, J.C. McKinnell, and D.C. Larbalestier, presented at the Applied Superconductivity Conference, San Francisco, 1988.

12) P.J. Lee, D.C. Larbalestier, and J.C. McKinnell, unpublished work, Applied Superconductivity Center, University of Wisconsin-Madison, U.S.A.

RECENT ACTIVITIES AT MIT IN THE APPLICATIONS OF ACOUSTIC EMISSION TECHNOLOGY FOR SUPERCONDUCTING MAGNETS

Y. Iwasa

Francis Bitter National Magnet Laboratory and Plasma Fusion Center
Massachusetts Institute of Technology, Cambridge MA 02139

ABSTRACT – Recent activities at MIT in the applications of acoustic emission (AE) technology are presented. Specifically these activities include: 1) acoustic-emission-technology based monitoring of and results from SSC dipole magnets, at both room temperature and 4.2 K; and 2) the analytical and experimental study of acoustic signals emitted by epoxy-impregnated adiabatic superconducting magnets. Our preliminary results from the SSC dipoles support the notion that the principal source of premature quenches in these dipoles is dissipative mechanical events taking place within the winding and that the performance of these dipoles is furthermore critically dependent on the 'mechanical stiffness' of the magnet structure. There appears to be plausible correlation between stiffness and room-temperature wave attenuation data on the one hand and between stiffness and 4.2-K performance on the other. The AE work on an epoxy-impregnated magnet indicates that even in these magnets most AE signals are generated by displacement events occurring within the winding and that the nonuniform temperature-induced thermal stresses can replace electromagnetic stresses as the sources of mechanical events.

Introduction

The source of AE signals is mechanical in that it is elastic energy released relatively quickly. That is, AE power $P_{ae}(t)$ is proportional to $V_{ae}^2(t)$, the square of the AE signal. Thus, $V_{ae}(t)$ is related to the rate of change of mechanical energy release, or mechanical power. Namely:

$$P_{ae}(t) \propto V_{ae}^2(t) \propto \frac{dE_{mech}(t)}{dt} = P_{mech}(t) \qquad (1)$$

Because of this basic equation, mechanical-power-releasing phenomena may be studied by observing AE signals. These include: 1) plastic deformation, *e.g.* twinning, dislocation motion; 2) cracking; 3) frictional motion; 4) phase transformation. Energized superconducting magnets are abundant sources of AE signals.

Up to now the most useful and successful applications of AE technology for superconducting magnets have been: 1) quench location triangulation in dipole magnets; and 2) quench mechanism identification, *e.g.* conductor-motion induced, epoxy-cracking induced, or short-sample-current induced, in epoxy-impregnated adiabatic magnets.

Currently we are engaged in the following two research projects that employ AE technology in ways different from the past practice: 1) effect of mechanical integrity on the performance of dipole magnets; and 2) analytical and experimental study of normal-zone-propagation-induced AE signals in epoxy-impregnated adiabatic solenoids. Results presented here are preliminary.

SSC Dipoles

The 16.6-m (magnetic length) dipoles are being developed collaboratively by groups at Brookhaven National Laboratory (BNL), Fermi National Accelerator Laboratory (Fermilab), and Lawrence Berkeley Laboratory (LBNL) for the proposed 20-TeV proton-proton superconducting super collider (SSC).[1] The main components of the single phase assembly cross section of SSC dipoles are the stainless steel beam tube, a two-layer coil assembly wound with niobium-titanium copper-stabilized cable, laminated stainless steel collars, laminated iron yoke, and a stainless-steel containment skin.

Objectives of the MIT Project on SSC Dipoles

The MIT project on SSC dipoles, which focuses more on long-term goals of understanding the behavior of SSC-like dipoles rather than short-term goals of solving the magnet performance problems at hand, has the following objectives:[2]

- Identify acoustic parameters which correlate with mechanical properties of the dipole measured at room temperature.

- Correlate mechanical properties of the dipole observed from acoustic measurements at room temperature with 4.2-K performance of dipoles.

- Correlate 4.2 K AE data from dipoles with 4.2 K performance of dipoles.

To achieve the above goals, two sets of measurements are being performed on some of the SSC dipoles, first at BNL where the dipoles are wound and assembled and then at Fermilab where the dipoles are placed in their cryostats and quench-current tests performed. At BNL, the attenuating AE wave amplitude along the dipole axial distance is measured at room temperature before the iron yokes are placed on the dipoles; at Fermilab, AE signals from AE sensors placed on the cryostat stainless steel shell are recorded as the dipoles undergo quench-current tests. The measurements during quench-current tests not only determine the axial distribution of quench locations but also record AE signals for spectral analysis.

Quench-Current Data of DD0014 Dipole

The eight AE sensors placed on the DD0014 dipole for quench-current tests are grouped in 3 and they are placed over three principal locations of the dipole: 1) one group of 4 sensors at the 'feed end' (FE) or the dipole end with the current terminals; 2) another group of 2 sensors at the 'middle' (MD) or the dipole midpoint; and 3) the last group of 2 sensors at the 'return end' (RE) or the far end of the dipole.

The quench results of DD0014 are summarized in Table 1. Runs 1 through 14 were at 4.3 K, while runs 15 through 19 were at 3.4 K. After the dipole was warmed up to room temperature, it was cooled down and tested 4 more times (runs 20~23). Note that the design current at 4.3 K of 7100 A was never achieved. Because of instrumentation problems not all the quenches could be located by AE signals alone; data from the pressure sensors, located at each end of the dipole, were also used for quench localization. From Table 1 it is quite evident that most quenches were initiated at the feed end (FE).

Table 1. Quench Currents and Locations of DD0014 SSC Dipole

Run	I_q (A)	location	Run	I_q (A)	location	Run	I_q (A)	location
1	6377	RE	9	6812	FE	17*	6827	FE
2	6563	FE	10	6729	FE	18*	6847	FE
3	6818	FE	11	6651	FE	19*	7199	FE
4	6759	RE	12	6818	FE	20	5703	FE
5	6793	RE	13	6778	FE	21	6070	FE
6	6666	FE	14	6832	FE	22	6681	FE
7	6778	RE	15*	7478	FE	23	6847	FE
8	6735	RE	16*	7170	FE			

* Operation at 3.4 K; all others at 4.3 K.

Frequency Data

To perform spectral analysis of AE signals, a 4-channel LeCroy signal processor was used for four selected AE sensors. Each channel has the capacity to store 128,000 sample points; with a sampling frequency of 1 MHz it was possible to record 4 AE signals over a duration of 128 ms. Although instrumentation problems prevented recording of frequency data for runs 3 through 12, Table 2 summarizes AE data for the other tests of the DD0014 dipole. Sensors ae1, ae2, ae3, and ae4 are located at the feed end; sensors ae5 and ae6 are at the middle; and sensors ae7 and ae8 are at the return end. The most significant point to note in Table 2 is that there is one dominant frequency, f_d, for each sensor location whether or not that particular AE-signal-inducing event was quench inducing or not. We believe each frequency represents the resonant frequency for that particular location of the dipole and as demonstrated below that the frequency may be related to the 'stiffness' of the dipole at that particular location. The stiffness, in turn, may be shown to be related to attenuation of AE waves through the dipole at that particular location.

Table 2. Dominant AE Signal Frequency f_d at Each Sensor Location

Sensor	Pos. (m)*	f_d (kHz)	Sensor	Pos. (m)†	f_d (kHz)
ae1	0.13	7.5	ae5	8.3*	7.5
ae2	0.25	8.5	ae6	0.94	$\sim$10
ae3	0.64	24.0	ae7	0.76	8.0
ae4	0.94	7.5	ae8	0.64	22.0

* Distance measured from the very end of the feed end.
† Distance measured from the very end of the return end.

Room-Temperature Attenuation Data

The AE wave attenuation data are being recorded at BNL prior to shipment of the dipoles to Fermilab. These data are taken at the three principal regions of the dipole: 1) feed end; 2) middle; and 3) return end. AE signals are generated by dropping a 5-mm diameter steel ball from a height of 10 mm from the dipole collar. The location of this AE sensor used as a signal receiver is fixed for a given region and the position of the AE source is moved along the dipole axis within the region. By using the same receiver and holding it stationary the effect of the axial variation of the receiver-collar coupling on measured AE signals is minimized.

Figure 1 shows three sets of attenuation data for the DD0014 dipole. The open-circles are attenuation taken with the collared dipole; the open-triangles are attenuation data taken directly from the winding before the dipole was collared; and the solid-triangles are attenuation data taken from the collar chunks. As in the earlier 1-m long dipoles,[2] the DD0014 data (open circles) may be in-terpreted in terms of two waves propagating from the source to the receiver, represented by the two sets of data shown by triangles. The first wave, represented by solid triangles, propagates axially through the collars, there-fore dominating when the receiver is close to the source but attenuating rapidly because the laminated collars are disconnected me-dia in the axial direction. The second wave (open triangles) propagates radially inward through the collars beneath the source and reaches the winding and then propagates ax-ially through the winding which is contin-uous in the axial direction and then prop-agates once again radially outward through the collars, reaching the receiver. At short distances from the source, therefore, the sig-nal reaching the receiver is predominantly of the first wave and at long distances it is of the second wave.

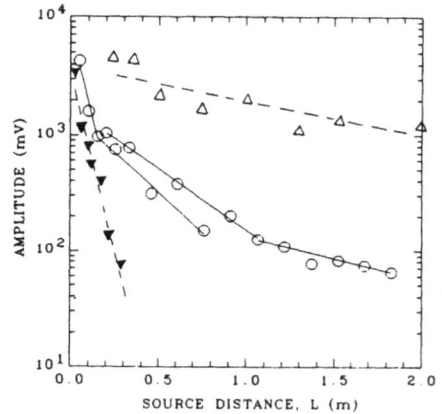

Fig. 1 Room-temperature AE wave attenu-ation data for DD0014 dipole (open circles). Open triangles are for the winding only; solid triangles are for the collar chunks only.

Wave Attenuation and Stiffness: The wave attenuation data are interpreted in terms of winding stiffness, because as will be shown later, quench current may be correlated to winding stiffness. There are three stiffnesses that must be considered in the dipole: 1) intrinsic winding stiffness, k_{wdg}; 2) intrinsic collar stiffness, k_{col}; and 3) effective winding stiffness, k_{eff}. k_{eff} is a combination of k_{wdg} and k_{col} and depends critically on how the winding and collars are interfaced. If they are separated, then $k_{eff} = k_{wdg}$; if they are intimately interfaced with 'zero' interface void, then $k_{eff} = k_{col}$. In real dipoles the interface falls between these two extreme cases.

The measured wave attenuation may also be thought of as an effective value, A_{eff}, which falls between the intrinsic collar attenuation, A_{col}, and the intrinsic winding atten-uation, A_{wdg}. As with k_{eff}, A_{eff} also depends critically on the collar-winding interface. Indeed, it can be shown that k_{eff} and A_{eff} are related as:

$$\frac{k_{eff} - k_{wdg}}{k_{col} - k_{wdg}} = \frac{A_{eff} - A_{col}}{A_{wdg} - A_{col}} \qquad (2)$$

Note that Eq. 2 reduces correctly in the two extreme cases. That is, as $k_{eff} \to k_{col}$, which is when the collars and winding are intimately interfaced, $A_{eff} \rightarrow A_{wdg}$ and when they are separated, then as $k_{eff} \rightarrow k_{wdg}$, $A_{eff} \rightarrow A_{col}$.

Stiffness, Wave Attenuation, and Quench Current: In a simple spring-mass system with frictional and electromagnetic forces that model the winding,[3,4] the incremental electromagnetic force ΔF_{em} necessary to move the winding by a slip distance Δx_s is given by:

$$\Delta F_{em} = k_{eff} \Delta x_s \tag{3}$$

It may also be shown that:

$$\Delta F_{em} \propto \sqrt{k_{eff}} \tag{4}$$

Since $F_{em} \propto I^2$, where I is the dipole current, we obtain the following relationships between ΔI and k_{eff} and ΔI and A_{eff}:

$$\Delta I \propto \sqrt{k_{eff}} \tag{5}$$

$$\propto \sqrt{1 - \frac{A_{eff}}{A_{col}}} \tag{6}$$

Equations 5 and 6 state that ΔI or an incremental current increase during the training sequence will be higher when the effective stiffness of the winding is greater or its effective attenuation is lower. Equations 5 and 6 imply that it may be possible to evaluate the dipole performance (quench current) in terms of the dipole effective stiffness, which in turn may be inferred from attenuation measurements. This conclusion relating dipole performance and attenuation data has been shown to be valid in 1-m long dipoles.[2] Table 3 summarizes the attenuation data and quench data for the DD0014 dipole. #Q refers to the number of quenches detected in the dipole's three principal regions.

Table 3. Attenuation Data vs. Quench Performance in DD0014 Dipole

	attenuation*	# Q
FE	74.3	18
MD	17.7	0
RE	36.9	5

* Ratio of AE signal V_{ae}, initial amplitude to amplitude at 20 cm.

Motion-Induced AE Signals in Epoxy-Impregnated Superconducting Magnets

In a quenching superconducting magnet two phenomena take place in the winding: 1) relaxation of magnetic stress as current decays; and 2) appearance of thermally induced stresses as a temperature distribution becomes nonuniform. Both generate acoustic signals. Our objectives for this research are as follows:

- Obtain quantitative relationships between V_{ae} and stress in the winding, both magnetic and thermal.

- Apply these relationships to a quantitative study of the quenching process.

368

AE Signals in Epoxy-Impregnated Superconducting Magnets

Since no significant amount of AE signals is generated when a material is in the elastic regime and the conductor in superconducting magnets is stressed only to within the elastic regime, there are only two important sources of AE signals in epoxy-impregnated superconducting magnets: 1) epoxy cracking; and 2) conductor motion. Of these two sources, cracking occurs predominantly in the early phase of operation when the magnet is first energized. Thereafter, as the winding is not completely rigid because of epoxy cracking, it produces mainly conductor-motion induced AE signals. The mass-spring model with frictional and electromagnetic forces referred to above in connection with the SSC dipoles is essentially applicable to epoxy-impregnated windings. Because of the epoxy that fills all the void space between conductors, the conductor's range of motion and the number of displaced sites are limited.

Figures 2 and 3 show oscillograms, each with magnet current and AE traces, obtained from an epoxy-impregnated magnet as it was energized and de-energized. In these tests, AE signals were first amplified ($\times 1000$), fed into an AC meter, and were recorded by a digital waveform recorder. When the current excursion was 100 A (Fig. 2), AE signals produced a trace with two humps, the first one corresponding to the charging sequence and the second one corresponding to the discharging sequence. These signals are the results of <u>dissipative</u> conductor motions taking place within the winding. Namely, the first hump is a result of Lorentz-force-induced displacement of movable conductor sites to new locations; the second hump corresponds to the return of these sites to the original locations. Each displacement event is dissipative and thus generates $V_{ae}(t)$ according to Eq. 1. $V_{ae}(t)$ is an exponentially decaying oscillatory signal.

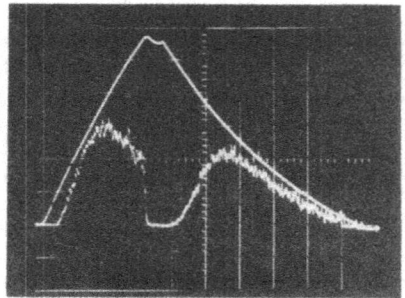

Fig. 2 Current and V_{met} vs. time traces when the magnet current was raised from zero to 100 A and returned to zero. V_{met} scale: 100 mV/div; time scale: 10 s/div.

In our experiment, an AC meter processes $V_{ae}(t)$: first it rectifies $V_{ae}(t)$; it then squares the rectified signal; next it integrates the squared rectified signal; finally it outputs $V_{met}(t)$ to a digital waveform recorder. If we assume that each displacement event, on the average, results in the same dissipative energy $< e_d >$ and that each event is well separated from each other, then the time integral of $V_{met}(t)$ should be proportional to the total number of events, N. Namely:

$$\int V_{met}dt \propto N < e_d > \qquad (7)$$

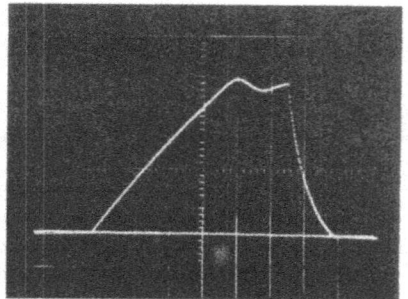

Fig. 3 Current and V_{met} vs. time traces when the magnet current was raised from zero to 15 A and returned to zero. V_{met} scale: 50 mV/div; time scale: 4 s/div.

Applying Eq. 7 for the 100-A case shown in Fig. 2, we obtain:

$$\int V_{met}dt \simeq 5.5 \text{ V s} \qquad \text{(sweep-up)}$$

$$\simeq 5.5 \text{ V s} \qquad \text{(sweep-down)}$$

That the integrals for sweep-up and sweep-down cases are identical is not surprising considering our conductor-motion model: the total number of displacement events, N, should be equal for both sweep-up and sweep-down sequences.

Figure 3 corresponds to the case with an energizing current too small to induce any displacement events. Here the maximum current was 15 A and during the entire sweep-up and sweep-down sequences, the frictional force was sufficient to prevent displacement events and absolutely no AE signals were recorded: the integral of Eq. 7 is thus obviously zero for each sequence.

Figure 4 shows an oscillogram of $V_{met}(t)$ (top trace) and $V_{ae}(t)$ (bottom trace) from a single event. $V_{ae}(t)$ lasts ~20 ms and the $V_{met}(t)$ integral for the single event is ~7 mV s.

Using this value to be typical, we estimate the total number of displacement events N to be ~800 for the sweep-up or sweep-down sequence of Fig. 2. This implies that in the 100-A sweep-down sequence, which took about 50 s, each event took place, on the average, at an interval of ~60 ms. (To record one isolated event, shown in Fig. 4, the magnet current was swept up at an extremely slow rate.)

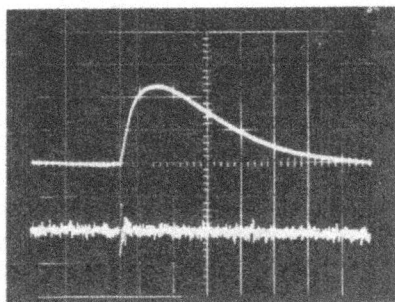

Fig. 4 V_{met} (top) and V_{ae} (bottom) vs. time traces for one displacement event. V_{met} scale: 10 mV/div; time scale: 0.1 s/div.

AE Signals from a Quenching Magnet

Figure 5 shows an oscillogram for the case in which the magnet was quenched at 100 A with a heater located at the magnet i.d. There are two interesting points to be made for V_{met} corresponding to the quench sequence: 1) the voltage integral appears significantly lower than that for the sweep-up and sweep-down modes of Fig. 2; 2) the AE signals persists over a long period, well after the current was reduced to zero and they appear to result from discrete events.

Voltage Integral: The V_{met} integral for the sweep-up is ~6 V s; the integral from 100 A to 0 is ~2 V s during quench and there is another ~1 V s during the 'post-quench' period. These integral values correspond to ~800 displacement events on the way up, ~300 events during quench, and another ~100 events during the post-quench period. We postulate a destructive interference

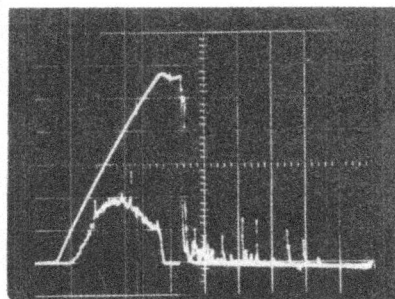

Fig. 5 Current and V_{met} vs. time traces when the magnet current was raised from zero to 100 A and quenched at 100 A. V_{met} scale: 200 mV/div; time scale: 4 s/div.

of AE signals during quench as the most likely cause for this much reduced number of events. This interference may occur if the spacing between events becomes comparable with or shorter than the duration of each signal. In the quench case shown in Fig. 5, the current dropped from 100 A to 0 A in $\sim$1 s; because $\sim$700 displacement events should have taken place, these events must have occurred at a rate of one event per $\sim$1 ms. Since each V_{ae} lasts $\sim$20 ms, it is very likely that under this much greater signal generation rate more than one event occurs simultaneously, resulting in interference.

AE Signals in Post-Quench Period: We postulate further that persistent AE signals that appear in the post-quench period are still generated by conductor displacement events. In the case shown in Fig. 5, $\sim$100 events occurred during this post-quench period. Quenching creates a nonuniform temperature distribution within the magnet, being hot in the i.d. and cooler radially towards the o.d. This nonuniform temperature distribution in turn creates thermal stresses within the winding in the same radially outward direction as that of electromagnetic stresses. Apparently these thermal stresses are sufficient to hold back some of the displaced sites even when the electromagnetic stresses are reduced to zero. The difference in temperature between the i.d. and o.d. of this magnet right after quenching is $\sim$30 K, sufficient to create thermal stresses equivalent to electromagnetic stresses corresponding to a central field of $\sim$4 T. The fact that the $V_{met}(t)$ trace during the post-quench period appears to consist of many discrete events also supports this postulate.

Conclusions

The preliminary results of the two ongoing programs at MIT, one with the SSC dipoles and the other with an epoxy-impregnated magnet, demonstrate that AE technology is applicable to the study of dissipative processes in superconducting magnets. For the SSC dipoles there appears to be correlation among room-temperature wave attenuation, stiffness, and 4.2-K performance. The two postulates presented here—destructive interference of AE signals and replacement of electromagnetic stresses by thermal stresses in an epoxy-impregnated superconducting magnet undergoing quench—require further investigation.

Acknowledgement

I wish to thank my graduate students for the work reported here: Segun Ige, whose thesis work is on the SSC dipoles; Tom Painter and Dimitris Zeritis for the work on the epoxy-impregnated magnet. I also express my thanks to Jim Strait of Fermilab and Peter Wanderer of BNL for their generous help in our work on SSC dipoles.

References

[1] See, for example, J. Strait, et al. Test of prototype SSC magnets, IEEE Trans. Magn. 24, 730 (1988).

[2] O.O. Ige, J. Strait, A.D. McInturff, and Y. Iwasa, Source location of quenches in SSC dipole magnets, IEEE Trans. Magn. 24, 1552 (1988).

[3] O. Tsukamoto and Y. Iwasa, Sources of acoustic emission in superconducting magnets, J. Appl. Phys. 54, 997 (1983).

[4] O.O. Ige, A.D. McInturff, and Y. Iwasa, Acoustic emission monitoring results from a Fermi dipole, Cryogenics 26, 131 (1986).

EVALUATION OF MATERIALS FOR SUPERCONDUCTING MAGNETS FOR FUSION AND OTHER APPLICATIONS

Toichi OKADA

Institute for Scientific and Industrial Research (ISIR)
Osaka University, 8-1, Mihogaoka, Ibaraki
Osaka, 567, Japan

Abstract - In this paper several examples of fusion oriented evaluation study for the superconductive magnet materials are reported. The strain and/or irradiation effects to A-15 superconductors have been recognized important in fusion magnet design. Particular attention has been paid to insulator, superconductor-supporting component. The development of radiation-resistant FRP has been well recognized as the key materials because of its radiation sensitivity.

Introduction

Superconducting magnets are considered as typical SERIES machine in the sense that magnetic field is produced by the transport current flowing through the coil conductor, and , therefore, soundness of the required current density must be kept under the operating condition all through current path. This means any local degradation might degradate the overall performance of the SC coil.

Another characteristics of the SC magnet is its COMPOSITE-NESS in terms of the combination of various types of materials which support the current flow through coil. These two factors make it difficult to establish high performance super-conducting magnets which must be operated under large electro-magnetic force and intense irradiation in such coil as fusion and/or accelerator magnets etc.

Supporting materials, therefore, as well as superconductor are equally important to establish overall performance of the magnet. In this paper several important topics on such items are reviewed.

Environmental Condition of Fusion Magnet Materials

Superconducting magnets for fusion are to be used under very severe conditions typically (i) very low temperatures, (ii) large electromagnetic forces and (iii) intense and high energy neutron irradiation.

Any component material must, therefore, keep their high performance under the different thermal contraction, under stresses against intense irradiation.

Superconducting Magnet Material

Any superconducting magnets mainly consist of following materials. (1) Superconductor, (2) Stabilizer, (3) Insulator, (4) Structural materials and (5) Coolant. In this paper two component materials, i.e., superconductor and insulator will be discussed from the view-point of material evaluation.

Strain Effects on A-15 Superconductor

The stress/strain effect of the superconductors seems mainly to arise from thermal contraction and hoop stresses. The hoop stresses σ on the conductor is given by the product of radius r of the coil, magnetic flux density B and current density J: $\sigma = J B r$. In case of large high field magnets with the bore radius larger than several meters the conductor itself cannot normally sustain the large electromagnetic force. A few techniques have alrealy been proposed for the support of the conductor such as wavy mode in large diurnal SMES magnet or D-shaped toroidal magnets for Tokamak fusion machine. It should, however, be stressed that finite stress or strain has to be shared by the conductor as well as spacer or insulators. Figure 1 gives strain effects on 'in situ' formed Nb_3Sn which show the high tolerance to the strain up to 1.6%. Improvement in strain tolerance in Nb_3Sn conductors made by other processes have greatly been made.

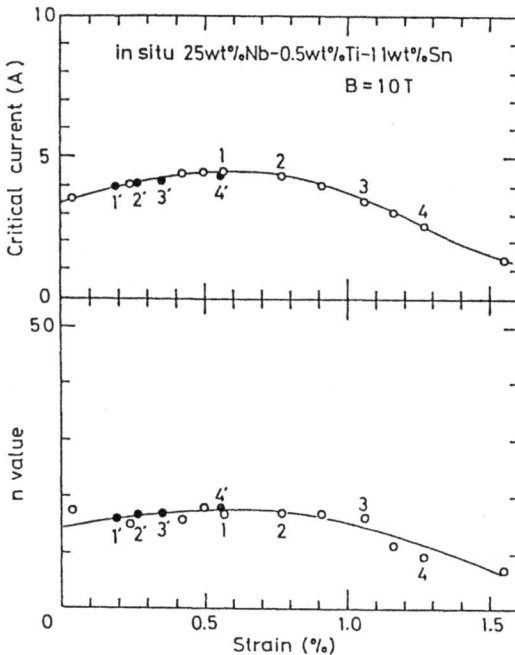

Fig.1.

Strain effects of in situ $(NbTi)_3Sn$.

Neutron Irradiation Effects

Table 1 and Figure 2 give the effect of reactor neutron irradiation on Nb₃Sn and (NbTi)₃Sn superconductors[1]. A slight harmful effects to neutron irradiation tolerance is recognized when Ti is added to increase high field performance of Jc(B).

Evaluation of Insulating Material for Fusion Magnet

Table 2 gives the material properties of FRP of which properties are measured at cryogenic temperatures and being evaluated.

Thermal and Mechanical Properties

Figure 3 shows the mechanical behaviors of typical glass fiber reinforced plastics (GFRP) which dramatically indicating that even at low temperatures GFRP has higher toughness compared with RT behavior. There are various ways of reinforcing resin with inorganic fibre which regulates specific performance of the material. These points will be discussed corresponding to special case in the following section.

Figure 4 shows the temperature dependence of the thermal conductivity in the fibre direction for advanced composite specimen being compared with 304 stainless steel from 300K to 4.2K. In Fig.5, the cryogenic performance of these material is given in the temperature range of LHeT to LNT. The SFRP and ALFRP are seen superior to GFRP or 304 stainless steel.

Table I Summary of material parameters for multi-filamentary (Nb,Ti)₃Sn wires

Compound	Filament dia. and number	Bronze ratio	Heat treatment
(Nb,Ti)₃Sn 0.3wt%Ti	5 µm 722	2.5	933 K 200 hr
Nb₃Sn	5 µm 745	2.0	973 K 120 hr

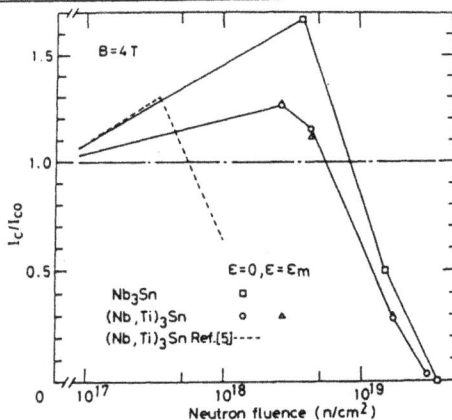

Fig.2.

Effect of neutron irradiation on the critical current.

Table 2. Property items of data
base for insulating
materials.

MECHANICAL PROPERTIES
Flexural, Tensile, Compressive
Impact Test (Charpy, Drop Weight, SHB)
Fatigue, Creep

THERMAL PROPERTIES
Conductivity, Contraction, Specific Heat

ELECTRIC PROPERTIES
Dielectric Constant, Loss, T.S.C
(Dielectric Strength)

NON DESTRUCTIVE TESTING
AE, Internal Friction

RADIATION DAMAGE
Mech.Prop., T.S.C.

VACUUM PROPERTIES
He gas permeation, Out gas analysis

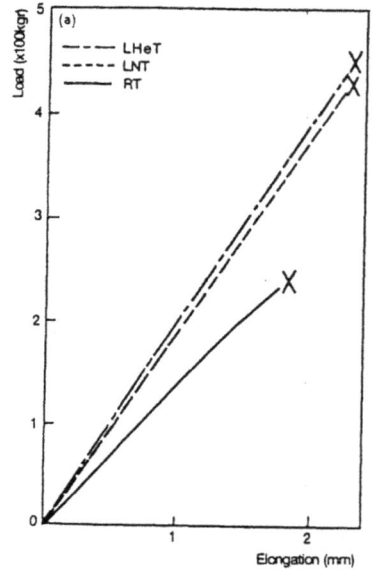

Fig.3. Typical load-elongation
curve for GFRP.

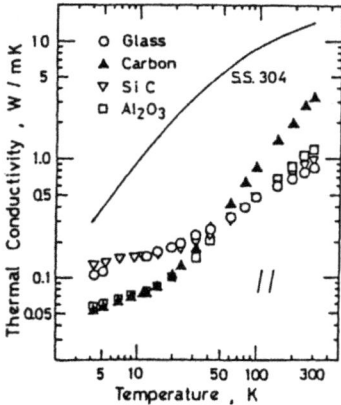

Fig. 4. Temperature dependence
of the thermal
conductivity in the
fiber direction.

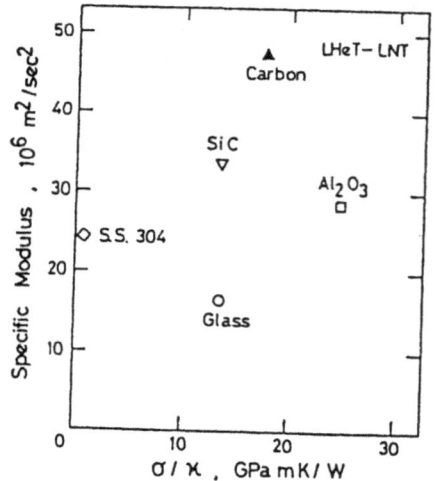

Fig.5. Specific modulus vs.
σ/κ between LHeT
and LNT.

Radiation Resistant Composite Material

   To prevent the degradation of ILSS, following two methods
should be proposed:
(1) Developing the radiation resistant matrix and/or finish
     agent of fibers.
(2) Designing the spacial arrangement of the reinforcement to
     get rid of the inter laminar area.
   In this work the latter method is adopted, that is, three
dimentional fabric reinforced composite materials (3D-GFRP) has
been designed and fabricated as the radiation resistant GFRP.
The radiation induced degradation of 3D-GFRP was studied and the
results were compared with those obtained in usual GFRP(2D-
GFRP).
   Figures 6 and 7 give the radiation degradation for 2D and
3D of interlaminar shear strength and nominal flexural strength,
respectively.
   Figure 8 demonstrates the configuration of three demension-
ally woven fibers.

Data Base for Cryogenic Composite Materials

Organic composite materials have not frequently been used for
cryogenic field because of insufficiency of reliable data base
for design.   In our laboratory the data base has been
constructed using only measured data which were obtained from
the same measuring apparatus.   The items now available have
been listed in Table 2.

Specific Irradiation Problems

14MeV Neutron Irradiation

   Guinan et al. give the typical results[2]of high-energy
neutron irradiation on the critical current of pure Nb_3Sn and
modified ones[$(NbTi)_3Sn$ and $(NbTa)_3Sn$] .   This type of
experiments are significant in two fold: (i) damage effectiveness of 14MeV neutron compared with ones with softer spectrum
in the fusion magnet region, (ii) preliminary data before
performing cryogenic irradiation up to the same fluence.

Comparative Study of Radiation Damage and Radioactivation

   The component materials in fusion magnet become more or
less radioactive after the operation of reactor for certain
length of time.   The radioactivation problem is also very
critical from practical stand-point, since fusion machine must
be subjected to hand on maintenance in the course of operation.
Development of low activation material has been well recognized;
a comparative study of the relative importance between radiation
effects and radioactivation is interesting.   Figure 9 gives the
summary of the results[3].

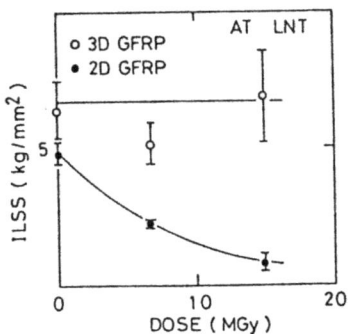

Fig. 6. Change of interlaminar
shear strength induced by low
temperature reactor irradiation
obtained at liquid nitrogen
temperature.

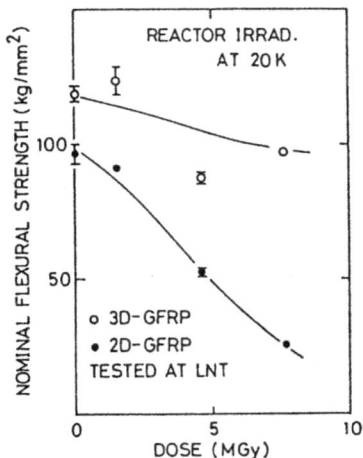

Fig.7. Effect of low temperature
reactor irradiation on nominal
flexural strength obtained at
liquid nitrogen temperature.

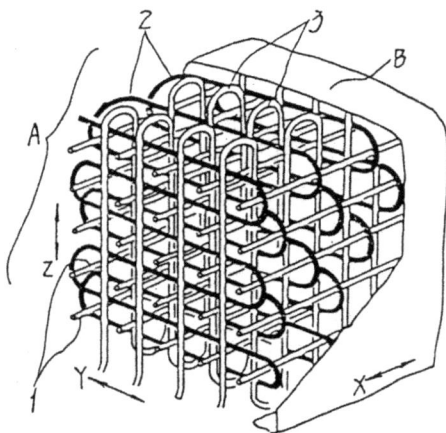

Fig.8. Three dimensionally
reinforced FRP.

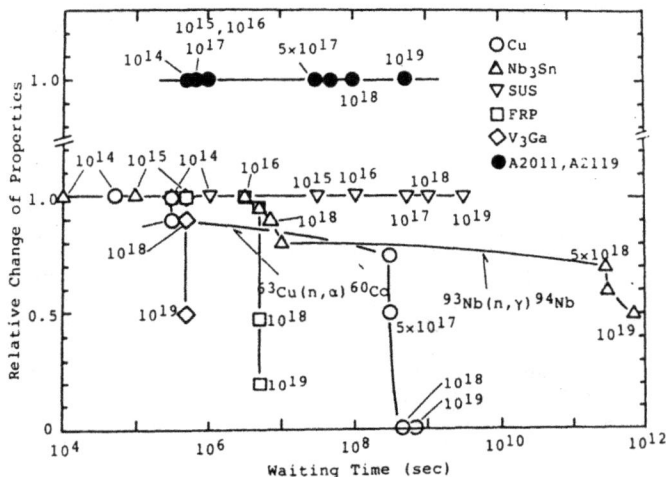

Fig. 9. Comparison of relative importance between degradation and activity on each component of fusion magnet induced by irradiation.

Conclusions

The design data base for fusion magnets is strongly insufficient. Those of organic insulating material are premature, particularly for cryogenic/irradiation condition. The methodology, however, to the development of radiation resistant materials is in progress.

References

1) K.Katagiri, K.Saito, T.Okada, H.Kodaka, and H.Yoshida, "Effect of Neutron Irradiation on Nb$_3$Sn Superconducting Composite", J.J.Appl.Phys.26 (1987) Suppl.26-3.
2) C.D.Henning, and B.G.Logan,"Overview of TIBER II— Evolution towardan Engineering Test Reactor", UCRL-96064 (1987).
3) T.Okada and S.Nishijima,"Comparative Study of Radiation Damage and Activation of Superconductive Magnet for Fusion Reactor", Adv.Cryogenic Eng. Vol.34, Prenum Press,p.p.917 - 924 (1986).

378

STUDY OF SUPERCONDUCTING MAGNET MATERIALS
FOR APPLICATIONS IN RADIATION ENVIRONMENTS

Harald W. WEBER

Atominstitut der Österreichischen Universitäten
Schüttelstraße 115, A-1020 Wien, Austria

Abstract - In the present report, experiments carried out over the past few years on the radiation-induced property changes of superconductors, insulators and stabilizer materials are reviewed. The main emphasis is placed on superconducting materials. In the first part, results are presented from 5 K irradiations of various NbTi-superconductors followed by thermal cycles to room temperature. The significantly different radiation response at low and high fields is ascribed to different flux pinning mechanisms. The high-field degradations of j_c ($\sim$ 20% at 5×10^{22} neutrons m^{-2}) are quantitatively related to the experimentally determined decrease of H_{c2} and T_c with neutron fluence. In the second part, results on Nb_3Sn and other high-field materials are presented. Nb_3Sn with small additions of Ti ($\sim$ 1.5 wt%) becomes useless at the low fluence level of $\sim 1\times10^{22}$ m^{-2}. On the other hand, recent data on sputtered NbN films show excellent properties even at a fluence level of 10^{23} m^{-2}, especially in the field range from 17 - 20 T. Although high-T_c superconductors are not considered as candidate materials for fusion magnets at present, brief reference is made to neutron-induced changes of critical current densities in single crystals of $YBa_2Cu_3O_{7-\delta}$. Finally, still existing material problems with insulators and stabilizer materials are pointed out.

Introduction

The application of superconductors in magnets designed for the magnetic plasma confinement in future fusion reactors imposes severe performance requirements on all the magnet components, in order to ensure safe and reliable operation over the whole plant lifetime of, say, 30 years. The most important issues, which have to be addressed in today's research, are (1) the strain and radiation tolerance of "commercial" high-field super-conductors, preferably studied under the operating conditions of the magnet including possible synergistic effects, (2) the strain and radiation-induced resistivity changes of the stabilizer material including the effects of thermal cycling to room temperature during plant shut-downs, (3) the performance of insulating materials, and (4) the stability of the whole composite forming the magnet windings under the same conditions.

In this contribution we will concentrate on property changes of various superconductors and the copper stabilizer induced by fast neutron irradiation, which have been studied by our group systematically during the past few years [1-9]. Strain and synergistic strain-radiation effects have been discussed recently by Okada et al. [10]. For work on insulators we refer to the preceeding paper in this volume [11].

Neutron damage considerations

The key issue for an evaluation of neutron irradiation experiments with regard to their relevance for the actual radiation environment at the magnet location of a fusion reactor, is the analysis of neutron damage as a function of neutron energy distribution. Since it has been well established that damage in superconductors is primarily caused by neutrons with energies exceeding 0.1 MeV (lower-energy neutrons which lead to Frenkel pair production and transmutations play only a minor role through their effect on the normal state resistivity), an assessment of neutron spectra in this energy range is essential. Furthermore, if the nature of defects does not change within the relevant range of neutron energies (0.1 $\leqslant E_n \leqslant$ 14 MeV), an attempt can be made to scale the damage effect on a certain quantity, e.g. the critical current density j_c, by the displacement energy cross section and the total energy transferred to each atom of the material ("damage energy", E_D) in the following way (cf., e.g., [5,12-14]:

$$<\sigma T> = \frac{\int \sigma(E) T(E) \left[d\Phi/dE\right] dE}{\int \left[d\Phi/dE\right] dE} \tag{1}$$

$$E_D = <\sigma T> \Phi t \tag{2}$$

where $\sigma(E)$ and $T(E)$ are the neutron scattering cross sections and the primary recoil energy distributions, Φ is the neutron flux density and t the total exposure time to the neutron flux Φ. Of course, calculations along

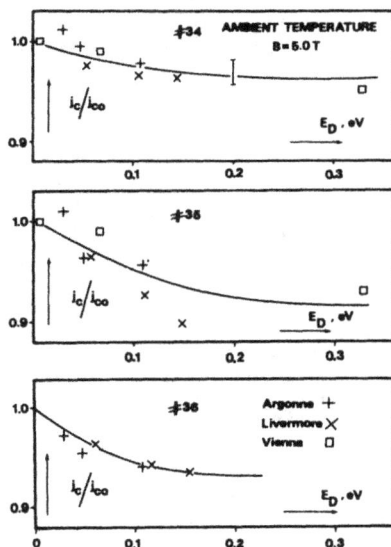

Fig. 1: Damage energy scaling of critical current densities in three multifilamentary NbTi-superconductors.

these lines require a detailed knowledge of material parameters, elaborate computer codes (which have been refined recently [14] to account for the compound nature of many superconductors) and information on the exact flux density distribution of the irradiation source.

In order to test this type of scaling and because of the availability of careful dosimetry studies of various neutron sources (cf., e.g., [15,4,5]), we have subjected identical pieces of NbTi-wire to ambient temperature irradiation in a reactor (TRIGA, Vienna), a spallation source (IPNS, Argonne) and the 14 MeV source RTNS-II in Livermore. The results on the critical current density j_c measured at 4.2 K and in a field of 5 T are shown in Fig. 1. Although the actual highest neutron fluences differ by nearly as much as a factor of 5, perfect scaling of the data with damage energy is observed. This result confirms firstly, that the type of defects introduced in the superconductor does not change with neutron energy within the energy range concerned (0.1 $\leqslant E_n \leqslant$ 14 MeV), and secondly, that any type of fusion reactor spectrum, e.g. STARFIRE [16], MARS [17], TIBER-2 [18] etc., can be used to evaluate the j_c-degradation from graphs like Fig. 1, if appropriately converted to damage energies according to Eqs. (1) and (2).

Superconductors

Niobium-Titanium Alloys

Radiation effects in NbTi superconductors are best understood at present. Extensive work on materials with different metallurgical microstructures [1-7,19] has established that superconductors with pre-dominant α-Ti precipitate pinning are affected least by neutron irradiation. This holds for ambient temperature irradiation, 5 K irradiation followed by thermal cycles to room temperature and 5 K in-situ experiments [19]. An example of this type of data is shown in Fig. 2a on the left-hand side, where the fractional change of critical current densities at 5 T is plotted versus neutron fluence.

However, if the data is taken at higher magnetic fields (B $\geqslant$ 7 T at 4.2 K), this clear correlation of j_c-degradation with microstructure disappears and a nearly uniform radiation response is obtained (Fig. 2a, right-hand side). This observation, which has been interpreted tentatively as being a consequence of a different pinning mechanism operative near H_{c2}, could be fully explained by a series of experiments on the radiation-induced change of upper critical fields H_{c2} and transition temperatures T_c [7]. As an example, we show an evaluation of H_{c2} based on the Schmucker theory of plastic flux line flow [20] near H_{c2} for the unirradiated and the irradiated samples in Fig. 2b. (It should be noted that the same field dependence of j_c, i.e. a linear $j_c^{1/2}$-B relationship, is also obtained within the recently proposed framework of a condensation energy pinning model [21]). In all investigated cases, H_{c2} was found to decrease by 2 - 6% with neutron irradiation up to a fluence of 3×10^{22} m^{-2} (E > 0.1 MeV). The physical origin of this result could be identified as being a small reduction of T_c with neutron fluence, as demonstrated by the excellent H_{c2}-T_c correlation shown in Fig. 2c.

Based on these results and assuming that only H_{c2} changes upon neutron irradiation in the equations for the pinning force at high fields, all pinning models result in the same expression for the fractional change of j_c:

$$\frac{j_c}{j_{c0}} = \frac{H_{c2,0}}{H_{c2}} \left[\frac{\mu_0 H_{c2} - B}{\mu_0 H_{c2,0} - B} \right]^2 \qquad (3)$$

An evaluation of eq. (3) in the field range from 8 to 10.5 T at 4.2 K proved highly successful and allows us to describe the experimental results on the j_c-degradation at high magnetic fields in general to an accuracy of ± 5%.

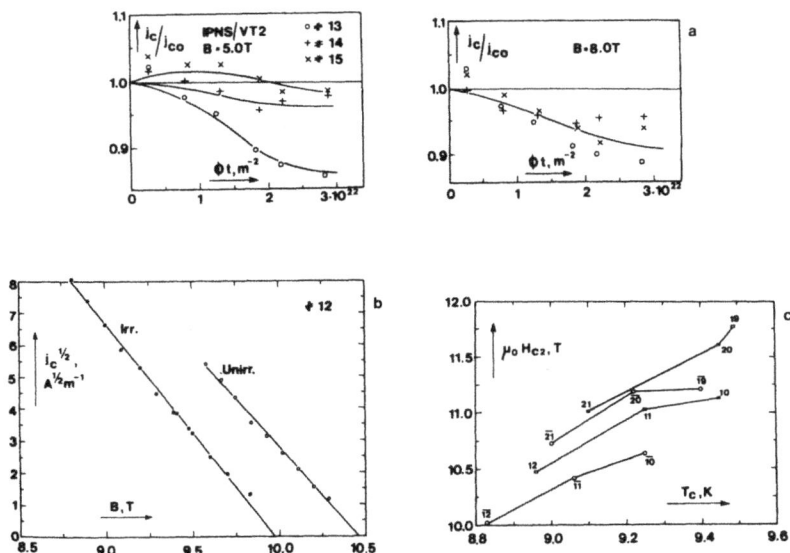

Fig. 2: Radiation damage (5 K irradiation plus thermal cycles to room temperature) in NbTi
   a) Fractional change of critical current densities with neutron fluence at 5 and 8 T. (The data refer to 3 NbTi-superconductors prepared under identical annealing and final cold working conditions, but differing in their Ti-content; o ... 42, + ... 49, x ... 54 wt% Ti).
   b) Schmucker extrapolation of H_{c2} before and after neutron irradiation to 3×10^{22} m^{-2} (E > 0.1 MeV).
   c) Correlation between H_{c2} and T_c in several NbTi-superconductors (sample numbers with a bar refer to the irradiated state).

Nb$_3$Sn, (NbTi)$_3$Sn, and NbN

The nature of radiation damage in commercial A15 superconducting wires has not been investigated in comparable detail. Most of the experiments have been done under ambient reactor temperature irradiation conditions [22,10] and have shown that j_c initially increases with neutron fluence, this effect being the more pronounced the higher the magnetic field is, and then precipitously drops to zero at fluence levels of the order 3×10^{23} m^{-2} (E $>$ 0.1 MeV), cf. Fig. 3. To explain this result, radiation-induced disorder leading to an enhancement of normal state resistivity ρ_n and, consequently, of the upper critical field H_{c2} has been invoked. This explanation has been supported more recently by low temperature irradiations followed by thermal cycles to room temperature, on various $(Nb_{1-x}Ti_x)_3Sn$ superconductors, where, starting from enhanced ρ_n- and H_{c2}-values due to the Ti-addition, a much faster degradation of j_c with neutron fluence has been observed [4,6], cf. Fig. 3. All these results suggest that the influence of the irradiation temperature should be negligible in these materials.

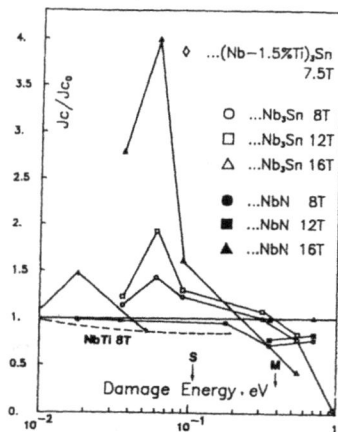

Fig. 3: Fractional change of critical current densities with damage energy for NbTi (Swiss LCT-conductor at 8 T), Nb$_3$Sn (from [22]), (NbTi)$_3$Sn and NbN.

An explicit experimental proof of this assumption was obtained only recently by a comparison of ambient and 12 K irradiations on the same conductors [23]. The results pertaining to the change of critical currents with 14 MeV-fluence are shown in Fig. 4. Although some scatter of the data is apparent, the general agreement between these two irradiation conditions is very good; even better agreement has been reported for the fluence dependence of the transition temperature and the upper critical fields, which were extrapolated from the j_c-data in the usual way.

From the point of view of a fundamental understanding of the damage process and, in particular, of the interplay between damage and the martensitic phase transformation [24], more experimental work is clearly needed. Similarly, more studies of damage energy scaling as well as more detailed studies of the effects of Ti-additions seem highly desirable.

With the quest for still higher magnetic fields to be provided by fusion magnets, radiation damage studies on "advanced" materials like NbN, the Chevrel phases and Nb$_3$(Al,Ge) have been initiated. In a first series of experiments, ambient reactor irradiation data on magnetron-sputtered NbN films have been obtained [8,9]. In addition to showing only a very small decrease of T_c (2 - 7%), these films proved to be extremely radiation-hard, especially regarding critical current densities in the field range from

15 - 20 T, under radiation levels of up to 10^{23} m^{-2} (cf. Fig. 3). Because of the observation of peak-effects in the irradiated materials near 19 T, increases of j_c even up to 80% were observed in some films, with the best j_c-value being ~ 3×10^8 A m^{-2} at 18 T.

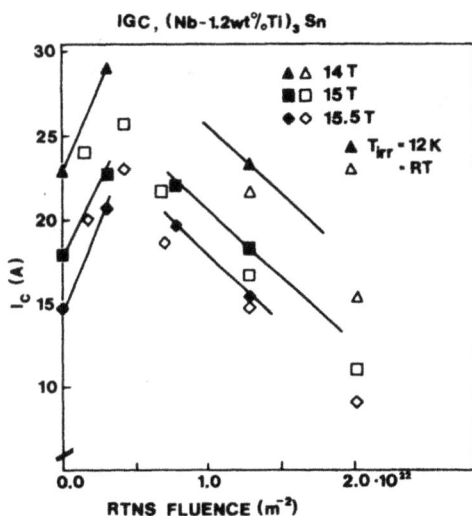

Fig. 4: Critical currents versus 14 MeV-fluence for an alloyed IGC conductor. The data refer to ambient and 12 K irradiation.

Single Crystals of YBa$_2$Cu$_3$O$_{7-\delta}$

Although not related to the main subject of this paper, a few remarks on neutron irradiation studies performed at Argonne National Laboratory on YBCO single crystals [25,26] will be made. The results refer to neutron fluences up to ~ 8×10^{21} m^{-2} (E > 0.1 MeV) and were evaluated from magnetization measurements in a 1 T-SQUID-magnetometer in terms of the Bean model.

The main conclusions of these experiments may be summarized as follows. The change of transition temperature T_c is still moderate in this fluence regime and follows a linear fluence dependence (-2.6 K per 10^{22} neutrons per m^2). In all cases, the critical current densities increase upon irradiation; at the same time, the intrinsic anisotropy is reduced. At 6 K, e.g., the critical current densities for the field applied parallel to the c-direction are ~ 44 times larger than for the field applied parallel to the basal plane. After irradiation, this ratio is decreased to ~ 9. The maximum j_c-enhancement occurs at 6 K for H || a, b (factor of 5), it still amounts to a factor of ~ 3 at 77 K for the same field orientation.

In summary, these results have shown that effective flux pinning centers can be introduced in high-T_c superconductors by neutron irradiation, but the nature of these pinning centers still needs to be clarified (cf. also the direct observation of defect production by electron irradiation [27]).

Stabilizing and insulating materials

In conjunction with the low temperature irradiation studies on NbTi discussed in a previous section, detailed investigations of the change of stabilizer (Cu) resistivity with neutron fluence including magneto-resistivity have been made [4]. These results, which included thermal cycles to room temperature simulating fusion reactor shut-downs after 2, 6, 9, 12.5, 15 and 20 years, were combined with in-situ low-fluence experiments by Klabunde et al. [28] to map out a complete simulation of stabilizer performance over the lifetime of the fusion magnet (Fig. 5). Although some extrapolation errors must be accounted for, the data pertaining to an operating field of 8 T clearly show that the present design limits allowing for a maximum increase of stabilizer resistivity by 25% will be exceeded very soon in the pre-annealing state. Since the inclusion of more annealing cycles cannot be reconciled with an economic operating schedule, clearly a problem has been identified which calls for further attention and design reconsiderations.

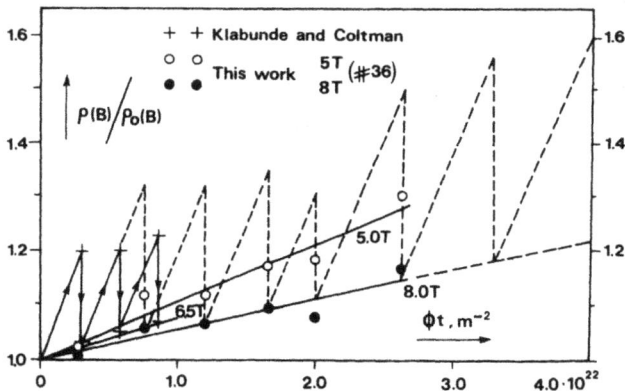

Fig. 5: Variation of Cu-resistivity with neutron fluence (low temperature irradiation plus thermal cycles to room temperature).

Concerning insulating materials, the situation is certainly still more serious. A few years ago, we reported on the complete destruction of composites made from glass-fiber reinforced epoxies, following reactor irradiation at 77 K to a neutron fluence of ~ 3×10^{22} m^{-2} and a γ-dose of 10^8 Gy [29]. However, the development of more radiation-resistant materials as well as of different types of reinforcement allowing for higher shear loads [11,30] is well under way and may provide acceptable solutions in the course of future test programs.

Conclusions

In summary, the neutron irradiation studies done on materials in view of their potential application in superconducting magnets for fusion reactors have provided the following results:

* Superconductors
  * NbTi
    - The degradation of j_c is, in general, small (20 - 30%).
    - Clear correlations of the j_c-degradation with microstructure are observed at low fields (B $\leqslant$ 7T at 4.2 K), whereas at high fields a more uniform radiation response prevails.
    - The physical processes involved in the high field range have been identified as follows: Due to a small radiation-induced decrease of T_c, the upper critical field H_{c2} is reduced, which accounts for the observed j_c-reduction in a quantitative way.
  * Nb_3Sn and $(NbTi)_3Sn$
    - The critical current densities are enhanced at low fluences (due to an increase of normal state resistivity) and degrade at higher fluences due to the significant reduction of T_c. Hence, the applicability of these materials depends on details of the field- and fluence requirements.
    - The influence of irradiation temperature on T_c, j_c and H_{c2} is insignificant.
    - Alloyed materials degrade at much lower neutron fluences.
  * NbN
    - Only small T_c-reductions are observed up to a fluence of 10^{23} m^{-2} (E > 0.1 MeV).
    - At high magnetic fields (B $\geqslant$ 15 T), j_c is either unaffected or even increases upon irradiation to the 10^{23} m^{-2} fluence range.
* Stabilizing and insulating materials
  * Cu-stabilizer: The resistivity increase of typical "magnet" copper at 8 T, including several thermal cycles to room temperature, has been found to exceed present design limits.
  * Insulators: Being presumably the most radiation-sensitive magnet component, more experimental work will be needed to test emerging primising concepts on improved radiation performance.

Acknowledgements

This work has benefitted from dedicated efforts of many graduate students and close cooperations with colleagues at Argonne and Lawrence Livermore National Laboratories, the University of Wisconsin and Brown Boveri Company in Switzerland. Partial support by the Bundesministerium für Wissenschaft und Forschung, Wien, and the U.S. Department of Energy is gratefully acknowledged.

References

1) F.Nardai, H.W.Weber, and R.K.Maix, "Neutron irradiation of a broad spectrum of NbTi superconductors", Cryogenics 21, 223-233 (1981).
2) H.W.Weber, F.Nardai, C.Schwinghammer, and R.K.Maix, "Neutron irradiation of NbTi with different flux pinning structures", Adv.Cryog.Eng. 28, 239-335 (1982).
3) H.W.Weber, "Neutron irradiation effects on alloy superconductors", J.Nucl.Mat. 108&109, 572-584 (1982).
4) H.W.Weber, "Irradiation damage in superconductors", Adv.Cryog.Eng. 32, 853-864 (1986).
5) P.A.Hahn, H.W.Weber, M.W.Guinan, R.C.Birtcher, B.S.Brown, and L.R.Greenwood, "Neutron irradiation of superconductors and damage energy

scaling of different neutron spectra", Adv.Cryog.Eng. 32, 865-872 (1986).

6) P.A.Hahn, H.Hoch, H.W.Weber, R.C.Birtcher, and B.S.Brown, "Simulation of fusion reactor conditions for superconducting magnet materials", J.Nucl.Mat. 141-143, 405-409 (1986).

7) H.W.Weber, W.Khier, M.Wacenovsky, and H.Hoch, "Radiation-induced changes of critical fields in NbTi superconductors", Adv.Cryog.Eng. 34, 1033-1039 (1988).

8) P.Gregshammer, H.W.Weber, R.T.Kampwirth, and K.E.Gray, "The effects of high-fluence neutron irradiation on the superconducting properties of magnetron sputtered NbN films", J.Appl.Phys. 64, 1301-1306 (1988).

9) H.W.Weber, P.Gregshammer, R.T.Kampwirth, and K.E.Gray, "High-fluence" neutron irradiation of superconducting NbN films", Proc. MRS Meeting on Adv. Materials, Tokyo, May 31 - June 3, 1988, in press.

10) T.Okada, M.Fukumoto, K.Katagiri, K.Saito, H.Kodaka, and H.Yoshida, "Effects of neutron irradiation on the critical current of bronze processed multifilamentary Nb_3Sn superconducting composites", J.Appl.Phys. 63, 4580-4585 (1988).

11) T.Okada, "Evaluation of materials for superconducting magnets for fusion and other applications", this volume, paper BD-1.

12) L.R.Greenwood, "Neutron source characterization and radiation damage calculations for material studies", J.Nucl.Mat. 108&109, 21-27 (1982).

13) L.R.Greenwood and R.K.Smither, "SPECTER: neutron damage calculations for materials irradiations", Argonne National Laboratory Report ANL/FPP/TM-197 (1985), unpublished.

14) L.R.Greenwood, "SPECOMP calculations of radiation damage in compounds", Proc. 6th ASTM-Euratom Symposium on Reactor Dosimetry, Jackson Hole, WY (1987), in press.

15) H.W.Weber, H.Böck, E.Unfried, and L.R.Greenwood, "Neutron dosimetry and damage calculations for the TRIGA Mark-II reactor in Vienna", J.Nucl.Mat. 137, 236-240 (1986).

16) C.C.Baker and M.A.Abdou, Eds., "STARFIRE - A Commercial Tokamak Fusion Power Plant Study", Argonne National Laboratory Report, ANL/FPP-80-1 (1980), unpublished.

17) M.Donohue and M.Price, Eds., "Mirror Advanced Reactor Study", Lawrence Livermore National Laboratory Report, UCRL-53480 (1984), unpublished.

18) J.D.Lee, Techn. Ed., "TIBER II/ETR. Final Design Report", Lawrence Livermore National Laboratory Report, UCID-21150 (1987), unpublished.

19) M.Söll, S.L.Wipf, and G.Vogel, "Change in critical current of superconducting NbTi by neutron irradiation ", IEEE Publication 72 CH 0682-5-TABSC, 434-439 (1972).

20) R.Schmucker, "The influence of plastic deformation of the flux line lattice on flux transport in hard superconductors", phys.stat.sol. b80, 89-97 (1977).

21) K.E.Gray, R.T.Kampwirth, J.M.Murdock, and D.W.Capone II, "Experimental study of the ultimate limit of flux pinning and critical currents in superconductors", Physica C152, 445-455 (1988).

22) C.L.Snead, jr., and D.M.Parkin, "Effect of neutron irradiation on the critical current of Nb_3Sn at high fields", Nucl.Technol. 29, 264 (1975).

23) P.A.Hahn, M.W.Guinan, L.T.Summers, and T.Okada: private communication, Sept. 1988.

24) M.A.Kirk, M.C.Baker, B.J.Kestel, H.W.Weber, and R.T.Kampwirth, "An HVEM study of displacement cascade damage in Nb_3Sn at 13 K", Proc. Int. Symp. on Effects of Radiation on Materials, Andover, June 27-29, 1988, in press.

25) A.Umezawa, G.W.Crabtree, J.Z.Liu, H.W.Weber, W.K.Kwok, L.H.Nunez, T.J.Moran, C.H.Sowers, and H.Claus, "Enhanced critical magnetization currents due to fast neutron irradiation in single-crystal $YBa_2Cu_3O_{7-\delta}$", Phys.Rev. B36, 7151-7154 (1987).

26) H.W.Weber, G.W.Crabtree, A.Umezawa, J.Z.Liu, W.L.Kwok, and W.K.Kwok, Proc. MRS Meeting on Adv. Materials, Tokyo, May 31-June 3, 1988, in press.

27) M.A.Kirk, M.C.Baker, J.Z.Liu, D.J.Lam, and H.W.Weber, "Electron irradiation effects in $YBa_2Cu_3O_{7-\delta}$ single crystals", in "High-T_c Superconductors", H.W.Weber, Ed., Plenum Press, New York (1988), p.59-65.

28) C.E.Klabunde and R.R.Coltman, jr., "The magnetoresistivity of copper irradiated at 4.4 K by spallation neutrons", U.S. Department of Energy Report, DOE/ER-0113/3 (1984), unpublished.

29) H.W.Weber, E.Kubasta, W.Steiner, H.Benz, and K.Nylund, "Low temperature neutron and gamma irradiation of glass fiber reinforced epoxies", J.Nucl.Mat. 115, 11-15 (1983).

30) J.Yasuda, T.Hirokawa, T.Uemura, Y.Iwasaki, S.Nishijima, T.Okada, H.Okuyama, and Y.A.Wang, "Cryogenic and radiation resistant properties of three dimensional fabric reinforced composite materials", this volume, paper BM-24.

TECHNICAL FEATURE OF POSSIBLE HIGH Tc SUPERCONDUCTING MAGNET

O. Tsukamoto

Faculty of Engineering, Yokohama National University
156 Tokiwadai, Hodogaya-Ku, Yokohama, JAPAN

Abstract - This paper discusses technical aspects of a possible high Tc superconducting (HTcSC) magnet, properties of magnet components, comparison to a conventional metal superconducting (CMSC) magnet, stability, quench protection and performances required to HTcSC wires.

Introduction

The discovery HTcSC materials with critical temperature reasonably far above the 77K boiling temperature of liquid nitrogen expanded expectations that many potential applications of superconductor become feasible. Of these applications, highly important are superconducting magnet applications, for example, fusion reactors, magnetically levitated trains, electric power generators, MRI etc.

The most remarkable advantage of HTcSC to CMSC is the drastic simplification of cryogenic system including refrigerator. At the present state, there are many R&D items to develop a practical HTcSC magnet wire and only short and very low current density wires have been developed. HTcSC may have disadvantages to CMSC, even when a practical HTcSC wire is developed. In some applications, those disadvantages may overcome the advantages.

In this situation, it is important to clarify the requirements to HTcSC material as a magnet wire and technical problems to use a HTcSC magnet wire.

Properties of Magnet Components

To discuss the differences in characteristics of HTcSC magnet cooled by LN_2 and CMSC magnet cooled by LHe, properties of magnet components are listed in Tab. 1.[1,2,3]

(a) Conductor materials

Properties of HTcSC are typical values of YBCO. The critical current density is most important property which is discussed later. Properties of CMSC are of NbTi. The most important difference in property of HTcSC and CMSC is the temperature margin which is ~10K for HTcSC and ~1K for CMSC. In the discussion of this paper, it is assumed that copper is used to stabilize HTcSC as well as CMSC. At 77K, the resistivity of copper is 10 times larger and the specific heat is 2000 times larger than the respective values at 4.2K. Allowable strain of HTcSC is less than that of Nb_3Sn (0.3%), while that of NbTi can be up to 0.6%.

(b) Coolants

As to the transfer heat flux of the coolants, it needs some discussions about which value should be selected for the calculation of cryostability, maximum nucleate boiling heat flux or minimum film boiling heat flux. This will be discussed later.

(c) Structural material

In large scale magnets, about 50% of the weight of the magnet consists of a structure which is required to support the electromagnetic force.[4] At 77K, the tensile strength of stainless steel is 20-30% smaller than that at 4.2K, which means that the cross section of the structural material in the winding will increase by these percentages, and that total weight of other magnet increases by 10-15% if weights of other components are same. HTcSC is much more sensitive to strain than CMSC, therefore, more amount of the structural material is demanded to keep the strain within the allowance.

Stability and Quench Protection

(a) Composite HTcSC wire
  Normal resistivity of HTcSC is about 10^{-5} Ωm and a bare HTcSC superconductor will burn out when it is quenched. Therefore, it is necessary to plate a low resistive normal metal on the HTcSC to reduce the normal resistivity of the wire. The HTcSC material has much higher specific heat and larger temperature margin than CMSC. There is an argument that a HTcSC wire can be much stable against external disturbance and might not need low-resistive metal plating. However, we should consider about the stability against microscopic defects which are intrinsically contained in any kind of superconducting materials. The MPZ (minimum propagating zone) of a superconducting wire can be calculated by the following equation.

$$\ell_m = \sqrt{\frac{2\,\kappa_s\,\Delta\,T}{J_c^2\,\rho_s}} \tag{1}$$

Where J_c is critical current density. Assuming $J_c = 2 \times 10^9 A/m^2$ and $\Delta T = 6.5K$, $l_m = 3\mu m$. This result means that there should not exist any defect longer than $3\mu m$ in the whole length of the wire, which is a too severe requirement, considering that the HTcSC is a very brittle material. By plating copper on the HTcSC, l_m can be easily increased to several mm.
  It is obvious from the above discussion that a wire using HTcSC material is desirable to be a composite wire plated by a low-resistive normal metal such as copper, aluminum or silver.

(b) Cryostatic stability
  The most reliable stability criterion is the cryostatic stability in which the wire can recover the superconducting state even when a long part of the wire becomes normal. There are calculations that evaluate the full stability condition of copper plated composite HTcSC wires by using the Stekly's parameter.[1,2] The results shows that the current density of the fully-stable wire cooled by LN_2 is roughly same as that of LHe cooling. However, those calculation neglects the temperature dependence of the resistivity of the normal metal. The figure 1 (a) shows the temperature dependences of the resistivity of copper and Fig. 1 (b) shows the heat transfer characteristics of LN_2. In the figure 2, relations between the heat generation at the normal zone and the heat transfer are shown. As is obvious from the figure, Stekly's stability criterion can not be applied to the HTcSC composite wire straightforwardly. Also obviously from Fig. 2, Maddock's equal-area theorem[5] can not be applied. If the wire current is selected to satisfy the equal-area theorem, the wire temperature where the heat generation of the wire is balanced with the heat transfer becomes too high and the wire will burn out when the wire current is not reduced after a normal zone appears. $\rho(T)$ linearly depends on T for T>200K and $q_{LN2}(T)$ is also for T=110-300K. From Fig. 1, $\rho(T)$ and $q_{LN2}(T)$ can be expressed by the following equations in those linearly-dependent regions.

$$\rho(T) = \alpha\,(T - T_a) \qquad \alpha = 6.7 \times 10^{-9}\ \Omega m/K\ ,\quad T_a = 49\ K \tag{2}$$

$$q_{LN2} = \beta (T - T_{\beta}) \qquad \beta = 10^2 \text{ W/m}^2\text{K}, \quad T_{\beta} = 30 \text{ K} \qquad (3)$$

To protect the wire from burning out, J should satisfy the following equation,

$$\frac{S \, \rho_c(T) \, J^2}{\lambda_c} < P \, q_{LN2}(T) \qquad (4)$$

where S is cross-sectional area of the wire, J is the current density of the wire and λc is the fraction of capper. For the equation to be satisfied,

$$\frac{S \, \alpha \, J^2}{\lambda_c} < P \, \beta \qquad (5)$$

For an example, when $\lambda c=1$, $S=1\text{cm}^2$ and $P=0.5\text{cm}$, $J<880\text{A/cm}^2$. This value is as small as a current density of water-cooled copper conductor. Obviously from the above discussion, to operate a HTcSC wire at a reasonably high current density, an active dumping of the energy stored in the magnet is required.

(c) Quench protection

To protect a magnet from damages caused by a quench, it is necessary to quickly detect the quench and dump the stored energy into the dump resister placed outside of the cryostat. Usually a quench is detected by measuring a terminal voltage of the magnet. A quench should be detected, at least, before the maximum temperature of the wire at the normal zone reaches to a maximum allowable temperature.[6] When a quench is detected, the voltage V_d across the normal zone originating from a point source and propagating one-dimensionally along the conductor is given by,

$$V_d = 2 \int_0^{t_d} \rho (T) \, J \, v_p \, dt \qquad (6)$$

where t_d is the time when the quench is detected and v_p is propagation velocity of the normal zone. In the above equation, λ_c is assumed to be 1. Assuming that v_p is constant and that $\rho_{av}(T)$ is the average resistivity of the wire in the normal zone, Eq. 6 is approximated by,

$$V_d = 2 \rho_{av}(T) \, J \, v_p \, t_d \qquad (7)$$

The maximum wire temperature T_{max} can be estimated by the following equation.

$$\rho (T) \, J^2 = C (T) \, \frac{dT}{dt} \qquad (8)$$

Modifying and integrating Eq. (8), t_d is given by,

$$t_d = Z(T_{max})/J^2 \qquad \text{where} \qquad Z(T_{max}) = \int_{77K}^{T_{max}} \frac{C(T)}{\rho(T)} \, dT \qquad (9)$$

Z(T) is given by Fig. 3. Assuming that the wire is adiabatic, v_p can be given by,

$$v_p = \frac{J}{C(77K)} \sqrt{\frac{\rho(77K) \, \kappa(77K)}{\Delta T}} \qquad (10)$$

Combining Eqs. (7), (9) and (10), V_d is given

$$V_d = \frac{2 Z(T_{max}) \, \rho_{av}(T) \sqrt{\rho(77K) \, \kappa(77K)/\Delta T}}{C(77K)} \qquad (11)$$

By this equation, the relation between V_d and T_{max} is obtained. A numerical example is given in Fig. 4. To obtain the result Fig. 4, $\rho_{av}(T)$ is approximated to be $(\rho(T_{max})+\rho(77K))/2$. Considering that it takes a time to dump the stored energy of the magnet after a quench is detected, T_{max} should be below 150K to protect the magnet winding from damages caused by thermal stresses. That is, a quench should be detected before V_d becomes 70mV. It should be noted that this small value of Vd is not changed by the size of the magnet and the current density of the wire. The above discussion indicates that the voltage measurement may not be appropriate as a quench detection method for a HTcSC magnet and that developments of other methods than the voltage measurement is necessary. Methods using optical fibers and acoustic emission technique are promising as the quench detection of the HTcSC magnet.[7,8]

Performances Required to HTcSC Material

(a) General requirements
For the HTcSC to be magnet conductor, following requirements should be satisfied:
(1) The conductors should have reasonably high critical current density at high magnetic field.
(2) The conductors should be long and flexible to be wound as a magnet.
(3) Characteristics of the conductors should be stable against heat cycle and long-term operation, and free from local defects.
(4) The conductors should not be damaged by the electromagnetic stress.

(b) Required Current Density
Of the requirements mentioned above, the requirement on the current density is the most critical issue for HTcSC magnet to be practical. Current densities required for the HTcSC wire to be practical are listed in Tab. 2 for various applications.[9] Even when HTcSC can be used to the Tokamak reactor, the current density required to the conductors are not much different from the values of CMSC, because the magnet system should be compact keeping the compatibility with the space for the plasma. To comply with field configuration and accuracy required to high-energy physics magnets even HTcSC magnets should have as high current density as of CMSC. If HTcSC magnets can be used to magnets for magnetically levitated trains, the magnet system can be lighter and more compact than CMSC system. For MRI, field accuracy is the first priority of the magnet performance and current density can be moderated to be 80A/mm^2. If current density of field winding of a superconducting generator is low, the diameter of the rotor increases and the efficiency of the generator decreases because of the increasing heat leak. Using HTcSC wire of the current density higher than 150A/mm^2, the efficiency of the generator could be higher than that of the CMSC generator because the efficiency of Liq. N_2 refrigerator is much higher than that of LHe one.
The values of the current density listed in Tab. 2 are over-all current densities of the composite wires consisting of low-resistive metal for stabilization and HTcSC materials. In practical magnet, operating current is 50%-80% of the wire critical current. Considering those factors, the critical current density required to HTcSC material itself is 3-4 times more than the values in Tab. 2.

Conclusions

From the discussions described above, following conclusions are obtained.
- In a large scale HTcSC magnet, amount of structural material increases significantly because strength of the stainless steel decreases at 77K compared to 4.2K, and HTcSC is very sensitive to strain, which requires more structural material.

- Current densities required to HTcSC material are comparable to that of NbTi in the magnets applied to Tokamak nuclear fusion reactors, high-energy physics experiments and superconducting generator.
- Cryostatic stability can not be applied to HTcSC wires because of too small conductor current density, and transient stability or adiabatic stability modes are required.
- To protect a HTcSC magnet from damages caused by a quench, tens of millivolts should be detected regardless of the current density.
- Application of HTcSC magnets to a large scale system seems to have many technical problems. Potential applications of HTcSC magnets are in the areas which use small or medium scale magnets.

Property	4.2K	77K
Conductor op. temp. range, ΔT (K) th. cond., κ_s (W/mk) heat cap., C_s (J/m) norm. resist., ρ_s (Ω/m) max. strain, (%)	NbTi ~1 0.1 0.5 6.5×10^{-7} 0.6	YBCO ~10 7 1000 2.0×10^{-6} 0.2
Copper th. cond., κ (W/mK) heat cap., C (J/m^3K) resistivity, ρ (Ω/m)	10^3 7.5×10^2 2×10^{-10}	500 1.5×10^6 2×10^{-9}

Coolant
boil. h. flux, (W/m^2)
  nucleate(max)
  film(min)
	LHe
	0.8×10^4
	0.3×10^4
		LN_2
		14×10^4
		0.8×10^4

Stainless steel tensile strength, (MPa)	SUS316	
	1450	1220

Tab. 1 Properties of magnet components

References

1. Y. Iwasa, IEEE Trans. Magn., Vol. 24, No. 2, 1988, p1211.
2. D. Ito, Cryogenic Eng. Vol. 22, No. 6, 1987, p383 (Japanese).
3. "Handbuch Kryotechnik" Arbeitskreis Kryotechnik der DPG, 1977.
4. R. J. Thome et al, Adv. Cry. Eng., Vol. 31, 1986, p341.
5. M. N. Wilson, "Superconducting Magnet", Oxford Scien. Pub, 1983.
6. Y. Iwasa et al, Cryogenics, Dec., 1980, p711.
7. O. Tsukamoto et al, Adv. Cry. Eng., Vol. 31, 1986, p259.
8. O. Tsukamoto et al, Adv. Cry. Eng., Vol. 31, 1986, p1269.
9. O. Tsukamoto, S. 9-5., Proc. Spring Meeting of J.I.E.E., 1988, (Japanese).

Application	Current density (A/mm^2)	Max. field (T)
- Fusion reactor	100~150	12
- Particle accelerator	800~1000	7
- Magnetic levitation	130	4
- MRI	80	3
- Superconducting generator	150	6

Tab. 2 Required currents

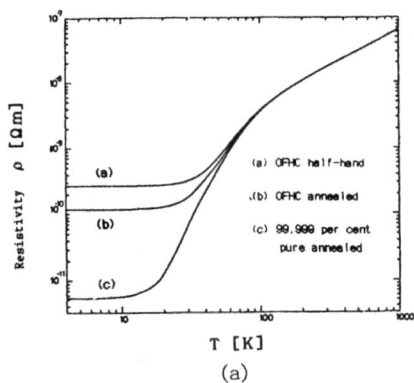

(a)

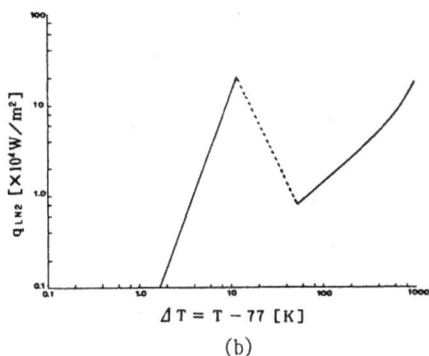

(b)

Fig. 1 (a)Temperature dependence of resistivity
of copper
(b)Heat transfer characteristics of LN$_2$

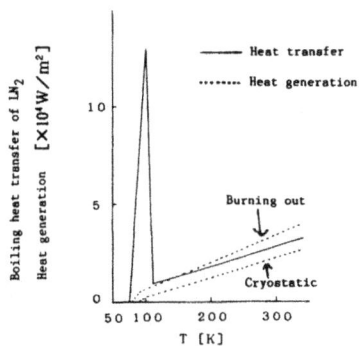

Fig. 3 T vs. Z(T) for copper

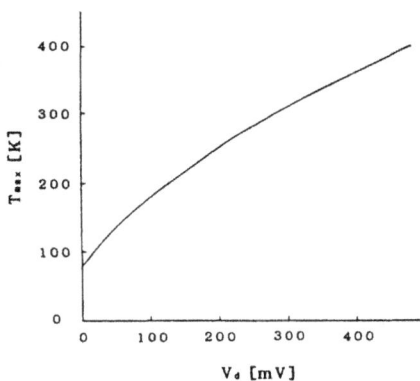

Fig. 2 Relation between heat transfer of LN$_2$ and
heat generation of wire

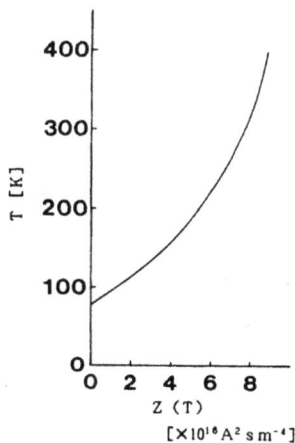

Fig. 4 Relation between voltage across normal
zone V$_d$ and maximum temperature of wire
T$_{max}$ in normal zone when a quench is
detected

THE CHARACTERISTICS OF HELIUM AS A COOLANT FOR SUPERCONDUCTING MAGNETS

T.H.K. Frederking*, J. Yamamoto**

*Univ. Calif. Los Angeles, CA 90024, U.S.A.
**Laboratory for Appl. Superconductivity, Osaka Univ.
2-1, Yamada-oka Suita, Osaka, 565, Japan

Abstract - Coolant properties of liquid Helium-4 are presented including interdependence of conductor quench domains and coolant heat transfer. A comparison of dynamic stability of He I and superfluid He II shows an enhanced temperature difference margin associated with triple-phase nucleate boiling in excess of limits expected from He II phase boundaries.

Introduction

Magnet energy range increases and high-transition temperature (T_c) superconductor developments have stimulated a reassessment of stability and of magnet coolant use. Starting from the He[4] phase diagram, and from MPZ lengths as quench parameters, various (selected) coolant modes are discussed including a comparison between He I two-phase nucleate boiling and triple-phase nucleate boiling of He II. We refer to early work in the magnet stability area[1] and to recent work in heat transfer-related stability.[2]

He[4] phase diagram and MPZ conditions

The He[4] phase diagram, shown schematically in Fig. 1, does not only have first order transitions but also extensions of higher order phase changes (shown as dashed lines) and superheat limits important for dynamic stability. The following notation is introduced in conjunction with the pressure(P)-temperature(T) diagram: He X (X=I, II); (subscripts sat and p denoting saturated liquid at vapor-liquid equilibrium and pressurized liquid respectively).

The coolant has to provide quench protection. For quench quantification, conductor properties are important. The minimum propagating zone (MPZ) is considered in the present context (Table I and Appendix A). Normal metal stabilizers, introduced by Laverick-Stekly,[1] prevent the short adiabatic length MPZ of bare conductor. Very long non-adiabatic lengths (MPZ > 10 cm) may exist in some parameter limits (Ito-Kubota[2]). Both extremes, the very short MPZ and the very long MPZ may endanger conductor operation, in particular in boiling liquid. In the non-adiabatic MPZ equation, the heat transfer coefficient plays a role. Therefore, a suitable coolant state, compatible with the MPZ - heat transfer requirements, ought to be selected for the particular conductor specifications imposed. Appendix A presents a simplified extension of non-adiabatic MPZ statements, e.g. Eckels.[2] It is noted that the original MPZ-introduction has features in common with nucleus behavior in first order phase transitions. This feature however is no longer pertinent for the non-adiabatic MPZ of steady heat transfer rates.

TABLE I : Characteristic properties of various MPZ results

Conductor type	conductor envelope	Parameters of influence	Remarks
bare	adiabatic	solid state properties	burnout danger no dia.effect
composite	adiabatic	geometry in-exerts influence	area ratio A_N/A_{sc} important
non-adiabatic composite conductor	heat trans-fer coeffic. h enters	liquid state;multiple steady states	optimization not always secured

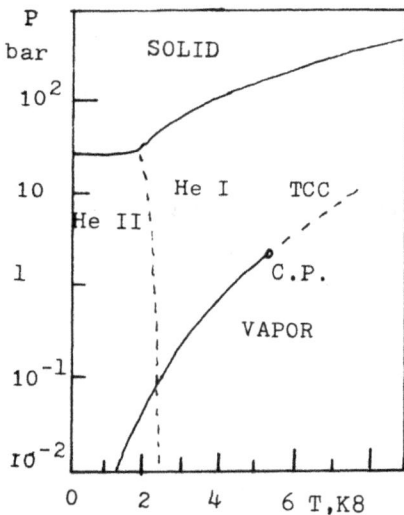

Fig. 1 Temperature-Pressure phase diagram* of He[4]; C.P. thermodynamic critical point; TCC transposed critical curve (=location of specific heat maxima); --- curves of higher order phase transitions without latent heat

(* schematically)

Figure 2 is a plot of MPZ examples. The very short adiabatic MPZ_a of bare NbTi is independent of the wire diameter. The composite conductor NbTi-Cu, with effective values (subscript eff) of the thermal conductivity (k) and of the electrical resistivity (o_e) is a weak function of the cross sectional area ratio (normal metal to superconductor) shown in Fig. 2. Optimum MPZ-values for dipoles, reported by Hassenzahl,[2] are rather close to an area ratio of unity. (The MPZ definition, used for dipoles, has a different prefactor, on the order of unity). The inset of Fig. 2 is an example of the non-adiabatic MPZ for boiling liquid He I baths. Boiling pool us is charac-terized by a long MPZ in the low current limit, to be discussed below.

The non-adiabatic MPZ equation contains the heat transfer coefficient (h). Therefore, this MPZ length is coupled to coolant convection modes, and in addition, the MPZ depends on the the liquid thermodynamics. In particular boiling baths, both the very short bare wire MPZ, and the very long MPZ, may lead to con-ductor deterioration and possible burnout danger. Examples are

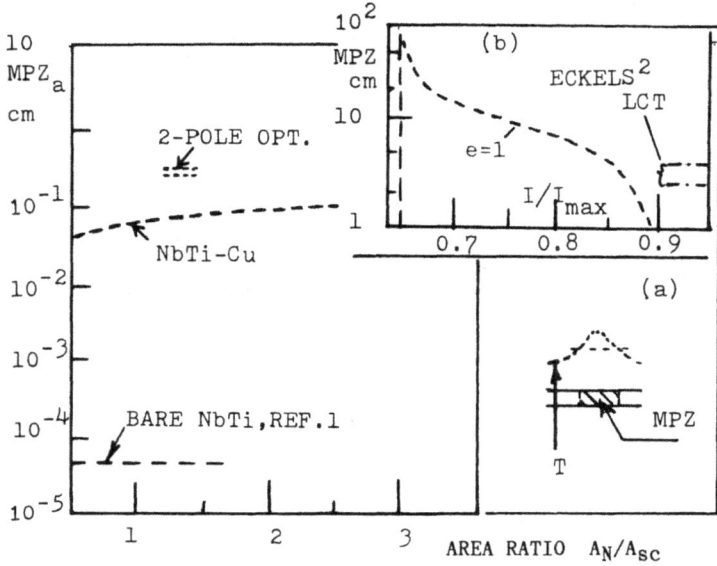

Fig. 2 Adiabatic MPZ_a as a function of area ratio (normal metal to superconductor); 2-POLE OPT=optimum of Hassenzahl[2]; Inset a: T-distribution associated with quench domain, schematically; Inset b: Non-adiabatic MPZ Ito-Kubota[2] for e=1 (cryostatics of EAC)

considered for illustration, referring to both He I and He II.

Examples of He I coolant characteristics

He I_p: Forced convection (forced flow) is assumed at pressures above the thermodynamic critical pressure of 2.3 bar. At large flow velocities, and at sufficiently high pressures, there is only a remote chance of a phase transition, the crossing of the transposed critical curve (TCC, Fig. 1). An example is the SMES simulation coil "SHETEM 2b" presented at this Symposium. The temperature margin is the difference between the superconductor transition temperature $T_c(B)$ which is a function of the magnetic field, and the "fluid bath" temperature T_B. There is an "energy" margin, given by the enthalpy difference (ΔH) over the magnet volume, from T_B to the current sharing temperature (T_{cs})

$$\Delta \dot{H} = V \int_{T_B}^{T_{cs}} (\rho \, c_p) \, dT \qquad (1)$$

([o c_p] specific heat per unit volume (V)). A more stringent criterion appears to be the power dissipated requiring a safe mass flow rate $\dot{m}$ for the T-margin available.

$$\Delta \dot{H} = \dot{m} \int c_p \, dT \qquad (2)$$

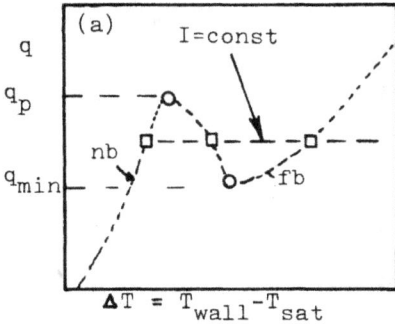

Fig. 3 Boiling curve, schematically;
3a Heat flux density q versus temperature difference, nb nucleate
boiling, fb film boiling; I=
const shown for 3 intersections
(squares) with the boiling curve.

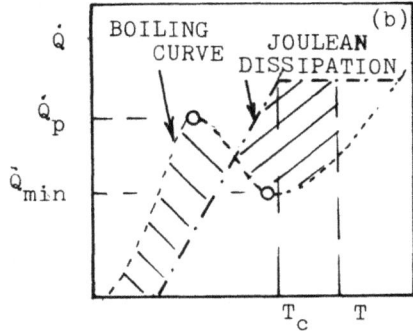

Fig. 3b Modified "equal-area-
criterion" (EAC) of
Kubota-Ito[2]; Power
$Q(T)$ is shown for the
heat removal of the
boiling curve.

The local condition at a quench domain is the heat flow rate (Q) resulting
from energy dissipation ($\dot{Q}$ = A·h·ΔT, A= fluid-wetted surface area, (Ah)
heating power per K). An MPZ restriction is possible, using a sufficient
mass flow rate, because of h proportional to m^n ; n in the range 0.8 < n < 1
depends on the coolant duct wall condition. Another parameter is the length
(L) of a coil section. The related "global" condition is written in a
simplified statement as

$$\int (d\dot{Q}/dx)dx \;\cong\; (\dot{Q}/L)L = \dot{m}\int c_p \; dT \qquad (3)$$

Thus, optimization is influenced but little by fluid phase transitions in
pressurized He I during forced flow.

   Natural convection at "1 g" (standard gravity) has much smaller flow
velocities associated with buoyancy (loss of forced flow scenario). In the
strong centrifugal fields of superconducting a.c. power station generators,
buoyancy associated with the order of magnitude of "1000 g" may be utilized
with benefits for stability. Once forced flow has been selected however, the
phase transition at the TCC (Figure 1) may exert a significant influence.

   He I_{sat}: In saturated liquid He I at vapor-liquid equilibrium, the
boiling curve is accessible readily, as in other saturated liquids, e.g.
liquid nitrogen. Historically, boiling He I baths have been preferred in
early developments of metal superconductor (MSC) magnets. However, the homogeneous liquid He I superheat limit is only about 0.2 K above the normal
boiling point (4.2 K). This constraint has caused considerable interest in
the "multiple steady states" (for constant current I) of the boiling curve,
shown schematically in Fig. 3. Its heat flux density (q) has the following
limits: "peak" value (q_p) at the maximum temperature difference $\Delta T = T_{wall}$ -
T_{sat} at "take-off" from the low energy nucleate boiling branch, and "minimum"
or "recovery" value (q_{min}) at the lowest T of the high energy branch of film
boiling. In this context several cryostatic stability criteria have been

applied. A special stabilization procedure, known as Maddock et al. "equal-area-criterion" (EAC)[1] has provided considerable insight into cryostatic, quasi-steady magnet stabilization. A recent numerical reassessment of EAC by Ito-Kubota[2] points out a limited useful range of MPZ versus electrical current (inset of Figure 2); [the normalized power, used as parameter"e", corresponds to cryostatics when "e" has the order of magnitude of unity]. Figure 3a indicates the recent EAC modification, including the cutoff T-condition.

The long MPZ length (Fig. 2) is to be avoided in the boiling bath for the following quench scenario: A high energy magnet, with a long MPZ in film boiling conditions, may "seek" locally to dissipate much energy in the high T domain. The result is a possible deterioration of conductor properties and integrity. In this context it is noted that the adiabatic propagation speed v_{ad} is on the order of $[k/(\rho \cdot c_p)]/ MPZ_a$. Unfavorably low speeds, tending toward cryostatic conditions, result when the MPZ_a is long, when the thermal diffusivity $[k/(\rho c_p)]$ is small, or when both conditions are imposed. For instance, low v_{ad} are predicted for high-T_c superconductors (HTSC) at liquid N_2 bath temperatures because of low thermal diffusivity (NBP: 77 K).

Superfluid liquid He II examples

He II_p: The choice of superfluid He II, at quasi-steady heat transfer conditions, introduces the thermal boundary resistance, known as Kapitza resistance. (During dynamic operation, KapitzaΔT contributions occur also in He I). Pressurization above the thermodynamic critical pressure (2.3 bar) is considered first.

Above the thermodynamic critical pressure, a severe quench produces only higher order phase transitions (from He II_p to He I_p and in He I_p from the low-T side of the TCC to its high-T side, Fig. 1). At high pressures, the selection of forced flow minimizes the influence of the Helium-4 phase transition dynamics. When forced flow is absent, the convection mechanisms have been characterized by the term "natural convection" in He I and by "zero net mass flow" (ZNMF) in He II.

Pressures near 1 atm and bath temperatures near 1.8 K have been found to be attractive for magnet stabilization: He II_p is easily pressure-controlled. The choice of T = 1.8 K is rather close to the maximum q-values associated with He II - He I phase changes. There is a subsequent first order transition, involving the latent heat of vaporization, when the pressure is below 2.3 bar. Therefore, dynamic stability limits, prior to a quench of NbTi, are associated with the peak values of transient "triple-phase nucleate boiling". He vapor is generated from a He I layer "sandwiched" between the conductor boundary and bulk He II.

He II_{sat}: The saturated liquid He[4] range of vapor-liquid equilibrium starts below 0.05 bar (= P_{sat}). Any vapor film-generating quench below this pressure tends to raise the composite temperature rapidly toward the high energy branch of the multiple steady states of the boiling curve. The He II "boiling curve" has a non-boiling branch at low T. This branch is dominated by Kapitza resistance (or by an overall Kapitza ΔT), illustrated in recent investigations[3-7].

Dynamic stability in both He I and He II produces Kapitza resistance-related temperature differences. He II_{sat} has been considered a required phase only in the auxiliary refrigeration subsystem for He II_p. An example of He II_{sat} as a magnet coolant is the "lambda" system, e.g. coolant use near 50 mbar.

Comparison of dynamic stability: He II_p versus He I at 1 bar

For a specified coolant duct geometry in a pool (free convection and ZNMF) the He II dynamics at take-off is characterized by significant improvements. For instance, for a step power input simulation of a quench, there is an order of magnitude increase in the take-off value (Q_t), at a specified time (t_t) when He I is replaced by He II_p at atmospheric pressure. In the superfluid He II_p the T-margin, prior to take-off extends from 1.8 K to T > 4.4 K (including the homogeneous liquid superheat) in the triple-phase nucleate boiling range. In He I alone, there is the usual two-phase nucleate boiling. The eventual T_t of the conductor at "take-off" depends on the coating chosen and Kapitza resistance contributions. For a thick coating and high Kapitza resistance, T_t may exceed T_c of NbTi. The comparison is illustrated by Figs. 4a and 4b. Fig. 4a presents 2-phase nucleate boiling near atmospheric pressure including recent data of Sakurai et al.[2], Giarratano et al.[3] and C. Schmidt.[4] Fig. 4b contains data of Gentile et al.[5] and of Chen - Van Sciver.[6] In both plots, the Lin et al. equation[7] is included. Nucleate

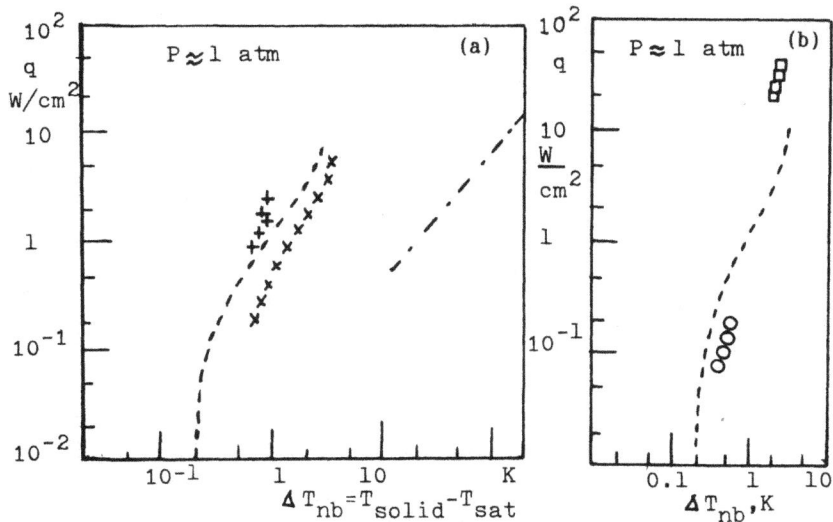

Fig. 4   Comparison of nucleate boiling characteristics:
   a. He I near normal boiling point at 1 atm;
      ++ Sakurai et at.[2]; x Giarratano-Schmidt[3,4]; ----- film boiling;
      --- Lin et al. equation[7]
   b. triple-phase nucleate boiling[10]; circles; Chen-Van Sciver;
      squares: Gentile et al.[5]

boiling modes appear to be consistent in both cases. There are substantial q_t range extensions in He II_p. - Further, it is mentioned that the He I data of Wilson[8] in ducts appear to be consistent with recent He II duct results of Filippov et al.[9]

Acknowledgments. One of us (THKF) acknowledges with appreciation support of JSPS, Prof. T. Ito, and of Osaka University. Further, Dr. Y. Kamioka's help is acknowledged with thanks.

References

1) M.N. Wilson, Superconducting Magnets, Oxford, Clarendon, 1983.
2) T.H.K. Frederking,, and T. Ito, Guest Ed., Joint Seminar, Cryogenics 1989, to be published.
3) P.J. Giarratano and N.V. Frederick, Adv. Cryog. Eng. 25, 513 (1980).
4) C. Schmidt, IEEE Trans. Magn., MAG-17, 738-741 (1981).
5) D. Gentile and W.V. Hassenzahl, Adv. Cryog. Eng. 25, 385-392 (1980).
6) Z. Chen and S.W. Van Sciver, Cryogenics 27, 635-640 (1987).
7) Li-He Lin and T.H.K. Frederking, Letters Heat Mass Transf. 9, 473-478 (1982).
8) M.N. Wilson, Proc. IIR Meetg. Boulder, 1966 (Suppl. Bull. IIR - Annexe 1966-5), 109-114 (1966).
9) Yu. Filippov et al., ICEC-12, Southampton, 1988.
10) S. Caspi and T.H.K. Frederking, Cryogenics 19, 513-516 (1979).

APPENDIX A: Simplified MPZ equations

Adiabatic length MPZ_a: Consider the adiabatic MPZ of a "point distur- bance".[1] The bare conductor result is[1]

$$MPZ_a = j_c^{-1} \left[(2k/\rho_e)(T_c-T_B) \right]^{1/2} \qquad (A.1)$$

(k thermal conductivity, j_c critical current density, T_c critical tempera- ture, T_B fluid "bath" temperature). Extending this result to the composite conductor (subscript cc), e.g. NbTi-Cu in Fig. 2, one may write for the composite

$$MPZ_{a,cc}/MPZ_a = \left[(k_{eff}/k)(\rho_e/\rho_{eff}) \right]^{1/2} \qquad (A.2)$$

It is noted for HTSC assessment that the adiabatic propagation speed (v_{ad}) is proportional to $k/(\rho c_p) / MPZ_a$.

Non-adiabatic MPZ: Various expressions have been given in the discussion of composites, e.g. Eckels[2]. A simplified form is written for a solid-to- fluid heat transfer coefficient (h) as

$$MPZ = j_N^{-2} (2/\rho_N) \Delta T \cdot h_{ax} \qquad (A.3)$$

h_{ax}=axial heat transfer coefficient $[k\, C_c\, h/A]^{1/2}$ where A is the cross sec- tional area and C_c the fluid-wetted circumference. The "high" -T modeling in a particular simple limit leads to a "high energy" film boiling MPZ with a Lorentz number (Lo) and normal state thermal conductivity and electrical current density (j_N).

This simplified MPZ for high T is written as

$$MPZ_{fb} = (k/j_N)/ Lo^{1/2} \qquad (A.4)$$

with $j_N A_N = j_{sc} A_{sc}$; subscript "sc" for the superconductor cross section.

PARAMETERS RELEVANT TO STRAIN DEPENDENCE OF CRITICAL CURRENT
IN PRACTICAL Nb$_3$Sn SUPERCONDUCTING WIRES

Kazumune KATAGIRI, Koji SAITO, Masashi OHGAMI, Toichi OKADA,
Akihiko NAGATA*, Koshichi NOTO* and Kazuo WATANABE*

Inst. Sci. Ind. Res., Osaka University
8-1 Mihogaoka, Ibaraki, Osaka, 567 Japan
*Inst. Maters. Res., Tohoku University
2-1 Katahira, Sendai, 980 Japan

Abstract - Some parameters relevant to the strain dependence of critical
current I_c were investigated with emphasis on their correlation with wire
constitution and heat treatment condition.   Examinations were made on three
kinds of practical Nb$_3$Sn superconducting wires which are in the course of
development.   Improvement of the strain characteristics in these wires
through proper choice of parameters is briefly discussed.

Introduction

Because of high sensitivity of critical current I_c to strain, the strain
effect of Nb$_3$Sn superconducting wire has been studied in order to realize
reliable performance of superconducting magnet at high fields (>10T).   The
strain dependence of I_c in the bronze processed multifilamentary superconducting wire is fairly well established [1,2].   It is characterized by 1)
the sensitivity of I_c on strain within the elastic region, 2) the peak value
of I_c and the corresponding strain ε_m, 3) the reversible strain limit ε_{irrev}
beyond which I_c data poins no longer fall on the curve on loading.
    Various process for fabrication of the wires have been developed.
Strain characteristics of the wires depend on the process.   In this paper,
the strain effect of some wires other than conventional bronze processed wire
is presented, and the controlling factors associated with wire constitution
and heat treatment are elucidated.

Experiment

Three kinds of wire, the details are to be described in the next section
were studied.   The strain dependence of I_c was measured at 6T or 15T (IMR,
Tohoku University) in liquid He using the loading apparatus [2].   The
criterion of I_c is 1 μV/cm and the accuracy of strain measurement is 0.05 %
strain.

Results and Discussion

Internally Reinforced Wire

In order to withstand huge electro-magnetic force as well as enable
winding process after reaction heat treatment, the composite wires with
reinforcement have been developed [3].   The wire examined is a bronze
processed multifilamentary (4 μm x 4620) composite with seven 405 stainless

steel reinforcing wires; one in the center and other 6 at corners of the composite inside the stabilizing copper sleeve. The wire, diameter being 1.2 mm, is heat treated at 923 K for 173 h.

The flow stress of the wire increased drastically as compared to that in conventional bronze processed wire. Figure 1 shows I_c vs. strain relation at 15T. Increase in strain torelance is seen in $\varepsilon_m = 0.4$ %, which is higher than 0.2 % in the conventional wire. The intrinsic reversible strain limit $\varepsilon_{0,irrev} = \varepsilon_{irrev} - \varepsilon_m$, however, is rather low (0.4 %) and the strain sensitivity of I_c normalized by its peak value I_{cm} is fairly high. These are suspected to be correlated with the serration around 0.9 % strain in the stress-strain curve and the micro-strain bursts preceding to it. Localized deformation in 405 steel (b.c.c. metal) through twinning and/or low temperature embrittlement deteriorate superconducting characteristics at that portion and leads to inferior strain dependence of I_c (Fig. 2 shows magnified brittle fracture surface in the wire). Similar behavior can also be anticipated in the case of reinforcement by f.c.c. stainless steel in which the serration by thermal instability during low temperature deformation is observed. Proper choice of reinforcing material and constitution which minimize direct transmission of local strain in the reinforcement to Nb_3Sn layer is required.

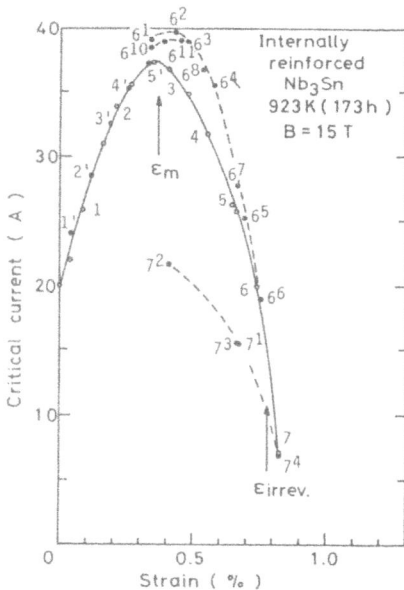

Fig. 1. I_c vs. strain relation of internally reinforced wire.

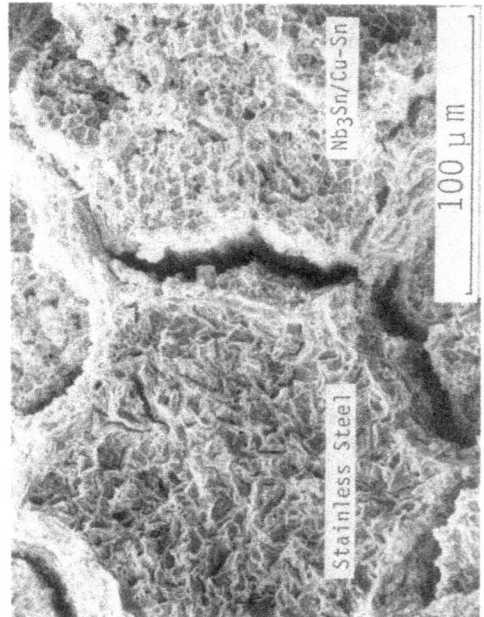

Fig. 2. Magnified view of fracture surface of internally reinforced wire.

Internal Diffusion Processed Wire

Because of ease of drawing and being free from limitation in Sn content, internal diffusion process is promissing [4]. The wire examined here is one in rather earlier stage of development, with 0.6 mm dia. consisted of 9100 x 1.7 μm filaments and single Sn core at the center of them.

The strain dependence of I_c is shown in Fig. 3. The strain sensitivity of the wire is higher and $\varepsilon_{0,irrev}$ (0.4 %) is smaller as compared with those of cohventional bronze processed wire ($\varepsilon_{0,irrev}$ = 0.6 %). Judging from the fractograph shown in Fig. 4, these inferior strain characteristics are attributable to the large Kirkendall voids, which induce concentrated strain around them in the wire. Recently, however, we have evidenced that excellent strain characteristics in the internal diffusion wire with controlled voids through heat treatment and also constitution with dispersed Sn cores [5].

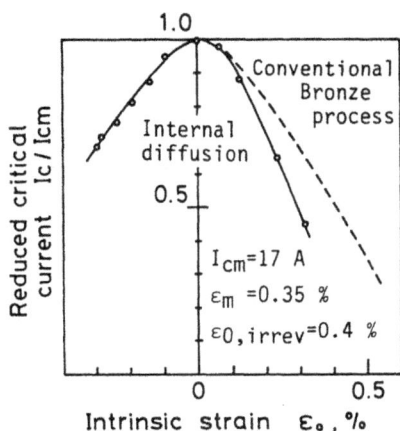

Fig. 3. Strain sensitivity of I_c in internal diffusion processed wire.

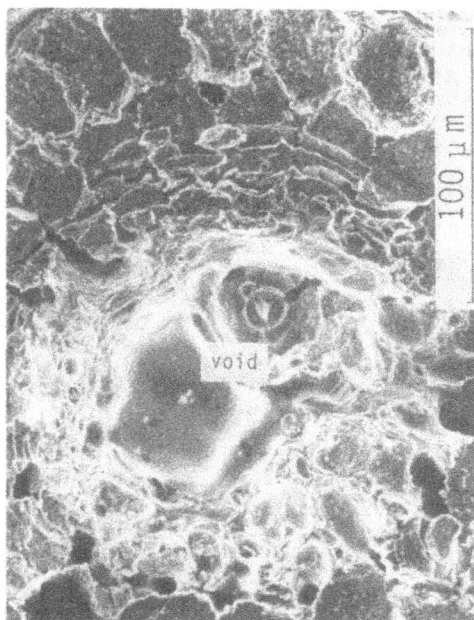

Fig. 4. Magnified view of fracture surface of internal diffusion processed wire.

External Diffusion "in situ" Processed Wire

External Sn diffusion "in situ" wires of 0.3 mm dia. with various Nb and Sn (electroplated) content [6] were examined. Heat treatment condition is

473 K (2 day) + 823 K (7 day). Figure 5 shows a fractograph of the wire (40 vol% Nb - 13 wt% Sn against Cu). According to an EPMA analysis, there exists a Sn gradient within the reacted area (outer ring) decreasing towards interior. Unreacted area manifested by dimple pattern is seen in the center of the fracture surface. The size of this area depends on initial Nb content as shown in Fig. 6. It is also a function of Sn content for a fixed Nb content. Figure 7 shows how critical current density J_c and characteristic strain values depend on Sn content ($\varepsilon_{irrev} = \varepsilon_f$ in this kinds of wire). The unreacted area fraction decreases J_c and increases the strain tolerance. There exist certain optimum Nb and Sn content in order to attain high J_c and good strain characteristics.

In the case of wires with high Nb and Sn content, improper heat treatment deteriorates both the I_c and the strain characteristics. Temperature rise beyond the melting point of Sn (510 K) in the stage of insufficient conversion to Cu-Sn intermetallic bilayer ($\eta + \varepsilon$), as well as that beyond the melting point of η layer (688 K) before full conversion to ε result in uneven Sn distribution in both longitudinal direction and radial direction of the wire [7]. Figure 8 exhibits deviation of reacted area in the fracture surface and the initiation of fracture occurred at the outside of the thicker region.

The tin balling is known as the extreme case of uneven Sn distribution along the wire. Deviation of Sn content from stoichiometry leads to degraded I_c and also to embrittlement. An example of fracture surface at a tin ball

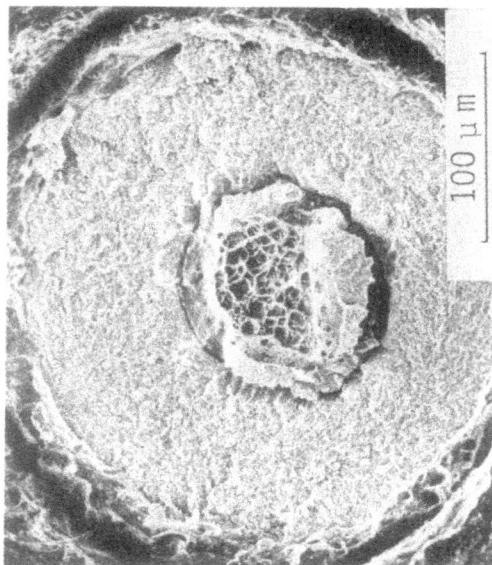

Fig. 5. Fracture surface of "in situ" wire.

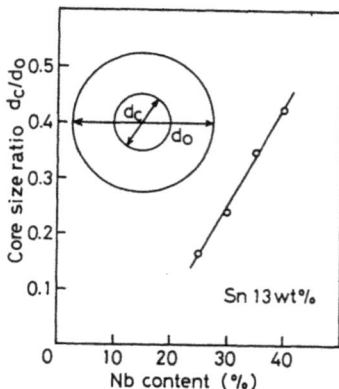

Fig. 6. Unreacted core size vs. Nb content in "in situ" wire.

is shown in Fig. 9.   Existence of free surface formed during the heat
treatment (arrow) suggests stress concentration around there.   These factors
will result in premature brittle fracture of the wire at small nominal
strain, 0.6 % for example.   Now, even the blistering which is more
troublesome than the tin balling in the external tin method has been
controlled through the proper design of Cu cladding thickness [7].

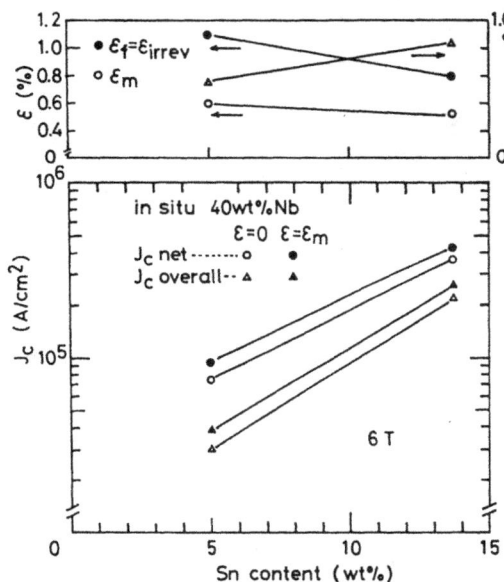

Fig. 7. J_c and characteristic strain values
vs. Sn content in
"in situ" wire.

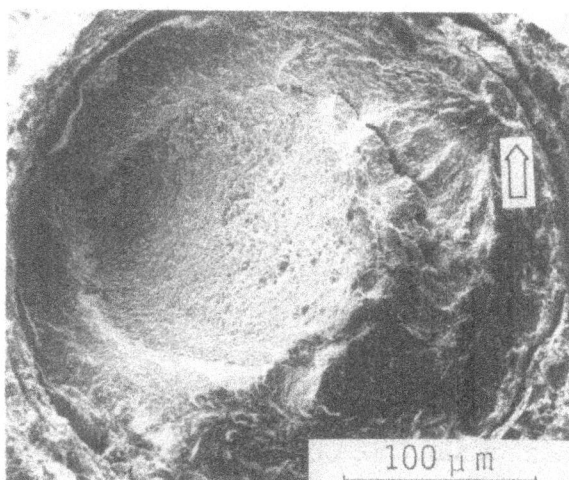

Fig. 8. Deviation of
reacted layer in
"in situ" wire.

406

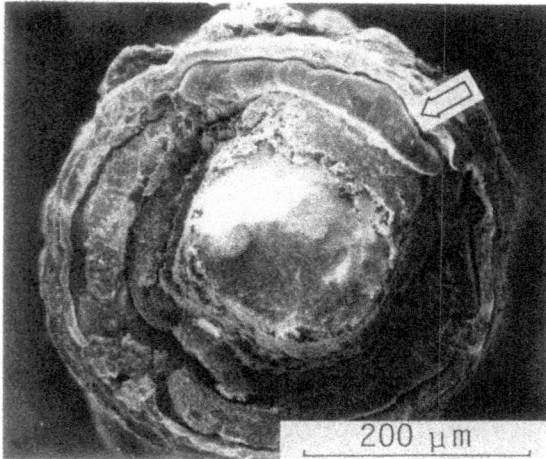

Fig. 9. Fracture surface of "in situ" wire at Sn ball.

Summary

Both Kirkendall voids and uneven Sn distribution in the case of internal and external Sn diffusion process are detrimental.   They as well as reinforcement can cause degradation of the wire through localized deformation and/or fracture.

The strain characteristics of the practical Nb_3Sn superconducting wires are improved through control of the wire constitution and heat treatment.

Acknowledgments - The authors would like to express their thanks to Prof. T. Anayama, Tohoku University, Dr. K. Yoshizaki, Mitsubishi Electric Co. and Dr. H. Takei, Sumitomo Electric Co., for providing the wires.   Staffs of High Field Lab. Supercond. Maters. and Low Temp. Center in IMR, Tohoku University as well as Low Temp. Center, Osaka University are acknowledged for their helpful assistance in experiment at high and low field experiment.   This study is partly supported by Grant in Aid for Sci. Res. No. 62050007, Ministry of Education, Japan.

References

1) J.W.Ekin, Cryogenics, 20, 611-624 (1980).
2) K.Katagiri, K.Saito, T.Okada, A.Ngata, and K.Noto, Adv. Cryog. Eng. Maters., 34, 531-538 (1988).
3) Papers in "Filamentary A15 Superconductors", ed. M.Suenaga, and A.F.Clark, (Plenum Press, NY, 1980).
4) M.Hashimoto, K.Yoshizaki and M.Tanaka, Proc. 5th Int. Cryog. Eng. Conf., Kyoto,   ed. K.Mendelssohn, (IPC Sci. Tech. Press, 1974), pp. 332-335.
5) K.Katagiri et al, to be published.
6) T.Anayama and A.Nagata, Proc. Int. Symp. Flux Pin. Electromagn. Prop. Superconds., Fukuoka, eds. T.Matsushita, K.Yamafuji and F.Irie, (Matsukuma Press, Fukuoka, 1985), pp. 75-82.
7) J.V.Verhoeven, C.C.Cheng and E.D.Gibson, Adv. Cryog. Eng. Maters., 32, (1986) 985-993.

DEVELOPMENT OF AC APPLICATIONS OF SUPERCONDUCTING WIRES

Kaname MATSUMOTO, Yasuzo TANAKA, and Katsuro OISHI

Yokohama R&D Laboratories, The Furukawa Electric Co.,Ltd.
2-4-3, Okano, Nishi-ku, Yokohama 220, Japan.

Abstract - We have been developing 50/60Hz use superconducting wire. For example, an 0.5μm NbTi multifilamentary superconducting wire has been developed and the practical test of epoxy-impregnated superconducting coils using the wire were performed. The capacities are 4 kVA, 100 kVA and others. Now, a 500 kVA disc-type coil has been tried and its current capacity is over 1300 A at 2 T. We analyzed a heat property in the winding of these coils and compared it with the results of the excitation test.

Introduction

The AC superconducting composite wires are very thin, typically 0.1-0.2 mm in diameter, for an improvement of cooling stabilization and for a reduction of coupling losses by a heavy twist pitch, so that current capacities of these wires are very small, for example, 5-20 A per elementary wire. However, a current capacity of conductor for a large scale superconducting transformer is required to be more than 1000 A at peak field of 0.5 T, and it is 10-30 kA at field of 2 T for an armature winding of a large scale fully superconducting generator [1],[2] . Therefore, it is necessary to make a large current capacity stranded cable with a lot of elementary wires, moreover, it is indispensable to fix every wire in this stranded cable with keeping good cooling condition so as not to move by Lorenz force.

In this case, an epoxy impregnation of the winding [1],[2],[3] is thought to be more effective than other methods, and if the cooling area of the winding is kept enough, the quench due to AC losses and/or the wire movements will not occur.

Using AC superconducting wire and epoxy impregnation technique, we have been trying to make large capacity coils operating at 50/60 Hz current mode. Firstly, we analyzed a temperature rise due to AC losses in the winding and discussed a cooling method. Secondary, the results of large capacity and large current coils are described. The goal of these coils is the application for AC machines.

Design of Epoxy Impregnation Coils

Usually, AC superconducting multifilamentary wire with around 0.5 μm NbTi fine fialments and a highly resistive CuNi matrix generates AC losses of 50-100 kW/m^3 at 1T, 4.2 K and 50 Hz. Therefore, the coil quenches if a cooling condition is not good. To design and make AC superconducting coils, it is indispensable to develop the techniques of keeping good cooling conditions and fixing windings simultaneously.

We introduce a design of a 100 kVA coil. This coil was developed by Central Resesarch Institute of Electric Power Industry (CRIEPI) and Furukawa and then analyzed at CRIEPI. The results were reported in 1988 [4] . A 100 kVA superconducting AC coil has a large cooling area by constructing cooling channels layer by layer in the winding and the coil is impregnated with epoxy for fixation of winding. The structure of the winding is shown in Fig.1. The excess epoxy is removed after impregnation so that all of winding is thought to be cooled enough by liquid helium through the heat conduction of the thin epoxy layers. The specifications of the wire and the coil are shown in Table 1. First of all, we carried out a heat analysis of the winding at AC operating mode. In longitudinal direction of the wire, the results were given by solving an one dimensional heat conductive equation[5]. Its model is shown in Fig.2. In the case of a 100 kVA coil, the wire is covered with the FRP(Fiber Reinforced Plastics) spacers, which construct the cooling channels, at interval of 3.4-10 mm so that those parts have a possibility of a temperature rise. We estimated a temperature rise in the winding from the calculated AC losses and got the value of 0.04 K assuming a good heat excahnge through the contact area with coolant. This efficiency is presumed only to depend on the differencies of temperature between the cable and the liquid helium in the cooling channels, too. This temperature rise is a very small and the winding is thought to have an adequate temperature margin compared with the critical temperature of NbTi.

In regard to a large current capcacity coil, the large size cable is needed so that it is more difficult to keep a good cooling condition because of the decreasing of the cooling area by cabling a lot of wires. In the case of a 500 kVA coil [6], we used a flat-shaped stranded cable for a coil winding in oder to improve the cooling condition. Parameters of the cable and the coil are given in Table 2. Using these cables, we made the coil with keeping cooling channels layer by layer and turn by turn in the winding. Schematic view of the winding is shown in Fig.3. This is a disc-type coil and it is wound by the same method as a 4 kVA disc-type coil [1]. We tried to analyze a heat distribution in the cable at AC mode by FEM (Fenite Element Method). For example, we show the one of the calculated results for a temperature rise in the cable in Fig.4. According to the increase of thickness of the epoxy layer, a temperature of the cable increases proportionally. Experimentally, the same results are pointed out using epoxy impregnated small coil [7]. This depends on both the AC losses and the ratio of the cooling area against the surface of the conductor, too. If there is the part of which has an excess epoxy in the winding and AC losses are large, the quench may occur there, especially, the part of which is covered with FRP spacer is a trigger to quench. The width of the cooling channel in axial direction of the coil is 2.0 mm and that in radial direction is 1.2 mm. Also, we selected the intervals between spacers and the length of the cable which are covered with spacers in oder not to raise the temperature in the winding by both the heat analysis and the experimental data [3].

100 kVA AC Superconducting Coil

The practical test of a 100 kVA superconducting AC coil was performed. The results of this coil at AC mode are shown in Fig.5. The excitation test was done using LC resonance method by CRIEPI [4]. The peak value of AC quench current reached critical current of short sample after third training and then its capacity attained about 100 kVA at 63.15 Hz. This means a very small temperature rise of the winding by the self field losses as same as the calculated result and the superiority of winding fixation by epoxy impregnation method.

Large Current Capacity AC Superconducting Coil

As a next stage, a 500 kVA disc-type coil has been developed. Its current capacity is more than 1300 A at 2 T which is the maximum field in the coil. A flat-shaped stranded cable shown in Fig.6 was used for coil winding. This cable are composed of 210 wires, a stainless steel tape for reinforcement and an epoxy-glass tape for insulation. Especially, a stainless steel tape was coated by a 10 μm thick copper layer for a heat sink in longitudinal direction of the cable. Firstly, the coil was excited up to 500 A at DC current mode because of the limit of power supply. Next, the AC excitation test was carried out at 50 Hz mode and total AC losses of the coil were measured by calorimetric method. The details were reported elsewhere [6] . The losses were about 10 times higher compared with the result of the small coil made of the same wire. This coil was quenched every time at the same current, its value was 380 A peak and then the maximum field in the winding was 0.7 T. The load line is shown in Fig.7. This phenomenon is thought to be caused owing to the eddy current in the copper layer on the stainless steel tape. The calculated maximum value of the eddy current loss in the winding is equivalent to 2700 kW/m^3 when converted to the losses of the superconducting wire. It is particularly interesting that this coil could keep the superconducting state under this high losses. If the eddy current loss is removed from the total loss of the coil, it is expected that the coil can be operated at the rated current.

Conclusion

Large current capacity AC superconducting coils have been developed and a 100 kVA coil using the epoxy impregnation technique could be operated stably up to the critical current of short sample wire. This result is in the good agreement with the calculated value from a heat analysis. On the other hand, the excitation of a 500 kVA disc-type coil with large curent capacity, over 1300 A at 2 T, was limited by the large eddy current losses in the copper layer on the reinforcement of stainless steel tape. This coil could be operated up to the rated current if the eddy current loss was removed.

Acknowledgement

The authors gratefully acknowledge Mr. Akita of Central Research Institute of Electric Power Industry for useful discussions, suggestions and giving a permission of quotation of data.

References

1) M.Yamamoto, T.Ikeda, Y.Tanaka, K.Matsumoto, T.Ishigohka, and O.Tsukamoto, "Development of 50 Hz Disc-Type Superconducting Coil", IEEE Trans.Mag.MAG-23, P.557, 1987
2) O.Tsukamoto, T.Ishigohka, M.Yamamoto, Y.Tanaka, and H.Kobayashi, "Characteristics of Epoxy-Impregnated AC Superconducting Winding", IEEE Trans.Mag.MAG-23, P1592, 1987
3) Y.Tanaka, K.Matsumoto, M.Yamamoto, O.Tsukamoto, and T.Ishigohka, "Effect of Composite Structure on AC loss of Superconducting Wire", IEEE Trans.Mag.MAG-23, p1588, 1987

4) S.Akita, and T.Ishikawa, "Characteristics of AC Superconducting Coils(part 1)", Koma Research Laboratory Rep. No.T87015, 1988
5) K.Matsumoto, Y.Yamada, and Y.Tanaka, "Development of AC Superconducting Wire", Workshop on Electric Instrument by the Institute of Electrical Engineers of Japan, SA-86-53, 1986
6) O.Tsukamoto, M.Yamamoto, T.Ishigohka, Y.Tanaka, T.Takao, and S.Torii, "Development of Large-Current Capacity Epoxy-Impregnated 50 Hz Superconducting Coil", presented at ICEC-12, 1988
7) O.Tsukamoto, H.Kobayashi, and S.Akita, "Stability of Epoxy-Impregnated AC Superconducting Winding", IEEE Trans.Mag.MAG-24, p1170, 1988

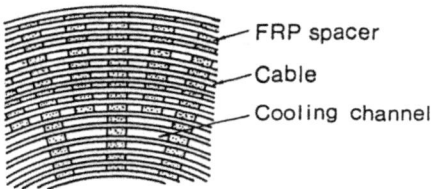

FRP spacer
Cable
Cooling channel

Fig.1 Structure of the winding of a 100 kVA coil.

Table 1   Parameters of a wire and a 100 kVA Coil

Wire	Wire Diameter(mm)	0.14
	Filament Diameter(μm)	0.5
	Twist Pitch(mm)	1.9
	Cu/ Cu-30%Ni/NbTi	1/3.5/1
	Insulation(μmt)	10(Polyester)
Cable	Number of Strands	7
	Cable Diameter(mm)	0.42
	Insulation(μmt)	60(Tetron)
Coil	Inner Diameter(mm)	45
	Outer Diameter(mm)	113
	Height(mm)	60
	Winding	Solenoid
	Number of Layers	20
	Number of Turns	1986
	Central Field Constant(T/A)	0.0256
	Inductance(H)	0.17

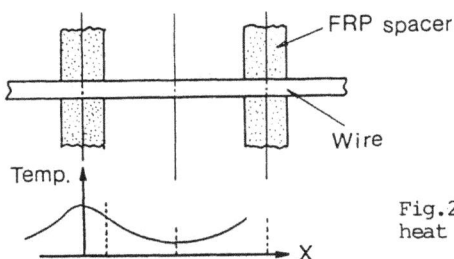

Fig.2 The model of one dimensional heat conductive equation.

Table 2 Parameters of a wire and a 500 kVA Coil

Wire	Wire Diameter(mm)	0.12
	Fialment Diameter(μm)	0.5
	Twist Pitch(mm)	1.4
	Cu/ Cu-10%Ni/NbTi	0.23/2.53/1
	Insulation(μmt)	10(Polyester)
Cable	Cable Size(mm2)	1.58X6.36
	Number of Wire	210
	Reinforcement(mm2)	0.74X5.52
	Insulation(μmt)	130
Coil	Inner Diameter(mm)	103
	Outer Diameter(mm)	196
	Height(mm)	96
	Winding	Disc-Type
	Number of Discs	12
	Number of Turns	195
	Central Field Constant(T/A)	0.0014
	Inductance(H)	0.0036

axial

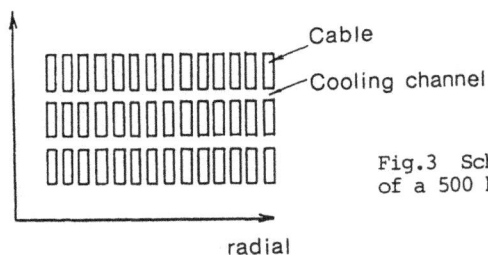

Fig.3 Schematic view of the winding of a 500 kVA coil.

radial

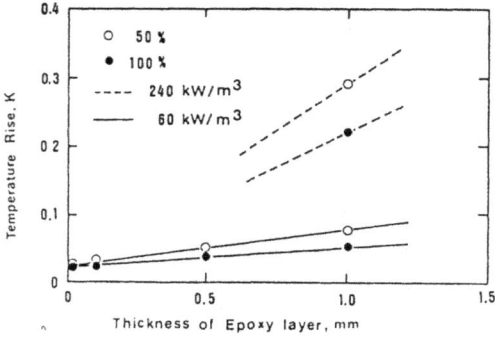

Fig.4 Calculated temperature rise of the cable against the thickness of the epoxy layer. ○, ● show a ratio of contact area with coolant against whole surface area of the cable.

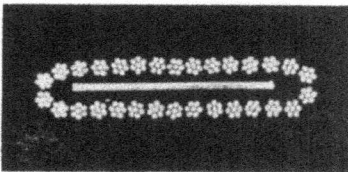

Fig.5 The load line of a 100 kVA coil. The number shows the sequence of the excitation 4).

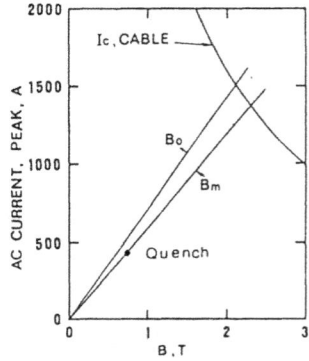

Fig.6 The cross section of the large current cable.

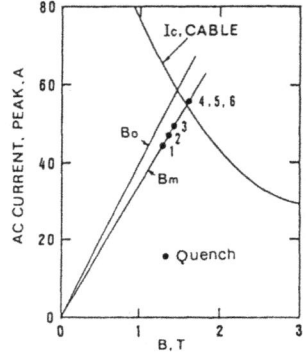

Fig.7 The load line of a 500 kVA coil.

EFFECT OF EPOXY CRACKS ON STABILITY OF IMPREGNATED WINDINGS

T.Yamashita, S.Nishijima, K.Takahata and T.Okada

ISIR Osaka University, Ibaraki, Osaka 567

Abstract-The change of the mechanical behavior and the superconducting instabilities of impregnated windings induced by epoxy cracking have been studied in order to establish the methodology of constructing the stable superconducting magnet.
The instabilities of superconducting windings induced by epoxy cracks are discussed together with the mechanical properties of the windings. The applicability of the results to the magnet construction is also mentioned.

Introduction

Training behavior and/or degradation, which impregnated windings often show, should be brought by the mechanical failures such as epoxy crackings or debondings[1]-[3]. In those impregnated windings, a number of cracks are introduced by thermal contraction, Lorentz force or thermal stress after quench until the winding achieves the expected performance. The large coil such as fusion magnet will be required to withstand for long term operation under high mechanical stress. The crack causes the change of thermal and mechanical property of the windings and has the possibility to degrade such magnets[3].
In this paper, the change of stability and the deformation of test coil induced by crack introduction was studied and the methodology of constructing the stable magnet was discussed.

Experimental

Test coil

The schematic illustration of the test coil is shown in Fig. 1. The conductor for the test coil is NbTi multifilamentary wire with copper to super ratio of 1.3. The diameter of the wire and filament are 0.5 mm and 0.042 mm, respectively. The twist pitch is 10 mm. The critical current is 150 A in 5 T and 100 A in 7 T. The test coil does not have end flanges so that shear stress are not induced.
The coil consists of two layers, inner and outer, as shown in Fig. 1. Between the layers silicon grease is spreaded so as to introduce the crack by energizing only outer layer. The volume fraction of superconducting wire is 48% in inner layer, and 65% in outer layer, respectively. The difference might induce shear stress over broad region in the test coil.
The test coil axis was positioned to coincide with the axis of back-up magnet. The direction of the transport current in the back-up magnet was determined to increase the field of the test coil. The peak self field constant of the test coil was 8.7×10^{-3} T/A and hence the designed maximum transport current is 160 A at back-up field of 3 T.
Three strain gauges were attached to the outer surface of the test coil

414

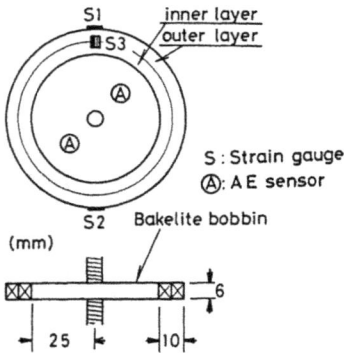

Fig.1   Overview of test coil
          arrangement

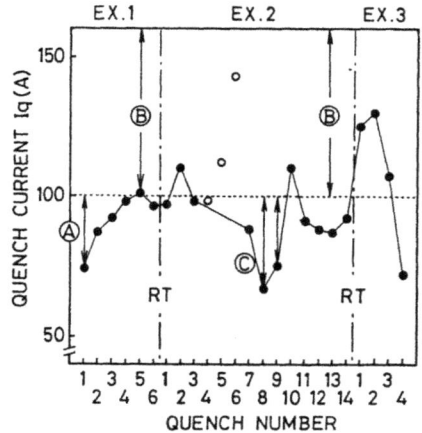

Fig.2   Quench current vs. quench
          number in test coil

to measure the deformation.   The strain gauges S1 and S2 are for
circumferential deformation and S3 for radial deformation.   A pair of AE
sensors were attached to the coil bobbin.   It was confirmed that the noises
from the support rod were negligible.

Experimental Procedure

    The test coil was energized in the back-up field of 3 T.   The energizing
rate was set at 2 A/sec.   Transport current, strains and AE signals were
measured simultaneously.   The following experiments were performed.
    1. Whole test coil was energized (named 'EX.1').
    2. After warming up to room temperature(RT), whole coil was energized.
       Then, only the outer layer of the test coil was energized in order
       to induce cracks between outer and inner layers.   After the crack
       introduction, whole coil was energized again (EX.2).
    3. After warming up to RT, whole coil was energized (EX.3).

Results

    Figure 2 shows the change of quench current against quench number.   The
closed circles present the quench current in whole coil energizing and the
open circles present those in outer layer energizing.
    Training in which quench current increases with quench number, appeared
on EX.1.   Degradation in which quench current does not achieve the designed
value (160 A at 3 T) was also seen since the quench current seems to be
saturated at about 100 A.
    In 1-3rd runs in EX.2 the coil showed quench current as high as maximum
Iq in EX.1 that is the test coil remembered the training in EX.1 even it was
warmed up to RT.   In the 4-6th runs in EX.2, outer layer was energized in
order to induce the crack.   The crack introduction was comfirmed by the

debonding of the strain gauge S3 in the 6th run.   In the 6th run, quench
current increased markedly, it was also confirmed that macro-crack had grown
up circumferentially and the outer layer was separated from the inner layer.
After the experiments, macro-crack was found by visual inspection on the
sectional view of the test coil.
    In the 7-14th runs in EX.2, whole test coil was energized again.   It was
found that macro-crack brought further instability to the coil and hence the
quench current came to be unstable showing between 70 and 110 A.   The
instability by the macro-crack remained even after the warming up to RT as
shown in 1-4th runs in EX.3.

    Figure 3-5 show the strain - transport current curves (ε-I curve)
obtained by strain gauge S1, S2 and S3, respectively.   The strain at 0 A
changed as quench number.   That might be caused by the introduction of
cracks.   The open and closed marks in Fig. 3-5 show the ε-I characteristics
before and after the macro-crack introduction in EX.2, respectively.
    The AE avarage voltage was also measured.   The typical feature are shown
in Fig. 6.   The burst of AE signal indicated by arrows in Fig. 6 were
observed reversibly that is Kaiser effect did not appear.   The sorces of AE
signals are reversible mechanical disturbances.   It was found that the
quench was caused by mechanical disturbances since the AE bursts were
detected just before the quench.

Discussion

Instabilities of Coil

    The instabilities of the test coil was discussed based on the data.   The
test coil showed both training and degradation before the macro-crack
introduction.   When the instability of the test coil is represented as the
difference between quench and designed current (160 A), the degree of the
instability in the training and degradation can be expressed as circle A and
B as shown in Fig. 1, respectively.
    The cause of the training in impregnated windings are thought to be the
irreversible mechanical disturbances such as micro-crack introduction.   The
cause of circle A, therefore, is thought to be the heat generation induced by
micro-crack.
    Degradation means that there are reversible disturbances in the test coil
in this case.   The AE signals also suggest the reversible mechanical
disturbances.   They should be originated from the friction at the
coil-bobbin interfaces.   The test coil was separated from its bobbin from
the 1st run, since the radial strain S3 showed compressive strain as seen in
Fig. 5(a).   The instability presented by the circle B would be brought by
the friction.

    The macro-crack brought another instability to the test coil and hence
the quench current was not settled.   It can be concluded that the
macro-crack decreases the quench current.   The reversible serration appeared
on the ε-I curve of S2 at approximately 50 A as seen in Fig. 4(b), after the
macro-crack introduction.   This is caused by the slip at the crack and which
should result in the low quench current.   The magnitude of heat generation
caused by the friction at the crack is evaluated as circle C in Fig. 2.

Deformation Behavior of Test Coil

    The deformation of the test coil is discussed based on the strain data.

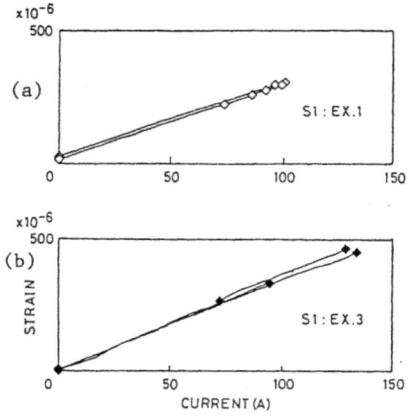

Fig.3 Transport current – strain curve of circumferential strain S1

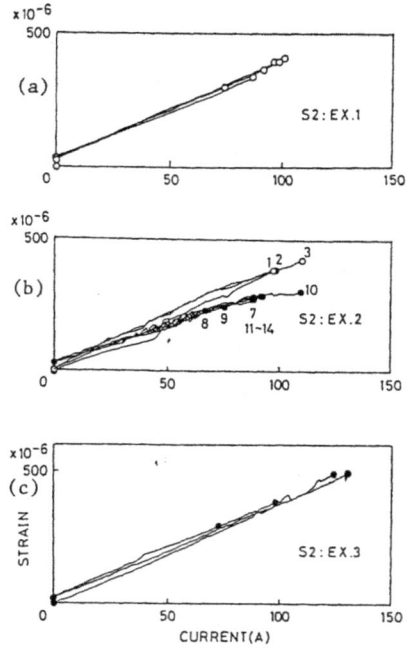

Fig.4 Transport current – strain curve of circumferential strain S2

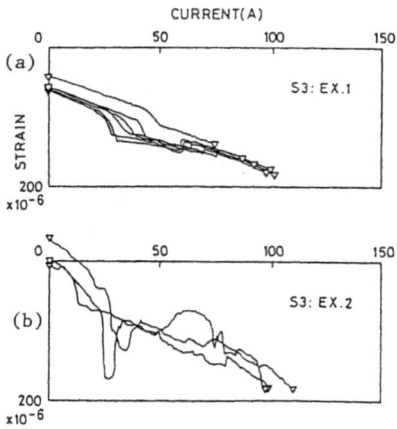

Fig.5 Transport current – strain curve of radial strain S3

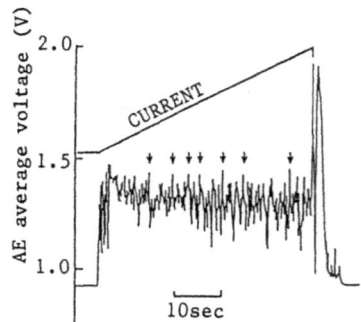

Fig.6 Typical AE average voltage data in energizing

Before the macro-crack introduction, circumferential strain S1 was smaller than S2 as seen in Fig. 3(a), 4(a).

It means that the deformation of the test coil was not uniform. Because of the ill-balanced deformation, the friction at coil-bobbin interfaces should occur. The irregular ε-I curve of S3 as seen in Fig. 5 might also be caused by the ill-balanced deformation. As shown in Fig. 7(a) the coil was enlarged circumferentially and compressed radically when the coil-bobbin adhesion is poor.

After the macro-crack introduction in EX.2, the decrease of strain and reversible serration appeared on the ε-I curve of the circumferential strain S2 as shown in Fig. 4(b). It is understood that the stress was not transferred from inner layer to outer layer at the crack face.

The crack might be opened when outer layer was energized at crack introduction. Following whole coil energizing decreases the crack gap and compresses the outer and inner layers independently as shown in Fig. 7(b). As a result, the circumferential strain decreases after the crack introduction.

These phenomena did not occur in EX.3 after warming up to RT as shown in Fig. 4(c) because the gap was closed by warming up due to the large thermal expansion of epoxy. The friction at the crack, however, remains and decreases the quench current though the macroscopic serration does not exist.

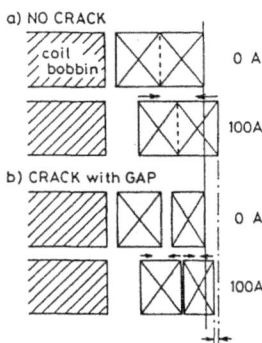

Fig.7   Model explaning decrease of circumferential strain S2 after macro-crack introduction

After the macro-crack introduction, circumferential strains S1 and S2 came to be equal as seen in Fig. 3(b), 4(c). It shows that the ill-balanced deformation of the test coil was recovered. The macro crack prevented the shear stress to transfer at crack faces. The relaxation of shear stress at the crack degrades the macroscopic rigidity of winding.

Application of Analysis

To investigate the location where the friction at the crack could be induced in an impregnated magnet, the stress analysis was made. Figure 8 shows the axisymmetric analysis model of impregnated magnet. In the analysis one macro-crack is introduced along the Z direction changing the position. The shear stress should cause shear crack. The friction could

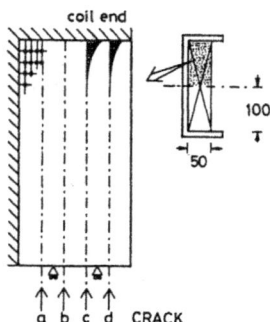

Fig.8   Analysis model of magnet

occur when the crack faces press each other and the shear stress is induced
there.   The stress analysis was made to know the crack position where the
compressive stress between the crack surfaces is induced.   When the
macro-crack c or d is introduced, the crack faces press each other though the
cracks open at a or b.   When the inner bobbin is debonded, any crack faces
press each other.   The compressive stress at the crack faces increases with
the radial position of the crack that is a < b < c < d.   Since the shear
stress is larger near the coil ends by the interaction between coil body and
coil flanges, the friction could occur at the ends of the outer layer of
winding (hatched area).

Conclusions

It was confirmed experimentally that macro-crack relaxes the shear stress
and causes the degradation of macroscopic rigidity.   It should be emphasized
that the crack introduction should be important problem for large coil.   The
crack causes the frictional heating when the surfaces are pressed each other.
It is reversible and make increase the instability of the magnet.   The
stress analysis indicates that the crack induced friction are apt to be
caused near the ends of outer layer of winding, where the margin is high.
It is, therefore, important for the stability of impregnated winding to
prevent inducing cracks by increasing the strength of impregnant even if the
conductivity is decreased.

References

[1] M. N. Wilson, "Some Basic Problem in Superconducting Magnet Design,"IEEE
    Trans. on Magnetics, MAG-17, 1815-1822 (1981).
[2] O. Tsukamoto and Y. Iwasa, "Epoxy Crackings in the Epoxy-Impregnated
    Superconducting Winding",IEEE Trans. on Magnetics, MAG-21, 377-379 (1985).
[3] T. Yamashita, S. Nishijima, et al. "Effect of Mechanical Disturbances on
    Stability of Impregnated Windings",IEEE Trans. on Magnetics, MT-10, 1186
    -1189 (1988).

DEFORMATION BEHAVIOR OF THREE DIMENSIONAL GLASS FABRIC REINFORCED PLASTIC
AT LOW TEMPERATURE

Yong An WANG, Shigehiro NISHIJIMA, Toichi OKADA,
Tsuguo UEMURA[+], Tetsuro HIROKAWA[+], and Jun YASUDA[+]

[+]ISIR, Osaka University, Ibaraki, Osaka, Japan,
[+]Shikishima Canvas Company, Ltd. Osaka, Japan,

Abstract - Deformation behavior of three dimensional glass fabric reinforced
plastic (3D-GFRP) has been studied to determine their applicability for
large scale superconducting magnet system as insulating and/or supporting
materials. An experamental study was carried out to clarify the
relationship between the construction of reinforcement and the mechanical
properties of the composites. The results were compared with those of the
usual GFRP (2D-GFRP). The tensile strength of 3D-GFRP does not depend on
the fibers in the thickness direction, but the Young's modulus does. It is
found that the fibers in thickness direction, which originate the
superiority of 3D-GFRP to 2D-GFRP, have important roles on the deformation
behavior in 3D-GFRP.

Introduction

In large scale superconducting magnets, a large quantities of
nonmetallic composite materials have been used as thermal insulating,
electrical insulating and suppoting materials[1-3]. Those materials are
usually reinforced by unidirectional fibers or by woven glass cloths. Such
composite materials have disadvantages such as low strength, low stiffness,
and large thermal contraction in the thickness direction. The low
interlaminar shear strength has also restricted the design of the
superconducting magnet systems[4].

To overcome the disadvantages, the three-dimensional fibric reinforced
composite (3D-GFRP) has been developed[5]. For the practical application,
the basic charateristics especially the mechanical behavior have to be
clarified because the selection standard of the composite material is to be
the mechanical characteristics at cryogenic temperature.

In this study, the tensile strength, Young's modulus and Poisson's ratio
were measured at room (RT) and liquid netrogen temperature (LNT). The
effects of fiber arrangement and glass content on the mechanical properties
were discussed.

Experimental

Specimens

The specimens used in this work were two types of 3D- and 2D-GFRP
containing E-glass fibers and epoxy resin. The 3D-GFRP is reinforced by a
three demensional fabric and the fibers in thickness direction run in the X

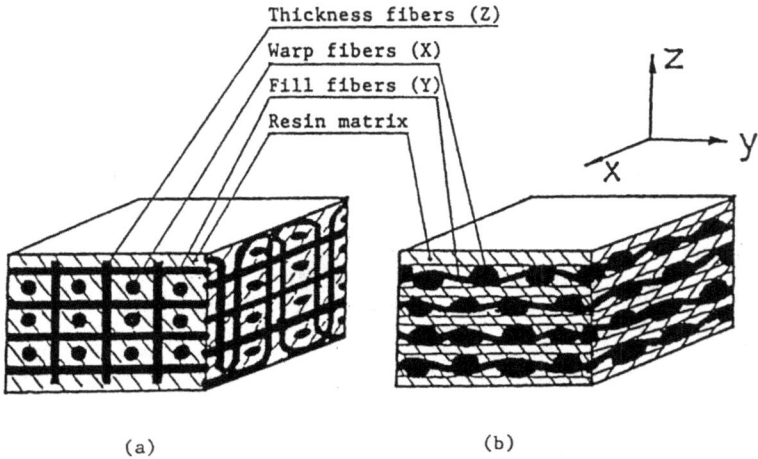

Fig. 1.   Arrangment of glass fiber in (a) 3D-GFRP
and (b) 2D-GFRP.

direction near the surface of the specimen (Fig. 1(a)).   The 2D-GFRP specimen
was reinforced by plain woven fabrics (Fig. 1(b)).   The specification of
specimens are given in Table 1.   Starch or epoxy was used as the sizing
material.

Mechanical Tests

Dumbbell shaped specimens were used in tensile test.   The Young's
modulus and Poisson's ratio were measured by strain gauges attached on the
surface of the specimens.   The measurements were made in X and Y direction
for Young's modulus and in X,Y, and Z direction for Poisson's ratio.   The
tests speed was set at 2.0mm/min. All mechanical measurements were
performed at LNT and RT.

Table1.   Specification of specimens

Name		H2	H2EP	C30	C30EP
Thickness (mm)		3.9	3.3	4.0	3.8
Fraction	X(%)	27	28	57	57
of glass fibers	Y(%)	59	61	43	43
in each direction	Z(%)	14	11	--	--
Glass content by volume	Vf(%)	59.1	67	59.4	62
Glass content by weight	Wf(%)	78	80	79	78
Density (g/cc)		1.79	2.14	1.95	2.08
Sizing material		starch	epoxy	starch	epoxy
Type		3D	3D	2D	2D

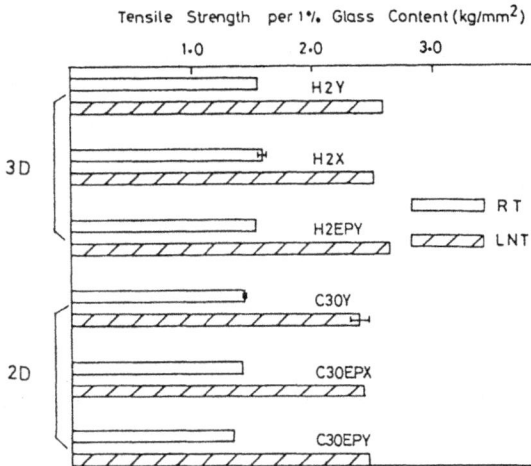

Fig. 2.   Tensile strength with 1 % glass content of
           3D- and 2D-GFRP obtained at RT and LNT.

Results and Discussion

Tensile Strength

To clarify the difference in the fracture behavior of the 2D- and
3D-GFRP specimen, the tensile strength with 1% glass content was calculated
and is presented in Fig. 2.   The values of the 3D-GFRP nearly equal to
those of 2D-GFRP.   It is clear that the strength in X or Y direction is not
affected by the fibers in the Z direction.

Young's Modulus

The Fig.3 shows Young's modulus with 1% glass content of 3D- and 2D-GFRP
obtained at RT and LNT.   The Young's modulus with 1 % glass content is
almost equal to each others in 2D-GFRP.   Those of 3D-GFRP in Y direction
nearly equal to those of the 2D-GFRP.   It means that Young's modulus of
3D-GFRP in Y direction is not affected by the fibers in the thickness
direction.   The Young's modulus in X direction, however, are larger than
those of 2D-GFR.   It would be originated from the contribution of the
fibers in thickness direction.

The effect of the fibers in thickness direction is to be discussed.
Since the fibers in thickness direction run in X direction near the
surface of the specimen, Young's moduli in X direction are devided by the
sum of glass content in X and Z direction that is (V_x+V_z) and are also
presented (with open markings) in Fig.3.   The values are equal to those of
2D-GFRP at RT and LNT, respectively.   It is found that the effects of
fibers in thickness direction to Young's modulus in X direction are almost
same as the fibers in other directions.

422

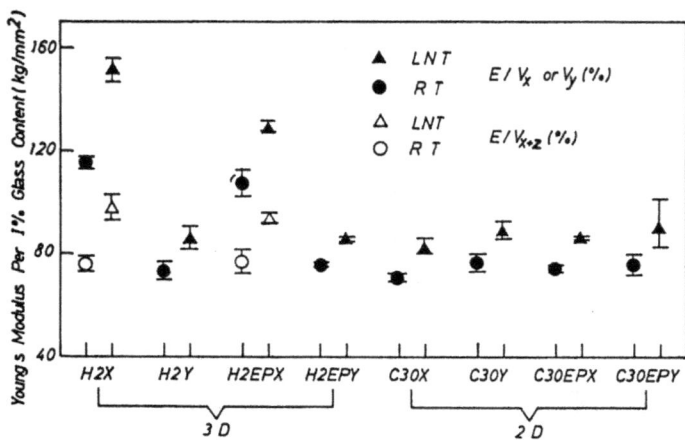

Fig. 3.   Young's modulus with 1% glass content of
          3D- and 2D-GFRP obtained at RT and LNT.

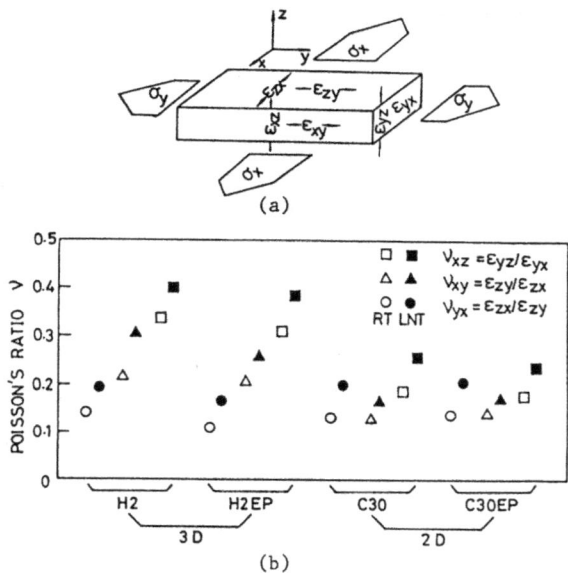

(a)

(b)

Fig.4.   (a) Definition of the deformation and
          (b) Poisson's ratio of 3D- and 2D-GFRP
          in each direction obtained RT and LNT.

Poisson's Ratio

In order to clarify the effect of the fibers in thickness direction on the deformation behavior, Poisson's ratios in each direction were measured at RT and LNT. The Fig. 4(a) shows the definition of the deformation in this work. The results are given in Fig. 4(b). The Poisson's ratios of 3D- and 2D-GFRP at LNT are larger than those at RT. The ratios in thickness direction of 3D-GFRP are also larger than those of 2D-GFRP. It could be considered that the fibers in thickness direction make the deformation in Z direction larger with "tighten effect".

Deformation Behavior

The difference of deformetion behavior between 2D- and 3D-GFRP were examined by meams of the Young's modulus and Poisson's ratio. The ratios between Poisson's ratio and Young's modulus in both X and Y direction were calculated and shown in Fig. 5.

When the plane stress condition was satisfied, following relationship has to be valid, though the equation is not sufficient condition for plane stress;

$$\nu_x/E_x = \nu_y/E_y$$

The 2D-GFRPs satisfy the equation both at RT and LNT. On the other hand 3D-GFRPs do not folow the equation as shown in Fig. 5. This would be the effect of the fibers in thickness direction.

The fibers in thickness direction do not affect on the fracture strength, but on the Young's modulus do.

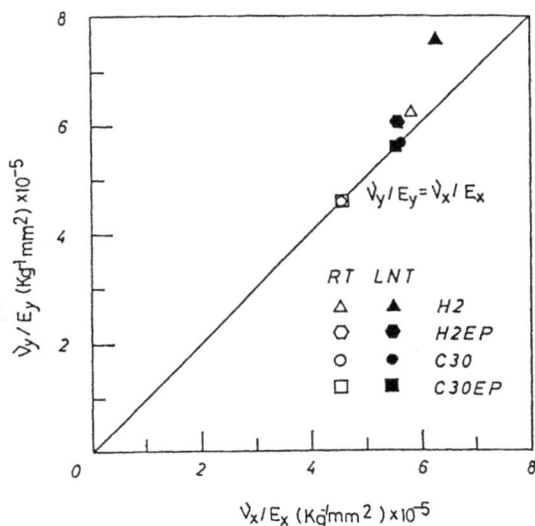

Fig. 5.   Deformation behavior of 3D- and 2D-GFRP.

Conclusion

Several mechanical properties of 3D-GFRP were measured and the results were compared with those of 2D-GFRP. Following conclusions were drawn.

1) Fibers in the thickness direction do not affect on the strength in other direction.
2) Young's modulus in X direction depends not only on the fibers in the X direction but also those in thickness direction.
3) Poisson's ratios of all specimen at LNT are larger than those at RT. The Poisson's ratios in thickness direction of 3D-GFRP are larger than those of 2D-GFRP, when the specimen was tensiled in X direction.
4) The deformation behavior of 3D-GFRP at small strain is different from that at large strain. This reflects the charcteristics of the fibers in thickness direction.

The mechanical behavior of 3D-GFRP was clarified in terms of experiments and the results strongly suggest the possibility of the cryogenic application of the 3D-GFRP.

Acknowledgment

This work was partly supported by Grant in Aid for Scientific Research No 63609511, Ministry of Education in Japan.

References

1) L. Coffman and J. C. Williams, "Room-temperature mechanical strength selection criteria of G-10 intended for cryogenic temperature", in Advances in Cryogenic Engineering--Materials, Vol. 26, (Plenum press, New York, 1980), pp. 245-251.
2) K. Koizumi, K. Yoshida, E. Tada, M. Nishi, M. Nagai, K. Kadotani, and N. Tada, "Mechanical properties of an insulator for the Japanese LCT coil", in Advances in cryogenic Engineering--Materials, Vol. 28, (Plenum press, New York, 1982), pp. 223-230.
3) G. Bogner, Present and future appliction of nonmetallic materials in cryogenic technology, in Nonmetallic Materials and Composites at Low Temperature 3, G. Hartwig and D. Evans. eds., (Plenum press, New York, 1984), pp. 209-214.
4) S. Nishijima, T. Okada, K. Miyata, and H. Yamaoka, "Radiation damage of composite materials at cryogenic temperatures", in Advances in Cryogenic Engineering--Materials, Vol. 34, (Plenum press, New York, 1988), pp. 35-42.
5) S. Nishijima, Y-A. Wang, T. Okada, T. Uemura, T. Hirokawa, and J. Yasuda, "Cryogenic properties of three-dimensional, glass-fabric-reinforced plastic", in Advances in Cryogenic Engineering--Materials, Vol. 34, (Plenum preess, New York, 1988) pp. 59-66.

THERMAL FRACTURE OF GLASSFIBER REINFORCED PLASTICS
AT LOW TEMPERATURES

Sei UEDA

Department of Technology, Faculty of Education
Shizuoka University, Shizuoka 422, Japan

and

Yasuhide SHINDO

Department of Mechanical Engineering II, Faculty of Engineering
Tohoku University, Sendai 980, Japan

Abstract - Large quantities of nonmetallic composites such as G10-CR and G11-CR are currently under study as candidate materials for superconducting magnet system in a fusion reactor at low temperatures. Here we study the thermal and mechanical response of G10-CR glass/epoxy laminates with a crack. Fourier transforms are used to reduce the problem to the solution of a pair of dual integral equations. The solution to the dual integral equations is expressed in terms of a Fredholm integral equation of the second kind. In the absence of the crack, thermal stresses at low temperatures are expressed in closed forms. Numerical results on the thermal stress intensity factor at different temperatures are obtained and are presented in a graphical form.

Introduction

Large quantities of nonmetallic composites will be used in superconducting magnet system in a fusion reactor at low temperatures[1]-[3]. Kriz[1] studied the influence of damage on the mechanical performance of G10-CR glass/epoxy laminates at low temperatures by finite element method.

In this paper, we investigate the thermal singular stresses of a crack in G10-CR glass/epoxy laminates. Fourier transforms are used to reduce the problem to the solution of a pair of dual integral equations. The solution to the dual integral equations is expressed in terms of a Fredholm integral equation of the second kind. In the absence of the crack, thermal stresses at low temperatures are expressed in closed forms. Numerical results on the thermal stress intensity factor at different temperatures are obtained and are presented in a graphical form.

Statement of Problem and Fundamental Equations

As shown in Figure 1, we consider the layered composite is made of a cracked layer I bonded between two layers II of different elastic properties. The system of rectangular Cartesian coordinates (x,y,z) is chosen with the crack of length $2a$ being centered along the x-axis and y being parallel to the interfaces. The layer I is reinforced along the z-axis having the Young's modulus E_{x1}, E_{y1}, E_{z1}, shear modulus G_{xy1}, Poisson's ratio $\nu_{xy1}, \nu_{yz1}, \nu_{zx1}$ and the linear thermal expansion α_1. The layer II is reinforced along the y-axis possessing the elastic and thermal constants $E_{x2}, E_{y2}, E_{z2}, G_{xy2}$

$\nu_{xy2}, \nu_{yz2}, \nu_{zx2}$ and α_2 . In the following, the subscripts x,y,z and 1,2 wi be used to refer to the directions of coordinates and the layers I,II.

The components of a stress tensor and a strain tensor in x,y and z directions are designated as $(\sigma_{xj}, \sigma_{yj}, \tau_{xyj})$ and $(\varepsilon_{xj}, \varepsilon_{yj}, \gamma_{xyj})$, respectively. Assuming plane strain, we have the following two dimensional strain and stress relationships for the orthotropic system:

$$\begin{bmatrix} \sigma_{xj} \\ \sigma_{yj} \\ \tau_{xyj} \end{bmatrix} = \begin{bmatrix} c_{1j}, & c_{3j}, & 0 \\ c_{3j}, & c_{2j}, & 0 \\ 0, & 0, & c_{4j} \end{bmatrix} \begin{bmatrix} \varepsilon_{xj} \\ \varepsilon_{yj} \\ \gamma_{xyj} \end{bmatrix} \quad (j=1,2) \text{---}(1)$$

where

$$\left.\begin{array}{l} c_{1j}=(1/E_{yj})[1/E_{zj}-\nu_{yzj}^2/E_{yj}]/H_j \\ c_{2j}=(1/E_{zj})[1/E_{xj}-\nu_{zxj}^2/E_{zj}]/H_j \\ c_{3j}=(1/E_{zj})[\nu_{zxj}\nu_{yzj}/E_{yj}-\nu_{xyj}/E_{xj}]/H_j \\ c_{4j}=G_{xyj} \\ H_j=1/(E_{xj}E_{yj}E_{zj})[1-2\nu_{xyj}\nu_{yzj}\nu_{zxj}- \\ E_{xj}\nu_{zxj}^2/E_{zj}-E_{zj}\nu_{yzj}^2/E_{yj}] \end{array}\right\} \text{---}(2)$$

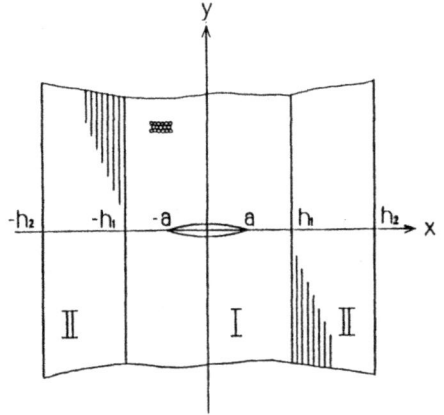

When the composite is cooled to liquid helium or liquid nitrogen temperatures, large internal stresses arise from the differential thermal contractions of dissimilar components. These thermally induced stresses and mechanical stresses cause the singular stresses at the crack tip. Applying the principle of superposition, the problem is separated into two parts. The first part is the thermal stresses problem with no crack while the second

Fig.1 Composite Model with a Crack

part is the singular stresses problem with the crack.

In x-y plane, the layer I is isotropic and the layer II is orthotropic. Then applying Fourier transform, we obtain the components of displacements and stresses as follows[4];

For the layer I,

$$u_{x1}(x,y)=-(2/\pi)\int_0^\infty[\{A_1(s)-(1-2\nu_1)B_1(s)\}\sinh(sx)+sxB_1(s)\cosh(sx)]\cos(sy)ds$$
$$-(2/\pi)\int_0^\infty(1-2\nu_1-sy)C_1(s)\exp(-sy)\sin(sx)ds \text{ ---}(3)$$
$$u_{y1}(x,y)=(2/\pi)\int_0^\infty[\{A_1(s)+2(1-\nu_1)B_1(s)\}\cosh(sx)+sxB_1(s)\sinh(sx)]\sin(sy)ds$$
$$+(2/\pi)\int_0^\infty(2-2\nu_1+sy)C_1(s)\exp(-sy)\cos(sx)ds \text{ ---}(4)$$
$$\sigma_{x1}(x,y)=-(4c_{41}/\pi)\int_0^\infty s[A_1(s)\cosh(sx)+sxB_1(s)\sinh(sx)]\cos(sy)ds$$
$$-(4c_{41}/\pi)\int_0^\infty s(1-sy)C_1(s)\exp(-sy)\cos(sx)ds \text{ ---}(5)$$
$$\sigma_{y1}(x,y)=(4c_{41}/\pi)\int_0^\infty s[\{A_1(s)+2B_1(s)\}\cosh(sx)+sxB_1(s)\sinh(sx)]\cos(sy)ds$$
$$-(4c_{41}/\pi)\int_0^\infty s(1+sy)C_1(s)\exp(-sy)\cos(sx)ds \text{ ---}(6)$$
$$\tau_{xy1}(x,y)=(4c_{41}/\pi)\int_0^\infty s[\{A_1(s)+B_1(s)\}\sinh(sx)+sxB_1(s)\cosh(sx)]\sin(sy)ds$$
$$-(4c_{41}/\pi)\int_0^\infty s^2yC_1(s)\exp(-sy)\sin(sx)ds \text{ ---}(7)$$

where

$$\nu_1=2c_{41}\nu_1'/E_1'$$
$$\nu_1'=E_1'(E_{y1}\nu_{xy1}/E_{x1}+\nu_{zx1}\nu_{yz1})/E_{y1}$$
$$E_1'=E_{x1}/(1-E_{x1}\nu_{zx1}^2/E_{z1})$$

For the layer II,

$$u_{x2}(x,y)=(2/\pi)\int_0^\infty[\{A_2(s)\sqrt{\beta_1}\exp(s\sqrt{\beta_1}x)+C_2(s)\sqrt{\beta_2}\exp(s\sqrt{\beta_2}x)\}\cos(sy)$$
$$-\{B_2(s)\cos(sy/\sqrt{\beta_1})+D_2(s)\cos(sy/\sqrt{\beta_2})\}\exp(-sx)]ds \text{ --------------------------}(8)$$

$$u_{y2}(x,y)=-(2/\pi)\int_0^\infty[\{\rho_1A_2(s)\exp(s\sqrt{\beta_1}x)+\rho_2C_2(s)\exp(s\sqrt{\beta_2}x)\}\sin(sy)$$
$$+\{\rho_1B_2(s)\sin(sy/\sqrt{\beta_1})/\sqrt{\beta_1}+\rho_2D_2(s)\sin(sy/\sqrt{\beta_2})/\sqrt{\beta_2}\}\exp(-sx)]ds \text{ --------------(9)}$$

$$\sigma_{x2}(x,y)=\{2(1+\rho_1)c_{42}/(\beta_1\pi)\}\int_0^\infty s[A_2(s)\beta_1\exp(s\sqrt{\beta_1}x)\cos(sy)+B_2(s)\exp(-sx)\cos(sy/\sqrt{\beta_1})]ds$$
$$+\{2(1+\rho_2)c_{42}/(\beta_2\pi)\}\int_0^\infty s[C_2(s)\beta_2\exp(s\sqrt{\beta_2}x)\cos(sy)+D_2(s)\exp(-sx)\cos(sy/\sqrt{\beta_2})]ds \text{ ------(10)}$$

$$\sigma_{y2}(x,y)=-\{2(1+\rho_1)c_{42}/\pi\}\int_0^\infty s[A_2(s)\beta_1\exp(s\sqrt{\beta_1}x)\cos(sy)+B_2(s)\exp(-sx)\cos(sy/\sqrt{\beta_1})]ds$$
$$-\{2(1+\rho_2)c_{42}/\pi\}\int_0^\infty s[C_2(s)\beta_2\exp(s\sqrt{\beta_2}x)\cos(sy)+D_2(s)\exp(-sx)\cos(sy/\sqrt{\beta_2})]ds \text{ ----------(11)}$$

$$\tau_{xy2}(x,y)=-\{2(1+\rho_1)c_{42}/\pi\}\int_0^\infty s[A_2(s)\sqrt{\beta_1}\exp(s\sqrt{\beta_1}x)\sin(sy)-B_2(s)\exp(-sx)\cos(sy/\sqrt{\beta_1})/\sqrt{\beta_1}]ds$$
$$-\{2(1+\rho_2)c_{42}/\pi\}\int_0^\infty s[C_2(s)\sqrt{\beta_2}\exp(s\sqrt{\beta_2}x)\sin(sy)-D_2(s)\exp(-sx)\sin(sy/\sqrt{\beta_2})/\sqrt{\beta_2}]ds \text{ --(12)}$$

where

$$\rho_j=(c_{12}\beta_j-c_{42})/(c_{32}+c_{42})$$

$\beta_j (j=1,2)$ are the roots of the characteristic equation.

$$c_{12}c_{42}\beta_j^2+(c_{32}^2+2c_{32}c_{42}-c_{12}c_{22})\beta_j+c_{22}c_{42}=0 \text{ --(13)}$$

And, $\Lambda_j(s), B_j(s), C_j(s)$ and $D_j(s)$ $(j=1,2)$ are the unknown function to be determined from the boundary conditions.

Internal Stresses Induced by Thermal Contraction

The boundary conditions for the no crack are

$$u_{y1}(h_1,y)+\alpha_1Ty=u_{y2}(h_1,y)+\alpha_2Ty \quad -\infty<y<\infty \text{ --(14)}$$
$$u_{x1}(h_1,y)=u_{x2}(h_1,y) \quad -\infty<y<\infty \text{ --(15)}$$
$$\sigma_{x1}(h_1,y)=\sigma_{x2}(h_1,y) \quad -\infty<y<\infty \text{ --(16)}$$
$$\tau_{xy1}(h_1,y)=\tau_{xy2}(h_1,y) \quad -\infty<y<\infty \text{ --(17)}$$
$$\sigma_{x2}(h_2,y)=0 \quad -\infty<y<\infty \text{ --(18)}$$
$$\tau_{xy2}(h_2,y)=0 \quad -\infty<y<\infty \text{ --(19)}$$

where α_j $(j=1,2)$ are the linear thermal expansion coefficient of layer I and II in y direction, and "T" denotes variation of temperatures. Considering the condition of no crack, the unknown function $C_1(s)$ is zero. We obtain the relations among $\Lambda_j(s), B_j(s)$ $(j=1,2)$, $C_2(s)$ and $D_2(s)$ from the boundary conditions (15)-(19). And we have the integral equation concerned with the unknown function $B_1(s)$ as follows;

$$(2/\pi)\int_0^\infty W(s)B_1(s)\sin(sy)ds=y \text{ --(20)}$$

where

$$W(s)=-[\{R_1'(s)+2(1-\nu_1')\}\cosh(sh_1)+sh_1\sinh(sh_1)+\rho_1R_2'(s)\exp(-s\sqrt{\beta_1}h_1)+\rho_2R_3'(s)\exp(-s\sqrt{\beta_2}h_1)$$
$$+\rho_1R_4'(s)\exp(s\sqrt{\beta_1}h_1)+\rho_2R_5'(s)\exp(s\sqrt{\beta_2}h_1)]/\{(\alpha_1-\alpha_2)T\} \text{ ----------------------------(21)}$$

$$R_1'(s)=P_2'(s)/P_1'(s)$$
$$R_2'(s)=\{\gamma_2'+\gamma_1'R_1'(s)\}/\{c_4(1+\rho_1)\gamma_5'\}$$
$$R_3'(s)=-\{\gamma_4'+\gamma_3'R_1'(s)\}/\{c_4(1+\rho_2)\gamma_5'\} \quad \text{----------(22)}$$
$$R_4'(s)=-[(1+\rho_1)(\sqrt{\beta_1}+\sqrt{\beta_2})\varepsilon_1R_2'(s)+2(1+\rho_2)\sqrt{\beta_2}\varepsilon_2R_3'(s)]/\{(1+\rho_1)(\sqrt{\beta_2}-\sqrt{\beta_1})\}$$
$$R_5'(s)=[2(1+\rho_1)\sqrt{\beta_1}\varepsilon_2R_2'(s)+(1+\rho_2)(\sqrt{\beta_1}+\sqrt{\beta_2})\varepsilon_3R_3'(s)]/\{(1+\rho_2)(\sqrt{\beta_2}-\sqrt{\beta_1})\}$$

$$P_1'(s)=\gamma_5'\sinh(sh_1)-(1/c_4)\{\sqrt{\beta_1}\delta_5\gamma_1'/(1+\rho_1)-\sqrt{\beta_2}\delta_6\gamma_3'/(1+\rho_2)\} \quad \text{---(23)}$$
$$P_2'(s)=\gamma_5'\{(1-2\nu_1')\sinh(sh_1)+sh_1\cosh(sh_1)\}-(1/c_4)\{\sqrt{\beta_1}\delta_5\gamma_2'/(1+\rho_1)-\sqrt{\beta_2}\delta_6\gamma_4'/(1+\rho_2)\}$$

$$\gamma_1'=2\delta_2\sinh(sh_1)+2\sqrt{\beta_2}\delta_4\cosh(sh_1)$$
$$\gamma_2'=2\delta_2\sinh(sh_1)+sh_1\delta_2\cosh(sh_1)+2\sqrt{\beta_2}\delta_4sh_1\sinh(sh_1)$$
$$\gamma_3'=2\delta_1\sinh(sh_1)+2\sqrt{\beta_1}\delta_3\cosh(sh_1) \quad \text{----------------------------------(24)}$$
$$\gamma_4'=2\delta_1\sinh(sh_1)+sh_1\delta_1\cosh(sh_1)+2\sqrt{\beta_1}\delta_3sh_1\sinh(sh_1)$$
$$\gamma_5'=\sqrt{\beta_1}\delta_2\delta_3-\sqrt{\beta_2}\delta_1\delta_4$$

$$\varepsilon_1=\exp(-2s\sqrt{\beta_1}h_2)$$
$$\varepsilon_2=\exp[-s(\sqrt{\beta_2}+\sqrt{\beta_1})h_2] \quad \text{----------------------------------(25)}$$
$$\varepsilon_3=\exp(-2s\sqrt{\beta_2}h_2)$$

$$\delta_1 = E_1 - F_1 E_2 + F_2 E_3$$
$$\delta_2 = E_4 - F_3 E_5 + F_1 E_6$$
$$\delta_3 = E_1 + F_1 E_2 - F_3 E_3$$
$$\delta_4 = E_4 + F_2 E_5 - F_1 E_6$$ ----------------------------------(26)
$$\delta_5 = E_1 + F_1 E_2 - (1 + \rho_1) F_3 E_3 / (1 + \rho_2)$$
$$\delta_6 = E_4 + (1 + \rho_2) F_2 E_5 / (1 + \rho_1) - F_1 E_6$$

$$E_1 = \exp[-s\sqrt{\beta_1} h_1]$$
$$E_2 = \exp[-s(2\sqrt{\beta_1} h_2 - \sqrt{\beta_1} h_1)]$$
$$E_3 = \exp[-s\{(\sqrt{\beta_2} + \sqrt{\beta_1}) h_2 - \sqrt{\beta_2} h_1\}]$$
$$E_4 = \exp[-s\sqrt{\beta_2} h_1]$$ ----------------------------------(27)
$$E_5 = \exp[-s\{(\sqrt{\beta_2} + \sqrt{\beta_1}) h_2 - \sqrt{\beta_1} h_1\}]$$
$$E_6 = \exp[-s(2\sqrt{\beta_2} h_2 - \sqrt{\beta_2} h_1)]$$

$$F_1 = (\sqrt{\beta_2} + \sqrt{\beta_1}) / (\sqrt{\beta_2} - \sqrt{\beta_1})$$
$$F_2 = 2\sqrt{\beta_1} / (\sqrt{\beta_2} - \sqrt{\beta_1})$$ ----------------------------------(28)
$$F_3 = 2\sqrt{\beta_2} / (\sqrt{\beta_2} - \sqrt{\beta_1})$$

$$c_4 = c_{42} / c_{41}$$ ----------------------------------(29)

The solution of equation (20) is

$$B_1(s) = (\pi/2)[\delta(s)/sW(s)]$$ ----------------------------------(30)

where $\delta(s)$ is delta function. Then stress component $\sigma_{y1}(x,0)$ along y=0 plane is

$$\sigma_{y1}(x,0) = 2c_{41}\{R_1'(0) + 2\}/W(0)$$ ----------------------------------(31)

$W(s)$ at the limit point s=0 is

$$W(0) = [-2(1-\nu_1') - \rho_1\{R_2'(0) + R_4'(0)\} - \rho_2\{R_3'(0) + R_5'(0)\}]/\{(\alpha_1 - \alpha_2)T\}$$ ----------------------------------(32)

And $R_j'(0)$ (j=1,5) are

$$R_1'(0) = 0$$
$$R_2'(0) = -2h_1/\{c_4(1+\rho_1)(\beta_2 - \beta_1)(h_2 - h_1)\}$$
$$R_3'(0) = 2h_1/\{c_4(1+\rho_2)(\beta_2 - \beta_1)(h_2 - h_1)\}$$ ----------------(33)
$$R_4'(0) = -[(1+\rho_1)(\sqrt{\beta_1} + \sqrt{\beta_2})R_2'(0) + 2(1+\rho_2)\sqrt{\beta_2}R_3'(0)]/\{(1+\rho_1)(\sqrt{\beta_2} - \sqrt{\beta_1})\}$$
$$R_5'(0) = [2(1+\rho_1)\sqrt{\beta_1}R_2'(0) + (1+\rho_2)(\sqrt{\beta_1} + \sqrt{\beta_2})R_3'(0)]/\{(1+\rho_2)(\sqrt{\beta_2} - \sqrt{\beta_1})\}$$

Finally, the stress component $\sigma_{y1}(x,0)$ is

$$\sigma_{y1}(x,0) = 4c_{41}/W(0)$$ ----------------------------------(34)

Singular Stresses at the Tip of the Crack

Assuming that the arbitrary load -p(x) operates on the crack surfaces, we have following boundary conditions:

$$\sigma_{y1}(x,0) = -p(x) \qquad -a < x < a$$ ----------------------------------(35)
$$u_{y1}(x,0) = 0 \qquad -h_1 < x < -a, \ a < x < h_1$$ ----------------------------------(36)
$$\tau_{xy1}(x,0) = 0 \qquad -h_1 < x < h_1$$ ----------------------------------(37)
$$\sigma_{x1}(h_1,y) = \sigma_{x2}(h_1,y) \qquad -\infty < y < \infty$$ ----------------------------------(38)
$$\tau_{xy1}(h_1,y) = \tau_{xy2}(h_1,y) \qquad -\infty < y < \infty$$ ----------------------------------(39)
$$u_{x1}(h_1,y) = u_{x2}(h_1,y) \qquad -\infty < y < \infty$$ ----------------------------------(40)
$$u_{y1}(h_1,y) = u_{y2}(h_1,y) \qquad -\infty < y < \infty$$ ----------------------------------(41)
$$\sigma_{x2}(h_2,y) = 0 \qquad -\infty < y < \infty$$ ----------------------------------(42)
$$\tau_{xy2}(h_2,y) = 0 \qquad -\infty < y < \infty$$ ----------------------------------(43)

Making use of mixed boundary conditions (35) and (36), we have the dual integral equations:

$$(4c_{41}/\pi)\int_0^\infty s[A_1(s)\cosh(sx) + \{2\cosh(sx) + sx\sinh(sx)\}B_1(s) - C_1(s)\cos(sx)]ds = -p(x) \quad 0 < x < a$$ ------------(44)
$$(2/\pi)\int_0^\infty (2-2\nu_1)C_1(s)\cos(sx)ds = 0 \qquad a < x < h_1$$ ------------(45)

The boundary conditions (37)-(43) will be satisfied provided that

$$A_1(s)=(2/\pi)[\{P_4(s)f_3(s,h_1)-P_2(s)f_4(s,h_1)\}/s+R_2(s)f_2(s,h_1)+R_3(s)f_1(s,h_1)]/R_1(s) \quad \text{------------------(46)}$$

$$B_1(s)=-(2/\pi)[\{P_3(s)f_3(s,h_1)-P_1(s)f_4(s,h_1)\}/s+R_4(s)f_2(s,h_1)+R_5(s)f_1(s,h_1)]/R_1(s) \quad \text{------------------(47)}$$

where

$$\left.\begin{array}{l} R_1(s)=P_1(s)P_4(s)-P_2(s)P_3(s) \\ R_2(s)=P_4(s)Q_1(s)-P_2(s)Q_3(s) \\ R_3(s)=P_4(s)Q_2(s)-P_2(s)Q_4(s) \\ R_4(s)=P_3(s)Q_1(s)-P_1(s)Q_3(s) \\ R_5(s)=P_3(s)Q_2(s)-P_1(s)Q_4(s) \end{array}\right\} \quad \text{------------------(48)}$$

$$\left.\begin{array}{l} P_1(s)=-2\cosh(sh_1)-(c_4/\gamma_5)[(1+\rho_1)\delta_1\gamma_1-(1+\rho_2)\delta_2\gamma_3] \\ P_2(s)=-2sh_1\sinh(sh_1)-(c_4/\gamma_5)[(1+\rho_1)\delta_1\gamma_2-(1+\rho_2)\delta_2\gamma_4] \\ P_3(s)=2\sinh(sh_1)-(c_4/\gamma_5)[(1+\rho_1)\sqrt{\beta_1}\delta_3\gamma_1-(1+\rho_2)\sqrt{\beta_2}\delta_4\gamma_3] \\ P_4(s)=2\{\sinh(sh_1)+sh_1\cosh(sh_1)\}-(c_4/\gamma_5)[(1+\rho_1)\sqrt{\beta_1}\delta_3\gamma_2-(1+\rho_2)\sqrt{\beta_2}\delta_4\gamma_4] \end{array}\right\} \quad \text{----------------(49)}$$

$$\left.\begin{array}{l} Q_1(s)=(c_4/\gamma_5)\{(1+\rho_1)\sqrt{\beta_2}\delta_1\delta_6-(1+\rho_2)\sqrt{\beta_1}\delta_2\delta_5\} \\ Q_2(s)=(c_4/\gamma_5)\{(1+\rho_1)\rho_2\delta_1\delta_8-(1+\rho_2)\rho_1\delta_2\delta_7\} \\ Q_3(s)=(c_4/\gamma_5)\{(1+\rho_1)\sqrt{\beta_1}\sqrt{\beta_2}\delta_3\delta_6-(1+\rho_2)\sqrt{\beta_1}\sqrt{\beta_2}\delta_4\delta_5\} \\ Q_4(s)=(c_4/\gamma_5)\{(1+\rho_1)\sqrt{\beta_1}\rho_1\delta_3\delta_8-(1+\rho_2)\sqrt{\beta_2}\rho_1\delta_4\delta_7\} \end{array}\right\} \quad \text{------------------(50)}$$

$$\left.\begin{array}{l} \gamma_1=\sqrt{\beta_2}\delta_6\cosh(sh_1)+\rho_2\delta_8\sinh(sh_1) \\ \gamma_2=\sqrt{\beta_2}\delta_6\{2(1-\nu_1)\cosh(sh_1)+sh_1\sinh(sh_1)\}-\rho_2\delta_8\{(1-2\nu_1)\sinh(sh_1)+sh_1\cosh(sh_1)\} \\ \gamma_3=\sqrt{\beta_1}\delta_5\cosh(sh_1)+\rho_1\delta_7\sinh(sh_1) \\ \gamma_4=\sqrt{\beta_1}\delta_5\{2(1-\nu_1)\cosh(sh_1)+sh_1\sinh(sh_1)\}-\rho_1\delta_7\{(1-2\nu_1)\sinh(sh_1)+sh_1\cosh(sh_1)\} \\ \gamma_5=\sqrt{\beta_1}\rho_2\delta_5\delta_8-\sqrt{\beta_2}\rho_1\delta_6\delta_7 \end{array}\right\} \quad \text{----------(51)}$$

$$\left.\begin{array}{l} \delta_7=E_1-F_1E_2+\rho_2(1+\rho_1)F_2E_3/\{\rho_1(1+\rho_2)\} \\ \delta_8=E_4-\rho_1(1+\rho_2)F_3E_5/\{\rho_2(1+\rho_1)\}+F_1E_6 \end{array}\right\} \quad \text{------------------(52)}$$

$$\left.\begin{array}{l} f_1(s,h_1)=\int_0^\infty \eta[(1-2\nu_1)/(\eta^2+s^2)-(\eta^2-s^2)/(\eta^2+s^2)^2]C_1(\eta)\sin(\eta h_1)d\eta \\ f_2(s,h_1)=2\int_0^\infty s[(1-\nu_1)/(\eta^2+s^2)-\eta^2/(\eta^2+s^2)^2]C_1(\eta)\cos(\eta h_1)d\eta \\ f_3(s,h_1)=2\int_0^\infty \eta^2[1/(\eta^2+s^2)-(\eta^2-s^2)/(\eta^2+s^2)^2]C_1(\eta)\cos(\eta h_1)d\eta \\ f_4(s,h_1)=4\int_0^\infty \eta^3s/(\eta^2+s^2)^2]C_1(\eta)\sin(\eta h_1)d\eta \end{array}\right\} \quad \text{------------------(53)}$$

The unknown $C_1(s)$ as given in [4] is

$$C_1(s)=(\pi a^2/4c_{41})\int_0^1 \sqrt{\zeta}\,\Gamma(\zeta)J_0(sa\zeta)d\zeta \quad \text{------------------------------------(54)}$$

where $J_0(\ )$ is the zero order Bessel function of the first kind and $\Gamma(\zeta)$ is the solution of the Fredholm integral equation of the second kind:

$$\Gamma(\zeta)+\int_0^1\Gamma(\xi)K(\xi,s)d\xi=(2\sqrt{\zeta}/\pi)\int_0^b p(au)/(\zeta^2-u^2)^{1/2}du \quad \text{------------------(55)}$$

where the kernel $K(\xi,\zeta)$ is given by

$$K(\xi,\zeta)=\sqrt{\zeta}\,\xi\int_0^\infty s\{s/R_1(s/a)\}[\{2F_b(s/a,\xi)-F_a(s/a,\xi)\}I_0(s\zeta)+s\zeta F_b(s/a,\xi)I_1(s\zeta)]ds \quad \text{------------(56)}$$

$$\left.\begin{array}{l} F_a(s/a,\xi)=P_4(s/a)F_3(s/a,\xi)-P_2(s/a)F_4(s/a,\xi)+R_2(s/a)F_2(s/a,\xi)+R_3(s/a)F_1(s/a,\xi) \\ F_b(s/a,\xi)=P_3(s/a)F_3(s/a,\xi)-P_1(s/a)F_4(s/a,\xi)+R_4(s/a)F_2(s/a,\xi)+R_5(s/a)F_1(s/a,\xi) \end{array}\right\} \quad \text{------------(57)}$$

$$\left.\begin{array}{l} F_1(s/a,\xi)=\{(sh_1/a-2\nu_1)I_0(s\xi)-s\xi I_1(s\xi)\}\exp(-sh_1/a) \\ F_2(s/a,\xi)=\{(3-2\nu_1-sh_1/a)I_0(s\xi)+s\xi I_1(s\xi)\}\exp(-sh_1/a) \\ F_3(s/a,\xi)=2\{(1-sh_1/a)I_0(s\xi)+s\xi I_1(s\xi)\}\exp(-sh_1/a) \\ F_4(s/a,\xi)=2\{(2-sh_1/a)I_0(s\xi)+s\xi I_1(s\xi)\}\exp(-sh_1/a) \end{array}\right\} \quad \text{------------------(58)}$$

where $I_0(\ )$ and $I_1(\ )$ are, respectively, the zero and first order modified Bessel function of the first kind.

For the case that the temperature change is T and the mechanical stress is σ_0, $p(x)=\sigma_0+4c_{41}/W(0)$, the stress intensity factor K_1 is

$$K_1=\lim_{x\to a^+}\sqrt{2(x-a)}\,\sigma_{y1}(x,0)=\sqrt{a}\,\Gamma(1) \quad \text{------------------------------------(59)}$$

Numerical Result and Discussion

We examine the effect of the temperature change and the length of the crack on the stress intensity factor K_1. Used elastic and thermal properties are listed in table 1 [1]. In table 1, "Vf" denotes volume fraction of glass fibers. Subscripts "l" and "t" indicate longitudinal and transverse directions to the fiber axis, respectively.

The stresses at the crack surfaces induced by mechanical load is 0.175GPa [1] and $h_2/h_1=2.0$. Temperatures of 395K(cure temperature at strain free state), 293K(room temperature), 77K(liquid nitrogen temperature) and 4K(liquid helium temperature) are chosen.

The normalized stress intensity factors

$$\overline{K}_1 = \Gamma(1)/\sigma_0$$

are plotted in figure 2 as a function of the normalized crack length a/h_1 for various values of the temperature. The thermal stresses on the crack surfaces are 0GPa at 395K, 0.023GPa at 293K, 0.073GPa at 77K and 0.089GPa at 4K and normalized stress intensity factors at $a/h_1=0$ are 1.0,1.131,1.417 and 1.513, respectively. In any case, as normalized crack length a/h_1 increases, the normalized stress intensity factors decrease.

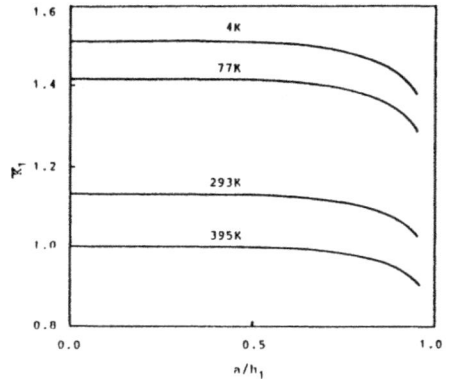

Fig.2 Thermal Stress Intensity Factors

Table 1. Elastic-Thermal Properties

Isotropic constituents	Young's modulus, E (GPa)		Shear modulus, G (GPa)		Poisson's ratio, ν	
Bisphenol A epoxy	7.16		2.768		0.293	
E glass fiber	82.22		32.55		0.263	
Transversely isotropic composite	E_l (GPa)	E_t (GPa)	G_{lt} (GPa)	G_{tt} (GPa)	ν_{lt}	ν_{tt}
Glass/epoxy (Vf=0.5)	44.69	17.88	6.804	6.890	0.276	0.297
Coefficients of thermal expansion	α_l 10^{-6} K^{-1}		α_t 10^{-6} K^{-1}			
Glass/epoxy (Vf=0.5)	5.963		23.39			
Elastic constants Material I	c_{11} (GPa)	c_{21} (GPa)	c_{31} (GPa)	c_{41} (GPa)		
	20.820	20.820	7.040	6.890		
Material II	c_{12} (GPa)	c_{22} (GPa)	c_{32} (GPa)	c_{42} (GPa)		
	20.820	48.928	7.682	6.804		

References

[1] R.D.Kriz, "Influence of Damage on Mechanical Properties of Woven Composites at Low Temperatures", Journal of Composites Technology & Research, Vol.7, No.2, 55-58 (1985)

[2] M.B.Kasen, G.R.MacDonald, D.H.Beekman and R.E.Schramm, "Mechanical, Electrical and Thermal Characterization of G10-CR and G11-CR Glass/Epoxy Laminates Between Room temperature and 4K", Advances in Cryogenic Engineering, Vol.26, 235-244 (1980)

[3] M.B.Kasen and R.E.Schramm, "Current Status of Standardized Nonmetallic Cryogenic Laminates", Advances in Cryogenic Engineering, Vol.28, 171-177 (1981)

[4] E.P.Chen and G.C.Sih, Mechanics of Fracture 6,33, Noordhoff, Legden (1981)

EVALUATION OF IRRADIATION EFFECT OF FRP BY AE METHOD

Tetsuya NISHIURA, Kazumune KATAGIRI, Shigehiro NISHIJIMA and Toichi OKADA

Inst. Sci. Ind. Res., Osaka University
8-1, Mihogaoka Ibaragi, Osaka 567, Japan

Abstract - Tensile tests of glass fiber reinforced plastics (FRP) were performed and acoustic emission (AE) was observed simultaneously to detect microscopic fractures.   The irradiation induced degradation of FRP was recognized by AE method, while the degradation of tensile strength did not appear remarkably with irradiation.

Introduction

Superconducting magnet constructed in a fusion reactor has been predicted to be irradiated up to the dose of 10^9–10^{10} rad by 14 MeV neutron and gamma ray.   An insulator and construction materials of SC magnet such as FRP and plastic films, therefore, exposed to the same irradiation dose, and are also subjected to high electromagnetic force.   Organic materials are so sensitive to the irradiation that many studies on effects of irradiation to the mechanical properties of resin and FRP have been conducted [1-4].   However, the microscopic studies such as crack initiation, propagation and/or debonding between resin and fibers are few [1-3].   We have tried to apply AE method to detect microscopic fracture processes at cryogenic temperature [5].   In this study, the applicability of AE method to the evaluation of irradiation effect on microscopic fracture in FRP is tested.

Experiment

Samples

A sample composed of 10 layers of glass cloths impregnated by epoxy resin [Epomik resin (Mitubishiyuka Co. Ltd.) 100 : hardener of Jefermin-D230 ,30 ; named Epomik] was prepared in our laboratory. Tensile specimens were cut into the form as shown in Fig.1 from the 2 mm thickness plate of the FRP.

Tensile tests

Tensile test was made by using an Instron type tensile machine (Shimadzu Co.Ltd.) with a cross head speed of 1.5 mm/min.   Tests were carried out at room and liquid nitrogen temperatures (RT and LNT).

Irradiation

The specimens were irradiated by γ-ray from Co 60 and by 20 MeV electron of LINAC (220 mA, 120 pps and pulse width 1.5 μs) in ISIR, Osaka University. The irraadiation of γ-ray was made in air at room temperature.   The electron irradiation was performed in cooling water of 283 K to avoid thermal destruction of specimens.

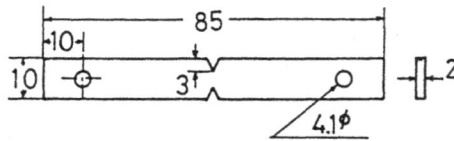

Fig.1. Tensile test specimen configuration.

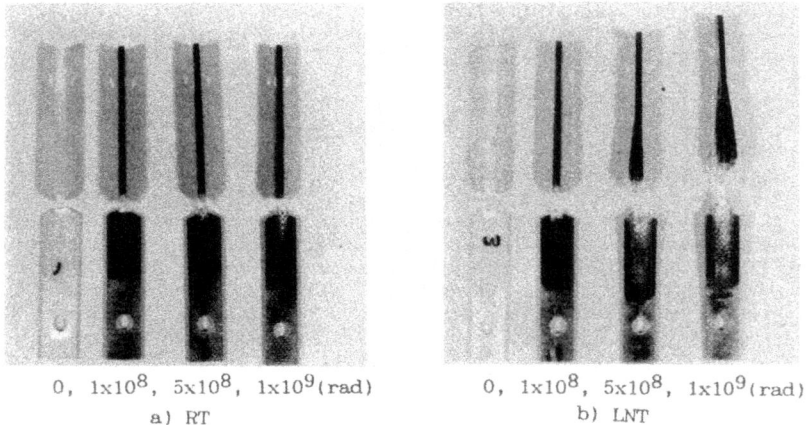

0, 1x10^8, 5x10^8, 1x10^9(rad)   0, 1x10^8, 5x10^8, 1x10^9(rad)

a) RT   b) LNT

Fig.2. Fracture profiles after the tensile tests of irradiated specimens.

AE measurement

AE signals were detected with a PZT transducer of 140 kHz of resonance frequency, which the signals were amplified to 70-80 dB by pre- and main-amplifiers, and the signals were counted and recorded on a multi-pen recorder during the tensile test.

Results

Observation of fracture profiles

Fracture profiles after the tensile tests of irradiated specimens are shown in Fig.2. The specimens are coloured from light brown to deep brown with increase of the dose. A whitening region and pulling out fibers are observed at the vicinity of breakage place. The area of the whitening region and the pulling out length of glass fibers increase with the radiation dose, and delaminations are also observed. These behaviors are more remarkable in specimens tested at LNT than those tested at RT. This whitening is believed to be formed by debonding of resin from the fibers and the pulling out of fibers is enhanced with debonding. Following the results, it is suggested that irradiation yields the degradation of adhesive force between resin and fibers and the increase of dose brings about the increase of the extent of degaradation. The debonding, the pulling out and the delamination are more easy to occur when the ratio of the strength parallel in fiber orientation to that perpedicular to it increases.   The ratio becomes larger with reducing temperature because strength of fibers becomes larger with decrease of temperature while strength of bonding does not.   This situation is more remarkable in irradiated samples in which the bonding strength has been degraded.

Tensile strength

The relation between tensile strength and irradiation dose is shown in Fig.3. At RT, the strength keeps almost constant with dose increase although slight decrease is observed at 1000 Mrad.   At LNT, the strength is also nearly constant like that at RT, except that the value is about 2.5 times larger than that at RT.   In the result of e-irradiation, the tensile strength does not exhibit remarakable degradation with the dose both at RT and LNT (Fig.3. b)).   In this study, no significant difference in irradiation effect between γ-ray and e was observed. We are now examining the quality effect of radiation sources in details.   The tensile strength is confirmed not to be affected by irradiation of these dose range.   This result indicates that degradation of resin does not make FRP degrade in these test direction. This means that macroscopic tensile strength of FRP is determined by fiber strength.

AE behavior

Activity of AE during tests is shown in Fig.4 as a relation between total AE counts and elongation.   The load is also drawn in the same figure. At RT, AE activity is almost independent on the dose except at 1000 Mrad where AE slightly increase in the early stage of loading as compared to that at lower dose.   Load-elongation curves exhibit no change with the dose at RT.   On the other hand, at LNT, AE rate and total AE counts drastically increase with radiation dose.   In the load-elongation curves at LNT, the

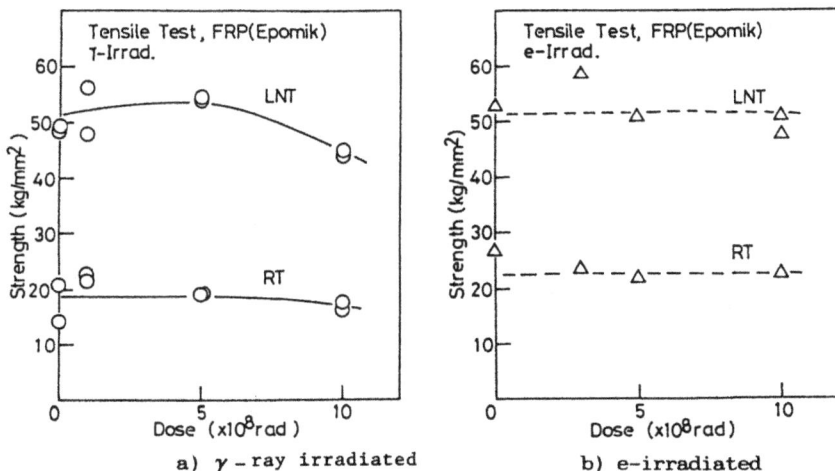

a) γ - ray irradiated   b) e-irradiated

Fig.3. Relation between tensile strength of FRP and irradiation dose.

434

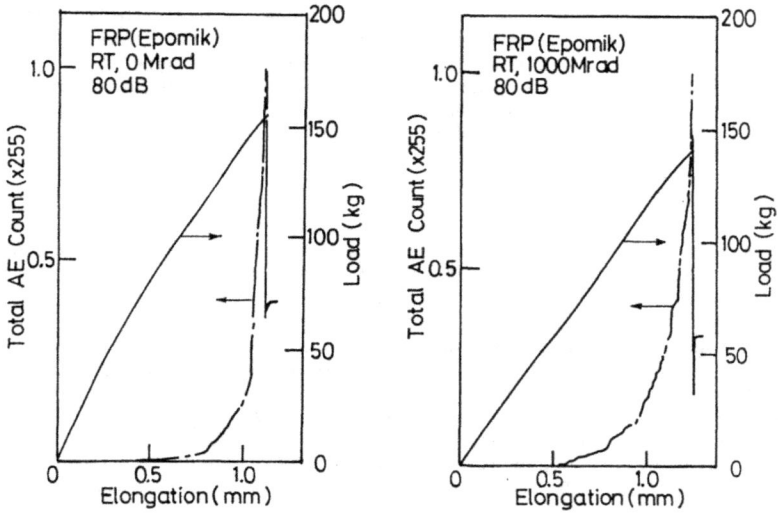

a) RT

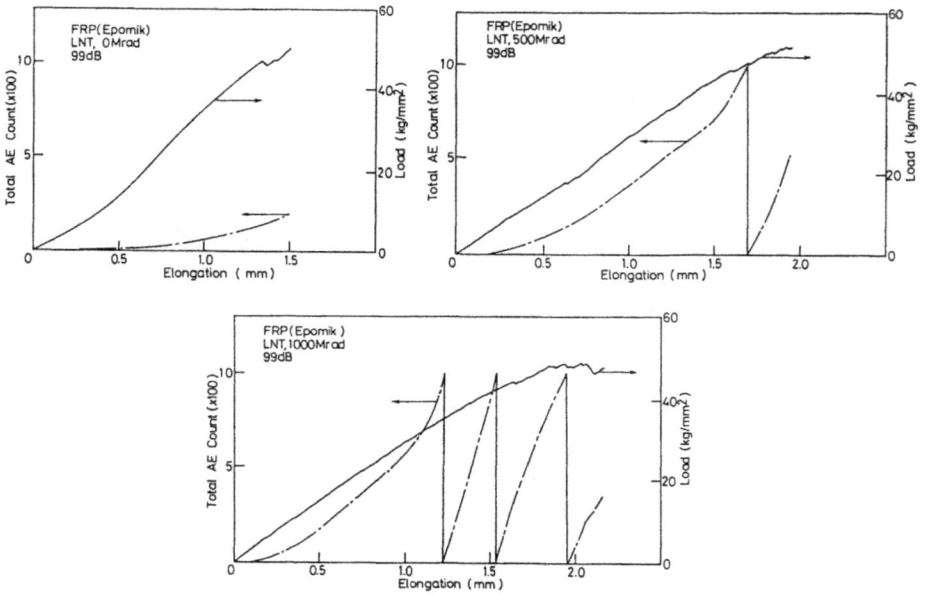

b) LNT

Fig.4. Activity of AE during tensile test of irradiated FRP.

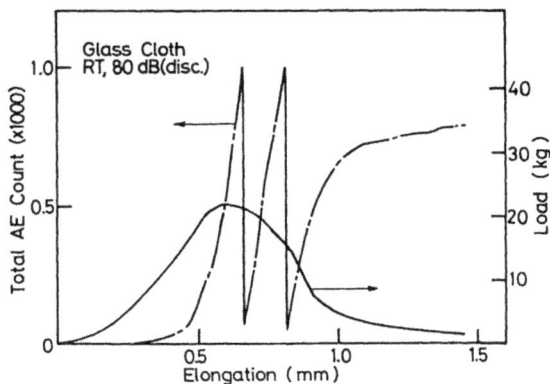

Fig.5. AE behavior during a tensile test of glass cloth.

elongation at breakage increases and the load for a certain elongation value
decreses with increase in dose.   AE behaviour during tensile test of glass
cloth alone is shown in Fig.5.   Total AE counts are remarkably many and
this suggests that AE activity of FRP irradiated to higher dose is similar
to that of glass cloth alone.

Discussion

    When the irradiation dose increased, total AE counts remarkably
increased during the tensile test at LNT, and the whitening region and
pulling out of fiber were also increased.   The coincidence between AE
counts and white region and/or pulling out suggests that they are correlated
each other.   White region is made as a result of micro-fracture such as
debonding between fiber and resin, and/or microscopic fracture of resin
along the fiber.   These fractures should release the strain energy and a
part of the energy should be converted to AE.
    The mechanism of these micro-fractures is macroscopically understood as
the change of the ratio of strength in the parallel direction to that
papendicular. Increasing the ratio, debonding and pull out occur more easily.
Irradiation of FRP degrades the matrix resin and the interface between
fiber and resin, which this situation results in reduction of bond strength.
    The picture of this micro-fracture is as followings.   In non-irradiated
FRP, the bonding strength of resin to fibers is so high that the breakage of
a single fiber does not result in the release of all the stress subjected
,because the fixed resin on fiber transfer the stress by shearing force.
On the other hand, in irradiated FRP, the strength of bond is so weakened
that breakage of fiber release the stress and makes debonding and pull out
of fibers.   These processes are further promoted with increase of
degradation of matrix resin, which processes activate AE.   AE method is
thereby applicable to evaluate irradiation induced degradaion of FRP.
    When the degradation of FRP progresses to its extream, the bonding resin
is perfectly separated, and, therefore, FRP is thought to be similar to
glass fiber cloth alone.   To confirm this, a tensile test of glass cloth is
made (Fig.5).   AE is remarkably detected in this test.   Consequently, in

this situation, because the resin is free, AE should not include the AE sources concerning resin fracture and debonding processes. The sources are attributed to friction between fibers themselves and to fiber break. Therefore, AE generated in the tensile test of irradiated FRP results from debonding, crack, friction during fiber pulling out and fiber breakage.

Conclusion

AE technique applied to evaluate the irradiation induced degradation in tensile properties of FRP and the following conclusions are obtained.
1. Tensile strength of FRP is not significantly degraded by irradiation up to 1000 Mrad.
2. Microscopic degradation such as debonding and microcracking, however, proceeded and was detected by AE technique.

References

1) C.E.Klabunde and R.R.Coltman, Jr., "Debonding of Epoxy from Glass in Irradiated Laminates", J. Nuclear Materials, 117, 345-350 (1983).
2) M.Hagiwara, A.Udagawa, S.Kawanishi, S.Egusa and N.Takeda, "Degradation Behavior of Fiber Reinforced Composites Under Irradiation by 3MeV Electrons", J. Nuclear Materials, 133&134, 810-814 (1985).
3) G.F.Hurley, J.D.fowler and D.L.Rohr, "Low dose cryogenic newtron irradiation effects in G-10CR", Cryogenics, August, 415-420 (1983).
4) D.Evans and J.T.Morgan, "A review of the effects of ionising radiation on plastic materials at low temperatures", Adv.Cry.Eng.-Mats., 28, 147-156 (1982).
5) T.Nishiura, K.Katagiri, S.Owaki and T.Okada, "Acoustic emission of composite materials in tensile tests at cryogenic temperatures", Cryogenics, June, 329-333 (1984).

DEVELOPMENT OF THREE DIMENSIONAL FABRIC REINFORCED PLASTICS

FOR CRYOGENIC APPLICATION

Hiroshi OKUYAMA, Shigehiro NISHIJIMA, Toichi OKADA
Tetsuro HIROKAWA[+], Jun YASUDA[+], Tsuguo UEMURA[+]
Shingo NAMBA[++]

[I]SIR Osaka University, Ibaraki, Osaka, Japan
[+]Shikishima Canvas Co., Ltd., Miyauchicho, Oomihachiman, Shiga, Japan
[++]University of Osaka Prefecture, Sakai, Osaka, Japan

ABSTRACT - The three dimensional fabric reinforced plastics (3D-GFRP) for cryogenic use have been developed making the selection of matrix, the optimization of 3D-fabrics and the investigation of fabrication processes. The epoxy matrix has been selected considering the durability against thermal shock and easiness of fabrication. The three dimensional fabrics have been constructed varying the orientation and the tension of glass fibers. The possible combination of epoxy matrix and 3D-fabrics were decided and then 3D-GFRP were fabricated. The cryogenic performance of the 3D-GFRP were examined. In this work, particular attention has been paid to thermal properties and they have been measured down to cryogenic temperatures. The thermal contraction of GFRP is ought to be small enough in practical cryogenic application. The usual GFRP (2D-GFRP) were also made using the identical components of 3D-GFRP and their dimensional stability were compared each other. It is found that concerning the thermal contraction 3D-GFRP shows much more isotropic characteristics compared with 2D-GFRP as expected. The thermal contraction of 3D-GFRP in thickness direction can be reduced as small as that of metals. By controlling the fabrication processes the thermal contraction can be designed in advance. The thermal conductivity of 3D-GFRP was also measured down to cryogenic temperature and indicated good thermal insulating performances. It is concluded that the 3D-GFRP for cryogenic use have been successfully developed.

Introduction

Glass fiber reinforced plastic (GFRP) materials have been used as insulating and/or structual materials in cryogenic apparatus[1]. The usual GFRP are reinforced in one or two directions with glass fibers. The composites have disadvantages such as low strength, low stiffness and large thermal contraction in the directions parpendicular to the fibers. Furthermore they show low interlaminar shear strength. For cryogenic use such demerits induce serious problems e.g. large thermal contraction degrades the rigidity of the magnet markedly. To improve these demerits three dimensional fabric reinforced composites have been developed[2]. In this paper the new composites and usual ones are called as 3D-GFRP and 2D-GFRP, respectively.

In this work, the thermal properties of 3D-GFRP were investigated aiming at practical cryogenic application. The matrix resin which affects the performances of GFRP were examined. The thermal contraction of GFRP were also measured down to liquid nitrogen temperature (LNT) varying the production methods. The thermal conductivity of 3D-GFRP and 2D-GFRP was measured down to liquid helium temperature (LHeT) and compared each other.

Experimentals

When the micro cracks are introduced to the composite materials in cooling down process, the rigidity of the composite materials is degraded[3]. The matrix resin is ought to show the durability against the thermal shock. In this work, 23 kinds of commercial available epoxy resins were examined in terms of the durability against the thermal shock. Samples of epoxy resin for thermal shock test, whose size were approximately 100mm x 50mm x 5mm, were put into liquid nitrogen. The crack density was measured after warming up to room temperature. The durability against the thermal shock was estimated by means of crack density. The thermal contraction down to liquid nitrogen temperature[4] and Vicker's hardness were also measured. The selected matrix resin was used for fabricating the composites.

Dimensional stability of the composites down to cryogenic temperature has been investigated. The specifications of fabricated samples were shown in Table 1. Not only 3D-GFRP but also 2D-GFRP were fabricated. Three types of 3D-GFRP were fabricated which were 3D-1, 3D-2 and 3D-3. The 3D-1 is usual 3D-GFRP. In 3D-2 the fibers in thickness direction were stretched at the fabrication. The 3D-3 was made by pressing the 3D-2 in the thickness direction.

The thermal contraction of the composites along X, Y and Z axis was also measured down to LNT.

The thermal conductivity was also measured for G-1 (3D-GFRP) and EL-405 (2D-GFRP) along X direction down to LHeT[5].

Table 1. Specification of specimens

Name	Type	Volume fraction of glass fibers in each direction (%)			Glass content by volume Vf (%)
		X	Y	Z	
C74	2D	57	43	--	63.0
C32		57	43	--	57.5
C24		57	43	--	52.9
C34		57	43	--	58.2
EL-405		50	50	--	49.9
H-3A	3D-1	33	39	28	43.0
H-3B		35	42	23	52.0
G-1		28	30	42	45.0
H2-4	3D-2	26	63	11	59.3
H3-3		36	50	14	52.1
H-4		23	71	6	60.8
H2-2		30	60	10	50.0
H-2	3D-3	30	60	10	59.1
H-2EP		30	61	9	67.0

Results and Discussion

The crack density was divided into 5 classes. The largest crack density was defined as class 5 and the relationship between the crack density and thermal contraction or Vicker's hardness was shown in Fig. 1 and Fig. 2, respectively. The general tendency between crack density and thermal contraction or Vicker's hardness can be observed. The thermal contraction decreases and Vicker's hardness increases, as the crack density decreases. It can be concluded that the screening of the epoxy resin matrix can be made using the tendency.

Using the selected resin matrix the 3D-GFRP and 2D-GFRP were fabricated. Figure 3 and Fig. 4 show the temperature dependence of the thermal contraction of 2D-GFRP and 3D-GFRP (3D-1 type), respectively. In regard to 2D-GFRP, thermal contraction in the thickness direction (Z direction) is approximately twice of that in the fiber directions (X and Y directions). On the contrary the 3D-GFRP shows the almost same thermal contraction in each direction. The 3D-GFRP shows isotropic properties compared with that of 2D-GFRP and is expected to show good compativility with metals whose thermal contraction down to cryogenic temperature is approximately 0.3%.

It was confirmed that any crack was not introduced to the composites after the thermal contraction measurements. It means that the suitable epoxy matrix selection was made.

The glass content dependence of thermal contraction down to LNT in 3D-GFRP is shown in Fig. 5. The glass content in Fig. 5 means the net glass content in each direction. In Fig. 5 the thermal contraction of 100% glass content shows thermal contraction of glass fiber itself and 0% that of epoxy resin which were measured. It was found to be possible to control the thermal contraction by changing the tension of glass fibers or pressing the composites.

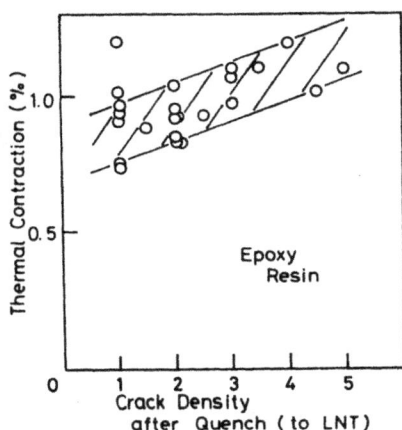

Fig. 1. Relation between thermal contraction and crack density.

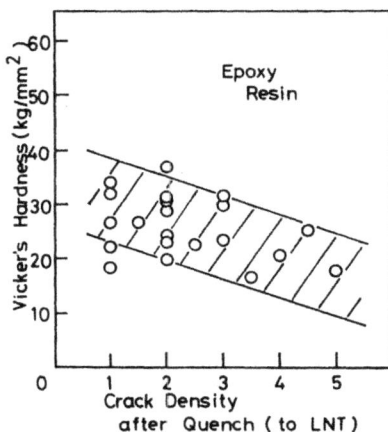

Fig. 2. Relation between Vicker's hardness and crack density.

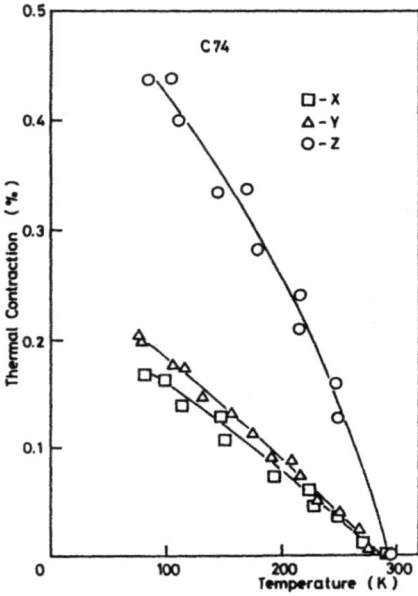

Fig. 3. Thermal contraction of
2D-GFRP.

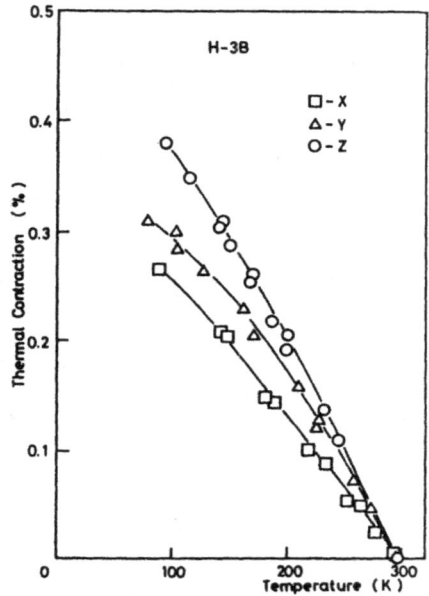

Fig. 4. Thermal contraction of
3D-GFRP(3D-1).

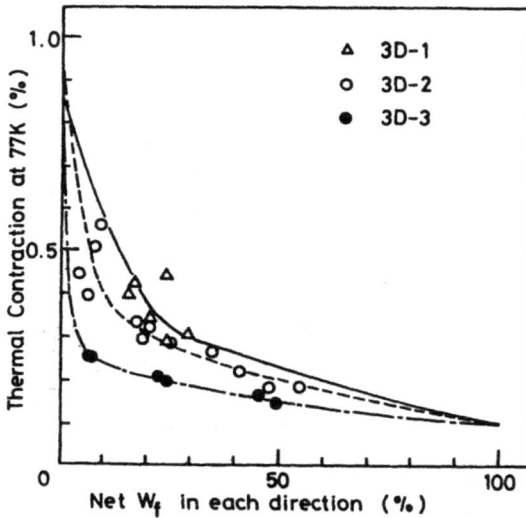

Fig. 5. Glass content dependence of thermal
contraction of 3D-GFRP.

The thermal expansion (contraction) coefficient was calculated and compared with the experimental data. In regard to the unidirectionally reinforced composites, the following law of mixture is valid:

$$\alpha_c = \frac{\alpha_f E_f V_f + \alpha_m E_m (1-V_f)}{E_f V_f + E_m (1-V_f)}$$

where α_f:thermal expansion coefficients of glass fiber
E_f:Young's modulus of glass fiber
α_m:thermal expansion coefficients of matrix resin
E_m:Young's modulus of matrix resin
V_f:glass content by volume.

The relationship between V_f and coefficient α_c is caluculated and shown in Fig. 6 by the solid line. The measured values of 3D-1 were plotted as circular markings. These values coincide with the law of mixture and hence this relationship can be used for design. On the other hand the values of 3D-3 (shown as triangular markings) are lower than the calculated ones. This would be originated from the fact that the fibers in other direction come to disturb the deformation with increasing the interactions between fibers. It would be concluded that the thermal contraction (coefficient) of 3D-GFRP is controlled by changing the fabrication process and glass content.

Figure 7 shows the temperature dependence of thermal conductivity of 3D-GFRP and 2D-GFRP, respectively.

It was found that the thermal conductivity of 3D-GFRP indicated good thermal insulating performances at cryogenic temperatures.

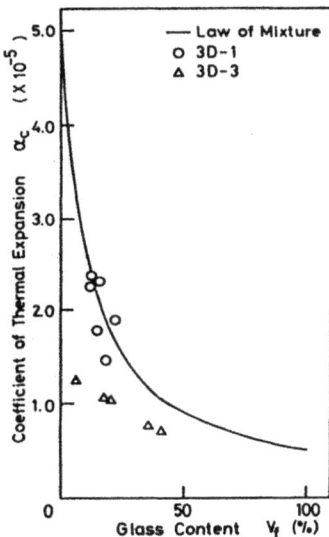

Fig. 6. Glass content dependence of coefficient of thermal expansion.

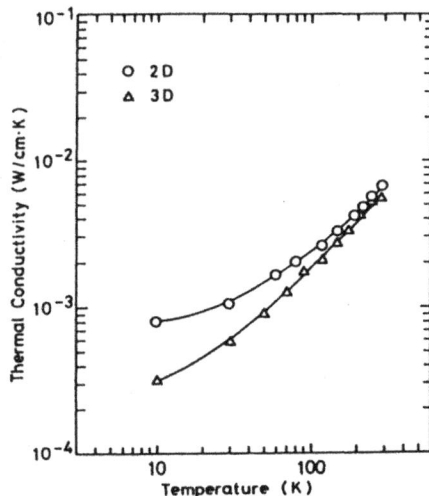

Fig. 7. Temparature dependence of thermal conductivuty of 3D-GFRP and 2D-GFRP.

442

Conclusion

The three dimensional fabric reinforced plastics have been developed aiming at the cryogenic applications.   Following conclusions have been drawn.
(1)It is possible to refer to the thermal contraction and Vicker's hardness in screening the resin matrix of the composites for cryogenic application.
(2)The 3D-GFRP has good dimensional stability and isotropy even at cryogenic temperature compared with 2D-GFRP.
(3)It is possible to control thermal contraction of 3D-GFRP by changing the tension of the fiber and/or pressing the composites.
(4)The thermal expansion coefficients of 3D-1 type (3D-GFRP) can approximatly obey the law of mixture.   Thermal contraction can be reduced by increasing the tension of fabrics and/or pressing the composites.   These effect would be originated from the fiber to fiber interactions.
(5)The thermal conductivity of 3D-GFRP shows good thermal insulating performances at cryogenic tempertures.
(6)It is concluded that the 3D-GFRP for cryogenic use has been successfully developed.

Acknowledgement

This work is partly supported by Grant in Aid for Scientific Reserch No.63609511, Ministry of Education in Japan.

References

1.   H. Nakajima, K. Yoshida, Y. Hattori, K. Koizumi, M. Oshikiri and S. Shimamoto, "Structual behavior of winding and superconductor under mechanical loading", IEEE Trans. on Magnetics, MAG-23, No.2, 1521-1530 (1987)
2.   S. Nishijima, Y. A. Wang, T. Okada, T. Uemura, T. Hirokawa and J. Yasuda, "Cryogenic properties of three-dimensional glass-fabric -reinforced plastic", in Advances in Cryogenic Engineering--Materials, vol.34, 59-66 (Plenum, New York, 1988).
3.   T. Okada, S. Nishijima, K. Matsushita and T. Okamoto, "Effect of interlaminar failure on dynamic young's modulus and internal friction in composites materials", in Advances in Cryogenic Engineering-- Materials, vol.34, 115-122 (Plenum, New York, 1988).
4.   S. Nakahara, T. Fujita, K. Sugiura, S.Nishijima, M. Takeno and T. Okada, "Two-dimensional thermal contraction of composites", in Advances in Cryogenic Engineering--Materials, vol.32, 209-215 (Plenum, New York 1986).
5.   M. Takeno, S. Nishijima, T. Okada, K. Fujita, Y. Tsuchida and Y.Kuraoka, "Thermal and mechanical properties of advanced composite materials at low temperatures", in Advances in Cryogenic Engineering--Materials, vol.32, 217-224 (Plenum, New York, 1986).

DATA BASE OF ORGANIC COMPOSITE MATERIALS FOR CRYOGENIC APPLICATION

Shigehiro NISHIJIMA and Toichi OKADA

Institute of Scientific and Industrial Research, Osaka University
8-1 Mihogaoka Ibaraki, Osaka, 567, Japan

Abstract - Data base of organic composite materials has been established in aiming at the practical application of these materials for cryogenic use. These materials have not used frequently for cryogenic used because of the lack of the data at cryogenic temperatures though they present the potentials for practical applications. In this work not only the data base but also the problems in obtaining the data have also been investigated. The measured properties are mechanical, thermal, electrical and vacuum behavior of organic and nonmetallic composite materials at cryogenic temperatures. The nondestructive testing and radiation damage have also been studied.

The recent activity of nonmetallic composite materials will be discussed referring the obtained data.

Introduction

Superconducting magnets are thought to be "series machines" in the sense that the total magnet system would not work when even one component of the magnet degrades the performances. The evaluation of the performance of the each component under the operated conditions, therefore, should be important. The establishment of the data base on the magnet components under various conditions are inevitable for practical application of superconducting magnets. Since the operation conditions varies with the design and/or the systems, the severest condition is to be supposed from the view point of establishing the data base. One of such superconducting magnets is ought to be the fusion magnets and hence the data base for the fusion magnets can also be used in various field other than fusion magnets.

The fusion magnets are to be operated at cryogenic temperature under high stressed conditions. The components should withstand the cryogenic temperature and high stresses. The radiation damages of the components have to be studied because the magnet is operated in radiation environments.

Among the components of the superconducting magnets, the component whose data accumulation is most urgent is insulating material especially organic composite material. The composites can be designed and fabricated arbitrarily with selecting and combining the materials or choosing the fabrication methods. The flexibility of designing the material, however, could be the demerits. The flexibility means the scatter of data and results in the delay of establishing the standard of data. It means that the data obtained by other researchers can not be used in specific application. The operating conditions of the magnets are created purposely and the experiments are made by the users and/or designers themselves. Furthermore, there is almost no irradiated data available.

In order to respond the demands the data base on organic composite materials has been established and the standardization of testing method has been done. The data base, which has been obtained so far, will be presented.

Data Base

The data base established is not the data accumulated from literature but those obtained in experimentals in our laboratory. Consequently it is necessary to develop the measuring systems for the data base. The systems developed are the apparatus for mechanical , thermal, electrical and vacuum properties. These systems can be operated from cryogenic temperatures to room temperature.

Concerning the mechanical properties flexural, tensile, compressive, impact (Charpy, Drop Weight, and Split-Hopkinson Bar test), and fatigue tests can be performed. The creep testing machine have just been installed and the creep test at cryogenic temperature is now performing.
Concerning thermal propertied, thermal conductivity, thermal contraction and specific heat of composite materials have been accumulated.
Dielectric loss measurement, thermally stimulated current have been performed as the electric properties. Such data give us the informations of the conditions of matrices in composite materials, especially the effect of curing conditions, irradiations and so on. The dielectric strength under stressed conditions at cryogenic temperature are now planning to perform.
As nondestructive testing, acoustic emission, internal friction and optical deformation measurement have been made. The acoustic emission and optical measurement during deformation have already technically been established and used as a monitoring techniques for FRP cryostat.
The effects of gamma, electron and neutron irradiation on mechanical and electrical properties have been investigated. The neutron irradiation experiments are performed in cooperation with Kyoto University Reactor Research Institute.
In table 1. the properties which are accumulated in the data base are shown.

Table 1. Material properties accumulated in the data base

Mechanical Property
> Strength (Flexural, Tensile, Compressive), Elastic Modulus (Young's modulus, Shear modulus), Impact strength (Charpy, Drop weight, SHB), Fatigue, Creep, Dynamic Young's modulus, Internal friction

Electric Property
> Dielectric constant, Dielectric loss tangent, Dielectric strength, Thermally stimulated current

Thermal Property
> Thermal contraction, Thermal conductivity, Specific heat

Vacuum Property
> Out gassing, Gas permeation

Irradiation effects
AE

Accumulated Data

The examples of accumulated data was shown in table 2, though they are written in Japanese. The data is concerned in the thermal contraction of the resins. These data are stored in personal computer and are possible to be obtained by means of floppy discs or printed matters. Hereafter the examples of accumulated data are presented in order to give the outline of the data base.

Table 2. Example of accumulated data on resins.

物 性	単 位	材 料	主 剤	硬 化 剤	混 合 率	DataLNT	メ ー カ ー
熱 収 縮	%	エ ポ キ シ	エピ゜コ‐ト828	エピ゜キュアDX-1	100/80/1	0.97	油 化 シェル
熱 収 縮	%	エ ポ キ シ	エピ゜コ‐ト828	エピ゜キュア113	100/33	0.759	油 化 シェル
熱 収 縮	%	エ ポ キ シ	エピ゜コ‐ト828	アクメックスH-84	100/50	0.926	油 化 シェル
熱 収 縮	%	エ ポ キ シ	アラルタ゛イトCY2アラルタ゛イトHY9		100/100/15	0.961	チバ゛ カ゛ イキ゛ ー
熱 収 縮	%	エ ポ キ シ	エピ゜コ‐ト828	アラルタ゛イトHT9	100/25	0.9399	油 化 シェル
熱 収 縮	%	エ ポ キ シ	XB-3052A	XB-3052B	100/38	0.86	チバ゛ カ゛ イキ゛
熱 収 縮	%	エ ポ キ シ	エピ゜コ‐ト828	カヤハ‐ト゛MCD	100/88/0.5	1.070	油 化 シェル
熱 収 縮	%	エ ポ キ シ	エピ゜コ‐ト828	リカシット゛DDSA	50/65/1	1.197	油 化 シェル
熱 収 縮	%	エ ポ キ シ	エボ゜ト‐トYC-1	HT972P	150/18	0.831	東 都 化 成
熱 収 縮	%	ビ゛スマレイミト゛	ハイボ゜リックN-3		100	0.737	三 井 石 油 化
熱 収 縮	%	エ ポ キ シ	ボ゜リセット2915ボ゜リセットW34		100/1/1	1.016	日 立 化 成
熱 収 縮	%	エ ポ キ シ	エボ゜ト‐トYC-2エボ゜ト‐トYC-2		100/80	0.952	東 都 化 成
熱 収 縮	%	エ ポ キ シ	エボ゜ト‐トLEX1エボ゜ト‐トLEX1		100/60	1.027	東 都 化 成

Mechanical Properties

Though the material usually comes to be hard and brittle as the temperature decreases, the organic composite materials especially glass fiber reinforced plastics (GFRP) are different from the usual tendency. Figure 1 shows the stress-strain curves obtained in tensile tests at room (RT), liquid nitrogen (LNT) and liquid helium temperature (LHeT) [1]. Not only the breaking stress but also breaking strain increases with decreasing temperature. The mechanical behavior at LHeT is almost same as that at LNT. This characteristic behavior is different from ordinal materials and is the advantageous characteristics. The behavior is also confirmed in flexural tests.

Figure 2 shows the results of Charpy impact tests of GFRP [2]. These results are obtained in the instrumented Charpy testing machine and hence the load-time diagrams in testing can be obtained. In the figure the diagrams obtained at RT, LNT and LHeT are presented. The absorbed energy is 53.9 kg cm/cm2 at RT, 73.4 at LNT and 79.5 at LHeT. The energy increases with decreasing temperature. The load-time diagrams obtained in Charpy impact tests show the similar tendency with temperature to those in static mechanical tests. The availability of GFRP at cryogenic temperatures was also confirmed by the impact test.

Thermal Properties

As the superconducting magnets are operated under cryogenic environments, the supporting materials are ought to show good thermal insulating properties and dimensional stability. The understanding of thermal properties is inevitable for real application of the materials. In Fig. 3 thermal contractions of the unidirectionally reinforced material in the fiber direction are presented [3]. They are organic composite materials reinforced by glass, carbon, silicon carbide and alumina fibers, respectively. The thermal contraction of metallic materials down to cryogenic temperatures is approximately 0.3% and hence the dimensional stability of the composite materials is good enough. The thermal contraction in the direction perpendicular to the fiber is approximately 0.7%. It is understood that we have to be careful about the anisotropy of the materials.

Figure 4 shows the temperature dependence of thermal conductivity of the identical materials as presented in Fig. 3 [3]. These are also reinforced unidirectionally. The thermal insulating capability of the materials is approximately two orders of magnitude higher than stainless steel. In the direction perpendicular to the fibers, thermal conductivity is smaller than that in fiber direction. It is confirmed that composite materials are suitable materials as supporting materials for cryogenic use.

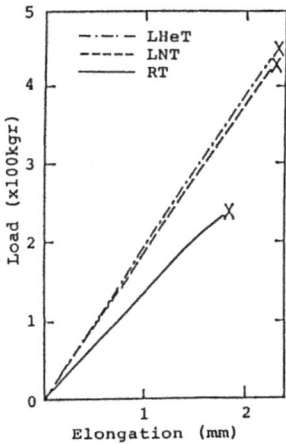

Fig.1. Stress-strain curves obtained at cryogenic temperatures.

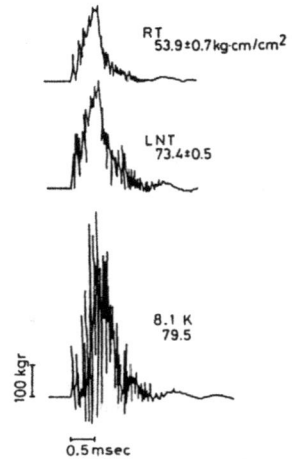

Fig.2. Load-time diagrams obtained at cryogenic temperatures.

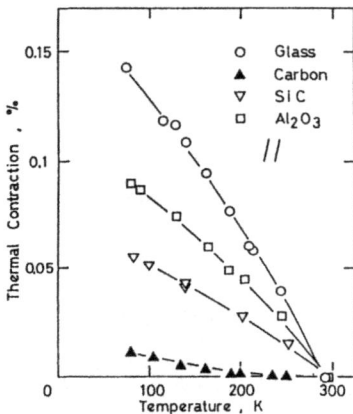

Fig.3. Thermal contraction of unidirectionally reinforced organic composite materials in fiber direction down to cryogenic temperatures.

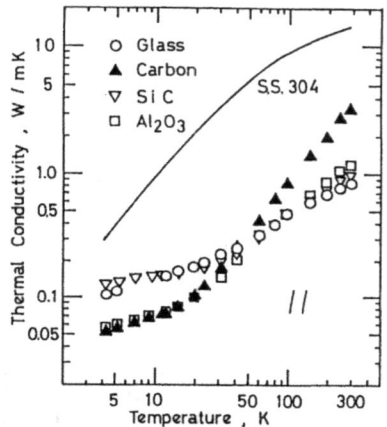

Fig.4. Thermal conductivity of unidirectionally reinforced organic composite materials in fiber direction down to cryogenic temperatures.

Electric Properties

Though the organic composite materials are used as an electric insulating materials, the material selection is often made based on the mechanical properties. In the superconducting magnet the induced voltage should be small and the dielectic strength of the composites increases with decreasing temperature. These are reasons why the electric properties are not the selection standards of the composite materials though they are used as electric insulating materials. The electric properties, therefore, can be thought as secondary properties in practical application. In this data base in order to check the soundness of the composite materials, the electrical properties were measured as the reference data. Figure 5 shows the temperature dependence of dielectric constants and dielectric loss tangent [4]. All of the samples used here are GFRP containing the different matrices. The different curing condition and different matrix induce the different temperature dependence and hence the evaluation of the soundness or the optimization of curing conditions can be made using the data.

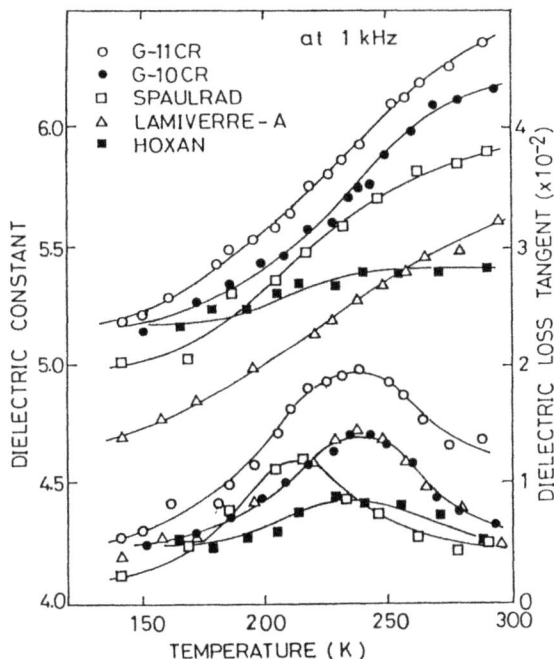

Fig.5. Temperature dependence of dielectric constant and dielectric loss tangent

Radiation Damage

Figure 6 shows the effects of irradiation on the interlaminar shear strength under three conditions: (1)RT gamma irradiation, RT test; (2) 20K neutron irradiation, LNT test; and (3) 360K neutron irradiation, RT test [5].

On the base of the neutron spectrum, a conversion factor was used to convert the neutron fluence into an absorbed dose (which is derived from the calculation of energy transfer). At an absorbed dose of 10 MGy, gamma irradiation causes only a 17% decrease in interlaminar shear strength, but neutron irradiation caused a 65% and 83% decrease in interlaminar shear strength at 360 and 20 K, respectively. The different effects of neutron irradiation are thought to cause this difference.

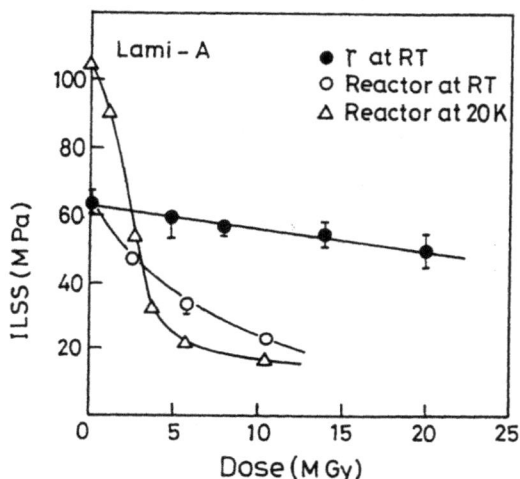

Fig.6. Change of interlaminar shear strength by irradiation.

Conclusion

A part of the established data base was introduced in this paper. The nondestructive testing method and evaluation method of radiation damage have already been established. Using the data base the real application of these materials are examined to the supporting members of cryostats. The promising radiation resistant organic composite materials are recently developed taking advantages of the data base. The data base will be available by means of floppy discs in near future.

References

[1] S.Nishijima, T.Nishiura, T.Okada et. al., "Thermally Stimulated Deformation in GFRP", Adv. Cryog. Eng., 34, 75-82 (1988).
[2] S.Nishijima, M.Takeno and T.Okada, "Impact Test of Reinforced Plastics at Low Temperatures", Adv. Cryog. Eng. 28, 261-270 (1982).
[3] M.Takeno, S.Nishijima, T.Okada et.al., "Thermal and Mechanical Properties of Advanced Composite Materials at Low Temperatures", Adv. Cryog. Eng. 32, 217-224 (1986).
[4] S.Nishijima, T.Okada and T.Hagihara, "Thermo-stimulated Current and Dielectric Loss in Composite Materials", Adv. Cryog. Eng. 32, 187-193 (1986).
[5] S.Nishijima, T.Okada, K.Miyata and H.Yamaoka, "Radiation Damage of Composite Materials at Cryogenic Temperatures", Adv. Cryog. Eng., 34, 35-42 (1988).

CRYOGENIC AND RADIATION RESISTANT PROPERTIES OF
THREE DIMENSIONAL FABRIC REINFORCED COMPOSITE MATERIALS

J.Yasuda, T.Hirokawa, T.Uemura, Y.Iwasaki,
S.Nishijima*,T.Okada*,H.Okuyama*,Y.A.Wang*

Shikishima Canvas Co.,Ltd., Miyauchicho,
Omihachiman, Shiga, Japan
*ISIR Osaka University, Ibaraki, Osaka, Japan

ABSTRACT- The insulating and/or structural materials for the fusion superconducting magnets are used under such strict environments as the cryogenic temperatures, high stresses and radiation environments. It is recognized that the usual laminated composite materials reinforced by glass clothes (2D-GFRP) are difficult to be used in such strict conditions. The three dimensional glass fabric reinforced composite materials (3D-GFRP) have high interlaminar shear strength due to the fibers in thickness direction. The cryogenic and radiation resistant properties of 3D-GFRP had been measured and the results were compared with those of the 2D-GFRP. The thermal contraction of 3D-GFRP in the thickness direction was smaller than that of 2D-GFRP at any temperature. The interlaminar shear strength and the nominal flexural strength of the 2D-GFRP considerably decrease when the 2D-GFRP is irradiated. Concerning the 3D-GFRP ,the interlaminar shear strength did not decrease and the nominal flexural strength showed little degradation. It was confirmed that the 3D-GFRP had high radiation resistant properties compared with the 2D-GFRP. It is concluded that the 3D-GFRP is suitable for the insulating and/or structural materials for fusion magnet.

INTRODUCTION

The organic composite materials reinforced by glass fibers (GFRP) show good thermal, electrical insulating and nonmagnetic properties at cryogenic temperatures.[1] These materials are promising as insulating and structural materials for cryogenic apparatus.[2]
The insulating materials for the fusion superconducting magnet, however, are required to show the excellent properties under the cryogenic temperatures, high stresses and radiation environments. The usual laminated composite materials [3][4] which are reinforced unidirectionally or by glass clothes (2D-GFRP) have some demerits in the practical application such as low strength, low stiffness and large thermal contraction in the thickness direction. Furthermore the irradiated 2D-GFRP comes to loose the adhesive strength of the matrix and to degrade mechanical properties.[5][6] The advanced composite materials, therefore, demerits of which are improved, should be developed.
It is expected that three dimensional glass fabric reinforced composite materials (3D-GFRP) have high shear strength and low thermal contraction at cryogenic temperatures. It has been known that 3D-GFRP has large shear strength due to the fibers in thickness direction and shows excellent properties compared with laminated composites [7]. The properties of these composites under the cryogenic temperatures and radiation environments have not been studied in detail.
In this paper, cryogenic and radiation resistant poperties of 3D-GFRP have been measured. In regard to the cryogenic properties, the thermal

contraction and compressive strength were measured at room temperature (RT) and liquid nitrogen temperature (LNT). The radiation resistant properties were evaluated by means of the interlaminar shear strength and flexural strength.

It is confirmed that the 3D-GFRP are promissing materials under these strict environments.

EXPERIMENTALS

Thermal Contraction Measurement and Compressive Tests

The specimens used in the tests were 4 types of FRP. Table 1 shows the specifications of the specimen. The reinforcement fibers and the resin-matrices used here were E-glass and epoxy, respectively. The 2D-GFRP was reinforced by plain woven fabrics. Two types of 3D-GFRP were fabricated and compared with G10CR and G11CR laminates produced by Spaulding Fibre Company, Inc,. Two kinds of sizing materials were used. These specimens were prepared by compression moulding method.

The thermal contraction [1][8] in the thickness direction was measured down to LNT.

The compressive strengths were measured in the thickness direction. The test speed was set at 1mm/min. The size of specimens was 5mm x 5mm.

Table 1   Specifications of specimens for thermal contraction measurement and compression tests

Name		PG3-1	PG4-1	G10CR	G11CR
Thickness (mm)		1.0	1.0	1.6	1.6
Volume fraction	X(%)	31	45	–	–
of glass fibers	Y(%)	53	51	–	–
	Z(%)	16	4	–	–
Glass content by volume	Vf(%)	54	57	46	54
Type		3D-GFRP	3D-GFRP	2D-GFRP	2D-GFRP

Radiation resistant tests

The specifications of examined specimens are shown in Table 2. The E-glass fiber and the epoxy resin were used. All of the FRP used were moulded under the identical conditions.

The neutron irradiation was performed at 20K in Kyoto University Research Reactor Institute and the mechanical tests were carried out at liquid nitrogen temperature without increasing the temperature. The tests made in this work were the interlaminar shear tests and the flexural tests. The interlaminar shear tests were performed in the Guillotine tests. The shape of the specimen is shown in Figure 1. The specimen for the flexural test was 40mm x 5mm x 4mm and the span length was 35mm. The test speed was set at 1mm/min..

Table 2 Specifications of specimens
for radiation resistant tests

Name		H2-2	P27
Volume fraction	X(%)	30	57
of glass fibers	Y(%)	60	43
in each direction	Z(%)	10	–
Glass content by volume	Vf(%)	50	47
Type		3D–GFRP	2D–GFRP

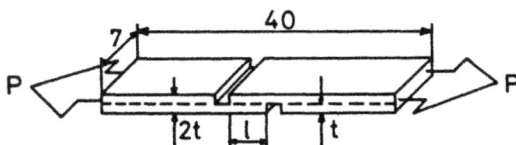

Fig.1 Guillotione test specimen
used in interlaminar shear test

RESULTS AND DISCUSSION

Thermal Contraction

Figure 2 shows the thermal contraction of 3D-GFRP and 2D-GFRP in the thickness direction. The thermal contraction of 3D-GFRP is smaller than that of 2D-GFRP at any temperature. It can be thought that the fibers in the thickness direction controll the thermal contraction effectively.

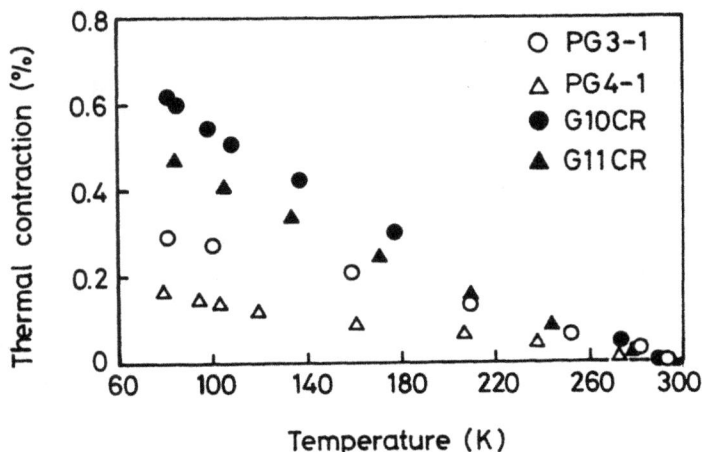

Fig.2 Thermal contraction of 3D–GFRP and 2D–GFRP

Among the 2D-GFRP the thermal contraction comes to be small with increasing the volume fraction of glass fibers as expected. Though the PG4-1 has smaller amount of the fibers in thickness direction than PG3-1, the thermal contraction of PG4-1 in thickness direction is smaller than that of PG3-1. The PG4-1 has almost maximum glass content estimated theoretically and hence the glass fibers in other directions in PG4-1 are thought to contact each other. This is the reason why the PG4-1 shows the smallest thermal contraction among the samples.

Compressive Strength

Table 3 shows the results of compressive strengths obtained at RT and LNT. The compressive strengths per 1 % glass content are also presented. At RT the compressive strength of 3D-GFRP was higher than those of 2D-GFRP. It means that the fibers in thickness direction affect on the compressive strength in thickness direction.

In the 2D-GFRP, G11CR which has higher glass content showed higher compressive strength than G10CR. In the 3D-GFRP, PG3-1 has more fibers in thickness direction than PG4-1. The fibers in the thickness direction affected the compressive strength and hence the compressive strength per 1% glass content of PG3-1 is larger than that of PG4-1. At LNT the compressive strength per 1% glass content comes to be equal. It means that at cryogenic temperatures the fibers in thickness direction does not affect on the compressive strength. The brittleness of the matrix might eliminate the effect of fibers in thickness direction due to the crack initiation at stress concentrated area. Further studies should be made to analyse the phenomenon.

Table 3 Compressive strength

	Name	PG3-1	PG4-1	G10CR	G11CR
R T	Strength (MPa)	677	675	445	545
	Strength per 1% glass content (MPa/%)	12.50	11.80	9.67	10.09
LNT	Strength (MPa)	1220	1146	940	1121
	Strength per 1% glass content (MPa/%)	20.70	20.10	20.40	20.76

Radiation Resistance

Figure 3 presents the change of interlaminar shear strength induced by low temperature reactor irradiation obtained at LNT. The 2D-GFRP shows approximately 85% degradation of interlaminar shear strength at total dose of 15 MGy. On the contrary the 3D-GFRP dose not show degradation up to the identical dose.

Figure 4 shows the change of nominal flexural strength induced by low temperature reactor irradiation. The nominal flexural strength was calculated by simple beam theory without considering the change of fracture mode. Concerning the 2D–GFRP the approximately 70% degradation of nominal flexural strength is brought by the 8 MGy irradiation. On the contrary less than 20% of degradation in 3D–GFRP is induced by 8 MGy.

It was confirmed that the 3D–GFRP had high radiation resistant properties compared with 2D–GFRP.

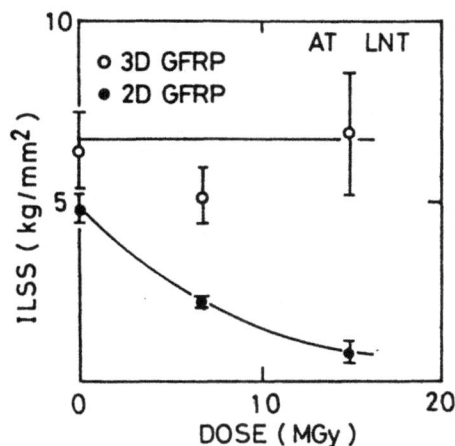

Fig.3   Change of interlaminar shear strength induced by reactor irradiation

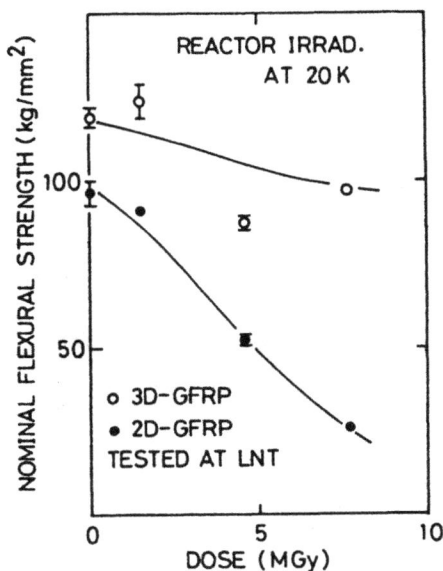

Fig.4 Change of nominal flexural strength induced by reactor irradiation

CONCLUSION

Some important properties of 3D–GFRP were measured aiming at the materials which were used under the strict environments. The following conclusion have been drawn.

(1) The 3D–GFRP have good dimensional stability at cryogenic temperature compared with 2D–GFRP.

(2) The 3D–GFRP shows high compressive strength compared with 2D–GFRP.

(3) The 3D–GFRP can show excellent mechanical properties under radiation environments and hence the 3D–GFRP was confirmed to be radiation resistant material.

REFERENCES

1) M.Takeno, S.Nishijima, T. Okada, K.Fujioka and Y.Kuraoka, "Thermal and mechanical properties of advanced composite materials at low temperature", in Advances in Cryogenic Engineering–Materials, vol.32,

(Plenum Press, New York 1986) pp.217-224.

2) A.Khalil and K.S.Han, "Mechanical and thermal properties of glass fiber-reinforced composites at cryogenic temperatures", in Advance in Cryogenic Engineering-Materials, vol.28, (Plenum Press, New York 1982) pp.243-252.

3) T.Okada, S.Nishijima, H.Yamaoka, K.Miyata, Y.Fujioka and K.Kuraoka, "Mechanical properties of unidirectionally reinforced composite materials", in Advances in Cryogenic Engineering-Materials, vol.32, (Plenum Press, New York 1986) pp.203-208.

4) S.Nishijima, H.Yamaoka, K.Miyata, Y.Tsuchida, K.Mizobuchi and Y.Kuraoka, "Mechanical properties of unidirectionally reinforced materials", in Nonmetallic Materials and Composites at Low Temperature, 3, (1986), pp.127-142.

5) T.Okada, S.Nishijima and H.Yamaoka, "Radiation Damage of Composite Material-Method and Evaluation" , in Advances in Cryogenic Engineering-Materials, vol.32 (Plenum Press, New York,1986) pp.145-151.

6) S.Nishijima and T.Okada, "Radiation Damage of Composite Materials at Cryogenic Temperature", CEC-ICMC 1987 St. Charles Illinois FY-5.

7) J.Yasuda, Y.Noguchi, T.Hirokawa and T.Tanamura, "Static and flexural fatigue strength of carbon fiber 3D fabric composites", in Reinforced Plastics, vol.34, No.1 (1988) pp.10-16

8) S.Nakahara, T.Fujita, K.Sukihara, S.Nishijima, M.Takeno, and T.Okada, "Two-dimensional thermal contraction of composite", in Advances in Cryogenic Engineering-Materials, vol.32, (Plenum Press, New York 1986) pp.209-215.

SELF-COMPRESSION OF THE MULTILAYER INSULATION
AROUND THE HORIZONTAL CYLINDRICAL SHIELD

Takao Ohmori*1, Ralph C. Niemann

Fermi National Accelerator Laboratory,
P.O.Box 500, Batavia, IL, 60510, USA*2

Abstract - Multilayer insulation (MLI) installed around the horizontal
cylindrical shield in the superconducting magnet cryostat is compressed at
the upper part of the shield due to the MLI weight. Pressure distribution in
the MLI is analyzed around the shield, and is related to the shield size and
the specific weight of the MLI. The cross section of MLI was examined with a
single plastic film around a horizontal cylinder and compared to the result of
the analysis. Thermal performance of four kinds of MLI are related to the
normalized compressive pressure by the specific weight of the MLI to discuss
the degradation of the performance by compression.

Introduction

Multilayer insulation in the superconducting magnet cryostat applied to
the high energy particle accelerator is usually installed around the
cylindrical thermal shield which is supported horizontally. The MLI is
compressed more due to MLI weight at the upper part of the shield and has
distributed layer density around the shield. The thermal performance of MLI
was tested by G. R. Cunnington[1] and N. Inai[2] with flat plate calorimeter and
reported that the heat flux through the MLI was increased by compression.
In the superconducting dipole magnets for the proposed Superconducting
Super Collider (SSC), the MLI blankets are installed around the very long
cylindrical thermal shields whose axis measures about 17 m. The degradation
of the MLI performance by the self-compression must be examined and compared
to the data obtained by the calorimeter of laboratory scale. These studies
are essential to design the SSC.
Detector solenoid magnets[3], for instance, TOPAZ in TRISTAN and CDF in
TeVatron, have so large diameter that the total weight of the MLI is huge. Is
the thermal performance of the MLI in such a case degraded more? It is the
intent of this paper to discuss how much the MLI is compressed by the weight
and how the compression affects the MLI performance.

Pressure Distribution Analysis

The object of this section is to derive the equations which describe the
pressure $p(\theta)$ and the tension $T(\theta)$ acting on the MLI around the horizontal
cylindrical shield (Figure 1). The azimuthal angle θ is measured from the top

*1 Visiting scientist from IHI Co., Ltd. 1 Shinnakahara-cho, Isogo-ku,
Yokohama, 235, Japan
*2 Operated by Universities Research Assn. Inc., under contract with the U.S.
Department of Energy

of the cylinder whose radius is r. In this analysis the MLI blanket is considered to be a single film which has the equivalent specific weight w (kg/m^2) of the MLI system and whose thickness is negligible. The friction between the blanket and the shield is also neglected. As the circumferential length S of the MLI blanket is practically longer than that of the cylinder (S>2πr), the blanket departs from the surface of the cylinder at the departure angle θ_d. The fabricated MLI blanket can be divided to two regions.

Region A; Upper part of the MLI blanket $(0<|\theta|<\theta_d)$
Region B; Lower part of the MLI blanket $(\theta_d<|\theta|<\pi)$

Balance of forces in Region A

Figure 1 shows the balance of forces acting on the circumferential element rdθ in Region A. We obtain the equations

$$T^*(\theta)\cos\frac{d\theta}{2} = \sin(\theta + \frac{d\theta}{2})d\theta + T^*(\theta + d\theta)\cos\frac{d\theta}{2} \tag{1}$$

$$P^*(\theta)d\theta = \cos(\theta + \frac{d\theta}{2})d\theta + T^*(\theta + d\theta)\sin\frac{d\theta}{2} + T^*(\theta)\sin\frac{d\theta}{2} \tag{2}$$

where $T^*(\theta) = T(\theta)/wr$, $P^*(\theta) = P(\theta)/w$. From eq. (1) and eq. (2), we obtain

$$T^*(\theta) - T^*(\theta_d) = \cos\theta - \cos\theta_d \quad, \tag{3}$$

$$P^*(\theta) = T^*(\theta) + \cos\theta \quad. \tag{4}$$

Balance of forces in Region B

Figure 1 shows the balance of forces acting on the circumferential element ds in Region B. We obtain the equations for tension T(x)

$$T(x + dx)\cos\varphi(x + dx) = T(x)\cos\varphi(x), \tag{5}$$

$$T(x + dx)\sin\varphi(x + dx) = T(x)\sin\varphi(x) + wds, \tag{6}$$

where ds = $\sqrt{1 + y'^2}$dx, y'(x) = tanφ(x) .
From eq. (5) and eq. (6), we obtain

$$T(x)\cos\varphi(x) = C \quad, \tag{7}$$

$$y''(x) = \frac{W}{C}\frac{ds}{dx} \quad. \tag{8}$$

Equation (8) gives the solution which describes the shape of catenary[4)]

$$y(x) = r\cos\theta_d + \frac{C}{W}\left\{\cosh(\frac{W}{C}x) - \cosh(\frac{W}{C}r\sin\theta_d)\right\}. \tag{9}$$

Boundary Condition

At the departure point P(rsinθ_d, rcosθ_d), the boundary condition for f(x) is given by y'(rsinθ_d) = -tanθ_d, hence the constant C is obtained

$$\frac{C}{wr} = -\cos\theta_d F(\theta_d), \quad F(\theta_d) = \frac{-\tan\theta_d}{\ln\{-\tan\theta_d + \sqrt{\tan^2\theta_d + 1}\}} \quad. \tag{10}$$

The Pressure Distribution of MLI in Region A

(7) and eq. (9) to eq. (3) and eq. (4), we obtain

$$P^*(\theta) = P^*_{max} - 2(1-\cos\theta),\qquad(11)$$

where the maximum pressure P^*_{max} is obtained at $\theta = 0$.

$$P^*_{max} = 2 - \cos\theta_d + F(\theta_d)\qquad(12)$$

Mean value of normalized pressure for $\theta_d = \pi$ is

$$P^*_{mean}(\theta_d = \pi) = \frac{1}{2\pi r}\int_{-\pi}^{\pi}P^*(\theta)rd\theta = 2.\qquad(13)$$

Circumference of the MLI

From eq. (8), we obtain the blanket length S_B in Region B

$$S_B = 2r\sin\theta_d F(\theta_d).\qquad(14)$$

The normalized circumference S^* and the excess circumference ratio Se are

$$S^* = \frac{S}{2\pi r} = \frac{\theta_d + F(\theta_d)\sin\theta_d}{\pi},\qquad(15)$$

$$Se = S^* - 1,\qquad(16)$$

where $S = S_A + S_B$, $S_A = 2r\theta_d$.
The vertical component of compressive force acting on MLI in Region A is integrated,

$$\int_{-\theta_d}^{\theta_d}p(\theta)\cos\theta rd\theta = 2wr(\theta_d + F(\theta_d)\sin\theta_d) = wS.\qquad(17)$$

Sag of the MLI

At the bottom of the shield, the sag H is defined as

$$H = -r-y(0).\qquad(18)$$

From eq. (10) and eq. (12), we obtain

$$y(0) = r\{\cos\theta_d - F(\theta_d)(1 + \cos\theta_d)\}.\qquad(19)$$

Combining eq. (18) and eq. (19), the normalized sag H^* (=H/r) is

$$H^* = (1 + \cos\theta_d)\{F(\theta_d)-1\}.\qquad(20)$$

Measurement of Departure Angle and Sag

In order to simulate the MLI on the shield, three kinds of single mylar films (Sample 1; w=0.171Pa, t=12μm^t, Sample 2; w=0.355Pa, t=25μm^t, Sample 3; w=1.42Pa, t=100μm^t) were wound around the horizontal aluminum cylinder which is 780 mm in diameter and 900 mm in length. The width of the film is 300 mm for each sample. For different excess circumference ratio S_e ranging from 10^{-3} to 0.2, the departure angle θ_d and the sag H^* were measured and are plotted in Fig. 2 and Fig. 3 to compare to the analytical value given by eq. (17) and eq. (21) respectively. The test results show good agreement with the analysis.

Results and Discussions

Maximum pressure P^*_{max} versus excess circumference ratio Se is plotted in Fig. 4. $P^*(\theta)$ and P^*_{max} do not depend on the size of the cylinder r for any MLI with different specific weight w. P^*_{max} does not exceed 4 for Se smaller than 0.12 and is almost 4 in this range. Fig. 4 shows that P^*_{max} exhibits no strong dependency to Se. Therefore, the MLI can be fabricated to keep P^*_{max} smaller than 4 without severe tolerance in the circumference of MLI.

The SSC magnet cryostat employs two thermal shields, at different temperature levels, whose diameter is about 500 mm. In the case of the MLI practically applied to the cryostat, excess circumference ratio is expected to stay around 0.03. Therefore, the circumference of the MLI for SSC cryostat is about 50 mm longer than that of the shield. The thermal shields in the SSC cryostat are, exactly saying, not round cylinders, but are flattened at the bottom. The azimuthal angles of the flattened part for each shield are 130° and larger than the departure angle 120° of MLIs.

Thermal Performance of MLI

As P^*_{max} is found to be smaller than 4, it is important to examine the heat flux through the MLI when the applied compressive pressure is 4w. G.R. Cunnington studied the heat flux through the MLI shown in Fig. 5 as a function of applied compressive pressure. These data were obtained in flat plate calorimeter for four types of MLI (listed in Table 1) between 278K and 77K, and also between 278K and 20K for each specimen. Normalizing the applied pressure by the specific weight of the specimen, heat flux versus P^* is plotted in Fig. 6. At compressive pressure value below $P^*=35$, the crinkled insulation exhibits the least increase of heat flux with compressive pressure. The heat flux ratio q_4/q_1 for $P^*=4$ and $P^*=1$ are listed in Table 1. The ratio for Sample 3 is expected to be 1.3. The heat flux q_1 is estimated by extrapolation of the q-P^* curve in Fig. 6.

Conclusion

Compression of MLI on the horizontal cylindrical shield due to the weight of MLI was analyzed. The maximum compressive pressure is smaller than 4w and does not depend on the size of the shield. The departure angle and the sag of single mylar film on a horizontal cylinder were measured to simulate the MLI and agree well to the analyzed value. Thermal performance data obtained by G.R. Cunnington are examined by the normalized compressive pressure. The degradation in thermal performance by the compression of 4w is estimated to be 30% to 120% for four MLI samples.

Acknowledgement

The authors wish to thank to Prof. H. Hirabayashi of KEK, Prof. K. Hijikata of Tokyo Institute of Technology and J.D. Gonczy of Fermilab for their helpful comments on this study. We also appreciate Dr. N. Inai of Toshiba Corp. for showing his data of MLI and valuable discussions and comments.

References

1) G.R. Cunnington, et al, NASA-CR 72605 (1971-4) p5-5.
2) N. Inai, Japan Soc. Mech. Eng., vol 44, no 385 (1978-9), p3116.
3) H. Hirabayashi, IEEE Trans. on Magnetics, vol. 24, No. 2, March 1988 p1256-p1259.
4) T. Hayashi and T. Mura, "Calculas of Variations", Corona Publishing Co., Ltd., Tokyo Japan (172), p36.
5) SSC Central Design Group, SSC-SR-2020, March 1986.

Table 1, Multilayer Insulation Specimens Tested by G.R. Cunnington, et al[1]

Sample	Reflective films and spacers	Specific weight, Pa	q_4/q_1
1	10 DAMs + 22 Silk Nets	2.07	1.5
2	10 DGMs + 22 Silk Nets	2.62	1.8
3	10 crinkled SAMs	0.83	1.3
4	10 DAMs + 11 Tissuglass	1.17	2.2

DAM; double aluminized mylar, DGM; double goldized mylar, SAM; single aluminized mylar

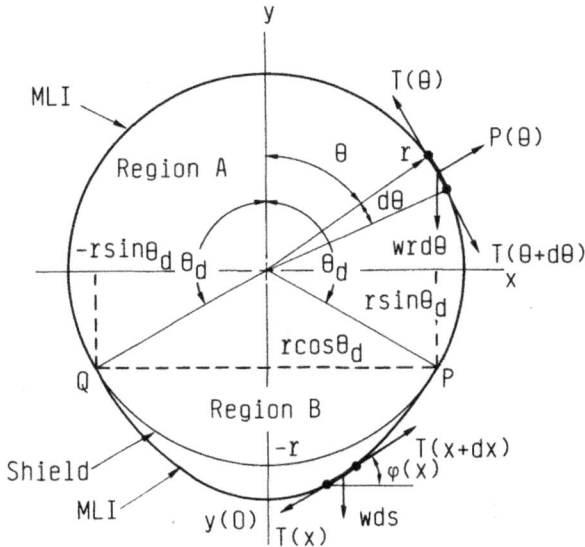

Fig. 1 Multilayer Insulation (MLI) around the Horizontal Cylindrical Shield

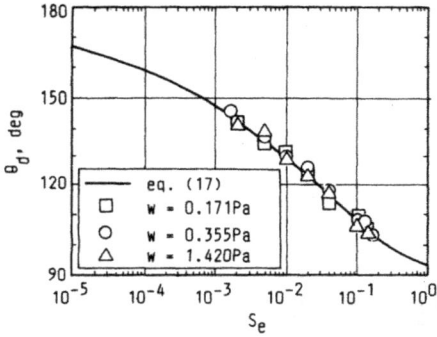

Fig. 2 Departure Angle of MLI

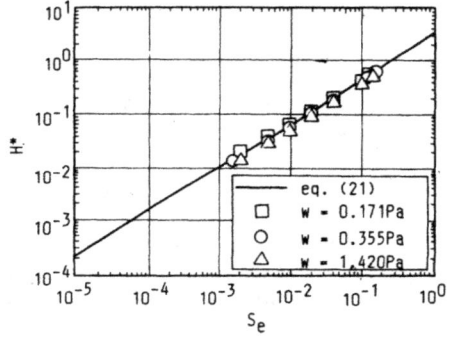

Fig. 3 Sag of MLI in Region B

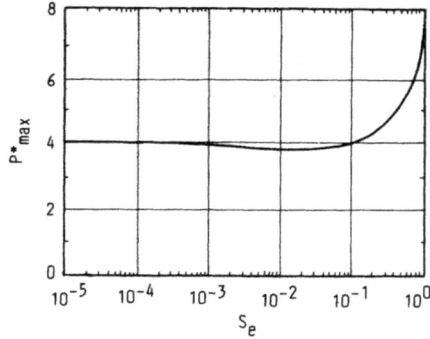

Fig. 4 Normalized Maximum Pressure in Region A

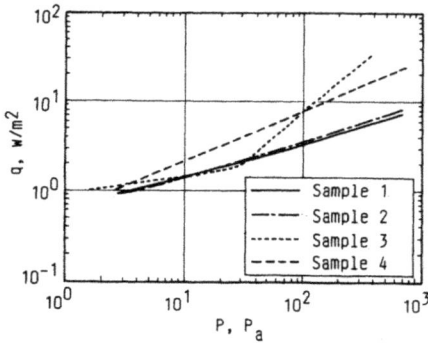

Fig. 5 Heat Flux through MLI Specimens as A Function of Applied Compressive Pressure[1]

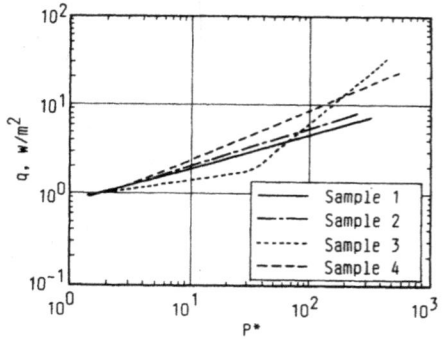

Fig. 6 Heat Flux through MLI Specimens as A Function of Normalized Compressive Pressure[1]

LOW-TEMPERATURE PROPERTIES OF HIGH PURITY COPPER

Kazuyoshi OHTA, Norio YAMAMOTO, Kazuhiko MOTOBA, Harumichi OKAMOTO

Central Research laboratory, Nippon Mining Co.,LTD
3-17-35 Nizominami, Toda-shi, Saitama, 335 Japan

Abstract - Electrical resistivity and thermal conductivity of a poly-crystalline high purity copper (HPC) were measured at various temperatures between 4.2K and 100K. The maximum value of the residual resistivity ratio (RRR) obtained was 9400 and the peak value of thermal conductivity was 302 W/cm·K at about 8K for some specimens of the high purity copper.

Introduction

The transport properties of metals at low temperature are controlled by the scattering of electrons caused by physical and chemical imperfections and by low energy phonons. Because of the lower impurities less than generally-used Oxygen Free High Conductivity Copper (OFHC) , HPC would have excellent transport properties at low temperatures.

Investigations on the change of electrical resistivity for polycrystalline HPC wire between 4.2K and 100K showed success in revealing some excellent properties when it was used at low temperatures.

A mean free path of conducting electrons in the HPC at 4.2K was determined by the experiment of the size effect due to scattering the electrons at the surface.

Thermal conductivity measurement of the HPC at low temperature was carried out at the TOKYO INSTITUTE OF TECHNOLOGY [1]. The results is also shown in this paper.

EXPERIMENTAL

The sample wire used in this experiment was made from high purity copper (HPC) ingot which was cast by Nippon Mining Co., LTD, mechanically treated, and then it was wound carefully around aceramics bobbin which was made from 4N almina to suppress the outgassing during the annealing. finally, the annealing was applied to these samples. Annealing conditions having a direct or indirect influence on the resistivity of copper at low temperature are (1) duration, (2) temperature and (3) cooling rate after annealing.

The annealing recognizes the crystal lattice and attenuates its physical defects. In paticular, it makes possible to reduce considerably the defects introduced through the mechanical treatment which the sample has been undergone.

In this experiment, all samples were annealed at 750°C for 4 hours under vacuum of about 10^{6} torr and cooled at the rate of 6°C per minutes.

The resistivity measurements were carried out with the four-point method. Avoltage tap was soldered to the HPC samples with commercial solder.

RESULTS

Resistivity measurement - The measured resistivity ρ is generally a function of temperature, and on approaching the absolute zero temperature it becomes to show aconstant residual value, well-known as the residual resistivity ρ_0. The quantity of ρ_0 arise from the presence of impurities, defect, and strain in the crystal lattice.

Figure 1 shows the results of resistivity measurement as a function of temperature. The experiments were performed for the two lots of HPC samples which contains different impurities, and the commercial pure copper (OFHC) sample. The contents of the impurities in these samples are listed in Table 1

The residual resistivity ratio (RRR) obtained these results are listed in Table 2 were 7400, 5700 and 105 for the samples of HPC(A), HPC(B) and OFHC, respectively.

The resistivity as a function of temperature, which arose in scattering the electrons by the thermal vibrations of the lattice, is given by the well-known Gruneisen-Bloch fomula for the resistivity of ideal metal, namely

$$\rho f(t) \backsim (T/\Theta)^5 \int_0^{\Theta/X} \frac{X^5}{(\exp(X)-1)(1-\exp(-X))} dX \qquad (1)$$

Where, T=absolute temperature, Θ=Debye temperature of the lattice and X= argument of the integrand ($=\hbar\nu/kT$). The total resistivity, ρ is devided into the two terms: temperature-independent term ρ_0 and temperature dependent term $\rho f(T)$. The following expression is well-known as Matthissen´s rule

$$\rho = \rho_0 + \rho f(T) \qquad (2)$$

In this paper, $\rho f(T)$ is apporoximately defined from expression (2).

Figure 2 shows the comparison between the temperature-dependent resistivity measurements and the theoretical curve(from equation (1): solid line) for the HPC(A) specimen.

Log $\rho f(T)$ vs log t plots are shown in Fig.3. The curve is separated into the two linear regions with a discontinuity in the slope at about 10K. Points from the linear region with above 10K were fitted by the least mean squares by the equation

$$\rho f(T) = A \times T^n \qquad \text{for } 10K < T < 60K \qquad (3)$$

The value of the constants, A and n, were listed in Table 3.

Size effect - Size effect, which may be very marked at low temperature in high purity metals, appear when the mean free path of electron "1" becomes of the same order of magnitude as the sample´s smallest dimmension "d".

The electrons then undergone additional interaction with the surface of the conductor, resulting in an incresse in resistivity.

A theoretical study on thin cylindrical wire was conducted by Nordheim[2] who derived the relationship (see ref.3):

$$\rho(d) = \rho(\infty) \times (1+\alpha l/d) \qquad (4)$$

Where, $\rho(d)$ is the resistivity of the wire of diameter "d", $\rho(\infty)$ the resistivity of sample of infinite medium and α a coefficient depending both on the type of reflection undergone by the electron and on the temperature. In practice, in most case $\alpha=1$ (case of diffuse reflection).

It should be noted that Noldheim´s relationship confirms in every case, to within 5% , the regorous expression of Ziman[4], if we take α=1. Thus, in practice, use will be made of the simple relationship which is sufficiently precise from the technical standpoint :

$$\rho(d)=\rho(\infty)+\rho(\infty)\times l/d \qquad\qquad (5)$$

The product $\rho(\infty)\times l$ is a characteristic parameter of the material and if the plot the curve $\rho(d)=f(d^{-1})$, we will obtain the straight line with slope of the value of $\rho(\infty)\times l$ intersecting the ρ-axis at the point $\rho(\infty)$.

The Table 4 give the result of measurements carried out on HPC wires having different diameters which are obtained by electrochemical polishing. The max. and min. diameter for each polished wire sample are also indicated in Table 4.   The average diameter corresponds to the arithmetical average value of three points of measurements made at points spaced regularly over the entire length of the sample.
The straight line of figure 1 plotted from Table 1 gives an average characteristic product $\rho(\infty)\times l=7.25\times10^{-10}$ Ωcm^2 and an average $\rho(\infty)=1.25\times10^{-10}$ Ωcm.   Therefore, the mean free path "l" and bulk sample resistivity $\rho(\infty)$ are determined from this method. In this way we obtained:

$$\rho(\infty)=1.25\times10^{-10}\ \Omega cm$$

$$l=0.58\ mm$$

giving a characteristic product:

$$\rho(\infty)\times l=7.25\times10^{-10}\ \Omega cm^2$$

Another property of HPC - Information on the thermal property of HPC at low temperature is important.   The measurements of the thermal conductivity of HPC were carried out at the TOKYO INSTITUTE of TECHNOLOGY[1], and Fig.5 shows the results of thermal conductivity is about 302 W/cm K at about 8K.

References

1) R.Li and T.Hashimoto : Thermal conductivity and electrical resistivity of very high-purity copper at low temperature. 13th ICEC (1988)

2) L.Nordheim : Act.Sient. et Ind.No.131,Paris (1934)

3) Electrical Properties of Copper at Low Temperature : INCRA Project No,114b

4) J.M.Ziman : Electrons and phonons. Oxford at The Clarendon Press,(1979) p463-p468

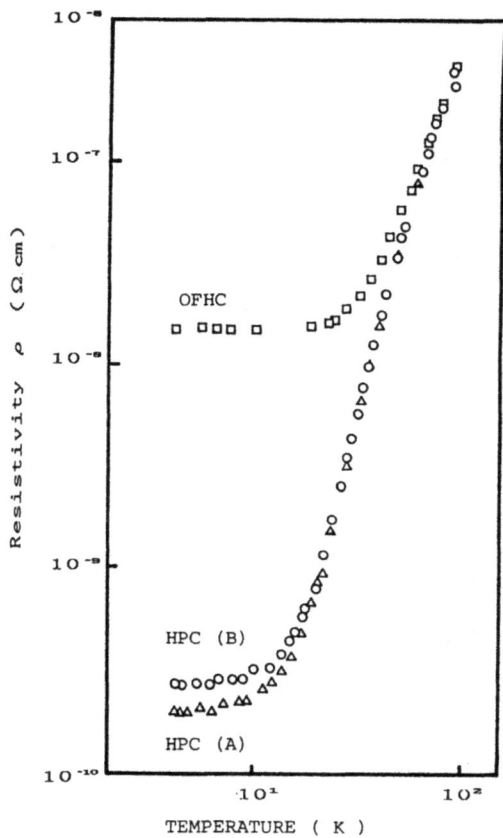

Fig.1 Resistivity variance as afunction
of temperature.

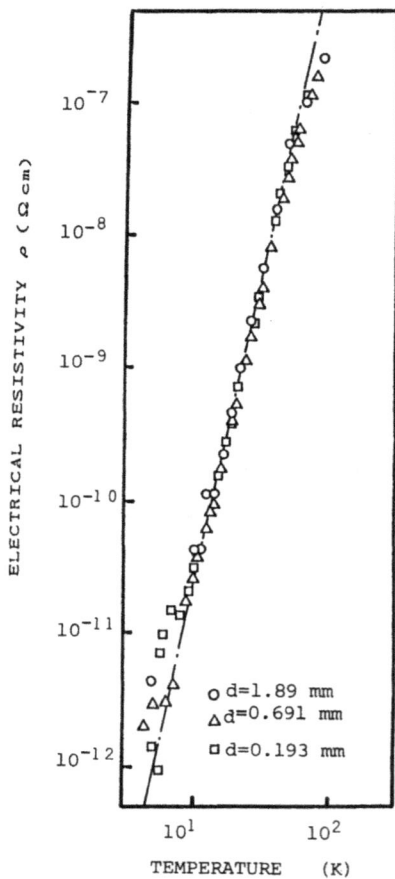

Fig.2 Temperature-dependent resistivity of HPC(A) below
100 K.

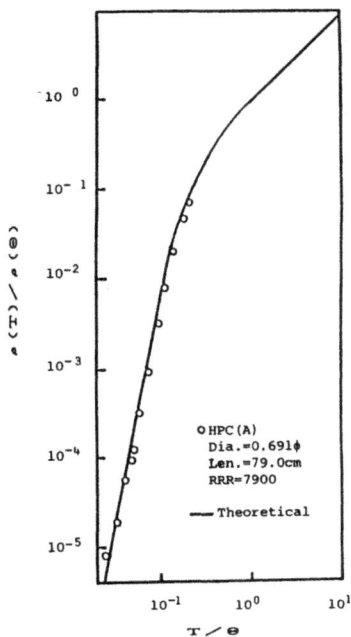

Fig.3 Comparison of the theoretical curve
with experimental data for the temperature-dependent resistivity.

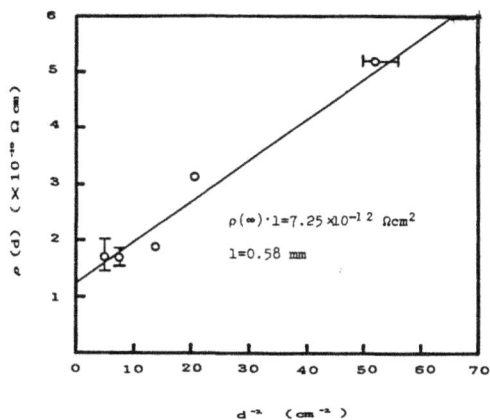

Fig.4 Electrical resistivity VS d^{-1}
plots.

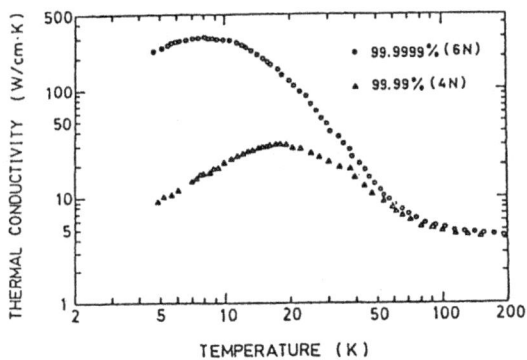

Fig.5 Thermal conductivity VS temperature curve.

Table 1 Impurities in specimens used. (ppm)

Specimen	Fe	S	Ag	Na	Ca	Cr	Ni	Mg	Al	Sb
HPC (A)	<0.05	0.06	0.10	<0.02	<0.04	<0.05	<0.07	<0.02	<0.05	-
HPC (B)	<0.05	<0.05	0.20	<0.02	<0.04	<0.05	<0.07	<0.02	<0.05	0.08
OFHC	3.0	5.9	14.1	<0.02	0.05	0.09	1.1	0.04	<0.05	0.25

Table 2 Value of diameter, residual resistivity and RRR of the indicated specimens.

Specimen	Diameter d (mm)	Residual resistivity ρ_0 ($\times 10^{-10}$ Ωcm)	RRR $\dfrac{\rho_{273\ K}}{\rho_{4.2\ K}}$
HPC (A)	1.023	2.0	7400
HPC (B)	1.029	2.7	5700
OFHC	0.917	150.6	105

Table 3 Values of the constant A and n in the equation(3), obtained from the experiment for the HPC(A) specimen.

Diameter d(mm)	$A \times 10^{-16}$	n
0.193	3.42	4.73±0.05
0.691	8.11	4.50±0.01
1.89	5.68	4.63±0.87

Table 4 Values of the minimum and maximum diameter. RRR and the residual resistivity for the HPC(A) specimens with the various diameters

Diameter d (mm)	d_{min} ~ d_{max} (mm)	$\dfrac{R_{273\ K}}{R_{4.2\ K}}$	Residual Resistivity $\rho(d)$ ($\times 10^{-10}$ Ωcm) at 4.2 K
0.193	0.178-0.201	2880	5.190
0.482	0.475-0.496	5760	3.128
0.691	0.685-0.695	7890	1.851
1.02	1.000-1.048	7340	1.60
1.89	—	9400	1.57

GFRP-STAINLESS STEEL HYBRID CRYOSTAT

T. Mori, K. Ikizawa, N. Takasu
Tsurumi Research Labs. NKK, Yokohama, Kanagawa 230

J. Ogata, Y. Yoshinaga
Kawasaki Research Labs. NKK, Kawasaki, Kanagawa 210

N. Yamada, H. Ohta
Advanced Material Div., NKK, Tiyoda, Tokyo 100

ABSTRACT

As an instrument to measure superconducting properties(Jc,Tc,Hc) by the magnetization method, a cryostat containing the magnet that generates an external magnetic field has been developed. To ensure thermal insulation ability and structural durability, this cryostat consists of a GFRP inner vessel and a stainless steel outer vessel. Various tests were carried out to verify the sufficient performance of this cryostat.

Introduction

The properties of a superconducting material are represented by critical current density (Jc), critical temperature (Tc), and critical magnetic field (Hc). These properties can be measured simultaneously by the use of the magnetization method. In this method, an external magnetic field, generated by a magnet, acts on a superconducting material. The intensity of a magnetic field caused to act can be generated, up to 2T, by the use of a liquid nitrogen cooled cooper coil, and, up to 8T, by a liquid helium immersed NbTi coil. In either case, the cryostat keeping the magnet in a refrigerant is necessary. Conventional cryostats have both the inner and outer vessels made of either metal or FRP, the former is excelling in structural durability and the latter in thermal insulation ability. In the measurement of superconducting properties, it is preferable that a cryostat has both structural durability and thermal insulation ability. For these reasons, a cryostat consisting of a GFRP inner vessel and a stainless steel outer vessel was fabricated, and various performance tests were carried out to develop a cryostat of hybrid construction.

Configuration of Cryostat

Construction

Fig.1 shows the construction of the cryostat. The inner vessel uses GFRP and the outer vessel and flange use 304 stainless steel. The inner vessel and the flange are bonded with an adhesive while the flange and the outer vessel, are bolted together. The space between the inner and outer vessels is filled with multilayer insulations.

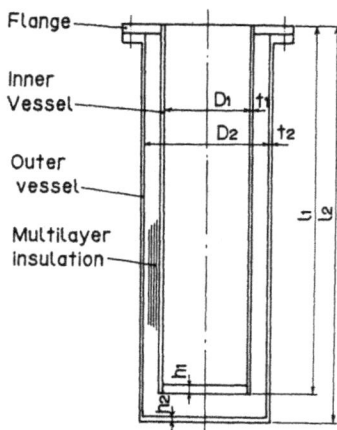

Fig.1. Schematic illustration of cryostat.

GFRP Inner Vessel

GFRP is expected to be resistant against low temperatures. The GFRP consisting of epoxy resin and E-glass fiber was used. As shown in Fig.1, the inner vessel consists of a cylinder and a bottom plate, both are bonded with an epoxy based adhesive. The cylinder is formed by the filament winding method, and the bottom plate is made by low pressure molding method. The thermal shock test of the inner vessel was conducted by injecting liquid nitrogen in it, and there were no failure. It was confirmed that they could be used at low temperature.

Filament winding angle

The mechanical and thermal properties of the GFRP pipe varies with winding angles. Fig.2 shows the winding angle and the fiber direction. Residual thermal stresses, σ_T and τ_{LT}, can be derived from the equations that follow:[1]

$$\sigma_T = 4l^2m^2(1-\nu_L\nu_T)G_{LT}E_L'E_T'(\alpha_L-\alpha_T)T/\Delta$$

$$\tau_{LT} = -2lm(l^2-m^2)(1-\nu_L\nu_T)G_{LT}E_L'E_T'(\alpha_L-\alpha_T)T/\Delta$$

$$\Delta = (l^2-m^2)^2(1-\nu_L\nu_T)E_L'E_T'+4l^2m^2G_{LT}(E_L'+E_T'+2\nu_TE_L')$$

$$E_L' = E_L/(1-\nu_L\nu_T), \quad E_T' = E_T/(1-\nu_L\nu_T)$$

$$l = \cos\theta, \quad m = \sin\theta$$

$$\tag{1}$$

where
σ_T : tensile stress of T direction
τ_{LT} : shearing stress in LT plane
E_L, E_T : Young's modulus of L and T direction
G_{LT} : modulus of shearing elasticity in LT plane
ν_L, ν_T : Poisson's ratio of L and T direction
T : temperature difference
θ : winding angle

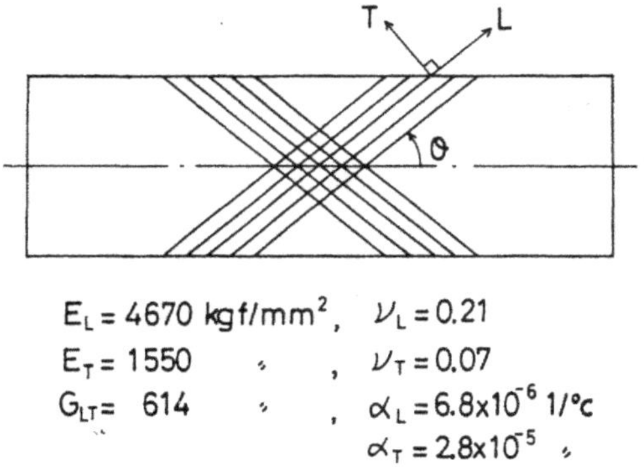
T
L
θ
E_L = 4670 kgf/mm², ν_L = 0.21
E_T = 1550 ″ , ν_T = 0.07
G_LT = 614 ″ , α_L = 6.8x10⁻⁶ 1/°C
α_T = 2.8x10⁻⁵ ″

Fig.2. Winding angle and fiber direction

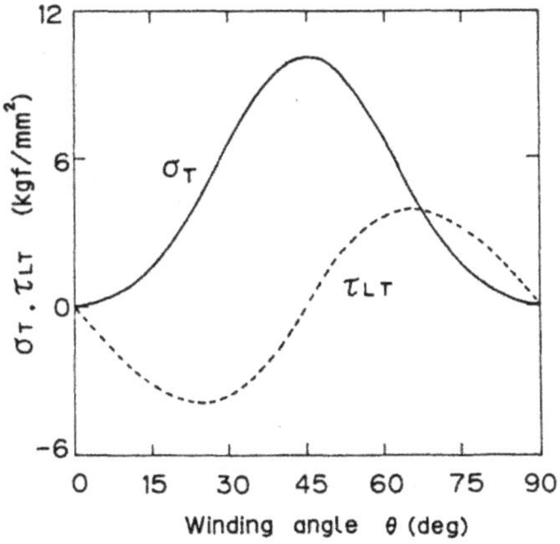
12
6
0
-6
σT
τLT
σT , τLT (kgf/mm²)
0 15 30 45 60 75 90
Winding angle θ (deg)

Fig.3. Relation between thermal stresses and winding angles.

Fig.3 shows the relationship between thermal stresses, generated during cooling, and winding angles when calculating eq.(1) by using the values shown on Fig.2. Temperature difference was assumed to be 400°C which was approximately equal to the difference between the curing temperature and the liquid helium boiling temperature. σ_T becomes maximum at 45° and τ_{LT} at 25° and 65°. Neither of these stresses occur at 0° and 90°. From the viewpoint of the axial and hoop strength, 55° is the degree of angle at which the filament winding of a pipe is generally performed, but 0° and 90° are preferable in point of thermal stress. However, the pipe is not reinforced in the direction intersecting 0° and 90°, and therefore its strength in that direction is smaller. Considering all these factors, the winding angles of 55° and 90° were selected.

Design of cryostat

Structural design[2),3)]

Table 1 shows the dimensions of the cryostat. The symbols in the table correspond to those shown on Fig. 1. The refrigerant is stored in the inner vessel under atmospheric pressure and the space between the inner and outer vessels is highly vacuum. So, the inner vessel is pressed at 1 atmospheric pressure from the inside, and the outer vessel from the outside. Hoop stress σ_t is the maximum stress generating in the cylinder, and assuming that pressure is p, inside diameter is D and plate thickness is t,

$$\sigma_t = pD/2t \tag{2}$$

Assuming that the thickness of the bottom plate is h, its maximum bending stress σ_m will be

$$\sigma_m = 3pD/16h^2 \tag{3}$$

Table 1. Dimensions of cryostat.   Table 2. Values of thermal design
                                                parameters.

D_1	180mm	D_2	261mm
t_1	5mm	t_2	3mm
h_1	20mm	h_2	10mm
l_1	840mm	l_2	900mm

ε	0.13	D	18cm
S	$2.5\times10^2 cm^2$	t	0.5cm
T_b	53K	T_f	300k
T_1	4.2K	l	60cm
Ω	2π	λ'	$2.6\times10^{-7} W/cm^2 K$
$\overline{\lambda}$	$2.6\times10^{-3} W/cmK$	S'	$1.7\times10^4 cm^2$
		T_h	300K

With the use of the values shown on Table 1,
  $\sigma t=0.18kgf/mm^2$, $\sigma m=0.00084kgf/mm^2$ for the inner vessel,
  $\sigma t=0.43kgf/mm^2$, $\sigma m=0.0048kgf/mm^2$ for the outer vessel,
both vessels have sufficient strength. In addition, it is necessary to
consider the bonding strength of the inner vessel and the buckling strength
of the outer vessel. The shearing stress in the bonded portion of the
inner vessel, τa, will be

$$\tau_a = pD/4h$$
$$(4)$$

and $\tau a=0.023kgf/mm^2$. And the buckling pressure of the outer vessel, pc,
will be

$$P_o = 2Et^3/(1-\nu^2)D^3$$
$$(5)$$

where E is elastic modulus and ν is Poisson's ratio. With $E=20000kgf/mm^2$ and
$\nu=0.3$, $pc=0.088kgf/mm^2$. Thus, both the bonding strength and the buckling
strength are sufficient.

Thermal design [2]

   Heat that penetrates into the cryostat can be divided into radiant
heat from the upper flange, Q_1, heat conducting along the wall of the inner
vessel, Q_2, and the radiant and conductive heat through the super
insulation space, Q_3, and each can be derived from the equations that
follow:

$$Q_1 = 5.67\times10^{-12} \varepsilon S(T_b^4-T_1^4) \Omega/2\pi$$
$$(6)$$

$$Q_2 = \overline{\lambda} \pi Dt(T_f-T_1)/l$$
$$(7)$$

$$Q_3 = \lambda'S'(T_h-T_1)$$
$$(8)$$

where ε : total emissivity
       S : surface area of flange
       Tb : temperature of baffle at the lowest end
       Tl : temperature of liquid helium
       Ω : effective solid angle
       $\overline{\lambda}$: average thermal conductivity of inner vessel
       D : radius of the inner vessel
       t : thickness of the inner vessel
       Tt : temperature of flange
       λ' : heat transfer coefficient inner and outer vessels
       S' : inner vessel's surface area being in contact with LHe
       Th : temperature of the outer vessel

The total amount of heat that penetrates into the cryostat will be 1.2W when calculating it by using the values shown on Table 2. The values used here are somewhat larger to allow for a safer evaluation, and therefore the actual total amount of penetrating heat will be considerably smaller.

Experimental fabrication of cryostat

Assembly

The FW(Filament Winding) pipe and the bottom plate are bonded together to form the GFRP inner vessel, and then the inner vessel is bonded to the flange. Subsequently, multilayer insulation, activated charcoal and molecular sheave are attached to the exterior surface of the inner vessel. After assembling the inner vessel and the outer vessel, the space between both vessels is evacuated to make a highly vacuum.

Performance tests

In order to examine the performance of the cryostat, helium leak tests and cold leak tests were carried out , and vacuum seal-off tests were carried out following the completion of the assembly. The results are demonstrated as follow:

Helium leak tests: To inspect that no leak occurs through the inner vessel, the outer vessel and the connections of them, helium leak tests were conducted. Any leak was not found.

Cold leak tests: Before and after liquid nitrogen was injected into the inner vessel, the degree of vacuum was measured. Fig. 4 shows an example of the results. Following the injection of liquid nitrogen, the degree of

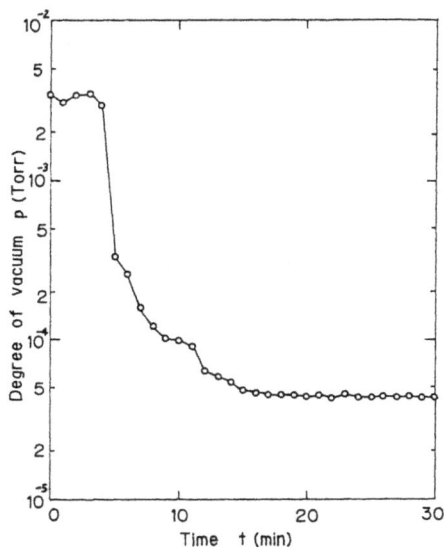

Fig.4. Results of cold leak test.

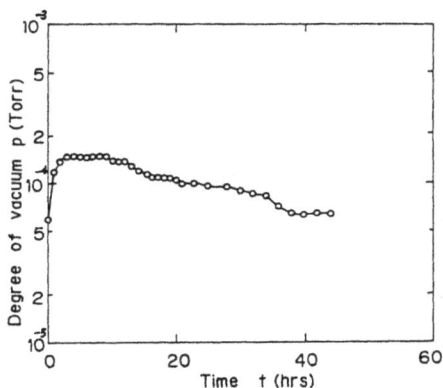

Fig.5. Results of vacuum seal-off test.

vacuum rose by two orders, and did not indicate any drop. Thus, any crack which becomes the cause of cold leak did not occur.

Vacuum seal-off test: After the assemble, liquid nitrogen was injected into the inner vessel and the vacuum exhaust valve was sealed off. Then, the change in the degree of vacuum measured and an example of the results is shown in Fig. 5. No vacuum deterioration is observed.

To ensure that the experimental cryostat has sufficient performance, as described above, the tests using liquid nitrogen were conducted. Any difference due to the winding angles (55° and 90°) of FW pipe was observed in the tests. With liquid nitrogen injected into the inner vessel, the degree of vacuum is an order of 10^{-5}-10^{-6} Torr, and thus the cryostat is considered as applicable to liquid helium. Actually it was confirmed that liquid helium could be stored in the cryostat. This proves that the experimental GFRP-stainless steel hybrid cryostat has sufficient performance for liquid helium. Currently, the boil off rate of liquid helium is being measured to obtain the accurate amount of penetrating heat in the cryostat.

Conclusions

The cryostat required for the measurement of the properties of a superconducting material by the magnetization method has been developed. The developed cryostat is of hybrid construction with its inner vessel made of GFRP, outer vessel made of stainless steel. It is excelling altogether, in thermal insulation ability and structural durability, and it is also useful for other cryogenic tests.

Acknowledgment

Prof. T. Okada and Dr. S. Nishijima, Institute of Science and Industry Research, Osaka University, are greatly acknowledged for their various instructions given in the development of this cryostat.

References

1) M. Uemura, et al.,"Thermal Expansion Coefficients and Residual Stresses in Filament-Winding CFRP Materials", Nippon Kouku Uchu Gakkaisi, vol.26, No.296, 471-479(1978).
2) Y. Kuraoka, et al.,"Development of the Large-Sized GFRP Dewar", Teionkougaku, vol.21, No.1, 44-50(1986).
3) K. Kadotani, "FRP Cryostat of Storing Liquid Helium", Koatugasu, vol.23, No.7, 351-358(1986).

CRYOGENIC MECHANICAL PROPERTIES OF A5083-SUS304L
EXPLOSIVELY BONDED JOINT

Yoshihiko MUKAI and Arata NISHIMURA

Faculty of Engineering. Osaka University
2-1. Yamadaoka Suita. Osaka. 565. Japan

Abstract   Tow types of the transition joint of alminium alloy and low
carbon austenitic stainless steel for cryogenic equipments were manufactured
by an explosively bonding method. and their cryogenic mechanical properties
were investigated. As the results. it was clarified that they had enough
strength even at 4.2K. and strength of these joints at cryogenic temperatures
was affected by bond interface strength together with those of materials used
and an inner calorification or a local reduction of area on bond interface.

Introduction

Many kinds of materials have been used for cryogenic equipments or structures. Although needs for joining of dissimilar metals often occur in some
engineering fields. it is not easy generally to weld dissimilar metals by fusion welding in strong dependence on assemblage of materials. As well-known,
however. the explosively bonding method can make the joining of dissimilar
metals possible. and a joinable combination of metals is very wide. So. this
method is considered to be useful for cryogenic transitionjoint.
    In this report. two types of the explosively bonded joints of A5083 and
SUS304L were manufactured to examine an applicability of the explosively bonding method for cryogenic equipments. One had commercial pure aluminum(A1100).
titanium(TP28C) and nickel(NNCP-0) layers as insert metals between A5083 and
SUS304L (5 layers explosively bonded joint. 5 LEBJ). and the other had Ag
layer only as a insert metal (3 layers explosively bonded joint. 3 LEBJ).
Their cryogenic mechanical properties and fracture modes were investigated.
Also. some factors affecting on their joint strength at cryogenic temperatures
were discussed.

Transition Joints and Experimental Procedure

    Chemical compositions and mechanical properties of materials used are
shown in Table 1 and Table 2. Material constructions and specimen locations in
each explosively bonded plate are described in Fig.1. In order to evaluate the
fracture stress on bond interface. specimens were machined out in the plate-
thickness direction from each bonded plate.
    Configurations of specimens are shown in Fig.2. A1100 or Ag layer was
placed on the center of a parallel part of specimens. Pipe and plate specimens
were prepared together with round bar specimen to investigate the effect of
specimen geometry on joint strength. Besides these specimens. 20mm and 12mm
diameter round bars were also tested [1]. Tensile tests were conducted at
4.2K. 77K and 293K and cross head speed of 8.3×10^{-3} mm/s was adopted usually.
    The transition joint studied here is a under-matching joint which has a
soft layer of A1100 or Ag. So. a parameter defined by following equation is
induced to present the degree of deformation restriction of a soft layer

$$X = h/t \qquad \text{------ (1)}$$

where h is the initial thickness of soft layer. t is the initial radius. wall thickness or thickness for round bar. pipe or plate specimen respectively. and X is called as a relative thickness. Range of X covering here is $0.08{\sim}1.20$ for 5 LEBJ. and $0.13{\sim}2.90$ for 3 LEBJ.

Results and Discussion

Cryogenic Strength of 5 Layers Explosively Bonded Joint

Relation between the tensile strength and the relative thickness in case of 5 LEBJ is shown in Fig.3. Dotted line shows a mean value of the strength of round bar specimens. Solid and open marks indicate a ductile fracture at Al100 and the bond interface fracture. respectively. At 293K and 77K. joint strength goes high according to decrease of X in each type of specimen. but when bond interface fracture occurs. the strength of pipe or plate specimen seems to be lower than that of round bar specimen. This is considered to be caused by the difference of stress condition on bond interface. At 4.2K. X region where bond interface fracture appears becomes wide and typical increment of strength is not observed. Also. since the scatter band is wide. the specimen configuration effect is not recognized clearly.

As the cause of this scatter seems to be a temperature rise resulted from a calorification in Al100 in plastical deformation process [2]. temperature is measured on specimen surface of Al100 by Au·Fe-chromel thermo-couple. Figure 4 shows it's results with stress-displacement curves. In a case that cross head speed is $8.3{\times}10^{-3}$mm/s (Fig.4(a)). there is a rapid temperature rise coinsident with a serration occurence. But when specimens are tested at fast deformation speed $(8.3{\times}10^{-1}$mm/s.Fig.4(b)). the temperature rises abruptly after apparent yielding. and no serration is observed. The maximum temperature measured is about 18K. and this is a good agreement with the fact that there is no serration when tensile test was conducted in 15K He gas environment.

From these results. it is considered that the temperature rise at Al100 caused by a concentration of deformation results in decrease of the bond interface strength in a certain X region. As X becomes very small. however. less than about 1.0. fracture occurs at A5083 which is the next low strength material. because the deformation restriction comes large and Al100 becomes hard to deform plastically .

Cryogenic Strength of 3 Layers Explosively Bonded Joint

Relation between the tensile strength and the relative thickness in case of 3 LEBJ is shown in Fig.5. Dotted line and marks are the same as those in Fig.4. and +mark shows A5083 fracture on Ag/A5083 interface (within 0.5 mm). At 293K. according to decrease of X. the fracture mode switched over from ductile fracture at Ag. SUS304L/Ag interface fracture. ductile fracture at A5083 on Ag/ A5083 interface to ductile fracture at A5083 away from bond interface. At 77K and 4.2K. the fracture at A5083 on bond interface occurs in wide region of X. Especially at 4.2K. joint strength becomes lower than Ag strength once. and it comes up again to A5083 strength in very small X region. And the effect of specimen configuration on the strength is not clearer than that in 5 LEBJ.

To investigate the strength drop at 4.2K. the reduction of area (RA) on Ag/A5083 interface was measured. It's results are shown in Fig.6. Dotted line shows RA of A5083 at each temperature. and the other marks are the same as those in Fig.5. It is realized that RA in which A5083 fracture on bond interface occurs is almost the same as that of A5083. From this result. it is con-

sidered that as restriction of deformation becomes high. RA on Ag/A5083 interface comes large. and when this value reaches to that of A5083. A5083 fracture
on bond interface is caused. But when the degree of restriction becomes higher. joint strength comes up to A5083 strength because of little RA.

Stress Distribution on Bond Interface

Results of stress conditions on bond interface calculated by axisymmetric
elasto-plastic FEM analysis are shown in Fig.7. The material constants shown
in Table 2 were used and the residual stress on explosively bonding and cooling were not considered because both of them had been less than 50 MPa [3].
On A1100/A5083 interface (Fig.7(a)). longitudinal stress (σ_z) in inner
part of round bar is higher than that in circumference. and this is clear especially in case of X=0.4. And on TP28C/A1100 interface (Fig.7(b)). σ_z in circumference is large and the distribution becomes flatter in case of X=0.4. As
shown in Fig.8. it is recognized that A1100/A5083 interface fracture (X=0.5.
Fig.8(a)) initiated on the center of bonded section. and TP28C/A1100 interface
fracture (Fig.8(b)) was observed on specimen surface.
From these results. it is expected that bond interface fracture initiates
on higher σ_z region. and as the distribution of σ_z on both interfaces becomes
flatter in the small X region. a fracture initiation position is hard to be
fixed. therfore. strength of bond interface fracture becomes high.

Cryogenic Strength of Dissimilar Joints Containing the Soft Layer

Schematic illustration of dissimilar joint containing a soft layer is described in Fig.9. This joint is supposed to be constructed by 3 materials. and
it is assumed that Mat.A is rigid. Mat.A/Mat.B interface dose not fracture.
and off-set stress of Mat.B is lower than that of Mat.C. When a static load is
applied on this joint. necking occurs at Mat.B. But it's degree depends on
relative thichness together with mechanical properties of Mat.B and Mat.C.
Four fracture types are observed in this study. The first is fracture of Mat.B
(Fracture section(FS) I). the second is Mat.B/Mat.C interface fracture (FS
II). the third is fracture of Mat.C on Mat.B/Mat.C interface (FS III). and the
forth is fracture at Mat.C away from bond interface (FS IV).
Schematic diagram of these fracture strength against relative thickness
is shown in Fig.10. Fracture strength on each fracture section becomes high
according to reduction of X except for Mat.B/Mat.C interface fracture strength
($\sigma_{gross}^{f.II}$). Since $\sigma_{gross}^{f.II}$ is affected by inner calorification of Mat.B at cryogenic temperatures. there is a region where it comes down. And the smallest
one among these strength for the given relative thickness becomes the joint
strength. Relative positions of these strength curves seem to depend on the
combination of materials. deformation rate and test temperature.

Conclusions

The mechanical and fracture properties of two types of explosively bonded
joint of A5083 and SUS304L were investigated in cryogenic temperatures. and
some factors affecting the joint strength were discussed. Main results are
summarized as follws:
1) Both types of transition joints have enough strength even at cryogenic
temperatures. but there is a region of relative thickness where the strength
becomes lower than that of soft material in each joint. The causes of this decrease were considered to be a inner calorification in A1100 for 5 LEBJ and
a extraordinary reduction of area in A5083 on bond interface for 3 LEBJ.

2) There are considered to be several fracture sections in these under-matching joints having bond interfaces, and the smallest one among the strength of soft layer, bond interface, hard material on bond interface, which depend on the relative thickness, becomes the joint strength.

Aknowledgments

Authors wish to thank members of the Low Temperature Center, Osaka University where the cryogenic mechanical testing has been performed. Thanks also to Asahi Chemical Industry Co.,Ltd for an offer of specimens.

References

1) Y.Mukai and A.Nishimura,"Cryogenic mechanical properties of A5083-SUS304L explosively bonded transition joint",Welding Metallurgy Committee of JWS, WM-1184-87 (1987)
2) Z.S.Basinski,"The instability of plastic flow of metals at very low temper-atures",Proc.Roy.Soc.,240.A,229-242 (1957)
3) Y.Mukai and A.Nishimura,"Estimation of Residual Stress and Thermal Stress at 4.2K of A5083-SUS304L Explosively Bonded Joint",Preprints of National Meeting of JWS,42,258-259 (1988)

Table 1 Chemical compositions of materials used

Material	Chemical composition (wt%)													
	Al	Fe	Cr	Ni	Mg	Si	Cu	Mn	Zn	Ti	Ag	C	S	P
A5083-O*	BAL.	0.21	0.12	-	4.55	0.13	0.03	0.64	0.01	0.01	-	-	-	-
A1100-H112*	99.15	0.58	TR.	-	TR.	0.10	0.15	TR.	TR.	0.02	-	-	-	-
TP28C*	-	0.03	-	-	-	-	-	-	-	BAL.	-	-	-	-
NNCP-O*	-	<0.01	-	99.20	-	0.22	<0.01	0.20	-	-	-	0.06	0.004	-
SUS304L*	-	BAL.	18.27	10.80	-	0.55	-	1.31	-	-	-	0.011	0.003	0.035
A5083-O**	BAL.	0.20	0.13	-	4.92	0.10	0.03	0.72	0.02	0.01	-	-	-	-
Ag**	-	-	-	-	-	-	-	-	-	-	99.99	-	-	-
SUS304L**	-	BAL.	18.19	10.17	-	0.51	-	1.48	-	-	-	0.013	0.002	0.034

* ; 5 layers explosively bonded material   ** ; 3 layers explosively bonded material

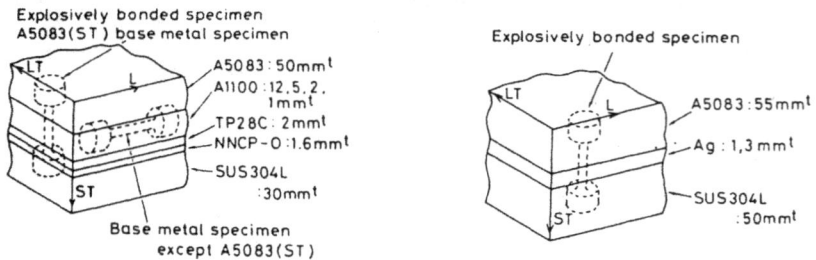

Explosively bonded specimen
A5083(ST) base metal specimen

A5083 : 50mm^t
A1100 : 12.5,2,1mm^t
TP28C : 2mm^t
NNCP-O : 1.6mm^t
SUS304L : 30mm^t

Base metal specimen
except A5083(ST)

Explosively bonded specimen

A5083 : 55mm^t
Ag : 1,3mm^t
SUS304L : 50mm^t

(a) 5 layers explosively bonded joint   (b) 3 layers explosively bonded joint

Fig.1 Specimen location in bonded plate

Table 2 Mechanical properties of materials used

Material	E (GPa)			$\sigma_{0.2}$ (MPa)			σu (MPa) *		
	293K	77K	4.2K	293K	77K	4.2K	293K	77K	4.2K
A5083(L)	71.9	80.6	83.3	249.3	295.1	294.2	347.6	459.0	603.2
A5083(ST)				217.3	253.1	278.2	304.6	402.7	412.0
A1100	69.0	81.7	82.7	120.1	147.8	166.6	134.3	214.7	380.1
Ag	64.5	73.9	75.5	105.9	119.2	122.7	181.5	308.4	396.4
TP28C	101.6	115.8	118.0	253.5	403.2	511.8	724.2	1086.9	1125.4
NNCP-0	197.3	214.1	216.8	258.9	308.9	380.9	-	-	-
SUS304L	196.1	210.9	231.1	382.8	352.3	373.1	628.1	1417.6	1651.8

Material	ϕ (%) *			H' (GPa) **			$E\ell$ (mm) ***		
	293K	77K	4.2K	293K	77K	4.2K	293K	77K	4.2K
A5083(L)	27.5	27.5	23.6	2.97	3.09	3.90	6.2	11.2	9.5
A5083(ST)	20.9	15.9	4.3	3.19	3.58	4.85	3.3	5.2	2.4
A1100	69.4	69.1	49.1	0.71	1.57	2.07	5.0	15.9	18.3
Ag	85.9	83.4	74.5	2.04	2.43	2.60	-	-	-
TP28C	61.7	65.0	32.7	1.66	4.43	8.44	1.5	1.7	1.5
NNCP-0	-	-	-	2.67	3.02	3.20	-	-	-
SUS304L	80.6	71.0	49.2	2.9	16.0	22.1	10.6	12.2	11.8

* ; Mean value   ** ; Strain hardening coefficient at e=1x10⁻²

*** ; G.L.= 40 (mm)   (L) ; L-direction   (ST) ; ST-direction

All data of A5083, A1100 and σu, ϕ, El of TP28C were obtained by round bar specimens taken from explosively bonded plate, and other data were obtained by plane plate specimens of virgin plate.

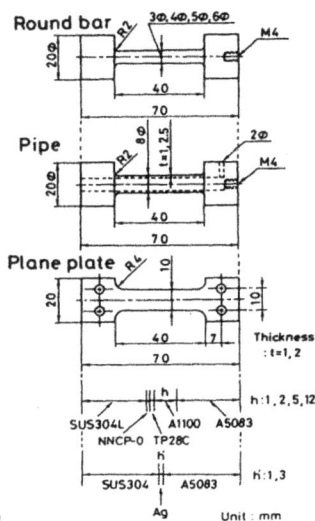

Fig.2 Test specimens

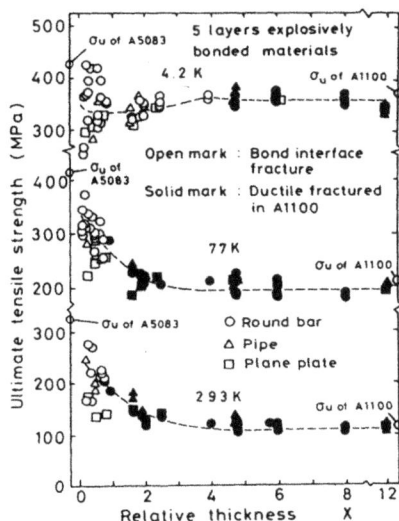

Fig.3 Relation between tensile strength and relative thickness in case of 5 layers explosively bonded joint

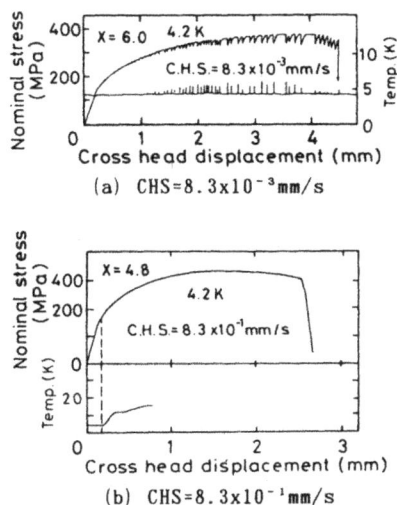

(a) CHS=8.3x10⁻³mm/s

(b) CHS=8.3x10⁻¹mm/s

Fig.4 Temperature rise and stress-displacement curve (5 LEBJ)

478

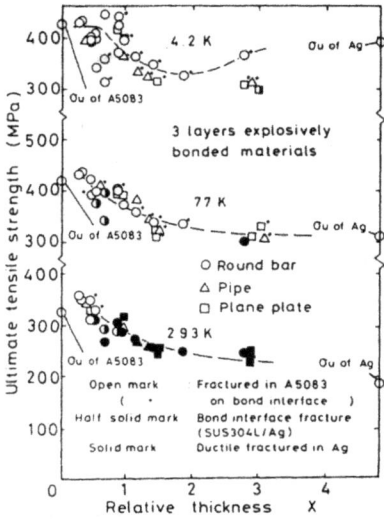

Fig.5 Relation between tensile strength
and relative thickness in case of
3 layers explosively bonded joint

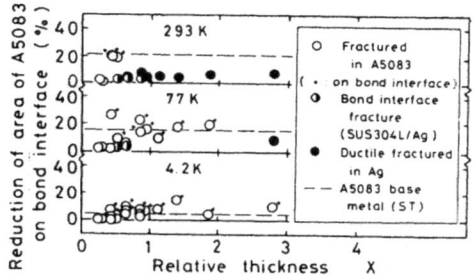

Fig.6 Reduction of area on Ag/A5083
bond interface (3 LEBJ)

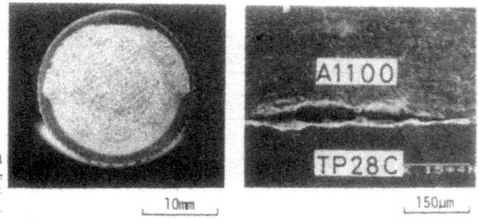

(a) 77K, X=0.5   (b) 4.2K, X=1.0

Fig.8 Examples of bond interface
fracture (5 LEBJ)

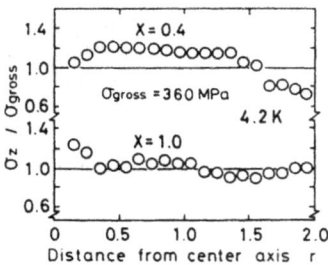

(a) A1100/A5083 interface

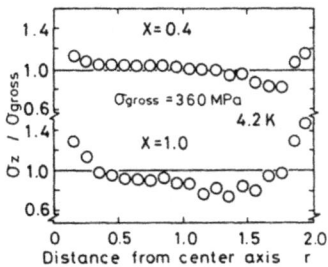

(b) TP28C/A1100 interface

Fig.7 Distribution of longitudinal
stress on bond interface
(Result of FEM analysis)

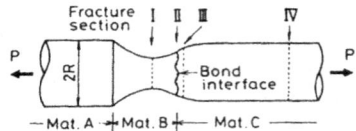

Fig.9 Schematic illustration of
deformation on bond interface
and fracture section

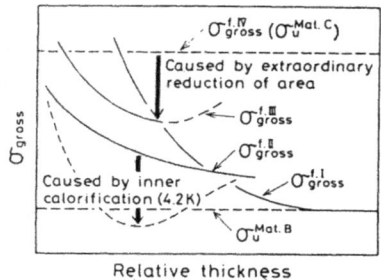

Fig.10 Schematic diagram of variation
in each fracture strength
against relative thickness

MECHANICAL PROPERTIES OF A LOW DENSITY Al-Li BASE ALLOY AT CRYOGENIC TEMPERATURE

Shigeoki SAJI[*], Ken YAMAMURA[**], Norio Furushiro and Shigenori HORI[*]

* Department of Materials Science and Engineering
  Faculty of Engineering, Osaka University
  2-1 Yamadaoka, Suita, Osaka 565, Japan
** Graduate Student, Osaka University

Abstract - The temperature dependence of the mechanical properties for an Al-Li-Cu-Mg-Zr alloy has been investigated by tensile tests in a temperature range between 7 and 293K. Toughness and elongation increased much as the test temperature decreased to about 20K, below which those decreased slightly. These remarkable increase is considered possibly to relate to the increase in the work-hardening exponent. The decrease below 20K is caused by the discontinuous deformation and the low value of the thermal conductivity. The effective concept to obtain a tough Al-Li alloy will be discussed briefly.

Introduction

An Al-Li base alloy has been expected to be one of possible materials in cryogenic service because of its low density, non-ferromagnetism and high rigidity. However, there is no report on its mechanical properties below 77K, except for the recent works by J. Glazer et al [1,2]. Even in their works, no experiment has been carried out between 4.2 and 77K and the explanation on temperature dependence of the toughness and ductility is still unclear.

Therefore, the objective of the present paper includes to make clear temperature dependence of the strength, ductility, toughness, work-hardening exponent or serrated flow for an Al-Li-Cu-Mg-Zr alloy after tensile tests at 7 to 293K, especially at every 10K between 7 and 70K. It is another objective to discuss important factors controlling ductility and toughness of the alloy.

Experimental

An Al-2.3%Li-1.2%Cu-0.67%Mg-0.14%Zr[*] alloy was cast using a flux. Contents of impurities were Si:<0.01, Fe:0.08, Mn:<0.01, Cr:<0.01, Zn:<0.01, Ti:0.02 in % and Na:3ppm. The ingot was homogenized at 793K in Ar, hot-forged and then hot-rolled from 30 to 10mm in thickness. Tensile specimens with the gauge section of 4mm in diameter and 22mm in length were machined from the plates as the stress axis was parallel to the rolling direction. The specimens were heated at 803K for 3.6ks, followed by quenching to iced water. Before the tensile test, some of these specimens were aged in an oil bath of 453K for 18ks and others 162ks, which correspond to under-aged (UA) and peak-aged(PA) condition, respectively [3].

---

* The symbol % refers to the weight percent.

Tensile tests were carried out at some temperatures between 7 and 293K at the initial strain rate of 4×10^{-4} s^{-1}, using an INSTRON machine. For tests below 130K, continuous flow cryostat [4] was used, in which the temperature could be maintained constant to ± 1K in each test by controlling the inflow rate of He gas and the heater current. Temperature of the specimen was monitored by the thermocouple of chromel and the Au-0.03%Fe alloy, which was cemented to its surface.

An optical microscope was used to estimate the grain size, the shape and the size of coarse inclusions before testing and slip bands on deformed specimens. Microstructures of aged and deformed specimens were observed by an transmission electron microscope and fracture surface by a scanning electron microscope.

Results

1) Microstructure of specimens

The grain structure before aging was characterized both by an elongated shape in the rolling direction (RD) and by a lamination in the normal direction to the rolling plane (ST) [3]. The mean size of grains in RD, LT or ST was 70-100, 50-70 or 10-15 μm. Inclusions of several microns in diameter were observed at grain boundaries.

Electron microstructures of UA and PA specimens show fine precipitates of $\delta'(Al_3Li)$ particles. Their sizes were 20 and 30nm for UA and PA, respectively. Composite particles [5] were also observed in places of aged specimens.

2) Temperature dependence of tensile properties

The ultimate tensile strength (UTS) increased with a decrease in test temperature above about 20K, while no remarkable change below 20K, as shown in Fig. 1. It is also seen that the difference between UTS and the proof stress in UA is bigger than that in PA, especially below 50K. This suggests that the amount of work hardening is more in UA than PA.

Figure 2 shows temperature dependence of the uniform elongation ε_u, which is defined as the amount of plastic deformation at the maximum load on the load-displacement curve. A maximum is seen in Fig. 2 at about 20K for both specimens. The maximum value of ε_u improved to 2.6 times as large as that at 293K. Values of ε_u at 293K for the present alloy is much smaller than those for A7075-T6 alloy, but rather larger below 70K [6]. On the other hand, the effect of testing temperature on the localized elongation ε_1 ($=\varepsilon_f-\varepsilon_u$, ε_f:the total elongation to fracture) is shown in Fig. 3. Values of ε_1 were much scattered, showing lower values around 20 to 70K. Such tendency of ε_1 is similar to that for A7075-T6, but extremely smaller values of $1/8 \sim 1/4$ [6].

Influence of temperature on the reduction of area is shown in Fig. 4. The reduction of area may take place before and after the necking. Values of ε_1 are never large, and its temperature dependence has analogy with ε_u. Therefore, it can be considered that the reduction of area occurred during uniform deformation. This is quite different from A7075-T6, in which 35$\sim$ 50% of the reduction in area was introduced after necking started.

Work done to fracture is one of indices to evaluate toughness of materials. The work per unit volume was calculated by the method described previously [6], and plotted as a function of the test temperature. The result is seen in Fig. 5. It is shown that toughness increases with a decrease in the temperature, showing a maximum at about 20$\sim$30K. The maximum value is three times of that at 293K. Change below 20K seems to be slight.

3) Work-hardening exponent

The stress-strain curve can be described often by the equation: $\sigma=K\varepsilon^n$, where σ:the true stress, ε:the true strain and K and n are work-hardening parameters [7]. The n value corresponds to ε_u under holding of this equation [7]. In the present study, the only n value was determined for each test. Temperature dependence of n is represented in Fig. 6, in which results are quite consistent with the tendency in Figs. 2, 4 and 5. This implies the existence of a close relationship between the uniform elongation, the reduction in area or the toughness and the n value.

4) Serrated flow

Serration was observed on the stress-strain curve of tests below 30K. Examples are shown in Fig. 7 for PA specimens. The amount of the stress drop was the larger, at the lower test temperature, the higher stress level and the higher strain. Temperature change corresponding to the stress drop was examined by continuous measurement of the temperature in the center part of the specimen deformed at 7K. Figure 8 shows an excellent correspondence between the temperature rise (a) and the stress drop (b). the maximum rise was achieved to 8K. It is also shown the amount of the temperature rise is not always higher at larger drop of the stress. This may be related to difference of the distance between the position of the thermocouple and the place where a localized deformation occurred [3,4,8].

Discussion

Much interests have been focused on the improvement of toughness and ductility for Al-Li base alloys. Webster[9] and Kobayashi et al [10] have been pointed out that the embrittlement at room temperature is possibly attributed to liquid phases at grain boundaries. The phase may be K, Na, having low melting point. Present results shows the more increase in the toughness and ductility even below 77K. Therefore, the suppression of the embrittlement due to the liquid phase is not the main reason for the improvement at low temperatures.

The increase in work-hardening exponent at lower temperature is another possible reason for the temperature dependence of toughness. The work-hardening behavior in FCC metals and alloys is much influenced by cross slip of dislocations. The frequency of the cross slip has been represented by the following equation:

$$\nu=\nu_0\exp(-Q_c/kT),$$

where Q_c is the activation energy of the cross slip, T:the absolute temperature, k:Boltzmann's constant and ν_0:the experimental constant. The equation means that the cross slip becomes difficult as the temperature decreases. This implies the increase in the work-hardening exponent incerases at lower temperatures. It is well known that the value of Q_c is high as the stacking fault energy of the matrix is low. For the Al-Li solid solution, the stacking fault energy will decrease with an increase in the solute atoms of lithium. This may be possible reason for higher values of the exponent in the UA- than the PA-specimen because of more solute atoms in its matrix. This implies that a significant method to obtain more improvement may be to add a solute atom, which lowers the stacking fault energy of the matrix effectively.

The decrease of toughness and ductility below 20K is attributable to the serrated deformation, which is resulted from the discontinuous deformation at very low temperatures [4,6,11]. Once discontinuous deformation takes place in this temperature range, it leads to the temperature rise localized in the deforming area because of extreme low value of the thermal conductivity, which is proportional to T^3. The local temperature rise lowers flow stress, then additional deformation proceeds while temperature falls to the test temperature. After the cooling, the specimen shows work-hardening again before the next discontinuous deformation. This procedure will make serrated flow at such very low temperatures.

Conclusions

The temperature dependence of the mechanical properties in the Al-Li-Cu-Mg-Zr alloy was made clear by the tensile test in a temperature range between 7 and 293K. Toughness and elongation increased much as the test temperature decreased to about 20K, below which those decreased slightly. These remarkable increase is considered to relate to the increase in the work-hardening exponent. The decrease below 20K is caused by the discontinuous deformation due to the low value of the thermal conductivity.

Acknowledgments

The tensile tests at low temperatures have been performed using the mechanical testing facilities of the Low Temperature Center of Osaka University.

References

1) J.Glazer, S.L.Verzasconi, E.N.C.Dalder, W.Yu, R.A.Eigh, R.O.Ritchi and J.W.Morris,Jr,"Cryogenic Mechanical Properties of Al-Cu-Li-Zr Alloy 2090", Adv. Cryo. Eng., 32, 397-404 (1986).
2) J.Glazer, S.L.Verzasconi, R.R.Sawtell and J.W.Morris,Jr, "Mechanical Behavior of Aluminum-Lithium Alloys at Cryogenic Temperatures", Met. Trans.A, 18A, 1695-1701 (1987).
3) S.Saji, K.Yamamura, N.Furushiro and S.Hori,"Mechanical Properties of an Al-Li-Cu-Mg-Zr Alloy at Very Low Temperatures", J. Jpn. Inst. of Light Met., in press (1988).
4) S.Saji,S.Senda and S.Hori, "Temperature Dependences of Mechanical Properties at Very Low Temperature in 5083 Aluminum Alloy", J. Jpn Inst. of Light Met., 37, 291-299 (1987).
5) N.Furushiro, S.Ishihara, D.Zhang and S.Hori, "Temperature Dependence of Tensile Properties of an Al-Li-Cu-Mg-Zr Alloy", J. Jpn. Inst. of Light Met., 36, 744-751 (1986).
6) S.Saji, K.Yasuhara and S.Hori, "Mechanical Properties of 7075 Aluminum Alloy at 6-130 K", Trans. JIM, 28, 773-780 (1987).
7) I.Gokyu and J.Kihara, "Work-hardening Behaviors of F.C.C. Polycrystalline Aggregates", J. Jpn. Soc. for Tech. of Plasticity, 9, 691-697 (1968).
8) Z.S.Basinski, "The Instability of Plastic Flow of Metals at Very Low Temperatures", Proc. Roy. Soc. London, 20, 229-242 (1957).
9) D.Webster, "Temperature Dependence of Toughness in Various Aluminum-Lithium Alloys", "Aluminum-Lithium Alloys III", edited by C.Baker, P.J.Gregson, S.J.Harris and C.J.Peel, The Inst. Met., London (1986) 602-609.
10) T. Kobayashi, M.Niinomi, K.Degawa, "Impact Toughness of Al-Li System

Alloys at Low Temperatures", J. Jpn. Inst. Met., 37, 816-823 (1987).

11) S.Saji, S.Senda and S.Hori, "Tensile Properties at Very Low Temperature in Aluminum Magnesium Alloys", "Strength of Metals and Alloys (ICSMA 7)",edited by H.J.McGueen, J.-P.Bailon, J.I.Dickson, J.J.Jonas and M.G.Akben, Pergamon Press, Oxford (1985) 471-476.

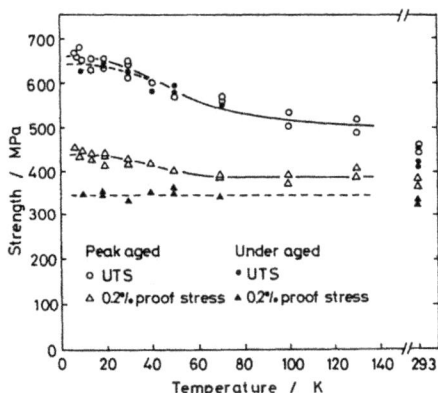

Fig.1. Temperature dependence of 0.2% proof stress and the ultimate tensile strength in the under-aged and the peak-aged specimens.

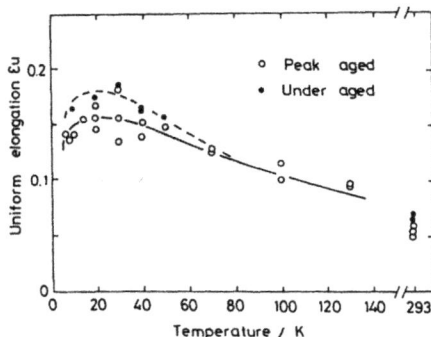

Fig.2. Temperature dependence of the true strain, ε_u corresponding to the uniform elongation in the under-aged and the peak-aged specimens.

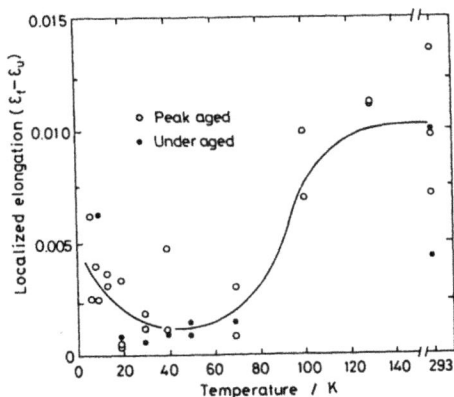

Fig.3. Temperature dependence of the true strain, $\varepsilon_f - \varepsilon_u$, corresponding to the localized elongation in the under-aged and the peak -aged specimens.

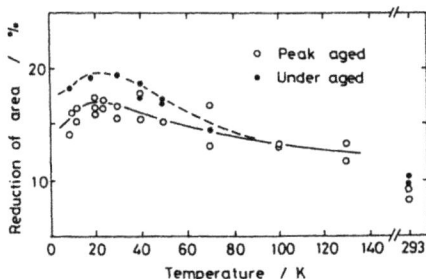

Fig.4. Temperature dependence of reduction of area in the under-aged and the peak-aged specimens.

484

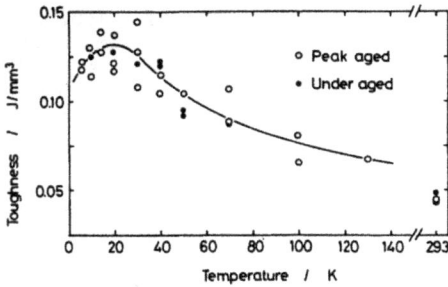

Fig.5. Temperature dependence of toughness; work done for fracture in the under-aged and the peak-aged specimens.

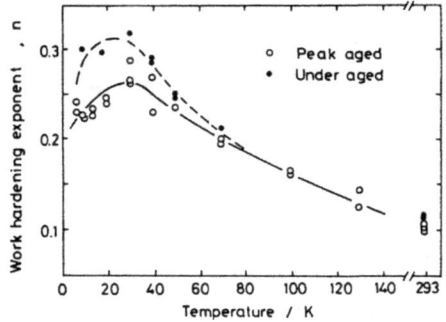

Fig.6. Temperature dependence of work-hardening exponent, n.

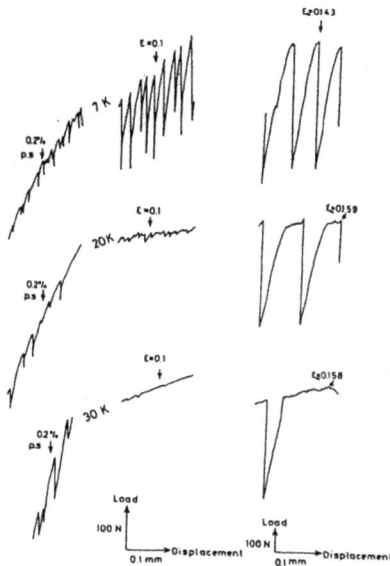

Fig.7. Typical load-displacement curves of the peak-aged specimens strained at the indicated temperatures. ε_u is the true strain corresponding to the ultimate tensile strength.

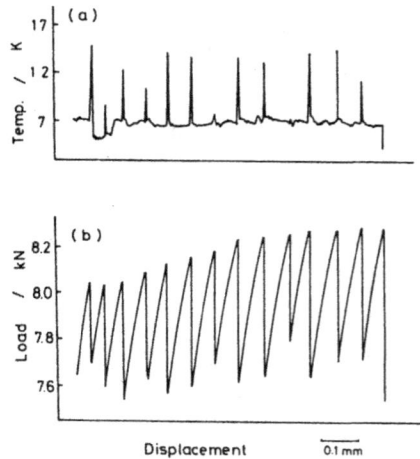

Fig.8. Correspondence between the temperature rises, (a) and the load drops, (b) during tensile test of the peak-aged specimen at 7K.

ADIABATIC DEFORMATION OF AUSTENITIC STAINLESS STEEL SHEET
NEAR ABSOLUTE ZERO

Koji SHIBATA*, Eiichi NAITO**, Hisaki SAKAMOTO***
and Kouzou FUJITA*

* The University of Tokyo, Dept. of Metallurgy and Materials
   Science, 7-3-1 Hongo, Bunkyo-ku, Tokyo 113
** Formerly Undergraduate School, Now, Matsushita Electric
   Industrial Ltd., Central Research Lab., 3-15 Naka-machi,
   Yagumo, Morikuchi-shi, Osaka
*** Graduate School, The University of Tokyo

Abstract - Deformation behavior of sheet specimens in liquid helium was
observed using austenitic steels.   Adiabatic deformation behavior could
be presented by computer simulation.   Specimen thickness affected the
onset strain of serrated deformation, tensile strength and total elongation.
In high strength steel, neckings inclined at about 45 degrees from the axial
direction of sheet specimens.

Introduction

Almost all metallic materials show serrated behavior in load-elongation
curves at very low temperatures near absolute zero[1].   This behavior is
known as serration at very low temperatures.   During the load drops in
this serration, unstable rapid deformation occurs locally in the specimen
together with a large increase in specimen temperature.   Hence, from a
practical viewpoint, it is important to examine the details of this phenomenon.   Many papers have been presented concerning the serrated deformation
of round bar specimens.   The serration of sheet materials, however, scarcely have been investigated.   Therefore, in the present work, the deformation behavior of sheet specimens were examined by using austenitic steels
in liquid helium.   Computer simulation of the serrated deformation was also
carried out in order to discuss the effects of the specimen thickness.

Experimental Procedure

High nitrogen high strength stainless steel and Fe-39%Ni binary steel
were used.   The former was a commercial YUS170 steel developed by Nippon
Steel Corporation[2].   The latter was melted for experiments by Kawasaki
Steel Corporation.   Chemical compositions are shown by Table 1.   Heat
treated plates were cut in the transverse direction and machined to specimens, of which gage length was 25mm and thickness was 2.0, 2.5 and 3.0mm.
Round bar specimens of 4mm diameter were also prepaired.   Geometry of
such specimens are shown in Fig.1.

Table 1.   Chemical compositions of steels (wt %).

steels	C	Si	Mn	Ni	Cr	Mo	N
YUS170	0.034	0.97	0.59	12.91	25.39	0.73	0.373
Fe-39Ni	0.001	-	-	39.5	-	-	0.0007

Tensile tests were carried out in liquid helium using an electro-hydraulically actuated servocontrolled machine under the constant crosshead speed condition.   Strain was measured with a strain gage extensometer attached to the specimen and the strain rate was 7.5×10^{-4} s^{-1} in the elastic deformation range.

Computer simulation was performed for Fe-39Ni steel, of which thermal properties at very low temperatures were available, by using a program similar to that developed by the present author et al[3-5] for roubd bar specimens.   Elements for calculation and a brief flow chart of the simulation are exhibited in Figs.2 and 3, respectively.

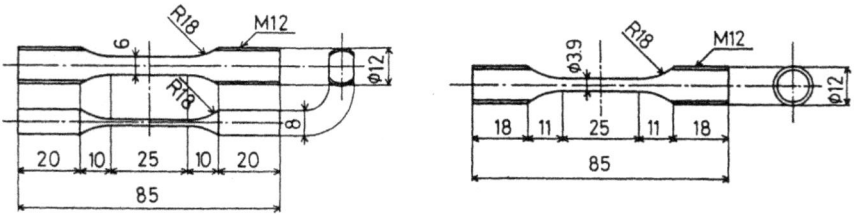

(a) sheet   (b) round bar
Fig.1 Geometry of specimens.

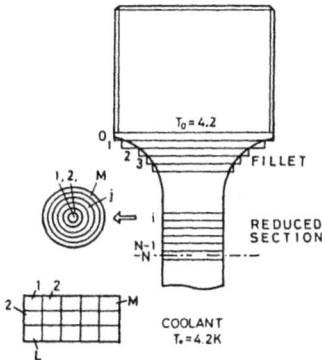

Fig.2. Elements for calculation.

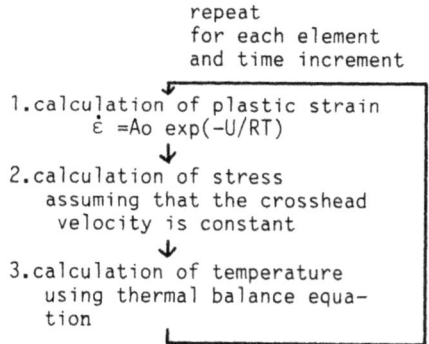

repeat
for each element
and time increment

1. calculation of plastic strain
   $\dot{\varepsilon}$ =Ao exp(-U/RT)

2. calculation of stress
   assuming that the crosshead
   velocity is constant

3. calculation of temperature
   using thermal balance equation

Fig.3. Brief flow chart of
calculation.

Results and discussion

Stress-elongation curves

The degree of load drops in YUS170 steel are remarkably larger than that in Fe-39Ni steel.   In comparison with round bar specimens, the serration was somewhat irregular in the sheet specimens as shown in Fig.4.

The effect of the specimen thickness on the degree of the serration, that is, the amount of load drops or the elongation accompanied with load drops, was not clear, but the degree showed a trend to reduce as the specimens were thinned.   For YUS170 steel, the onset strain of the serration was larger for thinner sheet specimens (Fig.5).   The effect of the specimen thickness on the onset strain of the serration is summarized in Fig.6.

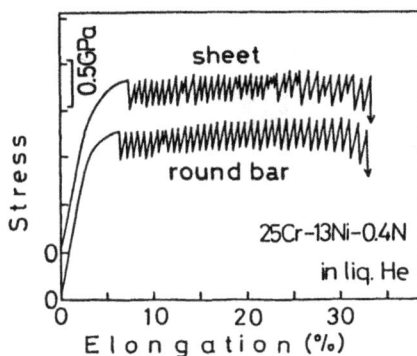

Fig.4. Stress-elongation curves for round bar (3.9mm in diameter) and sheet specimen (2mm in thickness).

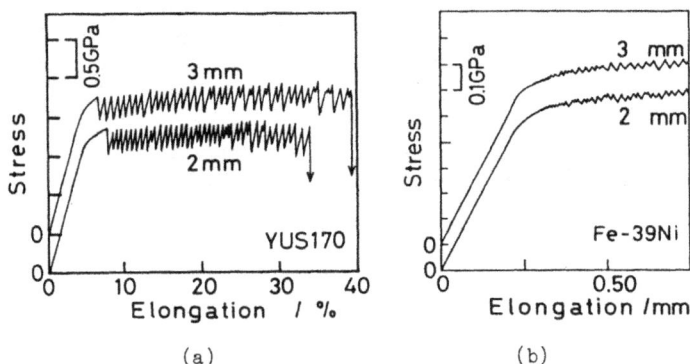

(a)   (b)

Fig.5.   Stress-elongation curves for sheet specimens.

Concerning Fe-39Ni steel, it was difficult to discuss the effect of the thickness on the onset strain only by using experimental data due to the smallness of load drops at the initial stage of the serrated deformation. Stress-elongation and -temperature curves obtained by   computer simulation are shown in Fig.7. The temperature is for the most inner element located at the center of the reduced section of the specimen.   Serrated behavior is presented by the simulation.

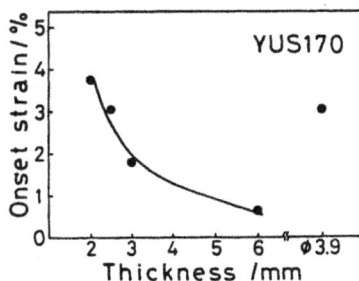

Fig.6. The effect of the specimen thickness on the onset strain.

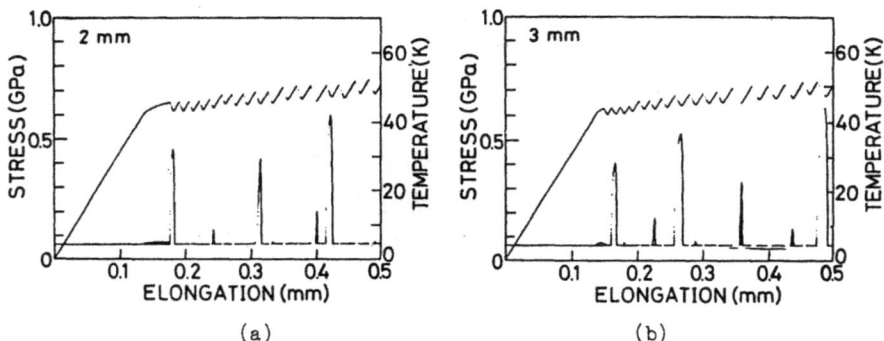

(a)   (b)

Fig.7. Stress-elongation and -temperature curves obtained by
       simulation.   Onset strain of the serration is larger for
       the thinner specimen.

Necking behavior

Multiple neckings were observed correspondingly to the load drops in
the serration in YUS170 steel as shown in Photo.1.   In this steel, necking
occurs at the angle of about 45 degrees from the axial direction of the
sheet specimen whereas at the angle of about 90 degrees in the round bar
specimen.   Figure 8 shows schematically the necking behavior of the YUS-
170 steel.   As shown in Fig.8 in sheet specimens, a newly occurred neck
used to intersect one or more pre-occurred necks where strain hardening has
taken place.   Such necking behavior is contributable to the irregurality in
the serration of the sheet specimen of this steel.

In the case of the Fe-39%Ni steel, neckings corresponding to the
load drops in the serration were not identified in both sheet and round bar
specimens, whereas the fracture took place with a necking propagated at the
angle of 90 degrees from the axial direction of the specimens.   The
obscurity of neckings corresponding to the serration is attributable to the
smallness of the necking, that is, smallness of the load drops in the
serration of this steel.

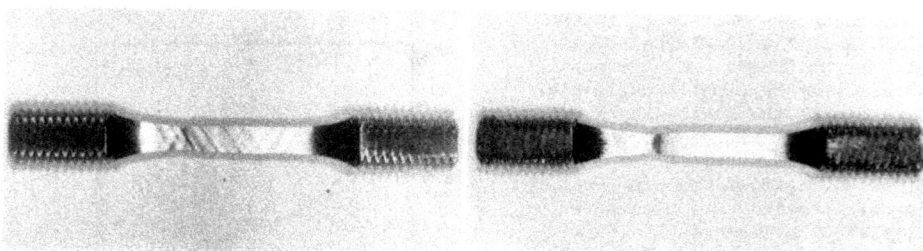

(a) YUS170 steel   (b) Fe-39Ni steel

Photo.1. Necking behavior of sheet specimens.

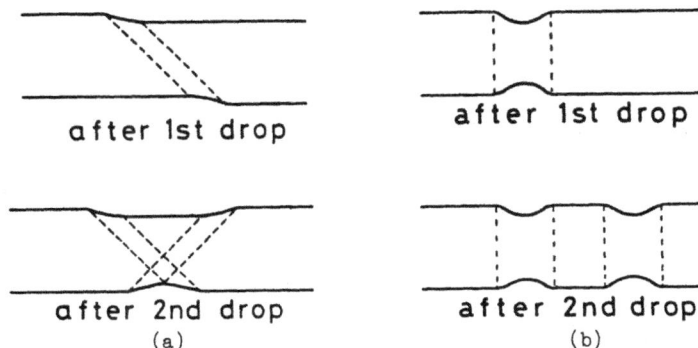

Fig.8. Schematic diagrams showing necking behavior of (a)sheet and (b)round bar specimens.

Effect of specimen thickness on tensile strength and total elongation

Total elongation was increased and tensile strength was decreased with an increase in the specimen thickness as shown in Figs.9 and 10. Symbols are representing the method for strain measurment. This effects of the specimen thickness on the ductility and the strength could be explained by the cooling ability of the specimen. That is to say, in the thinner specimen, the heat converted from to the plastic work can diffuse more easily. Therefore, the increase in the specimen temperature is more difficult to occur. In such a condition, the load drops occur at higher stress and this may increase strength of the thinner specimen. At the same time, the difficulty of the temperature rise reduces plastic deformation during the load drops in the serration. Hence, the total elongation may be smaller in the thinner specimen. In Fig.11, the relationship between cooling ability of the specimen and tensile strength. Cooling ability was evaluated as followings. First, the cooling curves of the most inner element at the center of the specimen were calculated as shown in Fig.11. Then, the mean cooling rate was obtained from the time for the temperature of the element to decrease downto the half value. The cooling rate was defined as cooling ability.

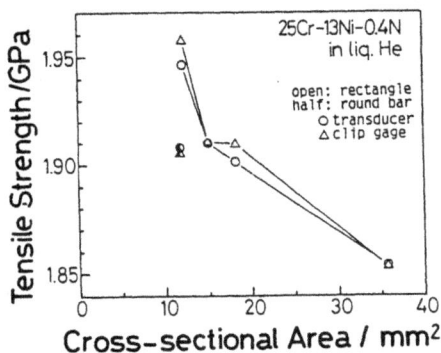

Fig.9. Effect of specimen thickness on tensile strength.

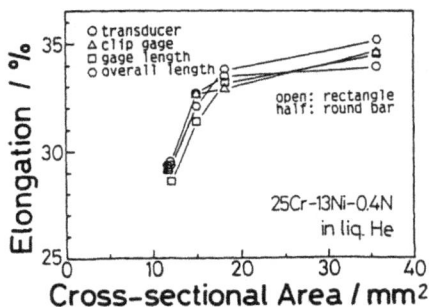

Fig.10. Effect of specimen thickness on total elongation.

490

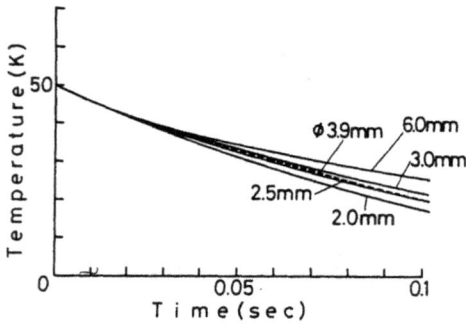

Fig.11. Cooling curve of the most
inner element of the specimen
(YUS170).

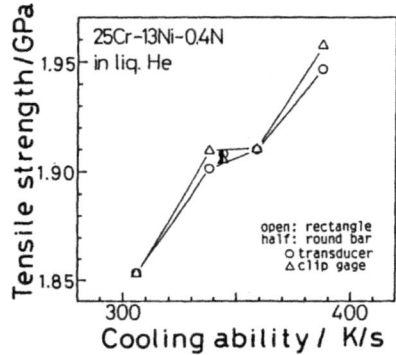

Fig.12. Effect of cooling ability
on tensile strength (YUS170).

Conclusion

The deformation behavior of sheet specimens of two austenitic steels
were observed in liquid helium.   Effect of the specimen thickness on the
onset strain of the serration, tensile strength and total elongation were
noticed. In high strength YUS170 steel, multiple neckings occurred at the
angle of about 45 degrees from the axial direction of the specimen, which
behavior could not be observed in round bar specimens. In Fe-39%Ni
steel,load drops in the serration were small and neckings accompanied with
the load drops were too small to be noticed.   The serrated deformation
behavior of Fe-39Ni steel could be presented by computer simulation.

This research was sponsored by the Special Cordination of Science and
Technology Agency of the Japan Government.   The athors are grateful to
Nippon Steel Corporation and Kawasaki Steel Corporation for preparing the
steels and Cryogenic Center of The University of Tokyo.

References

1) Z.S.Basinski : Proc.Roy. Soc. of LOndon, A240(1957), p.229
2) K.Suemune et al. : Advances in Cryogenic Engineering Materials, vol.34,
   ed. by A.F.Clark and R.P.Reed, Plenum , New York, (1987),p.123
3) K.Shibata and T.Fujita : Trans. ISIJ, 26(1986), p.1065
4) K.Shibata et al.:ibit, 28(1988), p.136
5) K.Shibata et al.: Advances in Cryogenic Engineering Materials, vol.34,
   ed. by A.F.Clark and R.P.Reed, Plenum , New York, (1987), p.217

EFFECTS OF Nb3Sn REACTION HEAT TREATMENT ON CRYOGENIC MECHANICAL PROPERTIES OF AUSTENITIC STEELS

Masao SHIMADA

Materials Research Laboratories, Kobe Steel, Ltd.
3-18, Wakinohamacho, 1-chome, Chuo-ku, Kobe, 651 Japan

Abstract - Effects of a Nb3Sn reaction heat treatment on cryogenic mechanical properties of Cr-Ni austenitic steels were investigated at low temperatures. Degradation of mechanical properties due to the heat treatment was observed. Their properties were closely connected to grain boundary precipitates and austenitic stability.

Introduction

As solution-treated (called solutioned hereafter) austenitic steels have been used as structural materials for superconducting magnets because they have a good combination of strength and ductility at low temperatures.[1] However, sheathing alloys for Nb3Sn superconductors will be reheated at a-round 700℃ for 50∼200 hours; Nb3Sn is formed by the diffusion reaction heat treatment with niobium and tin. This heat treatment causes sensitization, which makes carbides precipitate at the grain boundary. Therefore, this heat treatment (called aging hereafter) is supposed to degrade cryogenic mechanical properties of austenitic steels. But the effects of this aging on mechanical properties at 4K are not well studied and understood yet. In order to clarify its effects, the strength, ductility and fracture toughness of aged Fe-0.02C-18Cr-(10∼30)Ni alloys were investigated at cryogenic temperatures, especially at 4K.

Experiments

The tested steels were melted in a vacuum induction furnace. They were hot forged and hot rolled into 26 mm thick plates. They were solution heat-treated at 1050℃ for 1 hr. Half of each steel was aged at 700℃ for 75 hr, which simulated the reaction heat treatment of Nb3Sn.
The chemical compositions of the steels are tabulated in Table 1. All were low carbon steels. The nickel content ranged from 10 to 30%, which varied austenitic stability. Caluculated Md30[2], (the temperature at which 50% α' martensite is induced by 30% strain), is also shown in Table 1.
Tensile specimen were cut transverse to the rolling direction. Tensile tests were carried out at 4K, 77K and 300K. A strain rate of $8 \times 10^{-4} s^{-1}$ was

Table 1   Chemical compositions of the steels tested (mass%)

STEEL	C	Si	Mn	P	S	Cr	Ni	Md30[K]
N1	0.021	0.110	0.47	0.003	0.0014	17.69	10.02	258
N2	0.017	0.093	0.49	0.003	0.0014	18.12	12.60	177
N3	0.021	0.098	0.48	0.003	0.0009	18.04	15.03	108
N4	0.020	0.098	0.49	0.003	0.0010	17.34	20.35	-37
N5	0.021	0.110	0.48	0.003	0.0009	18.07	25.11	-185
N6	0.020	0.120	0.48	0.003	0.0010	18.11	30.16	-332

applied. A 4K test was conducted in a turret disk apparatus. [3]
25 mm thick compact tension specimens with a T-L orientation were used for
J_{IC} test at 4K. J_{IC} tests were conducted with a computer aided unloading
compliance method. [4] Fracture toughness K_{IC} was calculated from measured J_{IC}
by the following equation. [5]

$$K_{IC}^2 = J_{IC} \times E$$

where E is a Young's modulus.

Microstructures were observed through optical and electron microscopes.
Precipitates by aging were extracted chemically and investigated. Fracture
surfaces were also studied with a scanning electron microscope and permeability at the surface was measured.

Results

Precipitates

Photograph 1 shows grain boundary carbide in steel N1, N3 and N5. Each
aged steel showed intermittent distribution of grain boundary precipitates
but they were not observed in matrix. Electron diffraction and X-ray diffraction analysis revealed these precipitates were all Cr rich $M_{23}C_6$ carbides. An
increasing trend of carbides with nickel content was found .

Tensile Properties

Figure 1 shows 4K load-displacement curves of aged N1, N3 and N5 steels.
All steels showed serrated flow at 4K as shown in Fig. 1. The work hardening
rate decreases with nickel content. The difference between load-displacement
curves of as-solutioned and aged steels was not clear. Both intergranular and
intragranular fracture modes were observed in tension test sapecimens at 4K.
Figures 2 and 3 show yield and tensile strength plotted against nickel content. Open and solid circles stand for data of as-solutioned and aged steels,
respectively. There was no Ni content dependence of yield strength at 300K
but there was clear Ni content dependence at low temperatures. Minimum yield
strength existed around 15% Ni and it became more obvious in aged steels.
Tensile strength did not appear to be influenced by aging. But it decreased
as the nickel content incresed. Figure 4 shows Ni dependence of elongation at
each temperature. At 300K and 77K, aged steels showed almost the same curves
as solutioned ones, though aged ones had a little lower values. At 4K, the
elongation of aged ones have a peak of about 15% Ni content and showed a dif-

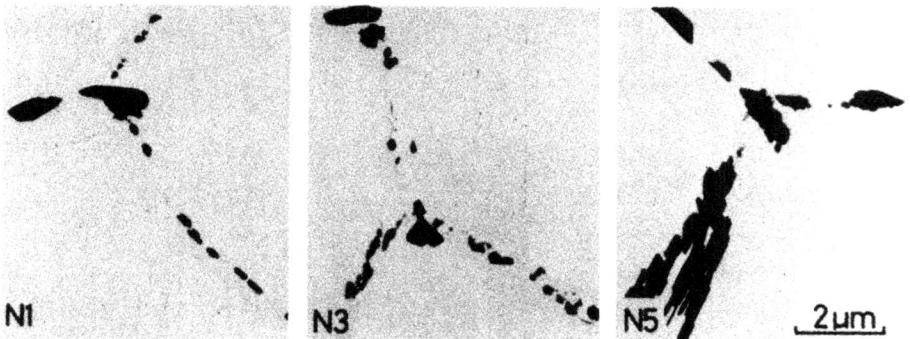

Photo. 1 Grain boundary carbides of the steels aged at 700℃ for 75 hours.

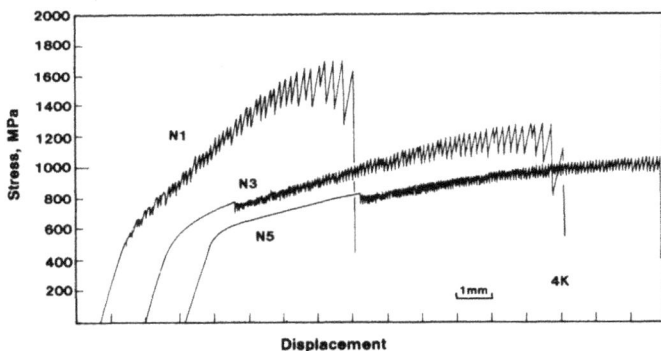

Fig. 1 Load-displacement curves of tension test at 4K for the aged steel.

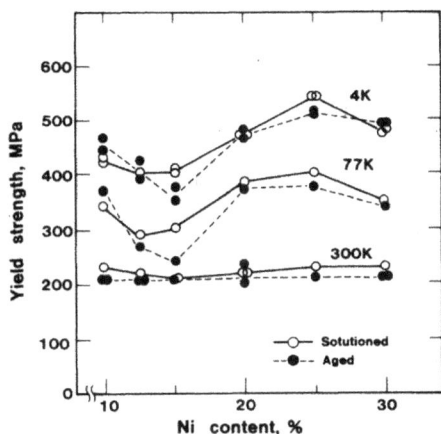

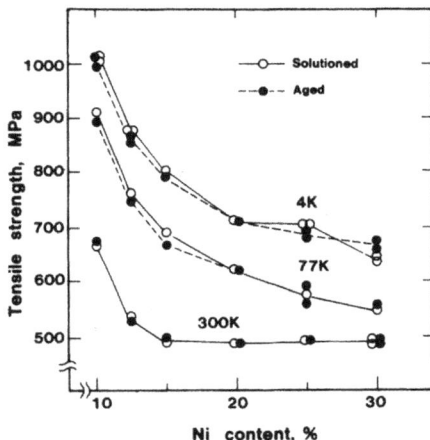

Fig. 2 Effect of Ni on yield strength.   Fig. 3 Effect of Ni on tensile strength.

ferent Ni dependence from that of solutioned ones.

Fracture Toughness

Figure 5 shows load-load line displacement curves in a J_{IC} test at 4K of the aged N1, N3 and N5 steels. Serration was observed but the initiation of N3 steel was retarded to a 3 mm displacement. Figure 6 shows Ni content depend-ence of K_{IC}. Valid K_{IC} could not be obtained for solutioned alloys with more than 12.5% Ni. J_{IC} of the aged N3 steel containing 15% Ni could not be meas-ured either because crack extended only 0.3 mm even if 5 mm of load line dis-placement was applied at 4K. Aging seemed to deteriorate 4K fracture tough-ness greatly. As one can seen, Ni content dependence curve was divided into two regions, namely high and low Ni regions with a border of around 15% Ni. The K_{IC} depends on Ni content strongly in the low Ni region but weakly in the high Ni region. Fractographs revealed that all aged steels were fractured in-

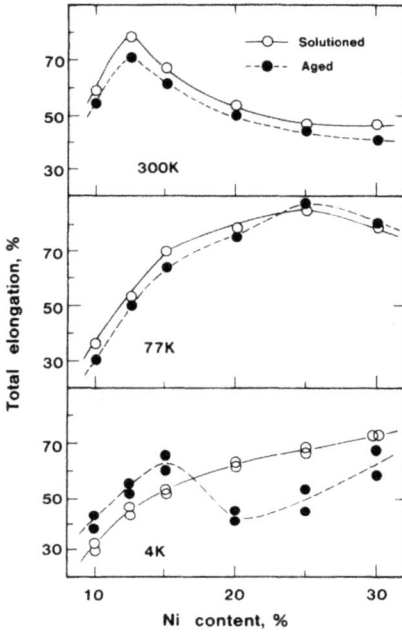

Fig. 4 Effect of Ni content
on total elongation.

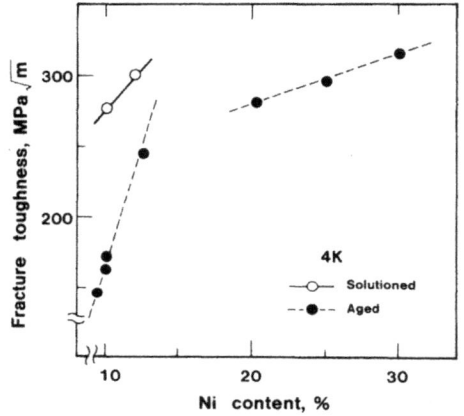

Fig. 6 Effect of Ni content on
4K K_{IC} of the aged steels.

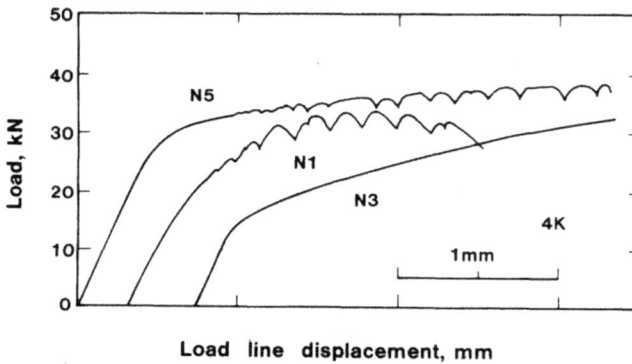

Fig. 5 Load-load line displacement curves of J_{IC} test at 4K
for the aged N1, N3 and N5 steels.

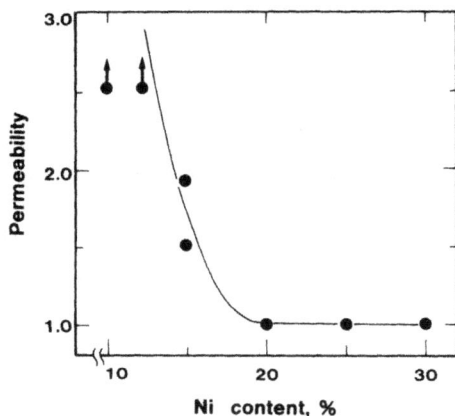

Fig. 7 Magnetic permeability at the fracture surfaces of the J_{IC} test.

tergranularly. Magnetic permeability measured at fractured surfaces are shown in Fig. 7. Both aged and solutioned steels gave the almost same results. Alloys belonging to the high Ni region appeared to be stable against α' martensitic transformation. Steels in the low Ni region had much higher permeability than 2.5, which was the maximum available value of the used meter.

Discussion

Cr rich $M_{23}C_6$ carbides precipitated along the grain boundaries during aging at 700 °C for 75 hr. This carbide leads to Cr depleted zone adjacent to grain boundary and leads to stress concentration around it. Stress concentration around the hard second phase will lead to easy void formation and will result in intergranular failure, especially during fracture toughness tests with a high tri-axiality of stress.

Judging from the Ni content dependence of mechanical properties in Fig. 2-4 and 6, the tested alloys are divided into two groups at about 15% Ni content. The first is lower Ni steel which is metastable austenite. The second is higher Ni steel which is stable austenite. The stable alloys have negative values of Md_{30} shown in Table 1. This classification is actually supported by the data shown in Fig. 7.

In metastable alloys, transformation induced plasticity (TRIP) was responsible for the lower yield strength because of easier martensitic deformation than disolocation gliding. Therefore the Cr depleted zone, which is more unstable than thematrix, gets weaker than the matrix. On the other hand, tensile strength was hardly affected by aging. As flow strength increases with an increase in the amount of transformed martensite, it is supposed that the overall amount of martensite is hardly affected by aging. Moreover it could be said that stress concentration at grain boundary precipitates along with the lower flow strength in the Cr depleted zone neighboring the grain boundaries degraded fracture toughness cooperatively because micro voids at precipitates gave a preferential crack path. Then 4K fracture toughness decreases rapidly with decrease in Ni content as shown in Fig. 6.

In the stable alloys, martensitic deformation does not take place but grain boundary precipitates act as stress concentration centers. Therefore

496

the grain boundaries become weaker than the matrix and the aged steels fail intergranularly in the fracture toughness tests at 4K though they have high fracture toughness.

N3 steel containing 15% Ni showed a distinctive behaviour. It lies between the higher and lower Ni regions. It shows high ductility, high toughness and low yield strength at 4K. Tamura et al reported that TRIP could keep an appropriate work hardening rate and relaxed stress concentration.[6]   A suitable amount of transformation proportional to a degree of stress concentration is supposed to be a cause for the high ductility and toughness. Electron microscopy revealed that there was no martensite in the N3 steel aged and cooled down to 4K but that α' martensite was distributed at intersections of ε martensite and/or stacking faults when deformed at 4K.   Magnetic permeability data in Fig. 7 indicates that the total amount of α' martensite is not large even at the fracture surface of N3 steel.   Accordingly, TRIP may play a very important part in mechanical properties of N3 steel at 4K.

It seems to be possible to improve fracture toughness of aged steels containing a number of grain boundary precipitates if stress concentration around precipitates in the crack tip region is relieved by TRIP.

Conclusions

(1) The aging caused Cr rich $M_{23}C_6$ carbide to precipitate at the grain boundary.   The precipitates increased with Ni content.
(2) Tensile strength was affected greatly by Ni content but not by aging because the amount of induced α' martensite corresponded to Ni content.
(3) The yield strength of metastable steel at 4K was lowered by induced martensitic transformation. This treand was enhanced by aging.
(4) Improvement of elongation was explained by TRIP in metastable austenitic steels.
(5) In the fracture toughness test, aging caused intergranular fracture and deteriorated K_{IC} regardless of Ni content, though K_{IC} increased with Ni content. Dependence of K_{IC} on Ni in metastable austenitic steels was much larger than in stable austenitic steels.
The N3 steel exhibited extremely high K_{IC} because TRIP relieved stress concentration effectively in the crack tip region.

References

1) F. R. Fickett: Adv. Cryog. Eng. 28, 1 (1982)
2) K. Nohara et al: Tetsu-to-Hagane 63 212 (1977)
3) T. Horiuchi et al : Proc. 5th Int. Cry. Eng. Conf. 465 (1974)
4) M. Shimada et al : Cryog. Eng. 21, 269 (1986)
5) ASTM Designation E813-81:1986 Annual Book of ASTM Standards. Section 3, Vol. 03. 01, (ASTM, Philadelphia) 768 (1986)
6) I. Tamura et al: 2nd Int. Conf. on Strength of Metals and Alloys. (ASM) 900 (1970)

CRYOGENIC STRENGTH OF LASER-WELDED 22 Mn STEEL

Motohiro NAKANO, Toshihiko KATAOKA, Keizo KISHIDA

Department of Precision Engineering, Fuculty of Engineering,
Osaka University, 2-1 Yamadaoka, Suita, Osaka 565, Japan

and Akira MATSUNAWA

Welding Research Institute, Osaka University
11-1 Mihogaoka, Ibaraki, Osaka 567, Japan

Abstract - The strengths of the laser-welded 22 Mn austenitic stainless steel are investigated at low temperatures, compared with the base metal and the electron-beam (EB)-welded joint. The tensile strengths of the laser-welded joint are equivalent to the base metal or the EB-welded joint. The fracture toughness seems to be slightly lower than the base metal. This degradation may be associated with the increase of oxygen content in the weld metal.

Introduction

The 22 Mn steel is a newly developed structural material [1] for the supercoducting magnet of the Fusion Experimental Reactor (FER) which has been designed by the Japan Atomic Energy Research Institute (JAERI).

The structural material was required to have the following properties[2];
(1) the yield stress and the fracture toughness higher than 1200 MPa and 200 MPa$\sqrt{m}$, respectively at 4K,
(2) the fatigue properties equivalent to SUS 316,
(3) the stable austenitic structure (paramagnetic property),
(4) the good weldability and machinability.

Weldability is the most important property for the material to construct a large magnet structure. It has been reported that the electron-beam (EB) welding is a good process for the 22 Mn steel of cryogenic use [3]. There is, however, a strong restriction in the EB welding that it is performed in a vacuum.

The present paper deals with the laser welding for the reason that it is performed in air and has similar welding characteristics to the EB in respect of the high energy density but the low total input energy.

Experimental Procedure

The material tested is an austenitic stainless steel whose chemical compositions are shown in Tab.1. The nitrogen content is increased instead of carbon for strengthening the austenitic matrix at cryogenic temperatures. High manganese content makes the steel tough and gives a stable austenitic structure.

Table 1. Chemical compositions of high Mn steel. (wt%)

C	Si	Mn	P	S	Ni	Cr	Mo	O	N
0.06	0.26	21.3	0.008	0.001	7.03	12.87	0.96	0.0018	0.222

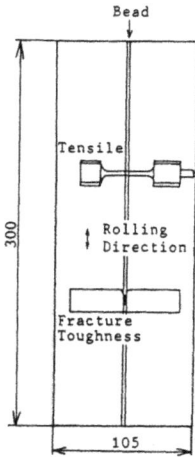

Fig.1. Specimens cut from welded plate.

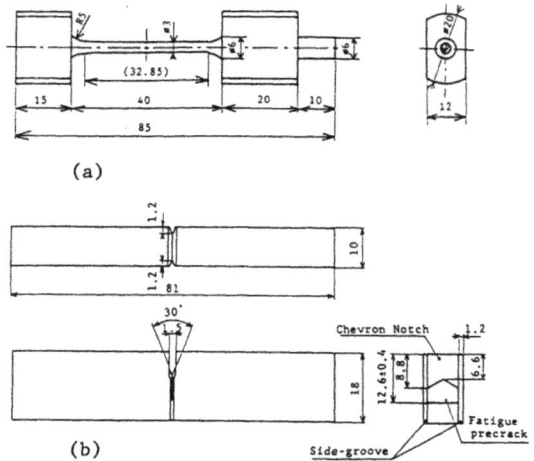

(a)

(b)

Fig.2. Specimens for tensile test (a) and fracture toughness test (b).

Table 2. Welding conditions.

Welding process		Laser			Electron-beam	
		L1	L2	L3	E1	E2
Heat input (kJ/m)		713	600	650	504	428
Welding speed (m/min)		0.8	1.0	1.2	0.5	0.8
Power (kW)		9.5	10.	13.	4.2	5.7
Shield gas		He (80 1/min)			Vacuum	
Plate thickness (mm)		12				
Welding position		Vertical				
Preheating		None				
Number of passes		1				
Groove preparation		Beam				

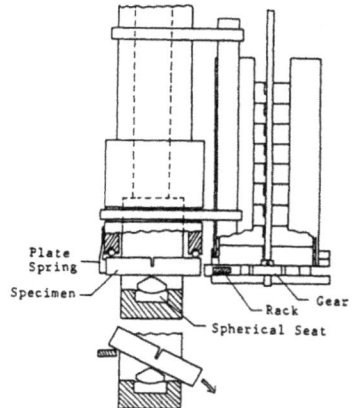

Fig.3. Bending apparatus for fracture test.

Specimens for tensile and fracture toughness tests were machined from 12mm thickness hot-rolled plate welded by the laser and electron-beam as shown in Fig.1. Figure 2 is the detail drawings of those specimens. Conditions of the laser and EB weldings are shown in Tab.2. They seemed to be good weldments judging from observations of beads formation on the plate.

Tensile tests were performed at room temperature, 77K and 4K using Instron 1125 testing machine under displacement controlled. Tests at 77K and 4K were carried out in liquid nitrogen and liquid helium, respectively, by the use of multi-specimen cryostat in which 8 pieces of specimens could be set. Crosshead speed of 0.5 mm/min was chosen as a standard. The crosshead speed is corresponding to the strain rate of 2×10^{-4} s^{-1} in the present test and is so slow that the temperature rise in the specimen can be suppressed at 4K test.

Fracture toughness tests were performed at 4K by using 3-point bending apparatus as shown in Fig.3. This apparatus has the storage unit for 10 specimens and the jig in order to set the new specimen on the proper location and remove the tested specimen.

Fatigue crack was introduced by a bending fatigue machine at the range of stress intensity factor less than 30 MPa$\sqrt{m}$. Since two grooves were made on the side surfaces of the specimen (Fig.2), such fatigue crack propagates exactly in the portion of weld metal.

Fracture toughness was estimated by the J-integral method in accordance with the ASTM standard E813. The bending load P was measured by the strain gage cemented on the support, and the load point displacement Δ by the linear variable differential transfomer (LVDT) attached between the loading bar and the support.

A series of tests were performed in which the specimens were deformed to some selected deflections so as to generate different crack growths Δa. After the tests all specimens were heat-tinted to mark the potion of crack tip. It was possible to measure the length of initial crack a_0 and the crack growth Δa for each test. The J-integrals were calculated from the load - load point displacement (P-Δ) curve using the following formula;

$$ J = \frac{2}{W-a_0} \int_0^\Delta (\frac{P}{B}) \, d\Delta \, , \tag{1} $$

where W is the width of the specimen (18mm), a_0 is the initial crack length (about 12.6mm) and B is the thickness of the specimen (7.6mm, bottom-to-bottom of side grooves).

The R curve was obtained from the data J and Δa for each test, and J_{IC} was determined as the J value corresponding to the intersection point of the regression line and the blunting line of $J=2\sigma_f \Delta a$, where σ_f is the mean value of the yield stress and the ultimate tensile stress. To convert J_{IC} to K_{IC} the following relationship was used,

$$ K_{IC} = \sqrt{E \, J_{IC}} \, , \tag{2} $$

where E is the Young's modulus of the specimen.

Experimental Results and Discussions

Chemical compositions of weld metals are shown in Tab.3 compared with those of the base metal. It is clear that there are no distinct change in the compositions except for the increase of oxygen content in the laser-welded metal.

Macro- and micro-structures of a cross section of the laser- and EB-welded joints are shown in Fig.4. The weld areas of both joints are considerably narrow compared with arc welding. The weld metal consists of the ascicular fine grains growing along the direction of heat flow during solidification. There seems to be neither hot cracking nor grain growth occasionally observed in the weldment of austenitic steels.

Figure 5 indicates the variations of hardness across the both welded joints at the 1/6-, 1/2- and 5/6-thickness locations. The increase of hardness is seen in the weld metal and in the heat affected zone (HAZ).

Table 3. Chemical compositions of weld metals. (wt%)

	C	Si	Mn	P	S	Ni	Cr	Mo	O	N
Base metal	0.06	0.26	21.3	0.008	0.001	7.03	12.87	0.96	0.0018	0.222
Laser	0.06	0.25	21.3	0.008	0.001	7.17	12.84	0.97	0.0028	0.238
Electron-beam	0.06	0.26	21.4	0.008	0.001	7.22	12.84	0.98	0.0016	0.218

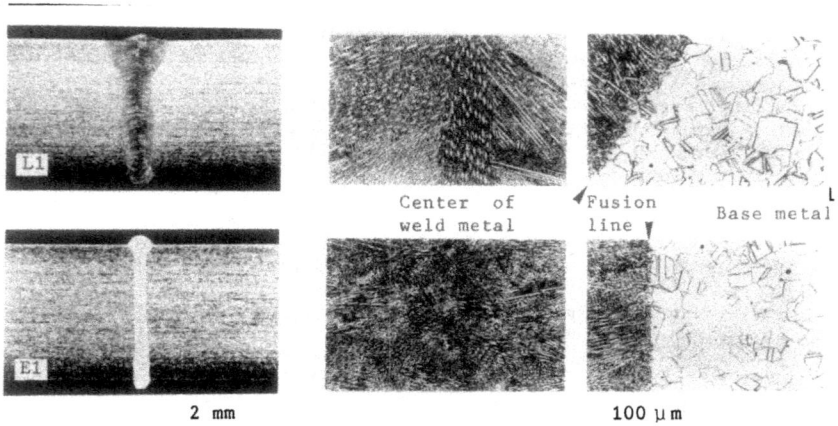

Fig.4. Macro- and micro-structures of laser-welded (upper)
and EB-welded (lower) joints.

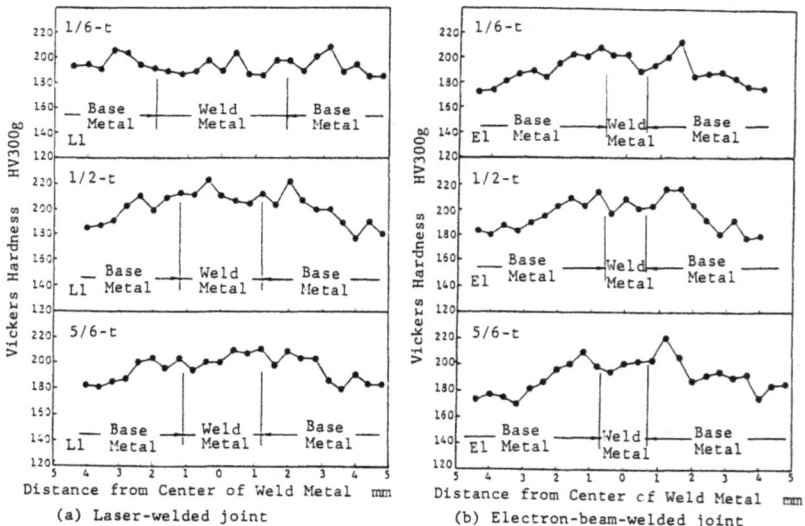

(a) Laser-welded joint   (b) Electron-beam-welded joint

Fig.5. Hardness distributions across welded joints.

Figure 6 shows the load-displacement curves of the laser- and EB-welded joints as well as the base metal. Any distinct difference is not seen among these curves. Yield stress, ultimate tensile stress, elongation and reduction of area are listed in Tab.4. The elongation of the laser-welded specimen seems to decrease at low temperature. This is caused by the fact that the plastic deformation is localized at the weld portion as is seen in Fig.7. It may be due to the ductility mismatch between the base metal and the weld metal in the case of laser welding. It cannot be inferred from the hardness measurement at room temperature as shown in Fig.5.

In Fig.8 the J-integral values are plotted against the corresponding crack growth Δa for the base metal, the laser- and EB-welded joints which were obtained from the 3-points bend tests at 4 K. In the figure the linear regression line for the data of the base metal determined by the least square method is shown. From the intersection of the regression line with the blunting line the J_{IC} of the 22 Mn steel was determined as 178 kJ/m^2, and converted to K_{IC} as 193 MPa$\sqrt{m}$. The J_{IC} for both the welded joints cannot be determined becuase of lack of the data, but it is presumed that the toughness of the laser-welded joint is slightly lower than the base metal and the EB-welded joint higher. The reason of the degradaion in the fracture toughness

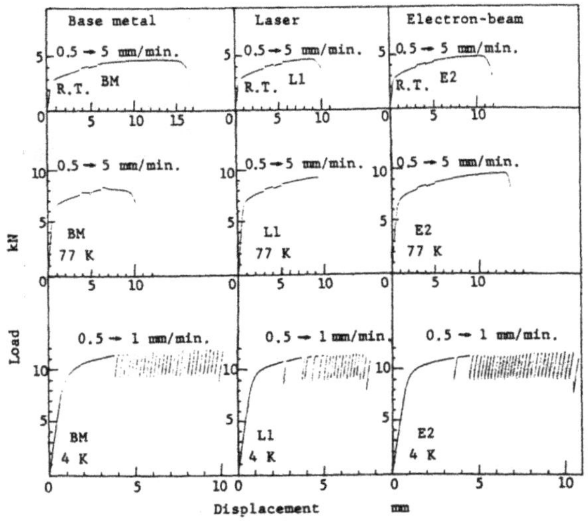

Fig.6. Load - displacement curves of the base metal, laser- and EB-welded joints at room temperature, 77K and 4K.

Table 4. Tensile properties.

		Test temp. (K)	Yield stress (MPa)	Tensile strength (MPa)	Elonga-tion (%)	Reduction of area (%)	Location of fracture
Base metal	BM	291	368	662	49	75	
		77	872	1172	30	50	
		4	1202	1634	30	39	
Laser welded joints	L1	284	381	659	30	72	Weld metal
		77	877	1319	24	39	Weld metal
		4	1231	1610	19	36	Weld metal
	L2	284	382	662	31	68	Weld metal
		77	880	1330	27	56	Weld metal
		4	1287	1646	19	48	Weld metal
	L3	284	381	668	31	71	Weld metal
		77	893	1341	25	47	Weld metal
		4	1231	1613	12	45	Weld metal
Electron-beam welded joints	E1	284	385	673	33	73	Weld metal
		77	880	1369	37	55	Weld metal
		4	1202	1652	31	45	Base metal
	E2	284	379	665	34	76	Weld metal
		77	840	1358	38	58	Weld metal
		4	1216	1646	28	40	Base metal

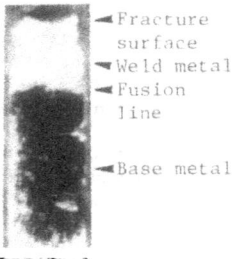

Fig.7. Longitudinal section of laser-welded joint specimen fractured by tensile deformation.

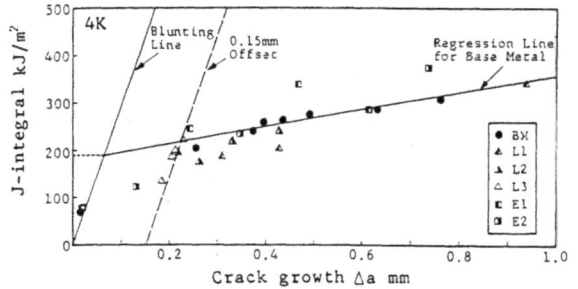

Fig.8. Relationship of J vs. Δa for base metal, laser- and EB-welded joints.

of the laser-welded joint is considered to be due to the increase of oxygen content in the weld metal. The oxide inclusions in the weld metal should control initiation, growth and coalescence of voids. It may be possible to prevent the intrusion of oxygen into the weld metal by some improvement of shield gas flow in the laser-welding process.

Concluding Remarks

The strength of the laser-welded joint of the 22 Mn stainless steel at low temperature was investigated compared with the base metal and the EB-welded joint. The following results were obtained;
(1) The laser-welded joint has equivalent tensile properties to those of the base metal and the EB-welded joint.
(2) The fracture toughness of the laser-welded joint is slightly lower than that of the base metal, while that of the EB-welded joint is slightly higher.
(3) It is important to prevent the intrusion of oxygen into the weld metal in the laser-welding process.
As a conclusion the laser-welding is a promising technique to construct a large structure of superconducting magnet in the field.

Acknowledgements

The authors wish to express their appreciation to Mr. S. Tone of Kobe steel Ltd. who contributed to this program. Further appreciation is expressed to Kobe Steel Ltd. for suppling the test material and Mitsubishi Electric Co. for the welding. The low temperature tests were performed using the mechanical testing facilities of the Low Temperature Center, Osaka University. This work was supported in part by the Grant-in-Aid for Fusion Research, the Ministry of Education, Science and Culture.

References

1) T.Horiuchi, R.Ogawa and M.Shimada, "Cryogenic Fe-Mn Austenitic Steels", Adv. Cryo. Eng. Mat., 32, 33-42 (1986).
2) Y.Yoshida, H.Nakajima, K.Koizumi, M.Shimada, Y.Sanada, Y.Takahashi, E.Tada, H.Tsuji and S.Shimamoto, "Development of Cryogenic Structural Materials for Tokamak Reactor", *Austenitic Steels at Low Temperature* , eds. R.P.Reed and T.Horiuchi, (Plenum press, New York, 1983) pp. 29-39.
3) S.Tone, M.Hiromatsu, T.Numata, T.Horiuchi, H.Nakajima and S.Shimamoto, "Cryogenic Properties of Electron-Beam Welded Joints in a 22Mn-13Cr-5Ni-0.22N Austenitic Stainless Steel", Adv. Cryo. Eng. Mat., 32, 89-96 (1986).

MAGNETIC FIELD EFFECT ON ELASTO-PLASTIC FRACTURE TOUGHNESS AT 4K

H. Yanagi*, A. Nyilas and W. Specking

*Maekawa Manufacturing Company
Engineering Research Laboratory
Aza-Okubo, Moriya-machi, Kitasooma-Gun
Ibaraki-Prefecture, Japan

Kernforschungszentrum Karlsruhe
Institute für Technische Physik
7500 Karlsruhe, FRG

Abstract - An experimental investigation has been carried out to obtain the possible magnetic field effect on elasto-Plastic fracture toughness at 4 K. The material tested was a stainless steel ∿AISI321 type(1.6903) frequently used for parts of superconducting machineries.

It was found that the critical value of the J_I Integral(JIc) value of this investigated material decreases with magnetic field. A ∿10% drop of the JIc value was determined at 10 Tesla. This reduction was thought to be derived from an additional stress of the internal ferromagnetic phase toward the axis of the applied magnetic field. The rotational movement of the strain induced martensite induces additional microstructural stresses.

Introduction

The fracture toughness JIc values are now widely used as a fracture parameter for characterization of elasto-plastic behaviors of materials. The structural materials of superconducting machineries at cryogenics are exposed to high magnetic fields. However, there is little knowledge about the influence of high magnetic fields on the JIc values. A recent report shows an effect of high magnetic fields on the JIc of stainless steel 304[1].

The main objective of this present paper is to find out the possible magnetic field influences on the critical elasto-plastic parameter. In addition, it is intended to give a microstructural explanation and establish a mechanism.

Experimental Procedure

Material and specimen

The material used for the present investigation was commercially available austenitic stainless steel 1.6903. The chemical composition of the material is shown in Table 1.

Modified compact tension(CT) specimens with Transverse-Longitudinal orientation were machined from a ~20mm thick plate material. Figure 1 shows the main dimension of the specimen.

Method of test

A special designed cryogenic test rig fitting in a 15 T split coil type superconducting magnet[2] was the loading unit. The bore size of the magnet was 30mm x 70mm. Figure 2 shows the schematic view of the test facilities under the high magnetic fields. The loading was carried out simply by a manual screw driven rod. The load was measured outside of the cryostat with a 120 KN piezo-crystal load cell.

The displacement measurement was carried out by two extensometers working with a full bridge strain gauge system. One extensometer was placed outside of the cryostat for the measurement of the load-line displacement, the other one was fixed directly on the specimen's front line.

For the detection of a crack-initiation, every specimen were instrumented with a strain gauge attached directly on the top of the specimens. The output signals of this strain gauge are directly correlated to the elastic bending of the upper half of the specimen, which acts as a beam. An increase of a/w ratio by crack advance results in a sudden slope change of load v.s. strain curve[3].

All specimens were fatigue-precracked at room temperature to a/w ratio of 0.64, where a and w are the crack length and the width of the specimen, respectively. Each specimen was loaded after magnetic field stabilization with an approximate displacement rate of ~10^{-2}mm/s. Load v.s. displacement and the direct strain gauge signals were recorded with a commercial x-y recorder.

After unloading and warm up, the specimens were hint tinted at ∿ 750 K for 1 hr. under atmospheric condition. After fracturing the specimens at 77 K in a liquid nitrogen bath, the fracture surfaces were observed by a stereo microscope to measure the fatigue precrack lengths a_F and physical crack length a_p.

<h2 style="text-align:center">Results and Discussion</h2>

The value of JIc was obtained by the equation

$$JIc = \eta_T \, \frac{A}{B \cdot b} + \eta_C \, \frac{P_M \cdot \Delta XM}{B \cdot b} \qquad (1)$$

where η_C = complementary energy coefficient
η_T = real energy coefficient
B = specimen thickness
b = ligament
A = area under load v.s. load-line displacement
P_M = maxiumun load
ΔXM = maxiumun load-line displacement

Table 2 gives the measured and evaluated values for each specimen. The JIc value of the specimen NO15 resulted to a higher value due to the small ratio of a/w 0.61. The plotting of JIc and magnetic field B values is given in Figure 3. In this figure a former obtained JIc datum is also included[4].

It was found from the Figure 3 that the JIc value decreased with increase of magnetic fields and that the reduction JIc revealed to be ∿10%.

This JIc reduction is attributed to the strain-induced martensite, a ferromagnetic phase at the crack tip. This microstructural transformation process generates an additional internal magnetic induced friction stress due to the deformation. A momentum resulting from the rotational force toward the axis of the applied magnetic field gives hence an additional stress.

The following relation of the stress intensity factor K_I can be predicated;

$$K_I = (\sigma_p + \sigma_m) \sqrt{\Pi} \ a \qquad (2)$$

where a = Crack length

σ_ℓ = Stress by mechanical load

σ_m = magnetic friction-induced stress at crack tip

During the loading of the specimen under magnetic field the load measurement comprises only the term σ_ℓ, which results in a pseudo K_i value. This K_i value is considerably lower than the K_I value determined without the magnetic fields. The fraction of the stress σ_m is by the common measurement technique not obtainable. Assuming this above given mechanism one can deduce the value of σ_m to be proportional to (K_I-K_i).

To verify this assumption, however JIc test must be proceeded using non magnetic transformation materials under magnetic fields.

Conclusions

4 K elasto-plastic fracture toughness tests of stainless steel 1.6903 under magnetic fields give the following finding:
- JIc values decreases by $\sim$ 10% with increase of magnetic fields to 10 Tesla.
- It was assumed that the JIc reduction was due to the additional internal magnetic induced frictional stress.
- Single-specimen method with use of strain gauge on the top of the specimen could be successfully applied for this tests.

Acknowledgment

The authors would like to thank Miss A. Kling, Mrss. H.P. Raber and S. Fischer for their contributions to this work and also would like to thank Dr. B. Obst for the valuable discussions. Last not least the authors thank Prof. Komarek and K.P. Jüngst for this support and encouragement.

References

1) Fukushima, E., et al., IUTAM symposium
2) Specking, W. and Flükiger, R., "A Compact 5 KN-Test Facility for Superconducting Conductors Carrying up to 1.5 KA in Magnetic Fields up to 14 T", Journal De Physique Colloque C1, supplement au NO.1, Tome 45, janvier 1984, page c1-79.

3) Nyilas, A. and Yanagi, H., "4 K fracture toughness investigation of 316 LN stainless steel plate and forging materials", to be published in Cryogenics.

4) Nyilas, A., "Kriterion für die Auslegung von austenitischen Werkstoffen im kryogenen Bereich", 03.03.04P16A, October 1986, Kernforschungszentrum Karlsruhe GmbH.

Table 1   Chemical composition of the austenitic stainless steel 1.6903 in wt. %.

C	Si	Mn	P	S	Cr	Ni	Mo	Ti	V	Co	Nb
0.07	0.691	1.29	0.031	0.015	17.26	10.94	0.315	0.433	0.049	0.196	0.017

Table 2   4 K results of J_{IC} tests of stainless steel 1.6903 under magnetic fields.

Specimen	Magnetic fields tesla	J_{IC} N/mm	a/w	mean J_{IC} N/mm	Remarks
NO 16	0	239	0.64	236	a_p = 0.166 mm
NO 6		233	0.64		
NO 9	5	223	0.63	220	
NO 8		217	0.64		
NO 15	10	221	0.61	213	
NO 2		205	0.64		a_p = 0.025 mm

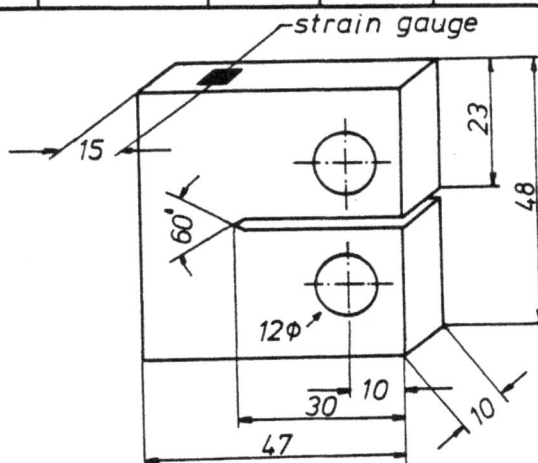

Fig. 1   Used compact tension specimen for the measurements.

508

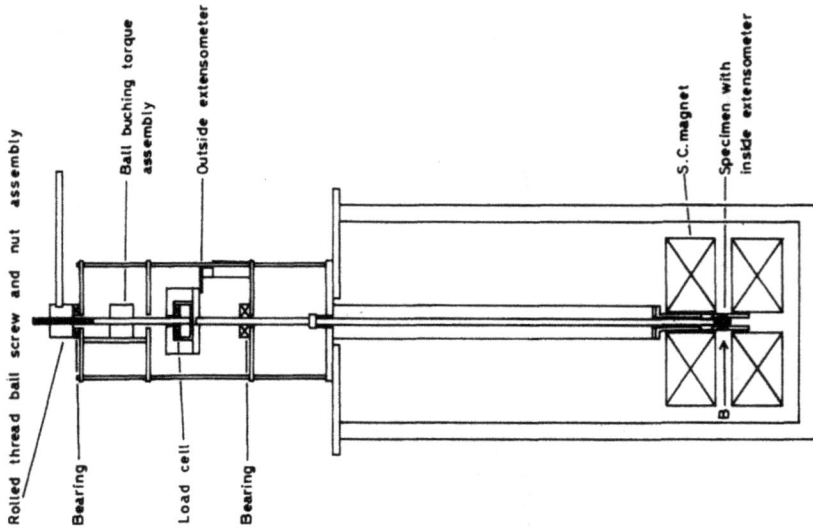

Fig. 3   4 K elasto-plastic toughness J_{IC} dependent on magnetic fields.

o——o 10mm thickness
△——△ 23mm thickness

J_{IC} (N/mm)

300

200

100

Magnetic fields (Tesla)

0   5   10

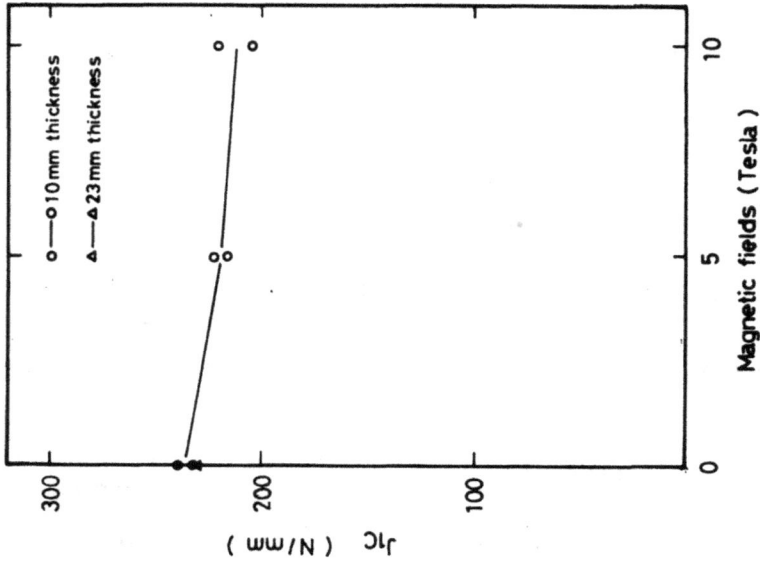

Fig. 2   The schematic view of the test facilities under magnetic fields.

Rolled thread ball screw and nut assembly
Ball buching torque assembly
Outside extensometer
Bearing
Load cell
Bearing
S.C. magnet
Specimen with inside extensometer
B

SHETEM PROJECT - SUPERCRITICAL HELIUM FORCE COOLED SUPERCONDUCTING TEST MAGNET

Junya YAMAMOTO, and Yoshishige MURAKAMI

Laboratory for Applied Superconductivity, Faculty of Engineering
Osaka University, Suita, Osaka 565, Japan

Abstract - Developing a new cooling technology of a superconducting magnet for a large scale system like SMES and fusion reactor, we have studied
supercritical helium force cooled superconducting magnet system. Refrigeration and coil requirements of the system are clarified not only for the steady
state operation (dc) but also for the pulsive operation (ac).

INTRODUCTION

Forced cooling by supercritical helium will be a powerful method for a
large scale superconducting magnet due to its high electrical insulation characteristics, high elasticity cooling channel, and small quantity of high density helium. For a large diameter and a high current coil, a forced cooled
type is easier than a bath cooled type in a production process. Several examples of application of the method to dc operated magnet can be seen like muon
channel[1] and bubble chamber[2] magnet, however its ac or pulsive application is
still under developing, because ac loss in a conductor degrades superconducting properties. Before applied to a pulse operation coil or/and a high current density coil, more fundamental properties have to be studied for designing the refrigerator specification, coil parameters, and safety devices.
Osaka University has started a project of SHETEM (Supercritical Helium cooled
superconducting TEst Magnet), which aims to get engineering data for pulsive
or ac use.

FLOW INSTABILITY OF SHE IN CURVED TUBING

Prior to the test coil experiment, we have studied the flow instability
of supercritical helium (SHE) in a test tube under pulsive heating[3]. A test
coil was a 3 m long stainless steel tube, 2.7 m in heating length, 2.88 mm in
inner diameter, 3.18mm in outer diameter, 100mm in coil diameter, and 9.6
turns. Ten carbon thermometers were mounted on the tube wall in distance of
300 mm. When the forced cooled superconductor is subjected to a thermal load,
it is generally considered that the temperature rise of a cooling channel will
be dominant at the end of the heated section in the flow direction, because
heated helium gas moves downward and the thermal input is constant along the
channel. However, the present experiment clarified that the maximum wall temperature did not appear at the end of stream, but the middle of the cooling
channel due to the flow instability. The point of maximum wall temperature
traveled down the heated channel with time. The position of the maximum temperature was strongly affected by the flow velocity and heating time.

TEST COIL PROJECT (SHETEM)

The coil is planning to be put in a FRP vacuum vessel in a 0.5 MJ pulsed
magnet.[5] Section of conductors and the coil specification are shown in Fig.

1. SHETEM 1 is for the purpose of a basic understanding of forced-cooling of a superconducting magnet and to establish SHE technology. SHETEM 2A is a real size solenoid coil, and is adequate for studying the relation between super-conducting properties and thermal and flow dynamic properties of coolant. SHETEM 2B is designed for experiments under pulsive bias field to get allow-able pulsive heat power and period of forced cooled magnet. SHETEM 3 is Nb$_3$Sn cable-in-conduit coil with an internal SHE generator, and is a most promising conductor design for pulsive or ac operation.

SHETEM 1

SHETEM 1, which was a ten turn coil composed of a NbTi hollow conductor with an FRP bobbin, revealed superconducting state recovery effect after quench[5] as shown in Fig. 2. This phenomena is due to the increase of heat transfer rate by fast SHE velocity after quench in narrow region.

DESIGN OF SHETEM 2A

SHETEM 2A[6] is a 215 turn solenoid coil. The conductor of SHETEM 2A is 87 m long and has a 3.0 mm ID cooling channel. There exists a FRP sheet (0.5 mm in thickness) between layers and the conductor is covered with Kapton tapes for electrical insulation. Then, transverse (turn-to-turn) heat conduction is very poor. Two heater wires are mounted on the whole conductor next to the superconductor. This configuration will be useful to study the effect of dc heat loads and ac losses on SHE in a hollow conductor. The proportional con-stant of the magnetic field at the center of the bore to a transport current is 1.55 x 10^{-3} T/A.

SHETEM 2A is put in a vacuum cryostat without bias magnetic field. Nor-mal transitions are detected by changes of electrical resistance in 6 sections along the conductor as shown in Fig. 3. Each section is composed of two layers of the solenoid coil. The length and the position of the section (X_n) are shown in Tab. 1, with the maximum magnetic field at each section and the critical temperature for 400 A. SHE passes through the coil from the inner-most layer. Rectangular thermal inputs were applied to the total length of the conductor, and the influence on temperature and mass flow rate of SHE was precisely observed. Two heating patterns were tested; one was a single pulse, and the other was repetitive pulses. The results suggested that (1) there was a minimum interval for periodic operation, and (2) there was a maximum ac load to keep the coil lower than a critical current.

OBSERVATION OF NORMAL TRANSITION

Figure 4 shows the typical results under the thermal input of 0.167 W/m. The initial conditions of SHE are as follows; mass flow rate 0.50 g/s, tem-perature at the inlet 5.85 K, pressure at the inlet 0.82 MPa, pressure drop 0.03 MPa, and background thermal load at the coil 2.0 W. The transport cur-rent was 400 A. As shown in Fig. 4, the thermal disturbance reduced the mass flow rate to 0.2 g/s due to the increase of flow impedance. The normal zone appeared at the time of 72 s after the beginning of heating. As the transport current and the heater current were decreased after all layers turned to the normal condition, heat dissipation in the normal zone reached 242 W. Conse-quently, the inverse flow of SHE happened at the inlet owing to the rapid pressure rise by the expansion of SHE, and the inlet temperature rose to 11 K. The normal zone appeared at X_4 (layers No.7 and No.8) and all layers turned to the normal in 6.6 s. It was found that this position was dependent on the heating intensity. Figure 5 shows in detail this dependence for the mass flow rate 0.8 g/s (open circle) and 0.5 g/s (closed circle). The axis of ordinates indicates the time of the normal transition (t_q) from the beginning of heating. The figure shows the time sequence of the quench propagation along the conductor against given dc thermal loads.

REQUIRED ENERGY FOR THE NORMAL TRANSITION

From these experiment, relation between the value of required energy, Q_j (q x t_q) (J/m) and the heating intensity q (W/m) is obtained as shown in Fig. 6, where

$$P = Q_j/V \int_{Ti}^{Tc} c_p \rho \, dT \qquad (1)$$

A denominator is an enthalpy change between the initial and the superconducting critical temperatures. If there is no temperature difference between the conductor and the helium coolant, P must be unity. The larger q/m value, the larger temperature difference, therefore P decreases like Fig. 8. Then as the necessary condition of the safety operation of the forced-cooled magnet, the eq. (2) should be satisfied.

$$Q_t < V \int_{Ti}^{Tc} c_p \rho \, dT \qquad (2)$$

where, Q_t is the thermal load given to the conductor during the transit time of SHE, t_s.

STABILITY AGAINST THE AC LOSS WITH SHETEM 2B COIL

The change of the mass flow rate by a rectangular heat input in SHETEM 2B coil was measured. During heating, the mass flow rate decreased by the expansion of SHE and the increase of the flow impedance, and recovered to the initial value in 10 s after heating. The decreasing speed of the mass flow rate is a linear function of the heating intensity[7].

The coil has four equal distance voltage taps. As shown in Fig. 8, a constant current (200A) was transported to SHETEM 2B and the bias magnetic field was changed repeatedly with the 0.5MJ pulsed magnet. The ac loss of SHETEM 2B for the different magnetic field sweep rate is measured by solo driven of the bias coil as shown in Table 2. The critical temperature of SHETEM 2B is 7.2 K at 200 A of I_t and 2.5 T of B. When the conditions of SHE were 5.3 K and 0.83 MPa, the right side of the eq. (2) is 8.83 J/m. From eq. (2), in case of dB/dt = 2.5 T/s and t_i = 1 s, it is expected that the normal zone appears after the fifth pulse. Figure 9 shows the bias magnetic field and the appearance of voltage due to the normal transition at X_3 in different magnetic field sweep rate. The numerals in the figure are the number of pulses. The mass flow rate of SHE was 1.0 g/s and t_s at the initial condition was about 41 s.

The normal transition was not observed before the fourth pulse for 2.5 T/s, it slightly appeared when the field was increasing, however, it turned to the superconducting state during the constant field. By the ac loss of the fifth pulse, the normal transition was observed while driving the magnet. The ohmic heating was about 8 J/m and it was considered that the temperature rise of SHE was about 2 K. In this case, the normal zone returned to the superconducting state owing to the increase of the critical temperature when the bias magnetic field became 0 T. At the sixth pulse, the superconducting state did not recover from the normal. In this way, the behavior of the superconductor should be considered with the relation between the thermal load by the ac loss and ohmic heating at the normal zone, and the change of the critical temperature according to the fluctuating magnetic field.

In Tab. 3, the summary of the experimental results of the different time interval are shown. The pulse number of the normal transition, N_p, agreed with that estimated by the eq. (2). With extending the time interval between pulses, the normal transition occurred at the down stream. This is due to decrease of integrated heat storage in the conductor.

CONCLUSION

1. Because of big difference of thermal capacity between the conductor and SHE at liquid helium temperature, temperature of the conductor is simply determined by SHE.
2. Temperature distribution of the conductor along the flow direction is determined by heat dissipation in the conductor and SHE initial condition like mass flow rate.
3. Quench is happened at the intersect point of the critical temperature of the conductor and SHE. The required energy for quench is obtained by equation (2).
4. Even pulse operation coil, the required energy for quench is the same as the dc operation. However the transit time of SHE determines the actual energy.
5. The high heating power generates the temperature difference between the conductor and SHE. The value of the heating intensity divided by the initial mass flow rate estimates the required energy.

The authors would like to thank Sumitomo Electric Industries, Ltd. and Kobe Steel, Ltd. for their technical supports. This work has been partially supported by Grant-in-Aid for Scientific Research from the Ministry of Education, Science and Culture.

REFERENCES

1) I. Horvath and G. Vecsey, Proc. MT-9, 174-177, 1985.
2) Morpurgo, Particle Accelerator 1, 255 (1970).
3) J. Yamamoto, K. Yamamuro, N. Ohuchi, and Y. Murakami, Adv. Cryogenic Engineering 31, 473-480 (1986).
4) Y. Murakami, J. Yamamoto, K. Kikuchi, N. Ohuchi, and Y. Inuishi, Adv. Cryogenic Engineering 29, 167-174 (1984).
5) N. Ohuchi, K. Yamamuro, J. Yamamoto, and Y. Murakami, Proc. ICEC 10, 398-402 (1984).
6) N. Ohuchi, J. Yamamoto, and Y. Murakami, IEEE Trans. MAG-23, 1539-1542 (1987).
7) N. Ohuchi, Y. Makida, J. Yamamoto, and Y. Murakami, Adv. Cryogenic Engineering, 159-166 (1988).

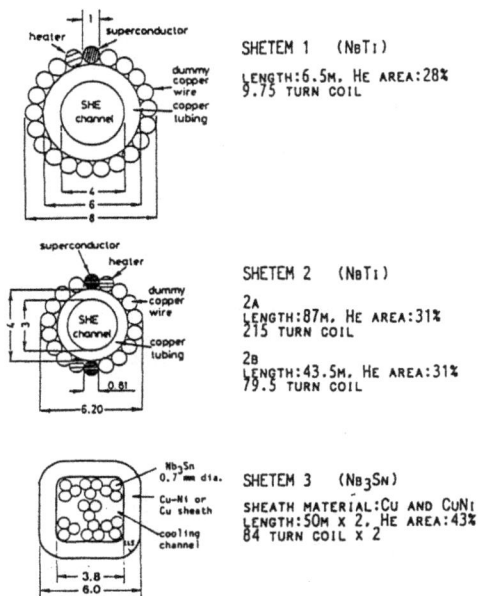

SHETEM 1 (NBTI)

LENGTH:6.5M. HE AREA:28%
9.75 TURN COIL

SHETEM 2 (NBTI)

2A
LENGTH:87M. HE AREA:31%
215 TURN COIL

2B
LENGTH:43.5M. HE AREA:31%
79.5 TURN COIL

SHETEM 3 (NB3SN)

SHEATH MATERIAL:CU AND CUNI
LENGTH:50M x 2. HE AREA:43%
84 TURN COIL x 2

Fig. 1. Conductor design of SHETEM.
SHETEM2B conductor has electrical tap
wire substituted to copper dummy wire.

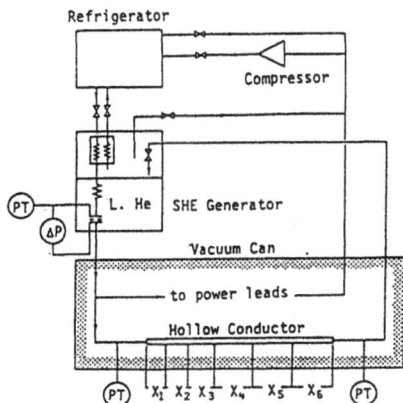

Fig. 3. Flow diagram of supercritical
helium force-cooled system. Test coil is
SHETEM2A.

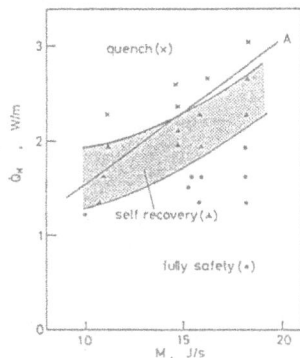

Fig. 2. Region of fully safety, self-
recovery, and quench of SHETEM1 coil.

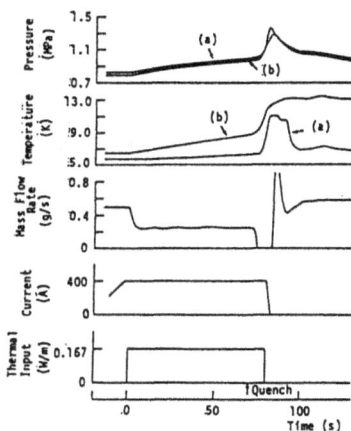

Fig. 4. Thermal condition of SHE at
quench. Initial mass flow rate: 0.5 g/s,
heating intensity: 0.167 W/m. (a) and
(b) are the inlet and the outlet of SHE,
respectively.

Table 1. Length and T_C of each section

layer No.		L(m)	B(T)	T_C(K)
X1	1,2	6.95	0.62	7.93
X2	3.4	10.1	0.52	8.16
X3	5,6	13.1	0.41	8.39
X4	7,8	16.0	0.31	8.62
X5	9,10	19.0	0.21	8.85
X6	11,12	21.9	0.10	9.08

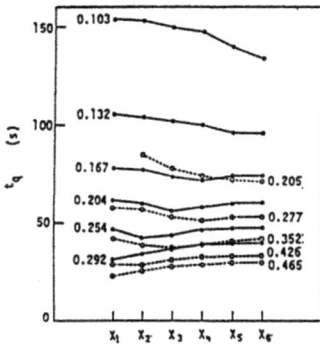

Fig. 5. Dependence of heating intensity on the position of the normal transition. o: 0.5 g/s, o: 0.8 g/s. The heating intensity (W/s) is shown.

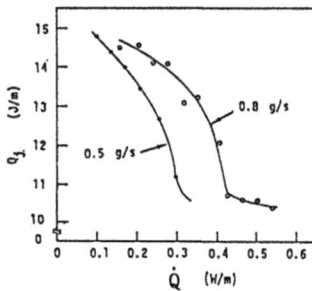

Fig. 6. Relation between the required energy for quench, Q_j, and heating intensity, $\dot{Q}$.

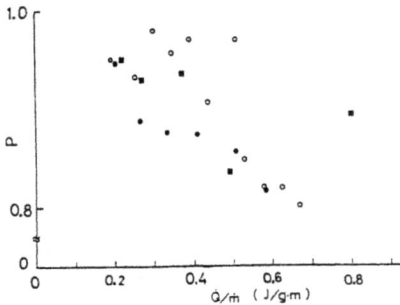

Fig. 7. Relation between P and $\dot{Q}/\dot{m}$ for the mass flow rate of 0.2 g/s(■), 0.5 g/s(●), and 0.8 g/s(o).

Table 2. AC loss of SHETEM 2B.

B	AC loss	heat load one pulse	t_p
T/s	W/m	J/m	s
1.25	0.23	0.91	4.00
2.50	0.91	1.82	2.00
3.75	1.82	2.43	1.33

t_p: heating duration for one pulse

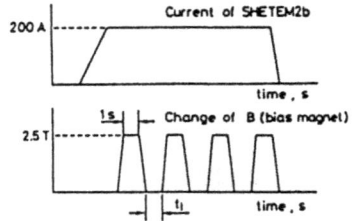

Fig. 8. The operating current of SHETEM2b and the bias magnetic field.

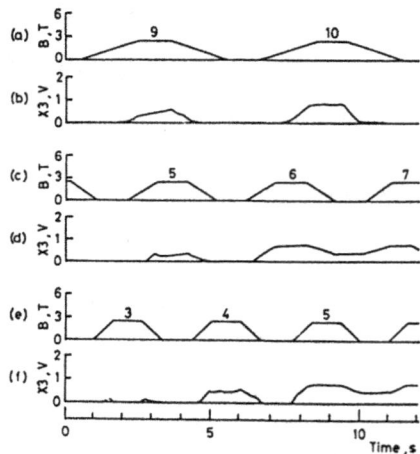

Fig. 9. The appearance of the normal zone at X_3 for the different sweep rate of the magnetic field. (a):1.25T/s, (c):2.5T/s, (f):3.75T/s.

Table 3. Normal transition region in different heating and cooling condition.

t_1(s)	X_1	X_2	X_3	X_4	N_p	
2	N	N	N	N	4	$\dot{B}$:3.75T/s
4	S	S	N	N	4	$\dot{m}_1$:1.0g/s
8	S	S	S	N	4	t_{s1}:41s
10	S	S	S	N	4	
1	N	N	N	N	5	
2	N	N	N	N	5	$\dot{B}$:2.50T/s
4	N	N	N	N	5	$\dot{m}_1$:0.8g/s
6	S	N	N	N	5	t_{s1}:51s
10	S	S	N	N	5	
12	S	S	S	N	5	

S:super, N:normal, t_1:pulse interval, N_p:No. of pulses, $\dot{m}_1$:mass flow rate at initial state, t_{s1}:transit time of SHE through coil.

VOLTAGE DISTRIBUTION WITHIN SUPERCONDUCTING COILS DURING QUENCH

Toshiharu TOMINAKA, Nobuhiro HARA, and Kunishige KURODA

Hitachi Research Laboratory, Hitachi, Ltd.
4026 Kuji-cho, Hitachi-shi, Ibaraki-ken, 319-12, Japan

Abstract - A computer program which can be applied to the calculation of a voltage distribution within superconducting coils during quench has been developed. Then, the calculation was compared with an experiment for a small superconducting solenoid, and the propriety of the calculation was discussed.

Introduction

In order to protect a superconducting magnet, it is necessary to estimate the maximum temperature and voltage of the coil during quench. If the adiabaticity is assumed in the calculation of the temperature rise, the time variation of coil current related to the maximum temperature is calculated only from the overall resistance of a normal zone. However, for the estimation of the maximum voltage, the voltage distribution within the coil from the resistance and inductance distribution along the superconducting wire must be calculated [1].

In this article, the quenching process in a superconducting magnet is calculated with the computer program "QUENCHM" which was developed on the basis of the program "QUENCH" [2]. Then, in "QUENCHM", the temperature distribution within the coil is calculated on the assumption that the normal zone is an ellipsoid, instead of a detail calculation of the thermal equation.

Experiment on a Small Superconducting Solenoid

The experiment about the quenching process of a small superconducting solenoid has been described elsewhere in detail [3]. In brief, several heaters, voltage terminals and thermocouple thermometers were attached within the solenoid as shown in Fig.1. The heaters H1-H9 attached within this solenoid were constructed from a manganin wire which is wound, extending over the length of about 10 mm along the superconducting wire. Then, these excited heaters can be considered to be point-like heat sources. The parameters of this superconducting solenoid and wire are listed in Table 1. The solenoid circuit is shown in Fig.2. The operating current in the experiment was chosen to be 162 A, and the dump resistor was 0.2 Ω.

After the heater was excited by the current of a pulse waveform, the voltage of the voltage terminals at each layer and the temperature at several locations within the solenoid were measured. Particularly for the cases in which the normal zone propagated from the heaters H2, H5 and H8, the calculation was compared with the experiment. The beginning and terminating times of normal propagation for each winding layer are plotted in Figs.3-5. In these figures, the circular markings mean the experimental results for the beginning of the normal propagation, and the square markings mean those at the termination.

In addition, the voltage distributions during these quenching processes are plotted for every 100 ms in Figs.6-8, corresponding to the quenching from heaters H2, H5 and H8, respectively. In these figures, the time origin was chosen to be the beginning time of the normal propagation at the layer with the excited heater, and the values of voltage were plotted for every one layer or two layers. In these measurements, the power supply was not switched off for at least 500 ms.

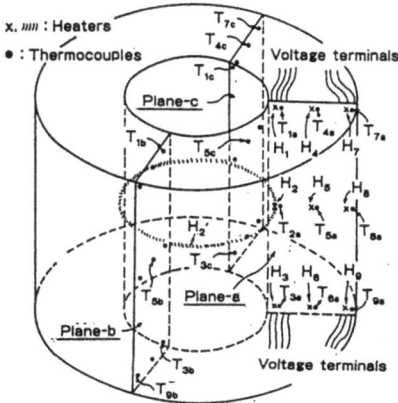

Fig.1 Arrangement of heaters, voltage terminals and thermocouples in the superconducting solenoid

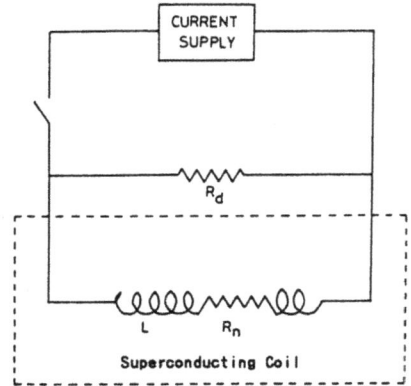

Fig.2 Electric circuit of the superconducting solenoid

Table 1 Specifications of superconducting wire and coil

Superconductor	Nb-Ti	Inner dia. (mm)	39.5
Filament dia. (μm)	50	Outer dia. (mm)	128
No. of filaments	42	Height (mm)	59
Cu/super ratio	7.6	No. of turns	1744
Overall wire dia. (mm)	1.0	No. of layers	46
		Inductance (mH)	126
		Quench current (A)	~ 165
		Central field (T)	3.67

Calculation and Comparison with Experiment

First of all, the propagation velocity of the normal zone was determined from the experiment. Secondly, with respect to the voltage distribution within the coil, the calculation was compared with the experiment.

The longitudinal velocity (the propagation velocity along the conductor) is much faster than the transverse velocity (the radial and axial velocities). Then, the propagation of a normal zone can be considered to be almost two dimensional and determined by the transverse velocity, on the assumption that the normal zone propagates with the shape of an ellipsoid. Furthermore, the beginning time of the normal propagation for each layer is determined only by the radial propagation velocity. In addition, the termination time of the normal propagation is determined by both the radial and axial propagation velocities. Therefore, both the radial and axial propagation velocities were determined so that the calculated beginning and terminating times fit the experimental results best. As a result, the calculated and experimental

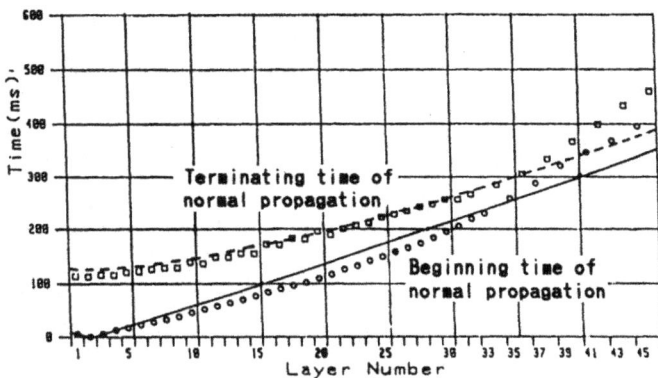

Fig. 3 Beginning and terminating times of normal propagation for each layer when heater H2 was excited

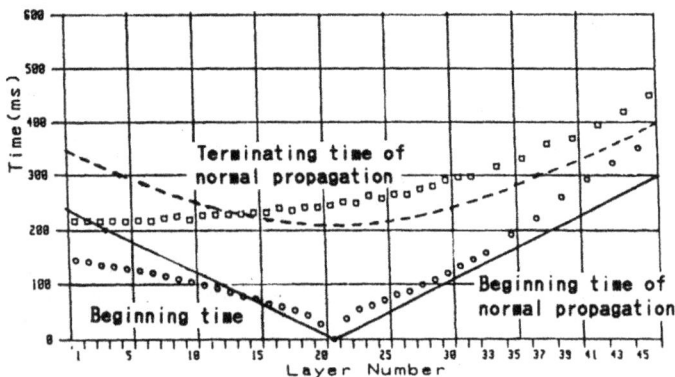

Fig. 4 Beginning and terminating times of normal propagation for each layer when heater H5 was excited

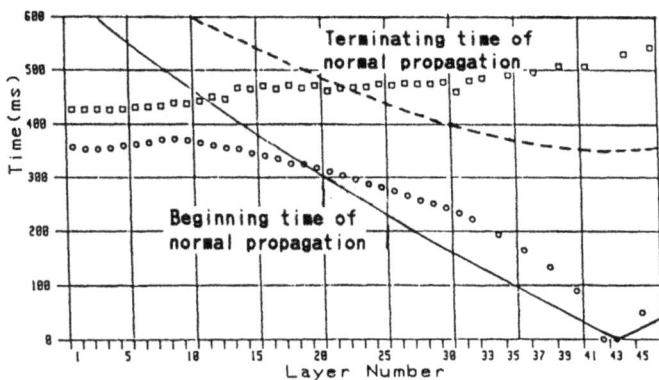

Fig. 5 Beginning and terminating times of normal propagation for each layer when heater H8 was excited

518

results are shown in Figs.3-5, corresponding to quenching from heaters H2, H5
and H8, respectively. In these figures, the solid curve means the calculated
result at the beginning of the normal propagation and the dashed curve means
that at the termination.

In this calculation, the propagation velocity was assumed to be
proportional to the current. In all cases shown in Figs.3-5, the initial
propagation velocity along the conductor is assumed to be 5.7 m/s. In the case
shown in Fig.3, the initial radial propagation velocity is calculated to be
0.13 m/s, and the initial axial propagation velocity is 0.23 m/s. In Fig.4,
the initial radial propagation velocity is 0.085 m/s, and the initial axial
propagation velocity is 0.14 m/s. And in Fig.5, these values are 0.074 m/s and
0.85 m/s, respectively.

In the case that quenching spreads from heater H2 located in the inner
layer, the calculation is comparatively consistent with the experiment as
shown in Fig.3. On the other hand, when quenching spreads from heater H8,
located in the outer layer, the calculation is not consistent with the
experiment as shown in Fig.5. This means that the normal zone initiated from
heater H2 can be assumed to be an ellipsoid, however, the normal zone from
heater H8 cannot be assumed so. In addition, when quenching spreads from
heater H5 located in the middle layer, the experimental radial propagation
velocity to the inner direction is faster than the outer one. However, the
calculational velocity is the same for both directions as shown in Fig.4.

The calculated results of the voltage distribution within coil are shown
in Figs.9-11, corresponding to quenching from heaters H2, H5 and H8,
respectively. In this calculation, with a fine division of the whole coil
conductor, the voltage along the conductor from the resistance and inductance
of the divided short conductors is calculated as follows:

$$V_j = V_0 + \sum_{i=1}^{j} \{r_i I + (l_i + m_i) dI/dt\},$$

where V_0 is the voltage at the termination point of the coil winding; r_i, the
resistance of each short conductor; l_i, the self-inductance; and m_i, the
mutual-inductance. The voltages at each mesh point are calculated from the
termination point to the beginning point of the coil winding. In these
figures, the current is assumed to be flowing from the beginning point to the
termination point. In addition, the voltage at the termination point V_0 is
assumed to be zero. Therefore, the voltage at the beginning point is equal to
the voltage drop of the dump resistor.

In these figures, the voltage values are plotted for every half turn.
But, the voltage is calculated for every 1/20 turn. Then these do not mean
that the fluctuation with the period of one layer on the voltage distribution
shown in Figs.9-11 conflicts with the experimental results, because the
experimental results are plotted for every one layer or two layers. In this
calculation, the value of the inductance was given for every layer. These
values were determined experimentally from the induced voltage during the
excitation.

In the case that quenching spreads from heater H2 located in the inner
layer, the calculation is qualitatively consistent with the experiment as
shown in Figs.6 and 9. On the other hand, when quenching spreads from heater
H5, located in the middle, or heater H8, located in the outer layer, the
calculation is not consistent with the experiment. This means that the
consistency between calculation and experiment on the voltage distribution is
strongly related to that on the propagation of the normal zone.

The calculation was also compared with measurements about the temperature
at 500 ms after quenching was initiated. In the case that quenching spread
from heater H2, the measured temperature at T2a shown in Fig.1 was about 32 K,

519

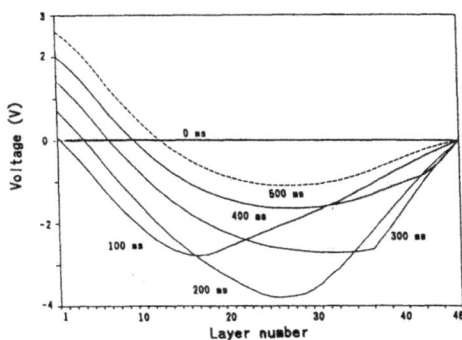

Fig. 6 Voltage distributions within coil
when heater H2 was excited (experiment)

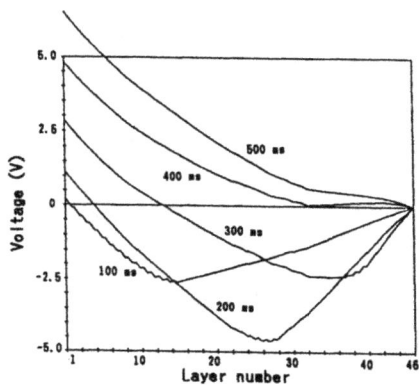

Fig. 9 Voltage distributions within coil
when heater H2 was excited (calculation)

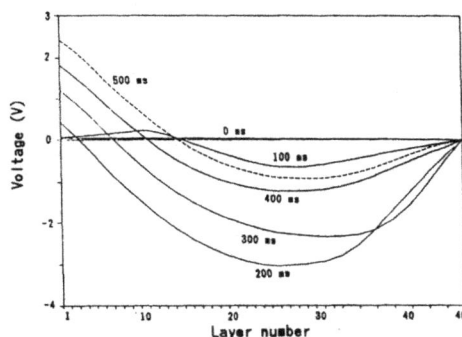

Fig. 7 Voltage distributions within coil
when heater H5 was excited (experiment)

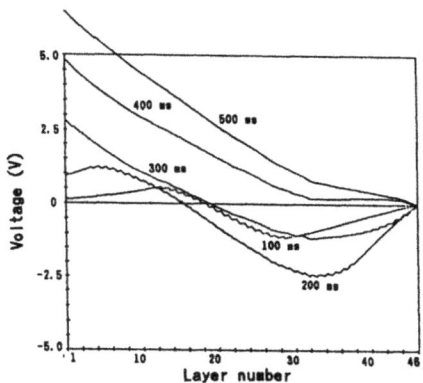

Fig. 10 Voltage distributions within coil
when heater H5 was excited (calculation)

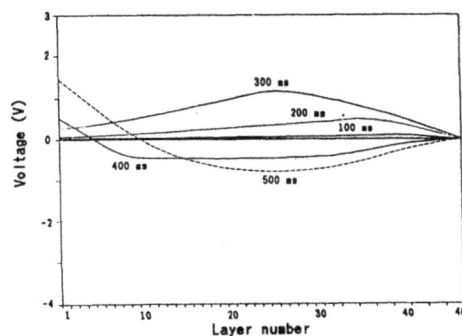

Fig. 8 Voltage distributions within coil
when heater H8 was excited (experiment)

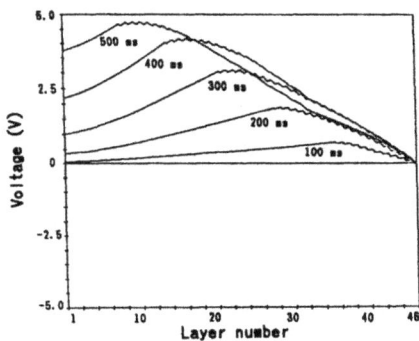

Fig. 11 Voltage distributions within coil
when heater H8 was excited (calculation)

and the calculated one was 28 K. When quenching spread from heater H5, the measured temperature at T5a was about 27 K, and the calculated one was 28 K. When quenching spread from heater H8, the measured temperature at T8a was about 19 K, and the calculated one was 29 K. In both cases of quenching from heater H2 and quenching from heater H5, the calculation is rather consistent with the experiment. However, for quenching from heater H8, the calculation is not so. It seems that the deviation is derived from the neglect of the magnetic field effect for the resistivity of copper. In the inner layer which is exposed to a strong magnetic field, the joule heat generation will be relatively large due to the large resistivity of copper. Conversely, in the outer layer exposed to a weak magnetic field, the joule heat generation will be relatively small.

From the comparison between calculated and experimental values of voltage, the calculated voltage at the beginning point of the coil winding which is equal to the voltage drop on the dump resistor is larger than the experimental one. This is derived from overestimation of the resistance of the normal zone. there are several causes for this overestimation of resistance. First of all, the magnet is cooled by liquid helium, but the temperature rise is calculated adiabatically. In addition, a calculational error is produced from the assumption that the conductor length per turn is the same for both the inner and outer layers. As a result, it seems that the approximation in this program is too crude, especially for a small size and low temperature rise of the coil.

Conclusion

With respect to the overall propagation of the normal zone and the voltage distribution within a superconducting coil, in the case that quenching spread from the inner layer of the solenoid, the calculation with the assumption that the normal zone was always an ellipsoid, was reasonably consistent with the experiment. However, in other cases, the calculation was not so. In most cases, quenching was initiated from the inner layer exposed to the strong magnetic field. Therefore, in this case, the voltage distribution could be calculated correctly with this program. The maximum voltage between the winding layers and between the coil and container could also be estimated from the voltage distribution. Then, it was expected that this program would be useful for safety estimations of a quenching coil.

References

1) J.Schultz, "Magnet Protection", MIT Plasma Fusion Summer Short Course, (Large Scale Applications of Superconductivity) 18-22 July (1983).
2) M.N.Wilson, "Superconducting Magnets", Oxford University Press (1983) p 218.
3) K.Kuroda et al, "Quench Simulation Analysis of a Superconducting Coil", Cryogenics, to be published.

DETECTION OF LOCAL NORMAL ZONE IN SUPERCONDUCTING MAGNET USING ULTRASONIC WAVE

Akira NINOMIYA*,Kazuyuki SAKANIWA*,Noboru KITAZAWA*
Takeshi ISHIGOHKA*,Hiromichi TOYODA*,Yakichi HIGO**

* Seikei University, 3-3-1, Kichijoji-Kita, Musashino 180 JAPAN
** Tokyo Institute of Technology, Nagatsuta, Yokohama 227 JAPAN

Abstract   We have been developing a new method to detect a quench of a superconducting magnet at its early stage. This method is a kind of non-destructive one which monitors a change of acoustic transfer function of a superconducting magnet induced by a local temperature rise or an epoxy crack etc..   Some experiments are carried out on a small epoxy   impregnated magnet. The detected acoustic signals are digitally processed by FFT analyzer and are quantitatively analyzed.   The experimental results show that a local temperature rise of about 2-3K can be detected by this method. And, some leading symptoms before quench were detected.

Introduction

In large scale superconducting magnets , a continuous monitoring and an early quench detection technique are essential. So far, the method to detect a quench has been mainly depending on the voltage detection. However, this method does not have enough sensitivity in the presence of electromagnetic noise.   Therefore, some novel methods other than a voltage detective one have been expected. We have been developing a new method to monitor and diagnose a superconducting magnet using ultrasonic wave [1-4].   This technique makes use of one pair of piezoelectric elements.   One piezoelectric element (driver) is placed on the top of the magnet and the other one (receiver) is attached to the bottom of the magnet. The injected ultrasonic wave from the driver travels through the magnet and is caught by the receiver. And, we have monitored the change of acoustic transfer function of the magnet induced by a local temperature rise or an epoxy crack. Some experiments on a small epoxy-impregnated superconducting magnet using an acoustic pseudo-random-noise with wide frequency range are carried out. In the experiment, the driver and the receiver signals are processed by a Fast-Fourier-Transform (FFT) analyzer. For a stable magnet state, a constant acoustic transfer function with enough reproducibility has been obtained. While, for any change of inner state, the transfer function shows obvious change. This phenomenon is confirmed for a local temperature rise of about 2-3K. And, some preceding symptoms which leads the quench were observed.

Experimental Procedure and Measuring system

The cross section and the specification of the experimental magnet are shown in Fig.1 and Table 1, respectively. The point heating heater made of manganin wire and thermo-couples(Au+0.07 at%Fe vs. Normal Silver) are located at the central part of the innermost winding. The driver and the receiver transducers are the wide band type piezo-electric elements. They are tightly fixed on the top and the bottom of the experimental magnet by sensor holders as shown in Fig.1.

The experiment was carried out immersing the whole assembly in liquid helium. In the steady state experiment, the magnet transport current and the heater current were kept constant. On the other hand, in the transient experiment, only the transport current was kept constant, and the heater was activated suddenly by the pulse discharge current from a capacitor bank. Also, the quench experiment without local heating was carried out. The receiver signals are amplified by 70dB. And at the same time, a multi channel data recorder is used to record the magnet voltage, the transport current, temperature, etc.. The measuring system is shown in Fig.2.

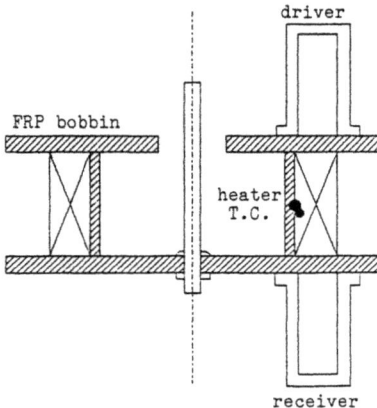

Fig.1. Section of magnet

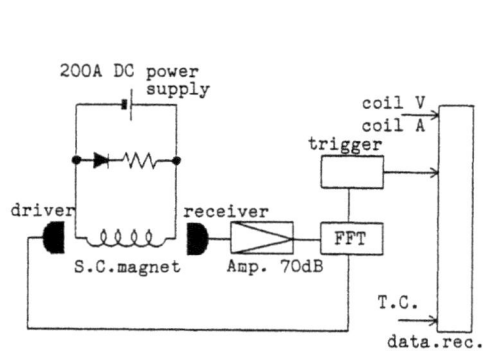

Fig.2. Measuring system

Table 1.   Parameters of magnet

```
-----------------------------------------------------
wire Nb-Ti
        diameter: 0.442mm ,filament no.: 564
        matrix  : Cu       ,matrix/SC   : 1.35/1
        twist   : 12.7mm  ,    Ic       : 122A at 5T
-----------------------------------------------------
No. of layers   26    , inductance   0.21H
inner dia.      60mm  , outer dia.   84mm
length          30mm  , material of bobbin  FRP
thermo-couple:(Au+0.07 at%Fe vs.Normal silver)
-----------------------------------------------------
sensor: piezoelectric element (wide band type)
-----------------------------------------------------
```

Experimental Results and Discussion

Steady state characteristics

Fig.3(a) shows the transfer function at I_m=0[A] (I_m:magnet current) and T=4.2[K]. On the other hand, Fig.3(b) shows the coherent function between the input and the output signals under the same condition. They are digitally processed through an averaging of 9 data to improve the S/N ratio. We selected the frequency range around 250-260kHz, because in this range both the sensitivity and the coherence are high. Fig.4(a) shows the ratio between the two data both at 0A, 4.2K transfer function before quench . On the other hand, Fig.4(b) shows the ratio of the data before quench to that after

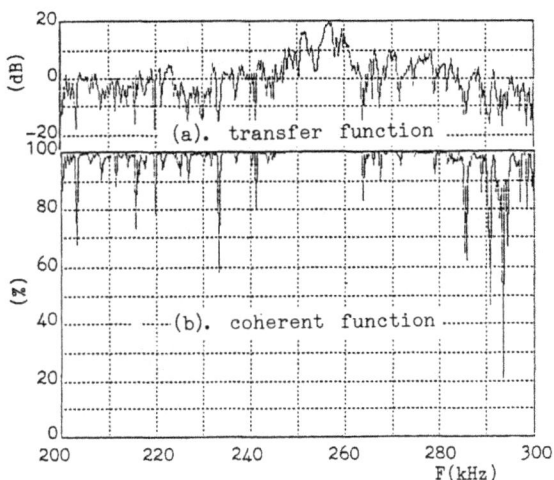

Fig.3. Transfer function,$G(\omega)$=Receiver(ω)/Driver(ω) and coherent function,I_m=0A,T=4.2K

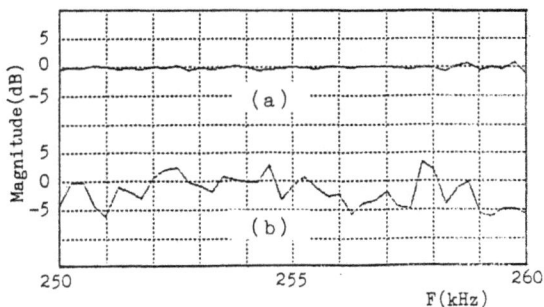

Fig.4. Comparison between the deviation of transfer functions before and after quench
(a). The ratio of two transfer functions before quench both at 0A,4.2K
(b). The ratio of transfer function before quench to that after quench

quench. These ratios show the change of the transfer function of the magnet itself. As can be seen in Fig.4, the transfer function does not change if the inner state of the magnet does not change. And also this method would be effective for an off-line diagnosis of superconducting magnets.

The quantitative change of the transfer function from the original one (obtained at 5A, 4.2K) due to the magnet current and the local temperature rise is shown in Fig.5. As shown in the figure, this method is sensitive to both the current rise and the temperature rise. However, the change due to the magnet current can be expected in advance. Therefore, if any deviation of the transfer function from the expected value was detected, it should be considered due to an anomalous disturbances like a local temperature rise or an epoxy crack, etc..

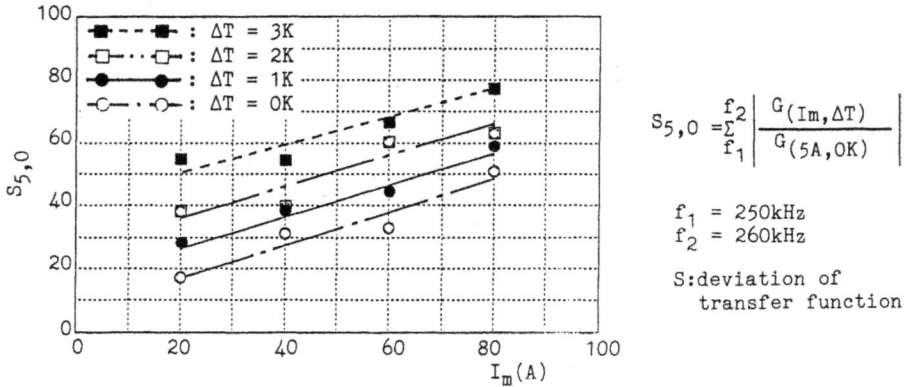

$$S_{5,0} = \sum_{f_1}^{f_2} \left| \frac{G(Im, \Delta T)}{G(5A, OK)} \right|$$

$f_1 = 250\text{kHz}$
$f_2 = 260\text{kHz}$

S:deviation of
transfer function

Fig.5. Deviation of transfer function at steady state

Transient characteristics

A transient characteristics at a quench due to the pulse spot heating is shown in Fig.6. Fig. 6(a),(b),(c),(d) show the magnet voltage, the magnet current, the heater current and the temperature of the spot, respectively. The deviation of the transfer function during the transient is shown in Fig.7. Fig.7(a) shows the deviation of the transfer function for the signals with the coherence higher than 97%. On the other hand, Fig.7(b) is the one for the signals with the coherence lower than 30%. We consider that Fig.7(a) shows the change of transfer function obtained mainly by the driving signals , and Fig.7(b) shows the effect of spontaneous AE signal other than the driving signal.

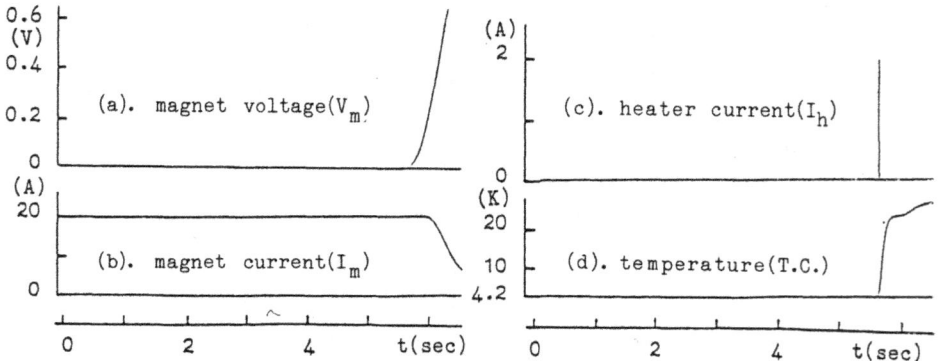

Fig.6. Transient characteristics by heater current

The quench characteristics due to the current increase is shown in Fig.8 and 9. As can be seen in the Fig.9(a) the deviation of the transfer function increases almost linearly with the magnet current. However, about 20 sec and 50 sec after the start of current rise, anomalous deviations from the expected line were observed. And in Fig.9(b), we can also find the sudden changes. We consider that some disturbances might have occurred in this instance.

525

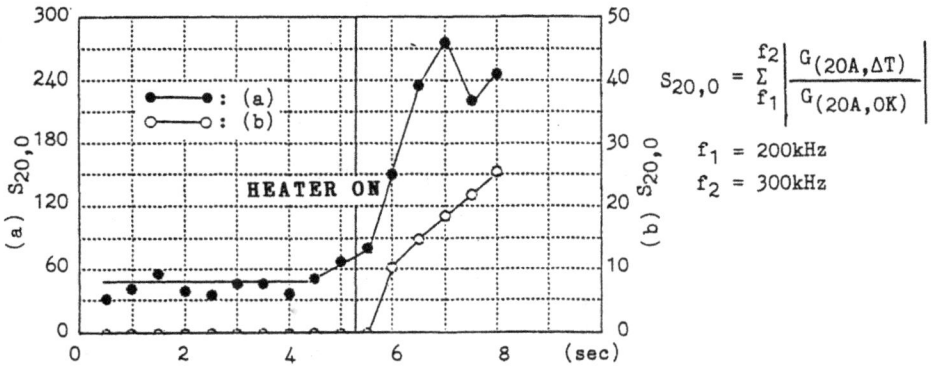

Fig.7. Deviation of transfer function at transient by
heater current
(a). the case for the signals with coherence higher than 97%
(b). the case for the signals with coherence lower than 30%

$$S_{20,0} = \sum_{f_1}^{f_2} \left| \frac{G_{(20A,\Delta T)}}{G_{(20A,OK)}} \right|$$

$f_1 = 200kHz$
$f_2 = 300kHz$

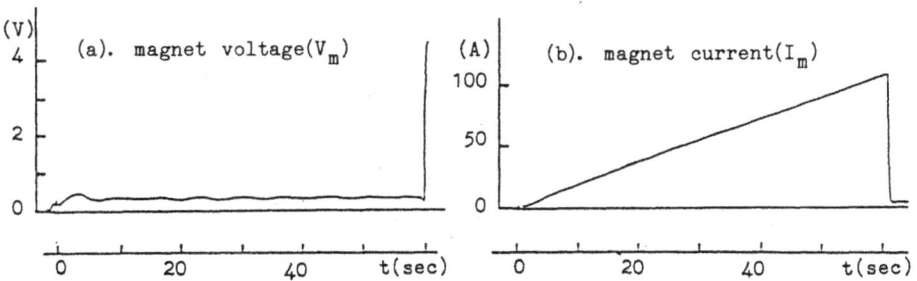

Fig.8. Transient characterictics during magnet current increase

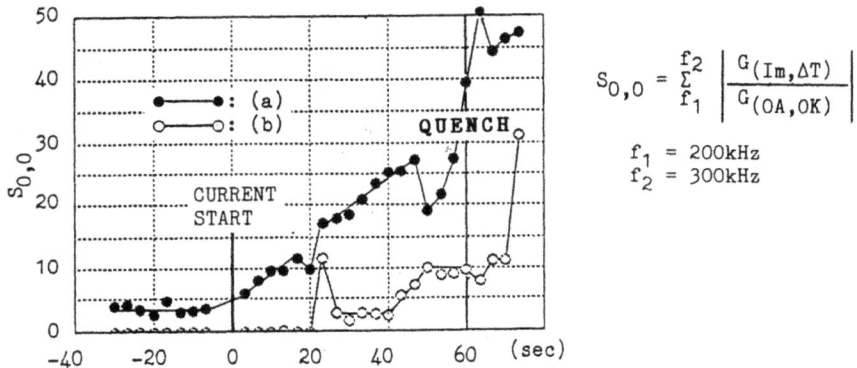

$$S_{0,0} = \sum_{f_1}^{f_2} \left| \frac{G_{(Im,\Delta T)}}{G_{(0A,OK)}} \right|$$

$f_1 = 200kHz$
$f_2 = 300kHz$

Fig.9. Deviation of transfer function during magnet current increase
(a). the case for the signals with coherence higher than 97%
(b). the case for the signals with coherence lower than 30%

Conclusion

We have offered a new method to detect a quench of a superconducting magnet. This method utilizes a change of an acoustic transfer function of the magnet. Conclusions are summarized as follows;

(1) The acoustic transfer function of a superconducting magnet changes by small heat input (about 0.1-0.3J) into the magnet, and the local temperature rise of 2-3K in the magnet can be detected.
(2) The transfer function does not change if the inner state of the magnet does not change.
(3) This technique would be effective for an early warning of quench by monitoring continuously the change of an acoustic transfer function of the magnet.
(4) This method would be also effective for an off-line diagnosis of the superconducting magnets.

There still exist many problems about the digital processing procedure etc.. However, we consider that fundamentally this technique would be effective for an early quench detection for large superconducting magnets.

Acknowledgment

This research is supported by Grant-in-Aid for Co-operative Research of the Ministry for Education, Science and Culture of Japan.
The authors are very grateful to Professor Emeritus N. Mizukami of Seikei University for many important suggestions.

References

[1] T.Ishigohka, O.Tsukamoto, and Y.Iwasa "Method to Detect a Temperature Rise in Superconducting Coils With Piezoelectric Sensors," Appl. Phys. Lett., Vol.43, No.3, pp.317-318, 1983.
[2] T.Ishigohka and A.Ninomiya, "Quench Detection by Acoustic Resonance," Cryogenic Engineering, Vol.22, No.1, pp.42-45, 1987.
[3] A.Ninomiya, T.Ishigohka, N.Mizukami, Y.Higo, "Monitoring of Superconducting Magnets Using Acoustic Resonance," IEEE Transactions on Magnetics, Vol.24, No.2, March 1988.
[4] A.Ninomiya, K.Sakaniwa, H.Kado, T.Ishigohka, Y.Higo "Quench Detection of Superconducting Magnets Using Ultrasonic Wave" 1988 Applied Superconductivity Conference, San Francisco, LC-9, Aug.21-25, 1988.

PROTECTION OF SUPERCONDUCTING MAGNETS USING ZnO ARRESTER

Takeshi ISHIGOHKA

Seikei University, Musashino 180 JAPAN

Abstract - A new protection method of superconducting magnet during quench is presented. The method utilizes ZnO arresters for dump resistor instead of conventional resistors. The ZnO arresters are used in power system to limit the overvoltages. By the introduction of ZnO arresters, both the discharge time and the heat dissipation in the superconducting magnet are reduced. The author carried out some preliminary experiments using small experimental magnets. The results show that ZnO arrester works as a dump resister for a superconducting magnet. And, it is confirmed that the discharge time of the magnet with ZnO arresters is much smaller than that with conventional dump resistors for the same maximum transient voltage.

Introduction

In large superconducting magnets, its protection during quench is very important. In a case of magnet quench, the stored energy has to be taken out from the magnet as soon as possible. For this purpose, the resistance of the dump resistor should be high. However, if it is selected too high, the over voltage which exceeds the insulation limit of the magnet may be generated. So, the resistance is limited to a certain value. As a result, the discharge time of the magnet current can not be made short enough. To improve such a situation, it would be effective to use a non-linear resistivity of ZnO arresters. So, the author has proposed a new quench protection method of a superconducting magnet using ZnO arresters instead of conventional dump resisters. ZnO arresters keep constant voltage (varistor voltage) when they are exposed to higher voltage than the threshold value. On the other hand, they exhibits almost infinite resistivity when lower voltage is imposed. As a result, ZnO arrester generates a constant voltage regardless of the current. So, it is considered to be an ideal dump resistor for superconducting magnets.

Theory

The new protection method of a superconducting magnet which utilizes ZnO arresters is shown in Figure 1(a). On the other hand, a typical quench protection circuit is shown in Figure 1(b).

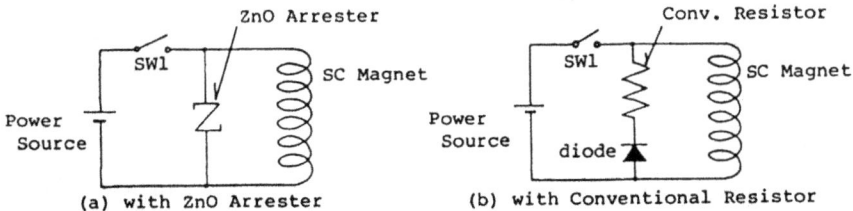

(a) with ZnO Arrester   (b) with Conventional Resistor
Fig.1 Protection Circuit of Superconducting Magnet

The voltage equations of both circuits when the switch SW1 is opened right after the quench become as follows;

(a) with ZnO arrester;

$$r_N i + L(di/dt) + V_v = 0 \qquad \qquad ...(1)$$

(b) with conventional resistor;

$$(r_N + R_d)i + L(di/dt) = 0 \qquad \qquad ...(2)$$

where

r_N : normal zone resistance of the magnet winding (time dependent)
L : inductance of the magnet
V_v : varistor voltage of the ZnO arrester (constant)
R_d : resistance of conventional dump resistor

The normal zone resistance of the magnet r_N is not constant, but it is time dependent. An example of measured resistance during quench is shown in Figure 2 [1]. It can be approximated as follows;

(a) for $t < t_q$

$$r_N = R_N(t/t_q)^2 \qquad \qquad ...(3)$$

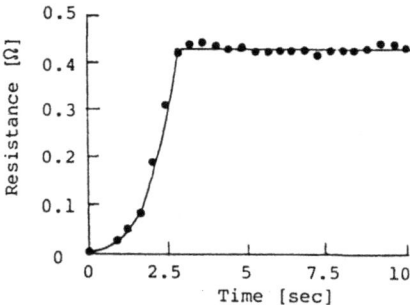

Fig.2 An Example of Normal Zone Resistance of SC Magnet [1]

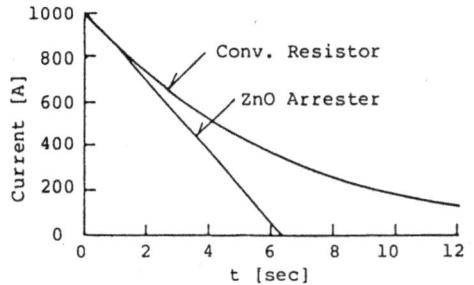

Fig.3 Coil Current

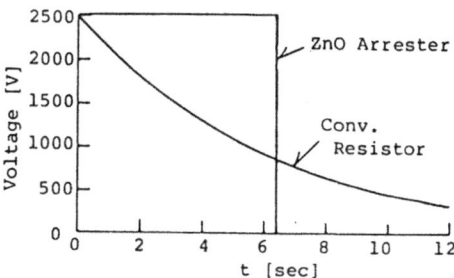

Fig.4 Terminal Voltage

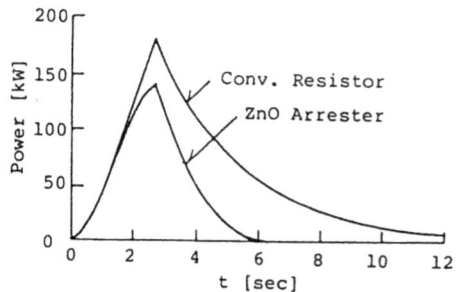

Fig.5 Dissipated Power in SC Coil

(b) for $t \geq t_q$

$$r_N = R_N \qquad \qquad ...(4)$$

where
    t_q : quench finishing time
    R_N : normal resistance of the coil (final value)

Equation (1) and (2) can be solved numerically. Taking the prototype coil of TORE SUPRA [1] as an example, the coil current, the terminal voltage, and the dissipated power in the coil, etc. are calculated. The results are shown in Figure 3, 4 and 5. The constants used for the calculation are summarized in Table 1. The maximum terminal voltages in both case are same. As shown in Figure 3 and 5, the discharge time and the total heat dissipation in the coil are much smaller in the case with ZnO arrester than in the case with conventional dump resistor.

Table 1   Constants of Coil Used for Calculation [1]

Inductance L [H]	16.75
Rated current I_0 [A]	1000
Stored energy E_0 [MJ]	8.37
Normal resistance R_N [ ]	0.42
Maximum terminal voltage V_{MAX}	2500*
Varistor voltage of ZnO arrester V_v [V]	2500**
Resistance of dump resistor R_d [ ]	2.5***
Quench finishing time t_q [sec]	2.7

*   This value is selected following the Euratom LCT coil [2].
**   The varistor voltage is same as the maximum terminal voltage.
*** $R_d = V_{MAX}/I_0$

The total heat dissipation in the coil during quench is a function of the maximum terminal voltage. The calculated values for both case are shown in Figure 6. As can be seen in Figure 6, the total dissipated heats in the coil decrease as the maximum terminal voltage increases. And, the effect of ZnO arrester becomes more remarkable when the higher terminal voltage is adopted. The effect of the ZnO arrester is shown in Table 2.

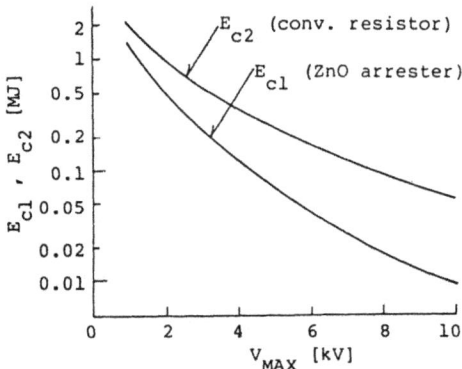

Fig.6 Energies Dissipated in SC Magnet

Table 2   Effect of ZnO Arrester

V_{MAX}[kV]	E_{c1}[MJ]*	E_{c2}[MJ]**	E_{c1}/E_{c2}
2	0.520	0.958	0.543
4	0.128	0.355	0.360
6	0.041	0.169	0.243
8	0.018	0.092	0.190
10	0.009	0.054	0.166

* E_{c1} is the energy dissipated inside the magnet when the ZnO arrester is used.
** E_{c2} is the energy dissipated inside the magnet when the conventional dump resistor is used.

Experimental Device

As a preliminary experiment to confirm the fundamental idea of this method, the fast discharge experiments were carried out using a small experimental superconducting magnet. The experimental circuit is one as shown in Figure 1(a). In this experiment, the switch SW1 is opened suddenly. During the experiment, the magnet does not quench. The experimental device is shown in Figure 7. As can be seen in Figure 7, the ZnO arrester is located close to the magnet, and the whole device is immersed in a liquid helium bath. The inductance of the magnet is 1.4H, and its stored energy is 2.2kJ.

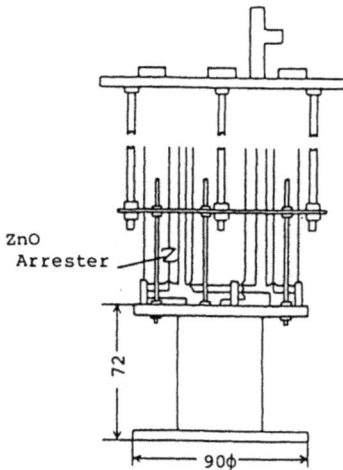

Fig.7   Experimental Device

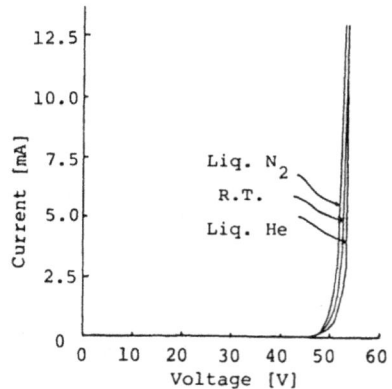

Fig.8   V-I Characteristic of ZnO Arrester

Experimental Results

Characteristics of ZnO Arrester

The characteristics of ZnO arrester at room, liquid nitrogen and liquid helium temperature are measured.   The ZnO arrester is Matsushita's M14JK470. Its varistor voltage and absorb energy in room temperature is 47V and 200J, respectively. The results are shown in Figure 8.   As can be seen in Figure 8, the characteristics at each temperature is close to each other. So, we can conclude that the ZnO arrester shows a sharp rising characteristics around 50V regardless of temperature.

Fast Discharge Experiment

As mentioned above, the fast discharge experiment was carried out using the experimental circuit shown in Figure 1(a). In the circuit, the switch SW1 was opened suddenly.   The transient oscillograms of the current and the voltage of the magnet are shown in Figure 9(a). On the other hand, the same experimental results in the circuit with conventional dump resistor are shown in Figure 9(b).

Comparing Figure 9(a) and (b) each other, we get that the current decays
almost linearly in the case with ZnO arrester, and that the transient voltage
is kept almost constant during quench. On the other hand, both the current
and the voltage decay exponentially in the case with conventional dump
resistor.   Therefore, the discharge time is much smaller in the case with ZnO,
arrester than that in the case with conventional dump resistor.

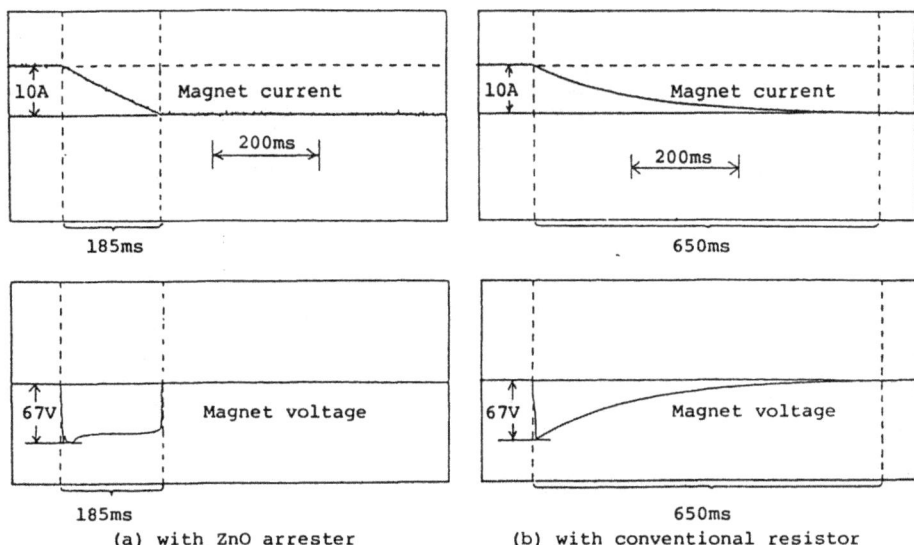

185ms   650ms
(a) with ZnO arrester   (b) with conventional resistor

Fig.9   Fast Discharge Characteristics of Superconducting Magnet

Discussions

Prospect for Large Scale Applications

For application of ZnO arresters to large scale magnets like fusion or
SMES, it is necessary to absorb high stored energy of the magnets more than
few MJ. For example, each LCT coil has a stored energy of about 100MJ. On the
other hand, the maximum absorb energy per unit volume of ZnO arrester is about
$400MJ/m^3$ in room temperature. From these figures, the required volume of ZnO
arrester for single LCT coil is estimated to be about $0.25m^3$. Therefore, it
is not so difficult to absorb this stored energy by ZnO arresters. Of course,
it is necessary to divide the discharged energy to many ZnO elements which are
connected in series and in parallel.

Suppression of Internal Overvoltage by ZnO Arresters

Other than as a dump resistor of superconducting magnet, the ZnO arrester
can be also used as an effective internal overvoltage protector. In this
case, a considerable number of ZnO arresters are connected to every
intermediate terminals of the magnet as shown in Figure 10. They should be
located in low temperature region to prevent the heat in-leak through the
leads. When enough number of ZnO elements are attached to the magnet, the

532

internal overvoltage would be suppressed perfectly. The difficulty of the
insulation design of superconducting magnets being reduced, the higher
operating voltage may be adopted. As a result, the design current can be
reduced. So, the heat leak into the low temperature region would be reduced
greatly.

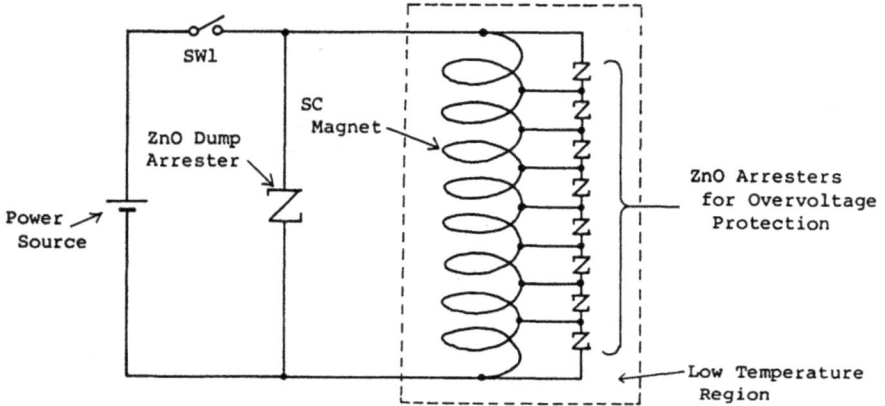

Fig.10 Overvoltage Protection Circuit Using ZnO Arresters

Conclusion

The conclusions are summarized as follows;

(1) It is confirmed that the ZnO arresters can be used as a dump resistor of
a superconducting magnet.
(2) The discharge time and the heat dissipation in a superconducting coil are
much smaller in the case with ZnO arrester than in the case with
conventional dump resistor for the same ceiling voltage.

By the introduction of ZnO arresters to the superconducting magnet
technology, the higher design voltage would become possible. However, further
extensive study should be done for the realization of the large scale
application.

Acknowledgement

This research is supported by Grant-in-Aid for Cooperative Research of
the Ministry for Education, Science and Culture of Japan.

References

[1] D.Ciazynski, C.Cure, J.L.Duchateau, J.Parain, P.Riband, B.Turck, "Quench
and Safety Tests on a Toroidal Field Coil of TORE SUPRA", IEEE Trans. on
Magn., MAG-24, No.3, pp.1567-1570 (1988).
[2] H.Tsuji, S.Shimamoto, A.Ulbricht, P.Komarek, H.Katheder, F.Wuchner,
G.Zahn, "Experimental Results of Domestic Testing of the Pool-Boiling
Cooled Japanese and the Forced-Flow Cooled Euratom LCT Coils",
Cryogenics, Vol.25, pp.539-551, (1985).

Section 3. Energy Applications

ENERGY STORAGE STUDY IN OSAKA UNIVERSITY

Yoshishige MURAKAMI

Faculty of Engineering, Osaka University
565 Suita Japan

Abstract - The small scale SMES which is oriented for power system stabilization is discussed. The energy storage 0.01 sec. with power 0.01-0.1 in per unit can stabilize power systems. Experimental results on the model system of long bulk transmission line verified the P(active)-Q(reactive) simultaneous control could stabilize the power flow oscillation and increased the transmitted power up to 50 % or so. Magnet technology and power electronics for The SMES are also discussed from the view points of the discovery of novel superconductivity.

1. Introduction

Through the study on SMES-Superconducting Magnet Energy Storage, we have recognized that the SMESs are classified according to the amount of storage such that: 1) 10^{13}J-, for peak shaving of the daytime power demand, 2)10^{10}J-, for power supply to the impulsive load and/or the storage for an intermediate sized load, 3)10^{6}J-, for the stabilization of power systems. This paper is mainly concerned with the SMESs of category 3).

The stability is the major issue of electrical power systems, and the strategies, methods and equipment have been developed for the stability, which have opened the abundant and diverse field of researches. However, we insist upon the distinguished features of SMES for stabilization. The system studies on various configuration of power systems have verified that the SMESs can stabilize systems in heavily loaded conditions as well as after violent disturbances by using the simultaneous control of active and reactive power[1]. It can even increase the transmitted power limited by instability in considerable amount, 50% or so.

2.SMES for Power System Stabilization

Capacities of Storage and Power of SMES: The basic configuration of SMES is composed of a superconducting magnet and a set of thyristor converters. For the control scheme of thyristor converters, we have proposed the P-Q simultaneous control in which the changing and discharging of active power P together with the reactive power Q of leading phase as well as lagging phase is controlled arbitrarily in the four quadrant P-Q domain, provided we have to use the thyristors of self turn of type such as GTOs[1]. We have verified this control scheme by the experiments in our laboratory.

Figure 1 shows the energy storage capacity with respect to the power capacity of converters. The dots plotted for the base power of 10,000 MVA in Fig. 1 show the converter powers which are in the range from several hundred to 1,000 MVA or from 0.01 to 0.1 in per unit are needed for the stabilization. The large converter powers which are the same order as for large scale SMESs are therefore needed. On the contrary, the small scale of stored energy, several hundred MJ or in the order of 0.01 seconds in per unit is sufficient.

Experimental Study[2]: Among electrical power systems, one that has a major load and a generator center separated by a long transmission line may experiences undamped and poorly synchronized power oscillations. For the experimental study of this sort of systems, we have developed a model system which is shown by Fig. 2. Parameters of the developed system together with the real scale parameters which are denoted in parentheses hereafter as for the base power of 2,000 MVA are given as: the generator rated power 10 KVA (2,000 MVA), the transmission line with rated voltage 460 V (500 KV) represented by the series and parallel connection of seven units of pi-figured lumped circuit, each unit of which corresponds to 40 Km. The SMES is composed of a pulse superconducting magnet of 500 KJ and of two sets of GTO thyristor bridges connected in series.

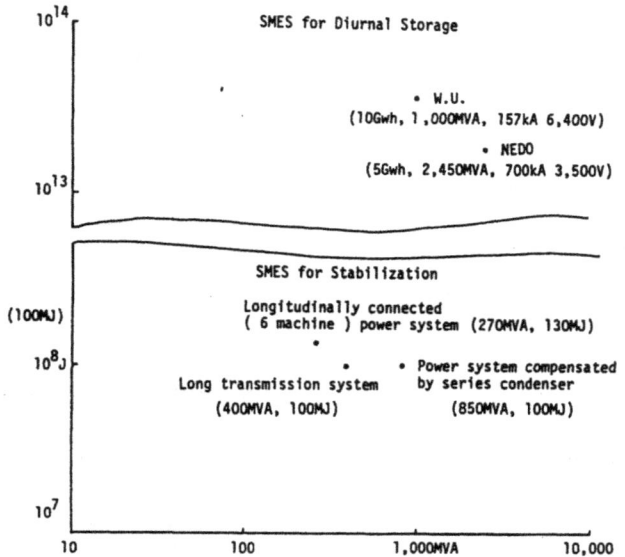

Fig. 1. Storage of SMES VS. converter capacity

Experimental Results:
The dynamics of the model power system after 3 phase 3 cycle short circuit were experimented, where the SMES located at the generator terminal, the length of transmission line 160 km, the initial condition of power flow 8kW, the generator terminal voltage 230 V. The simultaneous control of P and Q of SMES could damp out the oscillation in one and a half cycle of the oscillation, otherwise it would be sustained, where P and Q are controlled so as to follow the specified signals Ps and Qs given as:

$$Ps = -K_D \Delta \delta \qquad Qs = -K_V \Delta V.$$

Δ: Variation from steady state
δ: Torque angle of generator
Vs: Bus voltage where SMES located

The energy and power of SMES which were necessary for this stabilization were 500 J (100 MJ) and 2 KVA (400 MVA), respectively, which are consistent with the values given by Fig. 1 in order.

In Fig. 3, the occurrence of power flow oscillation with the gradual increase of the generator power up to 6.4 kW (1280 MW) is observed, for which the control by

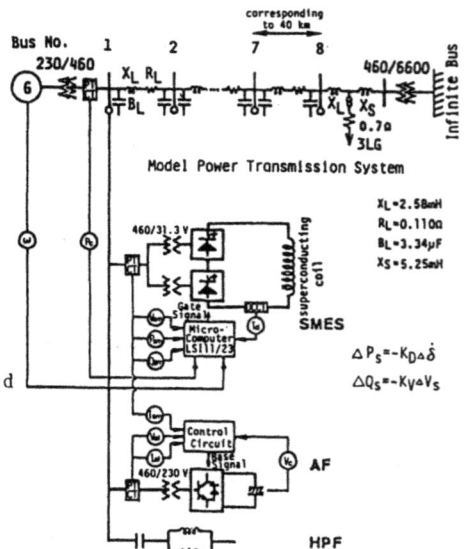

Fig. 2. Configuration of a model power transmission system and an SMES.

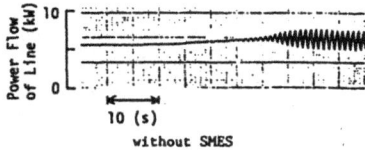

without SMES

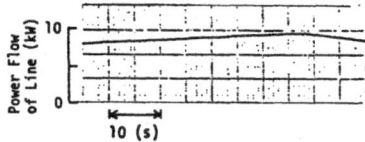

with control using active and reactive power

Fig. 3 Behavior of power oscillation with respect to the gradual increase in generator power output

SMES is not applied. This oscillation however can be suppressed up to 9.3 kW (1,800 MW) by using the P-Q control of SMES.

3.Superconducting Magnet

The superconducting magnet technology is developing continuously and even more indefinite on account of the discovery of novel superconductivity. Hence, we shall restrict our considerations to several view points : 1) the maximum/ minimum storage levels Emax/Emin compared to the average energy Eav, 2) the energy loss due to the ac loss of magnet and cryostat and that due to the thermal input to cryostat, and 3) stress and fatigue due to repeated drives.

Maximum/minimum Energy Levels: The increase of the ratio (Emax-Emin)/Eav specifies the coil design from dc coils to ac coils, which is accompanied with the increase of the converter power.

Energy Loss due to Thermal Input[3]: Although

Table 1. Parameters of the pulsed magnet

Coil bore diameter	288 mm
Winding inner diameter	310 mm
Winding outer diameter	494 mm
Winding length	256 mm
Rated current	1976 A
Conductor current density	168 A/mm²
Overall current density	79.9 A/mm²
Central field	5.0 T
Maximum field	6.1 T
Number of turns	952
Inductance	0.264 H
Stored energy	515 kJ

AC loss in cryostat, calculated value

POSITION		t(mm)	T(K)	ρ(10⁻⁸Ωm)	LOSS(W)	LOSS(W)
L.He.vessel①	side	2	4.2	49.0	221	253
	bottom	3	4.2	49.0	32	
Cu plate of radiation shield	side	3	78	0.24	~0	4092
	bottom	3	78	0.24	4092	
L.N₂.vessel②	inner	4	78	52.7	325	514
	outer	3	78	52.7	189	
outer vessel③	side	6	273	70.0	291	317
	bottom	6	273	70.0	26	
upper service flange④		48	273	70.0	0.3	0.3
						5176

t:thickness, T:temperature, and ρ:resistivity. Numbers in the position column correspond to ones in cryostat drawing.

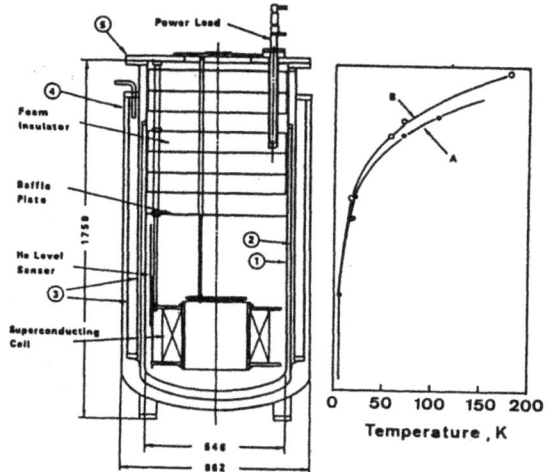

Fig. 4 Sectional view of cryostat and temperature distribution

the thermal input due to radiation can be shielded for cryostats presently available, that due to the thermal conductions through the power leads is inevitable. They were measured and analyzed for the cryostat whose sectional view is shown by Fig. 4, and also the parameters of the magnet are given by Table 1. The thermal input through the power leads amounted to 2.9 W which was estimated as for the calculated temperature distribution of curve B along the power lead. The estimated value of the thermal input due to radiation was 0.3 W. The summation of the two is very close to 2.98 W which corresponds to the measured helium evaporation of 4.2 1/hr for the helium level 10 cm above the uppermost flange of the magnet.

In order to decrease the main part of thermal input which is due to the temperature gradient along a power lead, we propose low temperature power electronics. By this development, the operation of a power converter in the same cold environment of a superconducting magnet becomes possible.

Ac Loss: In pulsive operations, the ac losses in a cryostat and in a magnet are added to the thermal input. The ac loss of the cryostat whose sectional is of metal type and shown by Fig. 4 amounts to 5,000 W for the sweep rate of 5 T/s by calculation. This value is 0.5% of the rated drive power of 1,000 kw, a part of which causes the evaporation of liquid nitrogen. The ac loss of the coil for one cycler of pulse drive measured by the evaporation of liquid helium becomes the same as the calculated amount 1500 J for the sweep rate 5 T/s both in increase to and decrease from the maximum 5 T as for the central field density. This loss is 0.3% of the rated stored energy 500 KJ. Based on these analytical and measured results, we recommend the cryostat which is made of GFRP(Glass Fiber Reinforced Plastic)for the pulsive usage.

Stress and Fatigue[3]: The electromagnetic forces calculated for the rated current of 2,000 A are 6.82/2.82 MN(695/288 ton) for the radial/axial directions. The maximum values of stresses in radial, circumferential and axial directions, $\sigma r(r,z)$, $\sigma \theta(r,z)$, and $\sigma z(r,z)$, where z=0 is the middle plane perpendicular to the magnet axis, are : $\sigma r(20(cm), 0(cm))$=8.5/7.8 MPa, $\sigma \theta(15.5,0)$=65/69 MPa and $\sigma z(15.5,0)$=27/26 MPa, respectively,in which the outer surface of the magnet are wound with/without GFRP bands. The maximum stress is far below the allowable stress of material. We have experimented however this coil for many purposes up to over three thousand cycles in pulse repetitions. At present, we observe the S-N transition in the current conduction of 2000 A, which might be caused due to the occurrence of looseness in the coil.

4.Future Study

Superconducting Magnet: The discovery of high temperature superconductivity stimulates the considerations on the improvements of magnet design which is based on the superconductors of much higher critical densities of current and field which have never existed.

We assumed the two cases of boundary conditions: 1)overall current density $j=0.1 \times 10^8 A/m^2$, maximum magnetic field Bmax≤7T, for conventional superconductors, Nb-Ti, 2) overall current density $j=0.1 \times 10^9 A/m^2$, Bmax≤30T.

We sought for the optimal design which minimize the coil volume with respect to the given energy E for these two conditions, and found the scaling low:

$$Vol=0.52E^{0.60} \text{ for condition 1),}$$

$$Vol=0.033E^{0.60} \text{ for condition 2),}$$

which means, for condition 2), the volume becomes 1/15 for the same E. The stress and strain distributions which are shown by Fig. 5 were analyzed based on the conditions : E=400MJ, $j=0.1 \times 10^9$ A/m^2, Bmax≤30T, Young's Modules =

1.0x1011 N/m2, Poisson's ratio = 0.3. The calculated maximum stress in circumferential direction is 1.45×10^9 N/m^2 which is far beyond the tensile strength of copper $1.2-1.7 \times 10^8$ N/m^2. This result suggests the impossibility of self sustained magnet for this case, and that even if we can obtain the superconductors of high quality in electromagnetic conditions, the mechanical stress needs the supporting materials which will increase the coil volume in total up to present state of magnet technology or the volume for condition 1).

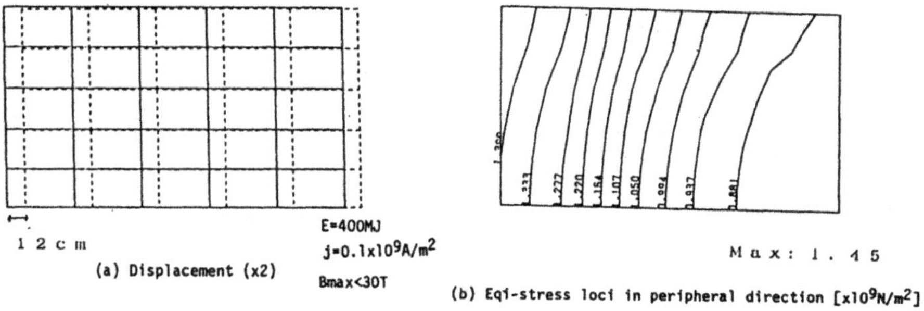

1. 1 2 c m

(a) Displacement (x2)

E=400MJ
$j=0.1 \times 10^9$ A/m^2
Bmax<30T

M a x : 1 . 4 5

(b) Eqi-stress loci in peripheral direction [$\times 10^9$ N/m^2]

Fig. 5 Analyses by finite element method

Low Temperature Power Electronics[4]: For the power electronics as the interface between a superconducting magnet and a power system, the following properties are required: 1)high density of integration of power switching devices, 2)low switching losses for high frequency operations, and 3) operation in cold environment with a superconducting magnet.

Item 1) is favorable for the requirement of comparatively large power with respect to the small storage of SMES for the stabilization purposes. The high frequency operation of PWM(Pulse Width Modulation) control which remove the problem of higher harmonics due to switching action requires property 2), and 3) is proposed to cope with the large thermal input due to the large temperature gradient. The discovery of high Tc superconductivity stimulates the development of low temperature power electronics in the liquid nitrogen temperature, in which semiconductors have high qualities.

Characteristics of Power MOSFETs in Cold Environment: The unipolar devices among which power MOSFETs represent the low loss current conduction and the low loss switchings in the liquid nitrogen temperature. The bipolar devices however have the problem of the freezeout of carriers in low temperatures. The current conduction of MOSFET is dominantly determined by the carrier concentration multiplied by the mobility of the conducting channel. The multiplicant which is the channel conductivity becomes maximum near the liquid nitrogen temperature.

The measured drain characteristics of an n-channel MOSFET whose configuration is integrated in three dimensions show the remarkable increase of drain current in the linear region at 80 degrees in Kelvin, which is shown by Fig. 6. By comparison of the drain current at 80 K with that at room temperature, we can observe the remarkable increase of the former in the linear region. This results in the low loss witching operations at 80 K. The average power loss for repeated on-off drives of this device becomes one fourth that of room temperature, which is shown with respect to switching frequency by Fig. 7. The increase of the average power loss with the increase of switching frequency suggests the loss due to switching trasients become in high frequency operations.

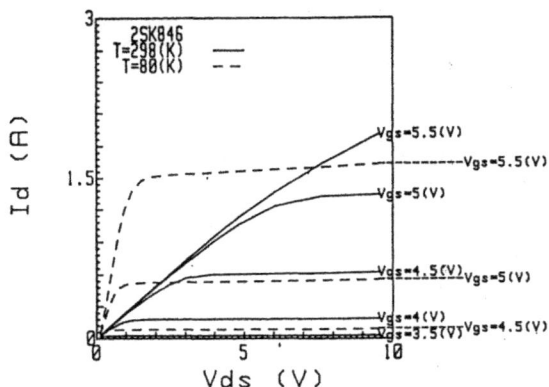

Fig. 6 Drain characteristics of a MOSFET

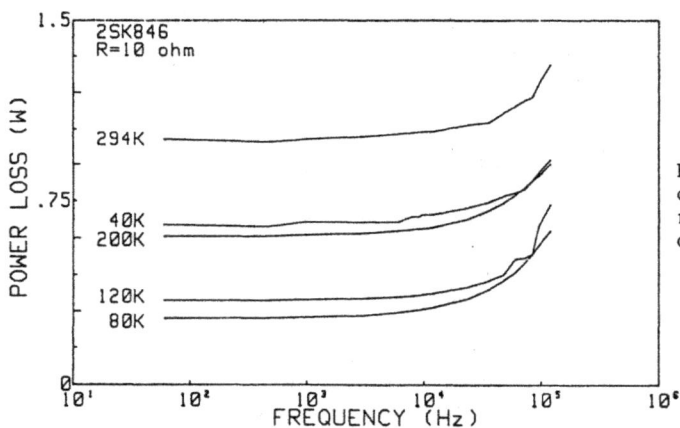

Fig. 7 Power loss of the MOSFET for repeated on-off drives

References

1) T.Ise, Y.Murakami, K.Tsuji, Simultaneous Active and Reactive Power control of Superconducting Magnet Energy Storage Using GTO converter, IEEE Trans. RWRD-1, 14-21 (1986)
2) Y.Mitani, T.Tsuji, Y.Murakami, Application of Superconducting Magnet Energy Storage to Improve Power System Dynamic Performance, IEEE Trans. PES winter meeting, 88 WM 191-9   (1988)
3) Y.Murakami et al, Experiments and Analysis of Thermal Characteristics and Stress/Strain Distributions of a 0.5MJ Pulsed Coil, Advances in Cryogenic Engineering, 29, 167-173(1984)
4) Y.Murakami, H.Kimura, Basic Discussion Low Temperature Power Electronics, Jpn. J. Appl. Phys., 26 Supplement 26-3(1987)

Acknowledgment

The author would like to express his sincere thanks to our colleagues, Professors J. Yamamoto, K. Tsuji, Drs. Ise, Mitani, Ohuchi and students of the Laboratory for Applied Superconductivity for their devotions to research works on SMES.

SMES DEVELOPMENTS AT THE UNIVERSITY OF WISCONSIN

R. W. BOOM, M. K. ABDELSALAM, Y. EYSSA, M. HILAL, X. HUANG,
G. E. MCINTOSH, and J. PFOTENHAUER

Applied Superconductivity Center, University of Wisconsin—Madison
1500 Johnson Drive
Madison, WI   53706 USA

Introduction

A long term SMES program in the Applied Superconductivity Center (ASC) has been continuously in progress at the University of Wisconsin since the 1970-71 invention of SMES by H. A. Peterson and R. W. Boom.[1] Seven major reports and 306 papers have been published by the Applied Superconductivity Center. Our first paper on SMES was "Superconductive Energy Storage for Power Systems" presented at INTERMAG-1972 in Kyoto, Japan.[1] In it we introduced H. A. Peterson's idea to charge and discharge storage coils with three-phase Graetz bridges, thereby achieving greater than 95% storage efficiency. SMES has been the major Applied Superconductivity Center activity; superconductive materials under D. C. Larbalestier and helium cryogenics under S. W. Van Sciver are the other two major center activities. The present principal interest in SMES stems from the US DNA-SDI program to build an Engineering Test Model (ETM) for utility and government use. This paper is a selected Wisconsin review of SMES design highlights and of some small scale SMES studies.

Electric Utility Usage of SMES

The appropriate SMES size for a given Utility Company is a function of the projected load curves for the next 50 years and of the economic planning for generation expansion. The opportunity for storage in the recent past years can be assessed by analyzing the past daily load curves to find the maximum SMES energy that could be used each day. In other words, if the generators deliver a constant power in a 24 h day then the energy stored at night would equal the energy discharged during the daytime. Figure 1 is a Wisconsin utility (1982) plot of the MWh of SMES vs. the number of days per year it would be fully used. The dashed lines show that 6500 MWh would be fully used 5 days/week all year and that 3000 MWh would be fully used 7 days/week. Future planning could build on this information.

One point not emphasized in the past is that the maximum power occurs at night. The power available in the daytime tends to be about twice as large as needed; usually it is planned that the energy charged at night in 8 h at higher power levels is discharged during the day during 15 h at lower power levels. Power levels vary continuously to match load with relatively fixed generation. It is difficult to rate an SMES in kW since the discharge is variable and always smaller than the bridge size purchased for charging.

A typical utility with E_{max} = 5500 MWh and E_o = 5000 MWh (charged & discharged) would have P_o = 1000 MW and V_{max} = 1.55 (P_o/I_{max}) for the most typical load curves. A conservative choice is to select V_{max} + 10% for the single maximum P_o of the year. Current I_{max} and V_{max} can be chosen from these relations.

Design Summary and Recent Trends for Large SMES

Modern SMES for electric utility use (Fig. 2) was established by the Wisconsin group in 1970. Research and development at the University of Wisconsin over the past 17 years has produced the following preferred SMES system design options and component developments:

- Efficient Graetz bridge energy conversion between ac (transmission line) and dc (superconductive solenoid).
- Double layer low aspect ratio solenoids mounted in surface trenches in soil or bedrock.
- Cryogenically stable composite (50-200 kA) conductors of NbTi and high purity Al cooled in a superfluid helium bath at 1.8 K for maximum helium heat transfer.
- Radial forces transmitted from the 1.8 K solenoid to the outer trench wall by reinforced epoxy insulating struts, which have heat intercepts at 28 K and 77 K.
- Axial forces balanced internally by aluminum alloy structure in the superfluid helium bath.
- Identification of long term "cryogenic stability" in the SMES open superfluid helium bath.
- Emergency energy discharge system to protect the storage coil by dissipating the stored energy as heat in the axial structure.
- Rippled solenoid and container designed to accommodate cooldown and warmup stresses.
- Or segmented non-rippled design to accommodate cooldown and warmup stresses.

Structure Design

The most recent SMES designs are two-layer low aspect ratio solenoids. Rippled and segmented non-rippled designs are described in references 2 and 3. The ripples are in the radial direction with nodes fastened by struts to the outer trench wall. Radial cooldown travel and the cooldown stresses are much smaller than would be for the large major radius circle. The segmented structure pieces are straight or slightly curved discontinuous pieces. During cooldown, the structural pieces slide on each other while the wall struts maintain windings in place. At the end of cooldown the telescoping structure is essentially unstressed and the conductors experience a one time cooldown strain of 0.43%.

Dynamic Stability

A new criteria for "dynamic stability" of large stabilized conductors has been developed at UW by considering the process of current diffusing from the superconductor strands into the adjacent aluminum. The steady state I^2R heating is assumed to be less than the helium heat removal, due to a large cross-section of aluminum and an adequate helium annulus. However, initially, current only diffuses into a small depth of aluminum and could take hundreds of milliseconds to reach steady state. As a result, the initial I^2R heating power can be very high. The combination of high

current, long diffusion length and time, and the very low specific heat of
low temperature aluminum may cause the conductor to be dynamically unstable
(unless enough surface cooling is provided) and results in a traveling
constant length normal zone. The heat flux transmitted along the aluminum
conductor coming from a normal zone triggers current sharing ahead of the
normal zone and causes current diffusion into the aluminum. The I^2R heating
is initially very high, drives the conductor temperature above the critical
temperature, and the conductor becomes completely resistive. While I^2R
eventually drops and the normal zone starts to recover, a new normal zone is
forming up at the end of the zone which moves down the conductor. We then
have a traveling normal zone that will sweep the whole conductor length,
warming the helium bath, and generating an unacceptably large heat load.
The above phenomena has been mentioned in the literature before without
consideration for the details of helium heat flux at the aluminum warm
surface.[4]

Dynamic stability is a stringent requirement on the perimeter and the
resistive quality of the aluminum in closest contact with the NbTi strands.
It is not related to the amount of helium next to the conductor. The
conditions under which a large stabilized conductor can be dynamically
stable have been identified. The critical condition for non-propagation is
that the (I^2R) peak during early diffusion should be less than the early
(QHe) helium heat removal at the surface. Both can be high surface heat
fluxes at short times of a few msec.

Protection (Internal Energy Dump)

All SMES designs should accommodate provisions for a rapid energy dump
in which the magnet can unload and warm to about 390 K in 10 to 15
minutes. Shorting switches are used to connect coil sections in parallel
before dumping the liquid helium. This concept results in uniform
temperature and voltage distribution during the current delay time.

Special Applications

We have designed several small and medium size SMES units for ground
and space use. Among these applications are cryoresistive aluminum hydrogen
cooled pulsed storage magnets and several superconductive solenoidal and
toroidal SMES coils for space applications.

Hydrogen Cooled Pulsed Cryoresistive Inductors

Cryoresistive magnetic storage can provide pulsed power for
accelerators, fusion devices, and electromagnetic launchers.
Electromagnetic launchers in space[5] could use available hydrogen as a
coolant. Only toroids are considered in order to limit external fields.
Figure 3 shows a proposed liquid hydrogen (LH_2) system for cooling
cryoresistive aluminum conductors. In one case the total system weight
including the cryogen coolant yields energy delivered per unit mass $\Delta E/M =$
40 J/g for $\Delta E = 500$ MJ for 1000 second pulses. The energy stored per unit
mass is 4 times larger than the energy delivered because of the requirement
for small current variation during the pulse: $I_{min}/I_{max} = 0.875$. High
modulus structural materials such as Kevlar-epoxy are required to limit

conductor strain and resistivity degradation. The reduction of total mass
is critically dependent on the aluminum final resistivity which is a
function of cycling strain and magnetic field. Eddy current losses during
pulsing are less for single layer coils. The higher losses of cabled
conductors or multilayer coils are caused by self shielding of each
conductor or layer from the ac fields from the other conductors. I^2R losses
for cable conductors are only small for wire sizes much smaller than the
skin depth. However, small wire sizes may have higher resistivity due to
size effects. The above 40 J/g is apparently superior to most other forms
of pulsed sources and is especially attractive if the exhausted H_2 is used
as fuel.

SMES Applications for Space Use

Cryogenics and superconducting magnets can be used in space for power
conditioning and base load energy storage, in addition to pulsed power
usage. The emphasis is to achieve large E/M, energy stored per unit mass.
We have introduced the pulsed shielded SMES in which the superconducting
coil is shielded by a parallel cryoresistive coil of equal number of
turns.[6] The shield turns are arranged so that the superconducting coil does
not experience ac currents or fields during the pulsing operation. For base
load energy storage we have designed a new toroidal configuration that fits
lifting requirements for space transportation.[7] The new configuration (Fig.
4) is called a thin (TD) shaped toroid where the height to width ratio, $\alpha =$
H/R_o-R_i, of the individual toroidal modules can be easily adjusted to fit
transportation vehicles. E/M values = 100-150 J/g are possible for large
units ($E_s > 10$ MWh). The materials are chosen to maximize E/M. The
conductor current density is 15 kA/cm^2 in the aluminum stabilizer around the
NbTi/Cu conductor. The tensile structure is Spectra 1000 and the
compressive structure is Boron/epoxy.

References

1) R.W. Boom and H.A. Peterson, "Superconductive Energy Storage for Power
Systems," IEEE Trans. on Mag., MAG-8, 3, 701-703 (1972) and H.A. Peterson
and R.W. Boom, "Superconductive Energy Storage for Power Systems," U.S.
Patent 4,122,512, October 24, 1976.

2) R.W. Boom, et al., Vols. I, II, III and IV (1974-1981) of the Wisconsin
Superconductive Energy Storage Project, UW-Madison; and R.W. Boom, et
al., "Cryogenic Aspects of Inductor-Converter Superconductive Magnetic
Energy Storage," Proc. Ninth Int. Cryo. Eng. Conf., Kobe (1982), 731-744.

3) R.W. Boom, et al., "Two-Layer Nonrippled Superconductive Magnetic Energy
Storage Systems," presented at the 1986 ICMC Conference, Berlin, West
Germany.

4) O. Christianson, "Normal Zone Evolutional Propagation in Cryogenically
Stable Conductors," Adv. in Cryo. Eng., 31 (1986).

5) Y.M. Eyssa, et al., "Design and Optimization of Hydrogen Cooled Pulsed
Storage Inductors for Electromagnetic Launchers," presented at the 4th
Symposium on Electromagnetic Launch Technology, Austin, TX (1988).

6) R.W. Moses and J.K. Ballou, "Inductive Shielding for Pulsed Energy
Storage," IEEE Trans. on Magnetics, MAG-11, 2 (1975).

7) X. Huang, et al., "Structure Optimization of Space Borne Toroidal
Magnets," presented at the 1988 Applied Superconductivity Conference, San
Francisco, CA.

WISCONSIN UTILITIES 1982

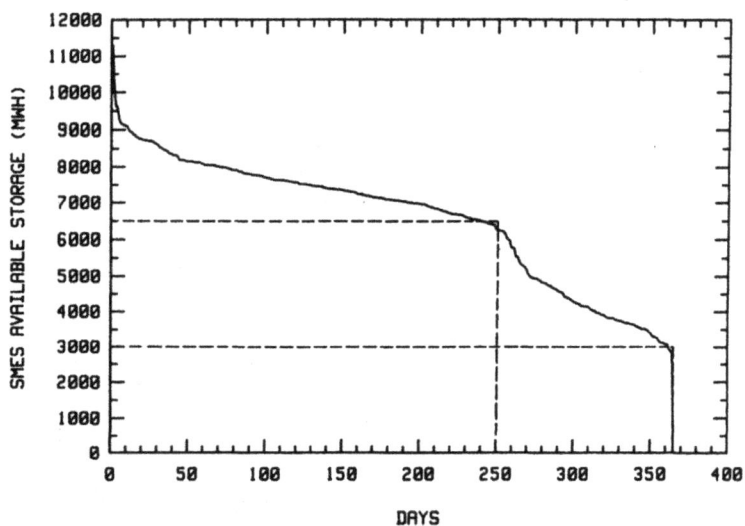

Fig.1.   SMES size duration curve vs. days of the year.

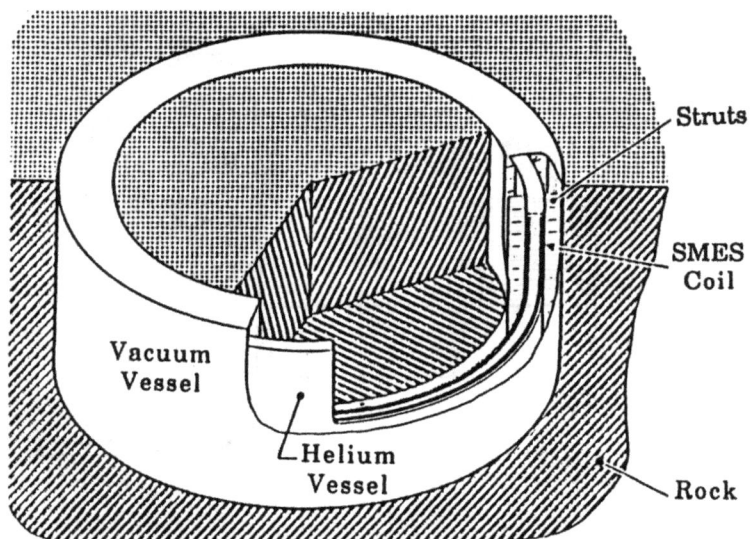

Fig.2.   Schematic SMES unit.

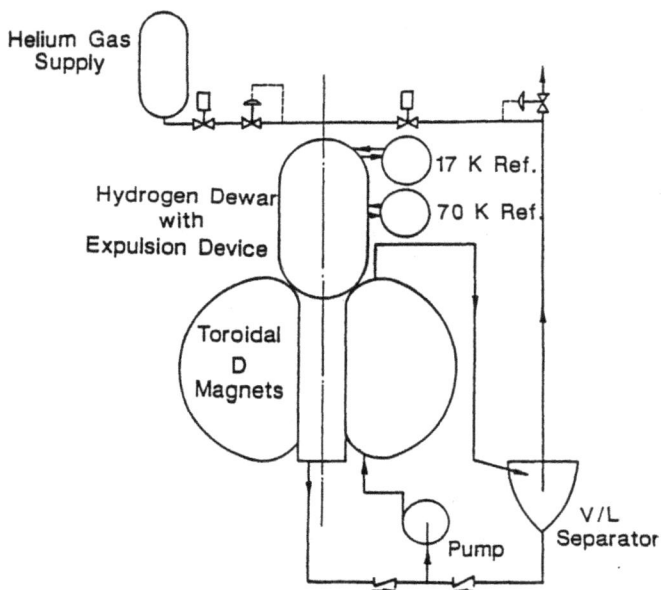

Fig.3. LH$_2$ cooling system for cryoresistive inductors.

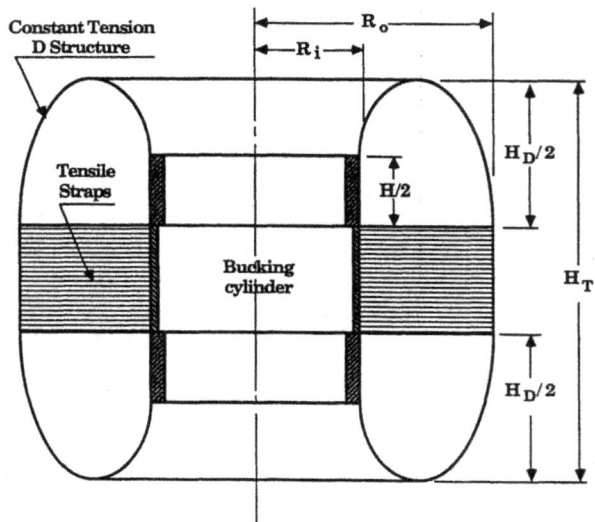

Fig.4. Thin D-shaped (TD) toroid.

RECENT TOPICS OF SMES RESEARCH

M. Masuda

Research Association of Superconducting Magnetic Energy
Storage (RASMES)
5-9-9, Tookoudai,Tuskuba

Abstract
Many applications of SMES have been discussed during a past
decade . They are roughly classified into three categories,
stabilization of utility network, load leveling and for
pulse load.
Research Association of SMES in Japan, RASMES conducted a
research for the load leveling that is directly oriented to
the saving energy management among SMES technologies. This
is to store the surplus energy at a bottom in load curve
and to deliver the electric power at its most peak in load
curve.
The unit, storage capacity of from several MWh to several
thousand MWh, have been discussed. They cover a range of
application from a local storage in a metropolitan sub-
center to a central storage in a major transmission
network. In connecting with this SMES, there are some
technological problems.
General description of the SMES technologies and some
topics on this matter that are the confinement of
electromagnetic forces, large current conductor problems,
environmental issues and economics for various scale are
discussed.

History of SMES Research
It is said that the opening of the SMES history is almost
just same as the history of superconductivity. The
discoverer of superconductivity, Onnes , also discovered
the very ingenious method to asses the zero resistance
characteristics of superconductivity which has never been
assured with any conventional means.
Figure 1 shows the basic circuit of Onnes's experiment.
Suppose the closed circuit formed with the inductance(L)
and the resistance(R), the circulating current decays
exponentially with the time determined by the inductance
and the resistance, L/R. When one wants to measure the very
low resistance of conductor such as superconductivity , the
time constant of current decay in this circuit can be a
good measure to tell the resistance.
The time constant of general coil made by conventional
conductors might be approximately 0.1 second at room
temperature no matter how its size and the configuration
are different. Larger coil has larger inductance and
resistance, smaller coil has smaller inductance and
resistance.
It is the famous story that Onnes performed two years
current flowing in a superconducting closed circuit. This

tells us the resistance of superconductivity might be one hundred millionth of the resistance of the conventional conductors.

In early 1970', University of Wisconsin proposed the energy storage using this permanent current characteristics of superconductivity[1]. They proposed the combination of the superconducting inductor and the thyristor converter to charge and discharge the energy to and from such inductors. They called such system as the I-C unit.

On the other hand, the National Laboratory for High Energy Physics in Japan has used successfully the thyristor converter to charge and discharge the conventional accelerator magnet. In middle of 1970's, they have thought to store the energy to the superconducting coil by this thyristor converter in place of accelerator magnet[2].

After such basic research carried out in Universities , National Laboratories in United States and Japan during past decade , the first remarkable attempt to asses the commercialization of SMES has started by EPRI in 1981[3]. The highly effective and attractive capital cost have been projected in this work. Independently in 1982 Japanese New Energy Development Organization (NEDO)[4] and Engineering Advancement Association (ENAA)[5] have carried out the conceptual design and evaluation of commercial size SMES . They came to the same conclusion as that of United States has done.

Table 1 shows the specifications of 5 GWh SMES designed in US and Japanese activities as mentioned above. The main parameters are not so different between two designs except some points that are the aspect ratio of coil, depth of the magnet from ground surface, temperature of liquid helium, the material of superconductivity. The agreement in major part between two designs, however, projected the consistency of such full size SMES. This will be discussed later.

In 1987, the enthusiasm for SMES technologies in Japan has inaugurated the Research Association of SMES which is organized with 40 major industries. They have already two years activities that are the evaluation of high Tc 5 GWh SMES, conceptual design of toroidal 20 MWh, safety analysis of SMES, investigation of environment issues and the analysis of the additional benefits for small SMES for stabilizing the existing network. Now they are proposing the construction of testing unit, 50 MJ, to the government. The recent and most striking activities in SMES is that two private firms in Unite States accepted the orders from Department of Defense in 1987 to develop the 20-50 MWh SMES for the power supply of the ground based laser weapon[6]. The initial phase that is the conceptual design of the unit is going to be finished by end of 1988 or beginning of 1989. The phase 2 that is the engineering and the construction of the Engineering Testing Model (ETM) will start consequently.

The activities in a research of SMES is now spread over in a world. France has also the idea of construction of

testing unit of 100 KWh. Korea started the research in
Korean Electrotechnical Laboratory from a couple of years
ago. Canada is thinking a SMES for railway business.
Finland designed 500 MWh unit with high Tc materials and
estimated the construction cost

Principle of SMES
Primarily the basic principle of modern SMES is the same as
Onnes's experiment as mentioned above already. However the
minor modification is given at the switching and the
battery. The modern circuit of the SMES is shown in Fig.
(b). The AC power from the commercial network is rectified
by the thyristor converter and charges the superconducting
coil to the maximum current which is determined by
superconducting and mechanical conditions . When the power
is required to draw back from the unit, the thyristor
converter is turn to the inverter mode. This is performed
by changing the timing of thyristor triggering.
We already have some plants of high voltage DC transmission
line in the world. In Japan, the network of Hokkaido
Electric Power Co. is linked to the network of Toohoku
Electric Power Co. through the DC transmission line laid
under the Tooya strait. The AC power in Hokkaido is
rectified to DC and once again converted to AC and to
connected to Toohoku line. This operation of DC
transmission is just same as the principle of SMES.
Therefore as far as the thyristor converter technologies is
concerned , it can be said that we have already high level
technologies.
Therefore the technological breakthrough, if exist, is
focused on the cryogenic region,including superconductivity
and its associates.

Technological Breakthrough
In a past research, the technological difficulties which
should be overcome have been extensively studied and were
made clear.
I like to point out three major difficulties among them,
confinement of electromagnetic forces, large current
conductor and the environment issues.

Confinement of Electromagnetic Forces
One difficulties is the confinement of huge
electromagnetic forces generated by the high field
superconducting coil at the cryogenic temperature.
In a common sense used to be in a superconducting
technology, such confinement of electromagnetic forces
should be performed by the structure made by
nonferromagnetic materials. However the theory of energy
science tell us that more than million tons of such
materials will be required for commerciasl size SMES. This
is the fatal economic disadvantages of SMES because the
cost of such materials surely exceed the cost of pumped
hydro which is the competitor of SMES. The solution has
been proposed. It is the using of the underground bed rock

to confine such forces.
Fig. 2 shows the artist rendition of 5 GWh unit which can
store the electric power of 1GW from Nuclear or Coal power
plant during 5 hours in midnight and deliver 5 GWh of
energy in daytime. The cryogenic structure which is
consisting of superconducting coil, liquid helium dewar,
thermally insulated strut to transmit the electromagnetic
forces to the bed rock , thermal insulation, vacuum chamber
and others are constructed in a deeper underground.
The effect of electromagnetic forces on the rock is
analyzed by finite element computation as shown in Fig.3.
Our conclusion on this issues is that we need deeper
underground and hard rock to bear the forces of high energy
density magnet.
Table 1 shows the rough specifications of 5 GWh unit
designed in United States and Japanese activities. The
major parameters of both design are not so different,
however, it has the minor differences between both designs.
The magnet configuration of the US design is low aspect
ratio and low energy density which means larger diameter
of solenoid coil. On the contrary the Japanese design is
small diameter and higher energy density because of
Japanese small useful land and fearing to spill the leakage
magnetic field widely for densely populated area.
Such design leads to the difference of bed rock structure.
United States constructs the coil in an open trench
excavated in soft rock or soil very near the ground
surface while Japan use deep underground hard rock such
like as the granite.

Conductor Current

The second technological breakthrough is the very large
current conductor in SMES. Why the large current should be
considered in SMES ?
The energy is stored in SMES as accordance with

$$E=1/2 \ LI^2 \quad , \quad P=V.I \qquad (1)$$

where E=stored energy, P=output power, L=inductance of
coil, I=coil current, V=terminal voltage of the coil.
The full size commercial SMES will generate the power of
1GW in day time. A coil design voltage of 10 KV is the
practical limit of superconducting coil. Therefore
according to Eq.1, the 1GW means 10 KV and 100 KA DC
current.
On the other hand, 5 GWh SMES discharges 90% of full
energy and still required to generate 1 GW even at remained
10 % of the energy in the superconducting coil. This means
that the current of superconducting coil decrease to 33 %
of maximum and the full power generation would be required
even at that moment. The design current, therefore, have to
be oversized by about of 3.0 times. Actually it might be
300 KA. The situation is shown in Fig.4. Anyway the SMES
needs the very large current.
On the hand, the conductor of SMES will be cryogenically

stabilized by tremendous amount of pure aluminum. This is
coming from the Steckly condition as shown in Eq.2.

$$hSA(Tc-To) > I^2 \rho \qquad (2)$$

in which h is the heat transfer coefficient, A is the
perimeter of stabilizer, S is the stabilizer cross section,
Tc is the critical temperature , To is the cooling bath
temperature, ρ is the resistivity of stabilizer and I is
the operating current.
Almost all of the SMES superconducting cable have been
designed under such criteria and the concept is called as
the full stabilization. We are assuming in this case,
however, that the current will transfer quickly from the
superconductor to the stabilizer at the normal transition
during a negligible short time compared with the time of
pulse disturbances such as the frictional move of the cable
in a magnetic environment.
Recently the slow current diffusion in a stabilizer has
been pointed out[8].The non-uniform current distribution and
the time constant of redistribution of the current is the
great concern in a high current superconductor. If the
current redistribution in the stabilizer can be a slow
process, a normal zone can propagate no matter how the
cryo-stable condition might be fulfilled.
The current diffusion time constant is given by,

$$t = k\, \mu_o/\rho \qquad (3)$$

where μ_o is the permeability of vacuum space and ρ is the
resistivity of stabilizer. The quantity of ρ/μ_o is the
magnetic diffusivity.
For pure aluminum stabilizer, for instance the RRR(residual
resistance ratio) is better than 1000 , the diffusion time
constant is very long something like longer than several
tenth milliseconds. This is an enough longer time to
stimulate the normal transition and promote the propagation
of normal zone.
If so, the superconducting cable carrying large current in
SMES needs more volume of stabilizer than defined by the
Steckly condition or other cable configuration than simple
monolithic has to be considered. The former requirement
would not be accepted because of economic disadvantages.
Thus the idea that the superconducting cables are
distributed with equally spaced in a round conductor or the
superconducting cables in conduit have been proposed for
200 KA cable[9].

Environmental Issues
One of the most difficulties encountered to commercialize
the SMES is the environmental issues. The larger SMES will
be located far from the load center because of its too big
capacity of storage or power. They have to be connected to
the high power transmission network. But the smaller SMES
is preferably located at the load center where is densely

populated or having the industrial area near the big city. Until today nobody can tell us the safety against the magnetic environment, however, it seems that there are three categories to classify the magnetic field strength for this problems. They are 200, 10 and 0.3 gausses. The first one is probably set just out of bound of the plant on the ground surface. The second may be adopted for sensitive instrument for magnetic field such as a pacemaker or something like that. The third is the geomagnetic field and may be impractical lowest limit for this discussion.

On the other hand, there are two configuration of the magnet, solenoid and torrid where the former is low cost to construct but more leakage of magnetic field in a surrounding area and the latter is safe for the magnetic field leakage but less economic.

For case of solenoid, Table 2 shows the radius and area requirement to attain various field level at the ground surface leaked from 10 GWh solenoid SMES buried underground in 150 m depth from surface..

We have two options to reduce the leakage field on the surface. One is increasing the burial depth of the coil and the other is the shielding coil constructed near the surface.

Fig. 5 gives the leakage field on the ground surface from 10 GWh solenoid SMES as a function of burial depth. It is clear that the depth play a major roll in reducing the magnetic field on the ground surface.

The shield coil or guard coil has been designed for full size SMES. The diameter of such shield coil is several times bigger than the main coil. The magnetic field strength out of the diameter at the shield coil is greatly reduced. The energy stored in the shield coil to cancel the leakage filed is approximately -1 MWh against the 10 GWh of main coil. The cost of shield coil including , conductor, dewar and structural support, is 4.3% of the same of main coil[10].

The rough conclusion is that if the field of everywhere on the ground is restricted less than 200 gausses, the problems can be solved by burying the magnet deeper underground . If the field is restricted less than 10 gausses, it needs the shield configuration, however, the cost of shield coil will not a main problem.

In a case of toroidal coil consisted of number of small coil, the magnetic filed is restricted inside the torrid and actually no leakage filed except very close to the outside of small coil.

The problem of toroidal coil is the cost. Table 3 shows the relative amount of superconducting materials per stored energy for various configuration of magnet[11]. This shows that the toroidal scheme requires almost 2 times amount of superconducting materials than solenoid. Moreover in small SMES which is to be sited in densely populated area, the cost of superconducting materials will play more dominant roll than larger coil. It can be said that the small toroidal coil have no ways to cut its construction cost.

Economics of Full Scale SMES

The largest applications of SMES is the load leveling of utility network. Therefore, the most economic concern has been put on this point.
The comprehensive studies on this problems have been first carried out in an activities of EPRI in United State from 1981. From 1982, Japanese NEDO also conducted the assessment work for three years. The United States and Japanese two independent works revealed very interesting results. Before these works, the reputation of SMES was that the nominal features of them is very attractive but they have the following doubt,
1) the efficiency of SMES is surely excellent, however, the refrigeration loss to keep the superconductivity in low temperature will deteriorate them.
2) judging from the current superconducting technologies such huge superconducting coil is only the matter of dream.
3) it is unlikely that the superconducting machines will meet the requirement of economics.
4) the cost of SMES can never compete with the existing pumped hydro forever
5) the natural resources particularly Nb and He will not permit the use of superconductivity in a society of utilities.
Almost all of such doubt came from the less knowledge on SMES technologies. It has been , however, cleared after two activities in Unite State and Japan.
Here the only economic discussion will be given.
The evaluation of SMES can be performed by comparing the value and the cost of SMES. The value is determined by the generation cost of power plant which is forced to construct if the SMES will not be used. This is compared with the gneration cost of the SMES. Thus the economic break even point are obtained as listed in table 4 . The figure covers 250,000 yen/KW to 450,000 yen/KW depending on which energy is stored in midnight and which power plant is set for comparison.
On the other hand, the construction capital of 5GWh SMES can be estimated. Table 5 shows the result of cost estimation. The cost can be compared with the BECC to asses the economics of full size commercial SMES. By this discussion, it is clear that the SMES has the highest commercialization potential.
The prospect of the economics of full size SMES has become clear to some extent, however, the small scale SMES is still in a strong doubt. Fig.6 shows the scale effect on the cost and BECC. This shows the higher unit cost and less economics of small SMES.
On the other hand, one of the attractive applications of small SMES, something like 20 MWh, is the energy storage for smart building or intelligence building which is fully computerized and so forbid a power failure even during a very short intervals. The SMES coil for such purpose is

customarily designed by the toroidal configuration to avoid
the exposure of the magnetic field to working people in
that building. This is shown in Fig. 7. The toroidal
coil is more expensive than solenoid as mentioned above.
One example of break even economic point is estimated to be
1.5 billion yen for a reference design of 6.5 MWh of 30
floors intelligent building .
This value can be compared with the cost as shown in Fig.
8. At the present moment the cost of small SMES is much
higher than the BECC. However reminding that the main part
of the cost of small SMES is superconducting materials ,
future improvement of productivity of superconducting
cable will enable the small SMES to have a chance to meet
the requirement of economics . A little bit larger unit,
20-200 MWh, can be economically used as the central storage
in a metropolitan sub-center supplying well stabilized
power to the complex of business buildings.

Conclusion
One of the applications of the SMES, load leveling has been
discussed. The possibilities to surpass the confrontation
against the difficulties in technologies, environmental
issues and economic conditions have also been shown.
The rough conclusion derived from the present study is that
the SMES for load leveling has no specifically mentioned
technological breakthrough. It has high possibilities to
commercialize not only for central storage in utilities
but also for local storage in the intelligent buildings or
in the metropolitan sub-centers.

Reference
1) Boom R.W. and Peterson H.A., IEEE Trans. on Magnetics,
   MAG-8, 701 (1972)
2) Masuda M. and Shintomi T., Cryogenics 17, 607 (1977)
3) "Conceptual Design and Cost of Superconducting Magnetic
   Energy Storage Plant", EPRI EM-3457 (1984)
4) "Investigation on a System of Superconducting Magnetic
   Energy storage"(in Japanese) NEDO-P-8408 (1984)
5) "Assesment of Superconducting Magnetic Energy Storage"
   (in Japanese) ENAA pu-1 (1984)
6) Loyd R.J. et al, presented at the Applied
   Supercondcutivity Conference (1988)
7) Eriksson J.T. et al, presented at the Applied
   Superconductivity Conference (1988)
8) Willig R.L., Proc. of 1978 MHD Magnet Design Conference
   249 (1978)
9) Walker D.L., et al, presented at the Applied
   Superconductivity Conference (1988)
10)"Evaluation of Environmental Control Technologies for
   Magnetic Fields", U.S. Department of Energy (1978)
11)Hassenzahl w., presented at the Applied
   Superconductivity Conference (1988)

Table 1   Major Parameters of 5 GWh SMES

		(Japan)	(USA)
Stored energy	(MWh)	5,000	5,500
Output power	(MW)	1,000	1,000
Aspect ratio		0.1	0.01
Coil diameter	(m)	400 376 400	1568
Coil height	(m)	10.8 18.0 10.8	15
Underground structure		3-tunnel	trench
Depth	(m)	150	15
Peak field	(T)	8.4	7.0
He temperature	(k)	4.2	1.8
Coil current	(kA)	707	765
No. of turns		330	112
Inductance	(H)	71.8	67.6
Conductor			
size	(cm)	21X10	13.5
material		Nb3Sn	NbTi
length	(km)	403	574
weight	(Ton)	25,000	24,000
Al structure	(Ton)	26,000	41,000
Excavation	(m^3)	5.4×10^5	5.3×10^5
Conv. efficiency		0.95	0.94
Cryogenic loss	(MW)	6.0	5.3
Overall effic.		0.90	0.91

Table 2   Radius of Field on Ground Surface to Attain Given Magnetic Field by Changing Burial Depth of SMES

Depth (m)	Magnetif Field(G)	Radius(m)
150	200	580
	10	1550
	0.3	4980
450	200	480
	10	1540
	0.3	4960
750	200	0
	10	1500
	0.3	4930
1050	200	0
	10	1390
	0.3	4870

Table 3 Relative Magnetization per Stored Energy
of Different Coil Configurations

Coil Configuration	Relative Ampere-meter/Joule
Thin Solenoid Best	1.0
Thin Solenoid Practical	1.51
Poloidal	1.12
Dipole	1.26
Toroidal	2.13

Table 4 Estimated BECC* of 5 GWh SMES

Principal charging energy	Type of standard plant	BECC(10^3yen/KW) (Availability) (10%)	(20%)
Nuclear	LNG**	334-406	364-436
	Oil	298-399	336-436
Coal Fired	LNG	290-361	306-378
	Oil	254-355	278-378

* BECC Break even captial cost
**LNG Liquid natural gas

Table 5 Estimated Cost of 5 GWh, 1 GW SMES
(100 million yen)

Magnet Parts
Conductor	561
Low temp.support	264
Themally insulated support	240
Thermal shield	39
Helium vessel	39
Vaccum chamber	94
Refrigerator	74
Liquid helium	3
Power conditioning	200
Control and protection	40
Excavation of tunnel	137
Assembly of coils	159
Direct construction cost	1850
Indirect cost (30% of total)	872
Contingency (10% of direct)	185
Grand total capital	2907

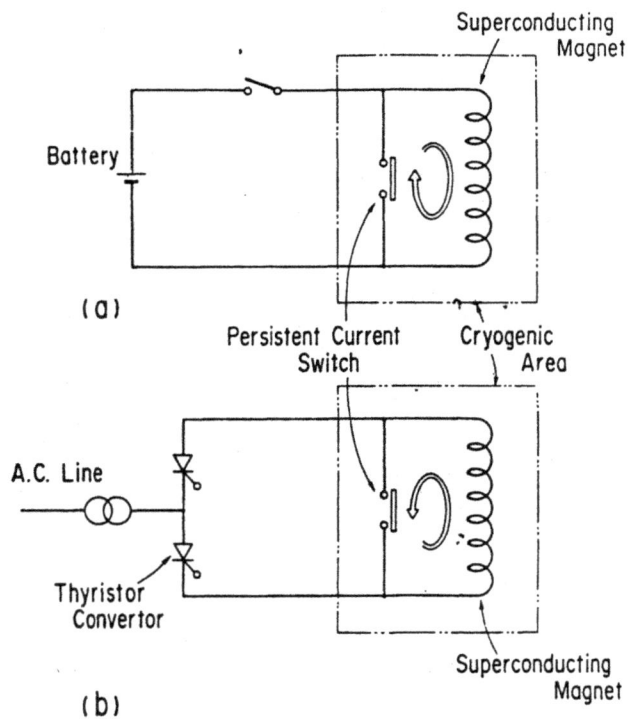

Fig.1 Principle of SMES, (a) Basic, (b) Modern

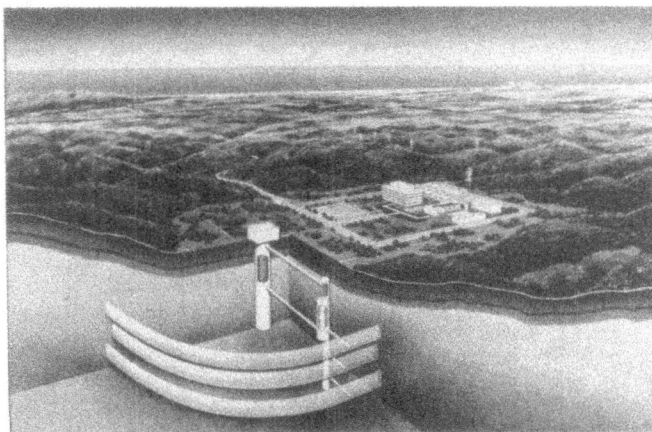

Fig.2 Artist rendition of 5 GWh SMES

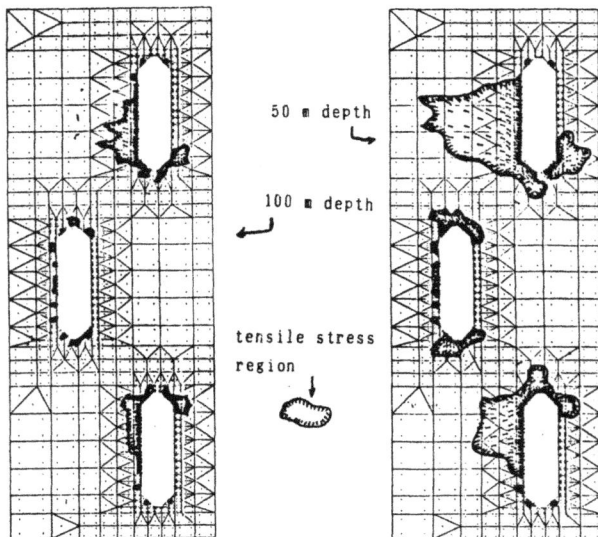

Fig.3 Computer analysis of stress in rock

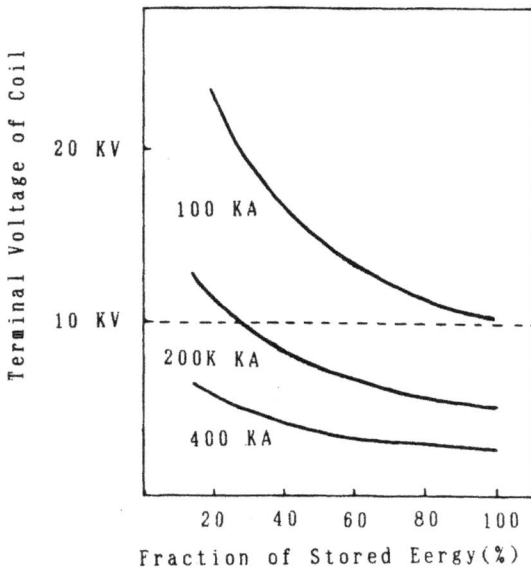

Fig.4 Required Terminal Voltage to Generate 1 GW depending
on a Fraction of Stored Energy in 5 GWh SMES. Design
Current is shown as Parameter.

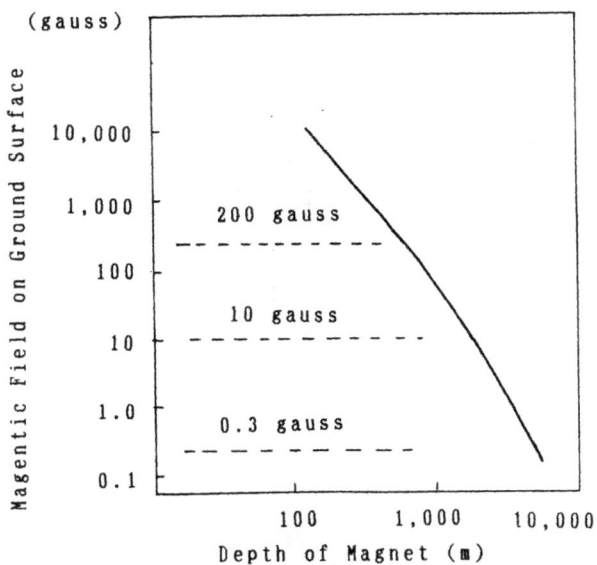

Fig. 5 Leakage Magnetic Field on a Center of Ground Surface
of 10 GWh SMES Depending on Burial Depth

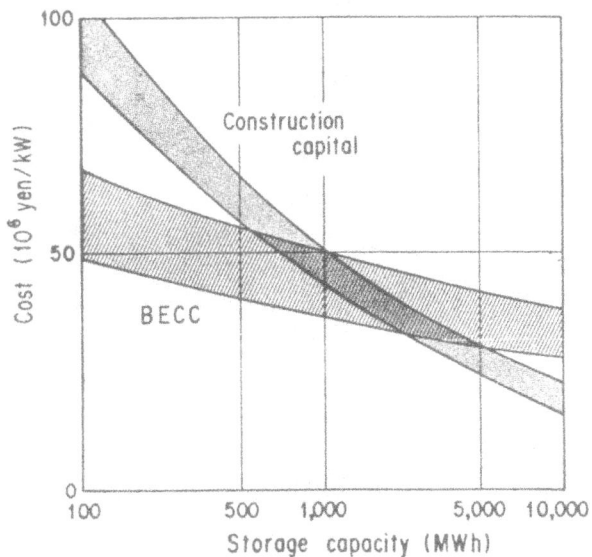

Fig. 6 Scale Effect of SMES Cost and Economic Break Even
Capital

Fig.7   Artist rendition of SMES for intelligent building

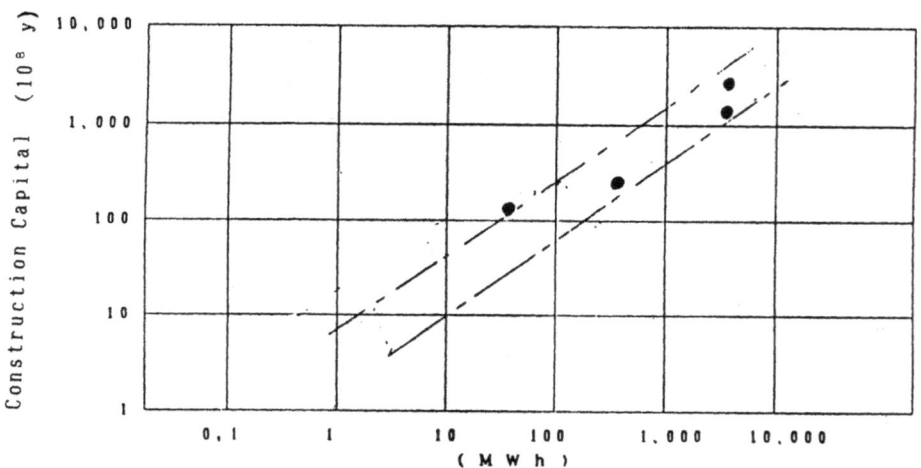

Fig.8   SMES cost of various scale

SUPERCONDUCTING SYNCHRONOUS GENERATOR AND
THE POWER SYSTEM CHARACTERISTICS

Tanzo Nitta and Takao Okada

Department of Engineering, Kyoto University
Kyoto, 606, Japan

ABSTRACT - The results of our studies on superconducting synchronous generator (SCG) are described in this paper. The standpoint of our studies on SCG is that the characteristics of SCG, especially the power system characteristics, must be understood for getting reliable power system apparates. Then experiments and simulation studies have been carried out by use of the small SCG (20kVA, 220V) as follows:
1) Two and three dimensional magnetic analyses for design, machine constants, machine characteristics and so on.
2) Experiments and simulation for steady and transient states of SCG in experimental single and two machine systems.
SCG with high response excitation, which is developed in collaboration with The Kansai Electric Power Co. Inc., Toshiba Corporation and Kyoto University, is introduced.

Introduction

Superconducting synchronous generators (SCG) have many advantages, such as high efficiency, light weight, small size, low synchronous impedance and so on. The low synchronous impedance improves power system stability. Therefore they are being developed and studied. In order to put SCG into practice, problems on SCG should be defined and solved. The fundamental configuration of SCG is already determined, which is discussed in the next chapter. In the process of developing highly reliable SCG, it is essential to understand the characteristics of SCG connected to power systems. It is the standpoint of our studies. For the purposes, we have carried out experimental and analytical studies by use of the small SCG(20kVA,220V). This paper describes the result of our studies.

We are also studying SCG with high response excitation in collaboration with The Kansai Electric Power Co. Inc., Toshiba Corporation and Kyoto University. The studies are introduced in this paper.

Superconducting synchronous generator

The configuration of SCG is already determined except some details, that is, a rotating superconducting field winding in a rotor of multiple cylinders and stationary normal conducting air gap armature winding.(1) The rotor consists of Liq.He bath, inner rotor, superconducting field winding, vacuum space, radiation shield, vacuum space and outer rotor. The stator consists of airgap armature winding and machine magnetic shield.

The different points among SCG's constructed or to be constructed are:(1)-(6)

Shape and supporting method of superconducting field winding.(racetrack, saddle;slots and wedge,binding, etc.)

Outer rotor damper. (one layer, three layer, winding damper, damper bar like conventional machine, etc.)

Radiation shield. (low resistance with screening and damping effect or high resistance without screening and damping effect)

Vacuum for thermal insulation (sealed off or continuous evacuation)

Inner to outer rotor attachment. (flexible support, rigid type, double bearing, etc.)

Structural material of rotor. (stainless steel, inconel, titanium alloy, A286, etc.)

Shape and supporting method of armature winding(diamond shape, helical coil, pancake coil, etc.; water cooling, oil cooling, etc.)

The 20 kVA SCG we used in our experiment as mentioned below is a vertical shaft type but the configuration is the same as those of other SCG's of horizontal shaft types.(9) The configuration of the 20 kVA SCG is shown in Fig.1

Since SCG has low impedance, then it increases power system stability. Furthermore when high response

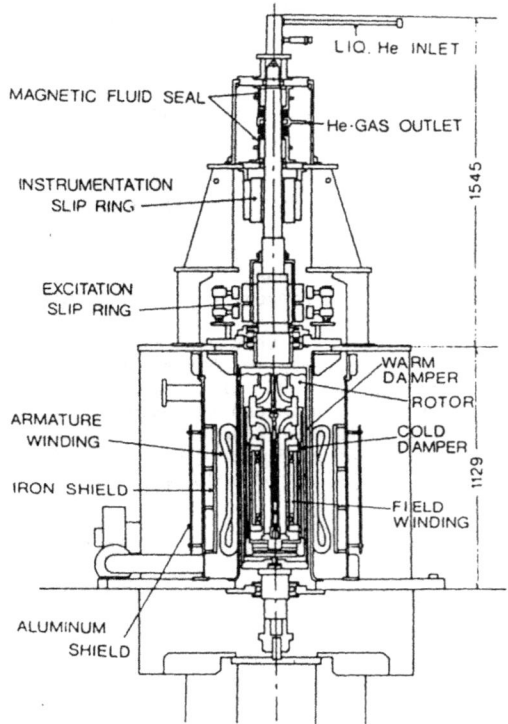

Fig.1 Configuration of 20 kVA SCG

excitation is possible in SCG, the power system stability could improve more.(6) SCG's with high response excitation will be constructed. In our researching group, an experimental 100 kVA SCG with high response excitation is under construction,(7) which is described bellow.

The development of superconducting wires with low losses under AC magnetic field allows to construct SCG with superconducting field and armature windings (full SCG). A 20kVA full SCG is under construction.(8)

Electrical characteristics of SCG

In order to put SCG into practice, problems on SCG should be defined and solved. Since SCG has advantages only when its capacity is rather large, the process of developing SCG is not the same as that of conventional one, where problems arising in increasing the capacities have been solved one by one. Then in the process the aid of theoretical approaches are necessary. The approaches must be reliable. It is difficult to define problems by theoretical approaches only. Then we have carried out the experiments and analyses by use of the 20kVA SCG.

Prior to understanding power system characteristics, the experiments and analyses for the electrical characteristics of SCG.

Experiments(9)

The distribution of flux density by the field current was measured. A three dimensional numerical analysis of magnetic field distributions was carried out as mentioned below. Good agreement between the measured and the calculated values is confirmed.

The basic electrical characteristics such as open circuit characteristics, short circuit characteristics, sudden three phase short circuit characteristics, machine constants and so on were obtained by the conventional test codes for synchronous machines.

Analyses

Two and three dimensional analyses for the electrical characteristics of SCG were performed.

From the results of the two dimensional analysis, The machine constants can be given by simple functions of the parameters such as the thicknesses and the conductivities of dampers, effective lengths of the field and armature windings, and so on.(10) The obtained machine constants were used in simulation studies on SCG connected to power systems as mentioned below.

In the three dimensional electro-magnetic analysis, the defference equations for the analysis expressed by vector potential are solved by SOR method. The distribution of flux density, no-load characteristics, the synchronous reactance and the self-inductance of the field winding were obtained numerically. The results agree well with those of the experiments.(11)

A three dimensional electro-magnetic analysis with due regard to eddy current in the cold and warm dampers was carried out. The three dimensional flux distribution and the eddy current distribution in the dampers at asynchronous states are obtained. From the calculations, the magnetic screening effects, the Joule heating, the torque, the radial stress of the dampers at the state could be obtained.(12)

Power system characteristics of SCG

The experimental and simulation studies on characteristics of SCG in power system were carried out.

Experiments

In the experimental system, the 20kVA SCG is connected to a reginoal power system (consider to be an infinite bus) through reactors (considered to be transmission lines). The experiment items are:
1) Steady state characteristics.(13)
2) Transient state characteristics.(14)
3) Steady state characteristics on parallel running with a conventional generator, which is connected to the terminal of SCG.(15)
4) Transient state characteristics on the parallel running.(16)

Steady state characteristics----The active-reactive power characteristics at the constant field current, the active power-load angle characteristics, the external characteristics and the field current-active power characteristics at the constant terminal voltage were obtained. The results indicate that 1) SCG has good property for the voltage regulation, and 2) the maximum power supplied by SCG without an AVR is not less than that by a conventional one with an ideal AVR.

Transient state characteristics---The experiment items are
1) Steady state stability test by a quasi-stationary increase of the generator output.
2) Disconnecting and reclosing of either of the transmission lines in parallel.
3) Transient stability test by three phase grounding fault with reclosing after clearing the fault.
    After SCG was going into loss of synchronism, the following procedure was implemented to prevent the rotating speed from increasing and the superconducting field winding from going into the normal state: 1)disconnecting switch between the SCG and the transmission line, 2)turning off the input power of the driving motor for SCG and 3) decreasing the field current of the SCG.   The procedure is necessary for protection system of power plants with SCG's.
    The experiment results are necessary for getting a reliable simulation code for transient behaviors of SCG.

Steady state characteristics on parallel running with a conventional generator---The experimental results show that since the SCG has a small synchronous reactance, it not only has a good stability but makes the whole system stable.

Transient state characteristics on parallel running---From the experimental results, it can be said that by installing SCG in power systems, the voltage stability and the power system stability can be improved in transient states as well as in steady states and the variation of armature current of SCG during a transient period is extreamly larger than that of the conventional one paralleled to the SCG.

Simulation studies

    In order to understand the transient behavior of SCG's, the transient analysis by a computer simulation was carried out for the above experiments.(17)   The equations describing the SCG and the conventional generator are so-called Park's Equations.   In case of the SCG, each inductance and the time constants are derived by the electro-magnetic analysis as mentioned above.
    In case of the conventional generator, these constants are obtained by the several tests.   The characteristics of the driver of the SCG and the conventional generator are taken into account, which affects the transient behaviors of the system.   The initial values when the system change (e.g. the system fault) are considered.   The simulation results agree well with the experimental results.   The torque and the Joule heating of the cold and warm dampers during the transient states could be calculated.

Superconducting generator with high response excitation(7)

    We are studying SCG with high response excitation (HSCG) through a detail design of an experimental 100 kVA HSCG in collaboration with The Kansai Electric Power Co. Inc., Toshiba Corporation and Kyoto University.

Problems for developing HSCG are defined.   Some solutions of the problems
are given.   The 100 kVA HSCG, whose pet name is "Hesper 1", will be
manufactured in 1988 and tested for power system characteristics, long-time
running test for reliability and so on in 1989-1990.   The specifications and
the configurations are shown in Table 1 and Fig.2, respectively.

Table 1. Specification of Hesper 1.

Filed winding	shape	racetrack
	outer diameter	327.6 mm
	max. axial length	500 mm
	insulation	epoxy impregnated
	superconducting wire	Cu/Cu-Ni/Nb-Ti
	cooling system	pool boiling
Rotor structure	room temperature damper	two layer damper(A286/Cu alloy)
	radiation shield	forced convection cooling type by gaseous He
	thermal insulation	continuous evacuation radiation shield(SUS316L)
	inner-to-outer rotor	flexible support
	attachment	(Inconel 718)
Stater		
Arnature winding	shape	diamond
	outer diameter	630 mm
	max. axial length	806 mm
	cooling system	air cooling
Machine shield		laminated silicon steel

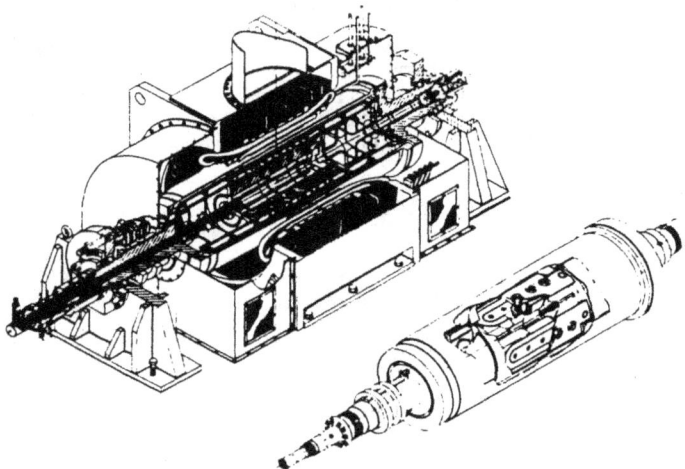

Fig.2 Configuration of Hesper 1.

Conclusion

A summary of the results of our studies on SCG is reported.   For further details, the reader should refer to our papers. We are studying characteristics and operations of power systems with applied power apparates. The Japanese 8-year project of research and development on superconducting generartors is started in 1988. We hope our studies will be useful for the project.

References

(1) J. L. Smith, Jr., "OVERVIEW OF THE DEVELOPMENT OF SUPERCONDUCTING SYNCHRONOUS GENERATORS", IEEE Trans. on Magnetics, vol.MAG-19,no.3 May 1983, pp.522-528.
(2) J. L. Sabrie and J. Goyer, "TECHNICAL OVERVIEW OF THE FRENCH PROGRAM", ibid. pp529-532.
(3) H. Fujino, "TECHNICAL OVERVIEW OF JAPANESE SUPERCONDUCTING GENERATOR DEVELOPMENT PROGRAM", ibid, pp.533-535.
(4) L. Intichar and D. Lambrecht,"TECHNICAL OVERVIEW OF THE GERMAN PROGRAM TO DEVELOP SUPERCONDUCTING AC GENERATORS", ibid. pp.536-540.
(5) I. A. Glebov and V. N. Shaktarin, "HIGH EFFICIENCY AND LOW CONSUMPTION MATERIAL ELECTRICAL GENERATORS", ibid. pp.541-544.
(6)CIGRE WG.11-05 Osaka meeting, "Development of Superconducting Generator and Material in Japan", March, 1988.
(7) T. Nitta, T. Okada, S. Hayashi, M. Tari and M. Kumagai, "SOME CONSIDERATION ON SUPERCONDUCTING GENERATOR WITH HIGH RESPONSE EXCITATION THROUGH DETAIL DESIGN OF EXPERIMENTAL 100 KVA GENERATOR", The Twelfth International Cryogenic Engineering Conference, 1988.
(8) Y. Brunet, P. Tixador, P. Vedrine, Y. Laumond and H. Nithard, "FULL SUPERCONDUCTING GENERATOR: STATOR AND ROTOR DESIGN", ibid.
(9) T. Okada, T. Nitta, T. Shintani, I. Muta, T. Ishigohka and H. Fujino, "THE BASIC TEST ON THE 20 KVA SUPERCONDUCTING SYNCHRONOUS GENERATOR", IEEE Trans. on Magnetics, vol.MAG-19,no.3 May 1983, pp.1043-1046.
(10) T. Shintani, T. Nitta and T.Okada, "EXPLICIT RELATIONS BETWEEN MACHINE CONSTANTS AND STRUCTURE-PARAMETERS OF SUPERCONDUCTING GENERATORS", JIEE vol.106-B, no.12, 1986, pp.1075-1082. (in Japanese)
(11) T. Okada, T. Nitta and T. Shintani, "THREE DIMENSIONAL MAGNETIC ANALYSIS ON SUPERCONDUCTING GENERATOR", Rotating Machine Seminar of IEEJ, RM-83-32, 1983. (in Japanese)
(12) T. Okada, T. Nitta, T. Shintani and Y. Akiyama, "THREE DIMENSIONAL MAGNETIC ANALYSIS ON SUPERCONDUCTING GENERATOR WITH DUE REGARD TO EDDY CURRENTS IN DAMPERS", Rotating Machine Seminar of IEEJ, RM-84-69, 1984. (in Japanese)
(13) T. Okada, T. Nitta and T. Shintani, "ON-LOAD TEST OF THE 20 KVA SUPERCONDUCTING GENERATOR", IEEE Trans. on Magnetics, vol.MAG-21, no.2, March 1985, pp.668-671.
(14) T. Okada, T. Nitta and T. Shintani, "TRANSIENT PERFORMANCE OF THE 20 KVA SUPERCONDUCTING GENERATOR IN POWER SYSTEM", IEEE Trans. on Magnetics, vol.MAG-23, no.2, March 1987, pp.1340-1343.
(15) T. Nitta and T. Okada, "PARALLEL RUNNING TEST OF THE 20 KVA SUPERCONDUCTING GENERATOR AND A CONVENTIONAL ONE", IEEE Trans. on Magnetics, vol.MAG-24, no.2, March 1988, pp.1340-1343.
(16) T. Nitta and T. Okada, "TRANSIENT CHARACTERISTICS OF PARALLEL RUNNING TEST OF THE 20 KVA SUPERCONDUCTING GENERATOR AND A CONVENTIONAL ONE", Applied Superconductivity Conference 1988.
(17) T. Nitta and T. Okada, "SIMULATION FOR ELECTRICAL CHARACTERISTICS OF A SUPERCONDUCTING GENERATOR CONNECTED TO A POWER SYSTEM", IEEE Trans. on Magnetics, vol.MAG-23, no.5, September 1987, pp.3551-3553.

AC SUPERCONDUCTIVITY APPLICATION TO ELECTRICAL MACHINES

Mitsuyoshi YAMAMOTO

Takushoku University
815-1 Tatemachi, Hachioji, Tokyo, 193, Japan

Abstract - As an example of the heavy current 50/60Hz application of superconductor for power electrical machines, a conceptual design of a 3phase-50Hz-1000MVA-275/66kV transformer has been carried out. Furthermore, the characteristics of a conceptually designed full superconducting generator has been investigated. In this study, the merits and the possibility of the superconducting AC electrical machines are reviewed.

Introduction

Recently, multifilamentary composite superconductors for 50/60Hz application have been developed. Thereupon, studies of electrical machines with AC superconducting windings have been started in many countries. In this paper, in the first place, configuration and characteristics of a heavy current AC superconductor under development are described. And, as an example of its application to electrical machines, a conceptual design of a transformer has been carried out. Also, the fundamental characteristics of a full superconducting generator has been studied. Through these studies, the construction, the loss characteristics, the merits on operation and the installation in site of the new machines have been reviewed in comparison with conventional ones.

Heavy Current AC Superconductor[1-3]

To realize the actual size AC power electrical machines, it is necessary to develop a superconductor with the current carrying capacity of several hundreds to several thousands amperes. For example, the rated current of both the primary and the secondary windings of a 3 phase-1000MVA-550/275kV auto-transformer is about 1kA, and that of a 1000MVA generator is about 5 to 20kA for terminal voltage of 30 to 120kV. Therefore, taking into account the use of parallel circuits, the development of a 1kA to 5kA class AC conductor would be necessary to realize the commercial superconducting machines. On the other hand, the AC loss in conductor has to be considered. There are several useful experimental data about the AC loss including ones obtained by the author's group. From these data, at present stage, it seems that the AC loss of $50kW/m^3$ at 1T and 50Hz is reasonable.

As the most practical selection for the heavy current AC superconductor configuration, the following two kinds of configurations are considered. One is a rectangular type as shown in Figure 1. And the other is a round cable type which has stranded elementary wires on a stainless center-core. In the case of the round cable, a number of wires are wounded over the large diameter core divided by insulation. However, this configuration would not be suitable for the winding of large machines. In view of this, we have been developing the rectangular type conductor. As the first step, we made a 3.6mH, 500A coil using this type conductor. As a matter of fact, the quenching current of this coil is 270A. This value is much lower than the critical current of the conductor estimated from that of a stranded wire. This seems to be because of an eddy current loss at the copper plated stainless steel tape for reinforcement. As the next step, an improved type conductor is under fabrication. The heavy current AC conductor with suitable configuration and characteristics for a large capacity electrical machine application will be realized in near future through these achievement.

A Conceptual Design of Superconducting Power Transformer

The superconducting transformer(SCT) has to be equipped with expensive superconductor, dewar, refrigerator etc.. Therefore, necessarily, its initial cost increases. On the other hand, the SCT has a great merit of high efficiency. If the effect of efficiency improvement exceeds that of the increase of the initial cost, the development of SCT will become reasonable. Besides, the SCT has a current limiting ability brought by quenching phenomena at the instance of short circuit fault. In addition to these, it has smaller floor area in site. Furthermore, it can be designed to have a large supply ability of reactive power by the adoption of a coreless configuration. However, needless to say, in these merits mentioned above, the decrease of loss would be most important.

In following paragraphs, the outline of the conceptual design is shown, and the meaning of the development and the image making of the SCT are reviewed.

(a) Rating

Type	Core type with disc windings				
Capacity	1000MVA				
Frequency	50Hz				
Pry winding	275/262,5/250kV	Impulse T.V.	1050kV	AC T.V.	460kV
Sry winding	66kV	Impulse T.V.	350kV	AC T.V.	140kV

(b) Fundamental Construction

Considering the efficiency of refrigerator, only windings should be contained in the dewar vessel. Meanwhile, the iron core leg should be located in a room temperature region which penetrates the winding bore. Both dewar and core are put into a steel tank filled with insulating oil as shown in Figure 2. As a result, the outer appearance of the SCT is not so different from conventional ones.

(c) Electrical Insulation System[4-6]

There are several useful experimental data as to the dielectric strength of liquid helium though they are obtained from small gap tests. Figure 3 shows the breakdown voltage of the boiling liquid helium as a function of the electrode spacing measured by R. J. Meats et al.. In this figure, also the data for the liquid helium given by B. Fallou et al.(CIGRE 15-04) are added. The data by Fallou et al.(1969) are close to the data of the gaseous helium breakdown voltage near the boiling point as shown in their CIGRE paper (15-04). They give the following experimental formula for the breakdown voltage of liquid helium as a function of electrode spacing.

AC   : $24d^{0.8}$ [kV]   for 2<d<15 [mm]
Impulse : $54d$ [kV]   for 2<d<6 [mm]

However, almost all of these data are measured in uniform fields.
Besides, we have to take note that these test voltages are rather too low
compared with test voltages of actual power transformers. So, it would be
yet early to discuss the full-scale insulation system of a large power
transformer on the base of these test data. However, at present, they seem to
be the useful data for the liquid helium insulation strength. Considering
this, in this paper, we employed above formula for the basis of the
insulation design.

The barrier insulation structure with multi insulating cylinders is put
to use for the insulation between high and low voltage windings which is
popular structure in commercial high voltage transformers. Each barrier
section is composed of 2mm thick cylinder and 8mm thick liquid helium, and
the total thickness of each barrier section becomes 10mm. As for the
dielectric constants, that of liquid helium is 1.05 and that of the barrier is
3.5 to 4.5. Therefore, considering the safety factor, it should be assumed
that all the voltage is imposed only on the helium gap. As for the main
insulation between the windings, the requirement for the AC test voltage being
more severe than that for the impulse one. Therefore, it would be enough to
consider only the requirement for the AC test voltage.

The breakdown voltage of the 8mm gap in helium is calculated to be
126kV(crest value) using the formula mentioned above. Therefore, that for 11
gaps is calculated to be 980kV(rms). Taking into account the correction
factor for non-uniform field (1.5) and the safety factor (1.4), the
withstand voltage of about 460kV can be obtained, which is just the same as
the AC T.V. for a 275kV winding. The details of this insulation design method
and that of other parts are mentioned in the reference[6].

(d) Core and Winding

Using the Richter's formula, the sectional area of the core is
calculated to be 14000cm^2. To get the %IX of 15% at the flux density of
1.72T, the height of the core leg is to be calculated 2.6m. From these data,
the weight of the core and the core loss become about 210 ton and 335kW,
respectively.

The fundamental configuration of the windings is a disc type, and a
rectangular type AC superconductor is adopted. The AC superconductor has, in
general, no or very small amount of copper stabilizer to reduce the eddy
current loss in the conductor. Therefore, it is necessary to obtain a good
cooling condition by exposing all of the four sides of the conductor directly
to liquid helium without any insulating covering. As shown in Figure 2, the
conductor and the windings are designed to get this condition.

The current density of the superconductor is about 200 to 300 times
larger than that of conventional conductor. But, the over all current density
of the winding is only several times larger than the conventional one because
of the presence of the cooling channels, etc..

The flux density of the main gap between two windings is calculated to be
about 0.4T. From these data, winding AC loss is calculated to be about 260W.

(e) Loss of Transformer

The calculated loss of the SCT is shown in Table 1 in comparison with
that of a conventional one. The efficiency of the refrigerator is estimated to
be 1/500 and the stray loss is assumed to be the same level of the
conventional one. As a result, the total loss of the SCT becomes about 1/3 of
that of the conventional one.

Generator with Superconducting Armature and Field Windings

Turbine generators with superconducting field windings are being developed in many countries. Generally, the year of its realization is considered to be in the next century. That means we have still enough time to develop the AC superconducting technique. Therefore, possibly, the full superconducting generator (FSCG) will be the main target of the developmental effort in that time. So, it would be appropriate to begin the preliminary study concerning the AC superconducting armature winding and the characteristics of FSCG. Considering this, it seems that the study of a 20kVA generator with superconducting armature winding by researchers of Alsthom is timely.

Generally, turbine generators must have the extremely high reliability in operation. It is required not to stop in any fault in the transmission system like three phases sudden short circuit fault. So, it is necessary to design the FSCG to be able to continue its operation without quench in any system fault.

In general, the superconducting armature winding has a high ampere-turns. As a result, it has high armature reaction. Therefore, the synchronous reactance Xd and the transient reactance Xd' becomes large. On the other hand, the FSCG will use the AC superconductor or the similar one also for the field winding on the rotor. This means that the FSCG has a high transient withstand capacity. So, the rotor can be designed with no or very week action damper, and as a result, the sub-transient reactance Xd" can be made high. A FSCG with following reactances is proposed bellow.

Xd   = slightly more than 200%
Xd' or Xd" = about 80-85%
      (This means that the sum of Xd' or Xd" and the %IX of the step-up transformer is about 100%.)

This generator can be operated without quench in any system fault, and the decrease of the stability of the transmission line is not so large.

If the more higher stability is required, it may be realized by the introduction of a SMES system.

We are also performing a conceptual design of a FSCG. According to the preliminary study, a considerable decrease of loss is expected.

Conclusion

Through this study on the application of AC superconductivity, it becomes clear that the total loss of the SCT can be reduced greatly. The construction of the SCT is relatively simple, so, the SCT will be one of the most promising targets in the field of superconducting AC power machines.

By the age of the realization of the SCG, the AC superconducting technique will be established. It may not be an overestimation to expect that almost all of the SCG will be manufactured as a FSCG in the future.

As mentioned above, the loss of electrical machines will decrease drastically by the introduction of AC superconducting windings. The development of these machines is feasible and very promising.

Acknowledgement

This research is supported by Grant-in-Aid for Co-operative Research of the Ministry for Education, Science and Culture of Japan.

References

1) A.Fevrier, "Latest News About Superconducting A.C. Machines", MT-10, Boston, pp.22-25 (1987).
2) M.Yamamoto et al., "Development of 50Hz Disc-Type Superconducting Coil" IEEE Trans., MAG-23, No.2, pp.557 (1987).
3) O.Tsukamoto et al., "Development of Large-current Capacity Epoxy-Impregnated 50Hz Superconducting Coil", ICEC-12, Southampton (1988).
4) R.J.Meats, "Pressurized-Helium Breakdown at Low Temperatures", Proc. IEEE Vol.119, No.6, pp.760-765 (1972).
5) B.Fallou et al., "Insulation Component for High Cryogenic Equipment", CIGRE PAPER 15-04 (1974).
6) M.Yamamoto et al., "High Voltage Technique Application to Superconductivity", 5th ISH Paper-93.04 Braunschweig (1987).

Table 1   Loss(kW) of 3 phase-50Hz-1GVA-275/262.5/250kV-66kV Transformer

	Conv.	Super.
Winding loss	1700	0.5
Input of Refrigerator	--	250
Stray load loss	250	250
core loss	380	335
total loss	2330	835

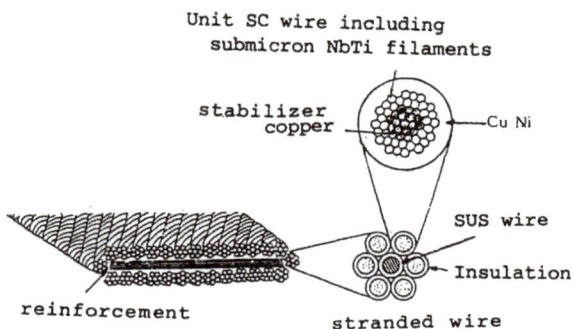

Fig. 1   An Example of heavy Current AC Superconductor.

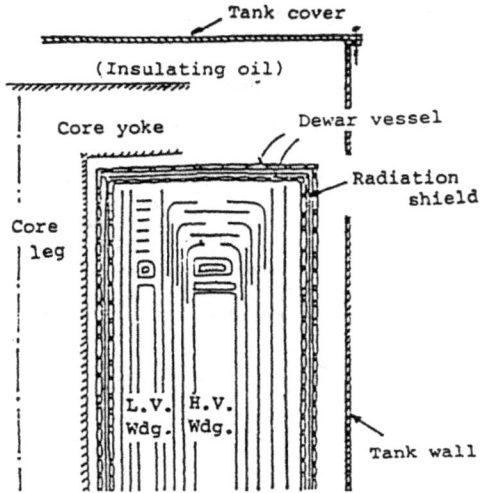

Fig. 2   Fundamental Structure of SCT.

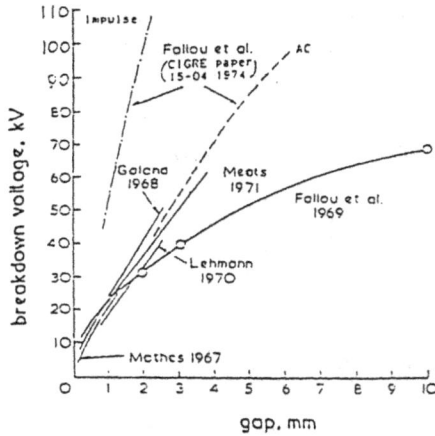

Fig. 3   Boiling Helium Breakdown Voltage
as Function of Electrode Spacing.

SUPERCONDUCTING MAGNET SYSTEMS FOR FUSION REACTORS

Albert ULBRICHT

Kernforschungszentrum Karlsruhe, Institut für Technische Physik,
D-7500 Karlsruhe, Postfach 3640, FRG

Abstract - The size of the plasma of the next generation of tokamak experiments require superconducting toroidal and poloidal magnet systems. In all tokamak designs around the world field levels needed will be about 12 T. Therefore A 15 (Nb$_3$Sn) superconductor technology is envisaged for this application. As a typical example for a development program the magnet technology program of the European community is described toward the superconducting magnet system for the Next European Torus (NET).

Introduction

Up to now the magnetic confinement for obtaining a controlled thermo-nuclear reaction is still the most promising way. The coming generation of machines for fusion research will reach a size where super-conducting magnet technology becomes stringent. The size of the plasma, increasing burning times for the plasma as soon as the break even point will be reached cannot be realized in an economic way by normal conducting magnets. Fusion technology people were aware of this task in the mid seventies when the first design of fusion reactors was proposed. For a mirror machine existing magnet technology was used for the construction of a 1.2 GJ s.c. magnet system. For the special topology of the magnetic field of the tokamak machine with pulsed field coils too, the state-of-art technology of this time was not immediately applicable. Under the leadership of the US-ERDA (now DOE), a technology experiment was initiated to develop magnets of such a size that one can scale up to a later reactor. This project called "Large Coil Task" (LCT) was running about more than ten years under the auspices of the International Energy Agency as an international collaboration of the USA, Japan, Switzerland and Euratom. Suitable conductors and magnets were designed, constructed and finally successfully operated for a test period of about two years. The operation of the magnet system at 9 T, the exposure of the magnets to poloidal field transients and the breakthrough of the forced flow cooled magnet technology were the highlights of this experiment. Stimulated by this success, the technical and economic necessity, the application of superconducting magnet systems is considered for future fusion devices with great confidence. One challenge was initiated by the next step of the national (NET, FER, TIBER, OTR) and international (ITER) fusion experiments under design now. Field levels up to 13 T stress the existing conductor technology of the A15 conductors. Sophisticated cable developments are running avoiding degradation effects as much as possible. Both coil systems for the toroidal and poloidal field will be pushed to its ultimate boundaries achievable with existing superconducting materials.

The tight relation of the poloidal field coils to a burning plasma in connection with the high AC loss requirements initiated the start of their development much later than those of the toroidal field. Projects for conductor and coil development are running in the Euratom Fusion Technology Program and in Japan. The most severe AC loss requirements originate from the withstanding of plasma disruptions and

574

control with field changes up to about 80 T/s. Model coils are under construction for the verification tests of the developed technology.

Last not least the high field requirement of fusion experiment pushed also the NbTi technology to higher field limits using 1.8 K cooling. This was realized in the French TORE SUPRA toroidal field coils. It is also considered as a possible backup solution for different coil systems of the next step tokamaks. Suitable cooling circuits are being investigated in laboratory scale.

Fusion Magnet Technology Program

The magnet systems which are needed for the next step of fusion machines represent a step partially beyond the present state-of-art in technology. The required field levels are considerably higher and thus, applicatin of the Nb_3Sn technology is needed. The capital costs of the magnet system are about one third of the complete machine. The complex geometrical arrangement requires a high availability and reliability for magnets as a basic component of the whole system.

The long term character of the required developments has been recognized already 10 - 15 years ago and there are early crown tasks which later became integrated in overall Fusion Magnet Technology Programs. These early tasks started more than ten years ago. The most of the early tasks were initiated by design studies of fusion reactors and conceptual designs of most generation devices. They contributed with the experience gained to the further steps of conductor and coil development and the experimental arrangements built up serve now in parts or completely as test facilities. Therefore in all countries which have the intention to build a next step machine development programs are running. The typical tasks of such a development program can be demonstrated by the Magnet Technology Program of the European Community (Fig. 1). The left hand side of the schema has to

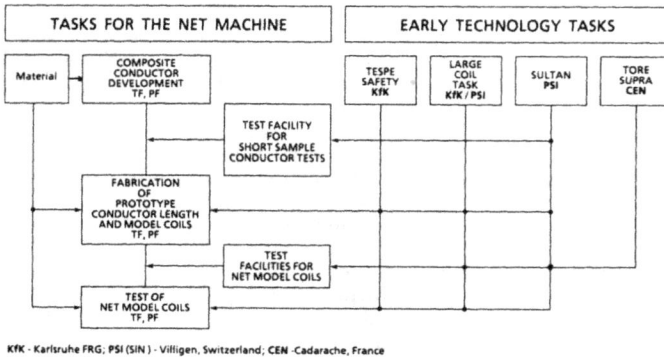

KfK - Karlsruhe FRG; PSI (SIN) - Villigen, Switzerland; CEN -Cadarache, France

Fig. 1. The scheme of the European Fusion Technology Program for the Next European Torus (NET) [19].

be applied for every conductor type which will be needed in the magnet system. Every step is costly. They have to be accompanied by basic programs to assure the success of the program [1]. The work is shared among the European superconducting magnet laboratories [2]. Similar programs are running in Japan, the Soviet Union and the United States, thereby the accentuation varies on this or that area.

The Outcome of the early Tasks in Euratom

TESPE:

The experiment was initiated as test bed for the investigation of basic investigations of toroidal magnet systems. The development of the conductor and coils needed similar engineering effort as for the LCT coil. The construction was handled by the laboratory (KfK/ITP) and only serial fabrication was handled by industry under the supervision of the laboratory. After the full torus was operated successfully the magnet system was used further on for experimental tests of safety related questions of superconducting magnet systems. There was investigated in detail :

- The buckling frequency of the system during the excitation of the torus, useable as preventive diagnostic before failures [3].
- Hot spot and quench propagation in a coil winding [4].
- Failure modes by loss of coolant, break down of insulation vacuum and short circuiting of one coil.
- Effects of arcing during the fast discharge [5].

The investigations were needed for the verification of computer code which raises the understanding and lighten the application of safety analysis procedures for larger magnet systems.

The Large Coil Task

The LCT was the first large step in the development towards toroidal magnet systems for tokamak machines. The international experiment ran over a decade of years with outstanding results not only in technical facts but also on the handling of the management of such large international experiments. The outcome was extensively evaluated and assessed [6]. The outcome for future magnet systems can be summarized as following:

- The scalability of the LCT magnets opens the possibility for the application of profen design principles for conductor and coil.
- the forced flow technology reduced the disturbance energy level so that conductor could be stably operated far outside the region of cryogenic stability. This lightens the integration of structural material in the conductor for reinforcement and resistive barriers for reduction of losses.
  The electrical integrity of the winding was outstanding.
- The mechanical stresses were in the range of expectations. Changing of boundary conditions change stress levels. A backlash free and pretensioned support will be indispensible for the long time operation of magnet systems.

Not only the conductors but also the coil designs of LCT were used as reference for the NET coils. The conductors needed modifications because some other design principles are required for Nb_3Sn.

It is intented that some of the LCT coils will be used as background coils in test facilities for model or prototype coils, fabricated from conductors developed for the next machines (NET, ITER, FER).

Sultan

The Sultan collaboration consists of three European laboratories PSI (before SIN)

Villigen, Switzerland, ECN; Petten, Netherland and ENEA, Frascati, Italy. The
Sultan facility was errected at the PSI. The general objective of the collaboration is
the development of forced flow cooled high field conductors and to provide for the
community a conductor test facility for full size NET conductors [6]. In the first stage
the test facility was built by PSI and the participants contributed with two NbTi
solenoids generating a 8 T field in 1,05 m free bore (Sultan I) [7]. The facility was used
for conductor tests of the next stage of Sultan. Three insert coils with Nb_3Sn
conductors in wind and react technique should increase the field to 12 T at 58 cm
accessable diameter (Sultan II) [8]. One of the insert coils is already tested
successfully. After this stage the Sultan facility will be changed now to a split coil
geometry with a gap of 10 cm and the 0.58 m free bore with 12 T (Sultan III)[9]. This
geometry is required for full size NET conductor short sample tests.

TORE SUPRA

The Tore Supra is the first European tokamak which has superconducting toroidal
field coils. In order to get a 4.5 T field on the torus axis the operation of the NbTi
superconducting coil at 1.8 K is neccessary because the field at the winding is then
about 9 T. The development of the s.c. coils started already in 1978 and construction
of the tokamak in 1982 [10]. The torus is cold since the beginning of 1988. Series of
promising tests and operations of the subsystems of the basic machine have been
started [11]. In April 88 the first plasma was ignited. The poloidal field coils are
copper coils. The investigation of long plasma discharges up to continuous plasma
discharges let it appear desirable to make also these coils superconducting in a later
stage of operation. In a collaboration between KfK-Karlsruhe and CEN Cadarache a
superconducting poloidal field coil is being developed. The coil has to withstand the
requirements coming from plasma disruption and plasma control. At first one
poloidal coil will be replaced by a superconducting one. If the results are satisfying
later on three other poloidal field coils will be exchanged by superconducting ones.

The use of superconducting toroidal and poloidal field coils in Tore Supra will give
useful information about the realibility and availability of superconducting coils. The
experience will be applied in the design of the NET machine.

Developments for the NET Superconducting Magnet System

Toroidal field (TF) coils: Three designs of the NET toroidal field conductor were
proposed by the European superconducting magnet laboratories. All three designs
were proposed in the "react and wind" technique. Besides this technique also a 40 kA
conductor in "wind and react" technique was proposed for the TF coils [17].

The basic idea of the "react and wind" technique is the usage of a reacted Nb_3Sn
Rutherford cable in Nb_3Sn technology: bronze, internal tin or external tin route are
represented in the three designs. The cable is sandwiched between two stabilizer
bars. The conductor components are soldered together and are enclosed in a stainless
steel jacket by laser beam welding (Fig. 2). The stabilizer bars have to have resistive
barriers in order to reduce losses caused by poloidal field transients. In the KfK
design flat Roebel transposed cables are proposed similar to the Euratom LCT
conductor design [12].

The conductors are designed for 16 kA rated current at a field of 12 T. They have
to carry hoop stresses of 140 MPa, radial stresses of -40 MPa and toroidal (axial)
stresses of -140 MPa. The conductors have to be designed that all stresses are
compatible with the axial and transverse degradation of Nb_3Sn conductors [13]. The
basic steps of the fabrication technique of such conductor types will be practiced in
subsize conductors or subsize similar conductors. They are tested in a coil or in
facilities under axial and transversal strain at the nominal field values. In the next

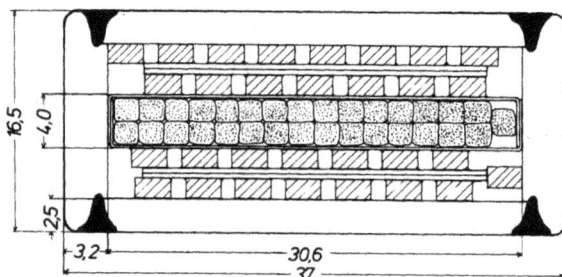

Fig. 2. The KfK design of NET TF conductor (16 kA, 12 T) in "react and wind" technique (Nb$_3$Sn bronze route, forced flow cooled by supercritical helium). The Nb$_3$Sn cable is sandwiched by two Cu stabilizer cables.

step of the development short samples of fulsize conductors will be tested in the modified Sultan facility [9]. This step will be followed by the fabrication of a prototype length which will be applied for maximal two different conductor designs. To verify the fabrication of the conductor and the fabrication technique of a coil it is planned to manufacture model test coils from this conductor pieces according to Fig. 1 and to test them in a most relevant manner. The most probable arrangement for the test of such coils is a so called TWIN configuration with the background field of one LCT coil in the TOSKA facility at KfK-Karlsruhe (Fig. 3) [14]. Field levels of about 11 T will be

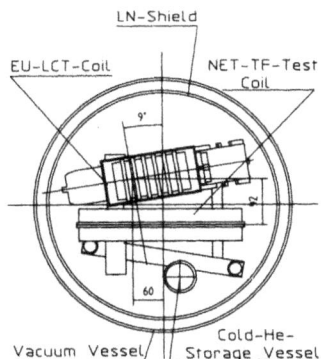

Fig. 3. Proposal TWIN configuration for NET TF model coil tests with the Euratom LCT coil operated as background coil at 1.8 K.

obtained in the model coil at 22 kA current. Some stress levels are lower (radial, shear and some are higher (toroidal, hoop). In a first step the field level of the Euratom LCT coil can be upgraded to 10 - 11 T at 1.8 K operation temperature. If a suitable reinforcement of the structure can be realized from the thermohydraulic and electrical point of view there are no reasons preventing the extension of the region of operation.

Poloidal field (PF) coils: About a decade later than for TF conductors the specific development of PF conductors were started. The specific requirements arising from plasma control and disruption need new designs. In KfK this development was started in collaboration with CEN Cadarache with the goal to replace one of the normal conducting PF coils of TORE SUPRA by a superconducting one. The reduction of losses needs a mixed matrix conductor and special insulation systems of the cable (Fig. 4). A dual cooling system was used for the removal of the losses within a nearly constant operation temperature. Two phase forced flow cooling was used for removal of the heat at nearly constant temperature. The stability of the conductor was assured by stagnant helium for transient heat transfer. After successful development the complete cable is now in the stage of production [16]. The cable jacket is fabricated from 4 quarter sections by laser beam welding. The conductor will be tested in a 3 m ø model coil in TOSKA. This conductor will be used in the later TORE SUPRA EHS coil probably with some modifications according to the test results. On the basis of this design CEN proposed a 40 kA conductor at 1.8 K for the central solenoid of NET, which should operate at rather high field so that as another solution a Nb_3Sn conductor becomes desirable too and is under development (CEN, ABB, LMI [17]).

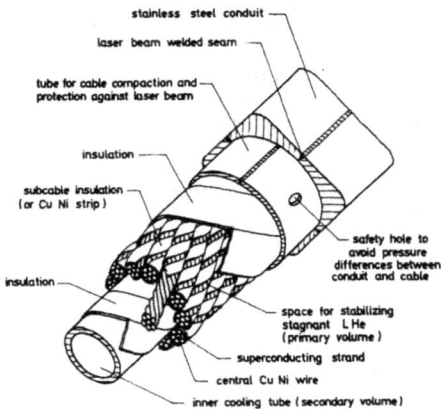

Fig. 4. Poloidal coil field conductor with a dual cooling system for stability (primary helium) and removal of losses (secondary helium).

The space limitations and the smaller size of the coil needs a coil fabrication in wind and react technique. Testing of such a type of conductor needs much more effort and is being in discussion within the European laboratories. Fast field change rates and high stored energies require a much more careful design of the coil considering the rules of the high voltage technique[18].

Conclusion

Looking forward the superconducting magnet technology in NbTi for fusion was well developed for some applications and is now able to replace copper magnets in all cases where the economy will be given. The operation at 1.8 K can increase the field level for NbTi coils up to 11 T. The higher field levels of the next generation of fusion machines need the development of A 15 superconductor cables and techniques of manufacturing windings from them within specified parameters and with sufficient live time. This is well proceeding up to large demonstration experiments within the fusion technology programs worldwide.

Acknowledgment

The author acknowledges the fruitful discussion with Professor Komarek during the preparation of this contribution.

Acknowlededgment

The author acknowledges the fruitful discussion with Professor Komarek during the preparation of this contribution.

References

1) P. Komarek, "Toward the availability of superconducting magnets for NET - the present KfK development effort. Nuclear Energy of Today and Tomorrow; 4th Internat. ENS/ANS Conf. and 9th Foratom Congress, Genève, CH, June 1-6, 1986; Transactions Bern: ENS, 1986. - Vol. 3 - S. 87 - 93

2) P. Komarek, Collaborative Superconductive Work in Europe, IEEE Transaction on Magnetics Vol-MAG-23 No. 2, March 1987, pp. 427

3) K.P. Jüngst et al., "First Results of the TESPE-S Magnet System Safety Experiments", Proc. 14th SOFT, Avignon 8-12 September 1988, France, page 1760

4) K.P. Jüngst, private communication

5) K.P. Jüngst et al."Arcing Experiments for Magnet Safety Investigations", 14th SOFT, Utrecht 19-23. September 1988

6) P.N. Haubenreich et al. "The IEA Large Coil Task, Development of Superconducting Toroidal Field magnets for Fusion Power". Fusion Engineering and Design 7, 1988

7) I. Horvath et al. "The forced flow high field test facility SULTAN", Journal de Physique Colloque CI, supplement au n° 1, Tome 45, Janvier 1984, page C1-93

8) J.D. Elen et al., "Upgrade of the Sultan superconducting test facility to 12 Tesla by three A-15 coils, Journal de Physique Colloque CI, supplement au n° 1, Tome 45, Janvier 1984, page C1-97

9) A. della Corte et al., The Sultan-III Project, Proc. 15 SOFT, Sept. 19-21, 1988, Utrecht, The Netherlands

10) R. Aymar, "The Tore Supra-Status report concerning the superconducting magnet after the qualifying development program IEEE Transaction on Magn. Vol. , No. 5, September 1981, p. 1911

11) R. Aymar, Commissioning and first operation of TORE SUPRA, Proc. 15th SOFT, 19-21, 1988, Utrecht, The Netherlands

12) R. Flükiger et al., "Status of the NET development of the KfK NET toroidal field conductor", Proc. 15th SOFT, Sept. 19-21, 1988, Utrecht, The Netherlands

13) W. Specking et al. "Effect of transverse compression on I_c of Nb3Sn" Adv. Cryo. Eng., Vol. 34 (1987), p. 569

14) A. Hofmann et al., Further use of the Euratom LCT coil, Proc. 15th SOFT, Sept. 19-21, 1988, Utrecht, The Netherlands

15) C. Schmidt, "Stability of poloidal field conductors: Test facility and subcable test results, Proc. ICEC 12, July 12-15, 1988, Southampton, England

16) U. Jeske, "Fabrication of a 15 kA NbTi-cable for the 150 T/s high ramp rate Polo model coil" Proc. 15th SOFT, Sept. 19-21, 1988, Utrecht, The Netherlands

17) J. Minervini et al., "Conductor designs for superconducting poloidal field coils of NET", Proc. 15th SOFT, Sept. 19-21, 1988, Utrecht, The Netherlands

18) G. Schenk, "High voltage insulation tests of cryogenic components for the superconducting model coil Polo, Proc. 15th SOFT, Sept. 19-21, 1988, Utrecht, The Netherlands

19) Next European Torus, fusion technology, July 1988, Volume 14, Number 1, p. 59

STATUS OF THE LARGE HELICAL PROJECT

Osamu MOTOJIMA

Planning Office for Institute of Fusion Plasma Science,
Nagoya University, Nagoya, Japan

Abstract - A large superconducting helical device (Heliotron Type) has
been designed for two years by Design Team (A Joint University Effort in
Japan). This will be a major experimental device for the new Toki
Institute (Institute of Fusion Plasma Science, Gifu Prefecture, Japan), which
will be founded in 1989 by Ministry of Monbusho. The specifications of the
present design option are : major radius R=4 m, coil minor radius a_h=0.96 m,
plasma minor radius a_p(plasma)=0.5-0.6 m, field period l=2 /m=10, magnetic
field B=4 T, plasma duration t=10 sec, heating power P_h=20 MW. The stored
magnetic energy of the superconducting coil is greater than 2 GJ. The construction is scheduled to be completed in 1995.

Introduction

After intensive design study of both super-conducting (SC) and normal-
conducting (NC) systems, the SC design is adopted[1-3]. This paper describes
the objective of this project and the conceptional design of the SC helical
device.

Objectives of the Project

The main objectives of this large helical system project are :

(a) To carry out transport study in a wide range of plasma condition and to
    achieve high $n\tau T$ plasma conditions extrapolatable to reactor plasmas.
(b) To realize high-beta plasmas with average beta value of 5%, which is
    required for reactor plasmas, and to understand the related plasma
    physics.
(c) To attain quasi-steady state operation and the related data base using
    helical divertors.
(d) To study the behavior of high-energy particles in this non-axisymmetric
    system, and to simulate alpha-particle behaviors.
(e) To increase the comprehensive understanding of toroidally confined
    plasmas by carrying out studies complementary to those in tokamaks.

There are two important engineering innovations:

(1) To build a superconducting helical system of which stored energy is more
    than 2 GJ.
(2) To develop new materials and particle control technique for the usage of
    first wall, divertor, and carbon wall.

There are three target plasma parameter regions which are shown in Table
1.

SC Machine Design

It is an important engineering innovation to build a superconducting helical device of which stored energy is larger than 2 GJ, largest in the world (Fig. 1). The construction is scheduled to be completed in 1995.

The device parameters are shown in Table 2. The major engineering items are as follows; (1) Superconducting Helical Coil, (2) Superconducting Poloidal Coil, (3) Vacuum Chamber, (4) Power Supply, (5) Control System, (6) Refrigeration System, and (7) Cooling System. Since the superconducting helical coil takes the primary part of the device, we have to pay the most of engineering efforts to make the practical view and engineering scope of this coil.

Three NbTi conductor designs are studied for helical coils;
(a) pool boiling,
(b) two-phase flow, and
(c) forced flow.
In order to keep the experimental flexibility, this helical coil is divided into 4 layers, each of which can be energized independently.

The flexibility of cross-sectional shaping by means of 3 pairs of poloidal coils are also taken into account. Moreover, the flexible long-pulsed operations are established by the SC coil system.

Aside from the merit of long pulsed operations, the SC device has other favorable features such as, (i) reduction of requirement for electric power supply, (ii) avoidance of X-ray production during start-up and shutdown of a discharge, (iii)avoidance of magnetic surface destruction due to the thermal coil deformation and induced current, and (iv) possibility of magnetic surface mapping at a high field.

The global arrangement of this large helical superconducting system is shown in Fig.2.

Schedule and R&D Items

Basic design of this helical device are carried out in FY 1988 to determine helical field period and winding law, and the detailed phase-1 design for the decision of the cooling system will be done in the beginning of FY 1989. After the decision of the main contractor for the machine construction, detailed phase-2 design will be conducted in FY 1990. The device contraction will be started in FY 1991 and completed in FY 1994. A first plasma is expected in FY 1995 (Table 3).

Before the start of the machine construction, 3-year R&D programs are arranged from FY 1988. The overall property of the SC helical system will be checked by the 1/5 reduced global model (Fig.3a). The technical arrangement for in-site coil winding will be tried by the partial model (Fig.3b), and the stress analysis will be done by the modular coil device (Fig.3c). Related to the poloidal coil design, the forced flow cooling test device and the conductor test machine will be constructed. The supercritical helium production system are also expected to be constructed in FY 1988-1989.

582

References

1. Design Group for the New Large Helical System Device, New Large Helical System Device (Outline of Design Proposal), March, 1988
2. O. Motojima, et al., The Design Study of The Large Superconducting Helical Experiment, Proceedings of Fusion Technology (1988).
3. K. Yamazaki et al., Comparative Design Study of Super- vs. Normal-Conducting Large Helical System, Proceedings of Fusion Technology (1988).

TABLE 1 Three Plasma Regimes

CASE 1 (high $n\tau T$) B = 4 T
  $<T_i>$ = 3 - 4 keV,
  $<n>$ = 10^{20} m^{-3},
  τ_E = 0.1 - 0.3 sec,
CASE 2 (hight T_i) B = 4 T
  $T_i(0)$ = 10 keV,
  $<n>$ = 2 x 10^{19} m^{-3},
  τE= 0.05 - 0.1 sec,
CASE 3 (high β) B = 1~2 T
  $<\beta>$ = 5 %,

TABLE 2 MACHINE PARAMETERS

Major Radius	4	m
Minor Coil Radius	0.96	m
Magnetic Field(plasma center)	4.0	T
(maximum)	8	T
Toroidal Field Period	10	
Helical Coil Current	8	MAT
Coil current Density	4.0 kA/cm^2	
Rotational Transform(center)	0.35	
(surface)	8.0	
Heating Power	20	MW

Helical Coil:	Nb-Ti wire
	50 mm thick coil can
Poloidal Coil:	Nb-Ti wire
	3 pairs, 6 MAT/coil
Vaccum Vessel:	dumbbell shape
	30 mm thick, SS 316
	100 ℃ baking
First Wall:	25 mm thick
	350 ℃ baking

TABLE 3
Schedule for SC Machine

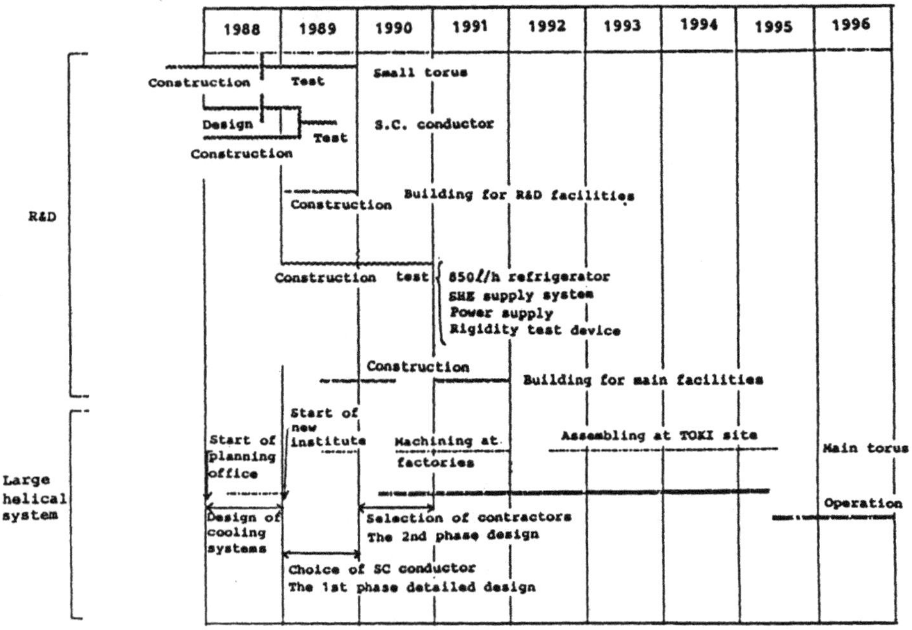

	1988	1989	1990	1991	1992	1993	1994	1995	1996

R&D

- Construction — Test — Small torus
- Design — Test — S.C. conductor
- Construction
- Construction — Building for R&D facilities
- Construction test { 850ℓ/h refrigerator / SHE supply system / Power supply / Rigidity test device
- Construction — Building for main facilities

Large helical system

- Start of planning office
- Start of new institute
- Machining at factories
- Assembling at TOKI site — Main torus
- Operation
- Design of cooling systems
- Selection of contractors / The 2nd phase design
- Choice of SC conductor / The 1st phase detailed design

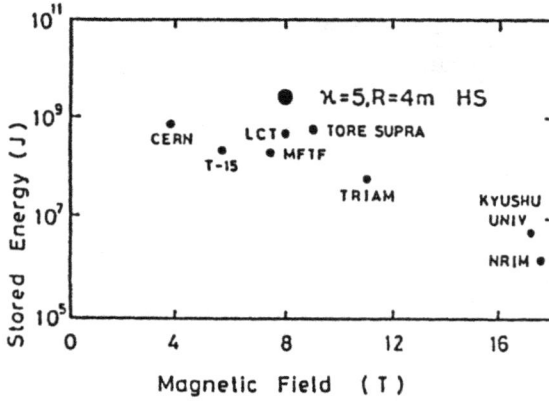

Fig.1

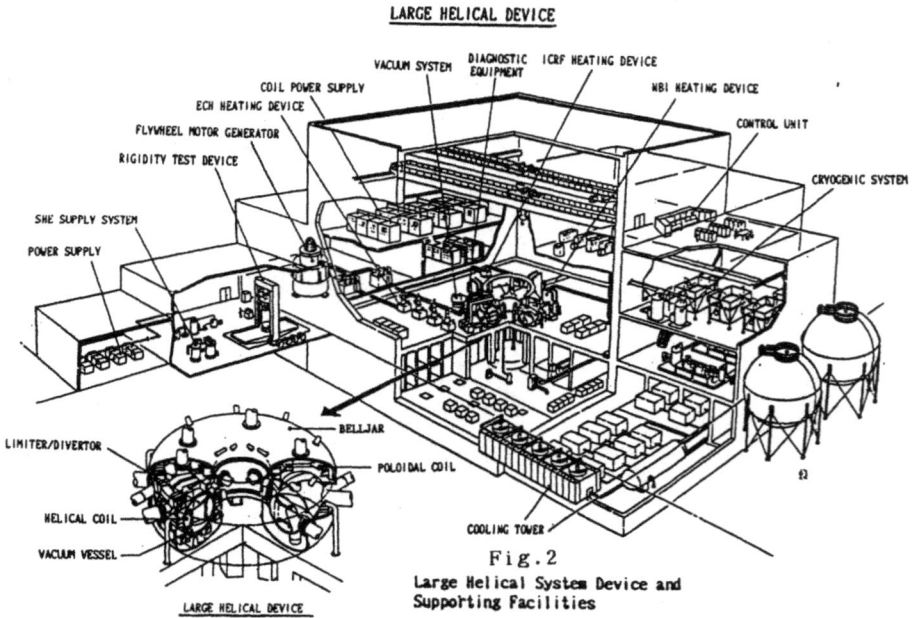

LARGE HELICAL DEVICE

Fig.2
Large Helical System Device and
Supporting Facilities

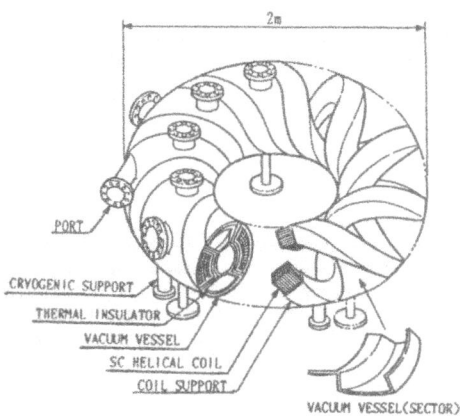

Fig.3a
1/5 reduced global R&D model

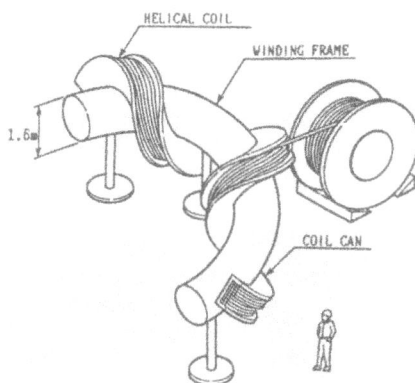

Fig.3b
Winding test partial R&D model

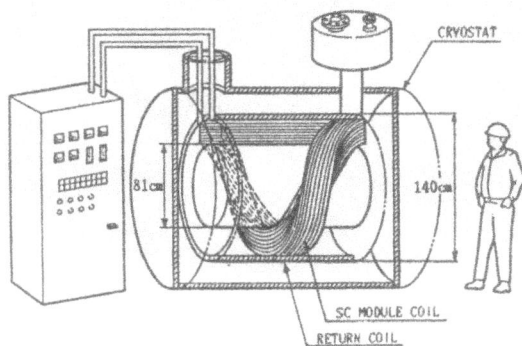

Fig.3c
Modular coil facility for stress test

586

DEVELOPMENT OF LARGE SUPERCONDUCTING MAGNETS
FOR FUSION APPLICATION AT JAERI

Kiyoshi OKUNO and Susumu SHIMAMOTO

Japan Atomic Energy Research Institute
Mukaiyama, Naka-machi, Naka-gun, Ibaraki-ken, 311-01, Japan

Abstract - This paper describes research, development, and demonstration work
of large superconducting magnets for a tokamak fusion reactor at JAERI. The
IEA-LCT project and the following project, the Proto Toroidal Coil Program,
are the major steps in the development of a toroidal magnet required for the
Fusion Experimental Reactor. As the development of a poloidal magnet, the
Demo Poloidal Coil Program has been progressing and three pulsed coils whose
stored energy is 40 MJ are now under construction.

Introduction

    Demonstration of fusion power by a tokamak reactor is one of the most
important missions for Japan Atomic Energy Research Institute (JAERI). Tokamak
is the most advanced type among several proposed fusion machines from the
viewpoint of plasma confinement, but the work requires the further advancement
of technologies over a wide field, and integration of technologies into a
system is also indispensable. These technologies must conquer the extreme
conditions such as large size, large forces, very high and very low

Fig. 1 Superconducting magnets for a tokamak fusion reactor.

temperatures, heavy heat load, nuclear radiation, and so on. While their requirements and complexity are fairly beyond our present status, steady progress is obtained in these area through research, development, and demonstration work. Superconducting magnet development is one of these examples. As the next generation machine after the critical plasma machine JT-60, design study of the Fusion Experimental Reactor (FER) is progressing at JAERI [1]. Superconducting magnet system for the FER is shown in Fig. 1. The height of toroidal magnets is about 15 m and peak magnetic filed is 12 T. The mean diameters of poloidal magnets are 3 m (ohmic heating coils) and 23 m (equivalent field coils), and peak field is 12 T. The following sections describe development work for these large superconducting magnets.

Toroidal Magnet Development

The Cluster Test Program

Toroidal magnets are the most important component in a tokamak machine because it is indispensable for plasma confinement. Therefore, toroidal magnet development started first at JAERI and the Cluster Test Program was initiated. In the program the Test Module Coil (TMC) was fabricated using Nb_3Sn superconductor with react and wind method. The inner and outer diameters of its winding are 0.6 m and 1.6 m, respectively. The program was successful and the coil achieved 12.2 T with the winding current density of 33 MA/m^2 without a quench [2]. Through the program the applicability of Nb_3Sn superconductor for large magnets was demonstrated and high field technology was established.

The Large Coil Task

In parallel with the Cluster Test program, JAERI performed the IEA Large Coil Task (LCT), which is a multinational collaboration to develop large superconducting toroidal magnets. Six coils provided by EURATOM, Switzerland, the United States, and Japan were installed and tested in the International Fusion Superconducting Magnet Test Facility (IFSMTF) at Oak Ridge National Laboratory (ORNL), as shown in Fig. 2. The LCT coils have D-shaped 2.5 x 3.5-m bore and are the largest superconducting coils up to date fabricated for

Fig. 2 Six LCT coils installed in the IFSMTF at ORNL.

fusion research. Each LCT coil has distinctive features in its design. Coils from Japan, U.S.-GD, and U.S.-GE have conductors cooled by pool-boiling helium and coils from EURATOM, Switzerland, and U.S.-Westinghouse have forced-flow cooling. Five coils use NbTi superconductor and the U.S.-Westinghous selected Nb_3Sn. Various experiments were performed at the design point (8 T) and beyond the design current and field [3]. The last test of the program was called "maximum-field torus test." All six coils reached the planned field of 9 T simultaneously, and a total stored energy of 944 MJ was achieved [4]. Results in the maximum-field torus test were summarized in Table 1.

Table 1 Achievements obtained in the maximum-field torus test.

Coil		US-GE	US-GD	Japan	Swiss	EURATOM	US-WH
Operating current	(A)	11,092	9,970	10,798	12,820	12,103	19,234
Maximum field	(T)	9.2	9.0	9.1	9.1	9.2	9.0
Centering force	(MN)	52.8	50.7	53.2	46.0	57.1	65.7
Out-of-plane force	(MN)	10.0	-0.2	1.6	-1.1	-11.2	1.2

The Proto Toroidal Coil Program

, Based on the above progress, JAERI started a new development work, the Proto Toroidal Coil Program [5]. Principal objectives of the program are as follows: (1) to develop high performance conductor (12 T, 30 kA, and forced-

Table 2 General requirements for FER toroidal magnet and achievements obtained in TMC and LCT projects.

		FER	TMC	LCT
Design magnetic field	(T)	12	10	8
Maximum operated field	(T)	-	12.2	9-9.2
Rated current	(kA)	30	6	10-18
Winding current density	(MA/m^2)	30-40	30[a]	20-30[a]
Maximum dump voltage	(kV)	20	1.0	2.5[b],0.5-1.0[c]
Environment				
Peak nuclear heating rate	(mW/cm^3)	1	-	2[b],55[c]
Maximum pulsed field		20T/s-200ms	-	0.14T/s-1s
Conductor				
Pulse loss time constant	(s)	< 0.1	> 10	0.3[f]
Stability margin	(mJ/cm^3)	50-100	cryostable	1000-1900[g] cryostable[c]
Hot spot temperature	(K)	100-150	< 40	< 100[f]
Superconductor		$(NbTi)_3Sn$	Nb_3Sn	$NbTi,Nb_3Sn$
Jc at design condition	(MA/m^2)	700	580	800-1,000[d]
Structural material		JCS^e	SS 304L, 316L	SS 304L,304LN, 316L,316LN,Al
Yield Strength	(MPa)	1,200	360	800[f]
Tensile strength	(MPa)	1,600	1,500	1,600[f]
Fracture Toughness	$(MPa\sqrt{m})$	200	-	250[f]
Cooling method		Forced flow	Pool boiling	Forced,Pool
Helium inlet temperature	(K)	4.0-4.5	4.2	3.8[b],4.3[c]
Helium inlet pressure	(Bar)	6-10	1	11-13[b],1.1[c]
Maximum inner pressure	(Bar)	50	2.4	15[b],3-5.5[c]

[a]At the design condition. [b]For forced-flow coils. [c]For pool-boiling coils. [d]In NbTi. [e]Japanese Cryogenic Steels. [f]For Japanese coil. [g]For WH coil.

flow cooled), (2) to develop manufacturing techniques of magnets for high rigidity and high voltage insulation, and (3) to fabricate and evaluate real conductor and coil as a prototype for the FER. While the FER design is not completely fixed, general requirements for conductor and magnets have been specified. The requirements are summarized and compared with the achievements obtained in the Cluster Test Program and the LCT project in Table 2. The electromagnetic properties of these requirements can be demonstrated by a small sample. However, mechanical and thermal properties have to be verified with a conductor or coil in a reasonable length and size. The design study of the FER indicated that the toroidal magnets will be exposed to much larger forces than those experienced in the LCT coils. Table 3 shows the comparison of these forces. Therefore,

mechanical investigation is the most important subject in the program. For this purpose, a test coil, named the Proto Toroidal Coil, will be fabricated. The coil has the shape of race truck with 3.0 x 5.1-m bore in order to have adequate stresses such as encountered in the FER. Most of the parameters of the coil, except coil size, shape, and ampere-turns, satisfy the general requirements indicated in Table 2.

Table 3   Comparison of forces between FER and LCT.

		FER	LCT
Size ratio		3	1
Stored energy	(GJ)	40	1
Centering force	(MN)	415	53
Vertical force	(MN)	195	50
Out-of-plane force	(MN)	-	26
Overturning moment (X-axis)	(MNm)	111	4

Poloidal Magnet Development

The Demo Poloidal Coil (DPC) program has been progressing for the development of poloidal magnets for the FER. Before starting the DPC program, JAERI had completed the development work of large-current NbTi pulsed conductors and fabricated 1.5-MJ pulsed magnet, the Poloidal Unit Pancake (PUP). The PUP has a 30-kA pool-cooled conductor and operated with 10 T/s at 7 T. Besides the PUP, JAERI fabricated and tested a 30-kA forced-flow conductor. These success are the technical background for the DPC. Figure 3 shows the major superconducting pulsed coils developed or being developed.

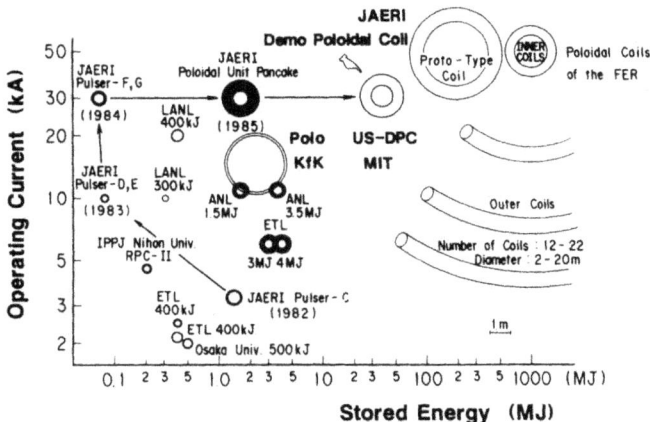

Fig. 3 Pulsed coil development for superconducting poloidal magnets

The Demo Poloidal Coil Program

In the DPC program, three pulsed coils, DPC-U1, -U2, and -EX, are being fabricated [6]. The coils will be installed and tested as shown in Fig. 4. In addition to the three coils, the U.S. MIT provides a test coil, which is indicated as the US-DPC in Fig. 3. The DPC-U1 and -U2 are forced-flow cooled NbTi coils whose total stored energy is 30 MJ. The purposes of the coils are to demonstrate reliable operation of large superconducting magnets in a rapid charge to 7 T in 1 s and to verify low ac losses. The DPC-EX and US-DPC use Nb_3Sn superconductor to demonstrate the applicability of Nb_3Sn for pulsed magnets. Their conductors, shown in Fig. 4, have thick stainless steel conduit to improve mechanical strength. The coupling loss time constant of a strand for the DPC-U1 and -U2 is much smaller than the specified value of 1 ms. This is achieved by arranging three CuNi barriers in the strand. Since the current distribution of poloidal magnets in a fusion reactor changes with time, several operation modes can be defined for the experiments of DPC. Two principal modes among them are mirror operation and cusp operation of the DPC-U1 and -U2 to full current. The former mode generates a 27-MN compressive force in the coils and the latter generates a 17-MN repulsive force. In order to support these forces, newly developed cryogenic stainless steels are employed for conduit material and coil structures. The conductor fabrication of the DPC-U1 and -U2 has been completed and winding work of their pancakes has started. The fabrication of the DPC-EX conductor is in progress.

For the experiments of the DPC, test facility (DPCF) has already been constructed and test operation was performed. A supercritical helium pump in the facility has succeeded in the operation of 500-g/s mass flow rate. Large current bus lines were connected to the JT-60 power supplies, which are powerful enough to charge the coils to 7 T in 1 s. Three pairs of 30-kA current leads were also tested and demonstrated a stable operation with the heat load of 1 W/kA. Thus the progress of the DPC program is favorable.

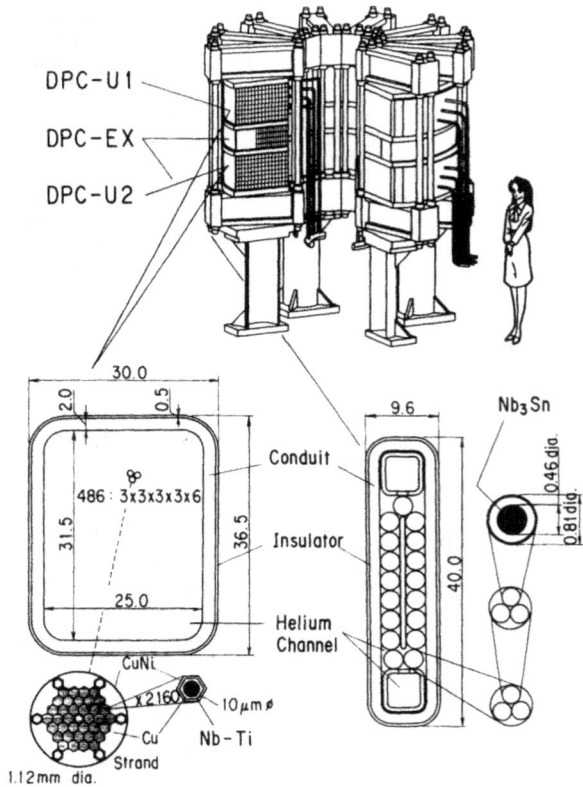

Fig. 4   Demo Poloidal Coils (DPC-U1, -U2, and -EX) and their conductors.

Fig. 5 Test facility for the Demo Poloidal Coils.

Concluding Remarks

Superconducting magnet technology for fusion has made remarkable progress in these ten years. The IEA-LCT project played an important role in the toroidal magnet development. A new program has started at JAERI aiming at the construction of a prototype coil. For poloidal magnets, the DPC program at JAERI is in progress and useful data will be obtained in a few years. These achievements promise to be useful for the construction of the FER.

Acknowledgments

The authors would like to thank Drs. S. Mori, M. Yoshikawa, and M. Tanaka for their continuing encouragement on the work above. Cooperation by all the members in the Superconducting Magnet Labo. at JAERI is greatly acknowledged.

References

[1] FER Design Team, "Conceptual Design Study of Fusion Experimental Reactor (FY87FER)-Summary Report-," JAERI-M 88-090.
[2] T. Ando et al., "Development of High Field Superconducting Coil for the Tokamak Fusion Machine in JAERI," Proc. of 11th Symp. on Fusion Eng. (1986) 991.
[3] K. Okuno et al., "Experimental Results on the Japanese LCT Coil in the IFSMTF: Pulsed Field Tests and Extended-condition Tests," IEEE Trans. Magn. MAG-24 No.2 (1988) 767.
[4] D. S. Beard, W. Klose, S. Shimamoto, and G. Vècsey, "The IEA Large Coil Task," Fusion Engineering and Design, Vol 7 Nos. 1 & 2 (1988).
[5] K. Koizumi et al., "Design of a Test Coil and the Test Facility for Proto Toroidal Coil Program," Proc. of 15th Symp. on Fusion Technology (1988).
[6] H. Tsuji et al., "Recent Progress in the Demo Poloidal Coil Program," Presented at the Applied Superconductivity Conference (1988).

DESIGN STUDY OF LARGE SCALE SMES

Takakazu SHINTOMI

National Laboratory for High Energy Physics
1-1, Oho, Tsukuba-shi, Ibaraki 305, Japan

and

Research Association of Superconducting Magnetic Energy Storage
Tokodai, Tsukuba-shi, Ibaraki 305-32, Japan

Abstract - The appropriate size of commercial SMES is estimated as 5 GWh in capacity and 1 GW in output power from the electricity demand in the year 2000. The diameter of superconducting coil in this scale is 400 - 1500 m depending on designs. By the optimization of the coil configuration and taking Japanese situations into consideration, a compact coil configuration whose diameter is 500 m is selected with a material of Nb_3Sn. The cost is also estimated and expected to be economical by comparison of the break even capital cost of a 5 GWh SMES unit. The possibility of application of the high T_C materials has been also considered.

Introduction

Superconducting Magnetic Energy Storage (SMES) is one of the most promising applications of superconductivity. A few works for assessment of its feasibility have been performed in Japan and the United States. In Japan the New Energy Development Organization (NEDO) carried out the conceptual design and the economic evaluation of a 5 GWh SMES unit from 1982 to 1984.[1] The Engineering Advancement Association (ENAA) did the cost estimation and technical feasibility for the same size from 1983 to 1985.[2] On the contrary, in the United States EPRI has performed the conceptual design and cost estimation of a SMES system of the same capacity.[3] The conclusions of those assessments are that SMES has no technical breakthrough and highly economic potential.

From 1987 the United States government started a project which intends to construct a SMES unit of about 20 MWh. In Japan, the Research Association of SMES was organized in 1986 with 40 members of private companies including utilities. The Association is now proposing a project of a 50 MJ SMES system to the government.

Optimization of Coil

Appropriate GWh and GW capacities of a practical SMES for load leveling have been determined by considering capacity of the power system, requirement of the energy storage plant, and an optimal operating strategy and economical benefit of SMES.[4] The stored energy of about 4.5 GWh and output power of 1 GW may be appropriate for load leveling in the year 2000. Supposing that the net available energy is 90 % of GWh capacity, the rated GWh capacity is evaluated as 5 GWh. Therefore, a SMES system of 5 GWh and 1 GW has been selected.

The SMES system depends on the scaling laws which are described with aspect ratio, stored energy and magnetic field. Using these laws, the coil configuration can be optimized in an economical point of view. The relative capital cost of the 5 GWh SMES plant has been calculated as shown in Fig. 1. The figure gives the minimum cost around 0.02 to 0.05 in the aspect ratio and 4 to 6 tesla in the maximum magnetic field at the midplane. The dotted line shows the limit of the coil height of 20 m.

Because of Japanese situation of small available area, a compact SMES plant may be preferable and the aspect ratio of 0.1 has been selected for the conceptual design.[1, 2] The coil should be divided into three segments because the coil height of 40 m is too high to be constructed. However, there were still difficulties to support each segment against vertical magnetic forces. We selected a single coil which has an aspect ratio of 0.04 for the next step.

The magnetic field of 5.9 tesla is selected around the maximum pinning of the superconductor. From these parameters the coil diameter and height are set to 500 m and 20 m, respectively.

The basic parameters of the coil are shown in Table 1.

Table 1 Basic parameters of the 5 GWh SMES system.

Capacity	5000	MWh
Output power	1000	MW
Coil Number	1	
Diameter	500	m
Height	20.04	m
Aspect ratio	0.04	
Current	500	kA
Layers	2	
Maximum field at midplane	5.9	T
Average magnetic pressure	3	MPa
Superconductor	Nb_3Sn	
Stabilizer	Pure Al	
Coil support structure	Al alloy	
Cooling	L He pool-boiling	
Thermal insulation	Thermal shield at 80 K	
Refrigerator power	3.5	MW

Design Features of 5 GWh SMES

Conductor

To meet an economic feasibility, the capital cost of SMES should be cheaper than Break Even Capital Cost (BECC). As a fraction of the conductor cost in the capital cost is relatively high, it is a key issue to select an economical superconducting material. In this point of view, we choose Nb_3Sn which may be expected to be less expensive than NbTi in future.

The conductor is designed to have the full stabilization and large cooling surfaces for saving pure aluminum. The conductor assembly consists of a superconducting strand cable, pure aluminum and high strength aluminum alloy, so called brick, as shown in Fig. 2.

Coil and Supporting Structure

The coil, which is designed to have a double-layer, is a soft structure and supported by the rock cavern through struts against radial expansion forces. The coil should stand also against thermal stresses at cooling-down. The ripple structure in Fig. 2 is effective to absorb thermal shrinkage. The rippleless coil has been designed as an alternative of the coil structure.[3] In this configuration the coil may stand against the electromagnetic forces as bending stresses in contrast with membrane stresses for the ripple structure. However, it has no margin against the thermal shrinkage and some supporting structures should be necessary inside the coil.

The stress analysis has been carried out for the thermal stresses and the electromagnetic forces to check the enough strength of the coil structures. One of the calculations is shown in Fig. 3.

For the material of the brick, we choose tentatively A7075 which is an excellent material among aluminum alloys. We, however, have not made our final choice because of the difficulty in welding.

Thermal Insulation

The struts which are made from fiber reinforced plastic are used between the coil and bedrock. The material should have properties of enough strength and good thermal insulation. The glass fiber is popular and available for them. An alternative is Al_2O_3 fiber which is stronger than the glass fiber.

As for radiation shields, only a thermal shield at 80 K is considered. A shield at 20 K is omitted because the contribution is not dominant. The omission brings the merits of the simple assembling and the cost reduction of the system.

Heat Loads and Refrigeration System

The heat loads caused by radiation, ac losses, conduction through the strut, power leads are shown in Table 2. As the result, the refrigeration power in room temperature is 3.5 MW which corresponds to the loss less than 4 % for the total efficiency in the daily operation of SMES.

Table 2   Cryogenic losses of 5 GWh SMES.

Source of heat load	Temperature(K)					Refrigeration power in room temperature
	4.2	10	13	20	80	
Conduction	0.5 kW			6.0 kW	68 kW	1.1 MW
Radiation	0.9				10	0.5
Ac losses	1.6					0.8
Transfer line	0.1			1.0		0.2
Power lead		0.1 kW	69 g/s			0.9
Total						**3.5**

In the design of the refrigeration system, the following conditions are taken into account: high reliability, large cooling mass in precooling even in the less stationary heat load, and cooling time shorter than 4000 hours. From the requirements the total system is composed of two identical subsystems. At the precooling, the two subsystems are operated simultaneously. One is operated and the other is rested as a spare at the stationary operation.

Cost Estimation

The initial cost estimation by NEDO and ENAA gave the capital cost of 5 GWh SMES as 290 B yen for the three-segmented coil. On the same basis of the estimation the cost may be reduced to 257 B yen by this revised design as shown in Table 3.

Table 3   Cost estimation of 5 GWh SMES (X 10^8 yen).

Materials	Conductor	535
	Brick	379
	Strut	112
	Vessels/Shield	104
	Refrigerator	64
	Power conditioning	200
	Control/Others	43
Construction	Trench	69
	Assembling	128
Contingency		164
Indirect		771
Total		**2569**

The BECC has been estimated using a cost-and-value method.[1] The values of highest and lowest BECC are summarized in Table 4. Our estimation of the capital cost lines almost on the lowest level of the BECC. Therefore, the SMES system in such scale is expected to have highly economic potential.

Table 4   BECC of a 5 GWh SMES system vs thermal power plant (X 10^3 yen/kW).

Principal charging energy	Type of standard plant	Availability (%)	
		16.8	14.3
Nuclear	Coal	352 - 490	346 - 482
	Oil	312 - 436	302 - 414
Coal fired	Coal	302 - 438	300 - 436
	Oil	270 - 384	258 - 372

The cost is still high in comparison with that of the United States.[3] The further effort should be paid to reduce the capital cost.

The capital cost mainly depends on the GWh capacity. The scaling is two-third of the stored energy. If it is assumed that the BECC also depends on the capacity, the economy of SMES for the stored energy will be given as shown in Fig. 4. The most optimistic case gives that a SMES system of a several hundred MWh will be feasible in an economic point of view.

Consideration of SMES with High T_C Superconductors

To consider the applicability of high T_C materials to SMES, it is assumed that the new materials are commercially available at the operating temperature of 77 K and has the same critical current density as Nb_3Sn.

The materials for the conductor have large heat capacity at the operating temperature in comparison with the liquid helium temperature. The coil may be tough against disturbances. Therefore, a large amount of the stabilizer may not be necessary for high T_C materials. On the contrary, the heat capacity is almost same as in case of the liquid helium temperature. As for the coil protection, necessary amount of materials should be required for the enough heat capacity.

As the coil is subjected with the same electromagnetic forces and structural materials are weaker at 77 K than at 4 K, more amount of structural material is necessary.

The thermal insulation is easier and simpler than that at 4 K. Some techniques for storage tanks of liquid natural gas may be available. One of the designs is shown in Fig. 5. Plates of 2 m square which are filled with evacuated perlite will be assembled to a plate of 4 m x 10 m at a factory, for example. In the design a vacuum vessel is not necessary. Therefore, the cryogenic system may be simpler and more reliable.

The cost reduction of the 5 GWh SMES system with the high T_C materials is expected as 15 to 20 % if they are available at the same cost as Nb_3Sn.

Development Schedule of Commercial SMES

The development schedule leading to a commercial SMES plant planned by the Association is to have steps of a pilot plant of 1 to 10 MWh and a demonstration plant of 100 to 500 MWh. The pilot plant will be around 100 m in diameter and intended for the technical feasibility. On the other hand, the demonstration plant may be operated to confirm the economic feasibility. Prior to the plants, a testing plant which has the capacity of 50 MJ and the diameter of 5 m has been planned to solve the basic technical problems.[5]

The superconducting coil of the commercial SMES plant is 10^4 times bigger than any existing coil systems. The most important problem is how to support the huge electromagnetic forces economically. The idea of using bedrock has been proposed to solve the problems. This is the main purpose of this plant to be constructed in bedrock for the first attempt.

596

Summary

The commercial size of SMES is designed to check the feasibilities in technical and economic point of view. The coil is 500 m in diameter and 20 m in height. The soft coil structure with ripple is adopted to absorb thermal shrinkage. The coil is supported with bedrock through struts which are made from fiber reinforced plastic against radial expansion forces. The structure is checked with the stress analysis using the finite element method. The cryogenic losses are estimated as less than 4 % of the total efficiency in the daily operation.

The capital cost of the plant is estimated as 257 B yen and lined almost on the lowest level of the BECC.

The commercial 5 GWh SMES plant has no technical breakthrough and highly economic potential. However, the great efforts on the technical development to scale up and the cost reduction should be paid. In this meaning the project of the testing plant which is intended to be installed in bedrock should be developed as the first step as soon as possible.

References

1) NEDO report, "Reports on the SMES System [III]," 1985 (in Japanese)
2) ENAA report, "Reports on the SMES System [III]," 1986 (in Japanese)
3) EPRI report, "Conceptual Design and Cost of a SMES Plant," EPRI EM-3457, 1984
   LASL report, "Design Improvements and Cost Reductions of a 5000 MWh SMES Plant, Part 2," LA-10668-MS, 1985
4) J. Hasegawa, et al., "Strategic Usage of SMES for Electric Power Demand in the Year 2000 in Japan," Proc. Int. Symp. and Workshop on Dynamic Benefits of Energy Storage Plant Operation, Boston, 1984, p. 284
5) Y. Hayakawa, et al., "Test Plant as the First Step Towards Commercialization of SMES for Utilities," IEEE Trans. on Magnetics, MAG-24, 1988, p. 887

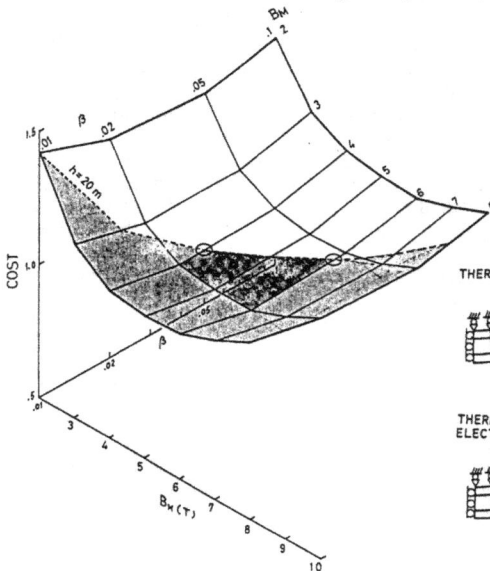

Fig. 1 The relative capital cost of a 5 GWh SMES system vs aspect ratio and maximum magnetic field at midplane.

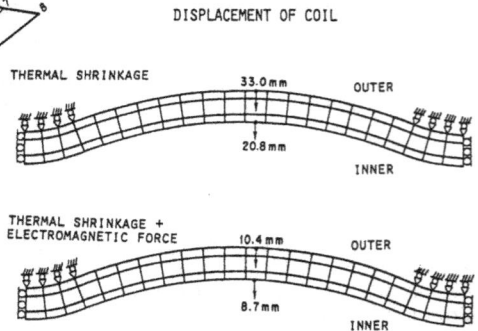

Fig. 3 Stress analysis of the coil structure of a 5 GWh SMES system.

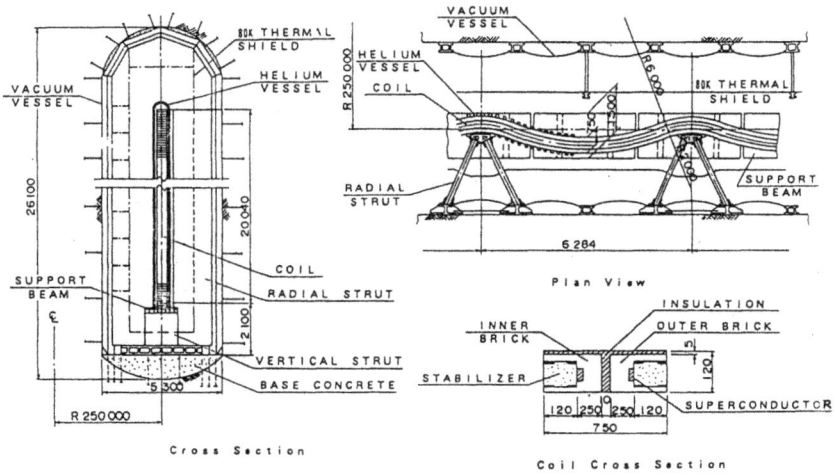

Fig. 2 Conceptual design of a 5 GWh SMES system.

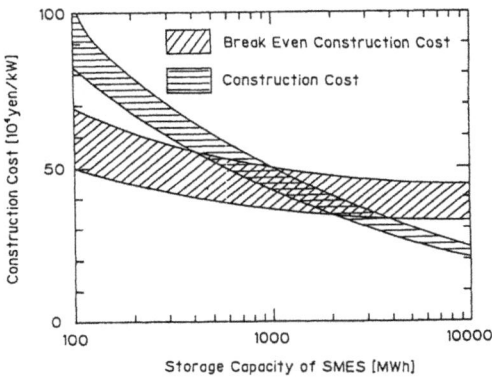

Fig. 4 The economy of SMES vs storage capacity.

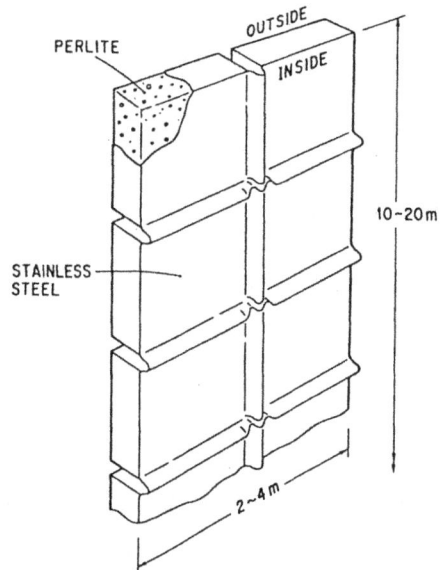

Fig. 5 One of the examples for thermal insulation with high T_C supercon-ductor.

CONCEPTUAL DESIGN STUDIES OF 5GWH SUPERCONDUCTING
MAGNETIC ENERGY STORAGE SYSTEM

Tsutomu MAEKAWA*, Hiroshi KUWAHARA*, Toshio FUKATSU*, Kyoji KASE*
Masatami IWAMOTO**, Tadatoshi YAMADA**, and Osamu OGINO**

  * Tokyo Electric Power Co. Inc., Minatoku, Tokyo, Japan
 ** Mitsubishi Electric Corp., Amagasaki, Hyogo, Japan

Abstract - On the assumption that high T_C superconductor will be
realized, the conceptual design studies of a 5GWh superconducting magnetic
energy storage system were made for the following four levels of superconducting temperature to survey the future feasibility; 1) liquid helium
temperature, 2) liquid hydrogen temperature, 3) liquid nitrogen temperature,
and 4) room temperature.

Introduction

Superconducting magnetic energy storage system (SMES) is one of the
possible energy storage systems for power utilities[1-2]. In the recent high
Tc superconducting fever, journalism has reported that superconducting
magnetic energy system is one of the most promising applications of superconductivity. At this moment, it is meaningful that we specialists of superconductivity, cryogenics and electric power engineering have made conceptual
design studies of a 5GWh superconducting magnetic energy storage system on
the assumption that high Tc superconductor will be practicable in future.
Four conceptual designs have been made for the following four levels of
superconducting temperature:
1) liquid helium temperature (4.2K)
2) liquid hydrogen temperature (20.3K)
3) liquid nitrogen temperature (77K), and
4) room temperature (300K).
The superconductor for liquid helium temperature is being commercially manufactured and widely used in high field magnets. However, the superconductors
for the latter three temperature levels are just assumed to be realized in
future.

5GWh SMES Using Liquid Helium Temperature Superconductor

Conceptual Design

The figure 1 shows the conceptual view drawing of the 5GWh SMES using
liquid helium temperature superconductor. The result of the conceptual design
is shown in the table 1. The huge superconducting coil with a diameter of
640m is installed in the bed rock in the depth of 100m. The radial electromagnetic expansion force is transmitted to and supoorted by the bed rock.
The averaged electromagnetic pressure is 10kg/cm². The bed rock which stands
the pressure of 10kg/cm² is widely found in Japan. The force supporting
column is made of fiber reinforced epoxy. The number of the column is 8400
for the support of the radial magnetic force and 850 for the support of the
weight.

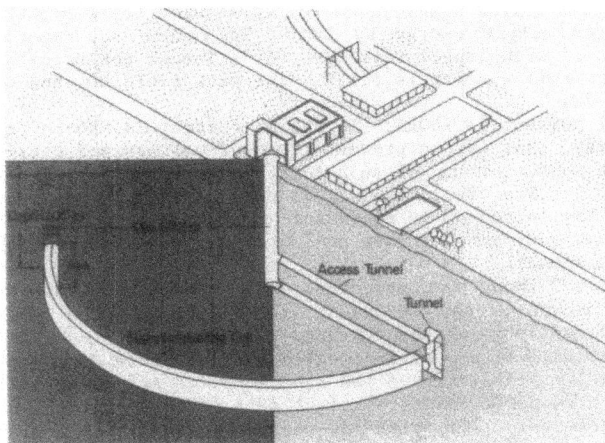

Fig. 1   Conceptual view of 5GWh SMES.
Coil diameter : 640m.
Depth   : 100m.

Table 1.   The summary of the conceptual design of 5GWh SMES (4.2K)

Maximum stored energy	5.5GWh
Charge and discharge energy	5GWh
Output power	1000MW × 5hr
Maximum current	700kA
Maximum voltage	4.52kV
Coil diameter	640 m
Coil length	32 m
Installation depth	100 m
Total weight	330000 tons
Cool down weight (4.2K)	235000 tons
Liquid helium	5800 m^3
Volum of cavern	850000 m^3
Power loss	522 MWh/day
Efficiency	90.6 %

600

The maximum stored magnetic energy is assumed to be 5.5GWh, being larger than the rated output energy by 10%. To reduce the magnetic pressure we designed a coil with aspect ratio β=0.05. Though the magnetic field at the center of the coil winding is 3.15T, the peak field at the end edge of the winding is 4.7T.

The coil current is 700kA. This high current is chosen to make the coil protection easy, that is, to prevent the over-voltage and over-heating during quench. The conductor is of so called bundle type[3]. The figure 2 shows the conductor. The liquid helium is circulated in the conduit of the conductor. The forced flow conductor of bundle type eliminates the helium vessel and the eddy current loss due to the electromagnetic short circuitry of the helium vessel.

The figure 3 shows the structure of the coil, the coil support and the thermal insulation in the cavern of the bed rock. The helium temperature region is thermally insulated by vacuum layer. The 80K thermal shield is thermally insulated by powder insulation with pressure of 1 atmosphere. The heat load and the refrigeration power consumption are shown in the table 2.

The electric power converter is of self commutating type using GTO (gate turn off) thyrister. The operation cycle is assumed to be 1 cycle a day; 5 hours for charging, 7 hours for holding, 5 hours for discharging, 7 hours for waiting. The loss of the electric power converter is 250MWh/day. The persistent current switch of mechanical type will be used to make a bypass circuitry of the coil current.

Fig. 2 Superconductor

Material : Nb-Ti
Current : 700kA.
Current density in strand : 60A/mm^2.

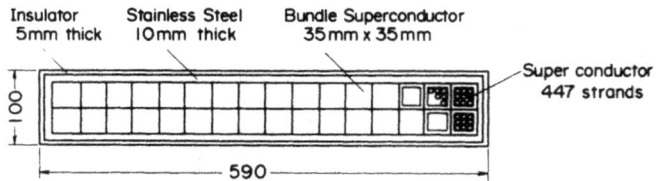

Fig. 3 Superconducting coil in the tunnel.

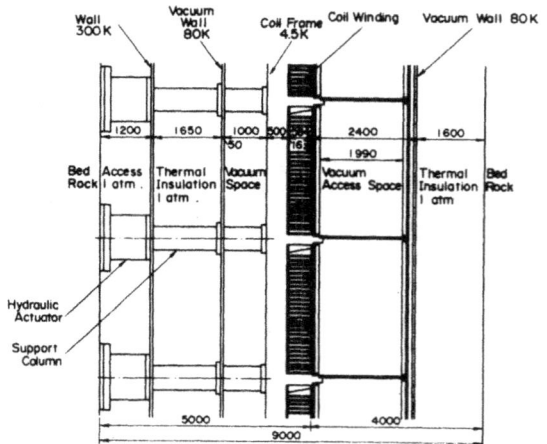

Table 2. The heat load and the refrigeration power of 5GWh SMES (4.2K)

Temperature			
4.2K	23.8 kW	7140 kW	(FOM=300)
Support column	6.3		
80K thermal shield	12.5		
Current lead	2.5		
Ac loss	0.3		
Others (10%)	2.2		
77K	509.9	4130 kW	(FOM=8.1)
Support column	33.4		
Powder insulation layer	430.1		
Others (10%)	46.4		
Total refrigeration power		11.27 MW	
Refrigeration power consumption / day		271.0 MWh/day	

Note : FOM = Figure of merit[4].

Efficiency and Cost Estimation

1) Initial cool down : After the completion of the consturction of the system, it takes 5.8 months to cool the whole coil down to 4.2K. The power consumption for the refrigeration is 94GWh. The cool down mass weight is 235000 tons for 4.2K and 63000 tons for 77K.

2) Efficiency : The system efficiency is 90.6%, which is significantly higher than the conventional pump-up power system of which the efficiency is around 70%.

3) Cost : The estimated cost in Yen (¥) of the total system is ¥3500×10⁸. This cost is higher than the present cost of the pump-up power system. By the way a 1000MW pump-up power system costs about ¥2000×10⁸.

In conclusion, the efficiency of the liquid helium temperature SMES is high and the construction cost is also high.

SMES Using High Tc Superconductor

Conceptual Design

1) Liquid hydrogen temperature SMES : The structural details of the liquid hydrogen temperature SMES are completely same with those of the liquid helium temperature SMES which were described in the previous section. That is, in the figure 3, the liquid helium (4.2K) is simply replaced by the liquid hydrogen (20.3K).

2) Liquid nitrogen temperature SMES : The superconductor is coolded by liquid nitrogen (77K). In reference to the structure shown in the figure 3, the vacuum thermal insulation layer between 4.2K and 80K is removed. The 80K

thermal shield is eliminated. Thus, the length of the suspension epoxy column is shortened and the width of the cavern is reduced correspondingly. It should be noted that the liquid nitrogen temperature SMES need not use the vacuum technology. The coil region (77K) is filled by the mixture of nitrogen and helium of 1 atmosphere and is thermally insulated by powder insulation with pressure of 1 atmosphere.

3) Room temperature SMES : The superconductor will be directly fixed on the side surface of the bed rock in the cavern . The vacuum and thermal insulation layers are completely removed. The axial magnetic compression force is supported by the bed rock in common with the radial magnetic expansion force. Thus, the structural material is significantly reduced. The width, thus, the volume of the tunnel is shortened.

Efficiency and Cost Estimation

The storage efficiency and the construction cost are estimated and the results are graphically summarized in the figure 4(a),(b) as a function of the conceptually designed operating temperature of the SMES. A big assumption is made in this estimation: The costs of the superconductor for the hydrogen temperature, the nitrogen temperature and the room temperature are assumed to be equal to the cost of the liquid helium temperayure superconductor.

Conclusion

(1) The efficiency of the liquid helium temperature will be significantly high (>90%) in comparison with a conventional pump-up power system. The construction cost, however, will be also high than that of the conventional pump-up power system.

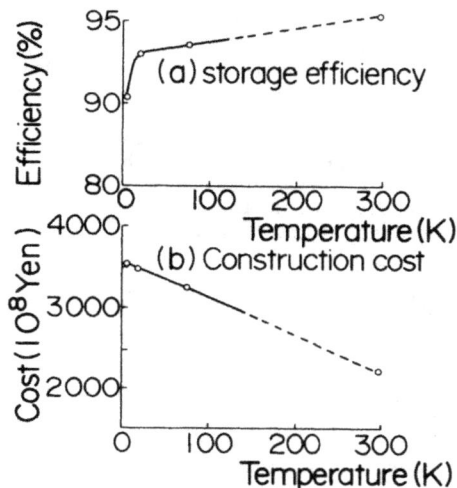

Fig. 4 The efficiency and the construction cost of 5GWh SMES
        as a function of assumed operating temperature of superconductor.

(2) If the liquid hydrogen temperature superconductor is applied to the SMES, the construction cost will be unchanged. The efficiency is improved because the hydrogen refrigerator need not use an inefficient Joule-Thomson valve. By the way, the hydrogen gas is not import goods in Japan. But the cost of hydrogen is presently not so cheap.

(3) If the liquid nitrogen temperature superconductor is applied to the SMES, the vacuum technology need not be used. However, the improvement of the construction cost is less than we have expected. The liquid nitrogen temperature superconductor is not so cost-effective as being presently expected.

(4) If the room temperature superconductor is applied to the SMES, the efficiency will be raised up to 95% and the construction cost will be significantly reduced ($¥2200 \times 10^8$ for 5GWh system).

In conclusion SMES will be feasible at least when the room superconductor becomes applicable. So we must continue the R & D efforts for SMES hereafter.

References

(1) K.Toyoda, T.Yamada, M.Iwamoto, Proceedings of the 3rd Meeting on Superconductive Energy Storage, Tsukuba, May 22 and 23, 1980, p18-21.
(2) T.Yamada, M.Iwamoto, Proceedings of International Symposium on Superconductive Energy Storage, Okaka, October 8-10, 1979, p177-181.
(3) K.Agatsuma, K.Kaiho, and K.Koyama, IEEE Transactions on Magnetics, Vol. MAG-19, No.3, May 1983, p382-385.
(4) T.R.Strobridge, Cryogenic Engineering, Vol.10, No.5, p185-189 (1975).

DESIGN STUDY OF 5GWh SMES

Toshihide NAKANO, Kazuyoshi HAYAKAWA, Masayuki Shimizu *

MITSUBISHI HEAVY INDUSTRIES, LTD.
KOBE SHIPYARD & MACHINERY WORKS
1-1, 1-chome, Wadasaki-cho, Hyogo-ku
KOBE, 652, JAPAN

* THE KANSAI ELECTRIC POWER COMPANY, INC.
11-20 Nakozi 3-chome, AMAGASAKI, 661, JAPAN

Abstract - The Superconducting Magnetic Energy Storage system (SMES) is expected to be a future electric power storage method instead of the pumped storage power generator. We have investigated on a commercial scale toroidal type of SMES, and compared it with solenoid type of SMES. Now, we came to the conclusion that the toroidal type was superior to the solenoid type on points of leakage magnetic field, production, maintenance and construction.
Thus the toroidal type of SMES can be considered to be a match for the solenoid type of SMES.

Introduction

The difference of electric power demand between day and night will be very large.

As the difference of power demand between peak load time and bottom load time will increase according to future industrial progress, it will become necessary to balance the load by planning to use the surplus power from the night for peak demand in the day. As one of methods for this purpose, SMES will be considered. There are 2 kinds of SMES i.e. the solenoid type and the toroidal type. We have investigated on the toroidal type of SMES, and have compared it with the solenoid type.

Conceptual design

The toroidal type of SMES will be very effective to reduce leakage magnetic field by confining the magnetic field in the coil. On the other hand, it is necessary to strengthen the coil structure, as load supporting by the solid rock can not be as effectively used in this case. Coils are installed 30m under the ground and 360 coil components are placed in a circle with a 260m radius. But the bedrock on the inside of the circle provides the necessary support against the centripetal force generated by the coils. Conceptual drawing of the 5GWh toroidal SMES main structure is shown in Fig. 1.

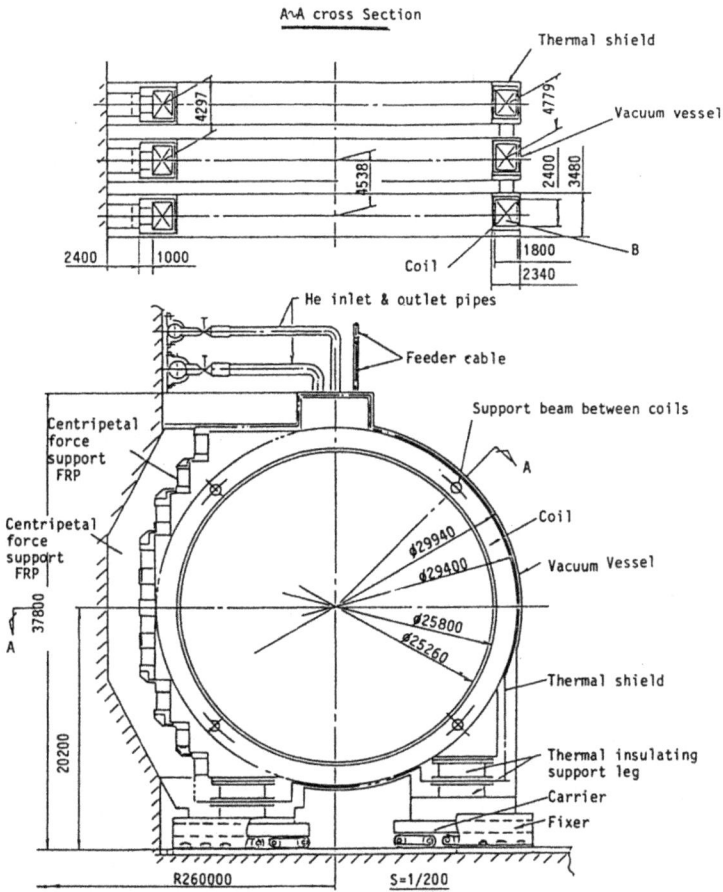

Fig.1   Conception of main structure

[1] Superconductiong coil

Conductor for the superconducting coil is made of 36mm×6mm□, Nb₃Sn/Cu molded strands for conductor, high pure aluminum for stabilizing material and
aluminum alloy for reinforcing material. A sectional drawing of coil is shown
in Fig.2

It consists of 8 lines×22 stages, every turn of 100mm×200mm. The specifications of the toroidal coil are summarized in Tab.1.

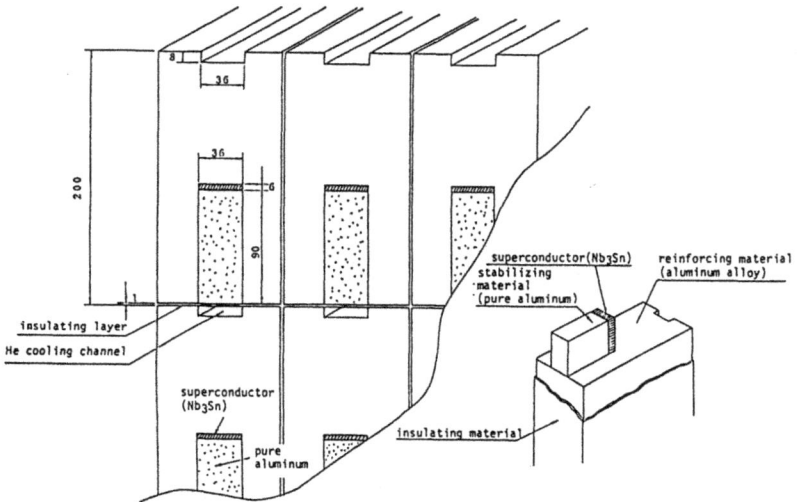

Fig.2 Superconducting coil cross section

Tab.1 The specifications of the toroidal coils

major radius	260	[m]
minor radius (coil radius)		
inner radius	13.0	[m]
outer radius	14.0	[m]
Coil thickness (in toroidal direction)	2.2	[m]
no. of turns	176	[turns]
no. of coils	360	[units]
inductance	$1.7×10^3$	[H]
average magnetic field at midplane	7.0	[T]
max. current (per coil)	$2.52×10^7$	[AT/coil]
(per turn)	$1.43×10^5$	[A/turn]

[2] Result of calculation

We have performed calculation of distribution of magnetic field, electromagnetic force and leakage magnetic field, according to this specification.
Results of calculation are shown in Fig.3, Fig.4, and Tab.2.

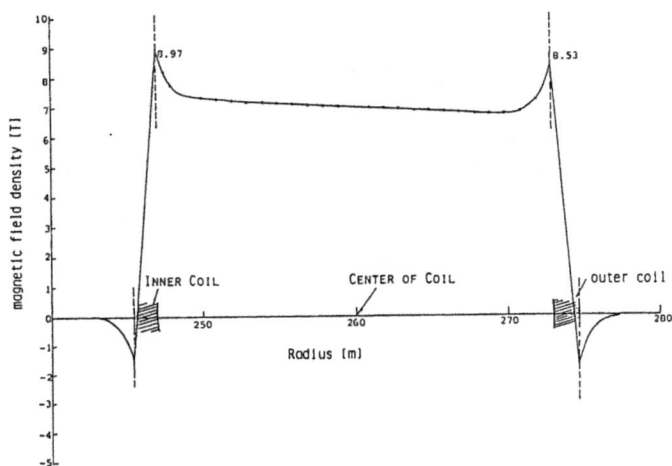

Fig.3 Radial distribution of magnetic field

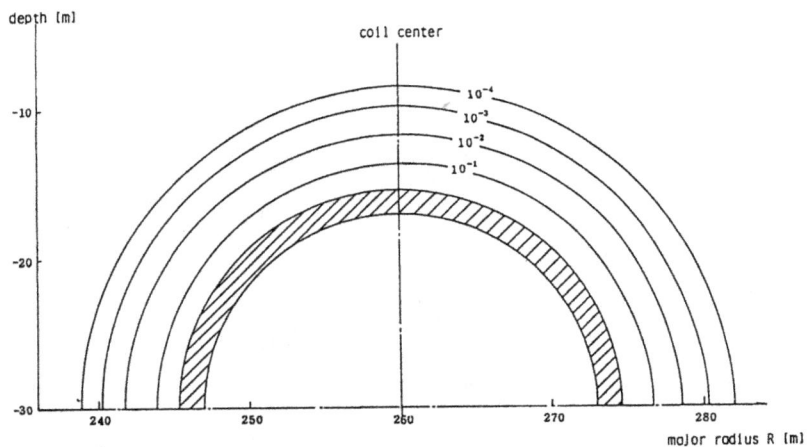

Fig.4 Leakage magnetic field near coil [T]

Tab.2 Electromagnetic forces of the toroidal coil (per coil)

centripetal force	FR	1.83×10^7	[kgf]
hoop force	Fr	7.76×10^8	[kgf]

[3] Thermal insulating support structure

The thermal insulating support structure for the toroidal type of SMES will consist of lower support legs which bear weight of coil, and structures which bear the centripetal force generated by electromagnetic force. These structures will be made of materials with small thermal conductivity (Aℓ-FRP [Fiber Reinforced Plastic]). Furthermore, on 80K thermal shild will be applied to provide insulation.

Centripetal force support structures, are provided to transfer the centripetal force to the bedrock and are installed between the coil and the vacuum Vessel.

As a result of structural strength analysis, it was determined that each structure was safe enough with respect to axial compression strength and buckling strength. Fig.5 shows the main structure of a coil.

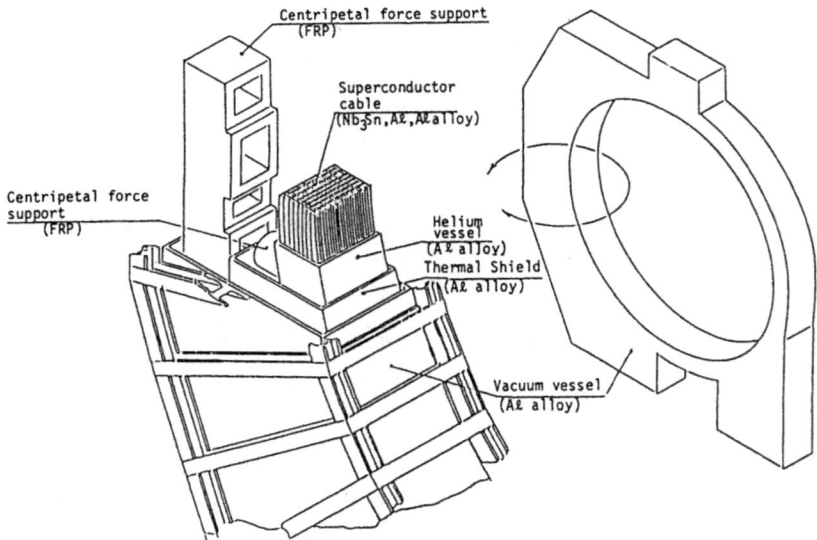

Fig.5   Cross section of centripetal force support structure

[4] Overall evaluation

The overall evaluation is shown in Tab.3

Tab.3 Overall evaluation

	leakage magnetic field	production	maintenance	construction			total
				site area	amount to be dug	cost	
toroidal SMES	magnetic field is conbined in the coil	factory production is possible	coil components are independent, so coil is exchangeable	200,000 m²	6,933,000 m³		O
	◎	O	O	O	△	△	
	×	△	△	△	O	◎	
solenoid SMES	scield coil is necessary	factory production is impossible (site production)	coil maintenance is almost impossible	800,000 m² (more than 200 Gauss)	362,000 m³		△

Remark ◎ : excellent O : good
△ : normal × : bad

Conclusion

We have investigated on the commercial scale toroidal type of SMES. And we came to the conclusion that the toroidal type was superior to the solenoid type, because low leakage magnetic field, fabricability, maintenance and installation.

But the toroidal type of SMES costs 3 times as much as the solenoid type of SMES. We find cost reduction is key point of the toroidal type of SMES relization.

References

1. M. Masuda "A Brief Report of Japanese SMES", Proceedings of ICEC-9 (1984)

2. M. Masuda "The Conceptual Design and Economic Evaluation of Utility Scale SMES"

3. NEDO, "Study on Storage System on Superconducting Energy [I], [II], [Ⅲ]"

4. T. Nakano "DESIGN STUDY OF A 5GWh SMES" Proceeding of 23rd IECEC Vol. 2

THE EFFECT OF SMES FOR THE SYSTEM STABILITY AND THE LOAD FREQUENCY CONTROL IN THE POWER SYSTEM

Kazuo MAEKAWA   Naoji HASHIMOTO
Engineering Research Association For Superconductive Generation Equipment
And Materials (14-10, Nishitenma 5, Kita-Ku, Osaka, 530 Japan)

Jun HASEGAWA   Hajime MIYAUCHI
Faculty of Engineering, Hokkaido Univ. Faculty of Engineering, Kyoto Univ.
(Kita 13, Nishi 8, Kita-Ku, Sapporo, (Yoshida-Hommachi, Sakyo-Ku, Kyoto,
Hokkaido, 060 Japan)   606 Japan)

Kiichiro TSUJI   Yasuharu OHSAWA
Faculty of Engineering, Univ. of Osaka Institute of Engineering Mechanics
(2-1 Yamadaoka Suita Osaka, 565 Japan) Univ. of Tsukuba(1-1, Tennoudai 1,
                                      Tsukuba, 305 Japan)

Noboru SUZUKI   Yukio ISHIGAKI
Hitachi Works, Hitachi, Ltd.(1-1, Saiwaicho 3, Hitachi, 317 Japan)

Abstract - This paper describes an effect of SMES for a stability and
a load frequency control in a power system. Computer simulations were
carried out to verify the stabilizing effect and the function of the load
frequency control of SMES. With these investigations, the effect of the
introduction of the SMES to the power system was verified.

Introduction

Superconducting magnetic energy storage (SMES) has good characteristics
for higher stability of a power system. In Japan, NEDO and Engineering
Promotion Association have executed FS from 1983 and have promoted conceptual design. We also have forwarded research regarding SMES. This paper
describes particularly the results of an examination with regard to the
effects of SMES on power system stability and load frequency control,
which was performed as part of the commissioned examination and research
under the Moonlight plan (Feasibility Examination and Research of the
Superconductive Generation related Equipment and Material Technologies).

Examination of Power System Stability

Sample System

A sample 8-machine system used for this examination is shown in Fig.1.
The machines No.1 and No.7 are assumed to be reduced from several generators
and the machine No.8 a pumped-hydro generator. A reference generator is
machine No.1.

Generators are assumed to be expressed in detail. All generators except
the machine No.8 are equipped with standard AVRs and governors. The machine
No.8 is equipped with an AVR and a governor for pumped-hydro generator
which are different from those of other generators.

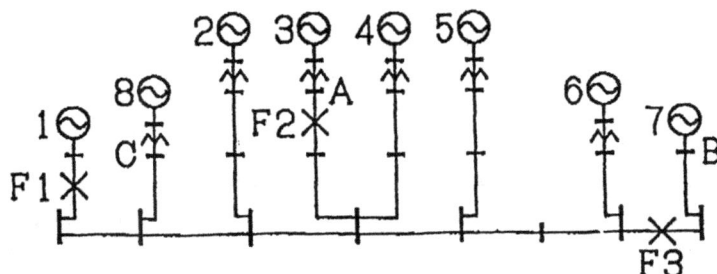

Fig.1 Sample System

SMES performs simultaneous control for both active power (P) and
reactive power (Q) which is expressed by the following equations.

$$\Delta Pd = \frac{s}{1+s} \frac{Kp}{1+sTp} \Delta\omega \qquad \Delta Qd = \frac{s}{1+s} \frac{Kq}{1+sTq} \Delta Vt$$

P, Q control characteristics of SMES are assumed to be semicircle on
the P, Q plane for simplification. When the power is deviated from
controllable range, P is controlled preferentially. Changes in coil
current of SMES are not taken into consideration. In addition, 80% of
maximum Q is compensated by a capacitor in order to control both leading
and lagging areas of Q.

Examination regarding Static Stability Limit

This section describes the static stability limit of the system by actually giving small disturbances.

For calculation, the output of the generators except the machine No.3 is kept to the original power flow condition at the peak time and the power of the machine No.3 is increased. The limit of stable power transmission is assumed as the static stability limit. The output of the machine No.3 in the original power flow condition is 2.24 p.u. As a disturbance to the system, 1 % of reactance is inserted in the point A at t = 0. SMES is assumed to be able to give the output required by the control without limitation.

If generator control systems such as AVR and governor are not taken into consideration, up to 3.8 p.u. of the output of the machine No.3 can be transmitted stably without SMES. When 3.9 p.u. is the output, the system becomes unstable. When SMES is installed to the bus A and control with SMES is executed, power transmission can be continued stably up to 5.2 p.u.

When all generators are provided with AVR control system, up to 4.5 p.u. of the output can be transmitted stably without SMES. When the control with SMES at the bus A is executed, up to 5.6 p.u. of the output can be transmitted stably.

The summary of the results above is Tab.1. This indicates that control with SMES increases the stability limit by 20 to 40%.

Table 1   Output Limit of the Machine No.3 (Stability Limit)

	Without SMES	With SMES
Without AVR	3.8 p.u.	5.2 p.u.
With AVR	4.5 p.u.	5.6 p.u.

Examination regarding Transient Stability

This section examines transient stability of the system. AVR and governor are taken into consideration as generator control system, the installation point of SMES is one of the three buses, A, B and C, and the converter capacity of the SMES is assumed to be 1 p.u.

3 φ short-circuit (~ 0.08 sec) in one of the 2 points (F2, F3) , both of which are the center of parallel two transmission lines, is assumed to be the fault given to the system. Power transmission with only one line

is continued after clearance of the fault and reclosure is not considered.

Qualitative comparative evaluation with the case without SMES control at the peak time is shown in Tab.2. The reason why the value of Kp varies with SMES installation point is that the gain is adjusted because of over response although Kp = 1 in standard.

Table 2 Transient Stability Improvement Effect with SMES and Required Capacity (peak time)

Failure point		F2			F3		
SMES installation point		A	B	C	A	B	C
	Kp	1.0	0.1	0.3	1.0	1.0	0.1
Effect	Machine No.3	O	—	—	—	—	—
	Machine No.7	O	—	—	O	O	—
	Machine No.8	O	—	—	—	O	—
max |Pd| p.u.		1.00	0.18	0.45	1.00	1.00	0.17
max |Qd| p.u.		0.80	0.06	0.09	0.80	0.80	0.04
Energy change		547	127	350	654	585	134

(Note) O : Effective   — : No significant effect

Tab.2 indicates that if SMES is installed to the bus A near the fault point F2, stabilization effect of SMES is extremely large. The reason why stabilization effect of SMES is not large in other cases is that the sample system itself is very stable and approximately 7° of oscillation of the machine No.7 is generated at most due to the fault at F3.

Load Follow-up Capability of SMES—Effect for Load Frequency Control

To inspect load follow-up capability of SMES, frequency variation characteristic and power output response characteristic of a generator against increase of load which is continued with ramp shape for three minutes are calculated.

Simulation Model and Assumed Condition for Calculation

The model used for simulation is shown in Fig.2. When AFC function is included in SMES, the integral value of frequency deviation is taken into consideration as input signal to SMES,though it is not shown in the Fig.2.

Assumed condition for calculation is as follows.

(1) System characteristic: Simulated as first-order lag

M = 1.7272 p.u. MWs/Hz (at peak time)

1.7289 p.u. MWs/Hz (at night)

D = 0.05 p.u. MW/Hz

(2) Governor and AFC channel of generator: Simulated faithfully in thermal power generator and pumped-hydro power generator

(3) SMES: Simulated as first-order lag (K_s = 1.0, T_s = 0.1 sec)

(Energy storage capacity: 9 MWh, Converter capacity: 300 MW)

(4) Load variation: Load variations with ramp shape of 400 MW/3 minutes and 600 MW/3 minutes are mainly considered.

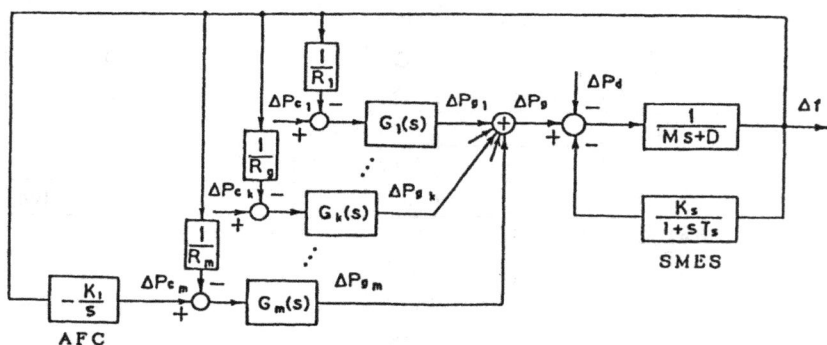

Fig.2 Simulation Model

Simulation Result

Tab.3 shows maximum frequency deviation (Hz) in various calculation cases. It indicates clearly that maximum frequency deviation is decreased effectively by SMES. Tab.4 shows approximate values of time for reducing frequency deviation within 0.03 Hz after excitation of frequency oscillation due to load variation in each calculation cases. It indicates that control characteristic is largely influenced by configuration of control system and gain setting and that fine tuning of control system is important.

Table 3 Principal Results of Simulation (maximum frequency deviation)

Load variation (MW/3m.)	Operation condition	Without SMES		With SMES(300MW)			
				P-Control		I-Control	
		Peak	Night	Peak	Night	Peak	Night
400	Governor free operation	0.266	0.183	0.068	0.049	0.068	0.049
	AFC with normal gain	0.222	0.168	0.049	0.045	0.047	0.045
	AFC with gain of 10 times	0.110	0.113	0.016	0.026	0.017	0.018
	AFC with gain of 20 times	0.086	0.110	0.013	0.023	—	0.017
	AFC with gain of 30 times	0.075	0.130	0.012	0.020	—	0.016
600	Governor free operation	0.484	0.275	0.229	0.142	0.228	0.142
	AFC with normal gain	0.436	0.251	0.207	0.134	0.198	0.133
	AFC with gain of 10 times	0.336	0.227	0.086	0.106	0.054	0.097
	AFC with gain of 20 times	0.321	0.226	0.046	0.084	0.037	0.069
	AFC with gain of 30 times	0.315	0.225	0.026	0.069	0.028	0.046

Table 4 Approximate Value of Time within 0.03 Hz (minute)

Load variation (MW/3m.)	Operation condition	Without SMES		With SMES(300MW)			
				P-Control		I-Control	
		Peak	Night	Peak	Night	Peak	Night
400	Governor free operation	*	*	*	*	>20	>20
	AFC with normal gain	16	25	5	8	4	8
	AFC with gain of 10 times	4	7	19	20	5	5
	AFC with gain of 20 times	14	12	10	11	—	4
	AFC with gain of 30 times	20	24	7	8	—	4
600	Governor free operation	*	*	*	*	*	*
	AFC with normal gain	*	34	*(6)	25	6	24
	AFC with gain of 10 times	*	16	*(4)	5	5	4
	AFC with gain of 20 times	*	37	*(3)	4	8	5
	AFC with gain of 30 times	*	>60	*	4	7	4

(Note) * means that stationary control error in frequency remains.

Summary and Acknowledgements

Examination on stability improvement by SMES is performed and improvements in static stability limit and transient stability have been found. In addition, effective supression of frequency deviation by introduction of SMES is indicated. Authors conclude that combination of SMES with governor free operation and AFC is valuable as a new method for supression of frequency variation and improvement of dynamic response.

Finally, authors would like to express their deep appreciation to Agency of Industrial Science and Technology of Ministry of International Trade and Industry (MITI) and other persons concerned for giving them the opportunity for writing this paper.

POWER SYSTEM STABILIZATION BY USING
SUPERCONDUCTING MAGNETIC ENERGY STORAGE

Yasunori MITANI[*], Yasuyuki KOWADA[**],
Kiichiro TSUJI[**], and Yoshishige MURAKAMI[**]

[*]Low Temperature Center, Osaka University
[**]Laboratory for Applied Superconductivity, Osaka University
2-1 Yamadaoka Suita Osaka, 565 JAPAN

Abstract -- This paper summarizes the results from an investigation concerning electrical power transmission system stabilization using superconducting magnetic energy storage (SMES).   Power system stabilizing control for a
one machine infinite bus system and a multi-machine power system is described.
The results show the effectiveness of SMES for power system stabilization
experimentally as well as numerically.

Introduction

Recent power systems have been characterized by long distance bulk power
transmission lines and longitudinal interconnections.   These structures may
cause an instability of power swing, that is, long term and undamped power
oscillations.[1]   At present, several stabilizing methods has been suggested,
such as the addition of a power system stabilizer (PSS) [2] to the generator
excitation system and control by static var compensators (SVC).[3]   This
paper is a summary of investigating the effectiveness of a new power system
stabilizer using SMES.[4-5]
SMES is capable of controlling active and reactive power simultaneously,
rapidly and smoothly at the bus where the SMES is placed.[6]   It has been
expected, therefore, that SMES is applicable to stabilize various kinds of
power systems effectively.[7]
This paper describes power system stabilizing control by SMES for undamped power oscillations in several types of power systems.   The analysis using a
long distance bulk power transmission system model shows the significance of
improved power system stability by SMES experimentally as well as numerically.
Another numerical study using the system with consideration to several kinds
of load characteristics demonstrates that SMES is always effective for power
system stabilization.   Furthermore, the stabilizing control scheme is applied
to a hydroelectric power generation system.   It demonstrates that SMES is
effective for controlling the instability of power swing in a multi-machine
system as well.

Active and Reactive
Power Control by SMES

Configuration of SMES

Figure 1 shows the fundamental
configuration of a SMES considered in
this paper.   The main components are a
superconducting coil and two sets of six
pulsed GTO Greatz bridge power converters in series.   The proper control of
the firing angles of these converters

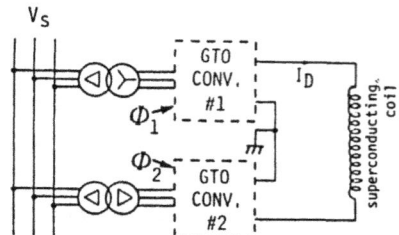

Fig. 1. Fundamental configuration
of SMES

makes it possible to control active and reactive power simultaneously, rapidly and smoothly in a four quadrant range at the bus where the SMES is located.

Controllability of Active and Reactive Power by SMES

Figure 2 shows the experimental waveforms of active and reactive power simultaneous control by the SMES connected to the artificial power transmission system.

In this experiment, we provided sinusoidal specified values of 0.5 Hz for the power controller of SMES. The results demonstrate that outputs of active and reactive power follow accurately the independent specified values.

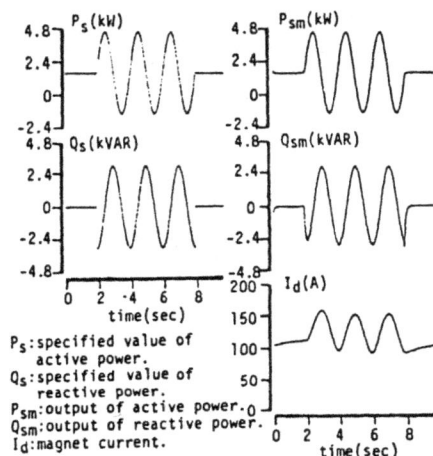

P_s:specified value of active power.
Q_s:specified value of reactive power.
P_{sm}:output of active power.
Q_{sm}:output of reactive power.
I_d:magnet current.

Fig. 2. Experimental results of active and reactive power control

Stabilizing Control of a Long Distance Bulk Power Transmission System

Configuration of a Power Transmission System

Figure 3 shows the configuration of the model power transmission system used for analysis . It is the most basic one machine infinite bus system model for a long distance bulk power transmission system. It corresponds to a system with a 2000 MVA power plant of turbine generators connected to a large power system through approximately 200 km long distance power transmission lines at 500 kV.

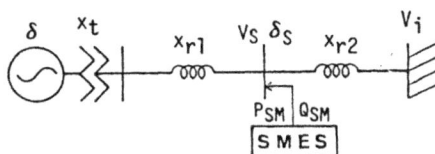

Fig. 3. Long distance bulk power transmission system with SMES

Power System Stabilizing Control Scheme

SMES is capable of controlling active and reactive power simultaneously to follow the specified active power (P_s) and reactive power (Q_s). In order to stabilize the power system, the following feedback control scheme is provided for P_s and Q_s.
$$\Delta \dot{P}_s=-K_D\Delta\omega, \quad Q_s=-K_v\Delta V_S$$
where Δ expresses the variable which represents the deviation from an operating point. ω is the angular velocity of the generator and V_S is the voltage at the bus with SMES.

Effectiveness of Power System Stabilization

Figure 4 shows the experimental results of power system stabilization in the artificial power transmission system rated at 10 kVA and 460 V. The assumed fault is a three line short circuit of 3 cycle duration. The location of SMES is the generator terminal. The result demonstrates that the simultaneous control of active and reactive power by SMES is capable of damping out both voltage and power swing quickly. The experimental waveforms show that the necessary capacity of the AC/DC converter is about 2 kVA and that the

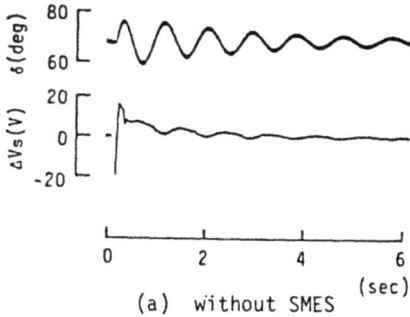

(a) without SMES

Fig. 4. Experimental results

energy used for power stabilization is about 500 J. These values correspond to 400 MVA and 100 MJ, respectively, in terms of a 2000 MVA real power system.

Figure 5 shows the behavior of power oscillation following the gradual increase in the generator power output. Without SMES control, the power system is able to transmit power up to only 6.4 kW. The active and reactive power control by SMES makes the power system stable up to 9.3 kW.

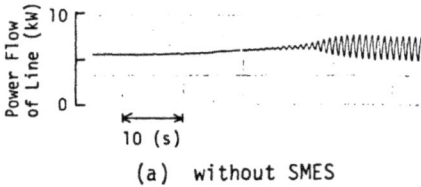

(b) with control using active and reactive power

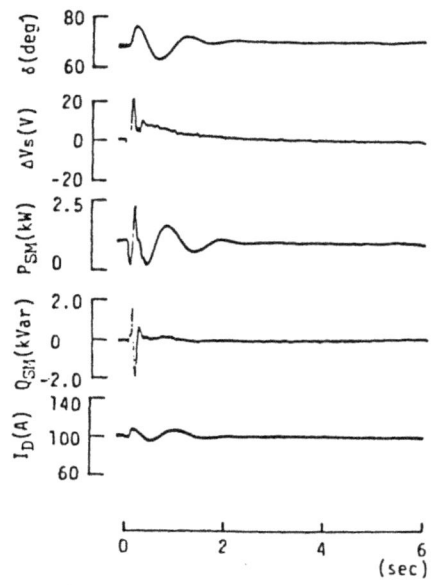

(a) without SMES

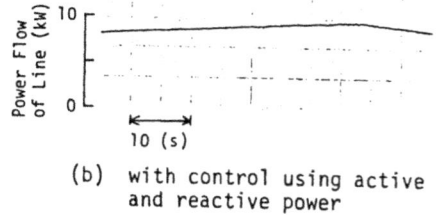

(b) with control using active and reactive power

Fig. 5. Behavior of power oscillation following the gradual increase in generator power output

Here, it should be supplementally remarked that the above results are all in good agreement with the numerical analysis by digital simulation.

Evaluation of the influence of load characteristics on the control by SMES

In order to obtain more detailed information concerning the significance of SMES, we prepared a system with consideration to load characteristics. Figure 6 shows the configuration of the model system with a local load on the transmission line. It is a system consisting of a 2000 MVA generator, a local load, 450 km long distance

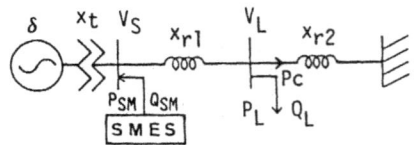

Fig. 6. Configuration of the model system with a local load

power transmission lines and an infinite bus. The following equations represent the characteristics of the load model.

$$P_L = P_{L0}V_L{}^a, \quad Q_L = Q_{L0}V_L{}^b$$

The power system stabilizing control scheme is the same as that in the previous discussion.

Here, the eigenvalues of the system state equation derived from the linearized mathematical model, are calculated for evaluating the system performance. Figure 7 shows the real part of the eigenvalue corresponding to the power swing mode for different load capacities. The results show that without SMES, load characteristics may cause the instability of power swing. In contrast with this, when the control by SMES is applied, the power system becomes stable independent of load status.

The digital simulation results shown in Fig.8 also demonstrate the effectiveness of SMES. The assumed disturbance is the gradual load cutoff at intervals of 10 seconds. The stabilizing control by SMES is so effective that the transmission capacity is increased significantly.

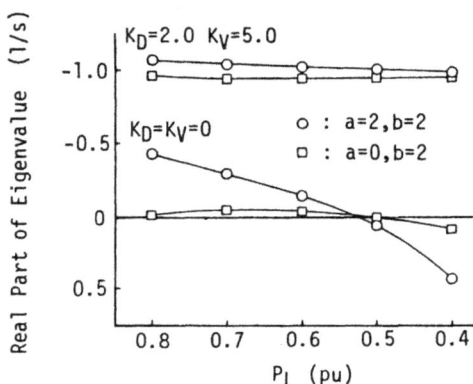

Fig. 7. Real part of the eigenvalue for different load status

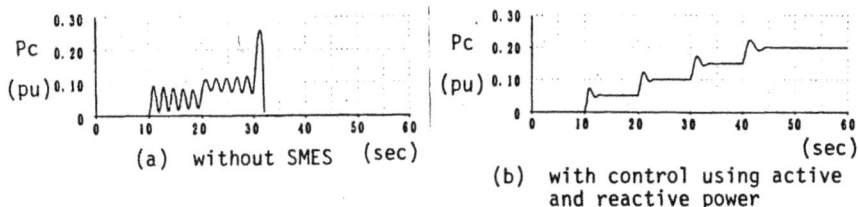

(a) without SMES (sec)

(b) with control using active and reactive power

Fig. 8. Results of digital simulation

Application of SMES to a Hydroelectric Power Multi-Machine System

Next, an application of SMES to stabilize a hydroelectric power multi-machine system is described. This system also has the instability of power swing associated with the long distance power transmission.

Configuration of the Model System

Figure 9 shows the configuration of the 26 generator hydroelectric power system model. The generated power of about 500 MVA is delivered downstream through long distance power transmission lines. The end of the transmission lines is connected to a large power system represented by the infinite generator (generator No.1).

Figure 10 shows the results of digital simulation. The assumed fault is a three line short circuit of 1 cycle duration at bus 3.

This system has an instability in the power swing of 1.3s duration where all generators oscillate together in phase. It equivalently looks like the instability in the one machine infinite bus system.

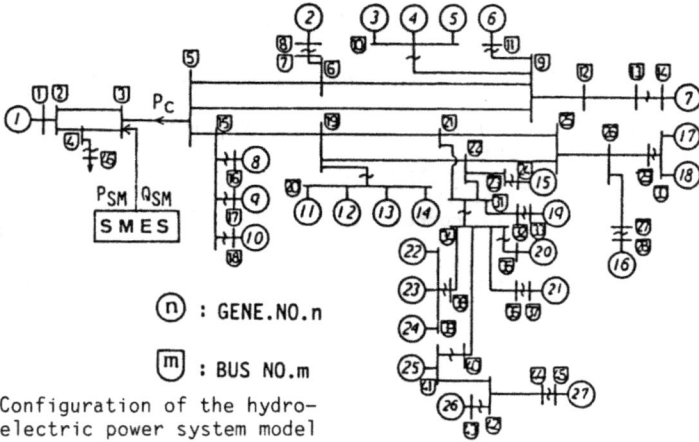

⊙ : GENE.NO.n

☐ : BUS NO.m

Fig. 9. Configuration of the hydroelectric power system model

Power System Stabilizing Scheme

Here, by assuming an equivalent
generator which gathers all inertia
of the generators together, we can
suggest the following power system
stabilizing scheme.

$$\Delta P_s = -K_D \Delta \omega, \quad \Delta Q_s = -K_v \Delta V_S$$

where ω is the angular velocity of
the equivalent generator and V_S is
the voltage at the bus with SMES.
In order to realize this control
scheme, however, a measurable signal
equivalent to ω is necessary.

Now, the most simplified torque
equation of the equivalent one
machine infinite bus system can be
represented as follows

$$Ms\Delta\omega = -\Delta P_c$$

where s is the Laplace operator and
P_c is the power flow through the
transmission line. We obtain,
therefore,

$$\Delta\omega = -(1/M)(\Delta P_c/s).$$

Consequently, the power system
stabilizing scheme for the
hydroelectric power system becomes

$$\Delta P_s = K_D \Delta P_c/(Ms), \Delta Q_s = -K_v \Delta V_S$$

in which all signals are measurable.

Results of the Power System Stabili-
zation

Figure 11 shows the results of
digital simulation. The location
of SMES is at bus 3. The simultaneous control of active and reactive
power is capable of damping out both
voltage and power swing quickly.

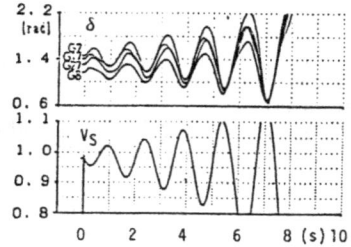

Fig.10. Results of digital simulation
(without SMES)

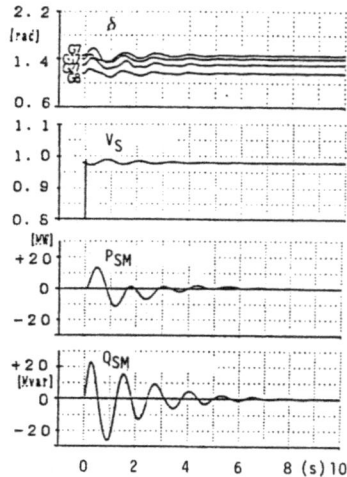

Fig.11. Results of digital simulation
(with SMES control)

It has nearly the same effectiveness as that for the one machine infinite bus system. The simulated waveforms show that the necessary capacity of the AC/DC converter is about 25 MVA and that the energy used for power stabilization is about 10 MJ.

Conclusion

(1) A Power system stabilizing control scheme using simultaneous control of active and reactive power by superconducting magnetic energy storage (SMES) has been presented as a powerful stabilizer for undamped power oscillation.
(2) The stabilizing effect of SMES for a long distance bulk power transmission system has been demonstrated numerically as well as experimentally by using the experimental SMES system and the artificial power transmission system installed at the Laboratory for Applied Superconductivity, Osaka University.
(3) The numerical study using the system with consideration to several kinds of load characteristics has demonstrated that the SMES is always effective for power system stabilization.
(4) Application of SMES to stabilize a hydroelectric power multi-machine system has been presented. The effectiveness of SMES for controlling the instability of power swing in the multi-machine system has also been confirmed.

Acknowledgment

This work was financially supported by Grant-in Aid for Scientific Research of the Ministry of Education, Science and Culture, Japan.
The authors would like to gratefully express their thanks to Mr. Y. Tanaka and Mr. T. Kaito of the Kansai Electric Power Co., Ltd. for their helpful comments and suggestions.

References

1) Y. Yu, Electric Power System Dynamics, London Academic Press, INC., 1983 pp.65-94.
2) F. P. de Mello, P. J. Nolan, T. F. Laskowski, J. M. Undrill, "Coordinated Application of Stabilizers in Multimachine Power Systems", ibid, Vol.PAS-99, p.892, May/June 1980.
3) F. Aboytes, G. Arroyo, G. Villa, "Application of Static Var Compensators in Longitudinal Power Systems", ibid, Vol.PAS-102, p.3460, October 1983.
4) Y. Mitani, Y. Murakami, K. Tsuji, "Experimental Study on Stabilization of Model Power Transmission System by Using Four Quadrant Active and Reactive Power Control by SMES", presented in 1986 Applied Superconductivity Conference (ASC 86) in Baltimore, 1986.
5) Y. Mitani, K. Tsuji, Y. Murakami, "Application of Superconducting Magnet Energy Storage to improve power system dynamic performance", presented at the IEEE PES winter meeting in New York, 88 WM 191-9, 1988.
6) T. Ise, Y. Murakami, K. Tsuji, "Simultaneous Active and Reactive Control of Superconducting Magnet Energy Storage Using GTO Converter", IEEE Trans. on Power Delivery, Vol.PWRD-1, p.143, 1986.
7) H. J. Boenig, J. F. Hauer, Commissioning Tests of the Bonneville Power Administration 30 MJ Superconducting Magnetic Energy Storage Unit", IEEE Trans. on Power Apparatus and Systems, Vol.PAS-104, p.302, 1985.

POWER CONDITIONING SYSTEM FOR
SUPERCONDUCTING MAGNETIC ENERGY STORAGE

Toshifumi ISE**, Yoshishige MURAKAMI*, and Kiichiro TSUJI*

*Laboratory for Applied Superconductivity, Osaka University
2-1, Yamadaoka Suita, Osaka, 565, Japan
**Nara National College of Technology
22, Yata, Yamatokouriyama, Nara, 639-11, Japan

Abstract - The studies of the power conditioning system for SMES (Superconducting Magnetic Energy Storage) at the Laboratory for Applied Superconductivity, Osaka University are summarized. Power conditioning system for 0.5MJ pulsed superconducting magnet consisting of line commutated type thyristor power converters(500V, 1000A, 500kVA, 2sets), forced commutated type power converters(300V, 300A, 90kVA, 2sets) using GTO thyristors, direct digital controller, sequence controller for protection and energy storage was developed and tested. The configuration and the performance of the system are presented.

Introduction

SMES is anticipated to play an important role in future power systems, not only for load leveling but also as a power system stabilizer[1-4]. Experimental SMES system was developed and tested at the Laboratory for Applied Superconductivity, Osaka University in order to investigate the charging and discharging control method, sequence control for protection and energy storage, power system stabilizing control, magnet technologies and cooling technologies[5]. The system mainly consists of 0.5MJ pulsed superconducting magnet, AC/DC power converters, direct digital controller and cooling system. This paper focuses on the power conditioning system of the developed SMES system.

Power Conditioning System for 0.5MJ Pulsed Superconducting Magnet

Main Circuit and Sequence Controller

Figure 1 shows the main circuit configuration of the power conditioning system for 0.5MJ pulsed superconducting magnet. Two sets of thyristor power converters can be connected in series or parallel. The switch S_{4a} is for series connection and S_{4b} is for parallel connection of the converters. Line commutated type converters using SCR's (SCR converters - 500V, 1000A, 500kVA, 2sets), and forced commutated type converters using GTO's (GTO converters - 300V, 300A, 90kVA, 2sets) are available. Energy dumping resistor is connected in series to the superconducting magnet. In case of emergency conditions such as a failure in control and quenching of the magnet, the high speed DC circuit breaker is opened and the stored energy is discharged rapidly to the resistor. The switches S_1, S_2 and S_3 are controlled by a programmable sequence controller. During the on/off control of the switches, the current path for magnet current must be maintained while avoiding the short circuit of AC power source. In order to satisfy this condition, through mode, in which a pair of thyristors connected in series is turned on, was introduced. The current path during through mode is shown in Fig.2.

Figure 3 shows experimental results of protecting operation. The protecting sequence was triggered by a manual push button. Figure 3(a) shows that the protecting operation began at around the circular area, and the stored energy was discharged to the resistor. Figure 3(b) shows detailed waveforms during protecting operation. First of all, the gate signals for through mode were given to the thyristors and the circuit became through mode. After 55msec from the beginning of through mode, the high speed DC circuit breaker S_1 was opened and the dumping resistor was inserted to the current path. After 55msec from opening switch S_1, the switch S_2 was turned on, and the magnet current flowed through dumping resistor and switch S_2. After this, switches S_{3a} and S_{3b} were opened. If the switch S_1 is kept on during above sequence control, energy storage is possible. During turning on switch S2, AC line voltage was observed in the voltage across the magnet but no SCR was damaged and the protecting operation was successfully performed.

Direct Digital Controller

Figure 4 shows the configuration of the direct digital controller of SMES. Active and reactive power and magnet current are detected by active and reactive power transducers and DCCT, respectively. These feedback signals are inputted to micro-computer (LSI-11/23) via 12-bit A/D converter with the sampling period 1/60sec (16.7msec). In the micro-computer, the calculation for compensation and determination of firing angles for thyristor converters are carried out within the sampling period. Many kinds of control algorithm for magnet current control and active and reactive power control were applied to this system.

Figure 5 is one of the excellent experimental results of magnet current control. In this experiment, magnet current was controlled with 4000A/sec charging and discharging rate and 10T/sec flux density changing rate. The control algorithm applied to this was LQI (Linear Quadratic Integral) with finite-time settling observer[6]. This ability was utilized for the measurement of magnet characteristics, especially AC losses of the magnet.

Active and Reactive Power Control Using GTO Converters

GTO Converter

Figure 6 shows the configuration of the main circuit of the developed GTO converters[7]. This GTO converter can operate at any firing angle from 0 to 360 degrees and can control both lagging and leading phase reactive power. In the circuit shown in Fig.6, the energy stored in the leakage inductance of the windings of the transformer is absorbed by the circuit which consists of diodes D_1-D_8 and capacitor C, and the charged energy is discharged to ac power system by line commutated converter which consists of thyristors S_1-S_6.

The possible region of active (P) and reactive (Q) power control varies with the combination of types of the converters as shown in Fig.7. In the case using line commutated converters, the region is restricted in the part of lagging phase. But in the case using two sets of GTO converters, the region becomes the wide domain between lagging and leading phase. The simultaneous independent four quadrant P-Q control is also possible by applying PWM control to the GTO converter. The PWM pattern with two pulses in one cycle of AC line voltage as shown in Fig.8 was applied to the GTO converter and P-Q simultaneous control was carried out by controlling the phase angle α and pulse width θ.

P-Q Simultaneous Control

The active and reactive power transducer was designed for the fast detection of power. Figure 9 shows the circuit configuration of the power transducer. The response time of this transducers is 50msec, which is faster than that of on the market. In this circuit, both three phase voltages and currents are transformed into two phase voltages and currents by matrix [C].

$$[C] = \sqrt{2/3} \begin{bmatrix} 1 & -1/2 & -1/2 \\ 0 & \sqrt{3}/2 & -\sqrt{3}/2 \end{bmatrix} \tag{1}$$

The calculated values p and q have ripples caused by the harmonics of the AC/DC converter. The filtered values P and Q are active power and reactive power, respectively. The response time for the power detection depends on the design of the low pass filters. Faster response will be available by using digital filters.

Figure 10 shows experimental results of P-Q simultaneous control using GTO converters. Independent firing angle control of the two GTO converters was applied to the results shown in Fig. 10(a), PWM control was applied to the results shown in Fig.10(b) and (c). In the case of PWM control, the average DC output voltages of the two converters are equal, so that both series and parallel connection of the converters are possible. Fig.10(b) is the results for series connection and Fig.10(c) is the result for parallel connection. Here, P_s and Q_s are the specified values for active and reactive power, and P and Q are the controlled outputs of active and reactive power, α_1 and α_2 are the firing angles of GTO converter, α and θ are the phase shifting angle and pulse width for PWM control, respectively. V_{sm} is the voltage across the superconducting magnet and I_d is the current through the superconducting magnet. The specified values P_s and Q_s were assumed to be a damped oscillation which would occur when SMES is used as a power system stabilizer. Both active and reactive power outputs followed accurately the specified values P_s and Q_s.

Figure 11 shows the result of examining the frequency characteristic of P-Q simultaneous control. Here, sinusoidal waveforms with the maximum amplitude of 10 kW and 10 kVAR were given for P_s and Q_s, respectively. From the amplitude of specified value V_{ps} and that of output value V_p, the gain $(=20\log(V_p/V_{ps}))$ was calculated. From this result, it can be concluded that both transfer functions from P_s to P, and from Q_s to Q are characterized as a first-order time lag with the time constant of 30 msec.

Conclusion

The studies of the power conditioning system for SMES at the Laboratory for Applied Superconductivity, Osaka University were summarized. The developed power conditioning system was installed in a model power transmission system with an active harmonic filter[8] and remarkable stabilizing effect of power system by SMES was verified. The experimental results of power system stabilizing control using SMES are presented in another paper[9]. Now the authors are testing a new power converter using SI (Static Induction) thyristors, of which switching losses are very low. High efficiency, low harmonic input current are expected from the power converter.

References

1) R.W.Boom, H.A.Peterson, and W.C.Young, "Wisconsin Energy Storage System Design" (1974).

2) T.Shintomi, M.Masuda, T.Ishikawa, S.Akita, T.Tanaka and H.Kaminosono, "Experimental Study of Power System Stabilization by Superconducting Magnetic Energy Storage", IEEE Trans. on Magn., MAG-19, 350 (1983).

3) H.J.Boenig, and J.F.Hauer,"Commissioning Tests of The Bonneville Power Administration 30 MJ Superconducting Magnetic Energy Storage Unit", IEEE Trans. on Power Apparatus and Systems, PAS-104, 302 (1985).

4) Y.Mitani, Y.Murakami, and K.Tsuji, "Experimental Study on Stabilization of Model Power Transmission System by Using Four Quadrant Active and Reactive Power Control", IEEE Trans. on Magn., MAG-23, No.2 (1987).

5) Y.Murakami, M.Sugita, H.Okuda, N.Ouchi, J.Yamamoto, T.Kinouchi, T.Okada, and Y.inuishi, "0.5MJ Pulsed Magnet with its Control and Cooling Systems", Proc. of 9-th International Cryogenic Engineering Conference (ICEC-9), Kobe, Japan, 130 (1981).

6) D.M.Auslander, Y.Takahashi, and M.Tomizuka, "Direct Digital Process Control: Practice and Algorithms for Microprocessor Application", Proc. of the IEEE, Vol.66, No.2, 199 (1978).

7) T. Ise, Y. Murakami, and K. Tsuji, "Simultaneous Active and Reactive Power Control of Superconducting Magnet Energy Storage Using GTO Converter", IEEE Trans. on Power Delivery, PWRD-1, 1, 143-150 (1986).

8) T.Ise, Y.Murakami, and K.tsuji, "Charging and Discharging Characteristics of SMES with Active Filter in Transmission System", IEEE Trans. on Magn., MAG-23, No.2, 545 (1987).

9) Y.Mitani, Y.Kowada, K.Tsuji, and Y.Murakami, "Power System Stabilization by Using Superconducting Magnet Energy Storage", presented at the Osaka Univ. International Symposium, Osaka, Japan, BE-5 (1988).

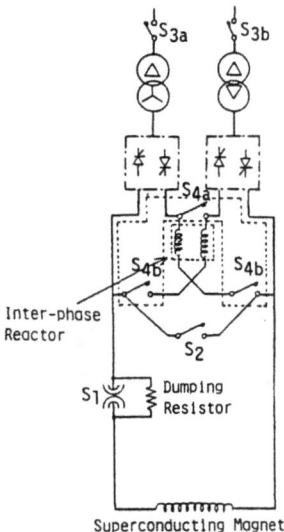

Fig.1 Main Circuit.

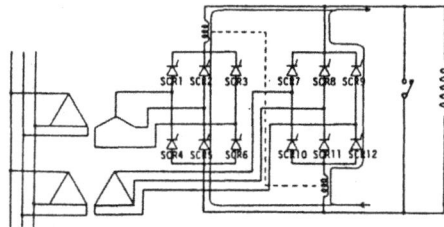

Fig.2 Through Mode.

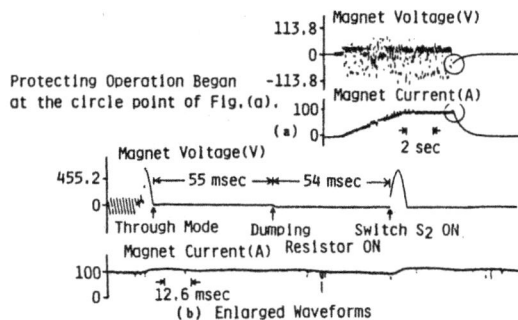

Fig.3 Experimental Results of Protecting Action.

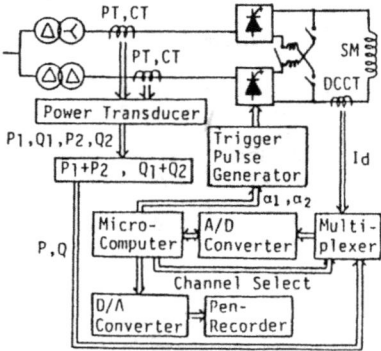

Fig.4 Direct Digital Controller.

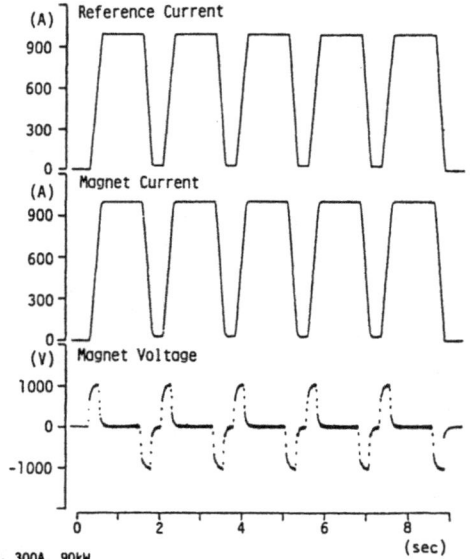

Fig.5 Experimental Result of Pulsed Charging and Discharging Control.

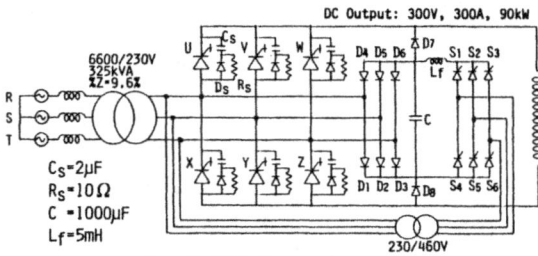

DC Output: 300V, 300A, 90kW

6600/230V
325kVA
%Z-9.6%

$C_S = 2\mu F$
$R_S = 10\Omega$
$C = 1000\mu F$
$L_f = 5mH$

230/460V

Fig.6 GTO Converter.

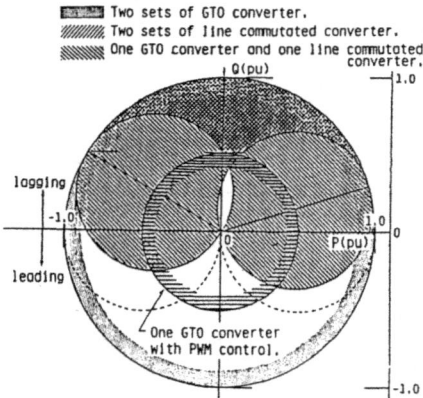

Two sets of GTO converter.
Two sets of line commutated converter.
One GTO converter and one line commutated converter.

One GTO converter with PWM control.

Fig.7 P-Q Controllable Region.

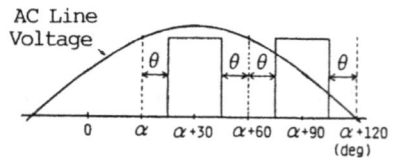

AC Line Voltage

Fig.8 PWM Pattern.

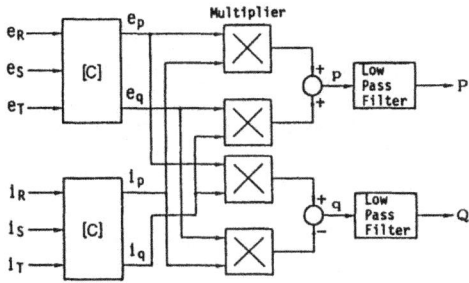

Fig.9 Power Transducer Circuit.

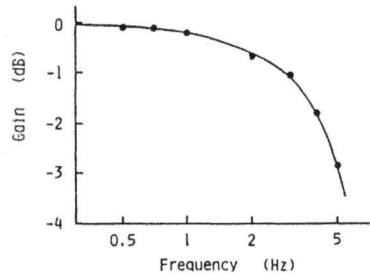

Fig.11 Frequency Characteristics of P-Q Control.

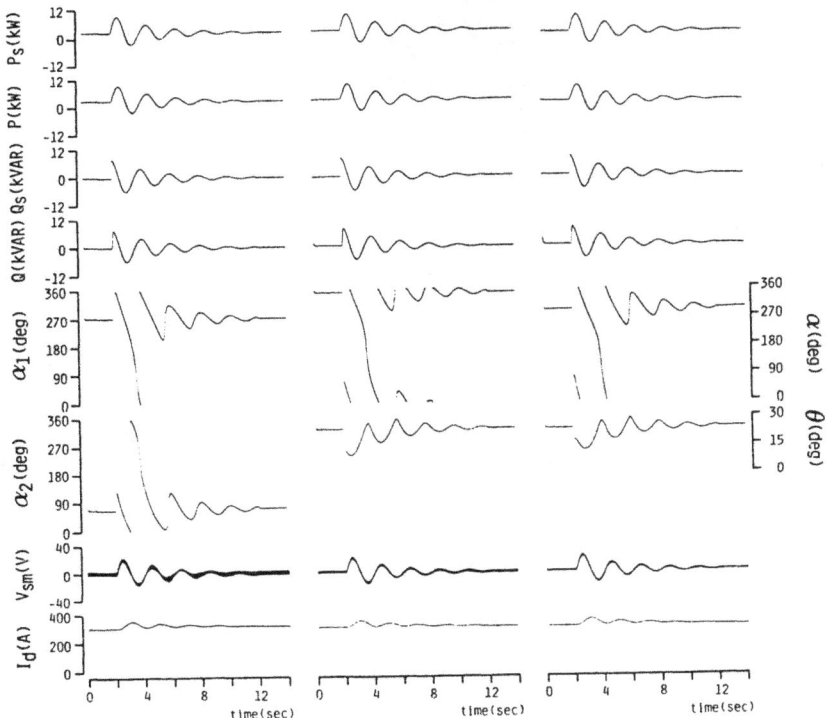

(a) Phase Control (b) PWM with series converters (c) PWM with parallel converters

Fig.10 Experimental Results of P-Q Simultaneous Control.

STUDIES ON SUPERCONDUCTING MAGNETIC ENERGY STORAGE
FOR PHOTOVOLTAIC POWER GENERATION

Yasuharu OHSAWA, Kazuhiko KATO, and Takuya HOMMA

Institute of Engineering Mechanics, University of Tsukuba
1-1-1, Tennodai, Tsukuba, Ibaraki, 305, Japan

Abstract – A circuit is proposed which incorporates a superconducting coil (SC) to a photovoltaic power generation system (PV system) for the energy storage, and its validity is examined via computer simulation of the operation. Also, the preliminary experiment is performed in order to confirm the operation of the proposed system.

Introduction

Since the output power of photovoltaic power generation systems (PV systems) is greatly influenced by the weather, some type of energy storage system is indispensable to the independent PV systems. The former is also advisable for the power-grid-connected PV systems from the viewpoint of the flexibility in operation and the efficiency of the PV system in case of the reverse power flow prohibited. Although lead-acid batteries have been generally used for that purpose, the widespread development of PV systems needs more efficient and easy-to-handle electrical energy storage device. Present study examines the possibility of application of a superconducting coil (SC) to the energy storage for PV system.

A solar cell should be operated at the maximum output point on the I-V characteristic curve with constant current and constant voltage in order that the solar energy be fully utilized, if the insolation is fixed. On the other hand, the superconducting coil carries a variable current according to the stored energy. Therefore, a current transforming circuit is to be put between the solar cell and the superconducting coil.

In this paper, we compared two possible current amplifying circuits, Ćuk converter circuit and step-down chopper circuit. The latter was chosen from the easiness and simpleness in the control. The overall operation characteristics of the proposed system was analyzed by computer simulation and the validity of the control strategy was examined. Furthermore, preliminary experiments were done with a normal-conducting coil and a chopper, which were at hand, in order to confirm the operation.

System Configuration

Solar Cell

The current-voltage (I - V) characteristics of a solar cell is described by Eq.1, and its equivalent circuit is shown in Fig.1[1].

$$I = I_p - I_o[\exp\{q(V - R_sI)/nkT\} - 1] - (V + R_sI)/R_{sh} \qquad (1)$$

where I_p : photocurrent
I_o : saturation current of the diode
n : diode quality factor
k : Boltzmann's constant ($= 1.38 \times 10^{-23}$ J/K)

T : absolute temperature (K)
q : charge on an electron
($ = 1.6 \times 10^{-19} $ C)
R_s : series resistance of electrode
and so on (Ω)
R_{sh} : shunt resistance due to defect
in PN junction (Ω)

Since R_s is small and R_{sh} is large, their influence can be neglected, and Eq.1 can be simplified as follows.

$$I = I_p - I_0 [\exp(\kappa V) - 1] \qquad (2)$$

where $\kappa = q/(nkT)$

Fig.1. Equivalent circuit of solar cell.

The I-V characteristics and the power-voltage (P - V) characteristics are shown in Fig.2, where I_{sc} denotes the short-circuit current, V_{oc} the open-circuit voltage, and I_m, V_m the current and the voltage at the maximum output power point (P_m). It is seen from this figure that solar cells should be operated near the P_m point from the viewpoint of the efficient utilization of solar energy. Although the maximum power point moves according to the insolation and the ambient temperature, it is assumed constant in this initial study.

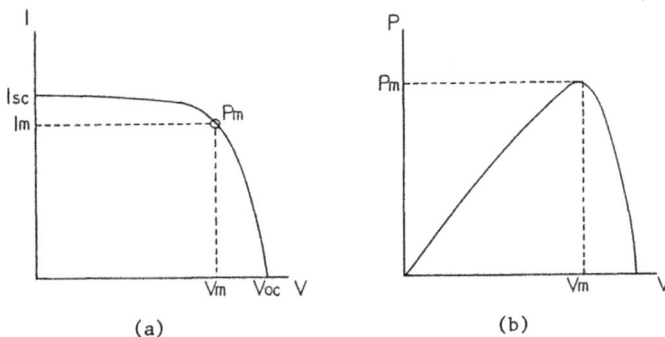

(a)   (b)

Fig.2. (a) I-V characteristics and (b) P-V characteristics of solar cell.

Ćuk Converter Circuit

Fig.3 shows the Ćuk converter circuit. When the chopper Ch is off in this circuit, the diode D is turned on, and the capacitor C is charged from the power supply. On the other hand, when the chopper is turned on, the diode becomes turned off, and a part of the energy stored in C is discharged to the load. If the currents I_1 and I_2 can be assumed constant during the short period of on-off of the chopper, from the condition that the time integral of the capacitor current should be zero in one cycle,

$$\int i_c dt = 0 \longrightarrow I_1 t_{off} = I_2 t_{on} \quad (3)$$

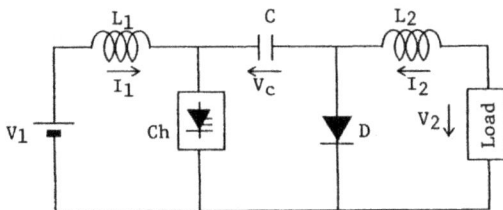

Fig.3. Ćuk converter circuit.

where t_{on}, t_{off} : time periods of chopper-on and chopper-off, respectively
Hence,

$$I_2/I_1 = t_{off}/t_{on} = (T - t_{on})/t_{on} = (1 - \alpha)/\alpha \qquad (4)$$

where α(conduction ratio) $= t_{on}/(t_{on} + t_{off}) = t_{on}/T$; $T = t_{on} + t_{off}$
As the current ratio is the reciprocal of the voltage ratio,

$$V_2/V_1 = \alpha/(1 - \alpha) \qquad (5)$$

The capacitor voltage V_c is given from the energy equation.

$$V_1 I_1 T = V_c I_1 t_{off} \longrightarrow V_c = V_1/(1 - \alpha) \qquad (6)$$

If C is initially charged greater than $2V_1$, this circuit works as a step-up chopper circuit ($V_2 > V_1$, $I_2 < I_1$). For example, when $\alpha = 0.8$, then $V_2/V_1 = 4$, $V_c = 5V_1$. From Eqs.4 and 6, however,

$$V_c = (I_1 + I_2)V_1/I_2 \qquad (7)$$

Hence, V_c is varried with I_2, even if V_1 is constant. In the circuit to be utilized here, the power supply and the load are replaced by the solar cell and the inverter circuit, respectively, and the coil L_2 becomes superconducting. Therefore, I_2 is varied according to the stored energy and the control of the chopper becomes complicated[2].

Step-Down Chopper

The circuit configuration of a step-down chopper is shown in Fig.4. In this circuit, the capacitor C is always charged from the power supply, while it is discharged only when the chopper is on. From the condition that the capacitor voltage be fixed,

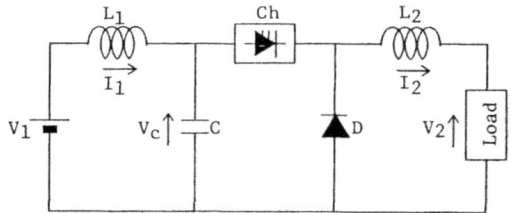

Fig.4. Step-down chopper circuit.

$$I_1(t_{on} + t_{off}) = I_2 t_{on} \longrightarrow I_2/I_1 = 1/\alpha \qquad (8)$$

$$V_2/V_1 = \alpha \qquad (9)$$

V_2 is always less than V_1 ($I_2 > I_1$). In the present circuit where the output of the solar cell is stored in the superconducting coil, this condition is usually satisfied. Although V_c is almost constant, it changes slightly; increases when Ch is on and decreases when Ch is off, resulting in a triangular variation, the mean value of which equals V_1 (Fig.5). When a solar cell is connected in stead of the constant voltage power supply, constant voltage control of the solar cell can be attained if Ch is controlled by the value of V_c.

In this study, we adopt the step-down chopper as a current amplifying circuit.

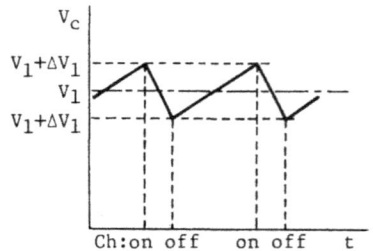

Fig.5. Variation of capacitor voltage.

Proposed Circuit Configuration

In Fig.6 shown is the overall circuit configuration of the superconducting magnetic energy storage for PV power generation system. The generated power is supplied to ac power grid through the externally commutated inverter.

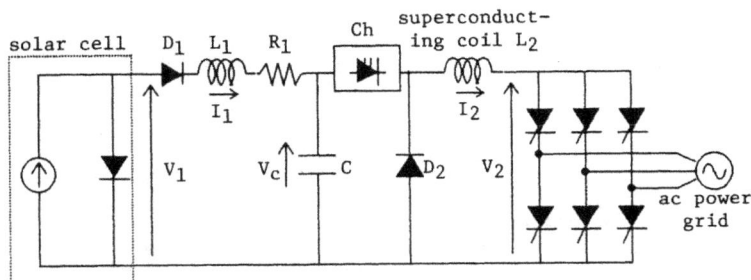

Fig.6. Circuit configuration of the proposed system.

The solar cell current is matched to the superconducting coil current by the step-down chopper. D_1 denotes the blocking diode.

Analysis of Proposed System

In the present section, the operation of the proposed system is analyzed by the computer simulation, and the validity of the control strategy is to be examined.

Circuit Equation

The equations of the circuit shown in Fig.6 are as follows.

chopper-off (charging)   chopper-on (discharging)

$$L_1(dI_1/dt) + R_1I_1 + V_c = V_1(I_1) \qquad L_1(dI_1/dt) + R_1I_1 + V_c = V_1(I_1)$$

$$L_2(dI_2/dt) = -V_2 \qquad (10) \qquad L_2(dI_2/dt) = V_c - V_2 \qquad (11)$$

$$C(dV_c/dt) = I_1 \qquad C(dV_c/dt) = I_1 - I_2$$

The voltage drops in the chopper and diodes are assumed negligible.

Control Strategy for Chopper

In the present study, the insolation is assumed constant and the control strategy for the chopper is considered in order to maintain the operating point of the solar cell at the maximum power point (V_m, I_m). The average capacitor voltage V_{co} should be equal to $(V_m - R_1I_m)$ in order that I_1 be equal to I_m. Therefore, we examine the control method, where Ch is turned on when V_c equals $V_{co}+\Delta V_c$, and is made off when V_c equals $V_{co}-\Delta V_c$. The constants of the solar cell and the circuit used for the simulation are tabulated in Tab.1. These values are taken from the apparatus for the experiment to be done in the near future. The operating voltage of the capacitor V_{co} is 28.74 [V]. Ch is made on at V_c=28.5 [V] and off at V_c=29.5 [V].

Table 1. Constants for simulation.

solar cell	$I_p(I_{sc})$	2.16	[A]
	I_o	1.726×10^{-5}	[A]
	κ	0.3101	[V^{-1}]
	V_{oc}	37.85	[V]
	I_m	1.952	[A]
	V_m	30.30	[V]
circuit	L_1	0.1	[H]
	R_1	0.8	[Ω]
	L_2	100	[H]
	C	5000	[μF]

Numerical Example

Numerical calculation was performed for the case that all of the output power from the solar cell is stored in the superconducting coil and there is no output to the ac power grid ($V_2 = 0$). The initial condition is such that $I_1(0) = 0[A]$, $I_2(0) = 10[A]$, and $V_c(0) = 0[V]$. The results are shown in Fig.7. At the early stage of the operation, V_c is small, Ch is kept off, and the output power of the solar cell is exclusively used for charging C. During that period I_1 becomes nearly equal to I_{sc}. As V_c becomes large, I_1 decreases. When V_c reaches $V_{co}+\Delta V_c$, charging of L_2 starts. It is seen from the figure that I_1 is kept almost constant (I_m). It takes 268 seconds for I_2 to vary from 10 [A] to 20 [A]. The stored energy is increased by $L_2(20^2-10^2)/2 = 15,000$ [J]. The output energy from the solar cell is $(V_m I_m - R_1 I_m^2) \times \Delta t = 15,033$ [J], which is approximately equal to 15,000 [J].

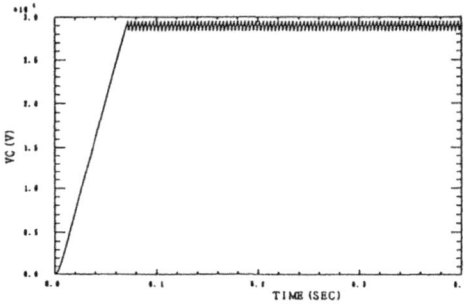

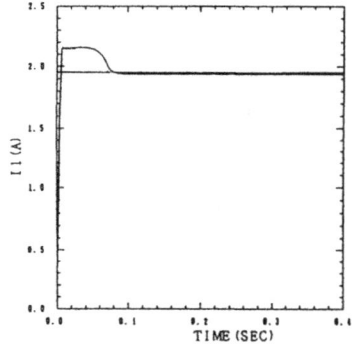

(a)

Fig.7. Results of the simulation
(a) capacitor voltage V_c,
(b) solar cell current I_1,
(c) SC current I_2.

(b)

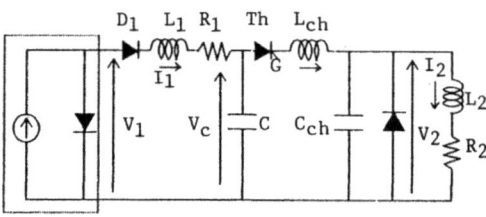

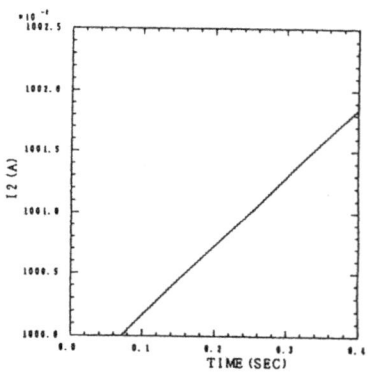

Fig.8. Configuration of the experimental circuit.

Table 2. Constants of the experimental circuit.

L_1	50	[mH]
L_2	150	[mH]
C	4700	[µF]
L_{ch}	500	[µH]
C_{ch}	15	[µF]

(c)

Preliminary Experiment

In order to verify the operation of the proposed system, preliminary experiments were done with a normal-conducting coil and a chopper at hand. The experimental circuit is shown in Fig.8. The chopper is of type which makes use of LC resonance (L_{ch} and C_{ch} in the figure) for the turn-off of the thyristor, so the on-period t_{on} is fixed to half of the resonance period. The off-period can be varied manually by changing the gate pulse frequency. The constants of the circuit are shown in Tab.2. The solar cell was connected with initial conditions $I_1 = I_2 = V_c = 0$, and the transients were recorded. The results are shown in Fig.9. It is seen from the figure that the solar cell current I_1 is equal to the short-circuit current I_{sc} since V_c initially equals zero. As V_c increases, I_1 decreases a little and I_2 increases. Since the coil L_2 is normal-conducting, I_2 increases when Ch is on, and decreases with the time constant L/R when Ch is off.

Fig.10 shows the time variation of each variable in the steady state. (The time-scale differs from that in Fig.9.) I_1 and V_1 are hardly varied. I_1 is about 70 [mA] while I_2 is 0.68 [A], which shows that current amplification is successfully done.

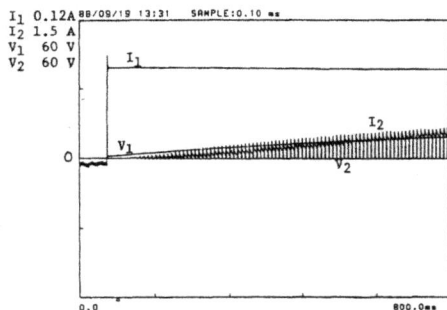

Fig.9. Transient performance of the experimental system.

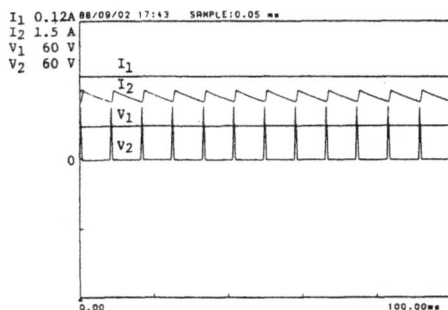

Fig.10. Steady-state performance of the experimental system.

Conclusion

A superconducting magnet energy storage system for PV power generation was proposed, which utilizes a step-down chopper circuit and a thyristor bridge converter circuit. Its operation was analyzed via the computer simulation and confirmed by the preliminary experiment. Experiments with a superconducting coil is scheduled in the near future. Also, the realistic system arrangement should be designed, because the voltage drops in the chopper circuit and/or the diode, which were neglected in this study, can harm the efficiency of the system. It can become problem, especially in small size systems.

References

1) Fonash, S. J., "Solar Cell Device Physics," Academic Press, New York, 1981.
2) Kato, K., "A Study on Superconducting Magnetic Energy Storage for Photovoltaic Power Generation System," Bachelor Thesis, University of Tsukuba, Tsukuba, 1988. (in Japaneses)

CONCEPTUAL DESIGN OF 200,000kW CLASS PILOT SUPERCONDUCTING GENERATOR
CARRIED OUT BY SUPER-GM (PART 1; LOW RESPONSE EXCITATION MACHINE)

Shiro KURIHARA[*1], Osami TSUKAMOTO[*2],
Kiyotaka UYEDA[*3], Shunichi HIROSE[*3], and Masaaki TANAKA[*3]
*1 AGENCY OF INDUSTRIAL SCIENCE AND TECHNOLOGY, MITI, *2 YOKOHAMA NATIONAL
UNIVERSITY, *3 ENGINEERING RESEARCH ASSOCIATION FOR SUPERCONDUCTIVE
GENERATION EQUIPMENT AND MATERIALS (SUPER-GM)
Umeda UN Bldg.,5-14-10 Nishitenma, Kita-Ku Osaka City, 530 JAPAN

Abstract - Conceptual designs of low response type 200,000 kW class
pilot superconducting generator were conducted and design specifications were
clarified. At the same time, research and development subjects of elementary
technology were reviewed, and actual policies and methods for the development
were examinated. An advanced type superconducting generator which was applied
Nb_3Sn supercondutor or AC superconductor were also studied.

Introduction

Feasibility studies on equipment and materials for superconducting
generator were conducted in 1985 and 1986 as the moonlight project sponsored
by the Agency of Industrial Science and Technology, Ministry of International
Trade and Industry.

Super-GM was established on Octorber 1, 1987. In its first year, 1987,
the conceptual design studies for superconducting generators, superconducting
materials and auxiliary equipment were conducted to clarify research and
development subjects for the national project to be started in 1988, and
research and development plan over an eight-year period has been established.

In this paper (Part 1), the conceptual designs of 200,000 kW class pilot
generator with a low response excitation system shown below and the
feasibility studies of advanced type superconducting generator mainly
described.

Outline of feasibility study on superconducting generator

The research and development plan for superconducting generators and the
process for achieving their practical application in this fiscal year
basically followed the proposals and assumptions presented in the feasibility
studies carried out in fiscal 1985 and 1986, that is, it was planned to carry
out careful research and development of element techniques. Those are for
trial fabricating and testing pilot machines (200,000 kW class) for technical
demonstration, and were supposed to subsequently introduce small capacity
practical demonstration machines (200,000 to 300,000 kW class) for further
development into large capacity machines.

The research of this fiscal year detailed the conceptual designs of
200,000 kW class pilot machines oriented for future superconducting
generators, based on the conceptual designs made in the research of fiscal
1985 and 1986. The 200,000 kW class pilot machines supposed in this fiscal
year are for thermal power generators (2-pole), and two models were examined:
low response excitation type superconducting generators (hereinafter called
the low response generators) and quick response excitation type
superconducting generator (hereinafter called the quick response generator).

The low response generators are intended to be applied for small to medium power plants in the suburbs of cities. They are provided with general merits of the superconducting generator such as enhanced efficiency and can contribute to the enhancement of system stability. The quick response generator is mainly applied to large capacity power plants, and is expected to be effective in enhancing system stability.

The low response generators are not required to be severely specified in the range and rates of current change and field change of superconducting field windings, since the exciter response control in the transient stage can be slow. On the other hand, to make the condition of disturbance (armature reaction) similarly less severe, the shield effect of the damper must be enhanced, and for this purpose, two dampers were decided to be used for room temperature and low temperature. Under these conditions, the effective magnetic flux of the field can be set at a high value. Thus since the AC loss of conductors is small, highly stable high current density windings can be realized. As a result, the synchronous reactance can be kept small.

The quick response generator required large current and field change range and high rate for excitation control. Considering the above, the damping characteristic of the damper must be improved, and only a room temperature damper was decided to be used, without using any low temperature damper. In this case, since the field change rate and superimposed AC field are large, conductors especially good in AC characteristics are required.

Both the low and quick response generators were selected, by carrying out a power system analysis parameter study for the synchronous reactance, and generally evaluating the effect of introduction into system such as the saving of operation cost and the increase of production cost.

The following states results of conceptual designs of 200,000 kW class pilot machines, the results of examination on the advanced superconducting generator using Nb_3Sn and AC superconducting wires.

Conceptual designs of 200,000 kW class low response pilot machines

The conceptual designs of almost the same level as made in fiscal 1985 and 1986 were examined with some items analyzed in more detail, to clarify the specifications and parameters of 200,000 kW class pilot machines. As a result, some prospect could be obtained for realization.

The preconditions and basic structures in the conceptual designs are approximately the same as those in the research of fiscal 1985 and 1986, except for the rated capacity. Since the structures of low temperature and room temperature dampers and the rotor thermal contraction are important element techniques, they were examined on both low response generators (A) and (B) individualy.

For these element techniques, the respective alternatives contain many unclear factors to be examined at present stage, and are not decisively different in features such as performance, ease of fabrication and economy. Therefore, selection cannot be made at the moment, and is thought to be accomplished when the specifications of the pilot machines are decided based on the results of research and development of element techniques (including various models) and comparative examination made in future.

Design parameters were obtained from conceptual designs which basic specifications are listed in Table 1, and the longitudinal sectional views and cross sectional views of 200,000 kW class pilot machines are shown in Figs. 1 to 4. Some of analytical approach were carried out which respectively exhibitted conductor temperature rises in system failures, transient torques in sudden shortcircuits, magnetic flux density distribution during forcing and density losses of low temperature structural materials.

636

The results of items especially examined in detail in this fiscal year are summarized below;

Table 1. Specifications of 200,000 MW Pilot Generator

Capacity	223	MVA
Voltage	20	kV
Current	6,438	A
Rotating speed	3,600	rpm

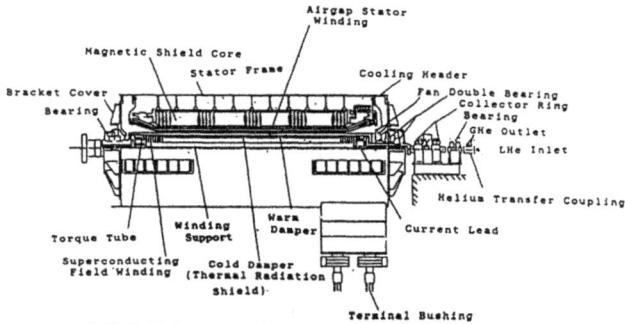

Fig. 1. Cross section of superconducting generator
(Low response excitation (A))

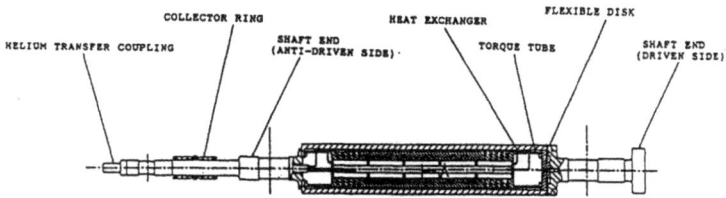

Fig. 2. Cross section of rotor
(Low response excitation (B))

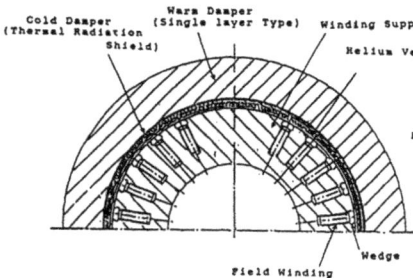

Fig. 3. Cross section of rotor
(Low response excitation (A))

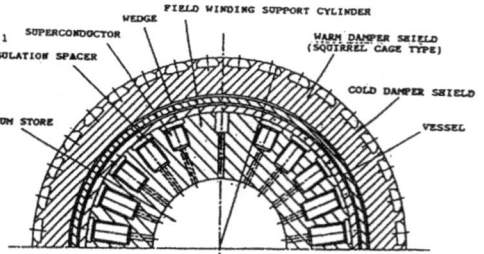

Fig. 4. Cross section of rotor
(Low response excitation (B))

Low response generator (A)

(1)For the room temperature damper,a single layer damper was selected considering ease of fabrication and high reliability, and for the low temperature damper, a three-layer damper was selected in view of strength.

(2)For prevention of thermal conduction of rotor, a double bearing structure was selected, in which the inner and outer cylinders are supported by separate bearings on the collector ring side.

(3)A high stabilized conductor design was made by using such as copper and aluminium as stabilizing materials, and filament configuration to show the applicability to generators.

(4)Also for the armature winding end, distributions of magnetic fluxes and elctromagnetic forces were calculated, and a structure was adopted for preventing the stress concentration at the armature coil caused by the thermal elongation and contraction of the straight portion of the winding, by adopting, at the winding end, a support structure with a sliding mechanism to allow the whole to move in the axial direction.(Fig.5)

Low response generator (B)

(1)For the room temperature damper, a squirrel cage type damper was selected, which contains conductive wedges in the slots formed in a highly strong non-magnetic cylinder. For the low temperature, a single layer damper using a highly strong aluminium alloy was selected. (Fig.6)

(2)For prevention of thermal conduction of rotor, a flexible disc structure was used, in which the inner and outer cylinders are flexibly connected in the axial direction and rigidly connected in the circumferential and radial directions.(Fig.2)

(3)Conductors of field windings were designed, considering three phase short circuit faults in the system (maximum current value 1.25 p.u.; maximum change rate 1.0 p.u./sec; rated current value 1.0 p.u.) based on the results of system analysis.(Fig.7)

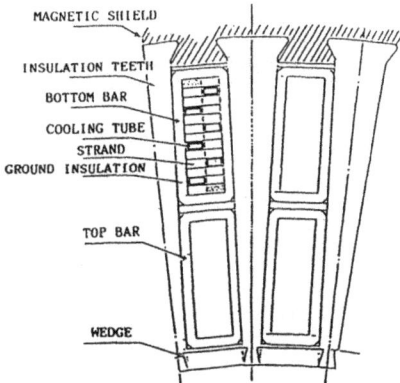

Fig. 5. Armature winding (in slot) support construction

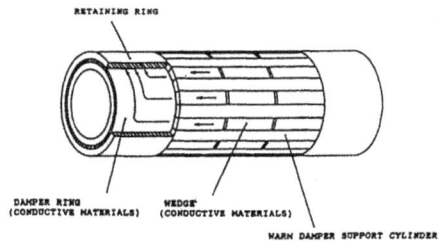

Fig. 6. Construction of warm damper shield (Squirrel cage type)

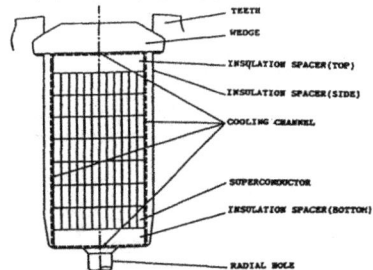

Fig. 7. Cross section of rotor slot

Examination of advanced superconducting generator

Advantages of applying Nb_3Sn wires on generators and their application methods were researched and examined, and subjects required for developing the Nb_3Sn generator were identified. An advantage clarified is that the current density is not less than 1.3 times that of NbTi, to decrease the rating of the generator (600,000 kW generator), and for application, two methods were examined; R&W (Nb_3Sn reaction heat treatment with subsequent winding) method and W&R (winding with subsequent Nb_3Sn reaction heat treatment)method.

For the fully superconducting generator with AC superconducting conductors applied, the results of the research of fiscal 1986 were reviewed, and development subjects were selected.

In table 2, the development subjects for the advanced superconducting generator are listed;

Table 2. Development subjects of advanced superconducting generator

Development subjects

Superconducting Generator applied Nb_3Sn to rotor

(Material)
1. Inprovement of strain characteristics for Nb_3Sn materials
2. Lower AC loss materials
3. Increase of critical current density
4. Development of winding technique in react & wind method
5. Development of insulation materials (Ceramic insulation, etc.) and construction materials on wind & react method. Adjastment of thermal expansion rate and retention of winding stiffness
6. Easier winding process (Simplification of winding tool. Shortening of winding time)
7. Decrease of material cost

(Generator construction)
1. Reduction of rotor diameter by applying technique of high current density and high magnetic field
2. Regulation of increasing core volume for magnetic shield with high magnetic field
3. Countermeasure of increasing fault current with low synchronous reactance
4. Others

Development subjects

Fully Superconducting Generator
1. AC superconductor with super low AC loss
2. Increase of mechanical stiffness in conductor
3. Development of large current capacity conductor
4. Helium cooling method under super critical condition
5. Supporting method and insulation method of armature air-gap winding
6. Quench protection
7. Others : Current limitter, vacuum shaft seal, etc.

Conclusion

The results obtained in this study are summerized as follows;

1)Conceptual designs of two kinds low response type superconducting generator were conducted in which design specifications were clarified.

2)Research and development subjects of elementary technology were reviewed, and actual policies and methods for the development were examined.

3)An advanced type superconducting generator which was applied Nb_3Sn or AC superconductors were studied.

Acknowledgement

This investigation and study were carried out as part of fiscal year 1987 Feasibility Study on Superconducting Machinary and Materials Technology Related to Electric Power Generation (Moonlight Project Promotion Office, The Agency of Industrial Science and Technology). The authors are much indebted to the members of Superconducting generator Subcommittee and other people concerned for their assistance and cooperation.

Reference

1) Feasibility study on superconducting machinary and materials technology related to electrical power generation (Superconducting generator) (1986,March).

2) Feasibility study on superconducting machinary and materials technology related to electrical power generation (Superconducting generator) (1987,March).

3) Feasibility study on superconducting machinary and materials technology related to electrical power generation (Superconducting generator) (1988).

640

CONCEPTUAL DESIGN OF 200,000kW CLASS PILOT SUPERCONDUCTING GENERATOR
CARRIED OUT BY SUPER-GM (PART 2; QUICK RESPONSE EXCITATION MACHINE)

Shiro KURIHARA[*1], Osami TSUKAMOTO[*2]
Toshio TANAKA[*3], Toshio KITAJIMA[*3] and Masakazu SUNADA[*3]

*1 AGENCY OF INDUSTRIAL SCIENCE AND TECHNOLOGY, MITI, *2 YOKOHAMA
NATIONAL UNIVERSITY, *3 ENGINEERING RESEARCH ASSOCIATION FOR
SUPERCONDUCTIVE GENERATION EQUIPMENT AND MATERIALS (SUPER-GM)
Umeda UN Bldg.,5-14-10 Nishitenma, Kita-Ku Osaka City, 530 JAPAN

Abstract - Feasibility studies on equipment and materials for
superconducting generator were carried out by Super-GM in fiscal 1987
as the framework of the moonlight project sponsored by the Agency of
Industrial Science and Technology, Ministry of International Trade
and Industry. Therein, conceptual design of a 200,000 kW class
superconducting generator with a quick response excitation system was
conducted. And research and development subjects of the
superconducting generator were clarified.

Introduction

As far as the research and development plan for superconducting
generator and the process for achieving the practical application are
concerned, it was planned to carry out careful research and
development of elemental technologies, thereafter to manufacture and
test the pilot generator (200,000 kW class) as the technical
demonstration one, and was supposed to subsequently introduce the
results into small capacity practical demonstration generator
(200,000 to 300,000 kW class) and furthermore into larger capacity
generator.
Two types of the 200,000 kW class generators of two poles for
thermal power plant were conceptually designed in more detail in
fiscal 1987, based on the conceptual design conducted in fiscal 1985
and 1986, which were the superconducting generators with low response
excitation system and with quick response excitation system.
In this paper (Part 2), the conceptual design of the 200,000 kW
class pilot generator with a quick response excitation system, the
results of investigation on low temperature structural materials and
research and development subjects of the superconducting generator
are mainly described. And also, the outline of research and
development plan in eight years is described.

Conceptual Design of 200,000 kW class
Quick Response Excitation Machine

In order to improve the power system stability and increase the
capacity of power transmission, the generator with quick response
excitation control system as well as low synchronous reactance is
effective. Thus, the generator with quick response excitation system
is intended to be applied to large capacity power plants with long
distance transmission line. High current and magnetic field changing

rate and speed in excitation control are required for that generator.

In order to realize it, the damping characteristic of the damper must be suppressed, and a warm damper is only used, without any low temperature damper. In this case, since the magnetic field changing speed and superimposed AC magnetic field increase, superconductors for windings with excellent AC characteristic are required.

Synchronous reactance was selected by carrying out a power system analysis parameter study, and generally evaluating the effect of introduction into power system such as stability, and the economic effect such as the saving of operation cost and the increase of production cost.

The conceptual design of almost the same level as performed in fiscal 1985 and 1986 were examined with some items analyzed in more detail, to clarify the specifications and parameters of the 200,000 kW class pilot machine.

The cross sections of the quick response excitation type superconducting generator and the rotor are shown in Figure 1 and Figure 2.

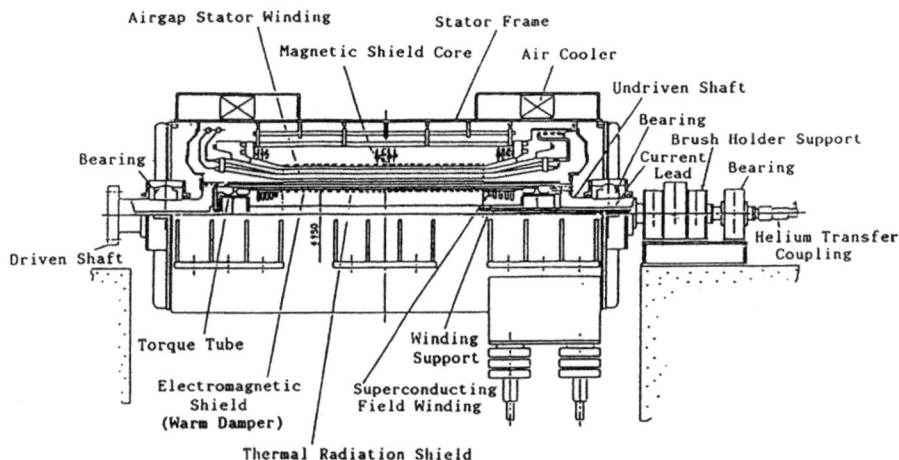

Fig.1. Cross section of superconducting generator.
(Quick Response excitation machine)

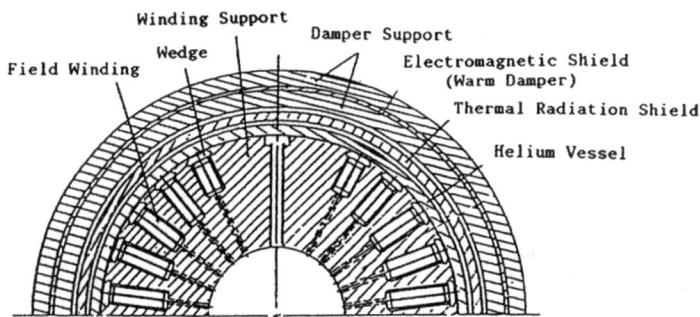

Fig.2. Cross section of rotor.

Some of analytical results in this conceptual design are as follows;
- temperature rise of the superconductor in power system faults
- transient torque in sudden short circuits
- magnetic flux distribution during forced excitation
- loss density of low temperature structural materials during forced excitation

And also, the results of items especially examined in detail in fiscal 1987 are summarized below.

1) The warm damper with three layered structure was applied to allow instant change of magnetic flux during forced excitation and to provide high mechanical strength. Furthermore, to optimize the initial open circuit time constant which contributes to the enhancement of power system stability, the structure and materials were reviewed.

2) In the design of field windings and structures, the excitation control responses for two values of ceiling voltages (300 p.u. and 400 p.u.) were taken into account. The former allows the same field current change rate as obtained in the research of fiscal 1986 and the latter allows the field current change rate obtained by the examination on the effect of transient stability in power system in fiscal 1987.

3) Optimum design for the field winding, multi-depth rotor slots and structural materials of cryogenic rotor was performed to reduce AC loss and stress.

Survey of Cryogenic Structural Materials

For cryogenic structural materials (metallic materials and organic composite materials), trends of research and development mainly on fatigue properties were surveyed, and requirment specifications were clarified by conceptual design. The following future subjects of research and development in application to 200,000 kW class pilot machines were found.

1) To identify material properties at cryogenic temperatures, and to establish a structural design standard based on them.

2) To develop techniques for testing and evaluating at cryogenic temperatures, and to establish a data base of them.

3) To identify fatique properties of weld zones and at complex (combination) loads, and to establish a technique for detecting fatigue and damage.

And so on.

As a part of examination of advanced superconducting generator, trends of research and development on the structural materials proposed for rotors suitable for W&R method of Nb_3Sn were examined.

Research and Development Subjects for The Eight-year National Project

The full-scale research and development that has already started since fiscal 1988 is carried out for "superconducting technology applied to electric power" for the purpose of practical use of various kinds of superconducting electric power apparatuses including superconducting generator.

The highest priority in the eight-year plan is to conduct

research and development of various elemental technologies required
to design and manufacture the 200,000 kW pilot superconducting
generator. Research and development subjects are shown in Table 1. It
is also important to develop high performance superconducting wires
and refrigeration systems applicable to various kinds of
superconducting electric power apparatuses including advanced
superconducting generator.

Two types of 70,000 kW class model generator rotors will be
manufactured and tested considering ease of testing and and element
technology scale rule, which will be a part of elemental technologies
for the 200,000 kW class pilot generator.

An intermediate evaluation of the overall research and
development must be performed for superconducting technology applied
to electric power. In addition, the evaluation and study of previous

Table 1. Research and development subjects
for superconducting generator.

Component	Research and development subjects
1.Metal-based Superconductor	1) High Current Density 2) AC Loss Reduction 3) Higher Stability 4) Higher Capacity 5) Electrical Insulating Method 6) Evaluation Technique
2.Field Winding	1) Winding Manufacturing 2) Supporting Method 3) Cooling at Transient Condition 4) Winding Insulation 5) Measuring Method
3.Multi-cylindrical Rotor	1) Vibrating Characteristics 2) Manufacturing and Assembling Technique 3) Manufacturing Technique of Structural Materials 4) Damper 5) Appropriate Structure of Whole Rotor
4.Cooling System	1) Efficient Cooling Method of Field Winding 2) Efficient Cooling Method of Rotor 3) Structural Materials' Cooling at Transient Condition 4) Reliability of Current Lead
5.Helium Transfer Coupling (HTC)	1) Gas Helium Sealing Method 2) Vacuum Sealing Method 3) Preventive Method of Heat Penetration to Liquid Helium 4) Reliability of HTC
6.Airgap Stator Winding	1) Winding Support 2) Stator Coil

research and development results, addition of new research and development items, and research modification must be checked and reviewed. In intermediate evaluation, a state of element technology research and development on the generators is grasped and addition of research and development of element technology for pilot generator is desirable whenever necessary.

Conclusion

The results obtained in this study are summerized as follows.

1)Conceptual design of quick response type superconducting generator was conducted in which design specification was clarified.

2)Research and development subjects for cryogenic structural materials in application to pilot machines were found.

3)Research and development subjects of elementary technology were reviewed, and actual policies and methods for the development were examined.

Acknowledgement

This investigation and study were carried out as a part of fiscal 1987 "Feasibility Study on Superconducting Machinary and Materials Technology Related to Electric Power Generation" (Moonlight Project Promotion Office, The Agency of Industrial Science and Technology). The authors are much indebted to the members of Superconducting Generator Subcommittee and other people concerned for their assistance and cooperation.

Reference

1) Feasibility study on superconducting machinary and materials technology related to electrical power generation (Superconducting generator) (1986,March).
2) Feasibility study on superconducting machinary and materials technology related to electrical power generation (Superconducting generator) (1987,March).
3) Feasibility study on superconducting machinary and materials technology related to electrical power generation (Superconducting generator) (1988).

EXPERIMENT OF BRUSHLESS FULLY SUPERCONDUCTING
GENERATOR WITH MAGNETIC FLUX PUMP

Itsuya MUTA, Hiroshi TSUKIZI, Tetsuya SASANO,
Yukio TSUTSUI and Tatsuya FURUKAWA

Department of Electrical Engineering, Saga University
1 Honjyo-machi, Saga-shi, 840 Japan

Abstract - Since the first success in the development of magnetic flux
-pumped brushlesss excitation system for superconducting AC generators in
1983[1], we have been building a testing machine to generate actual electric
power.
    The paper presents experimental machine system, and test results about per-
formances of the flux pump and output characteristics when operated as fully
superconducting brushless generator.

Introduction

    A common scheme of superconducting generators developed so far is well-
known to be of a rotating field type. However, they definitely require the
use of electric power leads, collector rings, brushes and so on for supplying
a current to the superconducting field coil from external source, which accom-
panies some problems of cooling efficiency being deteriorated due to external
heat leak and ohmic losses, and maintenance of mechanical contacts.
    To cope with this situation, the newly developed generator introduces, in
place of the collector ring system, a so-called "flux pump" system of indirectly feeding the field coil with an adjustable direct current. Hence, once
a current is fed into the circuit, the current becomes a "persistent current",
that is, keeping constant forever, but adjustable from external magnets.
    In addition, the output coil of the so-called "armature winding" is made
of superconducting wires with very low AC losses. Thus, we have developed a
so-called fully superconducting brushless generator.
    This time, because of mechanical constraints on its support system and
the low capacity of prime mover, only the output of 100 W was produced at the
speed of 113 rpm, but being equivalent to 1.7 kW at 60 Hz.

Scheme of Developed Machine

    Fig.1 shows the developed brushless fully superconducting ac generator
assembled in a cryosyat for the general purpose. Needless to say, this
facility is not specialized for the generator itself. However, the superconductive rotating field and staionary armature windings are provided in it.

Fully superconducting generator
    In Fig.2 is presented the schematic diagram of the hand-made generator,
the rotor of which is composed of 4-pole saddle-shaped superconducting field
coil with the straight section of 150 mm, and the outer diameter of about 100
mm, while the armature is composed of twelve spiral pancake superconducting
coils adhered to the FRP armature frame.
    The superconducting coil can be excited through the magnetic flux pump

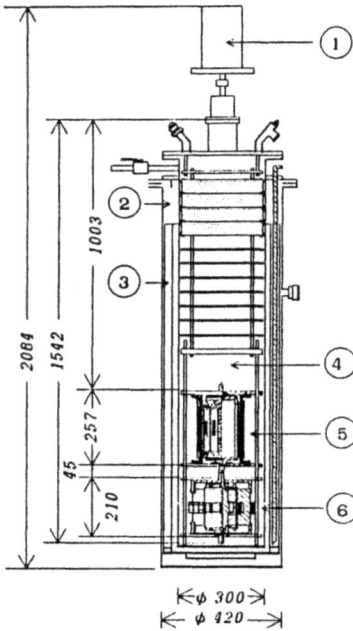

1 Driving motor
2 Vacuum space
3 LN2 jacket
4 LHe cryostat
5 Generator
6 Flux pump

Fig.1. General assembly of
        test facility.

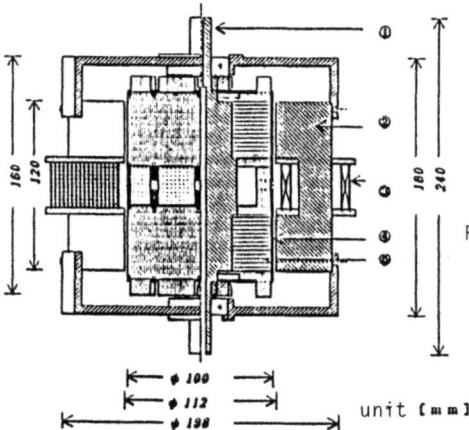

unit [mm]

1 Field winding
2 Armature winding
3 Search coil

(a) Whole assembly

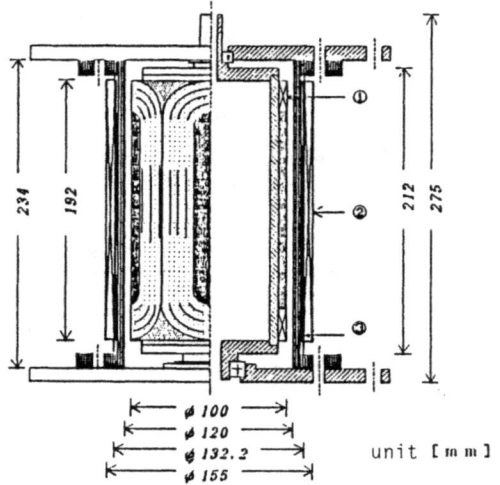

(b) Armature winding

Fig.2. Cross section of generator.

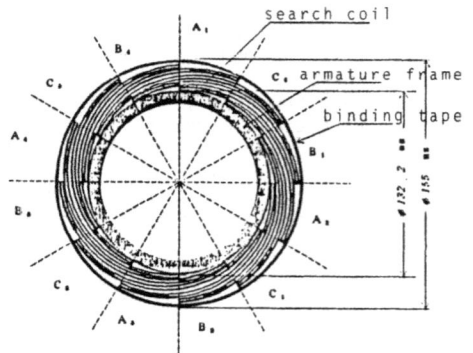

unit [mm]

1 Rotor shaft   4 Nb sheet
2 Magnet core   5 Back core
3 Exciting coils

Fig.3. Magnetic flux pump.

Table 1. Specification of the developed generator.

magnetic flux pump	excitation coil : 8-poles, 183 turn/pole, NbTi wire of ϕ0.535mm Nb cylinder : 20 μm thick, 60mm wide, 100mm dia.
field winding	4-poles, saddle-shaped, NbTi wire(ϕ0.385mm,ϕ0.375mm), No.of series turns 1420, Resistance 336[Ω] (at R.T.), self-inductance 67.5mH
armature winding	NbTi wire for AC applications of 0.209mm dia. (dia. of filament 0.93 μm, No.of filaments 15367, Composite ratio 1/0/2.3, twist pitch 0.98mm) Spiral pancake shaped coil, No.of turns 154/phase/pole, Resistance 414[Ω]/phase/pole (at R.T.), Self-inductance 26.0mH/phase, Mutual-inductance 22.78mH(phase), 40.69mH(line-to-line)

without the mechanical brush system. Main specifications of the generator are provided in Table 1.

Magnetic flux pump

Fig.3 shows the developed flux pump connected to the generator superconducting field coil. The magnetic flux pump is such a device as superconduting dc homopolar generator, for which one of the most practical and promissing versions, a type of rotating flux spots [2],[3] is adopted. The flux pump system developed here comprises two flux pump components connected in series, each of them having rotatable Nb strip cylinder of 6 cm wide, 20μm thick and 10 cm dia. as presented in Table 1. The current collected from the Nb strip cylinder is supplied to the generator field coils through NbTi superconducting wires. Flux pump excitation system is composed of 8 stationary unipolar electromagnets with soft steel cores, laminated silicon steel return core and superconducting excitation coils.

Experimental Results

For the present, because of low mechanical strength of the whole system and from point of view that the generator itself should be immersed in coolant of liquid helium, the revolving speed of the rotor must be limited to less than 400 rpm at the steady-state. /

Performances of the flux pump

Fig.4 shows a diagram of pumping-up process of the generator field current in the case of flux pump excitation current of Iex =13.8 A and the rotor speed of 400 rpm. From this test it has been recognized that the pumping-up current is almost linearly increasing with time and its maximum value or the quench current in this generator field system would be about 70 A. The pumping-up rates or the rate of increase of the field current against the revolving speed N [rpm] and the flux pump excition intensity Iex [A] are plotted in Fig.5 and 6, respectively. From these figures, it is clearly found

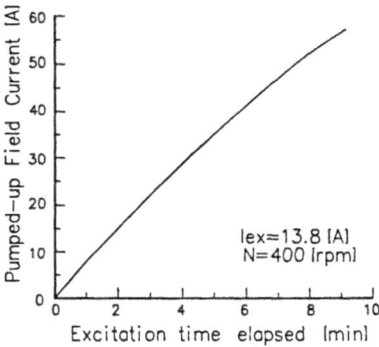

Fig.4. Example of flux pumping behavior.

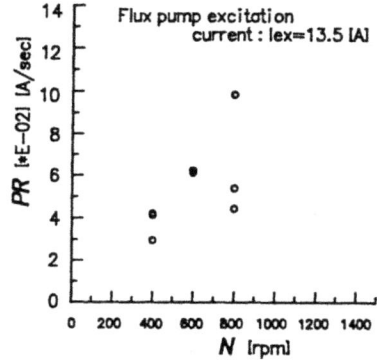

Fig. 5. Pumping rate of field current vs. revolving field velocity.

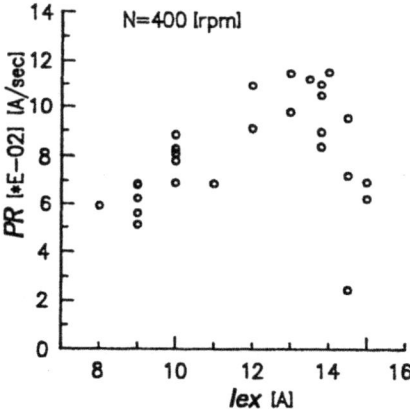

Fig. 6. Pumping rate of field current vs. flux pump excitation.

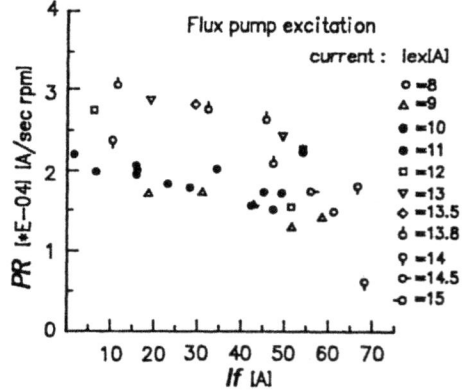

Fig. 7. Pumping rate of field current vs. pumped-up field current.

that the pumping-up rate approximately linearly increase in proportion to the revolving speed N, while it has the maximum value for the flux pump excitation current Iex.

Fig.7 for the pumped-up field current If [A] proves that the pumping-up rate normanized by both of the time and the revolving speed tends to decrease with the increse of pumped-up field current.

Characteristics of the generator

No-load terminal voltage of the brushless generator for the pumped-up field current was obtained at N=400 rpm and some other speeds. From such test results, no-load chgaracteristic curves for various speeds plotted in Fig.8 have been estimatted.

Fig.9 shows the waveform of terminal voltage under no-loading and loading conditions, which is proved to be almost symmetrical and sinusoidal.

Because of the small capacity of the driving motor, the rotor speed could not be kept constant this time. So, only the output power of 100 W at 3.18 Hz was generated, which is equivalent to around 1.7 kW at 60 Hz. In Table 2 are summarized test results.

Fig.10 gives phase voltage and output power curves at If=28.3 A and 50 Hz which are calculated by using machine constants presented in Table 1, in which the maximum output power will be predicted of around 1.1 kW.

Fig.11 shows a decremental waveform of the generator output voltage after

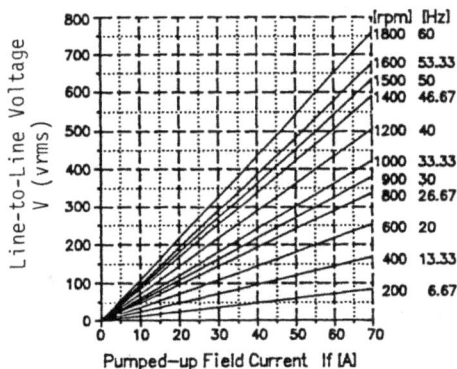

Fig.8. No-load characteristic curves of the generator.

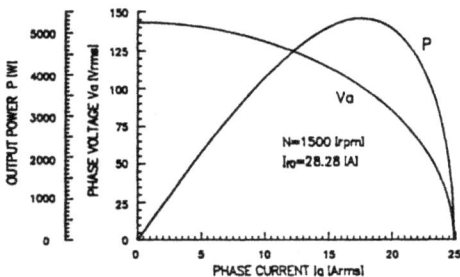

Fig.10. Predicted load characteristic curves.

at No-Load (Steady-state)   at Load (Steady-state)

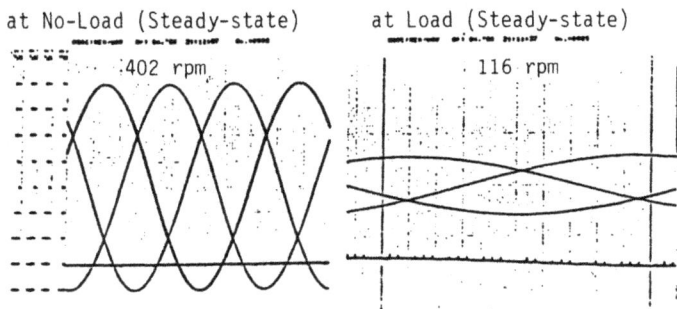

Fig. 9. Wave form of generator output voltage.

Table 2. Example of test results.

	No-load (steady-state)	Load (steady-state)	
terminal voltage	141.8	41.9	[v]
frequency	13.46	3.81	[Hz]
load resistance	--	57.51	[Ω]
armature current	--	0.73	[A]
field current	58.7	58.7	[A]
output power	--	91.64	[w]

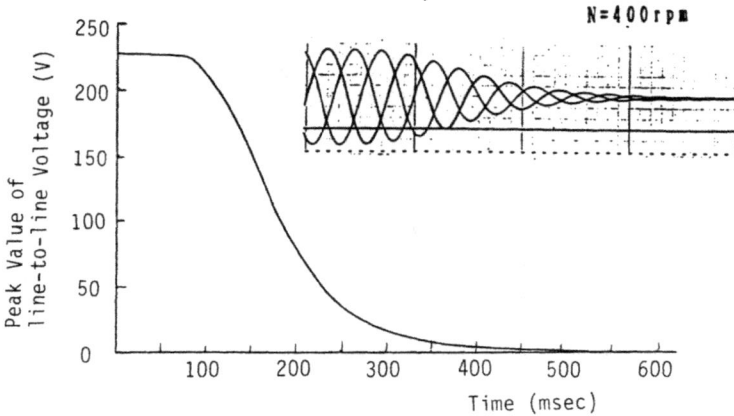

Fig.11. Wave form of decreasing terminal
voltage after quench.

a quench when the pumped-up generator field current is reaching the critical
value of around 70 A.

Conclusion

In general, from this study, it has been confirmed that the combination
system of the magnetic flux pump and the superconducting field coil could be
viable for the brushless excitation system of superconducting ac generators.
The results obtained here are summarized as follows :
1) The pumping-up rate of the current can be controlled by the intensity
of flux pump excitation, but have the maximum value for it.
2) The pumping-up rate of the flux pump decreases with the increase of
the magnitude of pumped-up current.
3) The maximum output power of the synchronous generator developed here
could be predicted to be about 18.6 kw at 50 Hz.
At present, the detail of performances of superconduting armature coil
could not be discussed yet because of the poor structure of the experimental
machine. Hence, we intend to improve the machine structure.

References

1) I.Muta, "Preliminary Experiments for Possibility of Brushless Superconducting Synchronous Generator Excited by Magnetic Flux Pump",Proc.of ICEM '84.,
Lausanne,Swiss., 1118-1121(1984).
2) I.Muta, H.Tanimizu, and E.Mukai,"Feasibility Test of Brushless Superconducting Synchronous Generator Excitated by Magnetic Flux Pump", Proc. of SCEC
'85, Shanghai, The People's Republic of China, 139-142(1985).
3) I.Muta, H.Tsukizi, E. Mukai and T.Furukawa, "Preliminary Experiment for
Brushless Supercondwuting Synchronous Generator with Magnetic Flux Pump",
Electric Energy Confernce '87, Adelaide, Australia, 301-305(1987).

ON THE EXCITATION CONTROL OF SUPERCONDUCTING SYNCHRONOUS GENERATOR CONNECTED TO ELECTRIC POWER GRID

Shinichi IMAI   Kenji MATSUURA

Faculty of Engineering, Osaka University
2-1 Yamadaoka Suita, 565 Osaka, Japan

Abstract -In this paper the authors propose an application of a thermoelectric generator(TEG) utilizing waste exhaust heat of power station to the exciter dc power source of a superconducting generator. The field voltage is controled by a d.c. direct converter. In the case of field winding quench, TEG can absorb the energy stored in the superconducting field winding.

Introduction

On the occasion of superconducting field winding quench the energy stored in it must be released as fast as possible. From the viewpount, the authors propose a new excitation system and its control for the synchronous generator with a superconducting field winding. The excitation system has a thermoelectric generator(TEG) as the dc power source and a switching circuit composed of a gate turn-off(GTO) thyristor bridge for the control of field voltage. TEG can utilize waste exhaust heat of the power station and absorb the released energy at the quench of the field winding. The proposed excitation control system operates as an automatic voltage regulator(AVR) in the normal operation in which no quench occurs in the superconducting field winding. Moreover it may be expected that TEG has effects of stabilizing power system transient swings and increasing transient stability power limit. Taking into consideration of these effects as well as the power absorbing function of TEG at the quench of field winding, an optimum design of TEG in the excitation system for the superconducting synchronous generator has been investigated.

System Description

Field Circuit of Generator

The proposed field circuit shown in Fig.1 consists of a thermoelectric generator(TEG), a gate turn-off(GTO) thyristor bridge, a superconducting field winding and a protection resistance and diode. For the purpose of automatic voltage regulation, the timing of switching of the GTO thyristors is done as a manner shown in Fig.2. When Ch_1 and Ch_2 are in on-state, $v_F = E_{th} - r_{th}i_F$ and when Ch_1 and Ch_2 are in off-state, $v_F = -r_p i_F$. The average of $v_F(\overline{v_F})$ is expressed as

$$\overline{V_F} = E_{th}\,\alpha - \{(r_{th} - r_p)\alpha + r_p\}\,i_F \qquad (1)$$

where $\alpha = T/T_p$. The nomenclature is listed in Table 1.

Note that the variation of i_F during the time interval of $T_p(s)$ can be neglected because the time constant of the field circuit is very large.

As TEG consists of many thermoelectric elements, the choice of E_{th} and r_{th} is easily realized by a properly connected series-parallel circuit of the elements. E_{th} and r_{th} can be determined as follows.

$$E_{th} - r_{th}\,i_{F0} = V_c \qquad (2)$$

Where i_{F0} is the steady state field current when both V_t and P_e are 1 pu and v_c is the ceiling voltage. The capacity of TEG, W_{th} is defined as the

652

Table1. List of symbols

Ch_1-Ch_4	GTO thyristor	T_m	Mechnical torque (pu)
D	Machine load damping coefficient	Tp	Switching period (s)
D_1	Diode	V_{ref}	Refrence terminal voltage of generator (pu)
e_q	Voltage behind transient reactance (pu)	V_t	Terminal voltage of generator (pu)
E_{th}	Effective EMF of TEG (pu)	V_∞	Infinite bus voltage (pu)
H	Inertia constant (s)	v_c	Ceiling voltage (pu)
i_d	d axis armature current (pu)	v_d	d axis armature voltage (pu)
i_F	Field current (pu)	v_F	Field voltage (pu)
i_q	q axis armature current (pu)	v_q	q axis armature voltage (pu)
K	Gain of control system (pu)	W_{th}	Capacity of TEG (pu)
k	$\sqrt{3/2}$ (pu)	x_d	d axis synchronous reactance (pu)
L_{AD}	Exciting inductance of d axis circuit (pu)	x'_d	d axis transient reactance (pu)
L_F	Field self inductance (pu)	x''_d	d axis subtransient reactance (pu)
P_e	Output power of generator (pu)	x_e	External reactance (pu)
r_F	Field resistance (pu)	x_l	Leakage reactance of generator (pu)
r_{th}	Effective resistance of TEG (pu)	x_{li}	Transmisson line reactance (pu)
T	Time constant of control system (s)	x_t	Leakage reactance of (pu)
T_{d0}	d axis transient open circuit time constant(s)	δ	Rotor angle (rad)
T_j	$2H\omega_B$ (pu)	ω	Angular frequency (pu)
		ω_B	Base frequency (rad/s)

Table 2 Superconducting generator data

Machine-rated MVA	1120MVA	r_F	0.6×10^{-6}pu
Rated kV	20kV	T'_{do}	1600sec
Rated PF	0.9	L_F	0.29pu
x_d	0.3pu	H	3.11sec
x'_d	0.2pu	Base MVA S_B	373.3MVA/phase
x''_d	0.13pu	Base kV V_B	15.01kV
x_l	0.13pu	Base field kV·V_{FB}	721.8kV

maximum power which is given by

$$W_{th} = E_{th}^2 / 4 r_{th} \tag{3}$$

By substituting equation (2) in equation (3), we have

$$W_{th} = (V_c + r_{th} i_{F0})^2 / 4 r_{th} \geq i_{F0} V_c \tag{4}$$

From equation (4) the minimum value of W_{th} is found to be $i_{F0}v_c$, and then

$$r_{th} = V_c / i_{F0} \tag{5}$$

$$E_{th} = 2 V_c \tag{6}$$

In order to prevent the field winding from dielectric breakdown, the voltage across r_p should be equal to 5kV for example when $i_F=1.5i_{F0}$. When i_F exceeds to $1.5i_{F0}$, Ch1 and Ch2 should be turned off. On the occasion of field winding quench the energy stored in it must be released as fast as possible. Therefore, the following control is applied to the system:

v_{F1}, the field voltage when Ch_1-Ch_4 are in off-state, is given by

$$v_{F1} = - r_p i_F \tag{7}$$

v_{F2}, the field voltage when both Ch_2 and Ch_3 are in on-state, is calculated from the relation,

$$v_{F2} = -r_p E_{th} / (r_{th} + r_p) - r_{th} r_p i_F / (r_{th} + r_p) \tag{8}$$

$|v_{F1}|$ and $|v_{F2}|$ are drawn in Fig.3 as a function of i_F. If $v_c>2.5$kV, on finding field winding quench, Ch_1 and Ch_2 should be turned off. Then if i_F decreases to become i_1, Ch_3 and Ch_4 should be turned on. If $v_c<2.5$kV, on finding field winding quench, Ch_1 and Ch_2 should be turned off. When i_F becomes to E_{th}/r_p, Ch_3 and Ch_4 should be turned on.

Power System

Dynamic computor simulation was performed on the power system shown in Fig.4. It is the one-machine infinite bus model of a power system with a superconducting synchronous generator connected through a transformer to two circuits of transmission line. The superconducting synchronous generator has the TEG-GTO switching excitation and control system described in the previous section. Generator data is shown in Table 2.

The transmission line impedance is expressed by jx_{1i} and the transformer leakage reactance of 0.15 pu is accounted for.

The governor is described by GE EHC governer model and the steam turbines are by tandem-compound single reheat steam.[1]

Result and Discussion

Effect of Excitation Control on Energy Release at Quench of Field Winding

The temperature rise of field winding at quench is proportional to Q, which is defined by $\int_0^t r_F i_F^2 dt$. Let Q_m be the maximum value of Q, then $Q_m = \int_0^{ts} r_F i_F^2 dt$, where ts is the time when i_F becomes to zero in the case that field winding quench has happened at $i_F=i_{F0}$ and $t=0$. Q_m was calculated as a function of v_c and the result is shown in Fig.5. From the figure, it is found that Q_m has a minimum value at $v_c=3.34$(kV)(v_{cs}). The reason why Q_m increases with increasing v_c in the range $v_c>3.34$(kV) is that TEG cannot be made to absorb the energy. This is because v_{F2} becomes larger than the dielectric breakdown voltage of the field winding. During the quench of field winding Q and v_F vary with respect to time as shown in Fig.6.

Effect of Excitation Control on Stabilizing Power System Transient Swings

The system describing equations of the generator, the transmission line, the control system, the steam turbines and the governor were linearlized to take the form,

$$\dot{X} = AX \tag{9}$$

$$X^t = [\Delta e'_{\xi} \ \Delta \omega \ \Delta \delta \ \Delta \alpha \ \Delta P_{GV} \ \Delta P_{M1} \ \Delta P_{M2} \ \Delta P_{M3}]$$

where PGV, PM1, PM2 and PM3 are output of the governor, the steam chest, the reheater and the crossover connection, respectively. From the eigenvalue analysis for the linearlized system eight eigenvalues were obtained. Note that the intial condition of the generator is $V_t = P_e = 1$(pu) and PF=0.988. The mechanical eigenmode is identified as follows. It may begin with the mechanical loop alone, which is a second-order system, and the undamped natural mechanical mode frequency ω_n is one of the mechanical mode. Relation between v_c and the real part of the mechanical mode eigenvalue is shown in Fig.7. The figure shows that the system is more stabilized by the larger value of v_c.

Figure 8 shows the block diagram of the system. Relation between $\Delta e'_{\xi}$ and $\Delta \delta$ is given by

$$\frac{\Delta e'_{\xi}}{\Delta \delta} = \frac{K A_2 A_6 A_8 + (A_2 A_{10} + A_3)(1+ST)}{(A_1 - A_2 A_9 + ST'_{d0})(1+ST) - K A_2 A_7 A_8} \triangleq G_{F\delta}(S) \tag{10}$$

The variations of $\Delta \delta$ and $\Delta e'_{\xi}$ during acceleration of the generator rotor are shown in Fig.9. The phase and the amplitude of $\Delta e'_{\xi}$ are decided by $G_{F\delta}(j\omega_n)$.

If the magnitude of e'_q increases, the system is stable since the power output of the generator increases and the speed of the rotor reduces. Here a new quantity S, a measure by which the stability of the system is evaluated, is defined by

$$S = \int_o^{\pi} \Delta e'_{\xi} \, d(\omega_n t) \tag{11}$$

The stability of the system will increase with increasing S. Relation between S and v_c for the system is shown in Fig.10. As compared with Fig.7 and Fig.10, it is confirmed that the system stability increases with increasing v_c. Figure 11 shows the variation of $\Delta \delta$ when a three phase grounding fault (3LG) happened at the sending end of #2 circuit of the transmission line shown in Fig.4, then the faulted circuit was tripped 4 cycles later and then reclosing was took place 60 cycles later. It is found that the case corresponding to $v_c = 3.34$kV(v_{cs})(Fig.11(b)) is more stable than the case corresponding to $v_c = 0.5$kV(Fig.11(a)). But the system needs more damping because it takes long time to stabilize the transient swings. For that reason a power system stabilizer(PSS) has been introduced and designed as follows.

The supplementary excitation control signal u_E expressed by eq.(12) is added to the input signal of AVR.

$$u_E = K_c \frac{S T_4}{1 + S T_4} \frac{1 + S T_1}{1 + S T_2} \Delta \omega \tag{12}$$

T_2 is chosen as 0.2(s) and T_1 as compensating the phase lag of AVR. K_c is chosen such that a damping of $0.02T_1$ is obtained. T_4 is 3.0(s). Figure 12 shows the transient swings of $\Delta \delta$ and ΔP_e for the system with PSS. It is found that the system with PSS is more stable and has better damping. Furthermore, it is found that the case (b) is also more stable than the case (a) for the system with PSS.

As a summary, the system stability is improved by increasing the magnitude of v_c regardless of PSS. But for the protection against dielectric breakdown of the field winding, r_c should be limitted within a proper value.

Relation between vc and the first swing transient stability power limit is shown in Fig.13. The first swing transient stability power limit increases also with increasing the magnitude of v_c as indicated in the figure.

Conclusion

We have shown, using a study of one machine infinite bus system, that during quench of the superconducting field winding of a synchronous generator a thermoelectric generator (TEG) for the exciter power source can absorb the energy stored in the field winding with the assistance of control of a d.c. direct converter composed of GTO thyristors.

The proposed TEG-GTO d.c. direct converter exciter system not only has compatibility with AVR in the normal state of power system, but also a remarkable effect on stabilizing transient swings with adopting a maximum permissible excitation ceiling voltage. It has been found, however, that there is an optimum excitation ceiling voltage to release as fast as possible the stored energy in a quenched superconducting field winding.

Reference

(1)Y.Sekine, "Analysis of power system transient, "Ohm Sha, Tokyo, Japan, 1984 (in Japanese).

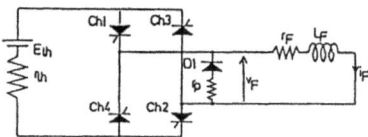

Fig.1   Field circuit

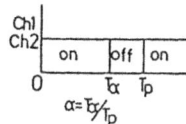

Fig.2   Timing of switching for the AVR operation.

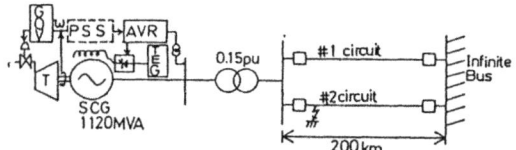

Fig.4   One machine to an infinite bus through a transmission line

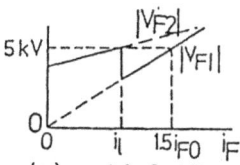

(a) $v_c > 2.5$ kV

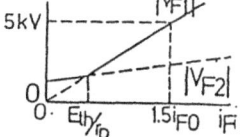

(b) $v_c < 2.5$ kV

Fig.3 $|V_{F1}|$, $|V_{F2}|$ versus i_F.

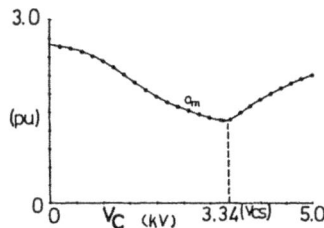

Fig.5   Relation between maximum of $Q(Q_m)$ and v_c.

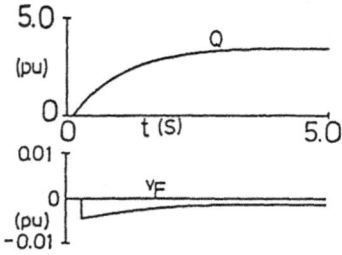

(a) $v_c = 0.5$ kV

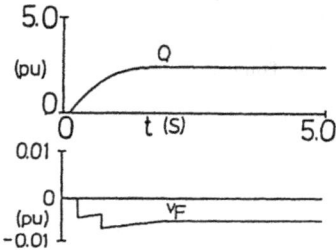

(b) $v_c = 3.34$ kV

Fig. 6　Variation of Q and v_F
during quench.

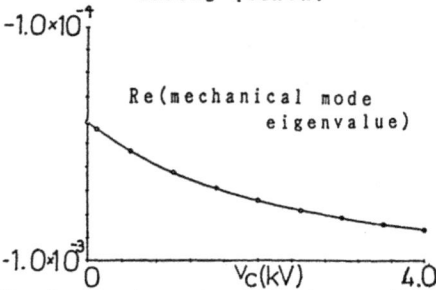

Fig. 7　Real part of mechanical
mode eigenvalue versus v_c.

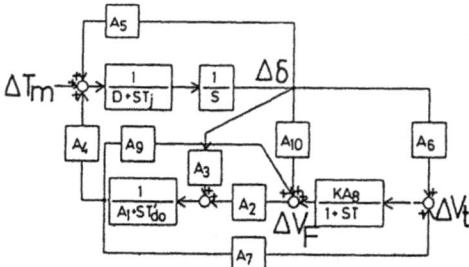

Fig. 8　Block diagram of system.

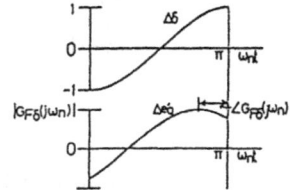

Fig. 9　Variation of $\Delta\delta$ and $\Delta e_q'$ during
acceleration of generator rotor

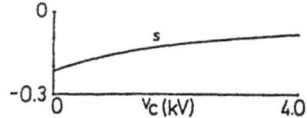

Fig. 10　S versus v_c

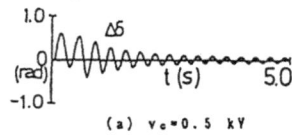

(a) $v_c = 0.5$ kV

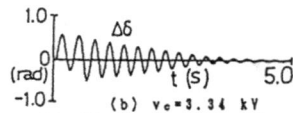

(b) $v_c = 3.34$ kV
Fig. 11　Variation of $\Delta\delta'$ for the
system without PSS.

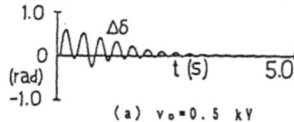

(a) $v_c = 0.5$ kV

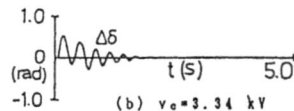

(b) $v_c = 3.34$ kV
Fig. 12　Variation of $\Delta\delta$ for the
system with PSS

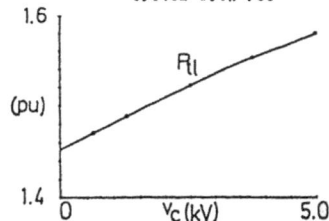

Fig. 13　Effect of v_c on the first
swing transient stability
power limit, P_{t1}.

DEVELOPMENT CONSIDERATIONS ON MULTIFILAMENTARY SUPERCONDUCTORS FOR SUPERCONDUCTING GENERATOR

Shiro KURIHARA[*1], Yoichi KIMURA[*2]
Shinichiro MEGURO[*3], Kiyotaka UYEDA[*3]
Takashi SAITOH[*3], Kenichi TAKAHASHI[*3], Hiroshi KUBOKAWA[*3]

[*1] AGENCY OF INDUSTRIAL SCIENCE AND TECHNOLOGY, MITI
[*2] ELECTROTECHNICAL LABORATORY, AIST
[*3] ENGINEERING RESEARCH ASSOCIATION FOR SUPERCONDUCTIVE GENERATION
EQUIPMENT AND MATERIALS (SUPER-GM)
Umeda UN Bldg.,5-14-10 Nishitenma, Kita-Ku Osaka City, 530 JAPAN

Abstract - Through the design study on multifilamentary
superconductors   for   superconducting   generator   equipped   with
rotational   field   windings,   inevitable   development   items   were
extracted for both NbTi and Nb_3Sn superconductors. The viewpoint to
realize the practical rotational machine was insisted in accordance
with the operational requirment for the superconducting generator.

Introduction

Feasibility studies on the equipment and materials for
superconducting generator were conducted in 1985 and 1986 within
the framework of the moon light project sponsored by The Agency of
Industrial Science and Technology, Ministry of International Trade
and   Industry.   The   Engineering   Research   Association   for
Superconductive Generation Equipment and Materials was established
on October 1, 1987. In its first year, 1987, the Association
conducted researches to identify the research and development tasks
over the 8-year national project to be started in 1988 and to
establish the research and development plan.
It is our common understanding that the development of
superconducting generator is coming to the stage of practical
machine development, completing the stage of verification of the
principles through the history of many research works for about 30
years in the world. [1] The practical superconducting machine
should compete successfully with the conventional machine relating
to reliability, efficiency, compactness, good stabilization and the
cost for power generation. In order to accomplish such advantages,
the benefit of superconductivity should be best incorporated in the
superconducting machine based on the optimum design and challenging
develpment program. As the result of this design study, technical
requirments   and   development   items   were   clarified   for   the
superconductors used in low response excitation machine [2], quick
response excitation machine [3] and advanced Nb_3Sn machine [2],
taking the operational conditions over the entire life of the
machine into account.

Operational Conditions of Field Windings

The field windings should be designed to withstand safely
under such operational conditions as shown below.
The major feature of field windings is that the superconductor
is operated under the rotational condition of very high speed

(3,000~3,600rpm), which may lead to unidentified phenomena that were not revealed in the non-rotating machine.

Stationary Operation

(1) Starting and stoppage
        Daily starting and stoppage
        Weak end stoppage
        Periodic inspection
(2) Rated operation
(3) Excitation control
        Low response   Quick response

Fault Endurance

(1) Short-circuiting   (2) Excessive rotation
(3) Quench propagation   (4) Vacuum leakage

Design Procedures of NbTi Superconductor for Field Winding

Typical design procedures of superconductor for field winding are shown in Fig.1. Operational conditions described above are taken into account as design conditions and design criteria of the procedures. Basic electrical requirment of field winding is expressed as the load curve in current-magnetic field configuration. An example of load curve is shown in Fig.2 for quick response machine. In our design study, the critical current of the superconductor is determined so as to maintain the temperature margin of 1K during the fastest change in magnetic field due to excitation control. The details of configurations of the superconductor are adjusted not to generate excessive temperature rise due to AC losses in the feed-back loop of the conductor design. Mechanical, thermal and geometrical requirments on the superconductor are imposed and checked in accordance with the detailed structural design of the generator. In order to apply these design procedures to the practical machine, the design criteria should stand on the long-term operation. However, strict numerical evaluation of superconductor's properties under such condition is not sufficiently available at present.

Development Considerations

When executing the design procedures of Fig.1, some difficulties are revealed because of the lack in experimental data which support the design criteria. This situation is most serious when the deterioration of the property due to the practical long-term operation is concerned. The development items of superconductor will be extracted by the careful considerations on design criteria. They are summarized in Table1. together with the explicit items which will be related to the realization of high-performance superconducting generator.

Another point of view should be payed to the standardization quality control and cost optimization at the developmental stage of proto-type machine. Superconducting generator will be specified by a new test cord which will include items originating from superconductivity. Reasonable procedures and its standardization should be developed to evaluate the electrical, mechanical and thermal properties of supercondutor. At the final stage of development, integrated manufacturing technology of superconductor should be developed before introducing superconducting generator into power plant.

Nb₃Sn Superconductor for Field Winding

It is usually believed that Nb_3Sn superconductor is not so reliable as to be applied to the windings of electrical power apparatus. However, the higher critical density and higher critical temperature are very attractive for the improvement of machine performance. The brittleness of material and high temperature heat treatment were the main hindrance of its application. Now, more brittle ceramic superconductors are under investigation including the application to power apparatus. It is worth while to take a new look at the Nb_3Sn for power application. It is needless to say that the adoption of Nb_3Sn leads to the fundamental design alternation of the machine.

React & Wind Method

A few percent of strain is usually applied to the superconductor when wound into field winding at the innermost corner. This winding strain is far beyond the acceptable limit for pre-reacted Nb_3Sn superconductor. One way to overcome this difficulty is to study the field winding configuration to allow more moderate curvature in the winding. Another way is to study the superconductor configuration to allow a few percent of strain, such as cabled configuration which consists of fine strands or improving the strain-sensitivity of Nb_3Sn itself by new technology of fabrication.

Wind & React Method

Using W&R method, Nb_3Sn will be relieved from suffering strain damage. On the contrary, structural material of the rotor encounters very high temperature for long time during the reaction heat treatment of Nb_3Sn. Moreover, considerable development work is necessary for reliable insulation material and method to fix the winding. Development of high-temperature-resistant structural material and at the same time the effort to lower the reaction temperature of Nb_3Sn are both necessary.

Anyway, developmental approach both from machine design and material improvement is essential to realize the superconducting generator with Nb_3Sn field windings.

Conclusion

As the results of the study, the necessity of the developmental work was recognized even for NbTi superconductor to realize the practical and reliable machine. Although there are many difficulties in realizing Nb_3Sn machine, the advantage of Nb_3Sn superconductor when used in the generator was identified for its higher critical current density and higher critical temperature.

References

[1] J.L.Smith, IEEE Trans. on Mag., MAG-19,3,522-528(1983).
[2] S.Kurihara et.al. to be presented in this Symposium BE-7
[3] S.Kurihara et.al. to be presented in this Symposium BE-8

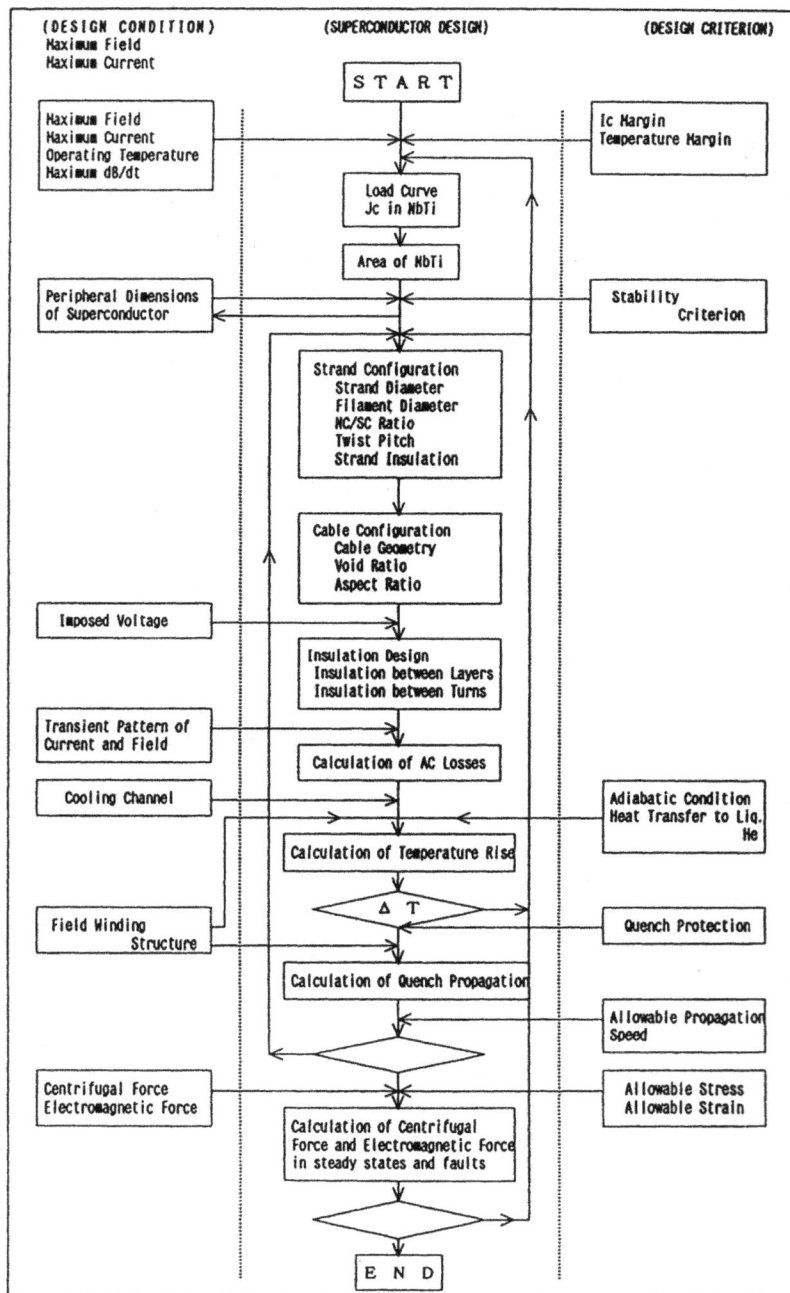

Fig. 1.   Design Procedures of Superconductor,
for Field Windings

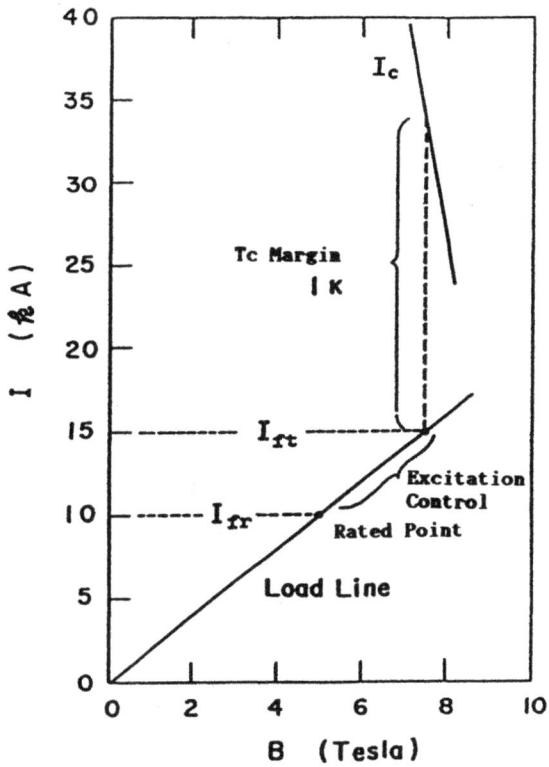

Fig. 2 An Load Curve for Field Winding

Table1. Development Items for NbTi Superconductor

	Contents of Development			Expected Merit				
	Improvement in Property	Long-Term Evaluation	Manufacturing technology	High Efficiency	Compact Geometry	Grid Stabilization	Cost Saving	Reliability
Jc	O	O		O	O		O	
AC Loss	O	O		O		O		O
Strength	O	O						O
Stability	O	O				O		O
Standardization			O					O
Quality Control			O					O
Cost Optimization			O				O	

CONCEPTUAL DESIGN
OF
275kV 300MVA SUPERCONDUCTING SHELL TYPE TRANSFORMER

S. HAYASHI, Y. SAIKAWA, T. KISHIDA,

THE KANSAI ELECTRIC POWER CO., INC.
11-20, Nakoji 3-chome, Amagasaki, Hyogo, 661 JAPAN

T. YAMADA, A. OHARA, M. MORITA, K. SHIONO*, T. KOHAN*, H. NAKAZAWA*

MITSUBISHI ELECTRIC CORP.
1-1, Tsukaguchi-honmachi 8-chome, Amagasaki, Hyogo, 661 JAPAN
* 651, Tenwa, Ako, 678-02 JAPAN

Abstract - A 275kV 300MVA superconducting shell type transformer was conceptually designed. The loss and the shipping weight of the superconducting transformer were compared with those of a conventional one. The loss is decreased to about 40 % at full load operation. The shipping weight is also decreased to about 40 %.

Introduction

Due to the development of ultra-fine filamentary superconductors, 50/60 Hz losses are greatly reduced and superconducting coils are prospective for AC applications [1-3]. The Westinghouse Electric Corporation group developed the a conceptual design for a 500/22kV 1000 MVA superconducting core type power transformer [4]. They concluded that an efficiency of the superconducting transformer was 99.85% and it had a weight advantage. Recently 220kVA and 72kVA superconducting core type transformers were constructed and tested in France and in Japan respectively [5-6].
This paper describes a conceptual design study for a 300MVA 275/77 kV shell type superconducting power transformer. The results are compared with the design of a conventional one. Advantages of a superconducting shell type power transformer become clear.

Features of conceptual design

Main features of the design are as follows which are introduced to reduce losses and a shipping weight.
-Low loss Nb3Sn superconducting cables are used : 6.6kW/m^3 at 0.1 T.
-Magnetic field in superconducting coils are decreased by increasing number of High-Low Coil groups : Less than 0.1 T.
-By the adoption of a copper machine design, a room temperature iron core with small cross section is used.
-Non metallic structural materials are used to reduce eddy current losses.
-Superconductors are designed to have very large critical current margin which prevents coils from quenching under fault conditions.
-Cryostats including superconducting coils are installed in the tank which has earth potential. SF6 gas is sealed in the tank for the insulation of high voltage bushings and cooling of a room temperature

iron core.
-The insulation system of superconducting coils is composed of solid insulators and liquid helium at 1 atmosphere.

Due to high critical temperature of Nb3Sn conductors, AC coils can be epoxy impregnated and solidified, and can be operated at the temperature slightly higher than 4.2K. This is a good method to prevent fine AC conductors from wire motions. Due to high critical current density of Nb3Sn conductors, it is possible to design coils which have enough current margin to carry fault current.

Conceptual design

The conceptual design of 275/77kV 300MVA shell type superconducting transformer is done under the conditions summarized in Table 1.

Table 1. Conditions for design

High voltage	275kV	300MVA
Middle voltage	77kV	300MVA
Low voltage	22kV	90MVA
Impedance		18%
Frequency		60Hz
Current density of a superconducting cable	about	120A/mm^2
Leakage flux density		0.1 T
AC losses of superconducting cables	6.6kW/m^3 at	0.1 T
Coefficients of performance of refrigerators		
liquid helium		1/500
liquid nitrogen		1/10

The conceptual view of a 300MVA superconducting shell type transformer is shown in Fig.1. Shell type transformers adopt an interleaved coil arrangement being easy to increase the number of H-L groups. Increasing the number of interleaving makes it possible to decrease the leakage magnetic flux density and losses in superconducting coils. The three phase superconducting transformer is installed in the tank containing SF6 gas and are connected at room temperature. Superconducting coils are fastened each other by outerbands to endure electromagnetic force. Figure 2. shows the cryostat structure. The material of the cryostat is GFRP. Superconducting coils are cooled by liquid helium. Liquid nitrogen is used for thermal shields.The heat leak of the cryostat is estimated and summarized in Table 2.

Table 2. Heat leak of cryostat

Temperature		77K	4.2K
Current leads	(W)	1260	30
Radiation	(W)	3970	36
Conduction	(W)	940	26
Delivery tubes	(W)	600	28
Total	(W)	6770	120

Heat leak from delivery tubes	77K : 10 W/m
	4.2K : 0.3W/m

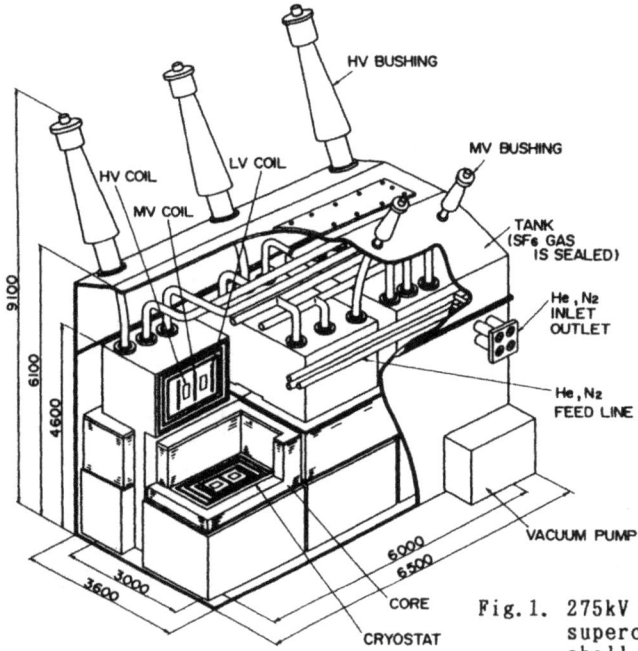

Fig.1. 275kV 300MVA superconducting shell type transformer.

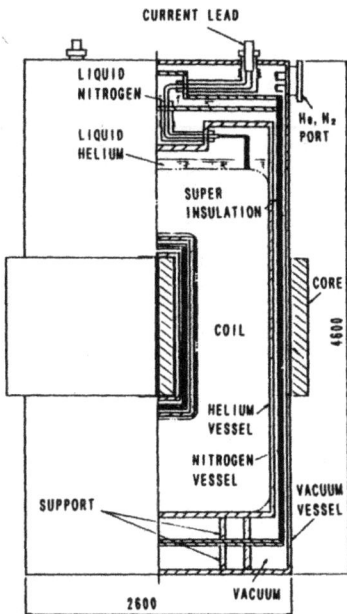

Fig.2. Cryostat structure

Figure 3. shows AC conductors. The diameter of a strand is 0.1mm. The filament size is of submicron. The conductors for 275kV and 77kV coils are composed of 3x7x6 strands and 7x7x9 strands respectively. These conductors are epoxy impregnated and solidified. AC losses of conductors are cooled by liquid helium through the epoxy resin and insulators around strands.

A 300MVA superconducting transformer is compared with a conventional one (Table 3.). A superconducting transformer has two obvious advantages: 1) The weight of a core and conductors are of very light-weight and the shipping weight is decreased to about 40% of the conventional one.

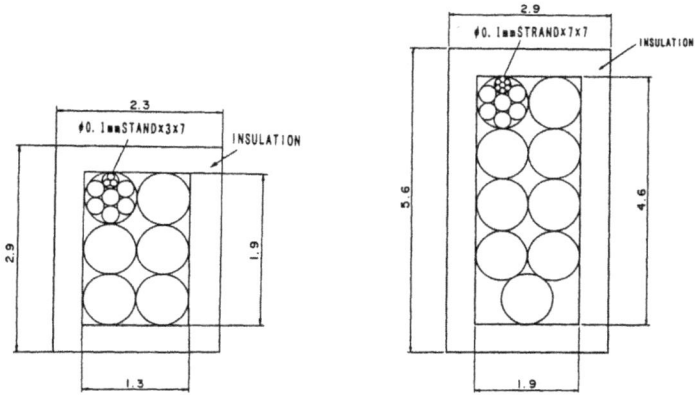

(a) Conductor for 275kV coils   (b) Conductor for 77kV coils
Fig. 3. Conductors

Table 3. Comparison of a 300MVA superconducting transformer with a conventional one.

Type		Superconducting	Conventional
Weight of materials			
Core	(kg)	22,000	83,000
Conductor	(kg)	4,000	30,000
Insulation	(ℓ)	8,000	12,000
Insulating oil	(ℓ)	—	44,000
Liquid helium	(ℓ)	12,000	—
Liquid nitrogen	(ℓ)	4,000	—
Shipping weight	(ton)	70	165
Number of H-L Coil groups		16	4
Loss			
No load loss	(kW)	40	160
Load loss *	(kW)	0.6	1,000
Auxiliary loss *	(kW)	450	40
Total loss	(kW)	490.6	1,200

* loss at full load

Conclusion

A 275kV 300MVA superconducting shell type transformer was conceptually designed. Advantages of a superconducting transformer are shown. The loss and the shipping weight of the superconducting transformer were compared with those of a conventional one. The loss is decreased to about 40 % at full load operation. The shipping weight is also decreased to about 40 %.

References

1) P.Dubots, et al.,"BEHAVIOUR OF MULTIFIRAMENTARY NB-TI CONDUCTORS WITH VERY FINE FILAMENTS UNDER A.C. MAGNETIC FIELDS",MT-8 Conference, 467-470, Grenoble Sept. 1983.
2) T.Ogasawara, et al.,"Recent Progress and Problems in the Development of Multifilamentary Composite Superconductors for Power Frequency Applications", CRYOGENIC ENGINEERING, Vol.22 No.2 76-91 (1987)
3) F.Sumiyoshi and K.Yamfuji, "SOME ELECTROMAGNETIC PROPERTIES OF MULTIFILAMENTARY SUPERCOMDUCTING WIRES",PROCEEDINGS OF INTERNATONAL SYMPOSIUM ON FLUX PINNING AND ELECTROMAGNETIC PROPERTIES IN SUPER-CONDUCTORS, 238-242, Fukuoka,Japan, Nov. 1985.
4) H.Riemersma, et al., "APPLICATION OF SUPERCONDUCTING TECHNOLOGY TO POWER TRANSFORMER", IEEE Trans. on Power Apparatus and Systems, Vol. PAS-100 No.7 3398-3405, July 1981.
5) A.Fevrier, et al.,"220 kVA SUPERCONDUCTING TRANSFORMER",ICEC-11, 474-478, BERLIN-WEST,GERMANY ,April 1986.
6) M.Iwakuma, et al.,"Fabrication and Preliminary Test of a 72kVA Superconducting Four-Winding Power Transformer",CRYOGENIC ENGINEERING Vol.22 No.6 354-361 (1987)

NB3SN SUPERCONDUCTING POWER TRANSMISSION CABLE

Noboru HIGUCHI, Naotake NATORI, Kazuaki ARAI,

Electrotechnical Laboratory
1-1-4 Umezono, Tsukuba-shi, Ibaraki, 305 JAPAN

and Tsutomu HOSHINO

Saga University
1 Honjo, Saga-shi, Saga, 840 JAPAN

Abstract - Characteristics of 10m-long Nb_3Sn superconducting power transmission cables are discussed. Cables are designed to be suitable for the power of 1 to 3 GW class and assembled using Nb_3Sn superconducting tapes. Current tests of those cables are carried out cooled with liquid or supercritical helium.

Details of cable design and experimental method are also described. Cable conductors are consisted of two layers of helically wound tapes, one for transport current and another for shielding. Electrical insulation is consisted of wrapped plastic tapes. Countermeasure against thermal contraction is one of the major factors in the design of this type of the cables, especially when superconducting materials applied are brittle.

The latest cable "N" is designed to suppress the stress inside the cable caused by thermal contraction, with a compromise between ideal design and restriction of cable assembly. The results of its current tests are also described, which proved the validity of the new design method.

Introduction

The annual increase of total energy consumption in Japan is slowing, such as 1-2% per year, however, the electric power demand in the biggest cities is showing different trend, because of high concentration of economic activities in those areas. Based on estimations on power demand in near future, it has been emphasized that the necessity of underground power transmission lines of large capacity, 1-3GVA, is going to be actualized.

In this situation, the advantages of superconducting power transmission lines, high power density, low losses, reasonable cooling distance etc., are of great significance to guarantee the transmission of a large block of power through densely populated areas, where big overhead lines can not be acceptable. On the other hand, results of economic evaluations performed so far indicate that superconducting transmission lines are not economically feasible, unless the capacities are over 5GVA or so.

As a result, it can be related that superconducting power transmission lines are superior candidates for power corridors, technically, but they need efforts to bring down the economical breaking point to the reasonable value, in order to bring the superiority into full operation. Therefore, the objective of our research is assigned on the fundamentals of superconducting cables suitable for the capacity of around 1-3GVA, with a rather simple structure, but accompanied with some technical difficulties.

Design and assembling of cables

Ac loss characteristics is one of the major factors to chose the material for conductors. Recently, NbTi wires with very low loss are available for 50-60Hz application, but the values are still too large to apply to transmission cables. Therefore, Nb_3Sn superconducting tape is still the first choice in spite of its brittleness, which may cause degradation of cable characteristics.

If oxide superconductors are available in future, Nb_3Sn tapes could be replaced with them without difficulty, since they are similar in the mechanical characteristics, and the same treatment to assemble cables may be applicable.

Cables are consisted of two layers of conductors, one for transport current, the other for shielding current to avoid eddy current loss in the cooling channels. They are helically wound with wrapped tape insulation between them, in order to make cables flexible. The lay angles of two helical superconductors are chosen to reduce the axial flux induced.[1] There exists another condition to be fulfilled, which is shown below as Eq. 1, to avoid damages which may be caused by the unbalanced thermal contraction inside the cable.

$$\alpha/\sin^2 \psi =const. \hspace{3cm} Eq. 1$$

α,ψ : coefficient of thermal expansion and lay angle

The latter condition is not consistent with the former one, and it is inevitable for the cables under consideration at present. This is the first problem to be solved to design the cable. The second one exists in the latter condition. It is not impossible to be fulfilled theoretically, but it requires to change the pitch continuously during the tape winding.

Compromising those problems, cable "N" was designed and assembled for current tests cooled with liquid helium. List of the cables tested so far and the details of the cable "N" are shown in Table 1 and 2 respectively.

Table 1. List of tested cables

cable	length(m)	O.D.(mm)	former	maximum current(A)	stabilizer
I	10	54.1	SUS rigid tube	6400	none
K	10	54.0	SUS helical coil	5500	none
L	10	57.0	SUS corrugated tube	5500	none
M	10	52.6	-	4500	yes
N	10	53.3	-	>4000	yes

To assemble the cable, SUS corrugated tube covered with SUS mesh was used as cable former, which is wrapped with tyvek tape for adjustment of diameter. Two layers of OFHC tapes were laid on the former as a stabilizer against fault current, which could be ten times as large as transport current in case of practical use. These OFHC tapes are preformed to round shape by rolling not to damage Nb_3Sn tapes above. Nb_3Sn tapes used were purchased, and laid as obtained. Four pairs of leads for loss measurement and four Au-Fe,Chromel thermocouples, of which cold junctions are set at the center space of the cable with two germanium thermosensors in total, were attached on those superconducting tapes.

Table 2. Dimensions of cable "N"

No.of layers	contents and materials	O.D.	winding direc. & pitch		tape size thick. x wid. (x pieces)
1	former,SUS corrugated tube with SUS mesh	27.0			
2-8	diameter adjuster,tyvek	28.7	LH	120	0.125 x 70
9	stabilizer,preformed OFHC tape	29.7	RH	250	0.3 x 6.5
10	-	30.2	LH	250	-
11	superconductor,Nb_3Sn tape	30.6	RH	250	0.12 x 6.45
12-22	insulation,tyvek	32.8	LH	112	0.125 x 70
23-28	-		RH	102	0.125 x 27 x 2
29-34	-		LH		-
35-40	-		RH		-
41-46	-		LH		-
47-52	-		RH		-
53-58	-		LH		-
59-64	-		RH		-
65-70	-		LH		-
71-76	-		RH		-
77-82	-	44.0	LH	64	-
83-88	-	45.0	LH	69	0.125 x 59
89	superconductor,Nb_3Sn tape	45.5	RH	250	0.17 x 5.0
90	stabilizer,OFHC tape	46.2	LH	250	0.3 x 6.5
91	-	47.3	RH	250	-
93	insulation(armor),tyvek	53.3			

(units in "mm")

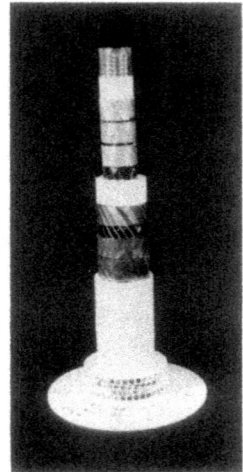

Fig. 1. Cable "N"

The material for electrical insulation is tyvek, but it does not mean this material is chosen as the final candidate. Because the purpose of the current test is to investigate the validity of the cable design scheme, consequently what we have to know concerning insulation materials is the mechanical characteristics. Since the cable is bounded for current test only, the maximum voltage to be applied is 15V. The design scheme will not be necessary to be changed when it is once established, except some parameters which may be modified when the material for insulation is decided finally.

The whole insulation layer is divided into 12 layers and each of the winding pitch is changed layer by layer, to make the lay angles close to the ideal value as shown in Table 1.

The next layer is a conductor consisted of Nb_3Sn superconducting tapes, which carries the shielding current. This conductor needs a

protection against fault current, also, and this is the role of the next
two layers, stabilizers. These are consisted of OFHC tapes same to the
stabilizers for transport current conductor. The final layer is a
protection armor, which guards the cable conductors from mechanical damages.
Fig.1 shows a model made from a cut piece of the cable "N".

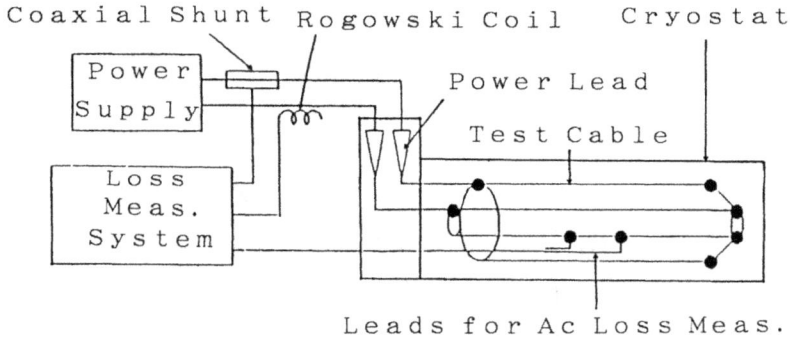

Fig. 2. Equipments for current test

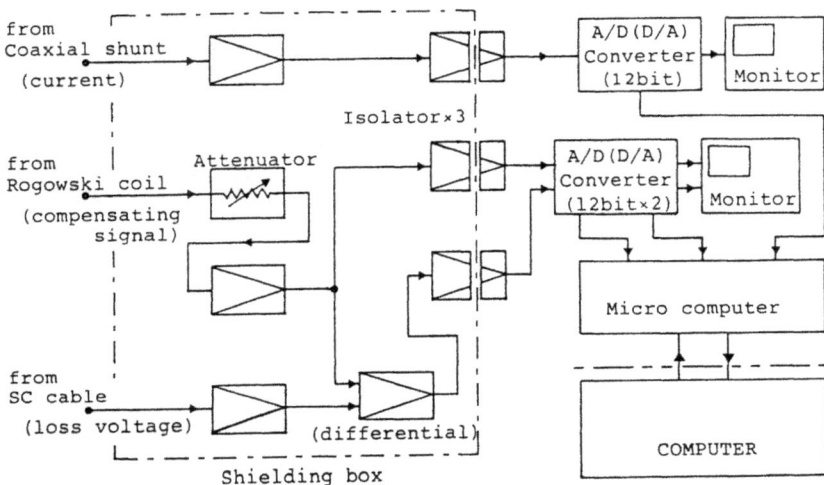

Fig. 3. Ac loss measurement system

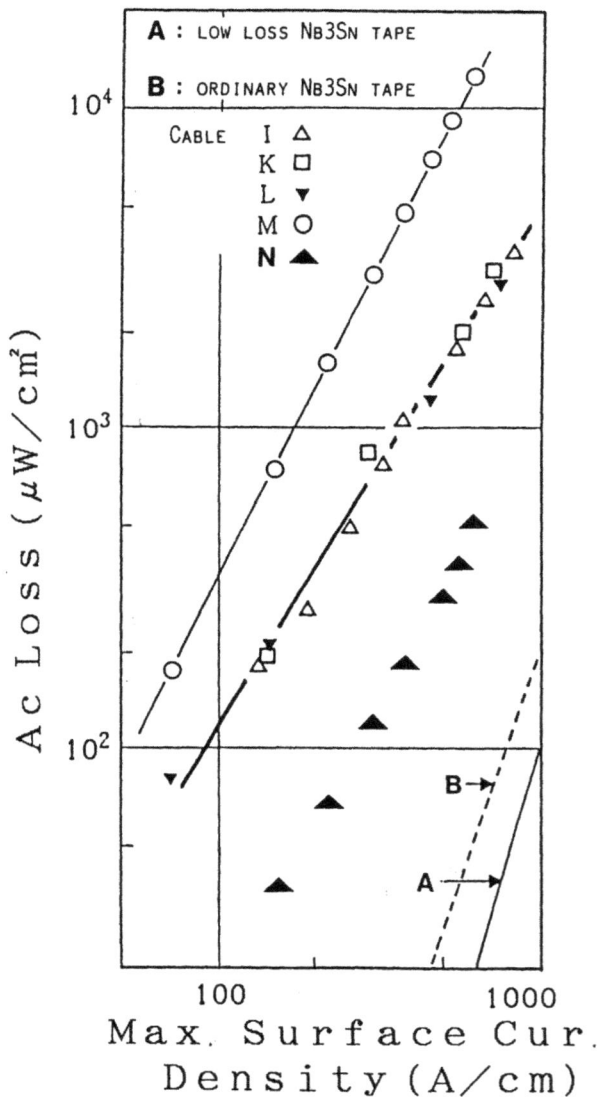

Fig. 4. Ac loss characteristics of cables

Experiments

The cable is 10m-long and short circuited at one end, and connected to power leads at the other end in the cryostat. The helium vessel is designed to endure the pressure over 10atm, and experiments can be performed with either liquid or supercritical helium. The power supply is a single phase transformer with the rating of 15V and 15kA. Fig. 2 and 3 show schematic diagrams of the equipments for current tests and loss measurement system, respectively.

Cable "N" was cooled with liquid helium, and ac loss was measured with the current up to 4000A. The result is plotted in fig. 4 together with the data of cables listed in Table 1. It shows that ac loss of cable "N" was successfully reduced compared with those of other cables, in case of which degradation of superconducting characteristics were found in the tapes taken from them after their disassembling. It may be concluded that the relief of stress in the cable did not affect superconducting characteristics of Nb_3Sn, not like other cables. Naturally, it needs the confirmation to bring to this conclusion. Current tests are planned to be repeated cooled with supercritical helium to obtain temperature characteristics, and later, the cable will be disassembled to take out samples of tapes to make it sure if the tapes are degraded,or not.

Conclusion

Ac loss characteristics of a superconducting cable are successfully improved compared with those of cables of similar design because of the relief of the internal stress probably, however, the loss in the cable is still about ten times larger than those of superconducting tape itself. Some possible reason have been considered, including edge effect of superconducting tapes in the cable, and further investigations will be necessary to establish technical feasibility of this type of cables.

References

1) Sutton,J. and Ward,D.A., "Design of flexible coaxial cores for ac sc cables", Cryogenics, vol.17,(1977)495-500
2) Forsyth,E.B. and Thomas,R.A., "Performance summary of the Brookhaven superconducting power transmission system", Cryogenics, vol.26,(1986)599-614
3) Higuchi,N. and Natori,N., "Nb_3Sn cables designed for ac current tests", Proc. of ICEC9,(1982)343-346

674

DEVELOPMENT AND TESTS OF EXTRUDED POLYETHYLENE INSULATED SUPERCONDUCTING CABLE

Masamitsu KOSAKI,   Masayuki NAGAO,   Yukio MIZUNO,
Kenji HORII* and Noriyuki SHIMIZU*

Toyohashi University of Technology, Toyohashi, Japan 440
* Nagoya University, Nagoya, Japan 464-01

Abstract   A superconducting power cable which has a structure similar to the conventional extruded polyethylene cable is proposed and the resulting laboratory size cable has been tested. The prominent feature of the design is to exploit the excellent electrical properties of polyethylene in the cryogenic temperatures and to separate the helium coolant from the electrical insulation. The cooling tests down to the liquid helium temperature, the current test up to 2.5kA and the voltage tests at the liquid nitrogen temperature have been carried out with fair success.

Introduction

The ever increasing electrical energy demand in the populated urban area requires high density reliable underground power cables instead of UHV overhead transmission lines with EMI and aesthetic problems. The superconducting cable is supposed to be the leading candidate among the future underground cables. Furthermore it has an indispensable role in the future superconducting power system including generator, transformer and energy storage apparatuses. The development activity of superconducting cable has been carried out at Brookhaven National Laboratory in U.S.A. where 100m cable has demonstrated distinguished capability[1]. However, its long term reliability especially in its electrical insulation and also its economic aspects need further improvement. The authors have long been proposing the idea of the extruded polyethylene insulation for superconducting cables[2]. The unique feature is the electrical insulation which takes full advantage of the excellent electrical properties of polyethylene at the cryogenic temperature region excluding liquid or supercritical helium.

In order to show the feasibility of the above proposal the superconducting cables with solid insulation were designed and fabricated. Low density or cross-linked polyethylene was chosen for the solid insulating material. This paper reports the development of superconducting cables and results of their cooling test down to the liquid helium temperature, current test up to 2.5kA and voltage test at the liquid nitrogen temperature.

Cable Design

The authors have been studying the thermal, electrical and mechanical properties of polyethylene(PE) for past several years[3,4] and drew conclusion that the superconducting cable insulated by extruded PE is feasible to be designed. As a following step, 15m ac superconducting cable test setup including two terminal bushings at ends was fabricated and installed by the cooperation of four cable manufacturing companies in Japan.

The structure of the cable in the first stage is illustrated in Fig.1 and the specifications are as follows: (a)voltage to ground, 20kV (35kV phase to phase); (b)transmitting current, 2kA; (c)cable length 15m, one piece; and (d)two terminal bushings.

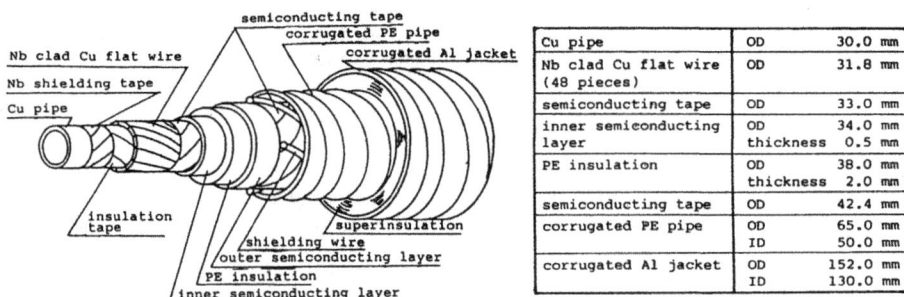

Cu pipe	OD	30.0 mm
Nb clad Cu flat wire (48 pieces)	OD	31.8 mm
semiconducting tape	OD	33.0 mm
inner semiconducting layer	OD thickness	34.0 mm 0.5 mm
PE insulation	OD thickness	38.0 mm 2.0 mm
semiconducting tape	OD	42.4 mm
corrugated PE pipe	OD ID	65.0 mm 50.0 mm
corrugated Al jacket	OD ID	152.0 mm 130.0 mm

Fig.1. Cable configuration.

Conductor Design

The superconductor is 10μm thick niobium which is clad over copper wire of 1.2 X 0.8mm. These 48 pieces of wires are helically wound around the copper core pipe. In order to eliminate the axial magnetic field on the cooper core pipe, a niobium tape of 10μm thick is helically wound between aforementioned main conductor and the core pipe in the opposite direction. The superconductors are indirectly cooled by the liquid helium in the core pipe. Since, there is no outer superconductor to shield magnetic field, eddy current loss is inevitably caused in the outer metallic parts like corrugated aluminum pipe. This is a common practice in the conventional cables. It can be seen that the conductor design is much simpler than other superconducting cables.

Electrical Insulation Design

The prominent feature of this cable lies in the electrical insulation composed of PE simultaneously extruded with both inner and outer semiconductor layers. Therefore liquid helium stays inside the core pipe solely as coolant and totally isolated from the electrical insulation. The idea is to exploit the distinguished electrical insulation capability of PE at cryogenic temperatures[5]. This structure enable the cable to be applied to any coolant depending on the superconductor used. The design stress is 10kV/mm. The most hazardous problem is the cracking due to the difference of thermal contraction of PE and conductor metals and also to cryogenic brittleness.

Thermal Insulation

The multi-layered superinsulation and vacuum constitute the thermal insulation and the liquid nitrogen thermal shield is not incorporated for the simple structure. The evacuation by rotary-diffusion pump sets attains to the order of 10^{-5}Torr. The corrugated PE pipe keeps the superinsulation layers in position and enhances the evacuation conductance. When the coolant flows through the core pipe, so-called cryopump effect can be expected.

Terminal Bushing

The normal conductor is made of copper braided wire which allows evaporated helium gas to flow through and to exchange heat generated by Joules' heating in itself effectively. The main electrical insulation consists of ethylenepropylene rubber(EPR) and vacuum. The surface flashover voltage along plastic and vacuum interface has been examined and found to have a sufficient withstand level in the present design. The thermal contraction of the cable at both ends is absorbed by the vacuum-tight bellows attached at the ends of the bushing.

Cooling and Voltage Tests

Cooling System

The cooling system shown in Fig.2 was used. The system consists of three parts: (1) cable test setup to be cooled; (2) controlled cooling apparatus using evaporated nitrogen gas; and (3) liquid helium supplier.

The cable setup (1) has two bushings which are named A and B. The bushing A is connected to (3) via a liquid helium (LHe) transfer tube. The bushing B is tied to (2) for the precooling down to about 100K. Nine pieces of temperature sensors measure the temperature profile of the cable. Six strain gauges attached on the outer semiconducting layer to monitor the axial and circumferential contraction of cable at both A and B ends and the center. Signals from these sensors and gauges are fed into the digital multimeter and then processed by the personal computer.

The cooling apparatus using evaporated liquid nitrogen (LN_2) gas was introduced to enable the slow and controlled cooling of the cable uniformly through 15m length. The LHe transfer tube is so designed that 20kVrms ac voltage can be applied during the LHe transfer. Therefore the double glass pipe sealed with vacuum is inserted in the central part of all metallic transfer tube. It was tested and found to have sufficient electrical insulation capability[6]. Evaporated He gas flows through the copper braided wire in the bushing for heat exchange and drives a gas flow meter. A LHe level sensor is set inside the core pipe of the B end to see if LHe fills the pipe all the way from the A end.

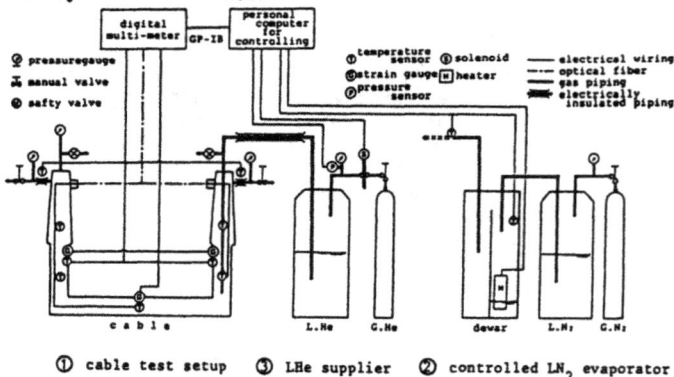

① cable test setup ③ LHe supplier ② controlled LN_2 evaporator

Fig.2. Cooling system.

Test Results

Though low density PE was first used as an electrical insulation of the cable, it turned out that cracking of the cable occurred around 170K during cooling. A new cable with 3.5mm thick cross-linked PE (XLPE) insulation was then developed taking the larger mechanical strength of XLPE into account.

The cooling was initiated by operating the controlled LN_2 evaporator and later by LHe supplied from the tank through the transfer tube. The temperature variation with time is shown in Fig.3. The cooling rate was set to 0.2K/min until about 100K. Though the B end, from which the cold nitrogen gas was introduced, was cooler than the other parts, it was confirmed that the temperature difference between the center and the A end of the cable was not large, showing that the slow cooling was carried out successfully as programmed. After the temperature of the core pipe reached 100K, the LHe transfer started. Soon the temperature of the A end, the LHe inlet showed 4.2K, as can be seen in Fig.3. Afterwards, the cold front advanced toward the B end.

During the cooling process of the cable, voltage test and measurement of loss tangent were done. The partial discharge experiments showed that the cable was free from partial discharges at the rated stress of 20 kVrms at 110K. The loss tangent of the cable was less than 10^{-4} at the liquid nitrogen temperature, which is a tolerable value for a superconducting cable.

Figure 4 shows the time variation of the contraction of XLPE due to the cooling. The contraction in the axial direction changed gradually and the difference between the ends and the center of the cable was very small. On the other hand, the contraction in the circumferential direction was relatively large. When the temperature of the cable reached about 40K, an anomalous sound was suddenly noted and the signals from the strain gauges

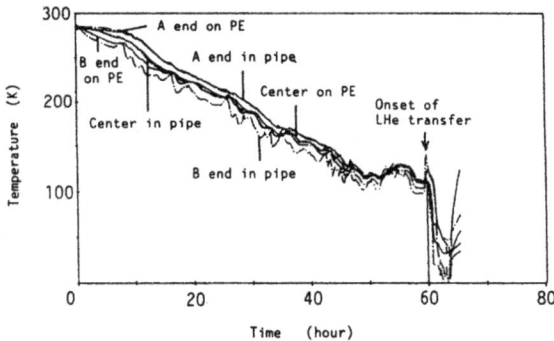

Fig.3. Variation of temperature during cooling.

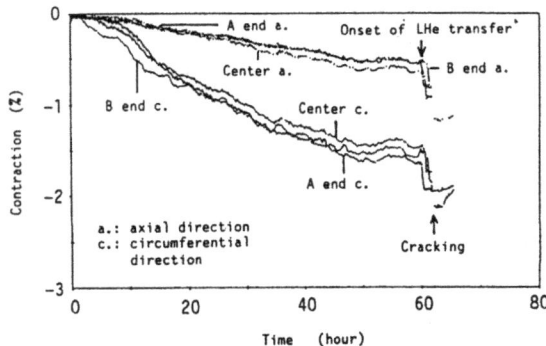

Fig.4. Variation of contraction of PE during cooling.

showed abnormal behavior. The cracking of the cable was confirmed by voltage application. The contraction in the axial direction was obviously small, which indicated that the residual stress in this direction had accumulated. Thus, the residual stress in the XLPE caused by the nonuniform thermal shrinkage eventually resulted in the cracking.

This cracking problem must be solved in view of the reliability during long time operation of the cable even if the cooling test were successful. Some attempts, such as the incorporation of the slipping layer between the metal conductors and the XLPE, and the straight installation of the cable, were unsuccessful in releasing the residual stress in the axial direction. One idea to overcome this problem is to introduce a loose contact of conductors and XLPE at room temperature which allows axial contraction at cooling. Another solution is to reduce the residual stress of the XLPE by increasing the thickness of it.

Although the cracking of the cable occurred at about 40K, this temperature is much lower than liquid nitrogen temperature, which is the operating temperature of the emerging ceramic superconductor. It is possible to say at this stage that the solid insulation design can be applied at least to the liquid nitrogen cooled cable.

Current Test

The current test was performed by the LC resonant circuit shown in Fig.5. Novel means was devised to feed large current to the cable. Namely, a distribution oil-filled transformer with a current capacity of 652A was directly cooled by LN_2 instead of oil. It has been experimentally tested that the current capacity can be increased by more than five times by this method.

At the first test the current was 1290A for fifteen minutes with steady

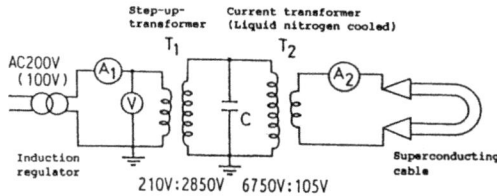

Fig.5. Current test circuit.

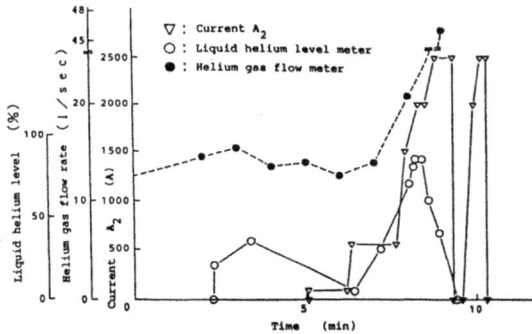

Fig.6. Results of current test.

supply of LHe. The rated current of 2000A was tested in the second trial successfully but when the current level was increased to 2500A sudden rise in the evaporation rate of LHe was observed as in Fig.6, apparently showing the quenching phenomenon (transition from superconducting to normal state).

It is difficult to explain this phenomenon but one probable reason of the quench may be due to the fact that the heat generated in the normal conductor in the bushing and inflow from the ambient could not be removed by boiling LHe at the junction of normal and super conductors. Another reason may lie in the enhancement of magnetic field at the edge of superconductors which caused the partial normal transition and the subsequent heat generation.

Conclusions

A superconducting power cable with extruded polyethylene insulation was designed, fabricated, installed and tested. The obtained results to date are as follows.
(1) The computer regulated cooling system was developed and slow cooling of the cable became possible.
(2) Liquid helium could fill the copper core pipe of 15m cable and the good thermal insulation was possible without liquid nitrogen thermal shielding layer.
(3) The loss tangent of the cable is less than 10^{-4} at the liquid nitrogen temperature.
(4) A liquid nitrogen cooled transformer with current up to 2500A was devised.
(5) The rated current of the cable 2000 A was tested with success.
(6) The voltage test of 20kVrms was performed with success at the liquid nitrogen temperature.

Acknowledgment

The authors are greatly indebted to The Furukawa Electric Co. Ltd., The Fujikura Cable Works Ltd., Showa Wire and Cable Co. Ltd. and Mitsubishi Cable Industry, Ltd. for their devoted cooperation in designing, fabrication and installation of the cable. Thanks are also to Chubu Electric Power Co. Inc., Nissin Electric Co. Ltd., Aichi Electric Co. Ltd. and Aichi Clock and Electric Co. for their support toward this project.

References

(1) E.B.Forsyth and R.T.Thomas, "Performances and Summary of the Brookhaven Superconducting Power Transmission System," Cryogenics, vol.26, pp.599-614, 1986.
(2) M.Kosaki and K.Horii, "A Design of Polyethylene Insulated SubGVA Superconducting Cable," presented at ICEC9, 146, Kobe, Japan 1982
(3) N.Shimizu et al., "Thermal Contraction and Cracking of Extruded Polyethylene Electrical Insulation at Cryogenic Temperature," Cryogenics, vol.26, pp.459-466, 1986.
(4) J.H.Hongoke et al., "The Dielectric Loss Tangent of Extruded Polyethylene Cables at Cryogenic Temperature," Trans. IEE of Japan, vol.105, pp.31-34, 1985.
(5) M.Kosaki et al., "Treeing Phenomena of Polyethylene at Cryogenic Temperature Region," Trans. IEE of Japan, vol.95-A, pp.292-299, 1975 in Japanese.
(6) Y.Mizuno et al., "Electrical Performance of Glass Insulated Transfer Tube of Cryogenic Liquid," Cryogenic Engineering, vol.23, pp.145-151, 1988 in Japanese.

NEW POWER GENERATION SYSTEMS BY SUPERCONDUCTING
ELECTRO-MAGNETIC PUMPING-UP DEVICE

Eiichi TADA, Tomomasa UEMURA, and Kensaku IMAICHI

Mechanical Eng., Faculty of Engineering Science, Osaka University
Machikaneyama-cho 1-1, Toyonaka, Osaka, JAPAN

Abstract - The electromagnetic thruster (EMT) has a poten tiality of the transportation system in the sea and the power generation system. The EMT propulsion unit is able to be used as the sea-water pump and the sea-water MHD generator.

Introduction

There are many kinds of methods of energy storage system; for exsample, superconductive magnetic energy storage system and pumping-up hydro-generation system. In this paper, we are studying a new pumping-up hydro-generation system with a superconducting electro-magnetic thruster(EMT). In Japan, the experimental EMT ship with two superconducting magnets will be sailed in 1990. Using the reverse principle of EMT, namely Fleming's right hand rule, it is possible to be used as sea-water MHD for power generation. An EMT is able to be used as the thrust pump and the MHD generator. In the sea-water MHD generator, sea-water flows with a velocity U across a magnetic field B, creating induced voltage of UxB and current density of $J = \sigma (E+UxB)$, where E is the electric field, and σ is the electric conductivity of sea-water.

The high conductivity is important for energy saving in order to obtain high energy efficiency. The electric conductivity of real sea-water is not so large, 5 (S/m). In our closed cylce system, the high salinity gains high conductivity on 25 (S/m).

EMT transfers electromagnetic energy direct into thrust and kinetic energy of fluid direct into electric energy without the need of moving parts such as propeller and shafting. This type of power generator has following advantages;

1) no moving parts;
2) simple construction and maintenance;
3) production of hydrogen gas which can be stored easily.

Basic Analysis

Generator characteristics

Under the magnetic field B (T), the sea-water is flowing in the duct, as shown in Fig. 1. The electric voltage and current are obtained by solving the Navier-Stokes's equation and the continous equation and Ohm's law equation. Now supposing the velocity of sea-water in the duct is constant, the velocity w (m/s) of sea-water in the duct is gained by the following equation;

$$w = Cv \sqrt{2g\,H} \qquad\qquad (1)$$

where H (m) is water head and Cv is coefficient of velocity.

The output voltage V (V) and current I (A) are obtained by the following relations;

$$V = E\,d = K\,w\,B\,d \qquad\qquad (2)$$

$$I = \sigma\,(1 - K)\,w\,B\,b\,l \qquad\qquad (3)$$

where b, d, and l is the length of generator duct as shown in Fig. 1, E (V/m) is electric field strength, and K is defined by

$$K = E\,/\,w\,B \qquad\qquad (4)$$

As the results, the output power Wg (W) is gained.

$$Wg = \sigma\,K\,(1 - K)\,w^2\,B^2\,b\,d\,l \qquad\qquad (5)$$

As the velocity w is constant in this type generator, it is possible to be gained the maximum output at K = 0.5.

$$Wgm = (1/4)\,\sigma\,w^2\,B^2\,b\,d\,l \qquad\qquad (6)$$

$$= (1/2) \; \sigma \; Cv^2 \; (\; bdl \;)(\; gH \;)^2 \; B^2 \qquad (7)$$

The kinetic energy Ww (W) of flowing sea-water is

$$Ww = (1/2) \rho \; w^3 \; b \; d \qquad (8)$$

where ρ is density of sea-water.
The power generator efficiency η g can be calculated.

$$\eta \; g = \frac{1}{2} \cdot \frac{1 \; B^2}{\rho \; w} \qquad (9)$$

Pump characteristics

When both the electric field E and the magnetic field B are acted to the sea-water, the sea-water is pressed by Lorentz's force. The water head of sea-water is obtained by solving the same equations for MHD analysis.
From momentum balance of fluid in EMT pump duct,

$$\rho \; g \; H \; b \; d \quad + \quad fr \quad = \quad F \qquad (10)$$

where F (N) is the total Lorentz's force and fr (N) is friction loss in the duct. The transforming factor η f is defined by the ratio of the momentum of sea-water to the Lorentz's force.

$$\eta f = \quad \rho \; g \; H \; b \; d \; / \; F \qquad (11)$$

The Lorentz's force F (N) can be calculated.

$$F = J B b d l \qquad (12)$$

where J (A/m^2) is the curret density.
The electrode voltage V (V) and current I (A) are gained.

$$V = J b d \qquad (13)$$

$$I = J d / \sigma \qquad (14)$$

The required electric power P (W) for sending the electric current into the sea-water is

$$P = I V = J^2 b d l / \sigma \qquad (15)$$

$$= (\rho g H)^2 b d / \sigma \eta_f^2 B^2 l \qquad (16)$$

The pumping efficiency η_p is calculated by the following equation;

$$\eta_p = 9800 Q H / P \qquad (17)$$

where Q (m^3/s) is flow mass of sea-water.

Calculation of 30000 KW

For the purpose of making EMT practical use, we will study the EMT power generator about 30000 KW. For this generator, the calculation conditions are the following;

$$H = 130 \qquad (m)$$

$$b \; = \; d \; = \; 0.717 \quad (m)$$

$$\sigma \; = \; 25 \qquad\qquad (S/m)$$

$$\rho \; = \; 1025 \qquad\quad (kg/\;m^3)$$

$$Q \; = \; 26 \qquad\quad\;\; (m^3/s)$$

$$Cv \; = \; 0.95$$

From EMT propulsive experimental results, the ηf can be given by the following;

$$\eta f \; = \; 0.98 \quad .$$

The calculated results are shown in Figs. 2 and 3. These figures say that it is possible to obtain the high efficiency when the magnetic field is 10 teslas.

Conclusion

We have studied the basic analysis for electromagnetic generator. It is ascertained that the present level of superconducting is sufficient in order to make the EMT generator with high efficiency practicable. We expect to a way of EMT generator practical use.

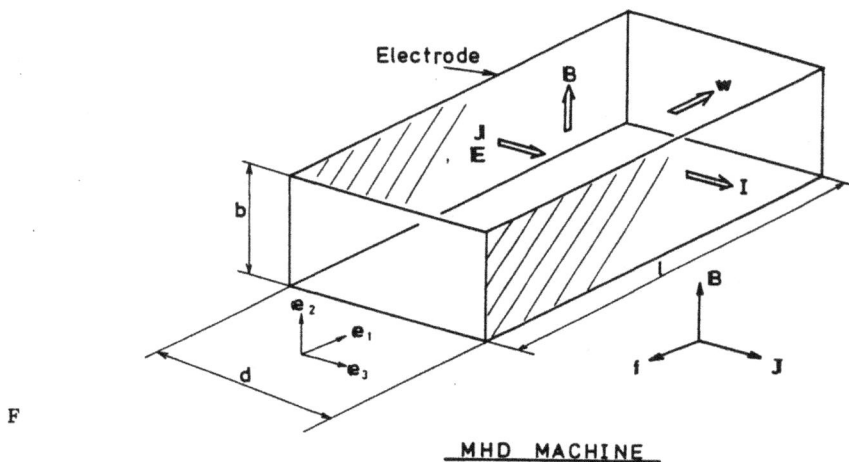

F

MHD MACHINE

Fig. 1   MHD generator

Fig. 2   Relation between duct length and magnetic field
         in case of generator

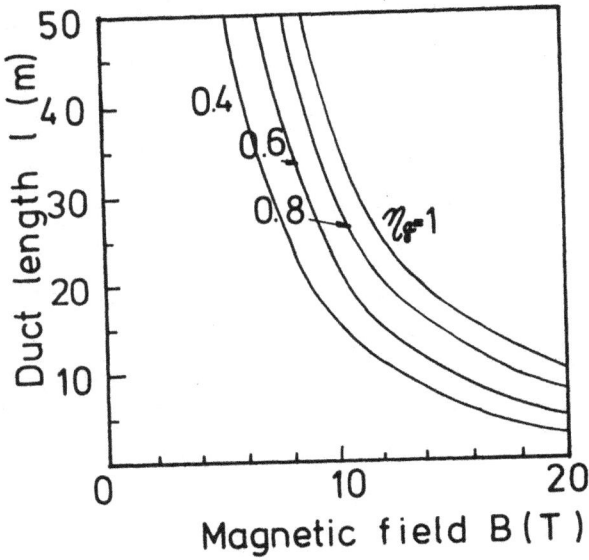

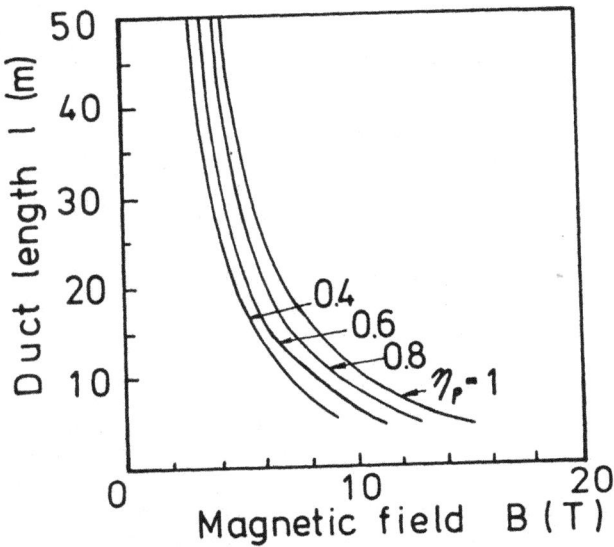

Fig. 3   Relation between duct length and magnetic field
         in case of pump

JAPANESE EXPERIMENTAL SHIP WITH THE SUPERCONDUCTING ELECTRO-MAGNETIC THRUSTER

*   *
Yohei SASAKAWA, Kensaku IMAICHI, Eiichi TADA, and Setsuo TAKEZAWA

Japan Foundation for Shipbuilding Advancement
Toranomon 1-15-16, Minatoku, Tokyo, JAPAN
* Faculty of Eng.Science, Osaka University
Machikaneyama-cho 1-1, Toyonaka, Osaka, JAPAN

Abstract - Superconducting Electro-Magnetic Thruster ship is attracting particular interest in the field of Superconducting technology. This technology creates the possibility of producing a lightweight superconducting magnet capable of generating a strong magnetic field enough to go ahead a ship. The JAFSA in Japan intends to operate sailing tests of the world's first prototype EMT experimental ship in 1990.

Principle and Development of EMT

The EMT propulsive unit with no screw and conventional machine is propelled by Lorentz's force generated by the interaction of a magnetic field made by superconducting magnet and electric current flowing through sea water from a pair of electrodes.

The EMT ships can be sailed by the jet of seawater moving astern like as water jet propulsion.

An EMT ship can be expected to have the following advantages:

1. Noise and vibration - free propulsion.
2. High controllability
3. High thrust efficiency
4. High thrust at bollard condition
5. Simple construction and maintenance free.

In 1968, Stewart Way[1] of Westinghouse Research Lab. built the ESM-1, the first model ship with an ordinary conducting coil, and tested it off the coast of California In 1973 and 79, Prof.Saji[2], A.Iwata, and E.Tada made two superconducting model ships, SEMD1 and ST-500. They studied the basic theory on EMT by theoretical and experimental investigations of SEMD-1 and ST-500. They have designed first practical full-scaled EMT icebreaker, ST-4000B.[3,4]

EMT Propulsion Systems and Project Program

The Japan Foundation for Shipbuilding Advancement (Chairman: Ryoichi Sasakawa) established the Research and Development Committee for Superconducting Electro Magnetic Thruster Ship (Chairman: Yohei Sasaskawa), whose Committee has two subcommittees; Hull Subcommittee (Chairman: Seizo Motora, Professor Emeritus of Tokyo University) and Equipment Subcommittee(Chairman: Kensaku Imaichi, Professor Emeritus of Osaka University).

In 1990, the Committee intends to conduct the first demonstration tests of experimental EMT ship. This ship will be laden with all necessary machinery and equipment. The experimental ship powered by EMT has the following characteristics:

Displacement	150 tonnes,
Equipment	100 tonnes,
Hull	50 tonnes,
Propulsive Thrust	8000 Newtons,
Engine	Inner Magnetic EMT,
Speed	8 knots,
Crew Complement	10 persons.

We made three Research and Development Dipole superconducting magnets in order to construct large dipole magnets of EMT propulsive experimental units. The total length of R&D-1 dipole magnet is 1.55 meters, straight part one meter, inner and outer diameter of this dipole 0.32 and 0.34 meters, single layer. The magnetic flux density of central part of sea-water duct is two teslas. The superconducting composite wire is consisting of NbTi filaments in a copper matrix and 550 filaments in a 0.775 milimeters diameter wire and 25 wires in a 10x15.7 milimeters retangular conductor.

We have studied the thrust experiments by R&D-1. The experimental results show that the thrust force transforming efficiency, defined by the ratio of the sea-water flow momentum to the Lorentz's force, is very high, so that this efficiency achieves over 95 percents. Because EMT propulsive unit has no mechanical devices which generate frictional loss, hydraulics loss and cavitation, and also electromagnetic force is the body force like as the gravity force.

Now we are making two R&D dipoles of double layer pancake winding, R&D-2, in order to obtain higher magnetic flux density of 4 teslas. Generally speaking, in straight-sided coils such as dipoles, the force problem is different from that in solenoids. It is more difficult because the conductor is unable to support the electromagnetic forces in tension. These forces must therefore be transmitted through the winding to external support structure. The R&D-1 dipole is very heavy because the electromagnetic force supporter is used a conventional collor structure, Fermi Lab., like as an accelerator dipole. We must study a new electromagnetic force support structure to lighten coil weights and to increase the intensity of magnetic field. We are now making two kinds of dipoles with new electromagnetic force support structures; one is by Al binding structure, the other by cast structure.

The experimental EMT ship has two EMT propulsion units. The propulsion unit is consist of a large superconducting magnet and a pair of electrode, and belongs to an inner magnetic field type EMT. In the inner type, the active electromagnetic region is restricted to the duct of ship's hull and the thrust is obtained by electromagnetically jetting water from the duct, which has the advantage that the magnetic field cannot leak out of the ship and that the thrust efficiency is very high.

The requirements of an EMT superconducting magnet are:
1. Light weight.
2. Stabilized against pitch and roll.
3. High magnetic flux density.
4. Full stabilization for heat generation.
5. Persistent Current mode.

The superconducting magnet of the experimental ship is made up six dipoles, shown in Figure 2, that are arranged as a round table. Each dipole, shown in Table 2, is 3.74 meters in length, double layer winding with Nb-Ti and PC mode. This magnet is 5 meters in length, 1.8 meters in diameter, 15 tonnes in weight, and the heat loss is about 10 watts at 4 K. Holding the cryostat at liquid Helium temperature, we are developing a small on-board Helium refrigerator with two micro turbo expanders of 6 milimeters in diameter and 600000 rpm.

In the d.c. EMT, a d.c. electric current is passed into the sea-water and an electrochemical reaction occurs on both surfaces of electrodes, producing hydrogen and chlorine gas from the anode and cathod respectively. The chlorine gas is dissolved in the sea-water. A new electrode material which generates oxygen gas instead of chlorine gas has been produced experimentally.

Before sailing a commercial EMT vessel, it is concluded that high magnetic flux density, about 20 teslas, is essential in order to obtain high thrust power efficiency, and the high electric conductivity of sea-water is important for the energy saving. Under the present magnet technical level, it is possible to obtain the magnetic flux density of 10 teslas, using Nb_3-Sn wire or Nb-Ti wire at 1.8 K and using the electromagnetic force support structure being developed by EMT project.

References

1). Way,S.   and Devin,C."Prospects   for   the   Electro-magntic
    Submarine", 67-432, AIAA(1967)
2). Iwata,A.,   Tada,E. and Saji,Y.,"Experimental and Theoretical
    Study of Superconducting Electro-magnetic ship Propulsion",
    5th Lips Propeller Symp.,2 (1983)
3). ·Tada,E.   and Saji, Y.,"The Prototype Bipolar Superconducting
    EMT --ST-10B", Proc.ICEC 10(1984)
4). Tada,E. and Saji,Y.,"Fundamental Design of a Superconducting
    EMT Icebreaker",Tras.IMarE,97, 6(1984)

Principle of propulsion

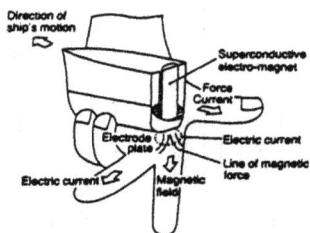

Figure 1. Principle of EMT Propulsion

690

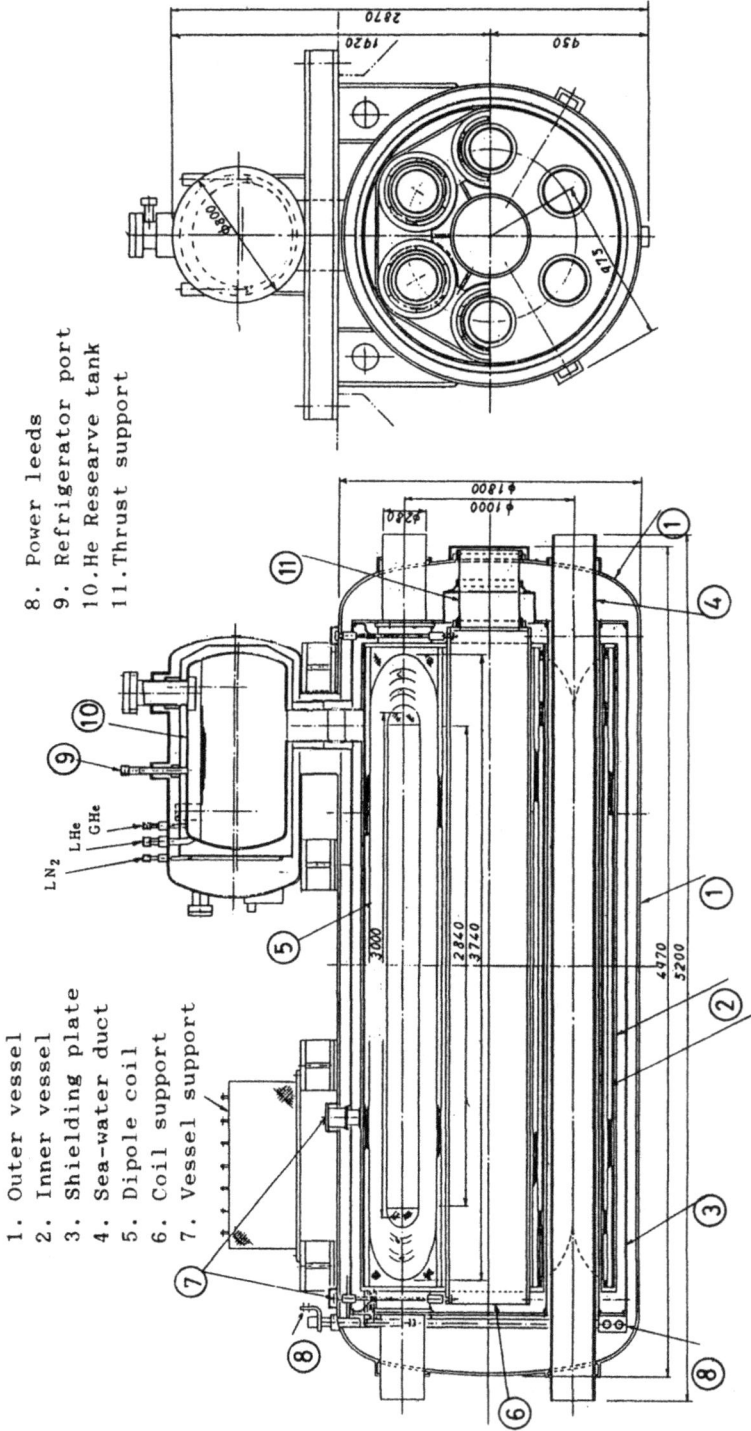

Figure 2. Superconducting Six Dipoles Magnet of EMT
Propulsion Unit

1. Outer vessel
2. Inner vessel
3. Shielding plate
4. Sea-water duct
5. Dipole coil
6. Coil support
7. Vessel support
8. Power leeds
9. Refrigerator port
10. He Reserave tank
11. Thrust support

Table 1. Specification for Superconducting Dipoles

One Dipole	
Length (m)	3.740
Straight part of length (m)	3.000
Outer diameter of winding (m)	0.40
Inner diameter of winding (m)	0.36
Number of turns	253
Magnetic flux density (T)	4.0
Conductor size (mm)	1.348/1.432 x 9.84
Angle of keystone	0.48
Wire diameter (mm)	0.772
Number of wire	26
Stabilized material	Cu
Material	Nb-Ti
Maximum current (A)	8500 at 6.5 T

Six Dipoles Unit	
Length of cryostat (m)	4.970
Diameter of cryostat (m)	1.8
Magnetic flux density (T)	
sea-water duct	4.0
maximum at wire	5.9
Total current (A)	3271
Current density (A/mm^2)	210
Self-inductance (H)	4.2
Stored energy (MJ)	22.5

EFFECT OF STRONG MAGNETIC FIELDS ON RELATIVE OXYGEN/CHLORINE EVOLUTION EFFICIENCIES IN MAGNETOELECTROLYSIS OF SODIUM CHLORIDE SOLUTIONS

M.Hiroi*[1], M.Muroya*[2], E.Tada*[3], and Y.Takemoto*[1]

*[1] Kobe University of Mercantile Marine
(5-1-1 Fukae-Minami, Higashinada, Kobe 658, Japan)
*[2] Faculty of Engineering, Osaka Electro-Communication University
(18-8 Hatsumachi, Neyagawa, Osaka 572, Japan)
*[3] Faculty of Engineering Science, Osaka University
(1-1 Machikaneyama, Toyonaka, 560, Japan)

Abstract - The effect of a magnetic field of 5T on anodic reactions in electrolysis of 3.5% NaCl solutions was investigated by measuring relative oxygen/chlorine evolution efficiencies at a Pt plate anode. No significant difference in the oxygen evolution efficiency between in magnetoelectrolysis at 5T and in the zero magnetic field was observed for an electrolysis time of 30 sec. When magnetoelectrolysis at 5T was carried out for 90 sec the oxygen evolution efficiency became higher by about 5% than that at 0T. The imposed magnetic field seems to have an effect of increasing the oxygen evolution efficiency, cancelling out the effect of the decrease in pH of the anolyte during electrolysis.

Introduction

The major practical advantage of electrolysis in magnetic fields is an increase in the mass transport rate as a consequence of the magnetohydrodynamic (MHD) effects [1]. A magnetic field, coupled with an electric field, generates the solution flow, improving the mass transport rates. Relatively little is known about the effect of magnetic fields on the electrode kinetics. Some investigators have reported experimental results suggesting that the imposed magnetic fields can affect electrode processes by promoting the orientation of reaction species in the diffusion layer or the electrical double layer[2,3].

In this work, the effect of a magnetic field of 5T on the oxygen/chlorine evolution rates at a Pt anode was investigated in magneto-electrolysis of 3.5% NaCl solutions of different pH. Whether or not strong magnetic fields would influence the oxygen evolution efficiency is of importance in connection with superconducting electromagnetic propulsion technology and the kinetics of electrode processes.

Experimental procedure

Sodium chloride solutions (3.5%) of pH 5.0 and 9.5 were electrolyzed at a current density of 3000 A/m² in the presence of a magnetic field of 5T. The volume of the anolyte used was 40 cm³. Pt plates were used as anode and cathode. The working area of the anode was $1.5 \sim 1.7$ cm^2. The amounts of chlorine evolved were determined by iodimetric titration to obtain Faradaic efficiencies for chlorine. Oxygen evolution was assumed to consume the balance of the anodic current.

The vertical magnetic field was produced from a superconducting magnet which could generate magnetic flux densities up to 5T at the center of cylindrical space (80 mm in diameter) in a cryostat (Fig.1).

The electrolytic cells used were two-compartment cells divided by a glass filter. One of them was H-type as shown in Fig.2 (a). With

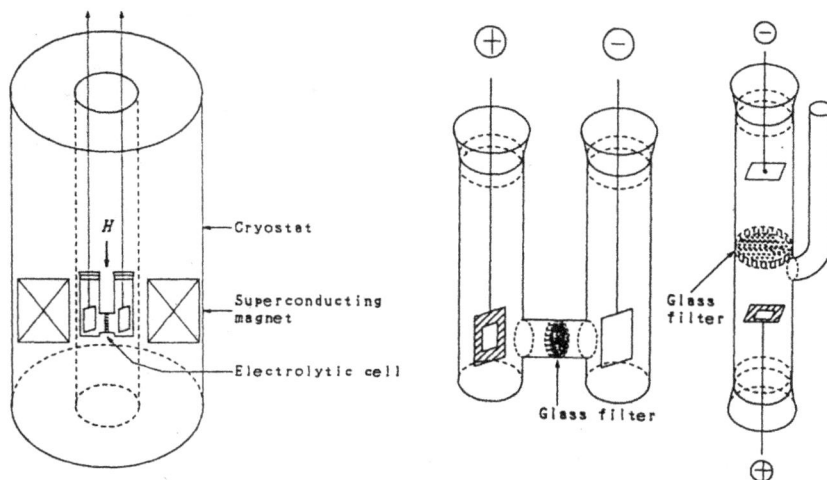

Fig.1 Schematic diagram of the equipment for magnetoelectrolysis.

(a) H-type cell (b) Cylindrical cell
Fig.2　Electrolytic cells used.

this cell, the effect of the magnetic field parallel to the anode surface and perpendicular to the electric field could be examined. The other type of cells used was cylindrical as shown in Fig.2(b). With this cell, the magnetic field could be imposed perpendicularly to the anode surface. The electrolysis was carried out at room temperature (26 $\sim$ 29 ℃).

Results and discussion

Oxygen evolution efficiencies in the cylindrical cell

The effect of the imposed magnetic field perpendicular to the anode surface on the oxygen evolution efficiency is shown in Figs.3 and 4 for 3.5% NaCl solutions of pH 5.0 and pH 9.5, respectively.

In the absence of the magnetic field the oxygen evolution efficiency decreased with electrolysis time. This decrease in the oxygen evolution efficiency is attributable to the decrease in pH of the anolyte due to both oxygen and chlorine evolution. When chloride solutions are electrolyzed both chlorine and oxygen normally evolve at Pt anodes. The oxygen evolution reaction generates H$^+$. In addition to this, any chlorine generated at the anode dissolves in the anolyte to generate H^+ as follows,

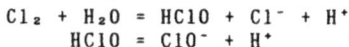

$$Cl_2 + H_2O = HClO + Cl^- + H^+$$
$$HClO = ClO^- + H^+$$

As the anolyte becomes more acidic, the thermodynamic potential for oxygen evolution becomes more anodic, a trend favoring chlorine evolution which is independent of pH. Therefore, as electrolysis proceeds the anolyte becomes more acidic and this causes the oxygen evolution rate to become lower[4].

In the presence of the imposed magnetic field the oxygen evolution efficiency increased or did not change significantly with time. No significant effect of the imposed magnetic field was observed at the initial stage of the electrolysis, but at 90 sec the effect became apparent, the oxygen evolution efficiency at 5T became higher by about 5% compared with the one without the imposed magnetic field.

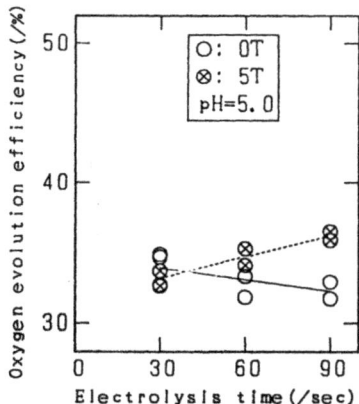

Fig.3 Oxygen evolution efficiency measured in the cylindrical cell. Electrolyte; 3.5% NaCl. Current density; 3000A/m².

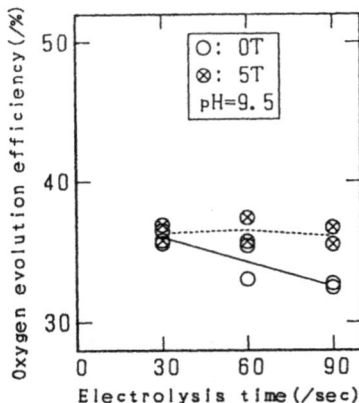

Fig.4 Oxygen evolution efficiency measured in the cylindrical cell. Electrolyte; 3.5% NaCl. Current density; 3000A/m².

Oxygen evolution efficiencies in the H-type cell

Figures 5 and 6 show the change in the oxygen evolution efficiency with time in electrolysis carried out in the H-type cell. No apparent magnetic field effect was observed for the electrolysis of 3.5% NaCl solution of pH 5.0. In the electrolysis of the solution of pH 9.5 a similar effect to that obtained in the cylindrical cell was observed.

It can be concluded from these data that the superimposed magnetic field increases the oxygen evolution efficiency. The effect was observed in both the cylindrical cell and the H-type cell. Therefore, this magnetic field effect cannot be explained in terms of only the MHD effect. In the H-type cell where magnetic field is

perpendicular to the electric field, the highest MHD effect is
expected, but in the cylindrical cell where the magnetic and the
electric fields are pallalel, the marked MHD effect is not expected.
We have no clear explanation for our experimental observations at the
present time. Although we can assume that the strong magnetic field
can promote to align reaction species or intermediates in the
electrical double layer on the Pt anode, such assumptions have to be
discussed after examining the magnetic field effect under various
experimental conditions.

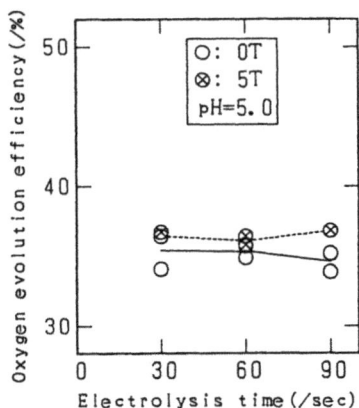

Fig.5 Oxygen evolution efficiency
measured in the H-type cell.
  Electrolyte; 3.5% NaCl,
  Current density; 3000A/m².

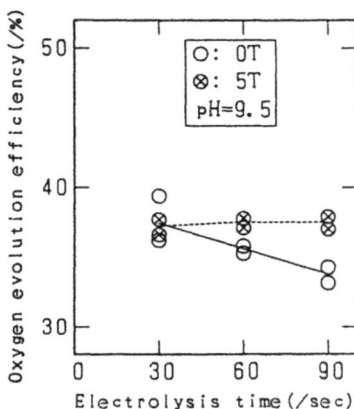

Fig.6 Oxygen evolution efficiency
measured in the H-type cell.
  Electrolyte; 3.5% NaCl,
  Current density; 3000A/m².

Conclusion

It has been found that the imposed magnetic field has an effect
of increasing the oxygen evolution rate at the Pt anode in
electrolysis of 3.5% NaCl solutions. This is an interesting
observation for the study of electrode kinetics, and this is favorable
for the electromagnetic thruster since it is desirable not to evolve
large volumes of chlorine from the thruster.

Acknowledgement

Financial support of a part of this work by Japan Foundation for
Shipbuilding Advancement is gratefully acknowledged. The authors wish
to express appreciation to Mr.H.Akazawa, Mr.S.Ohota, Mr.H.Konishi and
Mr.H.Maekawa of Osaka Electro-Communication University for kind
cooperation in the experiments.

References

1) T.Z.Fahidy. J.Appl.Electrochem., 13, 553 (1983).
2) F.Takahashi and Y.Sakai. Denki Kagaku. 49, 196 (1981).
3) F.Takahashi. K.Tomii and H.Takahashi. Electrochimica Acta. 31. 127 (1986).
4) J.E.Bennett. Int.J.Hydrogen Energy. 5. 401 (1980).

THE DEVELOPMENT OF ELECTRODES FOR USE IN THE SUPERCONDUCTING ELECTROMAGNETIC PROPULSION SHIP AND ITS EVALUATION

Masa-aki MUROYA[1], Masao HIROI[2], Soichi OGAWA[3], Eiichi TADA[4] and Yasuyuki TAKEMOTO[2]

[1] Osaka Electro-Communication University, 18-8, Hatsumachi, Neyagawa, Osaka, 572 Japan.
[2] Kobe University of Marcantile Marine, 5-1-1, Fukaeminami, Higashinada, Kobe, 658 Japan.
[3] Osaka Prefectural Industrial Research Institute, 2-1-53, Enokojima, Nishiku, Osaka, 550 Japan.
[4] Faculty of Engineering Science, Osaka University, 1-1, Machikaneyama, Toyonaka, 560 Japan.

Abstract - A screening test of the anode for use in the electromagnetic thruster was carried out. We have found some anode matrials which evolve oxygen at current efficiencies of higher than 85 %. Those are manganese oxide-coated DSA's. We are going to use one of them in the thruster. The influence of manganese oxide coating on the polarization curve was investigated to explain the increase in the oxygen evolution efficiency. The influence of flow rate of the electrolyte and of an imposed magnetic field of 5T on the concentration of hypochloric acid in the electrolyte was also investigated to know the electrode characteristics in the electromagnetic thruster.
-DSA is a registrated trademark of DST,SA.

Introduction

It is well known that the concept of electromagnetic thrust(EMT) has been proposed in 1961 by W. A. Rice, and he obtained the patent(1). The propulsion plants of this ship require, no propeller or other mechanical drive, and electromagnetic forces created by specifically Fleming's left-hand rule, which states that a magnetic field coupled with an electric field produces force. In 1966, the model MES-1 EMT-ship was built by S. Way et al(2). There was little expectation of running because the thrust power was too small. The thruster consisted of electrodes and electromagnet of which the flux density was about 150 gauss(0.015 T).

An EMT-ship(model SEMD-1) used a superconducting magnet has been built in 1976 by Y. Saji and his coworkers(3). This ship is designated as superconducting electromagnetic ship which is abbreviated as SC EMT-ship.

In 1979, the second model ship of SC EMT-ship, ST-500, has been built by Y. Saji et al, and was cruised at a spead of about one meter per second(4).

The construction of experimental SC EMT-ship, which have a displacement of 150 tons and the thrust power of about 8000 Newton, has started in 1985 by Japan Foundation for Shipbuilding Advancement. Investigators from various field are participated in the project. The thruster system consists of a superconducting magnet, electrodes, a helium refrigerator and others. Electrodes, a power souce and its control systems play an important role for the generation of power, the thrust efficiency, the spead control of run in the sea and forward and backward cruising.

This paper reports on the development of an electrode(anode) for use

Table I. Technological requirement for an anode

(1) Dimentionally stable in seawater electrolysis
(2) High electrical conductivity
(3) Homogeneity of the potential at the electrode/solution interface
(4) Highly catalytic activity for oxygen evolution
(5) Mechanical strength and long service life
(6) Simplicity, Availability and Low cost
(7) Capable for running ahead and astern
(8) Health safety

in this SC EMT-ship. The main requirement for the anode material to be utilized for SC EMT-ship are listed in Table I.

Experimental

The anode materials tested are listed in Table II. Pt/Ti and DSA_{O2} in group I were provided by Permelec Electrode Ltd, and Glassy Carbon(GC) was offered from Showa Denko Ltd. The anode materials of group II are modified

Table II. List of the anode species

Groups	Anode species
I	Pt/Ti DSA· DSA_{O2} GC·· Pt
II	MnO_x/DSA_{O2} MnO_x/GC MnO_2/PAN IrO_2/DSA···
	PbO_2/DSA $SnO_2/MnO_x/DSA$
III	ZrN/Pt/Ti TiN/Ti NbTiN/Ti Pt/Ti
	Ru/Pt/Ti

·Dimentionally Stable Anode
··Glassy Carbon
···J. M. Hinder, et al, J. Electrochem. Soc., <u>133</u>, 692(1986).

anode which were prepared by thermal decomposition, electrolysis or
sputtering methods in our laboratory. The anode matrials of group III were
prepared by plasma or reactive sputtering method.
    Test electrolysis was carried out in a H-type as is shown in Fig. 1(a).
The system used for electrolysis under continuous flow of electrolyte
solution is shown in Fig. 1(c). The cell was set in a inner house of
cryostat(Fig. 1,(b)), so that electrolysis could be carried out at an

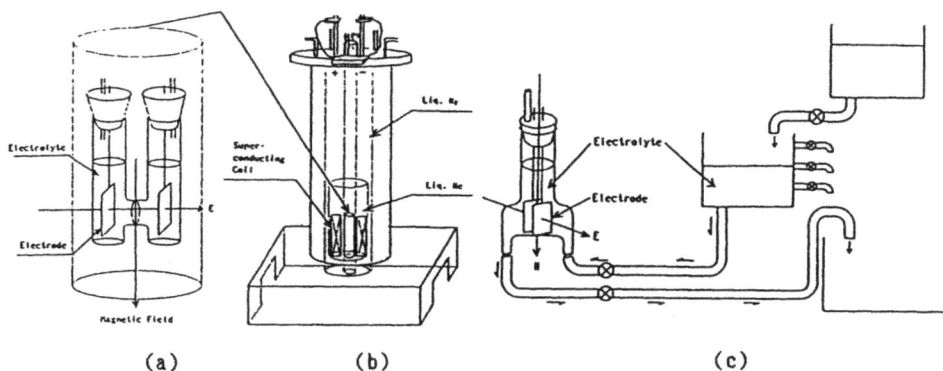

| (a) | (b) | (c) |

Fig. 1 Schematic diagram of the equipment for electrolysis, (a):H-type
    cell, (b):Cryostat, (c):Flow electrolysis system

composed magnetic field at 5 T. Flow rates of electrolyte solution were
1.6, 1.8, 2.6, and3.2 l/min within experimental error of about ±3%.
Electrolyte solution of 3.5 % NaCl solution was prepared from sodium
chloride of a reagent grade and distilled or ion-exchange water.
Electrolysis was carried out at a current density of 4000 A/m² or below.
    The oxygen evolution efficiency was estimated from a concentration of
hypochloric acid existing in electrolyte after electrolysis. The amounts
of hypochloric acid were determined.

Results and Discussion

    Some results of the oxygen evolution efficiency obtained with the
anodes listed in Table II are shown in Table III, together with that of
the results reported by J. E. Bennet(5). Low oxygen evolution effciencies
of less than 40 % and below were obtained at the anodes materials(a, b,
c, d, e, h, j, k, l and m in Table III) except for MnO_x/DSA and MnO_x/DSA$_{02}$
(f, g, and i). DSA and DSA$_{02}$ have the highest capability in regard to a
service life, which are estimated to be at least 10000 hours(6). From the
view point of long term stability, it is considered that DSA and DSA$_{02}$ can

Table III. Current efficiencies of oxygen evolution for
various electrodes. (Electrolyte: 3.5 % NaCl)

Electrodes	Oxygen evolution efficiency, % Current density: 10-30 A dm^{-2}					
	0	20	40	60	80	100
a DSA	—					
b DSA$_{o2}$		—				
c Pt/Ti			——			
d Pt			—			
e GC		—				
f MnO$_2$/DSA (Bennett)[1]						—
g MnO$_2$/DSA (Bennett method)[2]				——		
h MnO$_2$/PAN Ir/DSA (Hinder)[3]			—			
i MnO$_x$/DSA$_{o2}$						—
j SnO$_2$/MnO$_x$/DSA			—			
k MnO$_x$/GC				—		
l Pt/Ti (sputter)		——				
m Ru/Pt/Ti(sputter)			—			

[1] J. E. Bennett, Int. Hydrogen Energy, 5, 401(1980).
[2] Prepared in this work by Bennett method.
[3] J. M. Hinden, et al, J. Electrochem. Soc., 133, 692(1986).

be used to be a fundamental substrate material of the anode in the SC EMT
ship, but its surface is in need of modification to favor the evolution of
oxygen.

The MnO$_2$/DSA prepared by a method described by J. E. Bennet(5) shows
high oxygen evolution efficiencies as shown in Fig. 3. This electrode can
be used in the SC EMT-ship if its service life is long enough. However,
this anode lost its ability to evolve oxygen when it was dried before use.

As shown in Table III, an MnO$_x$/DSA$_{o2}$(EMT-AND1) anode, which was
prepared in our laboratory exhibited a high oxygen evolution efficiency,
and did not loose its capability even if dried before use.

Fig. 2 shows the polarization curves for the EMT-AND1 and DSA$_{o2}$ anodes.
It can be seen that the oxygen overvoltage of the EMT-AND1 is almost same
as that of DSA$_{o2}$. On the other hand, chlorine overvoltage of the
EMT-AND1 is high in comparison with that of DSA$_{o2}$. This indicates that the
oxygen evolution rate increased because the manganese oxide coatings cause
the chlorine evoluted to become higher. Since we are going to use this
anode materials, in the thruster, service life tests are now under way.

In order to know the amount of chlorine evolved, the concentration
of hypochloric acid containing in the electrolyte solution an flow out
from the cell, as shown in Fig. 1(c), was determined. Fig. 3 illustrates
the relation between the concentration of hypochloric acid and current
density of electrolysis when 3.5 % NaCl solution flowing the cell at a
rate of 2.6 l/min was electrolyzed by using anodes of these types. The

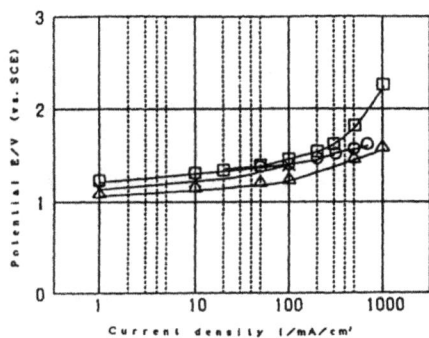

Fig. 2 Polarization curves of DSA
and MnO_x/DSA_{O2}.
O : 1N H_2SO_4, DSA   □ : 1N H_2SO_4, MnO_x/DSA_{O2}
△ : 1N HCl, DSA   × : 1N HCl, MnO_x/DSA_{O2}

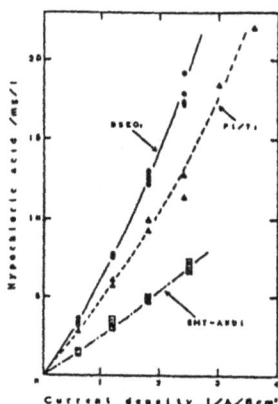

Fig. 3 Concentration of
hypochloric acid vs
current density

anode used were EMT-AND1, DSA_{O2} and Pt/Ti the cathode used was Pt/Ti.
The concentration of hypochloric acid in case of electrolysis using
EMT-AND1 increased with increasing current density, and the amount was
about 7 mg/l at 4000 A/m². The concentration of hypochloric acid evoluted
by EMT-AND1 anode was lower compared with that of DSA_{O2} or Pt/Ti anodes.
When the flow rate of the electrolyte solution was increased, the
hypochloric acid amounted to a little less than 1 mg/l.
    The influence of strong magnetic field, 5T, upon the characteristics of
EMT-AND1 was investigated by using a system which consisted of the cell,
the cryostat and the flow device of elctrolyte solution. The hypochloric
acid concentration in magnetoelectrolysis at 5T decreased by several
percentage compared with that of without field. The results of this
experiment were reported elsewhere in this symposium.

Conclusion

    We have identified some anode materials which evolve at high
efficiencies and evolve little amounts of chlorine in seawater
electrolysis. We have going to use these materials as anode in the
superconducting electromagnetic thruster which is being constructed.

702

Acknowledgement

Financial support of a part of this work by Japan Foundation for Shipbuilding Advancement is gratefully acknowledged. The authors wish to express their deep gratitude to President S. Nakagawa and Mr. K. Yamasaka, Permelec Electrode Ltd, for supplying electrodes for this reseach. The authers would like to thank Mr. H. Akazawa, Mr. S. Ohota, Mr. H. Konishi and Mr. H. Maekawa for experimental help to this reseach program.

References

1) W. A. Rice, U.S. Patent 2,997,013 (August 22,1961).
2) S. Way and C. Deblin, Prospects for the Electromagnetic Submarine, Paper 67-432, AIAA (1967).
3) Y. Saji, M.Kitano and A. Iwata, Adv. Cryo. Eng., 23, 159(1978).
4) A. Iwata, Y. Saji and S. Sato, Proc. ICEC 8, 775(1980).
5) J. E. Bennett, Int. J. Hydrogen Energy, 5, 401(1980).
6) S. Nakagawa, Soda and Chlorine, 38, 191(1987).

DEVELOPMENT OF SUPERCONDUCTING NETWORKS FOR A LARGE-SCALE MAGNETIC SHIELD

K. Takahata, S. Nishijima, and T. Okada

ISIR Osaka University, 8-1 Mihogaoka, Ibaraki, Osaka 567

Abstract-The calculations and experiments have been carried out on the superconducting shielding aiming at a practical use of superconducting networks for a large-scale magnetic shielding. Several types of disc-shaped networks were assembled with superconducting rings made of multifilamentary wires. The wires were connected electrically with Pb-Bi-Sn-Cd superconducting alloy. The shielding capabilities were calculated varying the diameter of rings. The calculation indicated that the highest capability can be obtained when the diameter of rings is half of that of disc. The calculation was confirmed by the experimentals using shielding discs of 40 mm diameter under uniform magnetic field of 0.5 T.

Introduction

When superconducting magnets are practically applied to various equipments, the leaked field must be reduced as low as possible. It causes the harmful effects on the various instruments or the human body. It is accepted that permissible field level to human body is five gauss [1]. This limit may disturb the practical application of the superconducting systems. The ferromagnetic shielding has been used in order to reduce the leaked field to the permissible level, though the whole weight of the system increases markedly. Superconducting shielding is expected to solve the problems.

The superconducting shield can be operated under higher field than the ferromagnetic shield. In case of superconducting shield, the field is concentrated at the end of the shield. The mechanism is different from that of ferromagnetic shield. For this reason, it may be difficult to confine the whole field of magnet with the superconducting shielding. For the practical use, it is advisable to use the combination of superconducting and ferromagnetic shield. The weight of ferromagnetic shield can be reduced remarkably with the combination. In this case, the shield is required to have following capabilities;
(1) not heavy
(2) adaptability to the field distributions
(3) control of the field concentration at certain space
(4) no disturbance inside the magnet.

Active superconducting shield will be put into practical use in near future [1]. It has, however, some demerits such as difficulty of the winding design, necessity of the power supply and so on. On the contrary, passive shield requires no complicated equipment and controls the field automatically if designed adequately. The remaining problem of the passive shield may be the superconductive stability.

In this series of works [2],[3], passive shield has been studied using networks with superconducting wires as shown in Fig. 1. The networks are expected to have higher stability against flux jumping compared with bulky superconductor. Here, the shielding capability of multi-ringed disc will be

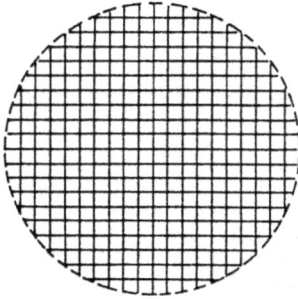

Fig. 1 Shielding networks.

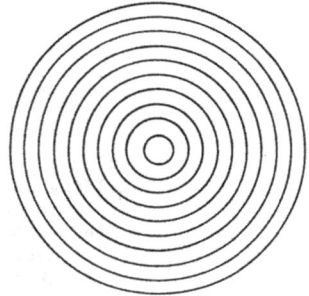

Fig. 2 Coaxial rings.

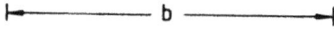

b

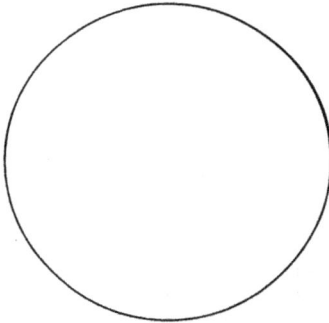

(a)

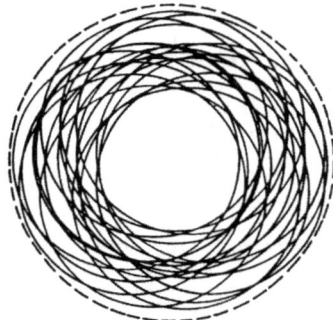

(b)

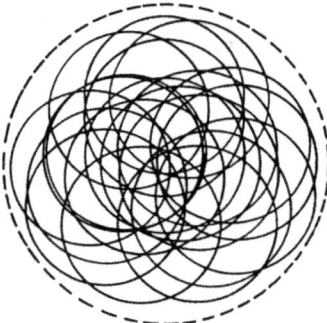

(c)

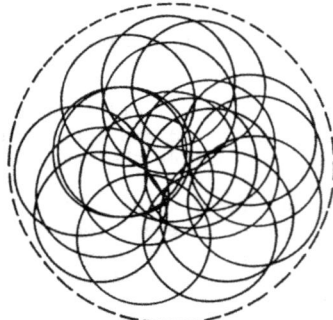

(d)

Fig. 3 Examples of multi-ringed disc with diameter "b". The a/b is (a)1, (b)0.7, (c)0.5 and (d)0.4.

discussed. The multi-ringed disc has a configuration as a simplified network and can be manufactured more easily. If the external field is symmetric with respect to the axis, coaxial rings, as in Fig. 2, are considered to have the same ability as the network. In case of unsymmetric field, rings must be arranged at random. In this paper, basic analyses will be made in order to design the disc shaped shield. The applicability of the multi-ringed disc to a large-scale shield was examined experimentally.

Basic Analyses

The rings of constant diameter "a" are placed at random inside of a circle of diameter "b" which is located on the plane, z = 0. Figure 3 shows the examples of multi-ringed disc of which the number of rings, N, is twenty. A hole exists at the center when a/b is larger than 0.5 as shown in Fig. 3 (b). The size of the hole comes to be larger as a/b increases and coincide with the circle at a/b is unity as shown in Fig. 3 (a). The fields made by the multi-ringed discs can be calculated as the total sum of those by the rings and written generally as

$$B = \mu_o I_s N/b \ f(a/b,z/b,r/b) \eqno(1)$$

where shielding currents are assumed to be constant, I_s. The f is calculated as a function of normalized ring diameter and location. The value of f becomes unity when the a/b is unity, z zero and r zero. This equation suggests that the shielding currents and the number of rings must be enlarged as the diameter of disc, b, increases. Considering the weight of shield, to increase the current is more desirable.

Figure 4 shows the value of f as a function of a/b at the center just above the surface. The z/b is fixed to be 0.01 considering the thickness of disc. The disc will indicate the shielding capability when a/b is larger than 0.3 and the highest at a/b = 0.5. The f value of coaxial rings, as in Fig. 2, is 3.6 at the same point. The random rings can compare with the coaxial rings when a/b is 0.5. Figure 5 shows the value of f as a function of r/b above the surface (z/b = 0.01). The field distributions on the r-axis have the peak except a/b = 0.5. It is also found that the field concentration at the edge (r/b = 0.5) is smallest when a/b is 0.5. The results suggest that it is desirable to make the diameter of rings half of that of the disc. For a large-scale shielding disc or plate, the diameter of rings must be designed adequately so that the field of shield corresponds to the external field distribution.

Experimentals

Fabrication of Rings

The superconducting wires used for the rings were multifilamentary Nb-Ti-Zr-Ta composites with copper matrix. The specifications of the wire are shown in Table 1. The wires were formed into a ring as illustrated in Fig. 6 and soldered with 25Pb-50Bi-12.5Sn-12.5Cd alloy (Wood's Metal Fusible) in order to joint them electrically. The Wood's Metal indicates superconductivity at 4.2 K. The field dependence of the critical current density of the Pb-Bi-Sn-Cd alloy was obtained by the magnetization measurement and is shown in Fig. 7. The upper critical field is above one tesla and it has higher value than usual tin-lead solder [3].

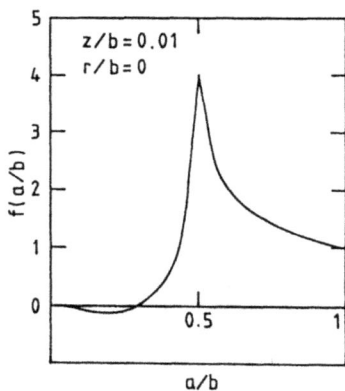

Fig. 4　Factor f as a function of
a/b at z/b=0.01 and r/b=0.

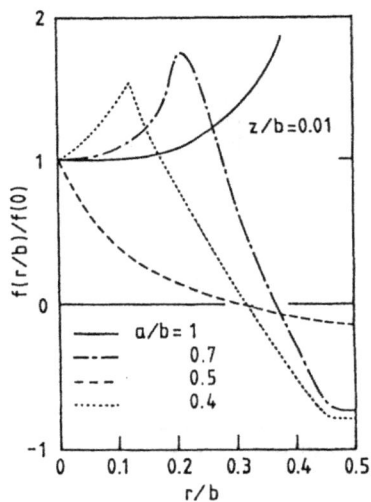

Fig. 5　Factor f as a function of
r/z at z/b=0.01.

Table 1　Specifications of the
multifilamentary superconductor

Conductor	Nb-Ti-Zr-Ta
Diameter	0.35 mm
Filament Diameter	0.03 mm
Number of Filaments	61
Cu/SC Ratio	1.3
Critical Current	80 A (at 5 T)
Twist Pitch	10 mm

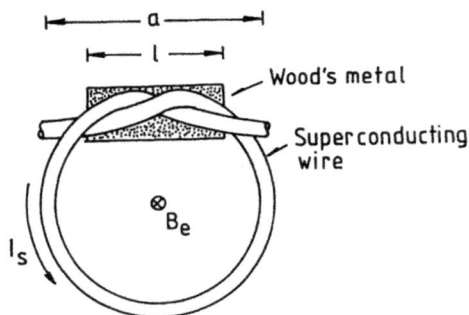

Fig. 6　Superconducting ring used for experiments.

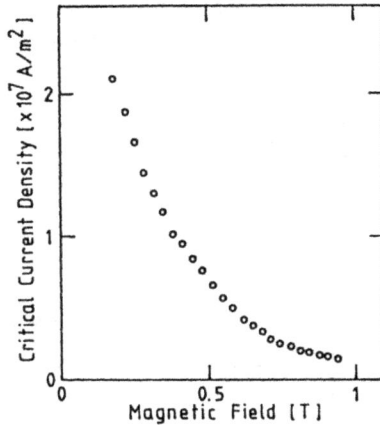

Fig. 7 Field dependence of the critical current density of Pb–Bi–Sn–Cd alloy. The critical current density was measured using a magnetization measurement.

40 mmφ Shield

Three types of shields of multi-ringed 40 mm diameter were fabricated in order to examine the effects of the number and diameter of rings. The diameter, a, was changed to be 10, 15 or 20 mm. The connected length, "l" in Fig. 6, was 7, 10 or 13 mm. The uniform field, B_{ext}, was applied to the multi-ringed shield and the field, B_{meas}, was measured at distance of 1 mm from the surface. The sweep rate of external field was 0.001 T/sec.

Figure 8 shows the N dependence of the shielding efficiency, defined as $(B_{ext}-B_{meas})/B_{ext}$, at the external field of 0.5 T. The efficiency of the shield with a = 20 mm was much larger than the others. That is consistent with the calculation. The efficiency increased linearly with increasing N. That is also consistent with Eq. (1) assuming the shielding current, Is, is constant. In case a = 20 mm, Is was obtained by the reverse calculation to be approximately 50 A. This value is lower than the critical current of superconducting wire. It suggests that the shielding capabilities were almost limited by the critical current density of the solder. In this case, the effective critical current of the ring may proportional to the connected length and may be written as

$$I_s = J_c \cdot l \cdot d \qquad (2)$$

where d is the diameter of wire. In case a = 20 mm, Jc will be approximately 1×10^7 A/m². This value is agree with that of the Pb–Bi–Sn–Cd alloy, as shown in Fig. 7. It is desirable to increase the connected length, but it cannot exceed the length of the circumference of a ring. The method of superconducting connection may be the most important subject if wires are used.

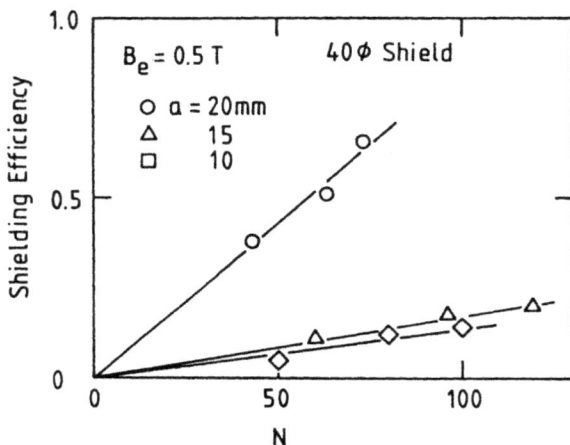

Fig. 8 Shielding efficiency as a function of the number of rings.

Conclusions

The shielding capabilities of superconducting multi-ringed disc were analyzed in order to design the shielding networks. The applicability to a large-scale shield was examined experimentally and the following conclusions are drawn.

(1) The multi-ringed shields show the highest performance when the diameter of rings is half of that of disc.
(2) If the diameters of rings are adequately designed, multi-ringed shield can control the field distribution.
(3) For a large-scale shielding network, the effective critical current of rings must be enlarged. If the superconducting composite wires are used, some devices are needed at the connected points.

References

[1] David. E. Andrews, "MAGNETIC RESONANCE IMAGING IN 1987," Adv. in Cryo. Engn., Vol. 33, pp. 1-7, 1988.

[2] S. Nishijima, K. Takahata, I. Miyamoto, T. Okada, S. Nakagawa and M. Yoshiwa, "MAGNETIC SHIELDING NETWORK WITH SUPERCONDUCTING WIRES," IEEE Trans. on Magn., Vol. MAG-23, No. 2, pp. 611-614, MARCH 1987.

[3] T. Okada, K. Takahata, S. Nishijima, S. Nakagawa and M. Yoshiwa, "MAGNETIC SHIELDING WITH SUPERCONDUCTING WIRES," IEEE Trans. on Magn., Vol. 24, No. 2, pp. 895-898, MARCH 1988.

MAGNETIC SHIELDING BY A TUBULAR SUPERCONDUCTING WINDING
IN PARALLEL AND TRANSVERSE FIELD

Masashi OHGAMI, Kazuya TAKAHATA, Shigehiro NISHIJIMA, and Toichi OKADA

ISIR, Osaka University, 8-1, Mihogaoka, Ibaraki, Osaka 567

Abstract- Magnetic shields with superconducting windings have been studied aiming at a practical application of superconducting shieldings. The tubular shields were made by windings using NbTi multifilamentary composite wires and impregnated with Wood's metal. The static fields parallel to the axis were applied to the shield and the field penetration was measured. The fields up to 2.0 T could be reduced less than 10^{-3} T at the center of the shield. To shield the vertical magnetic component to the axis, the superconducting wires were wound obliquely and examined. It was concluded that the magnetic shields with superconducting windings showed capability to shield not only the parallel but also the transverse fields to the axis. The stability of superconducting shield was also discussed.

Introduction

As superconducting magnet has been put into practical use, the equipment using high magnetic fields has been developed so far. Accompanying the wide use of superconducting magnet, high magnetic fields has brought the problems on the various instruments around the magnet. Magnetic shieldings with superconducting materials to replace ferromagnetic shields have been studied aiming at the high field shielding.

Some studies of superconducting shields have been performed in the form of plate[1], cylinders[2], [3] tapes[4] and thin films[5] . It becomes to be understood, however, as a serious problem that the bulky superconducting shields mentioned above have magnetic instability. In this work, superconducting shields using multifilamentary wires have been investigated and confirmed to show high stability. In the present paper, the authors have tried to shield parallel and transverse fields using four types of small superconducting shields. Furthermore, the stability of shields using multifilamentary(MF) wire were examined.

Samples and Experimentals

The superconducting wire used for shielding was multifilamentary Nb-Ti-Zr-Ta alloy with copper matrix. The specifications of the wire is shown in Table 1. Test shields were wound around the cylindrical bakelite bobbin and impregnated entirely with Wood's metal (Pb-Bi-Sn-Cd four elementary alloy) in order to connect electrically between wires. Two types of the magnetic shields were made. The specifications of the shields were shown in Table 2. The purpose of TEST SHIELD 1 is to shield parallel fields to the axis and that of TEST SHIELD 2 is transverse fields to the shield axis. In TEST SHIELD 1 (shown in Fig. 1 (a)) superconducting wires were wound spirally. TEST SHIELD 2 (shown in Fig. 1 (b)) were constructed to be wound wires obliquely at 45 deg. to the axis.

The experimental arrangements of the shield and the magnet were shown in Fig. 2. TEST SHIELDs were fixed in the center of the superconducting magnet.

Table 1 Specifications of the
multifilamentary superconductor

Conductor	Nb-Ti-Ta-Zr
Diameter	0.5 mm
Filament Diameter	0.042 mm
Number of Filament	61
Cu/SC Ratio	1.3
Critical Current	170 A (at 5T)
Twist Pitch	15 mm

Table 2 Specifications of TEST SHIELD

Name of shield	Inside dia. (mm)	Outside dia. (mm)	Length (mm)	Number of layers
TEST SHIELD 1	17	26	46	8
TEST SHIELD 2	19	23	73	6

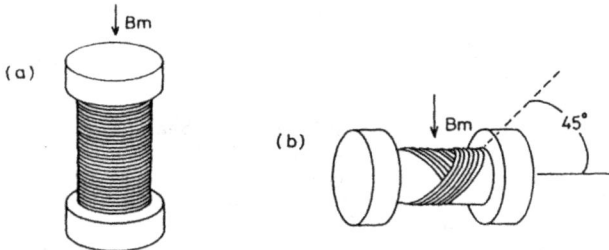

Fig. 1. Schematic illustrations of TEST SHIELD 1-2: (a) TEST SHIELD 1 and (b) TEST SHIELD 2.

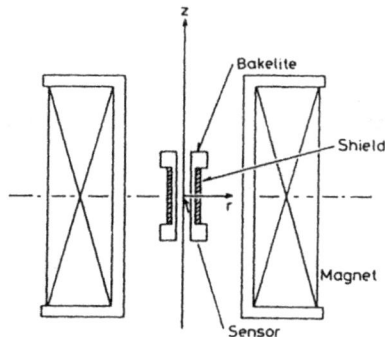

Fig. 2. Experimental setup of the magnetic shield and the superconducting magnet.

The parallel fields to the shield axis were applied by coinciding the magnet axis with the shield axis. This axis was defined as Z and the R vertical to Z axis to form a cylindrical coordinate. The center of shields was defined as the origin in calculation.

In the experiment, external magnetic fields (Bm) were applied to the shields at different sweep rate and the penetration of the magnetic fields (Bs) into the shields were measured using the magnetic sensor. The resolution of this sensor was 10^{-3} T at 4.2 K. The static fields were applied to the shields for a certain period (1500 sec) to examine the shielding efficiency in the static fields and the field penetration with time was measured.

Results and Discussions

Shielding efficiency in parallel field

The distribution of magnetic field along the Z axis both inside and outside of the shield was measured and shown in Fig. 3, when the static field of 2.0 T was applied to the TEST SHIELD 1. Broken line shows external field (Bm) obtained on the basis of calculation. Solid line shows the measured field distribution with the shield in the magnet. It was confirmed that the field of 2.0 T could be reduced less than 10^{-3} T in the center of shield without flux jump . Furthermore, the field was shielded completely in wide space in the shield.

In Fig. 4 time dependence of field penetration into the shield was shown when the static field of 2.0 T was applied to TEST SHIELD 1 for 1500 sec . In the field of 2.0 T, field penetration more than 10^{-3} T was not observed for 1500 sec into the shield and hence the field distribution did not change.

Shielding efficiency in transverse field

Figure 5 shows the result of applying the transverse field to

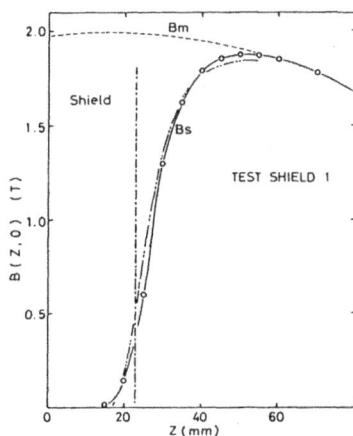

Fig. 3. Distribution of magnetic field along the Z axis in TEST SHIELD 1.

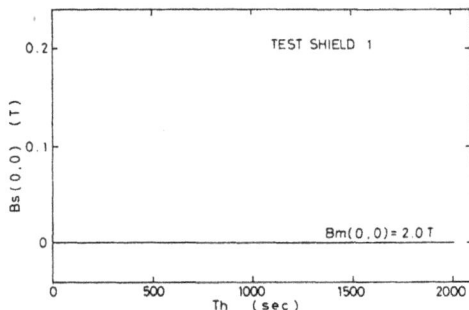

Fig. 4. Field penetration into the shield when the static field of 2.0 T was applied to TEST SHIELD 1 for 1500 sec.

TEST SHIELD 1. It is clear that the shield wound wire spirally has no shielding efficiency for transverse field. In order to shield not only parallel but also transverse fields, TEST SHIELD 2 were made with changing the wire arrangement. The transverse field was applied to TEST SHIELD 2 and the shielding efficiency was studied.

In Fig.6, shielding efficiency of TEST SHIELD 2 in the transverse field was shown. TEST SHIELD 2 had a shield capability of the transverse field up to 1.1 T. The field penetration into the shields was only three per cent of the external field Bm at field sweep rate of 1T/30sec. The field penetration was 5×10^{-3} T when the external field 1.0 T was applied to the shields for 1500 sec. This is caused by the fact that the superconducting contact between wires was insufficient.

From the results described above, it became clear that the transverse field could be shielded with obliquely wound wires. It is, however, important to get the sufficient superconducting contact between wires.

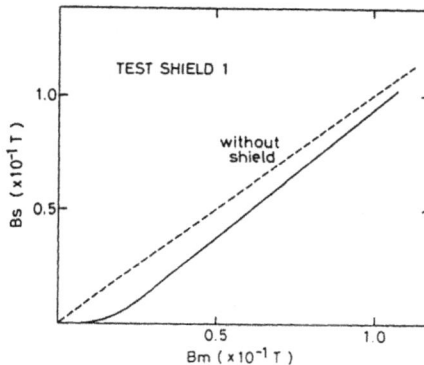

Fig. 5. Field penetration into the shield when the transverse field was applied to TEST SHIELD 1.

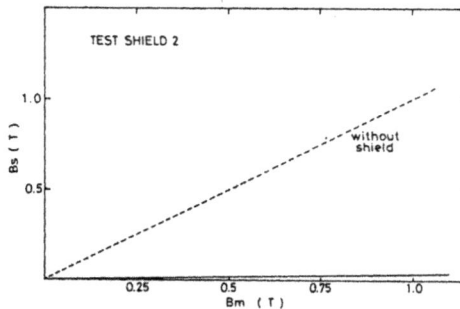

Fig. 6. Field penetration into the shield when the transverse field was applied to TEST SHIELD 2.

Stability of shieldings

The problem of magnetic shieldings with superconducting materials is magnetic instability induced by flux jump[3], [5]. In the shields using bulky superconducting materials, flux jumps are easily induced. In the experiments of TEST SHIELD 1-2, however, the field where flux jump occurs is high compared with that in bulky superconductor with same size. TEST SHIELD using superconducting MF wires is stable against flux jump and shows high shielding efficiency in high field of several tesla, because the hysteresis losses were reduced by rearrangement of current distribution with dynamic stabilization of the shields.

Calculation of fields and shielding current in magnetic shield

The distribution of shielding current in the shield corresponds to that of external field. TEST SHIELD 1 was divided into meshes and the distributions of shielding current and field in the shield were calculated . In the calculation, it has been assumed that the shielding current in wires flows at the critical current in the fields.

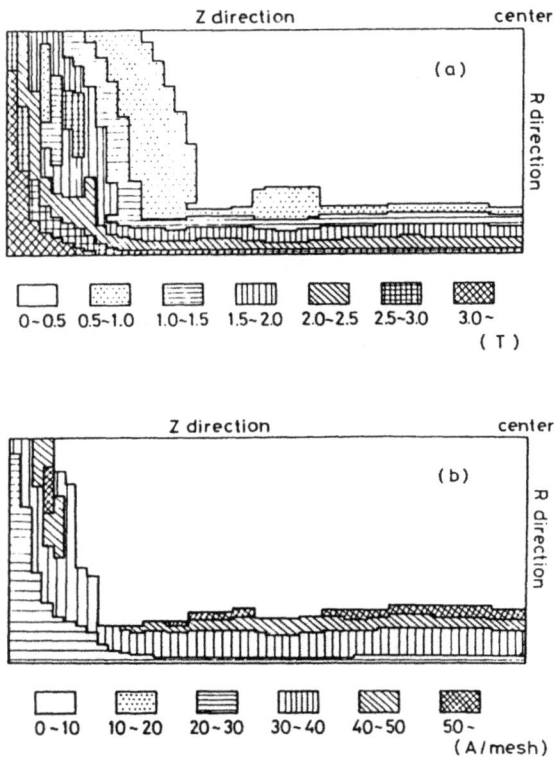

Fig. 7. (a)Distribution of field in the shield (b)Distribution of shielding current in the shield.

714

Figure 7 (a) shows the distribution of fields in the shield and Fig. 7 (b) shows that of shielding current. Both of them are shown in the 1/4 cross sections of shields. The field is concentrated near the edge of shields and more than 3.0 T is applied to the shield in external field of 2.0 T. The distribution of measured field agrees well with calculation as shown in Fig. 3 with dashed line. It is concluded that the field penetration into the shield was induced near the edge region by the concentration of field.

Conclusion

The shielding efficiency of parallel and transverse field were examined using two types of shields and possibility of magnetic shielding with superconducting MF wire were investigated. Following conclusions were drawn.

(1) The parallel field up to 2.0 T (maximum field of 3.0 T is applied to the edge of shield) were reduced less than 10^{-3} T for 1500 sec.

(2) The transverse field up to 1.1 T were shielded with 3 per cent penetration of external field by the shield wound wires obliquely. It is possible that the field penetration are reduced less than 10^{-3} T by keeping the superconducting contact between wires. It is confirmed that the shield wound wires obliquely could shield the high field in every direction.

(3) In the design of magnetic shieldings, the concentration of field near the edge of shield must be considered. The shields must be designed considering the shape, number of layers, the properties of wires, impregnating materials and superconducting contact between wires.

(4) The shields using superconducting MF wires can shield the high field stably.

References

1. I.Miyamoto, S.Nishijima, T.Okada, K.Yoshiwa and A.Iwata, "Study of Magnetic Shielding with Superconducting Materials," Proc. of MT-9, pp.841-844, 1985.

2. T.Okada, K.Takahata, S.Nishijima, S.Nakagawa and M.Yoshiwa, "Magnetic Shielding with Superconducting Wires," Proc. of MT-10, pp.895-898, 1988.

3. D.V.Gubser, S.A.Wolf, T.L.Francavilla, J.H.Classen and B.N.Das, "Multilayer Nb_3Sn Superconducting Shieldings," IEEE Trans. Magn., Vol.MAG-21, No.2, pp.320-323, 1985.

4. A.Shimizu and M.Inoue, "A Test of The Superconducting Shielding Tube Made of V_3Ga Tape," IEEE Trans. Magn., Vol.MAG-17, No.5, pp.2146-2149, 1981.

5. D.J.Frankel, "Model for Trapping and Shielding by Tubular Superconducting Samples in Transverse Fields," IEEE Trans. Magn., Vol.MAG-15, No.5, pp.1349-1353, 1979.

MAGNETIC FIELD SCREENING
WITH SUPERCONDUCTING NbTi-Cu MULTILAYER FILMS

Soichi OGAWA, Masaaki YOSHITAKE, Kazu NISHIGAKI[*]
Takao SUGIOKA[**], Masaru INOUE[**] and Yoshiro SAJI[**]

Osaka Pref. Industrial Research Institute
Kobe University of Merchantile Marine[*]
Koatsu Gas Kogyo Co.,Ltd.[**]

Abstract - We have studied the magnetic shielding effects (MSE) of NbTi single layer and NbTi-Cu multilayer films of various thickness in vertical field, as well as of designed hollow superconducting cylinders in transversal magnetic field to the axis of the cylinder. The total thickness (nd) dependence of MSE per unit thickness ΔBm/nd (ΔBm: maximum MSE, d: film thickness, n: number of NbTi films in a multilayer film) in the multilayer films increased exponentially with decreasing nd and increasing n in the region of nd>3μm and in any case of n=1,2,3. The relation between ΔBm/nd and nd can be expressed well by the following formula on the basis of these results;
$$\ln(\Delta Bm/nd) = a - \beta \ln(nd), \quad \alpha, \beta \text{ (constant)}.$$
These studies have developed for high magnetic shielding devices of hollow cylinders with superconducting NbTi-Cu multilayer films. The device provides a space with free from high magnetic field up to 0.8 T.
From these results, we conclude that it is possible to completely seal off or form a field free space against flux density of one tesla or more by laminating NbTi-Cu multilayer films of reasonable thickness.

Introduction

In many superconducting devices and applications such as SQUID and EMT using strong magnetic fields, it is necessary to shield the magnetic field efficiently. For this purpose, superconducting magnetic shield is required powerful mean. High magnetic shielding above a few teslas, however, has not been applied to the superconducting devices, and reports on this problem have scarcely been presented previously[1][2].

In recent year, it was shown that ΔBm/d(ΔBm:max. magnetic shielding effects(MSE), d: film thickness) increased by decreasing d and obtained maximum value at d=0.7μm. The value of MSE was more than 20 times of bulk NbTi plate on our study of magnetic shielding effects of superconducting NbTi thin film deposited by sputtering method on Al foil.[3]

In this work, We carry out basic study on MSE of NbTi single layer films and NbTi-Cu multilayer films of various thickness in vertical field, and of designed hollow superconducting cylinders in transversal magnetic field to the axis of the cylinder.

Experimental Procedure

Sample Preparation

NbTi (50at%Ti, Tc=7.8K) films were prepared by RF magnetron sputtering method. It is possible to sputter with relatively high deposition rate and to obtain high purity NbTi films. The film were deposited on Al foil(100 x 100 x 0.015mm^t) at Ar gas pressure in 10^{-3} Torr range with sputtering power of 300W while the substrate temperature was kept at R.T.. The film thickness ranged from 0.1 to 80μm with a deposition rate of 0.1μm/min. The crystal structure of NbTi films as defined by X-ray diffractmeter was cubic structure of the same as that of the NbTi target used for sputtering. The crystalline size of the films was a few tens of angstroms.

The hollow superconducting cylinders were constructed by piling disks of NbTi-Cu multilayer films and Al plates.

Measurement system

As shown in Fig.1 the core of the apparatus is composed of super-conducting solenoidal coil and cylindrical sample holder. A test film is inserted between pressing plates with three Hall probes and supported in the center of the coil so that the natural direction of magnetic shielding of coil and the surface of the film is in a right angle. The core of the apparatus is set in the cryostat by which test temperature is freely obtained between 4.2K and 1K. Measurements and data treatments are carried out automatically by a computer system.

Experimental Results and Discussions

Experimental data are put in order based on shielding magnetic field by superconducting planer film. We defined by eq.(1), because the effects of screened magnetic field is made remarkably as much as possible.

$$\Delta B = B_0 - Br \tag{1}$$

Where B_0 is natural magnetic flux density without screening NbTi film and Br is real magnetic flux density with the film.

Magnetic shielding effects of NbTi-Cu films

Figure 2 shows an example of measured data. Increasing the external magnetic field, the flux can not intrude into the shielding film. This means that the screened magnetic field ΔB is equal to the natural magnetic flux density B_0. Increasing the magnetic field from 0.1T, magnetic flux begins to intrude into the NbTi films. At 0.19T, the maximum shielding magnetic flux density ΔB of 0.136T is obtained. Increasing the magnetic field, the ΔB decreases gradually, but shielding effects of 0.015T is still remained at B_0=3T. Therefore, it is possible to shield high magnetic field of more than 3T by multilayer NbTi thin films. As magnetic fluxes intrude gradually from the edge rim of the NbTi film, the ΔB near the edge is much smaller than the one at the center of the film. In the case of the NbTi film, the value of ΔBm is independent of the exciting speed for the superconducting magnet.

Figure 3 shows the representative x-ray diffraction pattern of NbTi-Cu multilayers film(NbTi:2.9μm x 5, Cu:2μm x 5). As being seen the diffraction pattern,individual reflections from NbTi and Cu can be clearly detected. The facts show that alloying of each layer film does not occur.

Figure 4 shows relation of maximum MSE ($\Delta Bm/nd$) with total thickness of NbTi layer(nd) using the number of NbTi layer(n) of NbTi-Cu multilayers film as a parameter.

In the region of nd>3μm, $\Delta Bm/nd$ shows linear increase with decrease of nd in any case of n=1,2,3, and its relation conforms to the formula(2).

$$\ell n(\Delta Bm/nd) = \alpha - \beta \ell n(nd) \tag{2}$$
$$\alpha, \beta : \text{constant}$$

Figure 5 shows results of high magnetic field shielded by ten multilayer films and ten single layer films. In two cases, each total thickness of NbTi is 132 μm. From this figure, as ΔBm is 0.6T, it is known that a high magnetic field of 0.6T is shielded and as ΔB is 0.4T where B_0=1T, a high magnetic field of 1T can be shielded by three layers of this film having less than 0.3mm thickness.

Application to magnetic field shielding device (MFS device)

Table 1 shows the magnetic shielding effects of five hollow superconducting NbTi-Cu multilayer films. The hollow disk has an outer diameter(O.D.) of 46mm and an inner diameter(I.D.) ranging from 0 to 28mm by using the multilayer films which were composed of two NbTi layers and one Cu layer of 2μm thickness respectively. As the increasing I.D. of the disk from 0 to 28mm Δ Bm decreases from 0.10 to 0.02 T.

Table 2 shows the values of Δ Bm at the center (point 1 in Fig.7) of MFS device, forming by piling the hollow superconducting disks and Al disks alternatively(see Fig.6). As the increase of number of superconducting disks from 30 to 90 ΔBm increases from 0.31 to 0.45 T.

The height of the superconducting cylinder remains 32mm. As the number of the superconducting disks in a unit increases, Δ Bm becomes greater. As shown in table 2, the device of No.4 includes 180 of the superconducting disks (i.e. the NbTi layers are measured 0.72mm in thickness) providing more disks in one unit. The relation between a rate of magnetic shielding effects ($\Delta B/B_0$) and B_0 of the devices No.1 to No.4 are shown in Fig.8. The No.4 can completely shielded up to 0.8T(correspond to point A).

Figure 9 shows the result of shielding effects in which the height of the superconducting cylinder is increased from 30 to 60 and 90 mm while the number of the superconducting disks in one unit remains unchanged. The ΔBm_0(maximum complete magnetic shielding effects) of ambient magnetic field which can be interrupted by shielding completely at each of the positions on the cylinder is stated. As for a shielding area, an area in the cylinder having a height of 30mm covering from its center to a distance of 5mm can completely be shielded from the magnetic field of about 0.25 T. In the cylinder of 90mm in height, an area shielded from the magnetic field of 0.25 T extends from the center of the cylinder to a distance of 35mm.

In the results, a large area can completely be shielded from the magnetic field of a tesla or more, by increasing the number of the superconducting disks in one unit and also.

Conclusion

The magnetic shielding effects of NbTi-Cu multilayer films of various thickness in vertical field and of MFS devices in transversal field to the axis are summarized as follows,
1. The relation between ΔBm/nd and NbTi film thickness was represented as eq.(2). ΔBm/nd was raised with increasing n.
2. ΔBm of the sheet laminated multilayer films in total NbTi thickness of 132μm could be shielded the magnetic field of 0.6T.
3. The effective shield for high magnetic field is made possible by laminating superconducting NbTi-Cu multilayers film.
4. It is possible to completely shield up to 0.8 T or more with the MFS devise.

References

1. Simizu, A. and Inoue, M. 'A Test of the superconducting shielding tube made of V_3Ga type' IEEE. transaction on magnetics, vol.MAG-17, No.5 September (1981) pp.2146-2149.
2. Sato, S., Ikeuchi, M., Iwata, A., Saji, Y. and Kado, S., 'The Magnetic field screening with NbTi' proc. ICEC-9, (1982) pp.115-119.
3. Ogawa, S., Tada, E., Toda, H., Yositake, M., Sinpo, M. and Saji, Y. 'The Study of Superconducting NbTi films for Magnetic Shield' Proc. ICEC-11, (1986) pp.484-488.

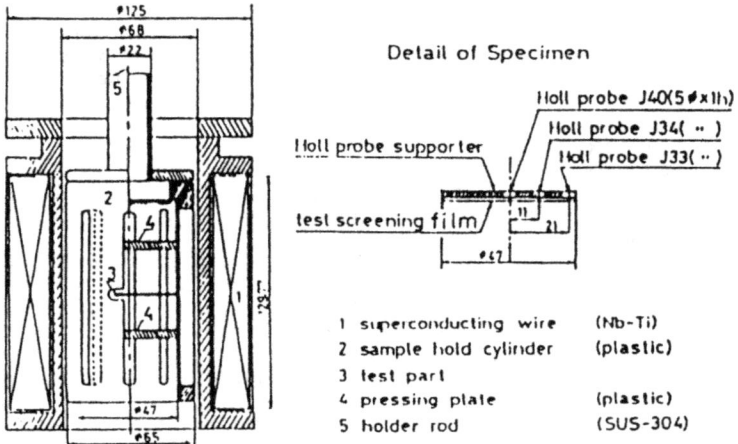

Detail of Specimen

Holl probe J40(5⌀×1h)
Holl probe J34(")
Holl probe J33(")

Holl probe supporter

test screening film

1 superconducting wire   (Nb-Ti)
2 sample hold cylinder   (plastic)
3 test part
4 pressing plate   (plastic)
5 holder rod   (SUS-304)

Fig.1. Cross-sectional view of core of apparatus.

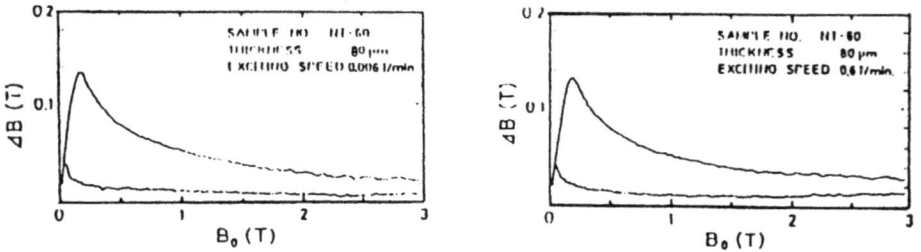

Fig.2. Shielding magnetic field versus magnetic flux density.

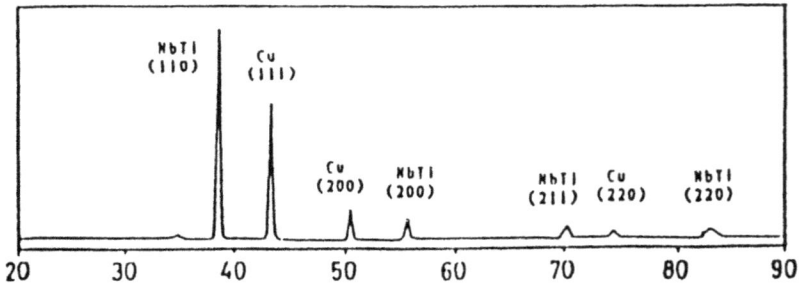

Fig.3. X-ray diffraction pattern of laminated superconducting NbTi-Cu films (NbTi 2.9Å x 5 layers).

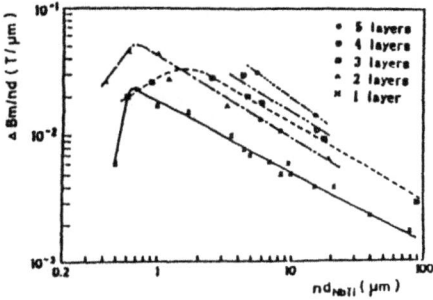

Fig.4.a. The thickness dependence of ΔBm/nd with the parameter of the number of NbTi layers for multi-layer sheets.

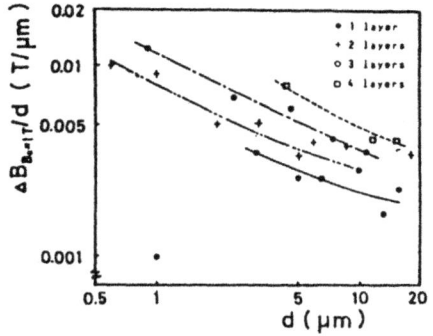

Fig.4.b. The thickness dependence of ΔBm/nd in Bo of 1 T.

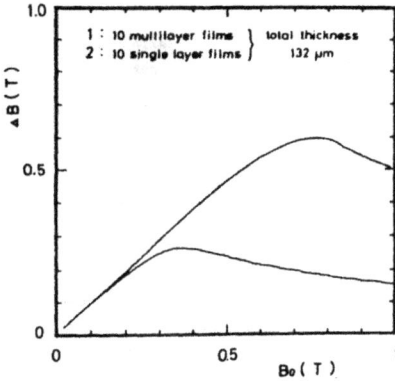

Fig.5. Magnetic shielding effect of multi-layer films or 10 single layer films with the same total thickness of NbTi layers

Table.1. Magnetic shielding effect of superconducting disks with a hole of different size.

O.D.(mm)	I.D.(mm)	ΔBm (Tesla)
46	-	0.101
46	15	0.040
46	20	0.031
46	23	0.026
46	28	0.020

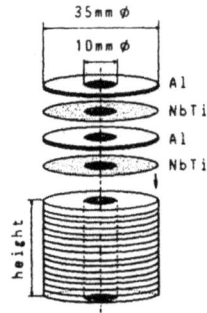

Fig.6. MFS device

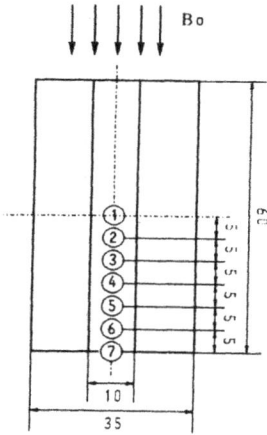

Fig.7. A cross section of the cylinder marked with measurement points. (sample No.5 in Table.2)

Table.2. Magnetic shielding effect of hollow superconducting cylinders.

NO	NbTi disk number	Aluminum disk (mm)		cylinder (mm)			maximum shielding effect
		thickness	number	O.D.	I.D.	length	ΔBm (Tesla)
1	30	1.0	31	35	10	32	0.308
2	60	0.5	61	35	10	32	0.448
3	90	0.5	61	35	10	32	0.569
4	180	0.5	31	35	10	20	0.883
5	60	1.0	61	35	10	62	0.333
6	90	1.0	91	35	10	93	0.338

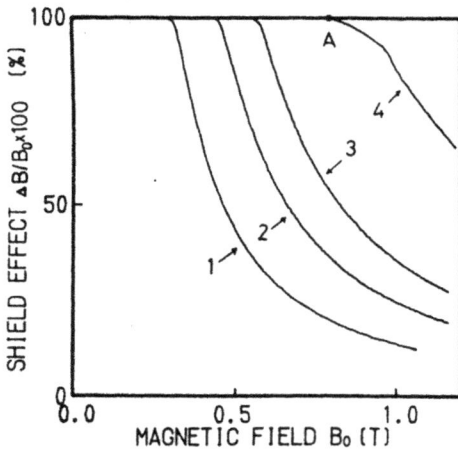

Fig.8. Magnetic shielding effect in the center of cylinders.

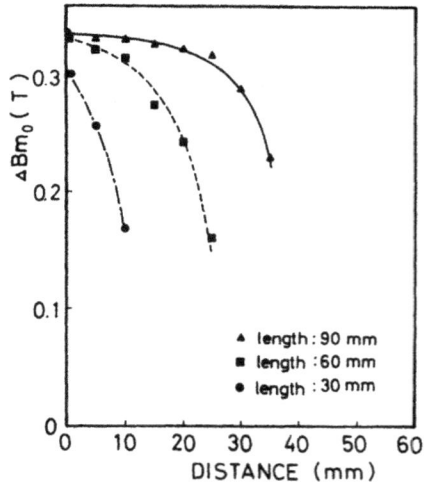

Fig.9. Magnetic shielding effect in superconducting cylinders of 30,60,90mm in height. (sample No.1,5 and 6)

Section 4. Applications for Accelerators and Measurements, Superconducting Electronics

SUPERCONDUCTING ACCELERATOR MAGNETS

Hiromi HIRABAYASHI

KEK National Laboratory for High Energy Physics
1-1, Oho, Tsukuba-shi, 305 Japan

Abstract

In the near future, a large number of high quality superconducting dipole and quadrupole magnets will be required for construction of the next generation multi-TeV high energy hadron accelerator-colliders. To establish the construction technology of such accelerator-colliders, extensive and world-wide R&D programs are now carrying out at several laboratories. In the first half of this paper the important issues in superconducting accelerator magnets such as cables, design, fabrication, testing and cryogenic system are discussed together with some details on coil cross-sectional current configurations, quality control of materials, quench protections, radiation heating and etc. In the latter part, the key technology in superconducting accelerator magnets will be summarized.

Introduction

At present superconducting accelerator magnets are mainly used in the high energy accelerators. To expand energy ranges more than one order of magnitude greater than those available at existing high energy physics facilities, extensive and world-wide R&D programs are now carrying out at several laboratories in America, Europe and Japan. Many physicists believe in the exploration to the microcosmos with multi-TeV accelerators will reveal new phenomena from fundamental material structures one thousand times smaller than the proton or neutron.

The most probable multi-TeV accelerator in the next decade will be high luminosity proton-proton collider with a large system of superconducting magnets. The constructions of multi-TeV superconducting proton-proton colliders such as SSC[1] and LHC[2] must be large undertakings. The technological facility of these colliders is based on the results of recent two decades of R&D of superconducting materials and magnets as well as the practical operational experience with Tevatron at the Fermilab in USA.[3]

In the following paragraphs, the issues of superconducting cables, coil designs, magnet fabrications, performance testings, cryogenic system and summary of key technology will be discussed in the order.

Superconducting cables

Superconducting cables are essential parts of accelerator magnets. Until early 1980's, it was generally believed that there were two choices of superconducting cable selection for the high field accelerator magnets.[4] One was the copper stabilized NbTi or NbTi (Ta) alloy cable and the other was the copper stabilized Nb_3Sn intermetalic compounds (A15) cable. The former is still applicable to high field superconducting dipole and quadrupole magnets, if it were cooled below 2 K and also is generally used in magnets up to several Tesla with 4 ~ 4.3 K cooling, and the latter is still unforeseenable for the accelerator application because of its brittleness and hard heat treatment in production process.

Today the accelerator magnet designer's interest is concentrated on the issue how to get higher coil current density (J_c) with enough mechanical toughness at the moderate cooling temperature such as 4.3 to 4.4 K. In this case copper stabilized Nb_3Sn superconducting cables could not be applicable because not for the critical temperature requirement (T_c, $[Nb_3Sn] > 10K$) but for the poor mechanical properties.

So far copper stabilized NbTi superconducting cable, the overall critical current density (without copper stabilizer) J_c is getting higher and higher in recent several years by the improvements of material homogeneity and cable fabrication process. The typical industrial J_c recently reached 2,850 to 3,000 A/mm^2 with several micron filaments at 4.2 K and 5 T. This value is about 50 % greater than that of the Tevatron construction age from 1979 to 1983.

Table 1. Comparison of superconducting accelerator cables and coils

Item S.C.Conductor	NbTi/Cu		Nb₃Sn/Cu
Operating Temperature	2 K	4.3 K	4.3 K
J_c without stabilizer			
at 5 T [A/mm^2]	4500*	2850	2500*
at 7 T	3500*	1700	1500
at 8 T	3000*	1100	1350
at 10 T	1400	200*	1200
Coil Current Density			
at 7 T [A/mm^2]	1400*	500	500
at 10 T	500	60*	400
Temperature margin in operation	small	medium	large
Mechanical Properties of cables	moderate, ductile	moderate, ductile	poor, brittle
Heat treatment	moderate	moderate	hard
Cryostat	complicated	simple	simple
Availability of cable	commercial	commercial	under development

* Estimated values.

The improved copper stabilized NbTi superconducting cables also could be used in the accelerator magnets cooled with superfuild helium. In this way at least 30 % increment of the J_c is expected. Table 1 summarize of the properties of available superconducting cables. Superconducting cable improvements have been carried out for many years at several places in the world. However most of them have been always in basic problems such as quality control in materials, process control in the wire and cable production. Even now the improvements are gradually proceeded in industry.

Magnet design

Superconducting accelerators magnets have transverse magnetic fields like electric power generators. To bend changed particles in the inside aperture, it is necessary that the field lines to be perpendicular to the long axis of the magnet. The coil windings must be saddle shaped as shown in Fig.1.

Figure 1. Saddle-shaped coil winding at KEK

At both ends of the coil winding, the field profiles are three-dimensional. It requires numerical calculations to estimate the field and careful attention should be paid to avoid excess enhancement of field flux density in the ends. At the center of a long coil winding, however it is permissible to use two-dimensional calculations for estimation of the field profile.

1) Ideal crosssections for dipole and quadrupole

The two-dimensional current configuration which gives rise to ideal dipole or quadrupole field, could be analytically determined. Starting from Beth's cylindrical current-sheet theorem with the complex field variable $B = B_y + iB_x$ in conjunction to the usual position variable $Z = x + iy$ and Cauchy-Riemann equations over the Z-plane as follows,

$$\frac{\partial B_x}{\partial x} + \frac{\partial B_y}{\partial y} = 0 \quad \text{and} \quad \frac{\partial B_y}{\partial x} + \frac{\partial B_x}{\partial y} = 0 \tag{1}$$

which are equivalent to Maxwell's equations in current-free regions[5].

Complex methods of field analysis may be extended to the interior of current carrying conductors by using a working variable $F(z) = B(z) - \left(\frac{1}{2}\right) \mu_0 J z^*$, where μ_0 is permeability, J is current density and z^* is the complex conjugate. It is trivial that the Cauchy-Riemann equations for $F(z)$ are equivalent to Maxwell's equations. With this variable, the fields inside and outside conductors carrying uniform current density may be determined. For example, the internal field of intersecting elliptical conductors may be given in the following equations.

$$B_{in} = \frac{\mu_0 J}{a+c} (ax - iby) \tag{2}$$

where a and b are the semi-axes of the intersecting ellipses in the x- and y- directions. When two intersecting ellipses carrying equal current density but opposite direction, the aperture field is given by the following equations

$$B_y = \frac{-\mu_0 J s a}{(a+c)} \quad \text{and } B_x = 0 \tag{3}$$

(a)
Dipole

(b)
Quadrupole

(c)

(d)

Small
Quadrupole

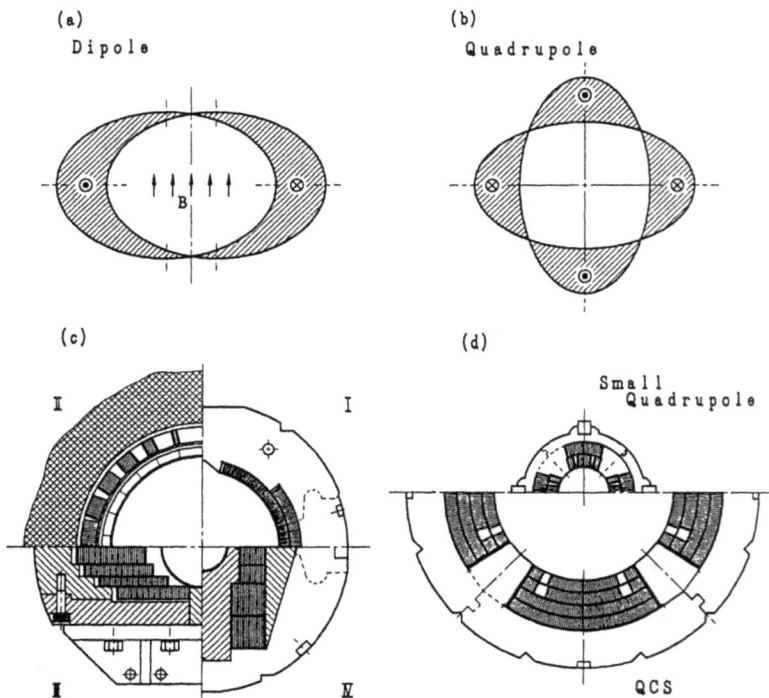

Figure 2. Ideal current crosssections for dipole and quadrupole and their practical approximations.

where s is the separation of the ellipse center points. This means that an ideal dipole field is generated in the aperture. If the ellipses are crossed quadrangle, a perfectly pure quadrupole field is given in the aperture. Figure 2 gives ideal cross-sections for dipole and quadrupole and their practical approximations.

2) Practical cross-sectional current configurations

Practical coil windings are usually approximations to the ideal shapes in cross-sections i.e. cos $(n/2)\theta$ or overlapping ellipses. In Figure 2(c), some practical cross-sectional examples are shown. For dipoles, the most popular approximation is in the first quadrant. This type of dipoles is called as double shell cos θ dipole and used in Tevatron, HERA[6], UNK[7] and others. The approximation in the second quadrant is one layer six-block cos θ winding and the ones in the 3rd and 4th have difficulties to accommodate the beam pipes at the ends.

So far quadrupole magnets, the typical crosssections are approximated cos 2θ windings as also shown in Figure 2.

3) Numerical field calculation

As mentioned previously, the field profile determination at the ends requires three-dimensional numerical calculations. However, the two dimensional field profile determination at the center of magnet with numerical computation is also very useful.

In practice, the field profile determination is possible with the real position data of every strand in coil cables. The field profile could be accurately estimated from the constituent of superconducting cable configuration in the magnets.[8]

Accelerator magnet production

When a large number of superconducting accelerator magnets is required for a multi-TeV accelerator, the production is usually proceeded as is in the following procedure.

(1) Cross-section and end design of magnets.
(2) Material selection for cables, insulators, non-magnetic supporting structure, magnetic shield, cryostat and etc.
(3) Developments of suitable superconducting cables.
(4) Qualified industrial cable production.
(5) Short magnet fabrication and performance testing.
(6) Full size prototype magnet fabrication and testing with cryogenics.
(7) Preproduction of magnets and system testing.
(8) Mass production of magnets for a real accelerator system.
(9) Accelerator system construction and test operation

The production process is indeed not a straight way but a series of many feedback loops from the latter processes to the former processes and it must be trial and error.

For instance if a short magnet test result is poor, a long magnet could not be fabricated and all previous processes should be carefully rechecked to shoot the reasons of poor characteristics. This means that if the problems such as insulation failure, heavy training and

728

insufficient field quality happen in a short magnet, one could not proceed to the next step. Today some short magnets seem to be acceptable for full size prototype magnet production.

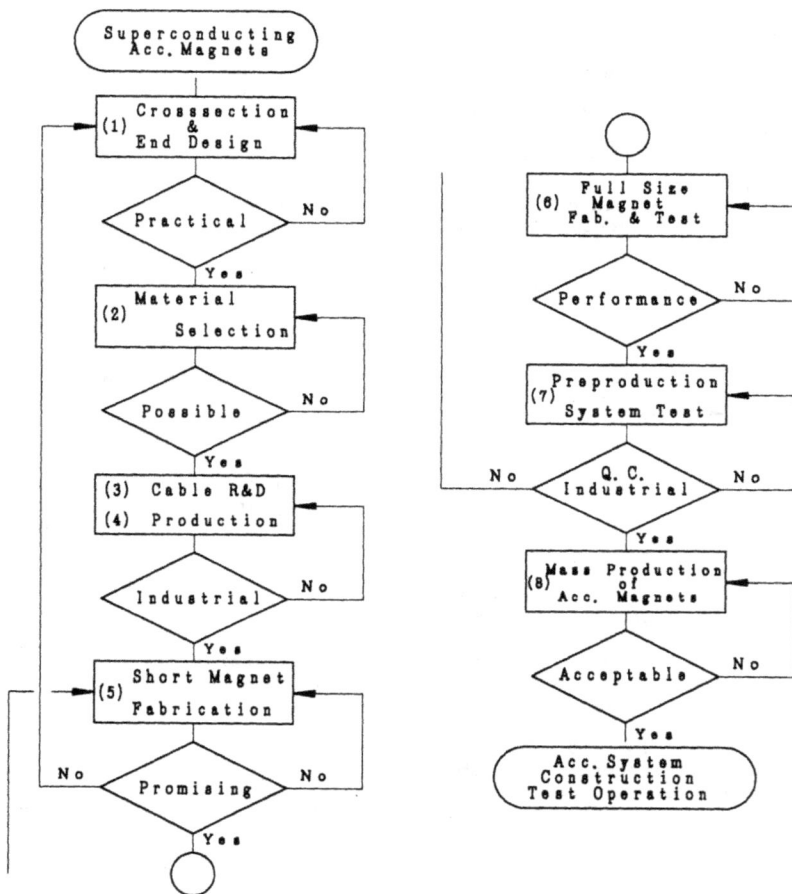

Figure 3. Production flow chart of superconducting accelerator magnets

Indeed the Tevatron was constructed with great effort of feedback loops and also trial and error. Presently the HERA magnets are in the preproduction stage (7) and the SSC magnets are under development of full size dipoles (6). Recently a LHC short magnet was made with collaboration of CERN and European industry[9]. Figure 3 shows the production flow chart of superconducting accelerator magnets.

Cryogenic system

Cryogenic system of a multi-TeV hadron accelerator-collider must be extensive to provide performance for various operation conditions such as cooldown, routine refrigeration, quench recovery and warm up of magnets. Such accelerator-collider must be too large in the diameter, therefore, it looks like long rather than circular. In this case a centralized cryogenic system is not practical and the cryogenic system should be divided in several units. Each unit could be operated independently and neighboring units should also be operatable cooperatively.

Heat load in cryogenic temperature is generated by synchrotron radiation, beam loss and static heat inleak. To get rid of heat load, the refrigeration powers at (2K), 4 K, 20 K and 80 K are required. The cold mass and helium inventory must be very big as several tens-thousand ton and a few million Nm^3, respectively.

Today helium refrigerator-liquefiers up to several kW at 4.2 K are commercially available. Problems are how to optimize and how to make efficient them for a accelerator-collider. One suggestion is to use subcooled single phase liquid helium as coolant for accelerator superconducting magnets. The stable operation with this coolant has been proved by the Tevatron experience.[10]

Summary of the key technology

Prior to the construction of the next generation hadron accelerator-collider, a lot of problems should be solved. Some of them look like difficult for the present technical level in the laboratories and related industries. The ancestor of superconducting accelerators is of course the Tevatron, therefore, the technical issues such as superconducting cable, field quality, end and iron effects, magnetic force, coil packing, quench protection training and system assembling, must be studied again with the Tevatron experience. However, the next accelerator-collider requires at least one order of magnitude larger number of superconducting magnets. It should be mentioned that many issues originate in the great difference of number of required magnets.

For example, if the magnets were produced by the same technology as the Tevatron, the accelerator operation loss time caused in magnets and cryogenics failures must be proportionally longer depending on the scale of accelerator-collider. It should be crucial problem not only for the accelerator-collider operation but also for physics experiments. It means a great reduction of useful machine time.

Before the mass production of magnets, we have to overcome the technical difficulties. Personally I belive that the key technology in the next generation hadron accelerator-colliders must be the up graded quality control in the magnet production, therefore, it requires the quality control system as is in the present automobile industries. Indeed several thousand accelerator magnets with a few hundred well qualified parts and performance are truly required for the next hadron accelerator collider.

The process for such superconducting accelerator magnet production shall be as follows.

(1) Qualified industrial superconducting cable production.
(2) Complete understanding on the full size magnet production method.
(3) Almost automatic magnet production with up-graded quality control.
(4) Arc cell magnet system tests with cryogenic.
(5) Accelerator-collider system installation and test operation.

Careful attention should be paid that the quantitative issues in high technology are often changed to the quality control problems. Now we are still remain in the first and second process for the next generation accelerator-collider.

There is not enough space to discuss on the other types of superconducting accelerator magnets, however, only a few words shall be given. For instance small superconducting accelerator magnets for compact synchrotron, cyclotron and etc., the problems should be solved individually depending on the characteristics of accelerators.

Acknowledgements

The author is grateful to the KEK staff working on accelerator superconductivity for their useful discussions on magnets and cryogenics and also thanks Mr. M. Ikeda of FEC for useful discussions on superconducting cable characteristics.

References

1) Conceptual Design of the Superconducting Super Collider, SSC-SR-2020, March 1986
2) G. Brianti and L. Burnod, "The Large Hadron Collider in the LEP Tunnel", Proc. of the ICFA Seminar on Future Perspectives in High Energy Physics, pp 179 - 199, October, 1987, BNL52114
3) H. T. Edwards, "The Tevatron Energy Doubler: A Supercoducting Accelerator", Ann. Rev. Nucl. Sci. Vol. 35, pp 605 - 659 (1985)
4) H. Hirabayashi, "High Field Superconducting Magnets for Particle Accelerators", IEEE Trans. on Nucl. Sci., Vol. NS-30, pp 3304 - 3308 (1983)
   K. Tsuchiya et al., "A prototype superconducting insertion quadrupole magnet of TRISTAN", Adv. in Cryog. Eng. Vol. 31, pp 173 - 180, (1986)
5) R. A. Beth, Proc. 6th Int. Conf. High Energy Accelerators, CEA, Cambridge, Massachusetts, p.387 (1967) and J. Appl. Phys. 40, 2445 (1969)
6) R. Kose, "Status of the HERA-Project", Proc. of the 1987 IEEE Particle Accelerator Conf., pp 29 - 33, March 16 - 19, Washington D.C. (1987)
7) A. I. Ageyev et al., "IHEP Accelerating Storage Complex (Status and Development)", Proc. of the 1987 IEEE Particle Accelerator Conf., pp 34 - 338 March 16 - 19, Washington D.C. (1987)
8) K. Ishibashi, et al., "Design Study on Future Accelerator Magnets Using Largely Keystoned Cables", Presented at (1988 Applied Superconductivity Conference at San Fransisco, Aug. 1988)
9) R. Perin, D. Leroy and G. Spigo, "The first Industry Made Model Magnet for the CERN Large Hadron Collider", (presented at 1988 Applied Superconductivity Conference at San Fransisco, Aug. 1988)
10) M. S. McAshan, "Refrigeration Plants for the SSC", SSC-129, May 1987

SUPERCONDUCTING MAGNET SYSTEMS FOR MRI

D.G. Hawksworth

Oxford Magnet Technology Ltd.,
Wharf Road, Eynsham, Oxon, England. OX8 1BP

Abstract - MRI is the first large scale commercial application of
superconductivity and has now achieved the status of a mature industry
with an annual turnover in the magnet industry alone in excess of $150M.
Conservative estimates put the investment of the medical industry in MRI
as a whole at more than a billion dollars.

In the nine years since shipment of the first superconducting whole
body imaging magnets of 0.3 Tesla field the standard product of the
industry has become a system of 1 metre bore and field strength 0.5 Tesla
to 1.5 Tesla although a significant number of magnet systems for
spectroscopy application have also been constructed with field strengths
of 2 Tesla and above. The highest field achieved to date in a whole body
system is 4 Tesla with a stored energy of 35 MJ.

In this paper the evolution of present day MRI magnets from small
bore but high field spectrometer magnets is reviewed and the direction of
future developments discussed.

Introduction

Although the principles of Magnetic Resonance Imaging (MRI) were
established by Lauterbur [1] in 1973 it was not until 1982 that the first
commercial units were introduced to the medical community at large.
However, since that time the growth of the MRI market has been rapid and
it is now the preferred diagnostic modality for imaging the head, neck and
spine. There are now more than 1200 MRI systems installed worldwide of
which approximately 85% utilise superconducting magnets. The ability of
the superconducting magnet industry to produce a reliable, cost effective
product has been a significant factor in the growth of this market.

Evolution of NMR and MRI magnets

High resolution NMR spectroscopy was first developed in the 1950's
and has proved to be a uniquely powerful enabling technique in many areas
of science, especially those concerned with the structure of complex
molecules. Indeed, the development of this technique has revolutionised
our understanding of these areas and has also expanded into medical
applications through MRI. Since the early 1970's when the first high
field NMR magnets were produced superconductivity has played an integral
part in the evolution of this technology. Some milestones in this process
are given in Figure 1. Note that the nomenclature of NMR refers to magnet
field in terms of equivalent frequency thus a 10 Tesla magnet is
equivalent to 42 MHz.

732

	NMR	MRI
1971	First Superconducting NMR Magnet (270 MHz/54mm bore)	
1973	First Multifilamentary NMR Magnet (360 MHz/54mm)	Lauterbur - spatial encoding. [1]
1977		0.1 Tesla Copper-wound MRI Magnet
1978	First High field NB3SN NMR Magnets (470 MHz/41mm)	0.2 Tesla Aluminium Wound MRI Magnet
1979		First superconducting MRI Magnets 0.3 Tesla (Hammersmith Hospital) 0.35 Tesla (U.C.S.F.)
1980	Introduce 500 MHz/51mm	
1981	First Horizontal MRS Magnet (85 MHz/310mm)	
1982		First 1.5 Tesla MRI Magnet
1983		First 2 Tesla MRI Magnet
1984	High Field MRS Magnet 200 MHz/330mm	
1986	Worlds Highest Field NMR Magnet 600 MHz/51mm	First 0.5 Tesla and 1 Tesla Active Shield Magnets
1987		First 4 Tesla MRS Magnets - 1 metre bore

Fig. 1 Milestones in the Evolution of NMR and MRI
magnet technology at OXFORD

The requirement of the original laboratory, high field, small bore NMR spectrometer magnet is for a persistent, homogeneous field and this in turn represents the technology base on which the MRI magnet of today is based.

Since their introduction to NMR in the early 1970's the continuous development of magnets for NMR application has proceeded along several parallel paths:
Firstly, the application of NMR spectrometry to systems of ever greater molecular complexity and importance has required the development of

magnets of increasing field strength since both the signal to noise ratio
and also spectral resolution increase with increasing field strength. The
world's highest field NMR magnet is presently a 600 MHz (14.7 Tesla) 51mm
bore system manufactured by Oxford Instruments. This magnet utilises both
NB3Sn and NbTi conductors.
Secondly, the advent of MRI has required the development of 1 metre bore
magnets of field strength 0.5 Tesla to 1.5 Tesla that are reliable and
cost effective, both in manufacture and performance. A signficant
requirement has also been that these products achieve the quality and
safety standards demanded by the medical industry.

A fact that has aided the development of superconducting magnets for
MRI is that the technology of homogeneous and persistent magnets was
already well established and that the scaling up from small bore NMR
spectrometer magnets to systems of 1 metre bore required no fundamental
advance in superconducting materials or magnet technology.

A comparison of industry standard specifications for MRI magnets in
1982 and 1988 is shown in Figure 2. From this it can be seen that while
there has been a gradual refinement of all aspects of the specification
over this period the only significant changes have occurred in the areas
of mobility and shielding. The reason for this relative stability of the
magnet specification is due to a reluctance of the MRI system manufacturer
to compromise on those elements of homogeneity, bore size and field
strength which impact image quality. This decision has a significant
impact on system cost. It is noticeable that in the two areas, mobility
and shielding, where significant changes have been made that this has been
the result of end user demand. It is probable that in the near future
end user pressure for reduced system cost will also result in changes in
other areas of the magnet specification. The trade-offs in such a
compromise are discussed below.

	1982	1988
Field	0.3 - 1.5T	0.5T - 2.35T (4.0T)
Bore	1.0m	0.9 - 1.2m
Homogeneity	40 ppm	10 - 20 ppm
Stability	<0.1 ppm/Hour	<0.1 ppm/Hour
Ramp Time	2 - 5 Hours	0.2 - 0.5 Hours
Helium Boil Off	0.5 Litres/hour	0.15* - 0.4 Litres/Hour
Nitrogen Boil Off	2.0 Litres/hour	0* - 1.0 Litres/Hour
Shielding	-	Room, Yoke, Active
Transportability		Mobile

*Two Stage Cryocooler

Fig. 2 Comparison of Industry Standard Magnet
Specifications in 1982 and 1988

The third area of NMR product development has resulted from the
desire to expand the application of NMR to involve in vivo experiments on
animals and humans. Magnets of field strength up to 4.7 Tesla and 330 mm
bore have been developed for animal experimentation and there are also
many installations in hospitals and research institutes worldwide with 1
metre bore 2 Tesla systems which are being used to undertake whole body
spectroscopy experiments. The highest field whole body MR system
constructed to date is of 4 Tesla field, representing a stored energy of
35 MJ. Two such magnets have been manufactured by OXFORD and are now
operative at the General Electric Research Laboratory in Schnectady, New
York and the Philips Research Institute in Hamburg, Germany. A third 4
Tesla has also been constructed by Siemens AG and is presently operative
in their own research laboratories. [2]

MRI System Configuration

The components of the MR system are shown in Figure 3; the main
elements of which are:

> magnet
> gradient system
> r-f coils

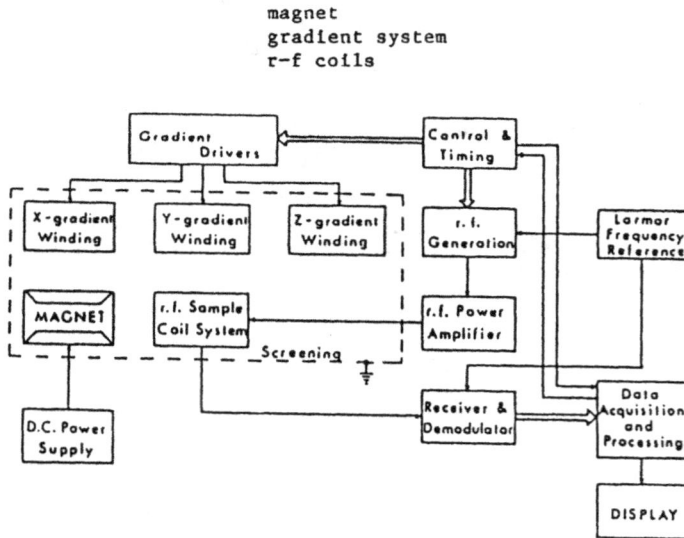

Figure 3: Components of the MRI system

The magnet provides a homogeneous and stable field over the volume to
be imaged.

The gradient system comprises an orthogonal coil set and associated
power supplies which generate pulsed linear magnetic field gradients over
the imaging volume. The switching of the field gradients with rise times
of order 1 millisecond provides the spatial encoding of the NMR signal and
the magnitude of the field gradient determines the image slice thickness.

The r-f coil(s) transmit signals to the sample and also detect the
re-radiated signal. Special or localised search coils are often used for
imaging different parts of the body.

Although the magnet specification has been relatively stable since
the introduction of MR, due at least in part to there being a limited
number of magnet manufacturers and the dominance of one supplier, OXFORD,
the gradient and r-f subsystems were quickly seized upon by the medical
equipment manufacturers as areas of proprietary technology and hence
product differentiation. However, while the R-F coil design is
independent of the magnet system, the gradient system performance is
critically determined by the magnet bore size and configuration. The
integration of the magnet and gradient system design must therefore
represent a significant opportunity in the near future for overall
optimisation of the MR system in terms of performance and cost.

MR Magnet Design

A cross-section through a typical MRI cryostat is shown in Figure 4.
The magnet and helium vessel are surrounded by two radiation shields, one
at approximatley 80K, cooled by a liquid nitrogen reservoir; the other at
about 40K cooled by heat exchange with the effluent helium gas. The outer
vacuum vessel (OVC) is constructed in accordance with international
pressure vessel regulations (ASME 8, BS5500, J.I.S). Inside the room
temperature bore of the cryostat is a resistive or passive (iron) shim
assembly which is used to correct the magnet homogeneity for manufacturing
tolerance errors or site environmental effects. Some manufacturers
utilise superconducting shim coils either to complement or in place of
resistive or passive shims.

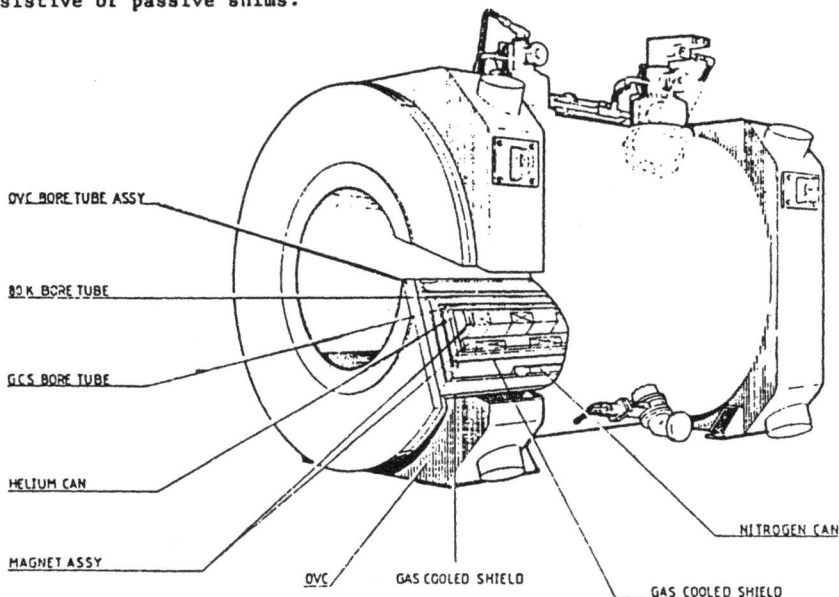

OVC BORE TUBE ASSY
80 K BORE TUBE
GCS BORE TUBE
HELIUM CAN
NITROGEN CAN
MAGNET ASSY
OVC   GAS COOLED SHIELD
GAS COOLED SHIELD

Figure 4:   Cross-section through an OXFORD Active Shield cryostat

The length of the system, of typically 2.3 metres, is dominated by the magnet, the dimensions of which are those necessary to achieve the desired level of homogeneity. The earliest magnet designs used 3 or 4 coil arrays but these were very quickly replaced by 6 coil systems during 1982 which are now the industry standard. In Figure 5 the theoretical homogeneity volume of 4 and 6 coil systems of 1 metre bore and 1.5 Tesla field is compared. While the 6 coil design in theory eliminates field error terms up to 12th order [3] and produces a spherical volume of homogeneity, a 4 coil array retains significant 6th and 8th order error terms and gives rise to an oblate spheroid homogeneity volume where the field quality degrades rapidly as one moves away from the midplane.

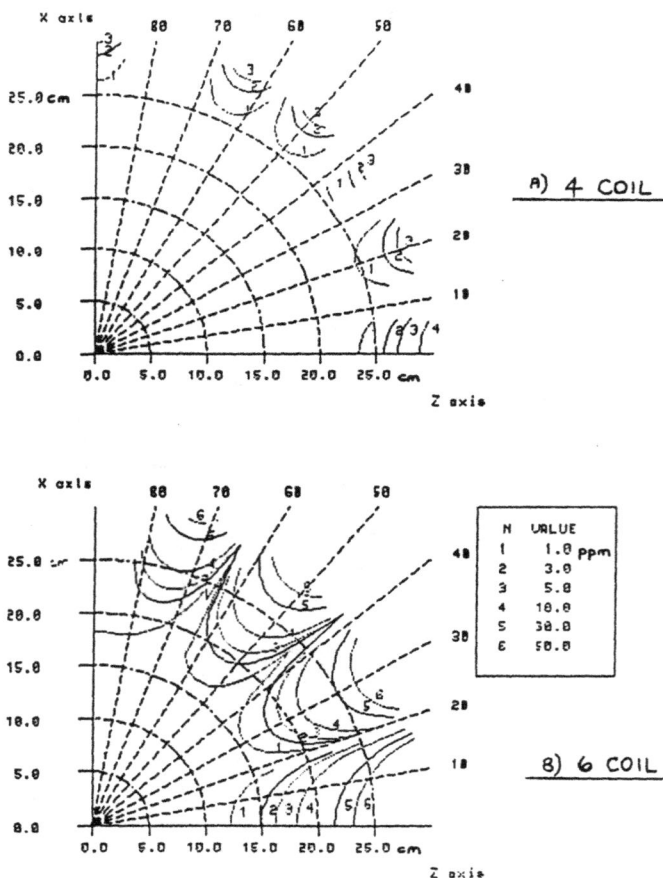

Figure 5: Theoretical homogeneity contours of 1.5 Tesla 1 metre bore magnets

In Figure 2 where the industry standard specifications of 1982 and 1988 are compared it is seen that the final magnet homogeneity specification has improved from 40 ppm on 50cm dsv to 10-20 ppm on 50 cm dsv over this period. This improvement reflects a refinement in magnet design, a greater understanding of manufacturing methods, an improvement in the definition of homogeneity (early definitions gave what would now be considered an optimistic result) and most significantly the development of improved shimming capabilities.

Passive iron shimming [4] is now routinely used to correct field inhomogeneities on magnets of all field strengths and is a uniquely powerful tool since it allows the correction, with relatively small amounts of iron, of high order field errors which cannot be corrected electrically. Thus where a magnet can be routinely shimmed to 15-20 ppm on 50 cm dsv using electrical shims only, if passive shims are used homogeneities of less than 10 ppm can be achieved. An added advantage of passive shimming is that it is a route to reduced cost since it can eliminate the need for expensive shim coil assemblies and power supplies.

Electrical shims, whether resistive or superconductive, are advantageous, however, where the magnet operating field is changed often or under circumstances where the environment is under-going dynamic changes e.g. mobile.

A consequence of the improved shimming method is that magnets of all field strengths at 1 metre bore can now achieve comparable levels of homgeneity (in ppm). This results in a situation where although even further improvements in the homogeneity of 2 Tesla systems are desirable since this will significantly enhance the signal to noise ratio of NMR spectra, the same improvement on a 0.5 Tesla system produces no real advantage and therefore it can be said to be ´overdesigned´ and hence by implication ´cost inefficient .

MR Mobility

In 1984 as a result of the adverse US regulatory climate [5] a significant market developed for a mobile MRI scanner which would travel between several hospital sites on a regular schedule. The market had already been primed and the technology developed by mobile X ray Computer Tomography (CT) devices so that the growth in this market sector was rapid. In the first year 35 mobile systems became operative and mobile MRI will account for 20% of all US MRI sales in 1988. The layout of a typical mobile MRI trailer is shown in Figure 6.

The mobile MRI market placed new demands on the magnet supplier for system robustness, low cryogen consumption and general user friendliness. These requirements have now become standard on all systems whether intended for the static or mobile market.

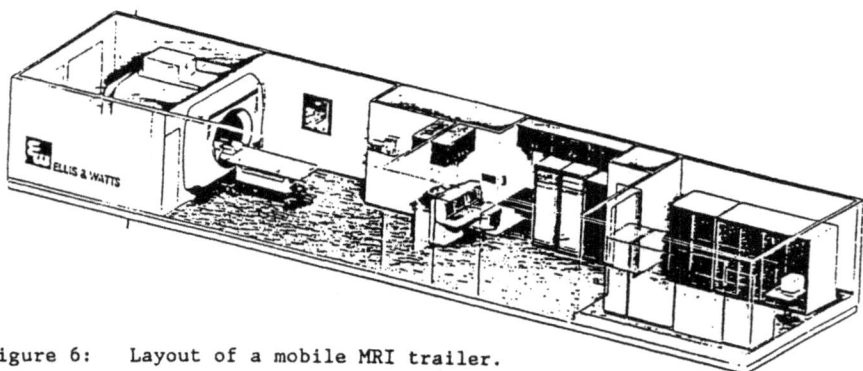

Figure 6:   Layout of a mobile MRI trailer.

In 1985 three magnet manufacturers OXFORD, Intermagnetics General
Corporation (IGC) and Applied Superconetics (ASC) competed for the Mobile
MRI market, OXFORD dominated the 1.0 Tesla and 1.5 Tesla market by virtue
of selling to General Electric and Siemens whereas IGC and ASC supplied
0.5 Tesla systems to Picker International, Technicare and Diasonics.   It
is perhaps strange that the mobile market, because of its emphasis on low
cryogen consumption, provided the first real test of refrigeration in the

market place.   The OXFORD cryostat utilised two stage cryocoolers; General
Electric adopted a Collins Cycle reliquifier and IGC evaluated a Gifford
McMahon cycle recondenser.   From the OXFORD perspective this experience
showed that two stage refrigeration is a viable and reliable product but
it is heavily dependant on the reliability of interactive system
components including chillers and electrics which must be adequately sized
and maintained.

The utility features developed for the mobile market included:

a.   automatic ramp to field so that imaging could start as soon as
      possible once the magnet arrived on site; for this development
      of diode protection circuitry was necessary in order to minimise
      liquid helium ramp loss.

b.   low loss permanently installed current leads were developed.

c.   ramp settling algorithms were developed in order to achieve
      adequate short term stability ($< 3ppm/hour$) of the magnet field.

d.   sophisticated magnet and shim power supplies were developed
      which allowed untrained staff to be used to operate the magnet.

The most recent development in the mobile market is the introduction
during 1987/88 of shielded magnets.   General Electric have introduced a
0.5 Tesla system, MR Max [6] where an Fe Yoke shield is integrated into
the cryostat OVC; the magnet is of 940mm bore and weighs 13 tonnes.
Simultaneously, OXFORD have introduced the 0.5 Tesla Active shield magnet
[7, 8] where the shielding is produced by superconducting cancellation
coils; this system has a 1 metre bore and weighs 9.4 tonnes.   Both the GE
MR Max and OXFORD Active Shield systems utilise a single two stage
cryocooler.

Refrigeration and MRI

Worldwide, refrigeration and particularly reliquification has so far played a minor role in static site MRI installations. In the USA where 70% of the MRI installed base lies, the reason for this is because direct cryogen costs contribute only a minor part to the cost of MRI scans which is instead dominated by fixed capital costs [9]. In Japan and Korea, however, the high local price of liquid helium has meant zero loss systems are more attractive although the relatively few superconducting systems installed to date has meant the market opportunity is small.

While there is now a relatively large installed base of mobile MRI systems (approximately 80) utilising two stage cryocoolers, the economic justification for these comes from the elimination of nitrogen boil off and hence the associated weekly service intervals, rather than the reduction in helium boil off which is typically a saving of 200 cc/hour or $10K/year.

The viability of zero loss systems is even more debatable:

1) Although Gifford McMahon systems have been successfully demonstrated on a mobile truck [10] and in a fixed site installation in Chiba, Japan, the latter for a period of 3 years, sales of such units have so far numbered less than ten. Due to a lack of cooling power at the 80K stage such units also require an auxiliary cryocooler to eliminate nitrogen boil off.

2) Collins Cycle reliquifiers introduced by General Electric on their mobile systems in 1985 although capable of eliminating nitrogen boil off have proven highly unreliable and have since been withdrawn as a product offering.

3) The most recent zero loss product offering is from Siemens A.G. which utilises a turbine reliquifier on their 2 Tesla 'Magnetom'. No data is yet available on the reliability of this product.

4) Although the Japanese market is predicted to grow rapidly in 1989/90 the price of liquid helium is falling and this will remove the economic justification for zero loss sytems. Since 1984, for example, the price of liquid helium in Japan has halved. Similarly in Korea where, although liquid prices are higher than in Japan, the lack of a local service support organisation means that yearly servicing and preventative maintenance costs are prohibitive.

In conclusion; unless helium prices rise significantly the future market for zero loss systems is small.

MR Magnet Shielding

One of the major considerations when siting an MRI system within a hospital is its interactions with the environment. Ideally one would like to place the MRI facility close to other imaging modalities such as X ray CT, in areas readily accessible to patients. However, Food and Drug Administration guidelines limit the maximum field for public access to 0.5mT, which for a 1 metre bore 0.5 Tesla system is some 9 metres from the magnet centre (for a 2 Tesla magnet 0.5mT is at 14.2 metres distance).

Furthermore, X ray CT scanner gantries must not be placed in fields exceeding 0.05 mT and there are similar limitations on other electronic and computer equipment. Clearly these factors pose significant problems and potential cost to the hospital administrator when siting an MRI system.

In most cases the response of the MRI system manufacturers to these problems has been to offer a room shield, where iron plates are placed in specific locations on the walls of the room in which the magnet is located, so as to provide limited shielding of selected areas. Siemens A.G., however, have marketed an iron yoke or ˆself shieldedˆ product which has given them significant product differentiaton. A variant of this

solution has also. recently been announced by General Electric who utilise an integral yoke as the cryostat OVC in their MR MAX product.

In 1986 OXFORD announced the Active Shield product which utilises superconducting cancellation coils to reduce the stray magnetic field. Since that time 50 0.5 Tesla and 10 1.0 Tesla systems have been manufactured. The principal advantage of Active Shield is the high shielding efficiency for low system weight. If the 0.1 mT contour were to be taken as the criterion rather than the standard 0.5 mT value then this difference would be even more pronounced. The excellent shielding efficiency of the Active Shield system is shown in Figures 7 and 8.

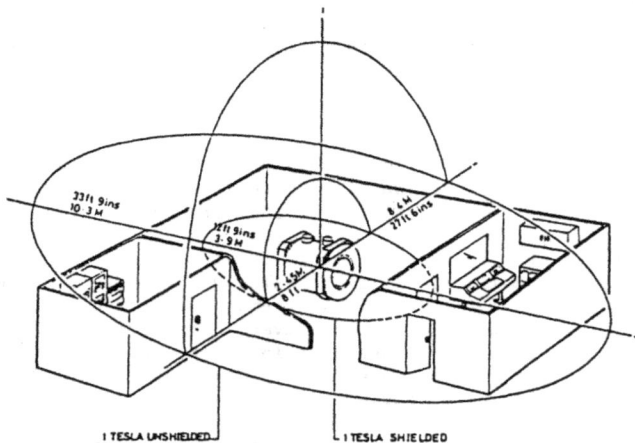

I TESLA UNSHIELDED   I TESLA SHIELDED

Figure 7:   0.5mT stray field contour for unshielded and Active Shielded 0.5 Tesla magnets superimposed upon a typical MRI site.

The relative merits of Room, Yoke and Active Shield are discussed in reference 8 which also considers the engineering issues associated with Active Shield.

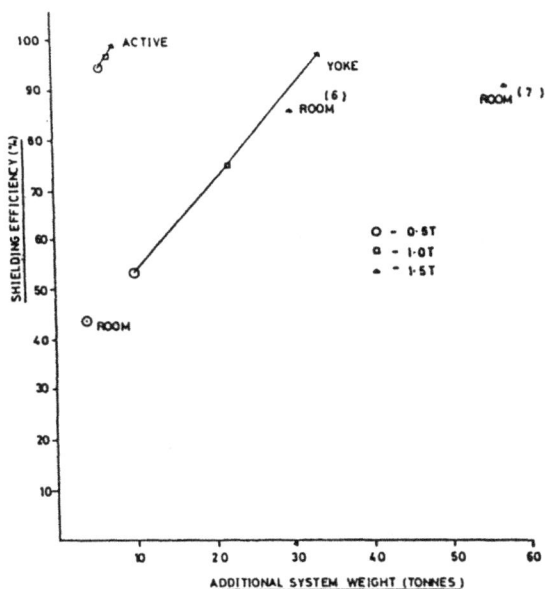

Figure 8:   Comparison of the Shielding Efficiency and weights of
Active and Passive Shielded magnets (ref. 8).

MR Market and New Developments

Although the growth of the MRI market has been rapid it has not
matched the phenomenal growth expectations of some tipsters who predicted
it would emulate the X ray CT market which doubled each year for four
years after its introduction in 1973. For example, in 1985 some estimates
of MRI system sales volume for 1988 were as high as 1200 units at a price
of $2M each; in practice actual installations are unlikely to exceed
500 - 550 units at an average price of $1.4M. Magnet prices in 1988 range
from $200K to $650K for field strengths 0.5 Tesla to 2 Tesla.

In figure 9 the segmentation of the MRI magnet market is shown for
1985 and 1990. The market share of resistive and permanent magnets is
expected to grow over this period reflecting the lower cost expectations
of the market and the growth of these magnets into new clinical areas
where siting difficulties are more acute e.g. emergency rooms or intensive
care units. Overall though it is expected that the dominance of the
superconducting magnet will continue to prevail on account of its higher
field, which gives improved signal to noise, and better temporal stability
and homogeneity.

742

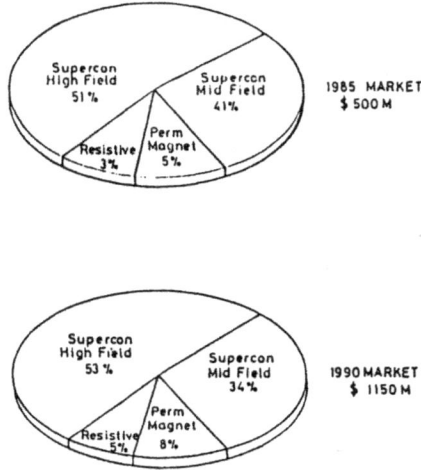

Figure 9:   Segmentation of the MRI market for 1985 and 1990.
(Courtesy of Frost and Sullivan).

There has been much debate within the Imaging industry since 1982 on the choice of the optimum field strength as it impacts perceived image quality. The argument made in favour of high field systems (1.5 Tesla) relative to their medium field counterparts (0.5 Tesla) is that as field strength is increased so better images result due to higher signal to noise, thinner sections and fast imaging. Hindsight shows that much of this debate was fuelled by the commercial marketing needs of General Electric since the relative merits of high field versus medium field imaging are not clear cut given the development of new technologies such as surface coils, and acquisition sequences, which for example, can increase S/N at midfield above those achieved at high field strength. Indeed, the initial protagonist, General Electric, has now introduced a midfield system. What is clear in the high field versus midfield argument is that there is a clear correlation of field strength versus cost, both in terms of the capital cost of the equipment and also its siting cost due to pollution of the environment by the magnet stray field.

The expectation of the market is therefore for systems of lower total cost and this can only be achieved by integrating all the cost elements: magnet, gradients, shielding, electronics and installation costs and searching for an optimum solution. There is negligible mileage from a magnet perspective, for example, in reducing cryogen consumption alone since this is a minor part of total scan costs which are dominated by capital costs [9]. The only effective means to reduce costs is to reduce the capital cost of the equipment or improve patient throughput.

The tradeoffs of cost against performance taking all the elements magnet, gradients etc. into account are complex. A 'wish list' specification would be that the system would be simultaneously of small dimensions, low weight and shielded and that there be no compromise over the imaging performance as judged by present day standards.

In practice the principal decision is in the trade off of magnet bore size versus increased gradient bore tube eddy current interaction.

In order to achieve economies in the magnet a reduction in bore size is desirable since this will both reduce conductor and cryostat vessel costs while at the same time reducing the magnet stray field and hence by implication shielding costs. Unfortunately the requirements of the pulsed gradient system are exactly counter to this since the reduction in bore size increases the magnitude of the eddy currents induced in the bore tube radiation shields and thus effectively reduces the gradient strength. The effect of reduced bore on the imaging experiment is thus to effectively reduce slick thickness/resolution and increase the eddy currents induced, time dependant, spatial variation in homogeneity. The sensitivity of reducing the magnet bore on gradient overshoot, which is a barometer of degree of compensation required to counter eddy current effects, is shown in Figure 10.

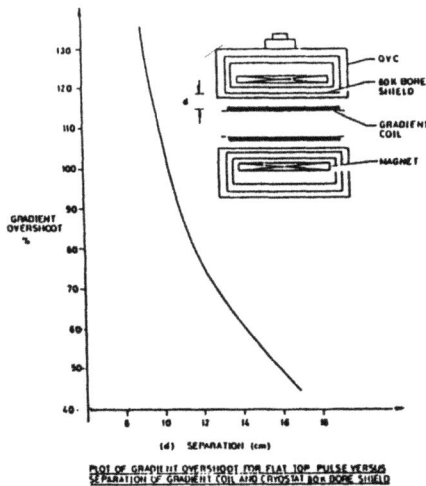

Figure 10: Plot of gradient overshoot versus separation of gradient coil and cryostat 80K bore shield.

Although the optimum system configuration is not clear, to the medical equipment manufacturer who solves this conundrum first the financial rewards will be great — who will it be?

References

1) P. Lauterbur
    Nature 242 190 (1973)
2) J. Vetter, G. Ries, T. Reichert "A 4 Tesla Superconducting Whole Body
    Magnet for M.R. Imaging and Spectroscopy" IEEE. Trans. Magn. 24, 2
    1285-1287 (1988)
3) M.W. Garrett, "Axially Symmetric Systems for Generating and Measuring
    Magnetic Fields" J. Appl. Phys. 22 1091-1107 (1951)
4) D.I. Hoult and D. Lee, "Shimming a Superconducting Nuclear Magnetic
    Resonance Imaging Magnet with Steel". Rev. Sci. Instrum. 56 (1),
    131-135 (1985).
5) R.E. Schwall, "MRI - Superconductivity in the Marketplace" IEEE
    Trans. Magn. MAG-232 1287-1293 (1987)
6) US Patent 4, 721, 934 Published 26.2.88.
7) US Patent 4, 587, 504 Published 6.5.86.
8) D.G. Hawksworth, I.L. McDougall, J.M. Bird, D. Black,
    "Considerations in the Design of MRI Magnets with Reduced Stray
    Fields" IEEE Trans Magn. MAG-23 2 1309-1314 (1987)
9) I.L. McDougall, N. Heiberg, K. White
    "Cryogenic Challenge from Imaging Magnets" Proceedings of the 11th
    Internation Cryogenic Engineering Conference, April 1986.
10) C.H. Rosner and G. Morrow
    "Superconducting Magnets for Magnetic Resonance Imaging Applications"
    IEEE Trans Magn. MAG-23 2 1294-1298 (1987)

DETERMINATION OF STRUCTURES OF PROTEINS IN SOLUTION BY USING A SUPERCONDUCTING MAGNET NMR INSTRUMENT

Yoshimasa KYOGOKU

Institute for Protein Reseach, Osaka University
Suita, Osaka 565, Japan

Introduction

NMR (nuclear magnetic resonance) is now widely used in the fields of biological sciences. Applications are classified into four fields: (i) structure analyses of biopolymers, (ii) in vivo NMR, (iii) solid state NMR of membranes and (iv) NMR-CT. In this article application in the field (i) will be described.

Molecule is composed of a number of atoms. Some type of nuclei like ^1H, ^{13}C and ^{15}N have magnetic moments. They are energetically equivalent where is no magnetic field. When they are placed in the field, however, energy levels split by the nuclear Zeeman effect and the nuclei populate them following the Boltzmann distribution law. When a radio frequency is applied, absorption takes place, if the irradiated frequency matches the difference between energy levels. This phenomenon is called nuclear magnetic resonance.

Requirements for NMR Instruments

Figure 1 is a plot of the Zeeman splitting against the strength of a magnetic field. Resonance frequency is simply

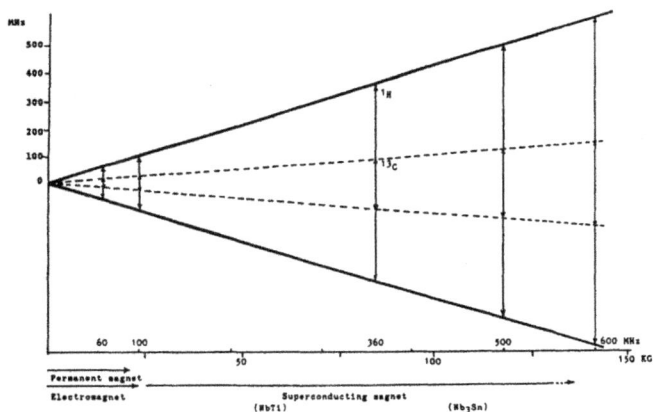

Fig.1. The relation between resonance frequency and the strength of magnetic field.

proportional to the strength of the magnetic field. The highest
magnetic field of the commercially available NMR instrument is
14.1 Tesler, i.e., 600 MHz at proton resonance. Advantage of the
higher-field instrument is good separation of overlapped
resonance signals. Each nucleus in a molecule experiences
different effective magnetic field due to the different shielding
effects by surrounding electrons, thus resonates at different
frequencies. For separation of such small differences, we
require NMR instruments quite severe field homogeneity and
stability. Guaranteed values for a 600 MHz NMR instrument are
given in Table 1.

Table 1. Guaranteed values for 600 MHz (14.1 T)
NMR spectrometer

Resolution	0.25Hz (5.6×10^{-2}mG) 4×10^{-10}
Homogeneity	(Line shape test) 10Hz at 5.5/1000
Stability	24Hz (5.5mG)/hour 4×10^{-8}

Structure Determination of Proteins by NMR

Now I would like to focus on the method to construct three
dimensional structure of a protein in solution by using NMR and
distance geometry algorithm. Structures of proteins are
generally determined for crystals by means of X-ray diffraction.
However, there is a strong demand to determine the structure of
proteins in solution where biological activity is functioned.
The process of the structure determination by NMR consists of
three steps: (i) signal assignments, (ii) estimation of inter-
proton distances from NOE (nuclear Overhauser effects), and (iii)
building of molecular model by using distance geometry
algorithms.[1]

An example of structure analyses will be presented on a
hormone peptide, α-human atrial natriuretic polypeptide, α-hANP.[2]
It consists of 28 amino acid residues, but contains one disulfide
bond forming a cyclic peptide backbone. It contains about 160

<pre>
          5   10
H–Ser–Leu–Arg–Arg–Ser–Ser–Cys–Phe–Gly–Gly–Arg–Met–Asp–Arg–
                            └─────────────────┐
 15   20   25
Ile–Gly–Ala–Gln–Ser–Gly–Leu–Gly–Cys–Asn–Ser–Phe–Arg–Tyr–OH
</pre>

Fig.2. Amino acid sequence of α-hANP.

protons (Figure 2). We observe the signals from these protons
and should assign them to individual protons. For that purpose
we employ two dimensional NMR. In 2D NMR multiple pulses are
applied and FID curves of two time variables, t_1 and t_2, are

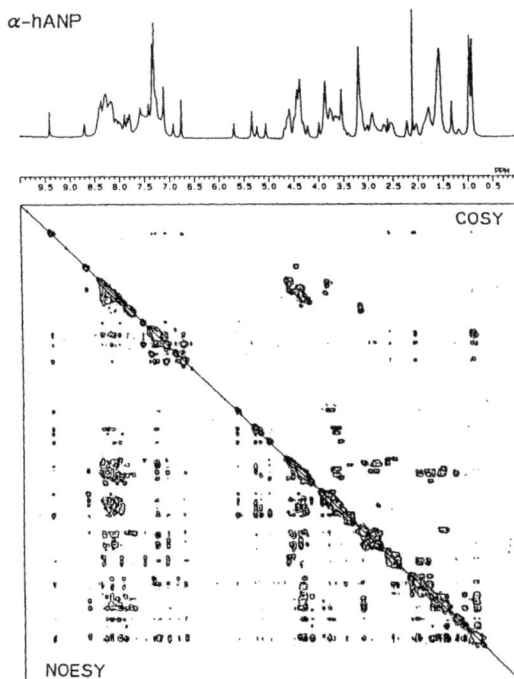

α-hANP

COSY

NOESY

Fig.3. Two dimensional
NMR spectrum of α-hANP.

collected, and are Fourier transformed into the functions with
two frequency variables. There are several variations of 2D NMR.
Upper right half of Figure 3 shows COSY type 2D spectrum of
α-hANP. A pattern of spin-spin coupling is typical for each type
of amino acid residue. Thus, for a simple peptide, most of peaks
can be assigned to each amino acid residue only from the pattern
of COSY spectrum. The lower half of the spectrum is NOESY type
spectrum where cross peaks correspond to the presence of magnetic
dipole-dipole interaction.

The intensity of the cross peaks in NOESY spectrum is called
NOE and is a function of inter-proton distances and a rotational
correlation time which indicates the extent of flexibility. If
we assume the correlation time τ_R is constant through a whole
molecule, then NOE is simply proportional to the inverse of the
sixth power of the inter-proton distance r.

$$NOE \propto A / r^6 \times f(\tau_R)$$

Then we translate the obtained NOE data into the inter-proton
distances following the above relation. Curve (a) in Figure 4
indicates the relation between r and NOE based on a rigid model
i. e., τ_R is constant. The other curve is based on a flexible
model where the distances change freely between the upper limit
and the lower limit, van der Waals contact.

Once the data of inter-proton distances are collected, we

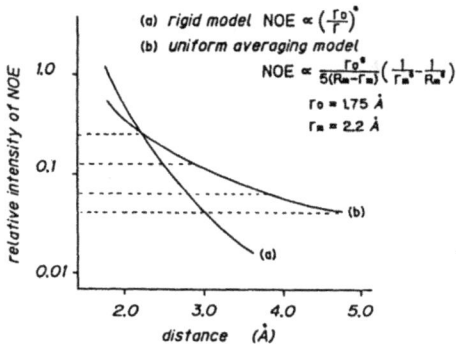

Fig.4. Dependence of NOE on interatomic H-H distances.

build up the model which satisfies the data of inter-proton distances. We will calculate the coordinates of the molecule based on the algorithm of distance geometry. There are several types of algorithms. We adopt a target function method where the variables are dihedral angles. We gave upper limits U_{ij} of inter-proton distances determined from NOE, and lower limits L_{ij} defined by the van der Waals contacts. By iteration we calculate r_{ij} to minimize the following target function.

$$T = \Sigma \frac{(U_{ij}^2 - r_{ij}^2)^2}{U_{ij}^2} + \Sigma \frac{\omega(L_{ij}^2 - r_{ij}^2)^2}{L_{ij}^2}$$

Figure 5 shows the superimpose of six backbone conformations which have smaller minimum of the target function out of one hundred trials. Their convergence seems not so good. However, when we look at the convergence of the local area they are fairly good. It means that NOE data of short range interactions are

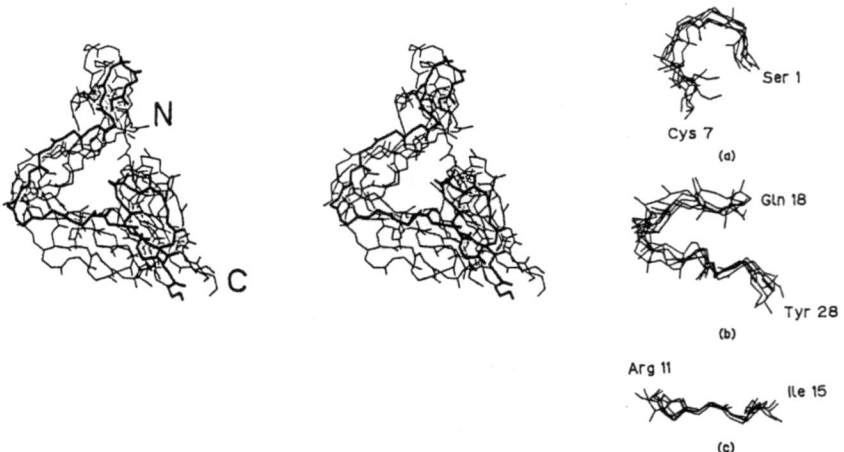

Fig.5. Computer drawings of the six conformers calculated for α-hANP.

available in a fairly large number, but there are not so many long range NOE data.

Now the method for structure determination of proteins in solution has been established. However, there are several problems and limitations. The upper limit of the molecular weight of molecules to which the method is applicable is about 15,000 because of severe overlap of signals and signal broadening. To avoid this problem there are several trials. It will be cleared, if a more powerful superconducting magnet NMR instrument like 700-1,000 MHz is developed. Another approach is a development of pulse techniques like three dimensional NMR which makes signal separation more evident.[3] The third and the most promising approach is the employment of isotope replacements. To eliminate unnecessary signals, deuterium substitution will be used and to get the signals of specific residues, ^{15}N and ^{13}C substitution will be done. The most desirable situation is the combination of the above three approaches.

References

1) K. Wuethrich "NMR of Proteins and Nucleic Acids", John Wiley-Interscience Publisher, New York (1986).
2) Y. Kobayashi, T. Ohkubo, Y. Kyogoku, S. Koyama, M. Kobayashi and N. Go., "The conformation of α-human atrial natriuretic polypeptide in solution", J. Biochem., 104, 322-325 (1988).
3) H. Oschkinat, C. Griesigner, P. J. Kraulis, O. W. Sorensen, R. R. Ernst, A. M. Gronenborn and G. M. Clore, Nature, 332, 374-476 (1988).

BIOMAGNETISM : MAGNETIC FIELDS PRODUCED BY HUMAN BODY

Yoshinori UCHIKAWA and Makoto KOTANI*

Department of Applied Electronic Engineering, Tokyo Denki University,
Hatoyama, Saitama, 350-03 Japan
* Department of Electronic Engineering, Tokyo Denki University,
2-2 Kandanishiki-cho, Chiyoda-ku, Tokyo, 101 Japan

Abstract - The computerized system with SQUID gradiometers for measuring
magnetic fields produced by the human body has been developed. By using this
system we measured the MCGs, including the MRGs related to the transient response of the retina, and the SEFs.

Introduction

Recent remarkable progress of neuromagnetism has been made it possible to
detect the exteremely weak magnetic fields produced by the excitation of the
His-Purkinje conduction system (1), the activity of the brain (2), and   the
transient response of the retina induced by the visual stimulation to the eye.
These magnetic recordings are called the high resolution magnetocardiograms (
HRMCGs), the magnetoencephalograms (MEGs), and the magnetoretinograms (MRGs),
respectivily. There is also the magnetooculograms (MOGs) which is associated
with the eye movement. This evolution is due to the fact that magnetometer
with SQUID and the computerized peripheral system for data acquisition and
signal processing have been improved.

This paper is to introduce the recent research on the biomagnetism having
been performed at our labolatory.

Configuration of measurement system

Magnetic fields produced by the human body have been measured in a nonmagnetically shielded room with a SQUID connected to a second order gradiometer
with a base line of 3.2 cm and a coil of 2.48 cm diameter. The block diagram
of the measurement system is shown in Fig.1. For data acquisition, signal

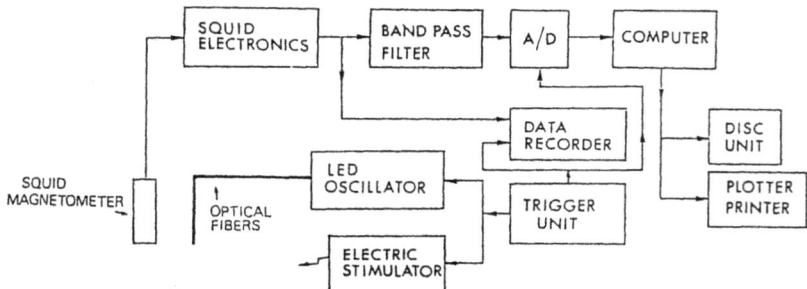

Fig.1 Block diagram of the measurement system of the biomagnetic fields.

processing, and the adjustment of the time parameter of a stimulation a micro computer with 12-bit A/D converter can control this system. All magnetic
data analyzed are stored in the hard disc and on the magnetic tape recorder.
In the MOG and the MRG measurement, in order to avoid the interference of
magnetic noise the optical fibers are used to transmit the indicating light
signals associated with the eye movement and the light stimulation of the
eye. As shown in Fig.1, this system is available to use not only for measurement of stationary fields like the MCG and the α - wave of the MEG but
also the evoked magnetic fields of the brain and the visual system which is
applied the signal averaging technique to get a signal-to-noise ratio.

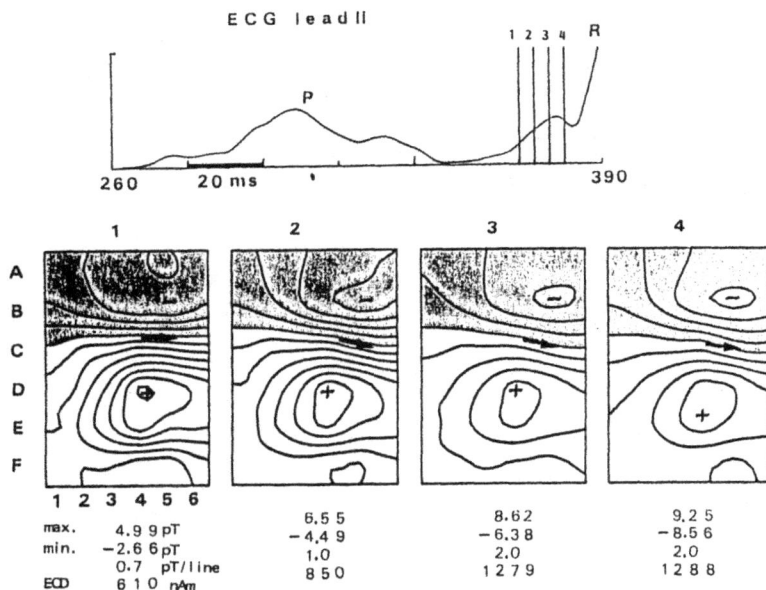

Fig.2 Isofield contour maps of the delta-wave of the WPW syndrome at
       four different instants during the PR segment. The arrows re-
       present the ECD.

High resolution magnetocardiogram (HRMCG)

Noninvasive localization of the preexciteation site in patients affected
by the Wolff-Parkinson-White (WPW) syndrome has been attemped on the basis
of the body surface mapping technique. Recently, Lorange et al.(3) proposed
a computer simulation of the WPW syndrome for electrocardiographic studies
using a modified Miller-Geselowitz Heart model. In this chapter we present
an example of the feasability of clinical application to estimate the preexcitation site of the WPW syndrome using high resolution magnetocardiographic
data.
In the healthy hearts the pacemaker signal coming from the sinus node is
delayed in the Atrio-Ventricular Node before passing to the His-Purkinje
conduction system. In the hearts showing the WPW syndrome, a leakage is

present between the atria and the ventricles. The atrial excitation passes to the accessaory pathway and produces a preexcitation. Consequently, the PR interval is reduced and a Delta wave appears. Fig.2 shows the isofield contour maps of eight succesive time instants which were obtained by measuring the MCG of a patient of the WPW syndrome. Solid arrows represent the location and direction of the equivalent current dipole (ECD) which is introduced in a meaning of the best fit between experimental and theoretical distribution. Comparing this result to normal subjects (1) it can be seen the great difference in the trace and locations of the ECD, the field pattern, and the intensities of magnetic fields. To understand and to interpret the MCGs of the WPW syndrome a three dimensional model of the ventricular excitation to be used to calculate the magnetic fields generated at the beginning of an ectopic excitation was developed (4). Using this model the computer simulation of the WPW syndrome suggested that in general in the initial stage of preexcitation a magnetic fields distribution with dipolar structure can be measured on the chest shown in Fig.2: this means that in the early part of the Delta wave it is a possible to localize the site of the earliest excitation using a simple ECD model. But the quardrupolar field pattern was also demonstrated: in this case a more complicated way is needed to describe the source structures. There are in agreement with the experimental results (5).

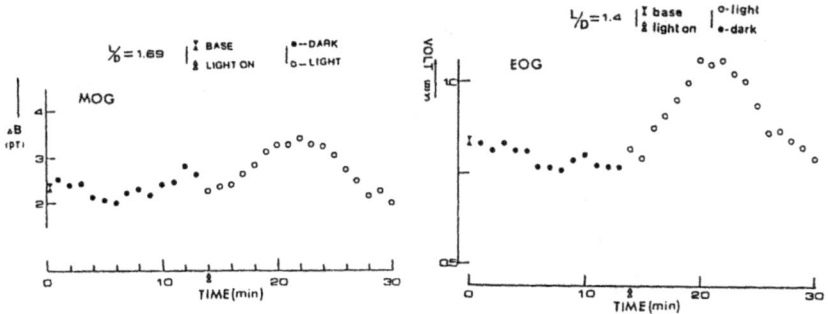

Fig.3 Light adaptation curve of the MOG and the EOG (normal).

Magnetooculogram (MOG) and Magnetoretinogram (MRG)

The strength of magnetic field, in general, can be observed on the order of 10 pT when the eye ball moves (6). The MOG associated with eye movement can be interpreted as magnetic fields produced by the current density sustained the electric potential difference across the retina and the pigment epithelium. The MOG has several advantages over the electrooculogram (EOG) in the noninvasive and contactless measurement.

Fig.3 shows the light adaptation curve of the EOG and the MOG to normal subject. The magnetic fields perpendicular to the frontal face were measured and the deflection angle of the eye movement was 30 degree horizontally. The dark trough (D) appeared at 7 minute after starting of adaptation. The light peak (L) appeared at 8 minute after a light on. The ratio of L/D was 1.69 in the MOG. A light adaptation curve of the EOG was also obtained from the same normal subject as in the MOG. Both the dark trough and the light peak appeared at close time courses as shown in the MOG and a L/D was 1.4.

Magnetic field of the MRG is so exteremely weak that it needs to avoid the magnetic noise and/or improve a signal-to-noise retio by averaging technique. The first measurement of the MRG induced by a flash was performed by a first order gradiometer by Katila and co-workers (7). Fig.4 shows the MRG induced by the light stimulation (flash: 12.4 J) with a second order gradio-meter in our wooden room, namely, nonmagnetically shielded room. This flash was generated by a stroboscope controled under electronics (see Fig.1). All magnetic fields of the MRG were measured at a distance of 3.5 cm from the surface of the bilateral head. Band pass filter was used in the range of 1 to 30 Hz and it was averaged of 100 samples. It can be seen a positive peak of the a-wave at latencies of around 13 msec and a negative peak of the b-wave at latencies of around 34 msec. These were a good agreement with the ERG recording obtained from normal subject (see the ERG of Fig.4). The amplitude between the a- and b-wave was 200 fT to 500 fT. It was found that in the MRG as well as the ERG the disappearence of the a-wave was observed when the intensity of the stimulation was decreased. Recently, we have discovered the off-response of the MRG by a simular method presented here(8). An aim of this study is to calculate the source localization of activity of the retina. From this viewpoint three dimensional current distribution in the retina will be investigated by measuring magnetic fields around the head and the face.

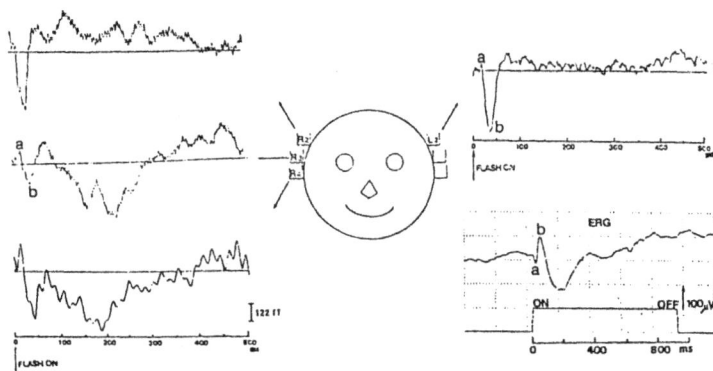

Fig.4   The MRGs and the ERG induced by stimulation of flash. Results of the MRGs are averaged of 100 samples. Illustration   shows the position and arrangement of the pick-up coil of SQUID gradiometer.

Neuromagnetism: Somatosensory evoked magnetic field (SEF)

Investigation of the human MEG has rapidly advanced in the several years due to the increase of researchers with the various motivations to study the human brain and with the development of the computerized magnetic signal processing system. In this chapter we present our neuromagnetic study of the human brain.

We have performed the magnetic mesurement at more than 25 different positions on the surface of the head in a wooden room with a SQUID connected to a
second order gradiometer (9). In the measurement of the SEFs the electric
current was applied to stimulate the median nerve of the right wrist with a
duration of 0.2 msec and the interval of 0.5 sec. All magnetic recordings
were averaged of more than 500 samples.

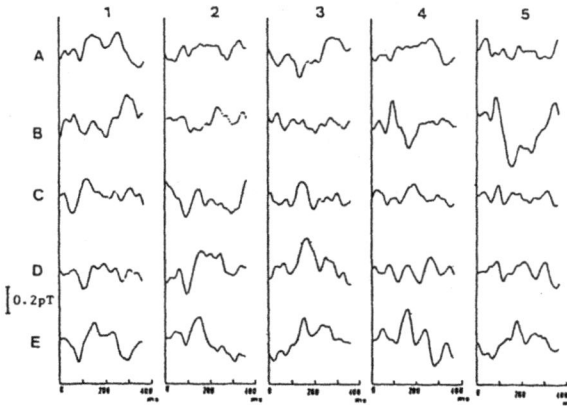

Fig.5 Averaged magnetic recordings of the SEFs obtained by
       elctric stimulation to median nerve of the right wrist
       displyed recording bandwidth 1 to 30 Hz.

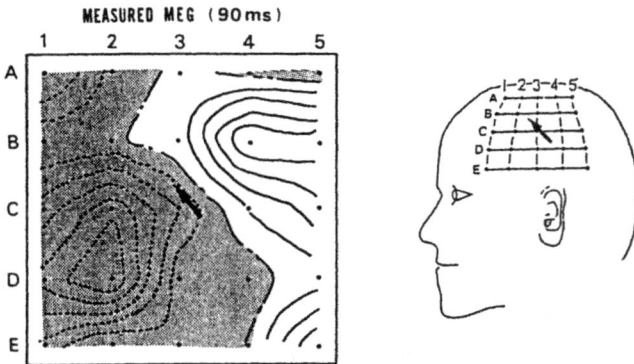

Fig.6 Location and direction of the ECD on the measuring grids calculated
       from the measured MEG at a latency of 90 msec. The calculated ECD
       is 1.53 nAm and the depth from the surface of the head is 2.5 cm.

Fig.5 shows the averaged magnetic recording of the SEF at the left hemisphere obtained by the electric stimulation to median nerve of the right
wrist displayed recording bandwidth 1 to 30 Hz. It can be seen that for
around 60 to 150 msec of latencies the upward (+) and/or the downward (-)
deflection were elicited, for instance, at position of D2, B4, respectivily(
see the measured position of Fig.6). In order to calculate the location of
the equivalent current source in the brain generating the magnetic field
distribution the ECD technique was introduced. Fig.6 shows the location of
the ECD on the measuring grids calculated from the measured MEGs (see Fig.5)
at a latency of 90 msec. The calculated ECD was located at the Rolandic
fissure of the left hemisphere where is sensory region related to the finger
and the direction of the ECD was perpendicular to the Rolandic fissure (see
Fig.6).

Conclusion

We have developed the computerized system with SQUID connected to second
order gradiometers for measuring magnetic fields produced by the human body.
By using this system, We measured the MCGs, the MOGs, MRGs related to the
transient response of the retina, and SEFs in a nonmagnetically shielded
room. In many laboratories a multi-channel SQUID magnetometer has been used
more and more. Research on the biomagnetism must be more quickly advanced
in the several years.

Acknowledgment

The authors wish to thank to Dr. Okuyama for his cooperation in the MOGs
and also wish to thank Mr. Adachi and Mr. Hasegawa for measuring at our
laboratory. The author's work has been facilitated by a grant of research
from Center for Research, Tokyo Denki University.

References

1) S.N.Erne, R.Fenici, H.D.Hahlbohm, H.Lehmann, Y.Uchikawa, in:Biomagnetism,
   ed. by H.Weinberg, G.Stroink and T.Katila, 132-136, Pergamon Press(1985).
2) S.Williamson, L.Kaufman and D.Nrenner, J. Appl. Phys., vol.50, 2418-2421
   (1979).
3) M.Lorange and R.M.Gulrajani, IEEE Trans. on Biomed. Eng. BME-33, 862-873
   (1986).
4) Yuchikawa and S.N.Erne, in: Biomagnetism'87, ed. by K.Atsumi, M.Kotani,
   S.Ueno, T.Katila, S.J.Williamson, 322-325, Tokyo Denki University Press
   (1988).
5) R.Fenici, M.Masselli, L.Loptz, G.Melillo, in: Biomagnetism'87, ed. by
   K.Atsumi, M.Kotani, S.Ueno, T.Katila, S.J.Williamson, 282-285, Tokyo
   Denki University press (1988).
6) Y.Uchikawa, A.Adachi, T.Hasegawa, N.Okuyama, M.Kotani, in:Biomagnetism'87,
   ed. by K.Atsumi, M.Kotani, S.Ueno, T.Katila, S.J.Williamson, 254-257,
   Tokyo Denki University Press (1988).
7) T.Katila,R.Maniewski, T.Poutanen, and T.Varpila, J.Appl. Phys., vol.52,
   2565-2571 (1981).
8) Y.Uchikawa, A.Adachi, T.Hasegawa, K.Aihara, M.Kotani, IEICE Jpn Tech.Rep.,
   vol.88, MBE-88-17, 49-56 (1988).
9) Y.Uchikawa, A.Adachi, T.Hasegawa, N.Okuyama, M.Kotani, Phys. Medicine &
   Biology, Vol.33, Suppl.1, Mp 16.4, 61 (1988).

JOSEPHSON DIGITAL CIRCUITS

Shinya HASUO

Fujitsu Laboratories Ltd., Atsugi
Fujitsu Limited
10-1, Morinosato-Wakamiya, Atsugi, 243-01, Japan

ABSTRACT-This paper describes recent progress on digital circuits using all niobium (Nb/AlOx/Nb) Josephson junctions. The niobium junctions have been applied to various circuits such as 8-bit shift registers, 16-bit ALUs (Arithmetic Logic Unit), and 4-bit microprocessors. We confirmed the high speed operation of less than 10 ps per gate on average for these circuits. We also developed a single-chip SQUID (Superconducting Quantum Interference Device), which is operated in a digital mode instead of analog mode.

INTRODUCTION

Performance of Josephson integrated circuits has been dramatically improved since niobium junctions were introduced[1-3]. Before niobium junctions, lead-alloy junctions were mainly used for the integrated circuits. Large scale integrated (LSI) circuits, however, seldom worked well. Almost all the reasons for these difficulties were originated from the unstable characteristics of lead-alloy junctions.

At the end of 1983, niobium junctions became available for use in integrated circuits and lead-alloy junctions were abandoned. After niobium junctions were introduced, various kinds of circuits operating much higher speeds than those using lead-alloy junctions were made. The higher speeds were due to the small scattering of the junction characteristics, and the inherent high performances of Josephson junctions became a reality.

At present, we make LSI-level circuits which include several thousands of junctions. Small scale Josephson computer operating at speeds of more than one order of magnitude faster than semiconductor computers has become a distinct possibility. In this paper, we describe our recent progress in Josephson digital circuits developed at our laboratory.

HIGH-SPEED GATE

Nb/AlOx/Nb junctions, whose excellent characteristics were demonstrated by Gurvitch et al.[4], and then further improved by Morohashi et al.[5],[6], are expected to be used in high-speed digital circuits and analog

applications.   The controllability, stability, uniformity, and
reproducibility gained through the use of Nb/AlOx/Nb are much better than
that from the use of lead-alloys.   We thus used the Nb/AlOx/Nb junction to
fabricate various digital circuits.

Cross section of the typical integrated circuit is shown in Fig. 1.
We compose various logic circuits using MVTL (Modified Variable Threshold
Logic) gate family.   The MVTL OR gate has an asymmetric interferometer and
a magnetically coupled control line.   The control current is injected into
the interferometer after magnetic coupling.   Figure 2 shows the equivalent
circuit of the MVTL OR gate.   Using the single junction J_3 and a resistor
R_i, the output current is isolated from the injected control current.   The
fastest gate speed obtained was 2.5 ps for the gate with a minimum junction
diameter of 1.5 μm.   The power consumption was 17 μW/gate[7].   This
gate is the fastest of all logic (including semiconductor) gates.   It must
be noted that Josephson logic gates can attain the gate delay of less than 10
ps/gate without using submicron process technology.

The AND gate is constructed with a single junction, and is always driven
by OR gate output signals.   This is because the AND gate cannot isolate the
output signal from the input signal.   Unit cells are combined using two OR
gates and an AND gate.   The gate delay of the unit cell was 16 ps for the
minimum junction diameter of 4 μm and 11.5 ps for a diameter of 2.5 μm.
We also desinged and tested a 2/3 MAJORITY gate and a TIMED INVERTER (TI).

DIGITAL CIRCUITS

Using the MVTL gate family, we fabricated various logic circuits to test
the high-speed operation of these gates.   They are a 16-bit ALU[1], an 8-bit
shift register[2], and a 4-bit microprocessor[3].   The performance of these
circuits is described here.   We also fabricated a single-chip SQUID
magnetometer[8].

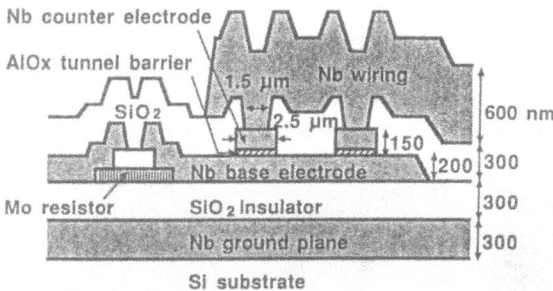

Fig.1   Cross section of an integrated circuit

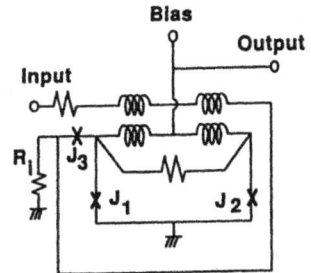

Fig.2   An equivalent circuit of an MVTL OR gate

758

A. 16-bit ALU

We fabricated a 16-bit ALU, which performs eight arithmetic and four logic functions. Figure 3 shows a photograph of the fabricated chip. The dimensions of the circuit size are 0.85 mm x 8.2 mm. There are 900 gates in the ALU, including the 36 gates needed to measure the critical path delay.

The critical path delay was measured to be 0.86 ns. The signal path during this operation covered 83 stages of the MVTL OR and AND gates. Since the propagation delay in the interconnecting lines on the signal path was calculated to be about 95 ps, the average gate delay was estimated to be 9.2 ps/gate. The total power consumption of the chip was 10.1 mW or an average of 11.3 μW/gate.

We also fabricated a 16-bit multiplier critical path model[9]. The model includes 828 MVTL gates, which are extracted along the critical path from the multiplier in order to estimate a multiplication time. The observed multiplication time was 1.1 ns.

B. 8-bit shift register

We designed an 8-bit shift register using MVTL gates. It is capable of SHIFT, LOAD, HOLD, and CLEAR functions. The fabricated chip contains 112 gates. The circuit dimensions are 1.1 mm x 2.1 mm. We confirmed that the 8-bit shift register operated correctly for all stages of all the control signals at an 80 μs clock. High-speed operation was tested. The SHIFT function was correctly operated up to a 2.3 GHz clock. The total power consumption was 1.8 mW.

We also developed a pseudorandom bit sequence generator[10]. The circuit is constructed with 9 stages of the one-bit shift register described above, and its output signal is fed back to the 5th stage through an exclusive-OR gate. Thus it can generate a pseudorandom number with a 511-bit sequence. We confirmed its correct operation up to 2.2 GHz.

C. 4-bit microprocessor

We have fabricated a 4-bit microprocessor[3]. This is the first instance to our knowledge, of application of Josephson devices to a microprocessor, so we wanted to verify the feasibility of the chip in

Fig.3 Photomicrograph of 16-bit ALU

comparison with a typical microprocessor constructed with semiconductor devices. We selected chip functions that were similar to those of the Am 2901 microprocessor made by Advanced Micro Devices Inc. This microprocessor has come to be regarded as the standard four-bit microprocessor slice.

It has a dual memory set which is used as a 16-word by 4-bit two-port RAM with a RAM shifter, an eight-function ALU, a Q register with a Q shifter, and several controllers. The total number of gates is 1841. A photomicrograph of the fabricated chip is shown in Fig. 4. The chip size is 5 mm x 5 mm. All functions and source combinations were confirmed at a clock frequency up to 100 MHz, the limit of the maximum clock of the word pattern generator. The operation along the critical path of the chip was tested using the high-speed pulse generator, and confirmed to operate correctly up to a clock frequency of 770 MHz. The gate power dissipation was 3.6 μW/gate, and the total power of the chip was 5 mW.

We verified that the Josephson microprocessor operated with a clock that was one order of magnitude faster, and consumed three orders of magnitude less power than semiconductor microprocessor. Performances of the AM2901 type microprocessors for three different materials are compared in Table 1.

D. A single-chip SQUID magnetometer

The SQUID magnetometer is a very high-sensitive magnetic sensor, that is expected to use as an image sensor for medical and other applications. We fabricated a single-chip SQUID magnetometer[8], which includes entire circuits such as the pickup coil, SQUID sensor, and feedback circuit. We introduced a digital feedback circuit and a superconducting storage loop. This made it possible to integrate the SQUID magnetometer into a single chip.

Table 1   Performance of
4-bit microprocessor

Device	Si [1]	GaAs [2]	Josephson
Maximum clock (MHz)	30	72	770
Power (W)	1.4	2.2	0.005

1) AMD, 1985 data book
2) Vitesse, 1987 GaAs IC Symposium

Fig.4 (left)   Photomicrograph
of a 4-bit microprocessor

The single-chip SQUID magnetometer requires only an AC bias and produces a digital output, with no peripherals, at room temperature. The output pulse can be processed by a digital processor or applied to a display instrument through a counter to directly monitor input magnetic field waveforms.

Figure 5 diagrams the circuit. The pickup coil transmits the magnetic flux to be measured to the SQUID sensor through coupling coils. The digital feedback circuit is fabricated using a superconducting storage loop and an interferometer as a write gate. The write gate receives a pulse sequence and writes a positive or negative flux quantum to the storage loop when a pulse arrives.

We fabricated the single-chip SQUID magnetometer and tested it. The magnetic flux coupled to the SQUID sensor was measured as low as 7×10^{-5} $\Phi_0/\sqrt{Hz}$, where Φ_0 is the flux quantum (2.07×10^{-15} Wb). This corresponds to a magnetic field of 4.7×10^{-12} $T/\sqrt{Hz}$, and the magnetic field gradient of $4.5 \times 10^{-9} T/m\sqrt{Hz}$ at the pickup coil. In our experiment, the sensitivity was believed to have been limited by environmental noise, not by device noise. Therefore we believe that the sensitivity can be further improved. In any case, this device is more sensitive than magnetocardiograms, which can only measure fields on the order of 10^{-11} T.

CONCLUSION

We described our recent progress on digital circuits with niobium Josephson junction. Progress has been rapid since we changed the junction material from the lead-alloy to niobium. We can operate LSI circuits with a few thousands gates. We verified in an LSI level circuit that the operating speed is more than one order of magnitude faster and the power consumption is more than two orders of magnitude smaller as compared with semiconductor circuits. Josephson memory circuit is also feasible up to 4 K bit with half a nanosecond access time [11]. As a result of our research, we feel that Josephson LSI with tens of thousands of junctions on a chip are

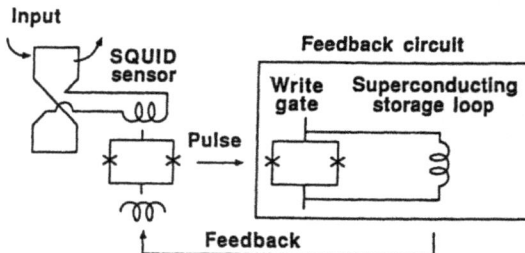

Fig.5   A circuit of the single-chip SQUID

feasible without any essential problems.   We also developed a single-chip SQUID magnetometer.   This will be widely used for medical and other applications.   Various kinds of digital superconducting circuits will break through new fields of superconducting electronics.

ACKNOWLEDGEMENT

The progress on Josephson digital analog devices described here is based on the work done by my colleagues, T. Imamura, N. Fujimaki, S. Morohashi, H. Tamura, H. Suzuki, H. Hoko, S. Kotani, A. Yoshida, and S. Ohara.   I want to express my sincere thanks for their efforts.
    The present research effort is part of the National Research and Development Program on "Scientific Computing System", conducted under a program set by the Agency of Industrial Science and Technology, Ministry of International Trade and Industry.

References

1) S. Kotani, N. Fujimaki, T. Imamura, and S. Hasuo, S., IEEE J. Solid State Circuits, 23, 2, pp.591-596 (April 1988).
2) N. Fujimaki, S. Kotani, T. Imamura, and S. Hasuo, IEEE J. Solid State Circuits, 22, 5, pp.886-891 (Oct. 1987).
3) S. Kotani, N. Fujimaki, T. Imamura, and S. Hasuo, S.,   Digest of Tech. Papers of 1988 International Solid-Circuit Conf. (ISSCC), San Francisco, 1988, pp.150-151 (Feb. 1988).
4) M. Gurvitch, M. A. Washington, and H. A. Huggens, Appl. Phys. Lett., 42, pp.472-474 (Mar. 1983).
5) S. Morohashi, F. Shinoki, A. Shoji, A. Aoyagi, and H. Hayakawa, Appl. Phys. Lett., 46, pp.1179-1181 (June 1985).
6) S. Morohashi, S. Hasuo, and T. Yamaoka, Appl. Phys. Lett., 48, pp.254-256 (Jan. 1986).
7) S. Kotani, T. Imamura, and S. Hasuo, Tech. Digest of International Electron Devices Meeting (Washington, D.C., 1987), pp.865-866 (Dec. 1987).
8) N. Fujimaki, H. Tamura, T. Imamura, and S. Hasuo, S., Digest o Tech. Papers of 1988 International Solid-State Conf. (ISSCC), San Francisco, 1988, pp.40-41 (Feb. 1988).
9) S. Kotani, N. Fujimaki, S. Morohashi, S. Ohara, and Hasuo,   IEEE J. Solid-State Circuits, SC-22, 1, pp.98-103   (Feb. 1987).
10) N. Fujimaki, T. Imamura, and S. Hasuo, IEEE J. Solid State Circuits, 23, 3, pp.852-858 (June 1988).
11) H. Suzuki, N. Fujimaki, H. Tamura, T. Imamura, and S. Hasuo, IEEE Trans. Magnetics, to be published.

A 4.7 T MAGNET FOR MAGNETIC RESONANCE IMAGING

H. MAEDA, M. URATA, T. WADA, T. YAZAWA and A. SATO

Toshiba Research and Development Center,4-1 Ukishima,
Kawasaki, Kanagawa, 210 JAPAN.

Abstract

A 0.3 m bore 4.7 T high homogeneity magnet, used for [31] P magnetic resonance imaging, has been constructed and installed in the Toshiba R & D Center. The magnets comprised of a set of solenoids, which attained designed 95 A current (1.75 MJ) without quenching. Diodes in the cryostat resulted in fast ramp rate, such as 3 T in 10 minutes. The field inhomogeneity for the bare magnet was 300 ppm/ 0.1 m dsv., which was finally reduced to 6 ppm/ 0.1 m dsv by shim coils. The field decay was less than 0.05 ppm/ h. The magnet system is currently used for ^{31}P magnetic resonance spectroscopic imaging experiments on living animals in the laboratory.

Introduction

Superconducting magnets for magnetic resonance imaging (MRI) or magnetic resonance spectroscopy (MRS) on ^{1}H generate a uniform and constant magnetic field of 0.5- 2.0 T[1]. A higher magnetic field, such as 4- 5 T , is necessary for MRI and MRS on ^{31}P, essential nuclei for energy metabolism of human body, while 1/ 15 in NMR sensitivity.

A 4.7 T superconducting MRI magnet, with a 0.3 m room temperature bore, has been manufactured and installed in Toshiba R & D Center to study ^{31}P chemical shift spectroscopic imaging. Design and test results are described in this paper.

Main Coil Design

The specifications for the MRI magnet are summarized in Table 1; a central magnetic field is 4.7 T, corresponding to a ^{1}H resonance frequency of 200 MHz, with a field uniformity of 1 ppm/ 0.1 m dsv. The field stability is less than 0.1 ppm/ hour.

Table 1 Specification of the MRI magnet system.

Central field	4.7 T
Room temperature bore	0.3 m
Field uniformity	1ppm/ 0.1 m dsv
Field stability with time	< 0. 1 ppm/ h

The circumferential vector potential, under an ideally uniform magnetic field in the plane perpendicular to the coil axis, is expressed as [2]

$$A_c = B_z \times (d/2.) / 2 ,$$

where A_c is the vector potential, B_z is the magnetic field and d is the diameter for an intersected circle between the plane and the sphere. Arrangement of solenoids, which gives such a vector potential distribution on the spheric surface, was obtained by the least squares method.

The magnet, thus obtained, consists of seven solenoids, as shown in Fig. 1, which give 2.4 ppm/ 0.1 m dsv field inhomogeneity: The error includes -2 ppm of uncorrectable Z^6 harmonics, while correctable Z^2 and Z^4 harmonics are less than a few ppm, respectively. Furthermore, the Z^8 harmonics reaches 0.07 ppm.

The field inhomogeneity arises from the inaccuracy of actual winding and structure [3]. The 0.1 mm radial position errors on coil # 3(#5) or #4 result in field inhomogeneity of 20-30 ppm, while the axial position errors of 0.1 mm on coil #3 (#5) or #2 (#6) are 30-31 ppm . Based on the calculation, more than 100 ppm is expected to appear for the actual magnet, even if the coil former tolerance was set to be ± 0.1 mm. A set of superconducting shim coils with a gradient of Z^n (n= 1, 2, 3) were thus adopted to correct such a field inhomogeneity.

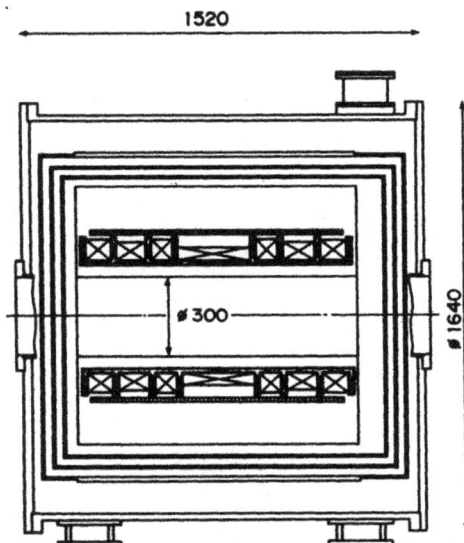

Figure 1.   Magnet and cryostat cross sectional view.

The magnet was wound by a Formvar coated NbTi conductor, with a copper/ superconductor ratio of 3. The maximum field in the coil winding at the operation current, 94.8 A, is 5.14 T; the current is 64 % on the coil load line, while the temperature margin is 2 K, corresponding to the minimum propagation zone, 12 mm in length[4].

An adiabatic solenoid tends to be quenched, if the inner winding layers are floated from the coil form during coil charging; a conductor moves easily by the electromagnetic force [5]. The winding tension which is sufficient to suppress the floating is numerically calculated by a multi- cylinder model[5]. Cumulative radial stress (S_r) distribution in the coil winding #2(#6) is shown in Fig. 2 for (a) coil winding with 100 MPa tension, (b) cool down and (c) coil energizing, respectively; the S_r 8 MPa remains at the operation current. The circumferential stress for the conductor is less than 100 MPa , which is less than the elastic limit for the NbTi conductor.

When a magnet is wound at 100 MPa winding tension, the winding form is pressed inward by the winding. The compressive stress for the coil form is 235 MPa, above the elastic limit of aluminum. Thus, a high manganese stainless steel was used for the coil form material, which has sufficiently high yield strength and is non-magnetic [6]. The stainless steel permeability, 1.0038, produces a field inhomogeneity of a few ppm.

The winding and substructure deformation values during (a) winding, (b) cooling down and (c) coil charging were calculated by both (i) multicylinder model and (ii) multipurpose finite element calculation program, NASTRAN/V64A. The calculated radial and axial displacement data are included in the initial dimension of the coil form. 70 % of the deformation is due to cooling down.

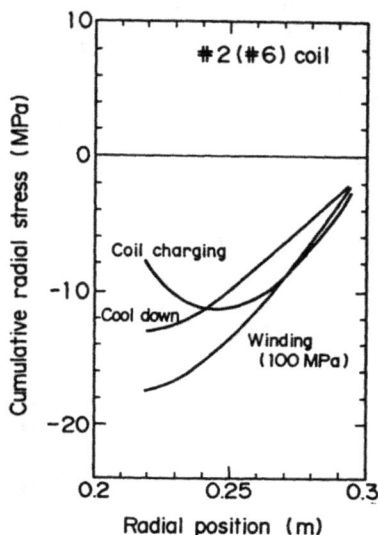

Figure 2. Cumulative radial stress distribution by winding, cooling down and coil energizing for solenoid #2 (#6).

Superconducting Shim Coil

Six sets of superconducting shim coils with a field gradient of Z, Z^2, Z^3, X, Y, ZY were prepared for the field correction. The design follows that suggested by Romeo and Hoult[7]. The coil used a 0.37 mm Formvar coated NbTi conductor with a copper/ superconductor ratio of 2.5. The coils were wound around a fiber-reinforced plastic tube. The field strength at the shim coil is 1.5 T; the maximum allowable current for the coil is 80 A, according to the coil load line. The current and the field gradient are summarized in Table 2.

Table 2 Shim coil current and field gradient.

Shim coil	Coil current	Field gradient
Z	10 A	± 100 ppm
Z^2	20 A	163 ppm
Z^3	10 A	± 10 ppm
X,Y	10 A	± 100 ppm
ZY	10 A	25 ppm

Coil Protection

Numerical calculations have been carried out to find the best protection scheme for the magnet system. The calculation assumed quench of the specified solenoids. Time dependence for the current and temperature after coil quenching were numerically calculated.

Particular features concerning the magnet protection are as follows:

(i) The magnet consists of seven separated solenoids, while its energy is as high as 1.8 MJ. It is probable that the quenching is bound to a single solenoid, which might result in excessive local heating and voltage. Thus, each of the solenoids was shunted by a resistor, which ensures separate dissipation of the coil energy. Note that this is the case, when the persistent-current switch (PCS) is " off", such as during coil charging.

(ii) The magnet is operated in the persistent mode, where current leads are removed from the cryostat. All of the coil energy is dissipated inside the cryostat in case of quench. Current decay after quench is slower in this case. The energy dissipation is concentrated on the shunt resistor of a quenched section, if the PCS remains "on" during current decay, especially when diodes were used in the circuit. The resistor volume, which assures the moderate heating, was defined based on the calculation.

(iii) Coupling between the main coil and the Z2 shim coil causes another problem in regard to protection[8];i.e. when the main coil is quenched , excessive current is induced in the shim coil. The current is dependent on the resistance, R_s, in the circuit(see Fig. 3). The coil and its resistor should be secured from quenching by an enhanced current. The R_s volume and the resistance was thus numerically obtained.

Magnet Fabrication

The main magnet was wound around a stainless steel coil form with a 100 MPa winding tension. In addition, both a copper alloy shunt resistor and a nonmagnetic phosphor bronze binder. The layer number is 60- 96, while the total turns number is 47,000. The coil outer diameter after being wound is 1-2 mm smaller than the calculated value; the error is due to the conductor tolerance and creep in the Formvar coating. They produce a field inhomogeneity of 20 ppm/ 0.1 mdsv. The inner diameter displacement agreed with the calculation results.

The shim coil was wound around a fiber-reinforced plastics tube, 0.63 m in inner diameter and 0.66 m in outer diameter. A rectangular flat coil, 1-3 mm in thickness, was bent around a bore tube to represent radial shim coils, such as X, Y, ZY.

The conductor/ conductor and conductor/ PCS were solid- diffusion welded at high temperature. Contact resistance was less than 10^{-10} Ω , which is sufficiently low to match the specification.

The magnet was fixed inside a cryostat, which has 3 thermal shielding stages and demountable power leads.

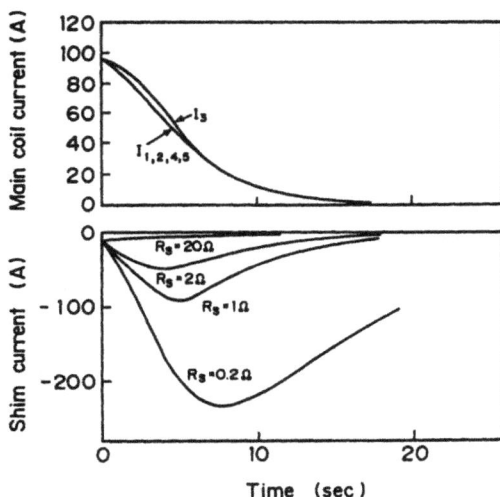

Figure 3. Shim coil current increase after a main coil was quenched.

The cryostat was made of Aluminum, whose outer diameter is 1.64 m while the length is 1.52 m. A two stage refrigerator , sometimes used for this kind of magnet system, was not utilized in the cryostat.

Coil energizing experiment

Quenching

The magnet has been energized to 95 A operation current, without any quenchings. The strain gauge on the coil form coincided with the calculation by NASTRAN. No saturation was observed on the strain, which demonstrates that the winding does not float from the coil form by electromagnetic force.

According to experimental results, the unfloated magnet shows the following characteristics[9]: Macroscopic wire motions are suppressed by winding tension, while premature quenchings still occur at above 80-85 % on the coil load line, due to a tiny slip event, such as a few μ m. Thus, the present magnet is stable, as it is operated at 64 % on the coil load line.

Fast Ramping

The magnet is expected to be ramped as fast as 1 A/ sec (0.06 T/ sec), as the current into shunt resistors is inhibited by diodes at 4.2 K. Results of quick ramping experiments are indicated in the following:
(i) Current ramp to 30 A(1.5 T), with a rate of 0.2 A/ sec (0.01 T/ sec). The coil voltage was 80 V, while the charging time is 2.5 min.
(ii) Current ramp to 61 A(3 T), with a ramp rate of 0.1 A/sec (0.005 T/ sec). Charging time is 10 minutes.

Field Uniformity

The field uniformity in 0.1 mdsv was measured by an NMR probe. Inhomogeneity was 250 ppm: (i) The axial gradient, such as Z^2, was -260 ppm, while the Z^4 gradient was 18 ppm. These values were much larger than was expected from the calculation. The Z^6 gradient, -2 ppm, coincided with the calculation. (ii) The radial gradient was about -30 ppm; X, Y, and YZ values were -10 ppm, respectively.

When the magnet was wound, a void gap was formed at the end of every layer. The field inhomogeneity produced by the disorder is calculated to be -289 ppm for Z^2 gradient, while it is 18 ppm for Z^4 gradient. It is suggested that the error is due to these gaps in the windings, as the calculated inhomogeneity agrees with the measurement.

The field inhomogeneity was reduced to 20 ppm/ 0.1 mdsv by superconducting shim coils, which were finally reduced to 6 ppm/ 0.1 mdsv by 14 sets of room temperature shim coils. The error comprised 4 ppm for Z^4 and -2 ppm of Z^6. The Z^4 shim power was limited by over-heating. Air cooling devices are being mounted, which will reduce the inhomogeneity to 1- 2 ppm/ 0.1 m in the future.

Field Decay

Field decays at 1 ppm/ h in the first 12 hours, which attains a low rate of < 0.05 ppm/h. The initial decay was suggested to be due to diffusion in its self field shielding current[10] in the magnet conductor.

Conclusions

The 4.7 T MRI magnet with a 0.3 m room temperature bore has been success-fully constructed and tested in Toshiba R & D Center. The magnet is being used for ^{31}P magnetic resonance spectroscopic imaging on living animals, which is necessary for medical application to a human body in the future.

References

(1) K. Roth, Proc. of the 11th International Cryog. Engineering Conf., pp 48, 1986.
(2) R. P. Feynman et al., *Lectures on Physics*, Addison- Wesley Publication Company, 1965.
(3) J. E. C. Williams, Proc. of the 9th International Cryog. Engineering Conf., pp 667, 1982.
(4) M. N. Wilson, *Superconducting magnets*, Clarendon Press Oxford, pp141, 1983.
(5) M. Urata and H. Maeda, IEEE Trans. MAG., MAG-23, pp 1596, 1987.
(6) H. Masumoto et al., Adv. in Cryog. Engineering Materials, Prenum Press, vol 30, pp 169, 1984.
(7) F. Romeo and D.I. Hoult, Magnetic Resonance in medicine., pp 44- 65, 1984.
(8) E. S. Bobrov et al., IEEE Trans. MAG-23, No. 2, pp 1303, 1987.
(9) M. Urata and H. Maeda, Presented at the Applied superconductivity Con-ference in San Francisco, 1988.
(10) S. Yamamoto and T. Yamada, IEEE Trans. MAG- 24, No. 2., pp 1292, 1988.

QUASIPARTICLE 90 GHz MIXER WITH Nb/AlOx/Nb ARRAY JUNCTIONS

Takashi NOGUCHI,   Tetsuya TAKAMI,   and Kohichi HAMANAKA

Central Research Laboratory, Mitsubishi Electric Corp.
1-1, Tsukaguchi-Honmachi 8-Chome, Amagasaki, Hyogo, 661 Japan

Abstract - A low noise SIS (Superconductor-Insulator-Superconductor) receiver for 90 GHz band using an array of four Nb/AlOx/Nb junctions has been made and tested at 94 GHz.   The junctions have a small subgap leakage current and a sharp current rise at a gap voltage of even 4.8 K. The best receiver noise temperature and conversion efficiency   of the mixer are 99 ± 4 K and 0.5 ± 0.02, respectively.

Introduction

Many SIS receivers have been made using Pb-alloy junctions, and it has been demonstrated that their noise temperatures are extremely low in the millimeter-wave region [1-4]. The Pb-alloy junctions, however, tend to degrade during thermal cycling between liquid He and room temperature, and their dc characteristics are fairly good at temperatures lower than 3 K. On the contrary, Nb/AlOx/Nb junctions are free from the degradation during thermal cycling and long-time storage in the atmosphere and have excellent dc characteristics even at 4-5 K. Hence, they are very suitable for the realization of a low-noise SIS receiver using a commercially available 4-K refrigerator to cool the receiver system. As far as we know, however, few works concerning the application of this type of junction to an SIS mixer have been reported [5,6]. Thus, we have made a low-noise 90 GHz-band SIS receiver using Nb/AlOx/Nb junctions. In this paper we describe the fabrication of Nb/AlOx/Nb junctions and the performance of a mixer/receiver with the junctions at 94 GHz.

Junction Fabrication

A Nb/AlOx/Nb tri-layer is formed onto a crystal quartz substrate (50 mm in diameter and 0.25 mm in thickness) using RF magnetron sputtering facilities. At first, the Nb lower electrode, including choke filters, is deposited to a thickness of 200 nm through photoresist mask. After deposition of the Nb lower electrode, Al film is deposited to a thickness of about 5 nm onto the Nb film subsequently without exposing the Nb film to the atmosphere. The AlOx barrier is formed by the thermal oxidation of the Al film with pure O_2. After completion of the oxidation, Nb upper electrode is deposited to a thickness of 100 nm. No intentional substrate heating is done during formation of the tri-layer structure. Once the tri-layer structure is completed, the photoresist mask is lifted off.

In the next step, the Nb upper layer is removed except for the junction area, which is protected with photoresist, by RIE (reactive ion etching) with a mixture of CF_4 and O_2. After the etching is completed, the area around the junctions is covered with a 400-nm thick SiO film to form an insulating layer between the lower and upper Nb electrodes. The SiO is deposited by a self-alignment lift-off technique with the same resist used in the RIE. Finally, a 1-μm thick Pb-alloy film is deposited through photoresist   mask in order to

769

(a)

(b)

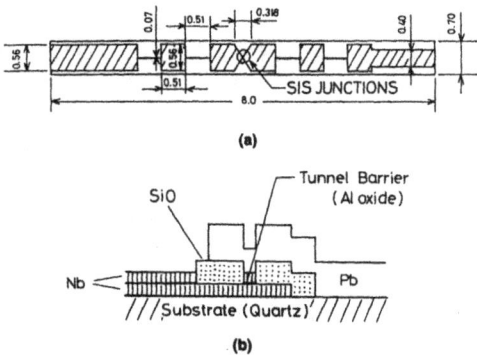

Fig. 1 (a) Deposited pattern of circuitry on the quartz substrate and (b) cross section of an SIS junction.

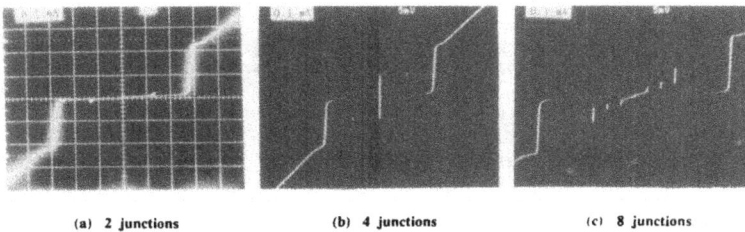

(a) 2 junctions   (b) 4 junctions   (c) 8 junctions

Fig. 2 dc I-V curves of series arrays of (a) two, (b) four, and (c) eight Nb/AlOx/Nb junctions.

interconnect isolated junctions.

After the final photoresist is lifted off, the quartz substrate is diced into 0.7×8 mm² chips. The circuit pattern on a chip made by the above process is illustrated in Fig. 1(a). In order to avoid saturation of a mixer by room-temperature radiation, a series array of 2, 4 or 8 junctions is fabricated on the chip. The designed area of each junction is 2.5×2.5 μm². A cross section of a junction is depicted in Fig. 1(b). The normal resistance of a series array of junctions can be adjusted both by changing the O_2 pressure at the oxidation of Al layer and by selecting an appropriate number of junctions in the array. The capacitance of a series array of junctions can be varied by changing the number of junctions in the arrays. The array junctions are located at the center of the chips and choke filters are placed on both sides of the array junctions.

In Fig. 2, typical dc I-V curves of series arrays of 2, 4 and 8 junctions at 4.2 K are shown. The dc I-V curves are not as good as those of Nb/AlOx/Nb junctions previously reported [5,7]. It is noted here that dc I-V curves of junctions fabricated on a Si substrate under the same conditions are better than those shown in Fig. 2. As the thermal conductivity of quartz is much lower than that of Si, the quartz substrate is heated up by plasma more easily than the Si substrate during the sputter-deposition of Nb electrodes. Thus, the degradation of the dc I-V curves of junctions on quartz are attributed mainly to the diffusion of Al into Nb lower electrode accompanied with the increase of substrate temperature [7]. Nevertheless, the dc I-V curves shown in Fig. 2 have both a subgap leakage current as low as and a current increase at the gap voltage as sharp as those of Pb-alloy junctions at temperatures much lower than 4.2 K, so we can expect that they

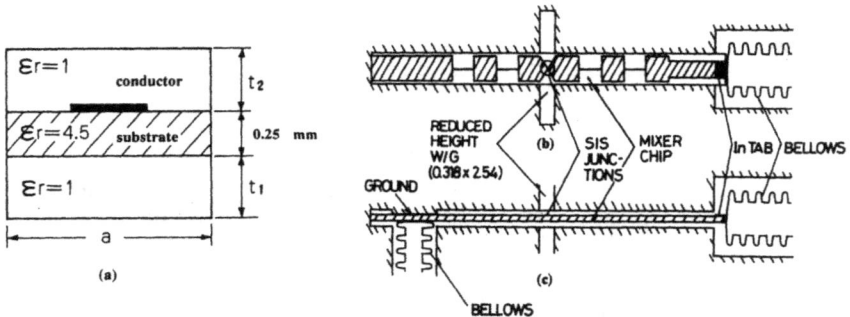

Fig. 3 (a) A cross-sectional view of a suspended strip-line and cutaway drawings of the mixer mount viewed (b) from the signal input port and (c) from the top.

would work well as high-performance mixers even at 4.2 K.

Receiver Configuration

The mixer chip shown in Fig. 1(a) is inserted into a narrow channel in a mixer mount made of brass. In this configuration, the conductor on the chip and inner walls of the channel form a suspended stripline as illustrated in Fig 3(a). By setting $t_1 = t_2 = 0.25$ mm, the lowest resonance frequency of the channel is estimated to be 137 GHz [8], which is much higher than the signal frequency (near 94 GHz), where t_1 and t_2 are upper and lower distances between the chip and walls of the channel. No resonance mode will be excited in the channel by the signal. A straight 1/4-reduced-height waveguide (0.318×2.54 mm^2) is grooved in the mixer mount. The inside surface of the channel and waveguide is gold-plated. The junctions are located at the cross point between the channel and the waveguide. Cutaway drawings of the mixer mount viewed from the signal input port and viewed from the top are shown in Figs. 3(b) and (c), respectively. One end of the stripline is joined with bellows, which is attached to a center conductor of an SMA connector, to bring out the IF from the mixer junctions through the choke filter. The other end of the stripline is grounded.

A movable contacting-type backshort with a choke structure is placed in the reduced-height waveguide behind the mixer chip. A waveguide T-junction with a movable short in the side waveguide is located at a distance of 2 mm ($\simeq \lambda_g/2$) toward the signal input, where λ_g is the guide wavelength. These two shorts work as tuners to adjust the source impedance of array junctions.

The LO (local oscillator) source used in the experiments is a 94-GHz Gunn oscillator. The LO and signal are combined by means of a 0.1-mm-thick teflon partial-reflector outside the cryostat and coupled into a pyramidal feedhorn through 0.5-mm-thick teflon vacuum window. The pyramidal feedhorn is followed by a standard WR-10 waveguide (1.27×2.54 mm^2). The transition from the standard waveguide to the reduced-height one in the mixer block is accomplished by means of a 3-section quarter-wave impedance transformer that is tightly bolted to the mixer mount. The mixer mount with the transformer is attached to a cold plate in the vacuum space of liquid He cryostat and the feedhorn and the standard waveguide inside the cryostat are thermally anchored to the cold plate.

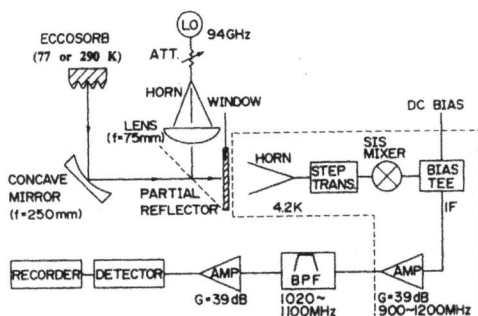

Fig. 4 Schematic diagram of the measuring equipment.

The first-stage IF amplifier is a cooled FET amplifier (Barkshire Technologies Inc., MODEL L-1.0-30) with a center frequency of 1.05 GHz and a 300-MHz bandwidth. A bias tee between the mixer and IF amplifier is used in order to provide the dc bias. A schematic diagram of the apparatus for measurements of mixer and receiver performance is shown in Fig. 4.

Receiver Performance

Receiver and mixer performance are determined from the I-V curves and measurements of IF output power with hot (290 K) and cold (77 K) Eccosorb loads placed in front of the partial-reflector. In Fig. 5 the data for a receiver with LO at a frequency of 94 GHz is shown. Curves (a) and (b) represent dc I-V curves with and without LO, respectively. As the gap voltage of curve (b) in Fig. 5 is slightly smaller than that measured at 4.2 K, the temperature of junctions in the mixer mount is higher than 4.2 K. Making a comparison between the gap voltages of curves in Fig.2 (b) and obtained at 4.2 K, the temperature of array junctions in the mixer mount is estimated to be in the range from 4.8 to 5 K. Here we assume that temperature dependence of the gap voltage obeys BCS theory, and the transition temperature of Nb film is 9 K. Very strong nonlinearity of I-V curve at the gap voltage is maintained even at nearly 5 K, so that no further decrease of bath temperature was made. On curve (a) in Fig. 5, several PAT (photon-assisted tunneling) steps are observed. The voltage separation of the steps are about 1.5 mV which is nearly equal to $4 \times (\hbar \omega_{LO}/e)$, where $\hbar$, ω_{LO}, and e are Planck's constant, angular frequency of the LO and electron charge, respectively. This fact indicates that 4 junctions are driven synchronously and are almost identical to each other. The normal resistance, R_N, of the array junctions shown in Fig. 5 is 55 Ω. The total capacitance of the array junctions is estimated to be 0.10 pF assuming that a specific capacitance of the Nb/AlOx/Nb junction is 6 $\mu F/cm^2$ [9], so that $\omega_{LO} R_N C$ product is estimated to be 3.2.

Curves (c) and (d) in Fig. 5 represent the IF output power in the band from 1.02 to 1.10 GHz as a function of voltage with hot and cold loads. The receiver noise temperature, T_{RX}, referred to its input is obtained from the expression

$$T_{RX} = (T_H - Y \times T_C)/(Y-1), \tag{1}$$

where T_H and T_C are the temperatures of hot and cold loads and Y is the

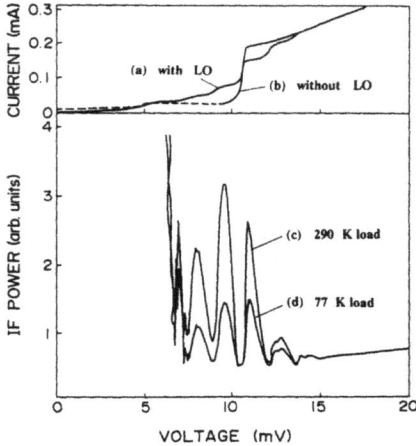

Fig. 5 dc I-V characteristics and IF output powers for a typical receiver with 94-GHz LO.

ratio of the corresponding IF output powers. When the array junctions are biased at a voltage in the middle of two adjacent PAT steps, the IF output power has a peak and the ratio Y reaches a maximum value. The largest value of Y among the maxima is 2.21 ± 0.03, which is obtained at a peak near 9.5 mV. Substituting $Y = 2.21 \pm 0.03$ into Eq. (1), we obtain $T_{RX} = 99 \pm 4$ K. As the signal-to-image ratio obtained in this case is greater than 8, the receiver is almost operating as an SSB (single sideband) system and the receiver noise temperature that we quote is approximately an SSB value.

Conversion efficiency of a mixer is calculated using the dc I-V curve and IF output power. It is shown that when a series array of N identical junctions is biased where the current is a linear function of voltage, the spectral power density of shot noise increases linearly at a rate of $e/2k_BN = 5.8/N$ K per mV [10], where k_B is Boltzmann's constant. In our series array of 4 junctions, not only the current but also the IF output power far above the gap increases linearly as a function of voltage. The 4 junctions in the array are almost identical to each other as mentioned above. Thus, the linear portion of the IF output power must have a slope of 1.45 K per mV referred to the IF amplifier input. Effective temperature of the IF output power referred to the IF amplifier can be calibrated using the value of slope. Conversion efficiency, G_c, is given as

$$G_c = \Delta T_0 / \Delta T_i, \tag{2}$$

where ΔT_i is the temperature difference between the hot and cold loads and ΔT_0 is the difference between the effective temperatures of the corresponding IF output powers referred to the IF amplifier. The largest value of ΔT_0 is 107 ± 5 K obtained at nearly 9.5 mV where the minimum receiver noise temperature is obtained. Substituting $\Delta T_0 = 107 \pm 5$ K and $\Delta T_i = (290-77)$ K into Eq.(2), we obtain $G_c = 0.5 \pm 0.02$.

In the same way, noise temperature of the IF amplifier, T_{IF}, referred to its input is calibrated to be 14 ± 0.5 K. Mixer noise temperature, T_M, is given by the expression

$$T_{RX} = T_M + T_{IF}/G_c. \tag{3}$$

Substituting $T_{RX} = 99$ K, $G_c = 0.5$ and $T_{IF} = 14$ K into Eq.(3), we obtain $T_M = 71$ K. The calculated mixer noise temperature is much larger than that theoretically predicted [11]. This is mainly due to the noise contribution of losses of signal power inside the cryostat and/or impedance mismatch between the signal and the array junctions. It will be improved by adjusting the signal input system and by using other array junctions with a different normal resistance and capacitance.

Conclusion

We have made a low-noise 90-GHz-band receiver using a series array of Nb/AlOx/Nb junctions. The Nb/AlOx/Nb junctions are of high quality even at 4-5 K. The best receiver noise temperature and mixer conversion efficiency we achieved are 99 ± 4 K and 0.5 ± 0.02, respectively. The performance of our receiver using Nb/AlOx/Nb junctions is comparable with the best performance of those using Pb-alloy junctions. Consequently, it will certainly be possible to make an actual low-noise receiver with Nb/AlOx/Nb junctions using a closed-cycle refrigerator to cool the system.

Acknowledgments

We wish to thank Dr. H. Ogawa of Nagoya University for many helpful discussions and suggestions and Dr. T. Nakayama and Dr. M. Ishii for their continuing support of this work.

References

1) P.L.Richards, T.M.Shen, R.E.Harris, and F.L.Lloyd, "Quasiparticle heterodyne mixing in SIS tunnel junctions", Appl. Phys. Lett., 34, 5, 345-347 (1979).
2) R.Blundel, H.Hein, K.H.Gundlach, and E.J.Blum, "An SIS receiver for the 3 mm wavelength range", Int. J. Infrared and Millimeter Waves, 3, 6, 793-799 (1982).
3) S.-K.Pan, M.J.Feldman, and A.R.Kerr, "Low-noise 115-GHz receiver using superconducting tunnel junctions", Appl. Phys. Lett., 43, 8, 786-788 (1983).
4) D.P.Woody, R.E.Miller, and M.J.Wengler, "85-115-GHz receivers for radio astronomy", IEEE Trans. Microwave Theory Tech., MTT-33, 2, 90-95 (1985).
5) J.Imatani, T.Kasuga, A.Sakamoto, H.Iwashita, and S.Kodaira, "A 100 GHz SIS mixer of Nb/Al-AlOx/Nb junctions", IEEE Trans. Magn., MAG-23, 2, 1263-1266 (1987).
6) A.W.Lichtenberger, C.P.McClay, R.J.Mattauch, M.J.Feldman, S.-K.Pan, and A.R.Kerr, "Fabrication of Nb/Al-Al₂O₃/Nb junctions with extremely low leakage currents", Abstracts of Applied Superconductivity Conference, EJ-7, 51 (1988).
7) H.A.Huggins and M.Gurvitch, "Preparation and characteristics of Nb/Al-oxide-Nb tunnel junctions", J. Appl. Phys. 57, 6, 2103-2109 (1985).
8) S.Lidholm, Research Report of Research Laboratory of Electronics and Onsla Space Observatory, Chalmers University of Technology, No. 131 (1977).
9) M.Gurvitch, M.A.Washington, and H.A.Huggins, "High quality refractory Josephson tunnel junctions utilizing thin aluminum layers", Appl. Phys. Lett., 42, 5, 472-474 (1983).
10) M.J.Feldman and S.Rudner, "Mixing with SIS arrays" in Reviews of Infrared and Millimeter Wave, Vol. 1, ed. by K.J.Button (Plenum, New York, 1983) 47-75.
11) J.R.Tucker, "Quantum limited detection in tunnel junction mixers", IEEE J. Quantum Electron., QE-15, 11, 1234-1258 (1979).

Superconducting mixers in the submillimeter region

K. Sakai, T. Fukushima, Y. Ichioka
Department of Applied Physics, Osaka University, Suita,
Osaka 565, Japan.

J. Inatani
Nobeyama Radio Observatory, National Astronomical Observatory,
Nobeyama, Nagano 384-13, Japan.

S. Kodaira
Electrical Engineering, Kisarazu National College of Technology,
Kisarazu, Chiba 292, Japan.

Abstract   The superconducting SIS mixers of Nb/AlOx/Nb have
been continuously studied, in order to improve the submillimeter
heterodyning. The mixer has two junctions in series and a printed
bow-tie antenna. They are connected through a matching circuit.
A fundamental mixing experiment was successfully made at 337.3 GHz.
In the course of the experiments an interesting incoherent/coherent
action of two junctions, depending on the incident power level, has
been observed. The result is also added.

Introduction

The importance of the SIS (Superconductor Insulator Superconductor)
mixer is increasing in the millimeter (mm) and the submillimeter
(submm) heterodyning. The SIS mixer for the mm region has been
developed to near the goal, while the submm mixer is still on its way
to development. The realization of the submm SIS is difficult because
of the inherent necessity of small junction area and the parasitic
junction capacitance. The authors have developed the submm SIS mixers
to use for submm astronomy at first and to use for other applications.
As the junction area is as small as 1.7 μm square and the thickness
of insulator is as thin as 15 Å, in addition to the normal sandwitch
type mixers, bridge type junctions were sometimes made, by breaking the
insulating layer. Both types have been used for the mixing experiments
at 337.3 GHz. A fundamental mixing and the incoherent/coherent feature
of array junctions have been observed. Those results will be reported.

Construction of SIS mixer

The mixer is mainly composed of a series-connected junctions and
a bow-tie antenna as may be seen from Fig. 1. The junction has a
structure of Nb/AlOx/Nb and two junctions are connected in series. Each
area is 1.7 μm square and the array junctions have a resistance of 50 Ω

and a capacitance of ~90 fF. The bow-tie antenna of 120° angle has a
pure resistance of 100 Ω. A quarter-wave (λ/4) impedance matching
circuit (Zt in Fig. 1(d)) connects the junctions and the antenna.
An inductance (Zc in Fig. 1 (d)) is connected to the junctions in
parallel in order to cancell the junction capacitance and to make the
junction impedance purely resistive. The Zsh is a λ/4 line to realize
a high frequency short circuit. The Zt, Zc and Zsh are microstrip lines
using one side of metal plane of the bow-tie antenna as a ground plane.
The junctions are fabricated on an under layer to protect the junctions
from the stress[1]. All components are fabricated on a quartz substrate.

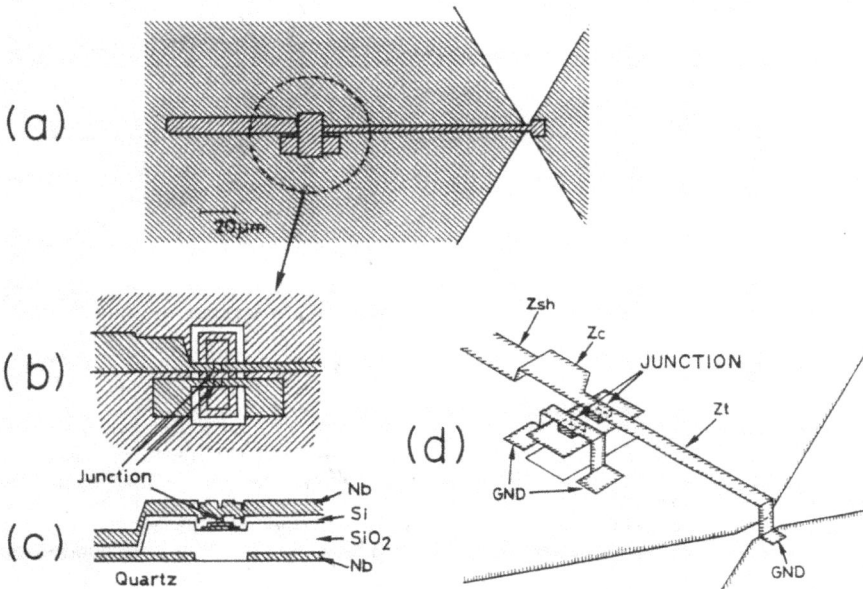

Fig. 1 Schematic illustration of Nb/AlOx/Nb submillimeter mixer with
a bow-tie antenna and matching ciruits. a) Top view of the mixer,
b) Expanded top view of the junction section, c) Cross-sectional
view of a junction, d) Three-dimensional picture of the mixer.

Experimental arrangement

The mixer has been mounted in a quasi-optical mixer mount as in
Fig. 2. The mixer is set up at the center of a hyperhemisphere
substrate lens (SL) made of quartz in such a way that the quartz
substrate of the mixer contacts to SL. The incident submm radiation
is converged by a TPX lens (L) and the SL, together with an external
TPX lens, in such a way that the incoming beam is focused as small as
possible. The mixer mount is set up in a liquid helium dewar, and the
mixer is cooled to 4.2 K.

An optically pumped submm laser (CH₂CF₂ 337.3 GHz) was used as a local oscillator(LO) and the frequency multiplier/klystron oscillator of 3rd harmonics was used as a RF signal source. Both beams were effectively combined by using a Martin-Puplett type diplexer. The whole arrangement of the experiment is shown in Fig. 3.

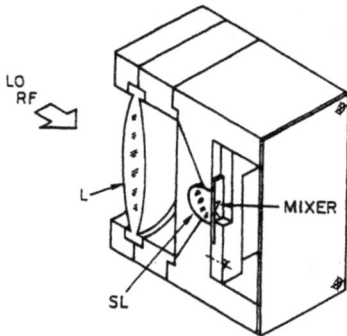

Fig. 2 Cross-sectional view of a quasi-optical mixer mount.

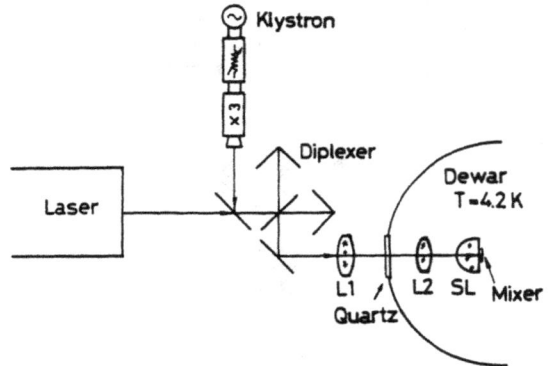

Fig. 3 Schematic of submillimeter fundamental mixing at 337.3 GHz.

Experimental results

The static I-V characteristics of the fabricated junctions are shown in Fig. 4(a) and (b). The Fig. 4(a) shows the one of original SIS tunnel junction and Fig. 4(b) is the one of short weak link (bridge type) Josephson junction. The gap voltage of the Nb is effectively 2.7 mV and, therefore, the present mixer with two junctions

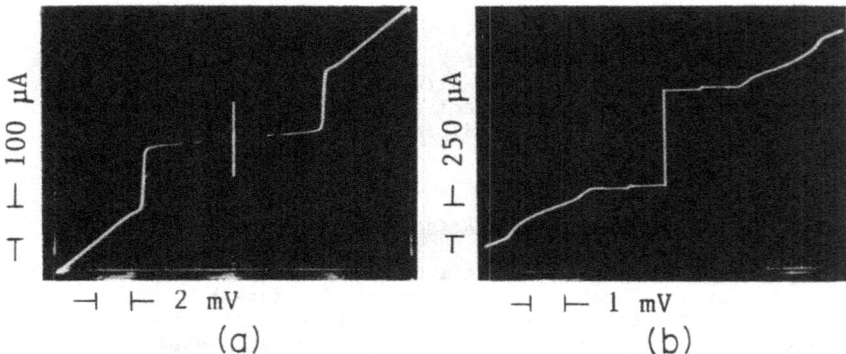

(a) (b)

Fig. 4 Static I-V characteristics of fabricated junctions. a) SIS tunnel junction, b) Short weak link Josephson junction.

in series shows a sharp gap structure at 5.4 mV as in Fig. 4(a), while
the short weak link junction showed a clear feature of Josephson
junction as in Fig. 4(b).

By using such mixers, fundamental mixing experiments were carried
out at 337.3 GHz. An intermediate frequency (IF) of 1.5 GHz was used
from the demand of the low noise amplifier. The result of mixing by a
SIS tunnel junction is shown in Fig. 5. The feature shows that the
fundamental mixing based on the quasi particle tunneling occurs clearly.
The width of broad peaks correspond to the one of PAT (Photon Assisted
Tunneling) step which is 2.78 mV in this case. Based on the feature, we
estimates the conversion loss of ~ 20 dB.

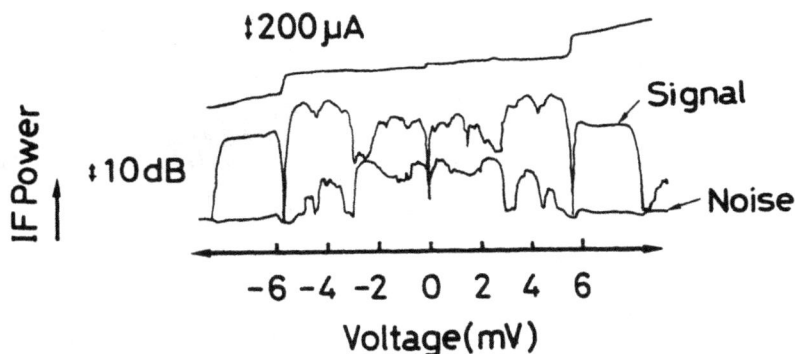

Fig. 5 IF power obtained from the fundamental mixing at 337.3 GHz.

The results of mixing experiments by a bridge type junction is
shown in Fig. 6. The feature shows that the array junctions act
independently when the incident LO power is weak (0.8 nW/junction
Fig. 6(c)) but they begin to act coherently on increasing the LO power
(1.7 nW/junction Fig. 6(b)) and act fully coherently above the LO
power (e.g. 2.8 nW/junction Fig. 6(a)). Such feature also has been
observed in SIS junction as may be seen from Fig. 5.

Conclusions

The superconducting SIS submm mixers of Nb/AlOx/Nb have been
fabricated. The mixer has two junctions in series, matching circuits
and a bow-tie antenna. By using the mixer a fundamental mixing
experiments were successfully carried out at 337.3 GHz. In the course
of the experiment, incoherent/coherent action of the series-connected
junctions was made clear.

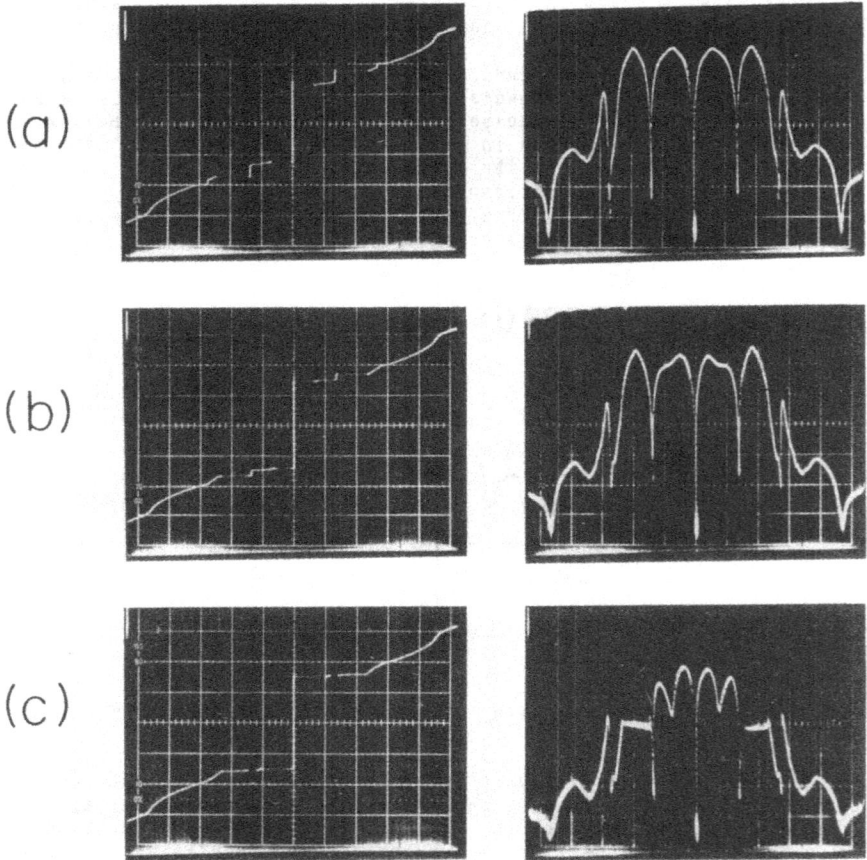

Fig. 6 Experimental results showing incoherent/coherent action of
series-connected short weak link Josephson junctions obtained from
the fundamental mixing at 337.3 GHz.   Left : I-V characteristics,
Vertical scale 250 μA/div, Horizontal scale 1 mV/div.
Right : IF power and applied vias voltage: Vertical scale 10 dB/div.,
Horizontal scale 1 mV/div. Feeded LO powers are (a) 2.8 nW/junction,
coherent action (b) 1.7 nW/junction, boundary of coherent/incoherent
action (c) 0.8 nW/junction, incoherent action.

Reference

1) M. Yuda, K. Kuroda and J. Nakano: Jpn. J. Appl. Phys. <u>26</u>, 1161
   (1987).

MULTI-CHANNEL SQUID SYSTEM AND APPLICATION TO BIOMAGNETISM

Hisao FURUKAWA, Toru KATAYAMA,
Satoshi FUJITA and Kimisuke SHIRAE

Faculty of Engineering Science, Osaka University
1-1, Machikaneyama Toyonaka, Osaka, 560, Japan

Abstract - We present three types of multi-channel SQUID amplifier with simple configuration, a developed rf type six channel SQUID system, and some experimental results of vector magnetocardiogram measurement using the system.

INTRODUCTION

The superconducting quantum interference device(SQUID) exhibits extreme sensitivity to magnetic field, and has been served as the main tool for the measurement of the weak magnetic signal such as biomagnetic field. In most of the magnetic field measurement reported to date, a single SQUID amplifier has been used. Recently, however, there are growing needs for the utilization of a multi-channel SQUID amplifier, especially in the investigation of the biomagnetism. The multi-channel SQUID amplifier can simultaneously measure the magnetic field at many positions on the body. This multi-point measurement can be expected to obtain more precise informations about the electric activity of organs.

Now, we have devised three methods for the multi-channelling of the SQUID amplifier, and developed a three directional SQUID gradiometer for vector magnetocardiogram(VMCG) measurement.[1,2] The developed SQUID system has a simple construction compared with the reported VMCG instruments[3,4,5], because of the intrinsic six channel rf SQUID amplifier use. Here, the principle of the multi-channelling methods, the developed six channel SQUID system, and some experimental results of the VMCG measurements using this SQUID system are described.

PRINCIPLE OF OPERATION OF MULTI-CHANNEL SQUID AMPLIFIERS

RF SQUID TYPE

As shown in Fig.1, N rf SQUID elements are coupled to one rf tank circuit. Rf voltage across the tank circuit, V_t is modulated by N input flux signals(ϕ_1, ϕ_2---ϕ_n) discriminated by separate audio frequencies(f_1, f_2---f_n). After rf detection of V_t, the varing parts ΔV_t of V_t can be expressed apporoximately as follows.

$$\Delta V_t = \left(\frac{K}{\sqrt{N}} \right) \sum_{n=1}^{N} \Phi_n \sin 2\pi f_n t$$

This voltage is applied to N phase sensitive detectors(PSD_1, PSD_2---PSD_n) driven by the same frequencies as the discriminating frequencies. The outputs of each PSD are fed back to the respective SQUID elements to obtain a well defined relation to the input signals.

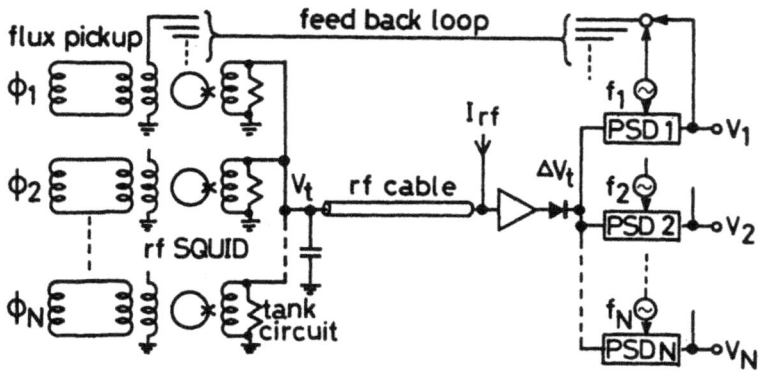

Fig.1 Multi-channel rf SQUID amplifier

DC SQUID TYPE

Similar method as rf SQUID can be applied to multi-channel dc SQUID amplifiers as shown in Fig.3.[6] It consists of two stage of dc SQUID elements, but all elements are dc biased by same current. In the first stage, N dc SQUID elements are driven by input flux signals($\phi_1, \phi_2 \text{---} \phi_n$) and discriminating frequencies($f_1, f_2 \text{---} f_n$). Voltage variations V_0 due to the input signals of each SQUID element are summed up by an inductance L coupled to the SQUID element of the second stage. We assume that all SQUID elements have identical characteristics. Then, current I through the inductance L can be expressed as follows.

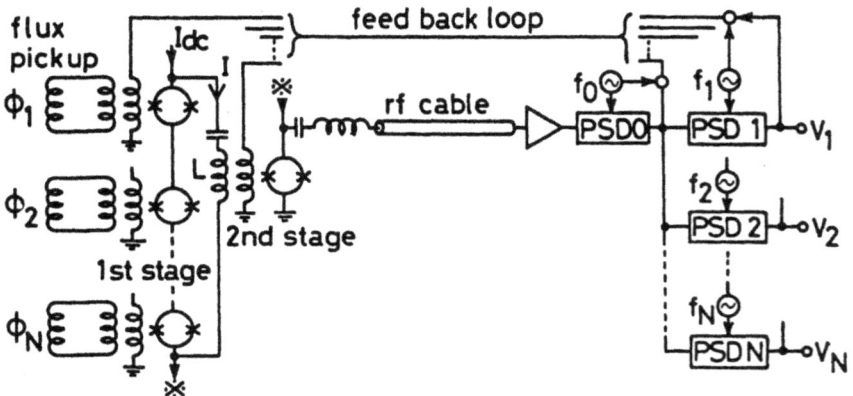

Fig. 2 Multi-channel dc SQUID amplifier

$$I = \sum_{n=1}^{N} \left(\frac{V_o}{Nr + j\omega_n L} \right) (\Phi_n \sin 2\pi f_n t)$$

where, r is output resistance of each SQUID element at the dc bias level. This summed up signals modulates the second stage SQUID element driven by much higher frequency than those of the first stage. Then, the modulated output voltage is applied to the phase sensitive detector PSD_0 through a resonance circuit and a preamplifier. Each component of N input signals of the PSD_0 output are separated by the same means as the rf SQUID type.

MULTIPLEXER TYPE

Above mentioned two multi-channel SQUID amplifiers, much simpler, of course, than the many single channel amplifier assemblies, requiers the same number of SQUID elements as the number of input signals. If the input flux signals can be multiplexed, a single SQUID amplifier will do. This multiplexing can be attained using the configuration shown in Fig.3. The multiplexing circuit switches sequencially the flux from the N superconductive closed circuits(first loops) to the second loop coupled to a single SQUID amplifier. The flux switching is accomplished by changing the coupling factor between the coil L_{n1} of the first loop and the coil L_{n2} of the second loop using the superconductive flux switchs(SFS). The SFS act as the superconducting enclosure for the coils, and made of a lower critical temperature material(such as Pb) than that of the coil material(such as NbN). So it is possible to transfer the SFS from super to normal state and vice versa, maintaining all coils in the superconducting state. The thermal radiation through the optical fibers are used for the transfer operations. In the superconducting state of the SFS, there is no coupling for two coils. But in the normal state, two coils tightly couples with each other, and the input flux signal is transferred to the SQUID element. Lower inductance of the second loop improves the S/N ratio because the flux

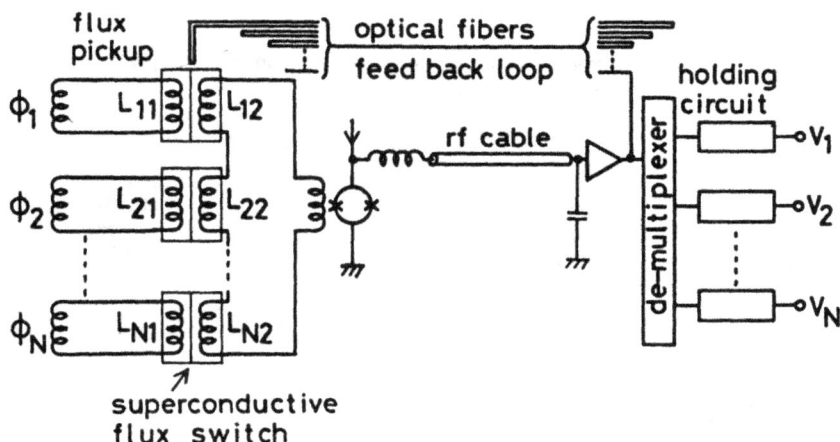

Fig.3 Multiplexer type multi-channel SQUID amplifier

transmittance rate increases. If the SFS couples tightly to the coils(L_{n1}, L_{n2}) except the sampled channel, the SFS makes short circuits for the coils. Therefore, the effective inductance of the second loop is apporoximately the sum of only two inductances coupled to the SQUID element and the switched channel.

CAPACITY OF THE INPUT NUMBER AND BAND WIDTH

In both rf and dc multi-channel SQUID amplifiers, the capacities of the input number are determined by the carrier frequency F, Q factor of the resonance circuit and requiered band width B of each channel. Practically available frequency range is $F/(2Q)$. The discriminating frequencies are selected in this range, but to avoid the interferences due to their harmonics, the recommended frequencies are between $F/(4Q)$ and $F/(2Q)$. If the crosstalk between channels less than one hundredth are requiered, the practical input number is $F / (400 Q B)$. Using the practical values of F=100 MHz, Q=100 and B=100 Hz, 25 channels is obtainable. In the multiplexer type, the input number is determined by the requiered band width and the transfer speed of the superconducting flux switchs. The transfer speed can be made as fast as a few tens microsec. When 250 samples per sec per channel is scheduled, amplifier with 80 channels is attainable.

COMPARISONS OF MULTI-CHANNEL SQUID AMPLIFIERS

Table 1 shows some compared performances for above mentioned three types amplifiers and ordinary one of multi-channel method. As shown, the required electronics of the conventional type is as many as the input number, then this leads to a bulky system, but it has widest band in four types. The rf SQUID type is inferior in the S/N ratio, but the fabrication of SQUID element is easier than that of the dc type. The dc SQUID type with two stage configuration is excellent for the flux sensitivity. On the simplisity of system, the multiplexer type is excellent because of one SQUID element use.

Table 1   Comparison of four type of SQUID amplifiers
for multipoints measurements

◎ excellent,   ○ good,   △ inferior,   ✕ not good

	conventional parallel amps	intrinsic multi-channel SQUID amplifiers		
		rf SQUID type	dc SQUID type	multiplexer type
inputs (N)	△	○	○	○
SQUIDs	N	N	N	1
rf cables	N	1	1	1
other wires	0 / 1	N	N+1	1+N(optical)
electronics	N	1	1	1
simplisity	✕	○	○	◎
S/N ratio	○	△	◎	○
band width	◎	○	○	△
interaction	△	○	○	○

SIX CHANNEL SQUID SYSTEM

Fig.4 shows rf type six channel SQUID system. Three second derivative gradiometers and three noise pickup coils are coupled to six rf SQUID elements(S_1~S_6). They are driven by a L–C resonant circuit, so the six input signals ride on a single rf carrier and are sent to the room temperature region by a single rf cable. As the signal in each channel is marked by amplitude modulation with the frequency assigned to the channel, signal separation is easily done by phase sensitive detector(PSD). PSD outputs P_1, P_2, and P_3 correspond to noise magnetic fields and P_4, P_5, and P_6 to the signals from three gradiometers. Properly weighted P_1~P_3 outputs are subtracted from P_4~P_6 outputs in the noise canceller block, producing noise suppressed outputs L, H, and U. Noise level of this system is about 3×10^{-4} $\phi_0/\sqrt{Hz}$ at 1 Hz in ordinary laboratory room. The system gain is about 1 V/nT, and the band width is from 0.1 to 12 Hz. Fluctuations of gain and band width between channels are less than 5 %. Cross talk between channels is less than 3 %.

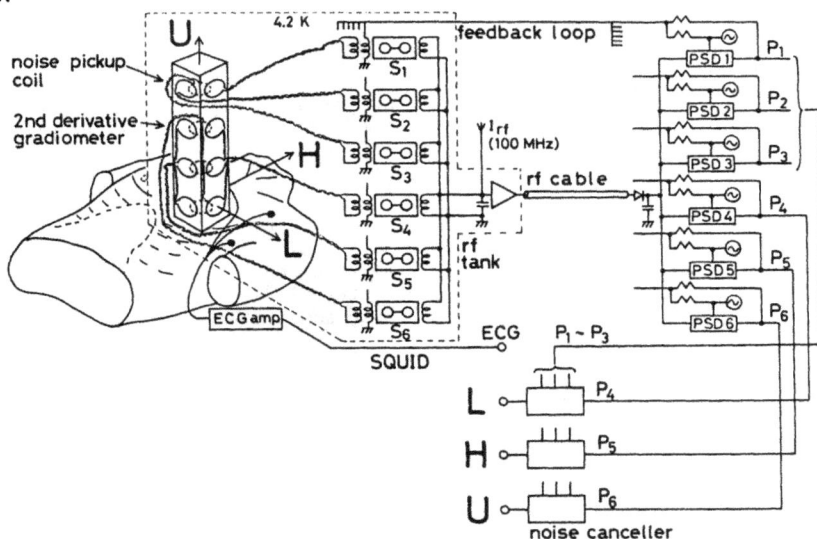

Fig.4   System diagram of a three dimensional SQUID gradiometer using six channel rf SQUID amplifier

MEASUREMENTS OF VMCG

As an application to the biomagnetism of the developed six channel SQUID system, the VMCG were measured in our laboratory room without magnetic shielding, but twenty times synchronous averaging was used to obtain clear results. We define an orthogonal coordinate system L, H, and U as shown in Fig.4. For the VMCG measurement, the subjects take the supine position on a wooden bed during measurement, and the cryogenic dewar tip is positioned as near as possible to the chest. Fig.5 shows an example of the wave forms obtained from the VMCG measurements. Using this example, three dimensional displays are shown in Fig.6. The circles show the vector top and the larger

784

radius means the vector turns to this side and the smaller radius to the other side. Overlap of the circles show the direction of the lapse of time.

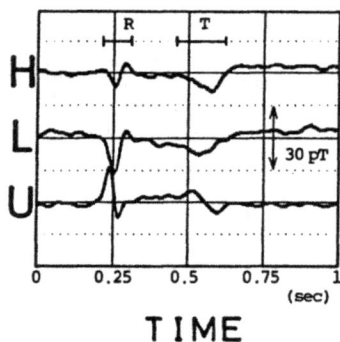

Fig.5 Wave forms of VMCG

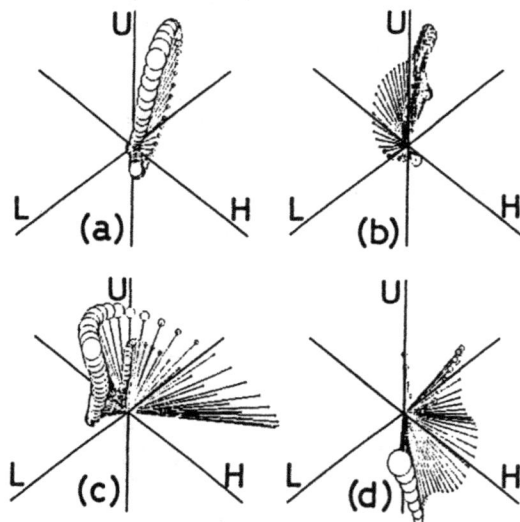

Fig.6 Vectorial representation : (a) QRS wave(interval R in Fig.5), (b) T wave (interval T), (c)(d) another example with same interval as (a) and (b)

SUMMARY

For the multi-channelling of the SQUID amplifier, three construction methods were devised. Using rf SQUID type, six channel SQUID system with three directional gradiometers were developed. The VMCGs were obtained in real time using the developed SQUID system at the ordinary laboratory, and the time variation of the measured VMCGs were displayed on a three dimensional screen.

REFERENCES

(1) K.Shirae et al   : CRYOGENICS, Vol.21, No.12, 707/710, 1981
(2) K.Shirae et al   : Intermag Conference digest, HE-07, April, 1987
(3) J.P.Wikswo and W.A.Fairbank : IEEE Tr. on MAGNETICS, MAG-13, 354, 1977
(4) J.A.V.Malmivuo and J.P.Wikswo : Proc. IEEE, 809, May, 1977
(5) W.H.Barry et al   : SCIENCE, Vol.198, 1162, 1977
(6) H.Furukawa et al : 1985 Ann.Conf. of Japan Soc. of Appl. Phys.,
                        2p-ZD-5, Oct. 1985

FLUX-FLOW NOISE IN TYPE-II SUPERCONDUCTOR THIN FILMS

Osamu OKAMURA, Junji SHIRAFUJI, and Naokatsu SANO[*]

Department of Electrical Engineering, Faculty of Engineering,
Osaka University, 2-1, Yamadaoka Suita, Osaka, 565, Japan
[*]Department of Physics, Faculty of Science, Kwansei-Gakuin
University, Uegahara Nishinomiya, 662, Japan

Abstract - Flux-flow noise spectra in thin film microbridges prepared from single and polycrystalline niobium (Nb) films grown by MBE and rf sputtering respectively have been measured. It is revealed that the flux-flow noise is governed by fluctuations of the penetration process at the microbridge periphery in the case of single crystalline samples, while pinning effects at grain boundaries dominate in polycrystalline thin film microbridges.

Introduction

When a direct current flows through a type-II superconductor under a magnetic field between the critical fields H_{c1} and H_{c2}, quantized fluxoids penetrating into the sample move in the direction transverse to both the current and the magnetic field under the influence of Lorentz force, inducing a voltage across the sample. This induced voltage is usually composed of dc and randomly fluctuating components. The noise component is due to fluctuations of flux-flow. If the entering and vanishing processes of fluxoids at the periphery occur periodically, the induced voltage is coherent and no noise component is generated. However, because the spontaneous penetration of fluxoids through the periphery is actually an activation process, the period of the penetration fluctuates randomly, causing voltage noise even in the ideal case where grain boundaries and impurity precipitation are not present in the sample. When the sample is prepared from polycrystalline films,the pinning effects on flux-flow motion are superimposed and dominate over the effect of random penetration process for voltage noise.

In previous experiments [1,2], noise power spectra generated by random flux-flow in narrow strip samples of type-I and II superconducting thin films have been measured and informations about pinning centers relating to grain boundaries have been obtained. An experiment on bulk samples of type-II superconductors [3] has also been performed. However, there has been no experiment on thin film microbridges. In this paper, noise power spectrum of superconducting microbridges prepared by MBE-grown single crystalline Nb thin films is studied experimentally and compared with that measured in rf sputtered polycrystalline thin film microbridges which contain numerous grain boundaries.

Simple Consideration of Flux-Flow Noise

When a fluxoid accompanying flux Φ travels across the bridge with a transit time τ_c, a voltage pulse of rectangular shape is generated [3,4]. If there is no pinning center in the sample, the voltage noise would be only because of fluctuations of the penetration of fluxoids into the bridge region. Therefore, the noise power spectrum $W_1(f)$ is analogous to the shot noise in a vacuum tube and given as

$$W_1(f) = 2\Phi V_{dc} [\frac{\sin(\pi f \tau_c)}{\pi f \tau_c}]^2 \tag{1}$$

where V_{dc} is the induced dc voltage across the bridge region. When pinning centers possibly due to grain boundaries exist in the bridge region, an additional voltage noise would be generated by an influence of pinning centers. Pinning centers interrupt the smooth motion of fluxoids to stop for a time and/or change the direction and velocity of their traverse. When it is assumed that the fluxoid motion is interrupted for vanishingly short time by randomly distributed pinning centers, the original rectangular voltage pulse of the duration τ_c is divided into subpulses with various durations τ_i [5]. By assuming noise power spectrum for each subpulse to be the same form as Eq.(1), the noise power spectrum $W_2(f)$ in this case is given by

$$W_2(f) = \int_{\tau_1}^{\tau_2} g(\tau_i) 2\Phi V_{dc} [\frac{\sin(\pi f \tau_i)}{\pi f \tau_i}]^2 d\tau_i \tag{2}$$

where $g(\tau_i)$ is a distribution function for τ_i.

Experimental Procedures

Single crystalline Nb films (thick 1000 Å) grown on sapphire substrates (R plane) by MBE method and polycrystalline films (thick 1000 Å) deposited on sapphire substrates (C plane) by rf sputtering were patterned into microbridges ($1 \times 1 \mu m^2$) by photolithography technique and plasma etching. When a direct current higher than a critical value I_c flows through the microbridge cooled down below the superconductivity transition temperature T_c, a dc voltage was induced. The voltage noise superposed on the dc voltage was measured by using a lock-in amplifier as a narrow-band ac voltmeter with a bandwidth of 1Hz. The noise power spectrum was sensitive to the dc current flowing through the bridge (bias current) and also to the ambient temperature. Dependence of bias current and ambient temperature on the noise power spectrum was measured to clarify further the mechanism of flux-flow noise.

Results and Discussion

Figure 1 shows the comparison of noise power spectra between single crystalline and polycrystalline thin film microbridges. A big difference between noise power spectra is evident.

In single crystalline samples the noise voltage would be

caused only by the influence of fluctuations of penetration process of fluxoids at the microbridge periphery, because there exist no grain boundaries (or pinning centers). The noise power spectrum should, therefore, follow Eq. (1). In Fig. 1 the experimental points for a single crystalline sample is fitted to Eq. (1) using adjustable parameters of τ_c=25ms, and Φ=2.59× 10^{-13}T.m^2. A good agreement is obtained in the frequency range between 10 and 40Hz.

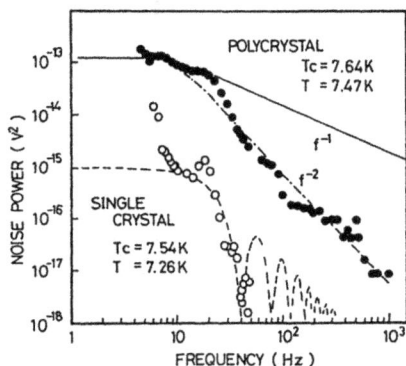

Fig. 1.
Noise power spectra in single crystalline and polycrystalline microbridges. The single crystalline microbridge was measured at T=7.26K and at the bias current I=2.26mA. The polycrystalline sample was measured at T=7.47K and I=1.50mA.

In polycrystalline microbridges, on the other hand, the noise voltage would be generated not only by fluctuations of the penetration period of fluxoids but also by the influence of pinning effects. The experimental noise power spectrum can be fitted to Eq. (2) when τ_1=10ms and τ_2=50ms are assumed. This indicates that fluxoids are pinned finite times at pinning centers due to grain boundaries.

Figure 2 shows the bias current dependence of the noise power spectrum in a single crystalline sample. It is seen that as the bias current is increased, the noise power increases. When Eq. (1) is fitted to the experimental curves, the value of the transit time τ_c , the total flux Φ , and the number of flux quanta in a fluxoid n(=Φ/Φ_0,Φ_0:flux quantum) are calculated as listed in Table 1. An increase in the Lorentz force with increasing bias current makes flux-flow velocity high, causing a decrease in τ_c. The size of bundle of flux Φ which penetrates into the bridge region increases as the bias current is increased, because an increased Lorentz force enhances the penetration of fluxoids.

Figure 3 shows the bias current dependence in a polycrystalline sample. The result is entirely opposite to those observed in the single crystalline microbridge (Fig. 2). As the bias current is increased, the noise power decreases. This can be understood by considering a dominated effect of pinning processes on flux-flow over fluctuations of the penetration process of fluxoids. When the bias current is increased, the increased Lorentz force reduces effects of pinning centers, causing a decrease in the noise power.

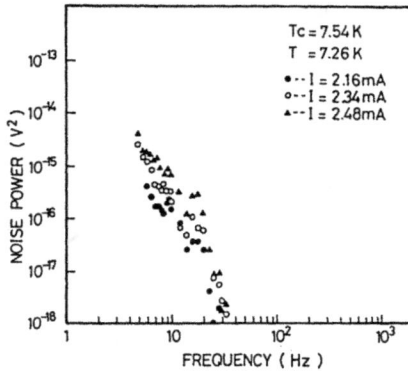

Fig. 2.
Bias current dependence of the noise power spectrum in a single crystalline microbridge.

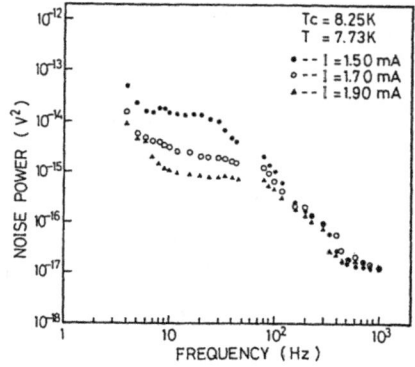

Fig. 3.
Bias current dependence of the noise power spectrum in a polycrystalline microbridge.

Table 1. Bias current dependence of τ_c, Φ, and n in a single crystalline sample at T=7.26K.

Bias Current I(mA)	Transit Time τ_c(ms)	Flux $\Phi(10^{-13}T.m^2)$	Number of Flux Quanta n(=Φ/Φ_0)
2.16	35	0.75	36.2
2.34	29	0.80	38.4
2.48	28	1.09	52.7

Figure 4 shows the ambient temperature dependence of the noise power spectrum in a single crystalline sample. As the temperature is lowered, the noise power decreases. When Eq. (1) is fitted to the experimental curves, the values of τ_c, Φ, and n are determined as shown in Table 2. A decrease in the temperature may decrease the frequency of the penetration period of fluxoids, causing a reduction of the noise power. The viscous resistance increases as the temperature is reduced; this causes an increase in τ_c with reducing temperature.

Figure 5 shows the temperature dependence of the noise power spectrum in a polycrystalline sample. As the temperature is lowered, the noise power increases; this behavior is the reverse of the result for the single crystalline sample. An influence of pinning centers on flux-flow increases with decreasing temperature, causing an increase in the noise power, because of increased pinning force and reduced velocity of flux-flow due to increased viscous resistance.

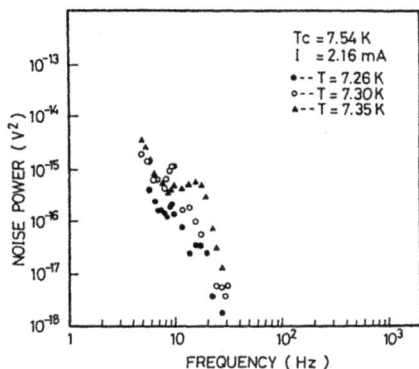

Fig. 4.
Ambient temperature dependence of the noise power spectrum in a single crystalline microbridge.

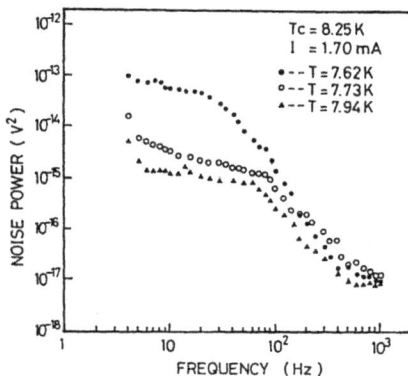

Fig. 5.
Ambient temperature dependence of the noise power spectrum in a polycrystalline microbridge.

Table 2. Ambient temperature dependence of τ_c, Φ, and n in a single crystalline sample at I=2.16mA.

Temperature T(K)	Transit Time τ_c(ms)	Flux $\Phi(10^{-13}$T.m$^2)$	Number of Flux Quanta n(=Φ/Φ_0)
7.26	35	0.75	36.2
7.30	32	1.70	82.1
7.35	30	2.86	138.2

Conclusions

Flux-flow noise in superconducting microbridges prepared by single crystalline films is mainly governed by fluctuations of the penetration period of fluxoids. On the other hand, in polycrystalline samples pinning effects at grain boundaries on flux-flow dominate fluctuations of the penetration period at the bridge in the noise power spectrum.

To clarify further mechanisms of flux-flow noise in type-II superconductor thin films, we are now going to measure effects of lattice defects introduced by ion implantation on noise power spectrum in Nb thin films. The similar measurement on high T_c superconductors may serve as a useful tool to get insight into the origin of pinning effects in these new materials.

Acknowledgement

The authors would like to express their thanks to Dr. T. Yotsuya of Osaka Prefectural Industrial Research Institute for his help in plasma etching.

References

1) C. M. Knoedler and R. F. Voss, Phys. Rev. 1326 (1982) 449
2) D. J. Van Ooijen and G. J. Van Gurp, Phys. Lett. 17 (1965) 230
3) G. J. Van Gurp, Phys. Rev. 166 (1968) 436
4) J. D. Thompson and W. C. H. Joiner, Phys. Rev. B20 (1979) 91
5) F.Habbal and W. C. H. Joiner, J. Low Temp. Phys. 28 (1977) 83

Section. Technical Data of Composite Materials

HIGH-STRENGTH AND HIGH-MODULUS POLYETHYLENE FIBER

Masako NAKAI, Hiroshi YASUDA and Ichiro YOSHIDA[*]

TOYOBO Research Institute, TOYOBO CO., Ltd.
2-1-1 Katata, Ohtsu, Shiga, 520-02, Japan
* Dyneema Japan Ltd.
2-8, Dozima Hama 2-chome, Kita-ku, Osaka 530, Japan

Abstract - High performance polyethylene (HPPE) fibers have been developed and their properties were examined. The characteristics of the fibers are reviewed.

Introduction

As one of the further improvement of the mechanical properties of man-made fibers, by orientation and chain extension, the gel spinning process was invented in Netherland and this process was applied to the ultra-high molecular weight polyethylene (UHMWPE) to produce the high performance polyethylene (HPPE) fibers.

Manufacturing Process of The Developed HPPE Fibers

As the results of many investigations for the improvement of the mechanical properties of the fiber of the flexible polymer chain, it was cleared that one of the ideal way to provide fiber with high strength and high modulus was to make fiber highly oriented by extending as much as possible alog the fiber axis. The technical points for this purpose were found to be existing in using ultra-high molecular weight polymer as high as possible, and applying the best condition that was suitable to carry out the super high extension.
We carried out many experiments, and we found that the super high extension can be obtained by drawing gel-like fiber with less entanglement content from UHMWPE solution. We called the process gel spinning ,because the fiber is spun and super drawn in gel state.
In this process, the very long molecules of UHMPE are dissolved in a volatile solvent and spun through a spinnerette. In the solution the molecules become disentangled and remain so after cooling in gel-like filaments. As the fiber is drawn, a very high level of macromolecular orientation is attained, and a fiber with a very high tenacity and modulus is obtained(Fig.1).
This fiber is characterized by a parallel orientation greater than 95% on a high level of crystallinity. This gives the developed HPPE fibers the unique properties. This fiber is now supplied under the name Dyneema SK60.

The Properties of The Developed HPPE Fibers

The fiber properties of the developed HPPE fiber are described below. These characteristics of the fiber are obtained by attaining the high level of orientation of UHMPE, making use of the structural flexibility of the polyethylene polymer chain.

1. High-strength and high-modulus

Table 1 shows the fiber properties of the developed HPPE fiber

comparing with other commercial high performance fibers.
On a weight-for-weight basis the newly developed fiber is the strongest
fiber on the market.
Its tensile strength is 2.7GPa which, combined with a density less than
1, gives a tenacity, or specific strength, of 30g/den. Modulus is also very
high; 87GPa and on a specific basis 1000 g/den.
Fig.2 and 3 compare this HPPE fiber with other high performance fibers.
Fig.2 gives specific strength versus specific modulus while Fig.3 is the
stress/strain diagram.

2. Impact strength

Fig.4 shows the impact strength of various fibers measured in
accordance with JIS 4013. Due to its high strength and modulus, the
developed HPPE fiber exhibits high energy to break at low elongation.

3. Abrasion resistance and flexlife property

Fig.5 shows the abrasion resistance of various fibers. The developed
HPPE fiber shows a good abrasion resistance. A good flexlife is also
expected due to a flexible chain material.

4. Loop strength and knot strength

Table 2 shows the loop strength/knot strength of various fibers.
The developed HPPE fiber has a high loop strength and a high knot
strength, because of its flexible chain of polyethylene.

5. Resistance to UV light, chemical resistance, and insensitivity to water and moisture

The developed HPPE fiber shows a good resistance to UV light , as
can be seen from Fig.6. The 1500 hrs. in this example corresponds to
about 2 years of outdoor exposure.
Chemical resistance is also good. In the pH range from 1 to 14, as shown
in Fig.7, the HPPE fiber shows good strength retention . The HPPE fiber
shows the good behavior in a range of solvents and chemicals.
The HPPE fiber is also expected to be very insensitive to water and
moisture due to its chemical structure.

6. Thermal stability

The HPPE fiber is polyethylene fiber with a high crystallinity.
Intrinsically, this fiber has a melting point between 145 and 155°C,
which means that use and processing should remain well below this
temperature. During processing, however, relatively high temperatures can
be used for a limited period without losing the properties of the fiber,
as can be seen from Fig.8.

7. Creep property

The properties of the developed HPPE fiber get better when the
temperature is lower. This holds especially with strength and creep. The
creep is sensitive to temperature (Arrhenius type of dependency),
reaching lower figures at lower temperatures, as shown in Fig.9. The
absolute figure for creep also strongly depends on the process and on the
grade of UHMWPE that is used to produce the fibers giving rise to
differences in the plateau creep rate of more than one decade.

8. Adhesive property

Due to the chemical nature of polyethylene, adhesion of the fibers to
matrix systems may be a problem. Of course there is no problem with the
adhesion of the fiber to regular polyethylene and very fine composites
have been made using this system. But with epoxy and other resins, it is
necessary to treat the HPPE fiber in a corona or plasma device to assure

HPPE fiber
Crystallinity >70%

Normal PE
Crystallinity <60%

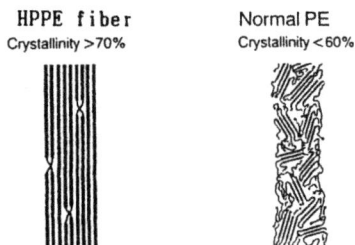

Fig.1 Crystalline models of HPPE fiber
and normal PE fiber

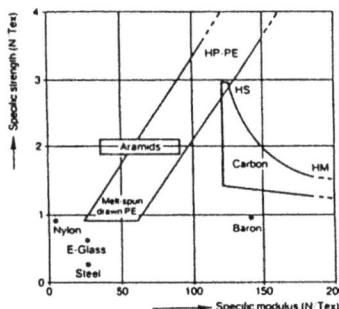

Fig.2 Specific strength vs specific modulus

Table 1 Fiber properties of HPPE fiber with other HP fibers

	HPPE fiber	Aramid 29	Aramid 49	Carbon HS	Carbon HM	E-Glass
density(g/cm)	0.97	1.44	1.45	1.78	1.85	2.55
tensile strength(GPa)	2.7	2.7	2.7	3.4	2.3	2.0
tenacity(N/tex)	2.65	1.9	1.9	1.9	1.2	0.8
tenacity(g/den)	30	22	22	22	14	9
modulus(GPa)	87	58	120	240	390	73
specific modulus(N/tex)	90	40	83	134	210	28
specific modulus(g/den)	1000	450	950	1500	2400	310
elongation at break(%)	3.5	3.7	1.9	1.4	0.5	2.0

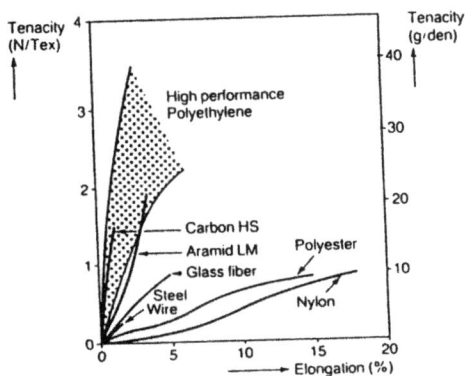

Fig.3 Stress/strain diagram

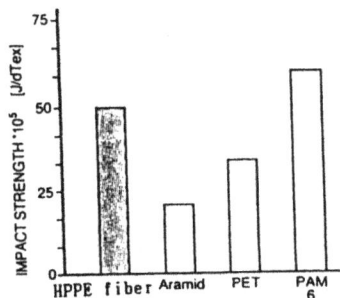

Fig.4 Impact strength of yarns

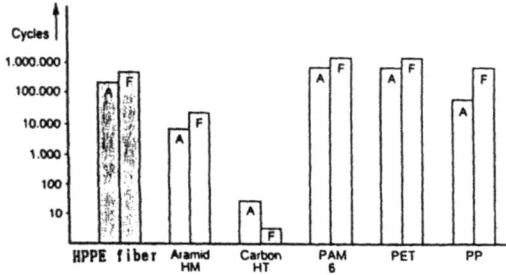

Fig.5 Abrasion resistance vs. Flexlife

Table 2 Loop strength vs. Knot strength

Material	Loopstrength N/tex	%	Knotstrength N/tex	%
HPPE fiber	1.3-2.0	40-65	1.1-1.7	35-55
Aramid	0.9-1.5	40-75	0.6-0.8	30-40
Carbon	0.01	~ 1	0	0
PET	0.6-0.7	70-75	0.4-0.5	50-60
PAM-6	0.6-0.7	70-75	0.5-0.6	60-65
PP	0.6-0.7	85-95	0.4-0.5	60-70

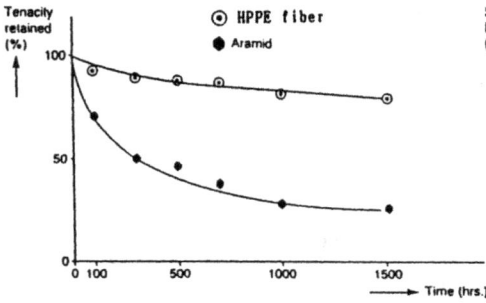

Fig.6 Light resistance (fadometer)

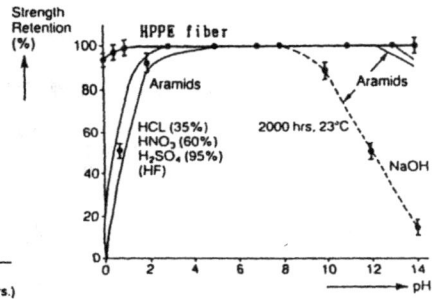

Fig.7 Chemical resistance

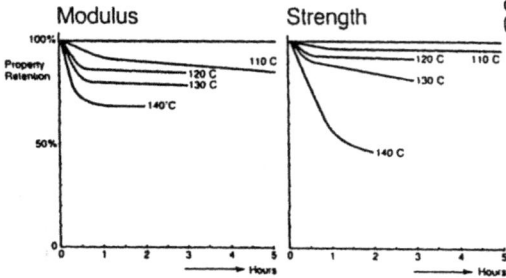

Fig.8 Short term thermal exposure

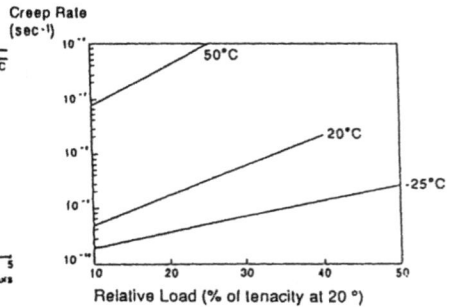

Fig.9 Creep of HPPE fiber

good adhesion. During this treatment the other properties of the HPPE fiber are not affected.

9. Other properties

The HPPE fiber is electrically nonconducting and transparent to X-ray and rader waves. This makes it possible to use the HDPE fiber to produce radomes, for example.

Conclusion

We have developed new high performance polyethylene fiber, from ultra-high molecular weight polyethylene, in the gel spinning process.
This fiber shows high-strength and high-modulus property, and on a weight-for-weight basis is the strongest fiber on the market.

Acknowlegements

The figures and tables are quoted from the following pamphlets 1) and 2).

References

1) Dyneema, Dyneema SK60 High strength/high modulus fiber PROPERTIES & APPLICATION (Pamphlet)
2) Dyneema, Dyneema SK60 High performance fiber in composite (Pamphlet)
3) H.Yasuda, Journal of the Textile Machinery Society of Japan, 40, P-261, 1987
4) R.Kirschbaum, H.Yasuda and E.H.M.van Gorp, 25th Int.Chemical Fiber Congress, 24-26 September, 1986, Dornbirn. Austria
5) H.Yasuda, Journal of The Society of Fiber Science and Technology,Japan, 43, P-139, 1987
6) K.F.M.G.J.Scholle et, SAMPE Symposium, June 1988, Milan

798

INNOVATIVE POLYIMIDE FILM
"UPILEX"

Kenichiro Yano

Polyimide Dept: Engineering Plastics Div. UBE INDUSTRIES, LTD.
ARK Mori Build. 12-32, Akasaka 1-chome, Minato-ku, Tokyo 107, Japan

Abstract - UPILEX is a completely new type of polyimide film with a
unique chemical structure. Developed by Ube Industries using
proprietary technology, UPILEX features outstanding properties over a
wide range of temperatures, and offers the following advantages over
previously available polyimide film.

* Ultra-High Heat resistance
* Excellent Cryogenic Properties
* High Tensile Strength and Modulus
* Excellent Radiation Resistance (γ-ray, Electron, neutron)
* Excellent Weather Resistance (Ultraviolet)
* Superior Dimensional Stability
* Excellent Chemical Resistance
* Low Water Absorption
* Low Gas Permeability

Thanks for these outstanding features, UPILEX has been used in the
electronics and other leading-edge Industries.

Introduction

Polyimide is synthesized by a polycondensation reaction between
an aromatic dianhydride and an aromatic diamine. UPILEX proves unique
when compared with well-known conventional polyimide film because UPILEX
employs BPDA as a monomer.

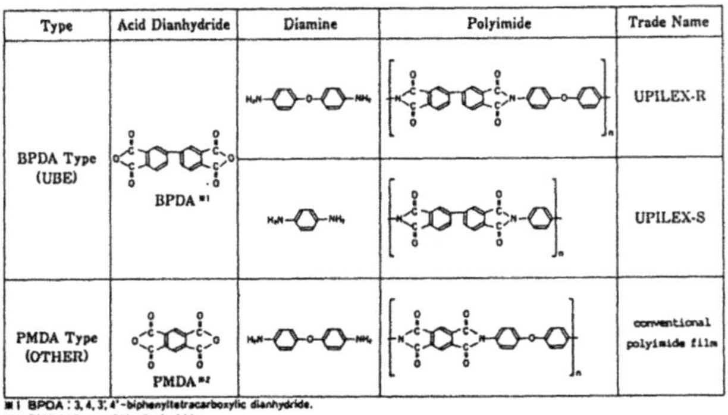

Type	Acid Dianhydride	Diamine	Polyimide	Trade Name
BPDA Type (UBE)	BPDA *1			UPILEX-R
				UPILEX-S
PMDA Type (OTHER)	PMDA *2			conventional polyimide film

*1 BPDA : 3, 4, 3′, 4′-biphenylltetracarboxylic dianhydride.
*2 PMDA: Pyromellitic dianhydride.

Fig.1. Chemical Structure of Aromatic Polyimide

UPILEX is available in two types. UPILEX-R is a standard polyimide film. UPILEX-S, with even more advanced specifications, is ideal for use in high added-value products.

Typical Properties of UPILEX

The characteristics of these grades are shown on Table 1. and Fig.2. through Fig.9.

Table-1   Typical Properties of Polyimide Film

Property	Unit	UPILEX-R	UPILEX-S	Other	Test Method (Test Condition)
Density	g/cm³	1.39	1.47	1.42	ASTM D1505-63T
Mechanical Properties					
Tensile Strength	kg/mm²	25	40	20	ASTM D882-64T (25°C)
Tensile Elongation	%	130	30	75	ASTM D882-64T (25°C)
Tensile Modulus	kg/mm²	380	900	330	ASTM D882-64T (25°C)
Thermal Properties					
Melting Point	°C	None	None	None	
Glass Transition Temperature	°C	285	None	None	
Thermal Decomposition Temperature	°C	520	560	500	Temperature Rise in 3°C/min.
Continuous Service Temperature	°C	270	290	250	Heat Treatment; 20,000 hrs.
Flammability		UL94V-0	UL94VTM-0	UL94V-0	UL94 File No.48133
Heat Shrinkage	%	0.18	0.07	0.25	JIS C2318 (250°C×2 hrs.)
Coefficient of Linear Thermal Expansion	cm/cm/°C	2.8×10^{-5} (20~250°C)	1.5×10^{-5} (20~400°C)	2.5×10^{-5} (20~250°C)	Micro Linear Thermal Expansion Tester Temperature Rise in 5°C/min.
Thermal Conductivity	cal/cm·s·k	5.6×10^{-4}	6.8×10^{-4}	4.1×10^{-4}	TC-3000 (Raser method)
Electrical Properties					
Dielectric Strength	kV/25μm	7.0	6.8	6.9	ASTM D149 (25°C, 50Hz)
Dielectric Constant		3.5	3.5	3.6	ASTM D150 (10³Hz)
Dissipation Factor		0.0014	0.0013	0.0020	ASTM D150 (10³Hz)
Environmental Resistance					
Water Absorption	%	1.3	1.2	2.1	ASTM D570 (Immersion in water: 23°C×24 hrs.)
Hydrolytic Stability (Residual Elongation) pH=1.0	%	90	85	30	ASTM D882 (100°C×2 weeks)
pH=10.0	%	95	85	55	ASTM D882 (100°C×4 days)
Alkali Resistance (Residual Elongation)	%	80	60	Soluble	ASTM D882 (25°C×5 days)
Gas Permeability H₂O	g/m²/25μm	22	1.7	47	ASTM E96 (38°C×RH90%×24 hrs.)
O₂	ml/m²/25μm	100	0.8	180	ASTM D134 (30°C×1 atm×24 hrs.)
Outdoor Weathering Resistance (Elongation)	%	85	25	3	Exposure: OSAKA (JAPAN), Jul., 1986 − Jan., 1987 (214 days)
γ-ray Irradiation Resistance (Residual Elongation)	%	80	—	6	Japan Atomic Energy Research Institute, Co-60 γ-ray 1,500Mrad. (Dose rate: 10⁶rad/hr.), O₂ Pressure (7 atm)

*   High Tensile Strength and Modulus

UPILEX-R's tensile strength and elongation are 25kgf/mm² and 130% respectively at room temperature, making it a very durable film. UPILEX-S, on the other hand, displays superior characteristics not only at room temperature, but at high temperature as well, with 22kgf/mm² of tensil strength and 350kgf/mm² of tensile modulus at 300°C.

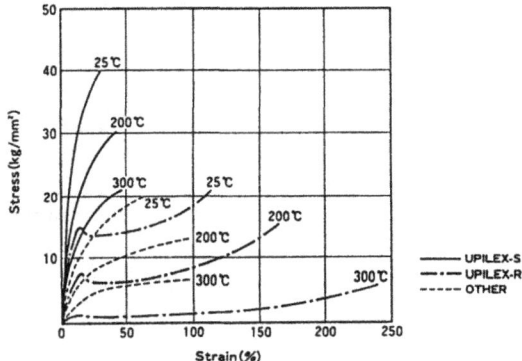

Fig.2. Tensile Stress-Strain Curves

*   Ultra-High Heat Resistance

UPILEX has no melting point. Thermal decomposition dose not start until over 600°C for UPILEX-S higher than any other organic compound. With a glass transition temperature of over 500°C, this new film exhibits superior physical heat resistance. Both UPILEX-R and S can be used continuously at temperatures of 250°C and above.

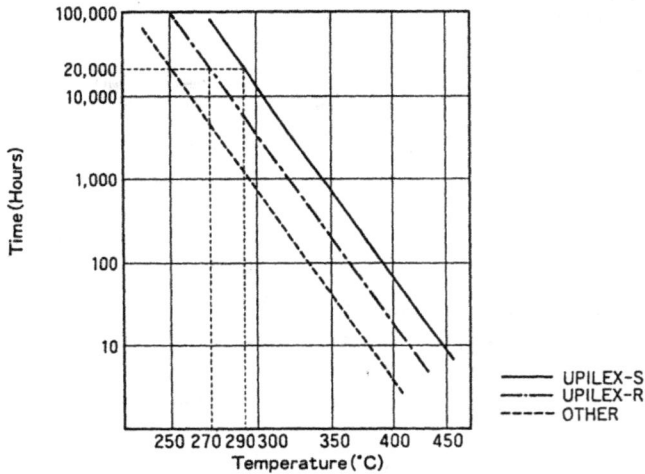

Fig.3. Temperature to 50% Reduction in Tensile Strength

*   Excellent Cryogenic Properties

UPILEX also features outstanding cryogenic resistance. Fig.4. shows the results measured at the Institute for Solid State Physics , Tokyo University.

Property	Unit	Film Temperature	UPILEX-50R	UPILEX-50S	OTHER
Tensile Strength	kgf/mm²	25°C	21	35	18
		−196°C	27	47	37
		−269°C	33	53	40
Tensile Elongation	%	25°C	81	16	50
		−196°C	63	7	20~38
		−269°C	2~60	6	2~20
Tensile Modulus	kgf/mm²	25°C	390	840	300
		−196°C	510	1,180	570
		−269°C	550	1,200	590

Test Condition ⎧ Gauge Length : 20mm
⎨ Crosshead Speed : 0.5mm/min.
⎩ Strain Speed : 4.2×10^{-4} S^{-1}

Fig.4. Cryogenic Mechanical Properties

* Excellent Radiation Resistance (*1)

Compared with other films, polyimide films are highly resistant to irradiation. And,among polyimide films, UPILEX is especially resistant, with a life several times longer than conventional polyimide film exposed to the same amounts of irradiation. Figures 5 and 6 show the results of tests conducted at the Japan Atomic Energy Reseaech Institute.

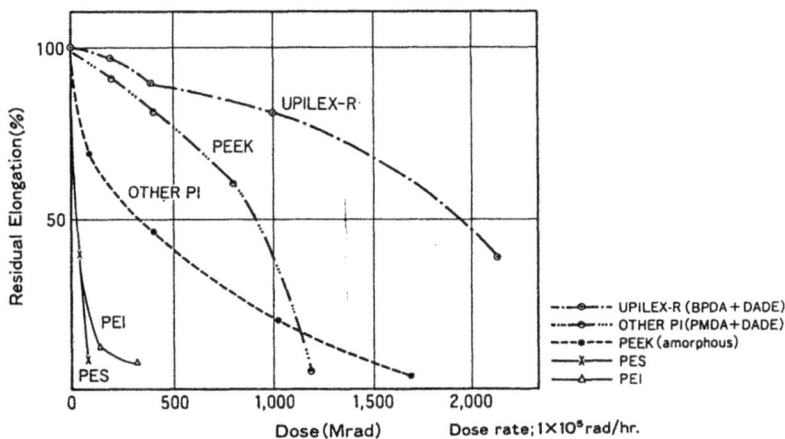

Fig.5. Relation of Residual Elongation and Dose (Co-60 𝛾-ray) under 0.7 MPa Oxgen Pressure

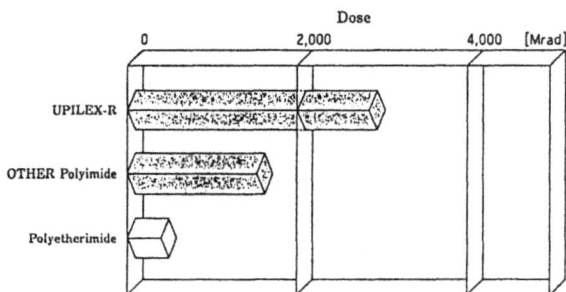

Fig.6. Electron Doses to 50% Reduction in Elongation

802

* Superior Dimensional Stability

UPILEX-S has dimensional stability far in advance of currently available products. Linear expansion, heat shrinkage and hygroscopic expansion are all extremely small.

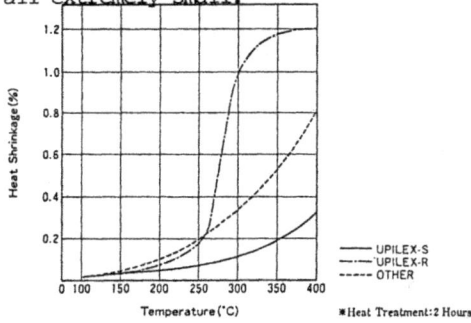

Fig.7. Heat Shrinkage

* Excellent Chemical Resistance

UPILEX is insoluble in all types of organic solvents and is resistant to virtually all chemicals. Further, unlike other polyimide films, UPILEX dose not hydrolyze readily or dissolve in alkalis.

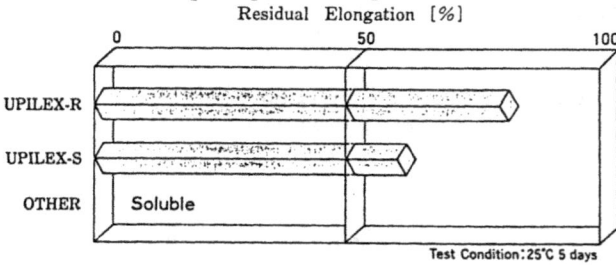

Fig.8. Residual Elongation after Immersion in 10% NaOH

* Low Gas Permeability

In particular, UPILEX-S exhibits excellent gas barrier properties.

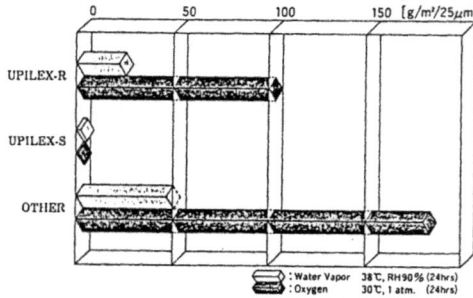

Fig.9. Gas Permeability

Some Typical Applications

* Superconducting wire and coil Insulation
* Super Insulation Film
* Vacuum Bagging Film (Autoclave Forming)

Furthermore Ube Industries, Ltd. can provide Metallized UPILEX, UPILEX-BOARD,U-VARNISH, UPIMOL (SHAPE) and POLYIMIDE-FIBER.

UPILEX-BOARD

Dimension

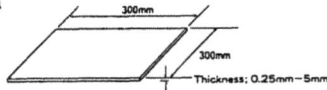

300mm
300mm
Thickness: 0.25mm~5mm

Table-2 Typical Properties of UPILEX-BOARD

Property	Temperature	−40℃	25℃	150℃
Tensile Strength	(kg/mm²)	19	21	11
Tensile Modulus	(kg/mm²)	450	440	260
Tensile Elongation	(%)	30	55	65
Flexural Strength	(kg/mm²)	24	20	12
Flexural Modulus	(kg/mm²)	470	430	300

Ube Industries,Ltd. has succeeded in styling boards from polyimide, something that was previously considered impossible. Boards can be manufactured in thickness ranging from 0.25 to 5 mm. In addition to possessing superior specific tensile strength that is about 2.5 times higher than steel, this tough material also features excellent impact and fatigue resistance. UPILEX-BOARD has similar basic properties to UPILEX as it is of the same molecular structure as UPILEX-R.

Conclusion

As mentioned above, UPILEX has surpassing properties compared with well-known conventional polyimide film because UPILEX employs biphenyl-tetracarboxylic dianhydride (BPDA) as a monomer. UBE believes that UPILEX will exhibit its highest capabilities paticularly under the severe environment.

Reference

*1) T.Sasuga and M.Hagiwara ; Papers of Technical Meeting on Electrical Insulating Materials,1984,EIM-84-132,IEE JAPAN

805